P9-CCJ-555

Practical Handbook of
Marine Science

Edited by
Michael J. Kennish, Ph.D.
Senior Environmental Scientist
GPU Nuclear Corporation
Forked River, New Jersey
and
Visiting Professor
Fisheries and Aquaculture TEX Center
Cook College
Rutgers University
New Brunswick, New Jersey

CRC Press
Boca Raton Ann Arbor Boston

Library of Congress Cataloging-in-Publication Data

Kennish, Michael J.
CRC practical handbook of marine science / edited by Michael J. Kennish
p. cm.
Bibliography: p.
Includes index.
ISBN 0-8493-3700-3
1. Oceanography. 2. Marine biology. I. Title. II. Title:
Practical handbook of marine science.
GC11.2.K46 1989
551.46—dc19

88-21003
CIP

This book represents information obtained from authentic and highly regarded sources. Reprinted material is quoted with permission, and sources are indicated. A wide variety of references are listed. Every reasonable effort has been made to give reliable data and information, but the author and the publisher cannot assume responsibility for the validity of all materials or for the consequences of their use.

Direct all inquiries to CRC Press, Inc., 2000 Corporate Blvd., N. W., Boca Raton, Florida, 33431.

Second Printing, 1990
International Standard Book Number 0-8493-3700-3

Library of Congress Card Number 88-21003
Printed in the United States

PREFACE

Contemporary research in marine science involves the multidisciplinary research efforts of marine biologists, marine chemists, marine geologists, and physical oceanographers. Significant findings during the past 3 decades in the major scientific disciplines have greatly advanced our understanding of the oceanic environment. The theory of plate tectonics and the concept of seafloor spreading, the complex chemosynthesis-based benthic communities flourishing at deep-sea hydrothermal vents, the close coupling of Pacific Ocean currents and overlying air masses in the development of El Niño, and the chemical transformations associated with the energy pathways in marine ecosystems provide examples.

The *CRC Practical Handbook of Marine Science* represents a systematic collection of selective physical, chemical, and biological reference data on the ocean. Arranged in a multisectional format, this book presents fundamental research material in expository, illustrative, and tabular forms. Discussions are provided on recent discoveries and current knowledge on the sea. It serves as a companion volume to four other handbooks published by CRC Press, including the *CRC Handbook of Marine Science* (Volumes I and II edited by F. G. Walton Smith and F. A. Kalber) and the *CRC Handbook of Marine Science, Compounds from Marine Organisms* (Volumes I and II edited by Joseph T. Baker and Vreni Murphy). Consisting of a compilation of the most useful oceanographic data derived from these four volumes, the *CRC Practical Handbook of Marine Science* focuses on nine subject areas: (1) air-sea interaction; (2) chemical oceanography; (3) physical oceanography; (4) marine geology; (5) ocean engineering; (6) phytoplankton; (7) primary productivity; (8) zooplankton; and (9) compounds from marine organisms. As such, it constitutes a useful, single-volume source of information covering all of the major disciplines as they relate to the ocean. Contributors to the *Handbook* are Joseph T. Baker (Section 9), Kurt Bostrom (Section 4), Harry DeFerrari (Section 5), Cesare Emiliani (Section 4), Carl Holm (Section 5), Donald Hood (Section 2), Frederick A. Kalber (Sections 6 to 8), Eric B. Kraus (Section 1), Vreni Murphy (Section 9), Richard M. Pytkowicz (Section 2), Joseph Reid (Section 3), and F. G. Walton Smith (Sections 6 to 8).

Although this publication has been designed as a reference for marine scientists, it should also be of value to administrators and other professionals dealing in some way with the management of marine resources and the investigation of problems pertaining to the marine realm. Its primary objective, however, is to amass data that will have appeal and utility for practitioners and students of oceanography and related disciplines. The scope of this task, of course, has grown immensely commensurate with the rapidly expanding data base in oceanography.

I want to thank the many scientist, colleagues, and friends who have interacted with me during the preparation of this volume. An expression of gratitude is extended to the managers of GPU Nuclear Corporation, particularly D. J. Cafaro and J. J. Vouglitois of the Environmental Controls Department, for their support and encouragement of this work. At Rutgers University, I am expecially grateful to R. E. Loveland and R. A. Lutz who have continually inspired me to pursue marine research. A specific ackowledgment is given to the marine scientists at the Fisheries and Aquaculture TEX Center of Cook College at Rutgers University involved in the assessment of the coastal waters of New Jersey. I also ackowledge the efficiency and guidance of the Editorial Department of CRC Press, especially Amy G. Skallerup, who coordinated the production of this book, as well as James Brody and Paul Gottehrer. In addition, I thank the authors and publishers responsible for granting permission to reproduce some of the material comprising this publication. Finally, I am most appreciative of my wife, Jo-Ann, and sons, Shawn and Michael, for recognizing the importance of my commitment to complete this *Handbook*. For me, even marine science remains far secondary to the love and well being of these three people.

THE EDITOR

Michael J. Kennish, Ph.D., is a Senior Environmental Scientist at GPU Nuclear Corporation, Forked River, New Jersey and a Visiting Professor at Cook College, Rutgers University, New Brunswick, New Jersey.

He received is B.A., M.S., and Ph.D. degrees in Geology from Rutgers University in 1972, 1974, and 1977, respectively.

Dr. Kennish's professional affiliations include the American Fisheries Society, American Geophysical Union, American Institute of Physics, American Society of Limnology and Oceanography, Estuarine and Brackish-Water Sciences Association, Estuarine Research Federation, New England Estuarine Research Society, Atlantic Estuarine Research Society, Southeastern Estuarine Research Society, Gulf Estuarine Research Society, Pacific Estuarine Research Society, National Shellfisheries Association, New Jersey Academy of Science, and Sigma Xi. He is also a member of the Advisory Board of the Fisheries and Aquaculture TEX Center of Rutgers University, overseeing the development of fisheries and shellfisheries in marine waters of New Jersey.

Although maintaining research interests in broad areas of marine ecology and marine geology, Dr. Kennish has been most actively involved with investigations of anthropogenic effects on estuarine ecosystems. He is the author of *Ecology of Estuaries*, published by CRC Press, and the editor of *Ecology of Barnegat Bay, New Jersey*, published by Springer-Verlag. In addition to these two books, Dr. Kennish has published articles in scientific journals and presented papers at numerous conferences. Currently, he is the co-editor of the journal, *Reviews in Aquatic Sciences*. His biographical profile appears in *Who's Who in Frontiers of Science and Technology*.

TABLE OF CONTENTS

Introduction

INTRODUCTION

SIGNIFICANCE OF MARINE SCIENCE

The study of marine science is necessarily a multidisciplinary endeavor, encompassing the efforts of many workers in the biological, chemical, geological, and physical sciences. The field concerned with research on the oceans — oceanography — has united biology, ecology, chemistry, geochemistry, geology, geophysics, physics, and meteorology, as well as other disciplines in investigations of the marine environment. During the past 2 decades, much information has been collected on the vast oceanic ecosystem, helping to delineate the processes operating in the water column and on the seafloor. Advances continue to be made on the circulation of the oceans, chemical reactions occurring in marine waters, and the biota occupying numerous habitats. Some of the most profound scientific discoveries in the recent past have involved geological oceanography and the elucidation of the structure of the ocean bottom. The confirmation of seafloor spreading has been of paramount importance in understanding the concept of continental drift and in formulating the theory of global plate tectonics which represents a unifying model in geology.[1] Since 1950, extensive exploration of the seafloor has unraveled new findings on deep-sea trenches, submarine volcanic and seismic activity, and mid-ocean ridges — a 75,000-km-long mountain system encircling the earth in the ocean basins.

The discovery of hydrothermal vents on the seafloor exemplifies the interdisciplinary nature of contemporary research in marine science. These hot-water vents, initially observed at a depth of 2500 m on the Galapagos Rift, support dense colonies of exotic organisms (clams, crabs, and giant tube worms) via chemosynthetic primary production.[2] While diving at the Galapagos Rift in the submersible ALVIN, geologists and chemists first recovered specimens of the bivalves *Bathymodiolus thermophilus* Kenk and Wilson[3] and *Calyptogena magnifica* Boss and Turner,[4] the limpet *Neomphalus fretterae* McLean,[5] the brachyuran *Bythograea thermydron* Williams,[6] the galatheid *Munidopsis subsquamosa* Henderson, and the tube-dwelling worm *Riftia pachyptila* Jones.[7] Biologists later visited this site on an expedition in 1979.[8] Subsequently, vent biota were found at other intermediate-to-fast-spreading oceanic centers in the eastern Pacific Ocean, including the East Pacific Rise, the Guaymas Basin of the Gulf of California, and the Juan de Fuca Ridge, and at slow-spreading oceanic centers in the rift valley of the mid-Atlantic ridge.[9,10] A summary of the biology of these unique vents has been reported elsewhere.[11-18]

Two types of seafloor hot springs can be differentiated based on the temperature of the emanating fluids. At the Galapagos Island rift zone, springs flow with a maximum temperature of approximately 16°C. Mineralized chimneys release solutions of much higher temperature (about 380°C) at the East Pacific Rise near Baja California.[1] Temperature appears to be a major determinant of faunal abundances at the vents.[10]

Chimney-like edifices and massive sulfides precipitated by the venting solutions have attracted the attention of geochemists and mineralogists. Rona et al.[9] discerned that sulfides constituted 66% by weight of dredged material removed from black-smoker-type hydrothermal vents at the mid-Atlantic ridge near latitude 26°08′N. At the Explorer Ridge in the northeast Pacific, sulfide samples taken from active chimneys or from mounds adjacent to active vents were characterized by high zinc:copper and zinc:iron ratios.[10] The temperature of the spewing hydrothermal fluids has been related to the sulfide mineralogy encountered in localized areas. Hence, with high temperature fluids of about 350°C, assemblages of minerals deposited at the chimneys or mounds may consist of chalcopyrite ($CuFeS_2$) bornite-chalcopyrite (Cu_5FeS_4-Cu_2S), in association with small concentrations of wurtzite ($(ZnFe)S$), pyrite (FeS_2), silicates such as talc ($Mg_3Si_4O_{10}(OH)_2$) and chrysolite ($Mg_3Si_2O_5(OH)_4$), and sulfate (anhydrite). At lower temperatures of approximately 260°C, minerals commonly noted are wurtzite, chalcopyrite, barite ($BaSO_4$), cubanite ($CuFe_2$), and anhydrite ($CaSO_4$), with wurtzite typically being dominant. Marked differences exist between the chemistry and mineralogy of active and inactive vents at the same locale.[19]

The inorganic chemistry of the hot-spring solutions has been discussed by Edmond et al.,[20-22] Edmond and Von Damm,[23] and Von Damm et al.[24,25] Grassle (p. 314)[14] synthesized the results of their work. "A combination of leaching from the basalts, and reduction of sulfate and bicarbonate in the seawater, produces hydrothermal fluid emanating from vents that is rich in biologically important compounds and ions such as H_2S, CO_2, NH_4^+, H^+, Fe^{2+}, Mn^{2+}, Cu^{2+}, Zn^{2+}, Ca^{2+}, and SiO_2.[22,23] Compared with seawater, hydrothermal fluid contains little or no sulfate and substantial amounts of hydrogen sulfide. Calcium, barium, silicates, alkaline earths, and trace metals are greatly elevated relative to sea water, and magnesium is absent."

Comita et al.[26] assessed the concentrations of particulate and dissolved organic carbon in vent waters. Clearly, the chemistry of the hydrothermal fluids and the ore-forming processes at the vent sites generate new minerals which have great potential value to mankind. The influence of these hydrothermal systems on the overall chemistry of the oceans, in turn, can be significant, because every 8 million years, the entire ocean cycles through them.[1]

Other areas of scientific inquiry of the sea have progressed rapidly as well over the two or three past decades. In physical oceanography, for instance, fronts and warm-core rings have become a focal point of many investigations. Fronts, defined as boundaries between horizontally juxtaposed water masses of dissimilar properties,[27] have been studied intensely as a result of their enhanced biological productivity and value of marine living resources. Le Févre[28] details the relevance of fronts to marine life. Six types of fronts are recognized: (1) river plume fronts; (2) upwelling fronts; (3) shallow sea fronts; (4) shelf break fronts; (5) fronts at the edge of major western boundary currents; and (6) fronts of planetary scale removed from major ocean boundaries.[29,30] Characteristics of frontal systems — complex circulation patterns and transport — involve surface convergence (purportedly responsible for accumulation of surface-living organisms), residual circulations parallel to the fronts, and cross-frontal exchanges through eddy motions.[28]

Warm-core rings develop in the North Atlantic Ocean between the North American continental shelf and the Gulf Stream in the slope water region.[31] Having a diameter of about 100 to 200 km, the warm-core rings form when a northward meander of the Gulf Stream closes, thereby enclosing a core of warm Gulf Stream or Sargasso Sea water.[32] By this process, the core rings are isolated from the Gulf Stream, and they subsequently move southwestward through the slope water region. The core rings, with a mean longevity of 6 months, account for the transport of streamers of continental shelf water into the slope water region adjacent to the Gulf Stream. In addition, they provide a means for the exchange of biological and chemical constituents across the Gulf Stream into the slope water region.[32,33] Comprehensive biological investigations have recently been undertaken in warm-core Gulf Stream rings.[34-37]

Ocean circulation is inextricably linked to the atmosphere.[38,39] Winds and density differences which drive circulation in the ocean largely depend on atmospheric conditions. The oceans, meanwhile, serve as an energy source for the atmosphere. The strong coupling between these two entities is evident in heat budgets formulated for oceanic and coastal areas.[40,41] The principal source of heat flux across the sea surface — solar radiation — enters directly or via reflection and scattering by the atmosphere and clouds. The loss of heat from the sea surface to the atmosphere occurs by means of longer wavelength back radiation and evaporation. In addition to heat fluxes through the sea surface, heat exchanges by advection or eddy diffusion contribute to the distribution of temperature in an oceanic region.[42] Advective changes in the heat content of various areas due to major ocean currents can exert a marked influence on land climates.[39] Such is the case for the Gulf Stream and Kuroshio Current, which transport substantial volumes of warm equatorial waters to high latitudes, affecting the climate of adjacent continental masses. The El Niño represents another example of changes in ocean currents (in the eastern Pacific along the Peru-Chile coast) that have a regional or global impact on land climates.[43] The high heat capacity of the oceans, in general, is a major factor in climate regulation.

Meteorological forcing mechanisms generate local and regional wave and current patterns in the ocean. Some effects are quite apparent, such as responses of waves to wind direction and velocity;[44] however, others remain subtle. The application of a new generation of sophisticated models has allayed various problems of air-sea interaction,[45-47] but additional observations and mathematical treatments must be pursued. The introduction of satellite vehicles and remote sensing devices promises to greatly facilitate the acquisition of data on surface and internal waves, sea-surface temperature and topography, mesoscale eddies and upwelling, coastal currents and frontal systems, as well as other phenomena in physical oceanography.[48,49]

The ocean is used by man for a wide range of purposes, including food resources, energy products (e.g., natural gas and oil), minerals, chemicals, construction materials, communications, and transportation. Another vital function of the ocean involves its utilization as a waste repository. Fishes, whales, crustaceans, molluscs, and seaweed are exploited renewable marine resources of great value to humans. The world fisheries seem to be approaching a maximum value which may be less than 1×10^8 MT/year.[50] By more effective management of the living marine resources, however, some experts believe that the fisheries yield could be doubled.[51] The bulk of the fisheries harvest is derived from estuaries and continental shelf waters (approaching 50% of the total) within 160 km of the shoreline and from highly productive upwelling zones (the remaining 50%).[52] Pelagic waters of the open ocean contribute only a small fraction to the world catch. Mariculture, or aquaculture, yields additional food from the ocean and could be a critical nutritional source in future years.

Some marine organisms (e.g., macroscopic algae) constitute more than just a food source. These seaweeds are valuable as fertilizer and fodder. They also have useful applications in medicine and paper production.[53]

As in the case of fisheries, most natural gas, oil, and construction materials (i.e., sand, gravel, and shell) derived from marine environments are obtained nearshore from estuaries and continental shelves. For example, estimates of undiscovered natural gas and oil equal about 70% for continental shelves and shallow marginal ocean basins and 23% for continental slopes.[39] Phosphorite nodules also exist in highest concentrations on the continental shelves; they serve as a source of fertilizer. On the deep-ocean floor, massive sulfide deposits (at spreading ocean centers formed by hydrothermal vent processes) and manganese nodules comprise potentially valuable sources of copper, nickel, cobalt, manganese, and other metals.

Seawater itself supplies a number of chemical elements for industrial products (e.g., bromine, iodine, magnesium, and potassium). Large evaporating ponds have been constructed to yield salt from evaporating seawater. In addition, the desalination of seawater provides freshwater, especially in arid regions of the world.

Nearshore coastal environments have rapidly become major depositories for anthropogenic wastes, notably dredged material, sewage sludge, and industrial and municipal effluents.[54,55] Degraded environments often lie in proximity to large metropolitan areas where disposal has resulted in contamination by heavy metals, organic compounds of synthetic (xenobiotic) and natural origin, organic carbon, nutrients, and pathogens.[52] Insidious chemical pollutants (e.g., polychlorinated biphenyls or PCBs) accumulate in the tissues of certain marine organisms, and are transferred through food chains in the sea. Biological magnification tends to amplify the chemical effects on living systems.[56] Recently, concerns of plastics and other debris in the ocean have been the subject of conferences and symposia.[57] Oil spills originating from tankers and offshore drilling platforms periodically impose severe constraints on marine communities. Of more global consequence is radioactive waste impinging on the oceans as fallout from the testing of nuclear weapons[58] and from unusual anthropogenic events (e.g., the Chernobyl nuclear plant accident). A local, albeit significant source of radioactive waste occurs at nuclear fuel reprocessing plants.[59] Waste disposal is not confined to coastal zones, but has been documented in the deep sea as well.[60]

The many, varied, and increasingly complex problems arising from anthropogenic use of the ocean have raised man's consciousness regarding the pressing need to protect marine ecosystems. As the human population expands and complex industrial technologies develop, it is anticipated that more areas of the ocean, especially nearshore habitats, will become repositories for waste materials. Consequently, new disposal strategies and waste management techniques must be devised to mitigate the effects of contaminants on the abiotic and biotic environment of the sea. Obviously, this will be a difficult task. Societal demands on the marine hydrosphere should continue to grow into the 21st century, which may threaten to compromise the structure and function of component ecosystems as well as their aesthetic value. These demands will require the integrated effort of marine scientists from many disciplines working in concert to achieve the long-term goal of preserving the integrity of the ocean.

LITERATURE REVIEW

Marine science comprises a comprehensive field of study that embodies marine biology, marine chemistry, marine geology, and physical oceanography, as well as related disciplines such as ocean engineering and the atmospheric sciences. Its scope is extensive, covering the natural phenomena of estuaries, harbors, lagoons, shallow seas, continental shelves, continental rises, abyssal regions, and mid-ocean ridges. Since World War II, the burgeoning literature in this field has contributed greatly to the advancement of contemporary civilization, providing valuable insight into major areas of scientific inquiry related to the marine realm.

The formation of new scientific societies concerned with investigations of the sea has enabled scientists to disseminate knowledge rapidly by communicating results of basic and applied research at conferences and symposia. Examples of organizations dedicated to estuarine science include the Estuarine Research Federation (ERF), established in the U.S. in 1971, and five regional, associated societies: (1) the New England Estuarine Research Society (NEERS); (2) the Atlantic Estuarine Research Society (AERS); (3) the Southeastern Estuarine Research Society (SEERS); (4) the Gulf Estuarine Research Society (GERS); and (5) the Pacific Estuarine Research Society (PERS). Founded in England in 1971, the Estuarine and Brackish-Water Sciences Association also deals with estuarine ecosystems, together with other aquatic environments exclusive of freshwater and the open ocean. Two noteworthy scientific bodies which promote research on the open ocean as well as the nearshore are the American Society of Limnology and Oceanography (ASLO), with headquarters in Grafton, Wisconsin, and the Oceanic Society. The Deep-Sea Biological Society has interests largely confined to the abyssal regions and mid-ocean ridges. The Oceanic Sciences Section of the American Geophysical Union and specific sections of the American Chemical Society, the Ecological Society of America, the Geological Society of America, and the American Meteorological Society cover a broad spectrum of oceanographic matters. The main objectives of each of these scientific organizations are not only to encourage the publication of research findings but also to facilitate informal exchange of information between its members and to foster cooperation among specialists in the field.

Many journals relevant to marine science have been published over the years. Some of these journals center around a single discipline of study, whereas others stress an interdisciplinary approach. Journals which have served as a principal basis of instruction and reference for marine scientists are:

- *Annual Reviews in Oceanography and Marine Biology*
- *Australian Journal of Marine and Freshwater Research*
- *Bulletin of Marine Science*
- *Continental Shelf Research*
- *Reviews in Aquatic Sciences*
- *Deep-Sea Research Part A — Oceanographic Research Papers*
- *Estuaries*
- *Estuarine, Coastal and Shelf Science*
- *Geochimica et Cosmochimica Acta*
- *Helgolander Meeresuntersuchungen*
- *Hydrobiologia*
- *Initial Reports of the Deep Sea Drilling Project*
- *Journal of Atmospheric and Oceanic Technology*
- *Journal of Atmospheric Sciences*
- *Journal of Coastal Research*
- *Journal du Conseil, Conseil International pour l'Exploration de la Mer*
- *Journal of Experimental Marine Biology and Ecology*
- *Journal of Geophysical Research — Oceans*
- *Journal of Marine Research*
- *Journal of the Marine Biological Association of the United Kingdom*
- *Journal of Physical Oceanography*
- *Limnology and Oceanography*
- *Marine and Petroleum Geology*
- *Marine Behaviour and Physiology*
- *Marine Biology*
- *Marine Biology Letters*
- *Marine Chemistry*
- *Marine Ecology — Progress Series*
- *Marine Environmental Research*
- *Marine Fisheries Review*
- *Marine Geodesy*
- *Marine Geology*
- *Marine Geophysical Researches*
- *Marine Geotechnology*
- *Marine Mammal Science*
- *Marine Policy*
- *Marine Pollution Bulletin*
- *Marine Technology Society Journal*
- *Netherlands Journal of Sea Research*
- *Ocean Management*
- *Ocean Science and Engineering*
- *Oceanography and Marine Biology*
- *Oceanologica Acta*
- *Oceanus*
- *Progress in Oceanography*
- *Publication of the Institute of Marine Science, University of Texas*

In addition to these journals, many excellent books have been published on various aspects of marine science during the past decade. A number of reference volumes incorporate outstanding state-of-the-art summaries of well-defined topics. Textbooks in this area, however, tend to emphasize principles and to treat a wider range of subjects at the expense of all-encompassing detail. Only a few texts released each year attempt to cover an entire field of inquiry in a coherent fashion. Because of their broad scope, these ambitious works generally attract a readership that seeks an overall synthesis of the field. Useful books assessing principles and processes in marine biology include those of Mann,[30] Dawes,[53] Nybakken,[56] Thurman and Webber,[59] Longhurst,[61] Levinton,[62] Ketchum,[63] Rowe,[64] Hobbie and Williams,[65] Valiela,[66] and Baker and Wolff.[67] Recent texts in marine chemistry[68-72] compile data and draw conclusions on a diversity of subjects ranging from chemical oceanography to aquatic toxicology. Anderson,[18] Fanning and Manheim,[73] Kennett,[74] Emery and Uchupi,[75] and Nairn et al.[76] give well-organized overviews on marine geology. The volume by Emery and Uchupi[75] represents an unprecedented, encyclopedic presentation on the marine geology of the Atlantic Ocean. Books in physical oceanography — by Gross,[39] Bowden,[42] Beer,[77] Csanady,[78] Pickard and Emery,[79] and Deacon,[80] — offer lucid qualitative and quantitative descriptions of coastal and offshore waters.

PLAN OF THIS VOLUME

The dynamic nature of the ocean is strongly coupled to atmospheric conditions. Air-sea interaction not only affects the chemistry of the ocean via gaseous exchange, but also influences physical parameters through meteorological forcing mechanisms such as wind stress. Moreover, the biology of marine ecosystems is sensitive to input and output processes between the atmosphere and sea surface. Section 1 contains data on the chemical composition of the atmosphere as well as wind regimes and dynamic tables. The topic of solar radiation also is treated in this section. Geodetic, astronomic, and hygrometric tables provide additional useful data.

Section 2 deals with marine chemistry. Dissolved inorganic constituents of seawater may be broadly subdivided into the major and minor components, trace metals, and nutrient elements. The major dissolved constituents of seawater (e.g., sodium, chlorine, calcium, magnesium, potassium, and sulfate) typically act conservatively, their concentrations being constant except at ocean boundaries due to input or output processes, including dilution by river runoff, anthropogenic activity, or precipitation. The minor dissolved constituents, in contrast, frequently behave nonconservatively; their concentrations tend to vary in response to biological processes and chemical reactions. The trace metals (e.g., cadmium, copper, lead, and nickel) may be toxic to marine organisms even at low concentrations. Of paramount importance to the organic production of the oceans are the nutrient elements (e.g., nitrogen, phosphorus, and silicon) which represent the chief limiting elements to phytoplankton growth. This section presents data on the major and minor constituents of seawater. The solubilities of selected gases and minerals comprise a number of tables. Other parameters investigated are the density, viscosity, and conductivity of seawater.

Section 3 covers physical oceanography. A brief account is given on the major ocean circulation patterns, both surface circulation and thermohaline deep circulation. The section also considers temperature, salinity, and density distributions in the oceans and illustrates surface currents.

In the past 2 decades, the theory of plate tectonics has revolutionized the field of marine geology. Major structural features of the seafloor (e.g., mid-ocean ridges, transform faults, and deep-sea trenches) are being reassessed in respect to this unifying theory. Section 4 focuses on various aspects of marine geology, yielding data on the location, size, and depth of deep-sea trenches, submarine canyons, and other topographic features of the seafloor. In addition, heat flow measurements are recorded for deep-sea sediments of the Atlantic Ocean.

Concepts in ocean engineering are found in Section 5. Seven subject areas have been delineated: (1) materials for marine applications; (2) ropes, chains, and shackles; (3) marine power sources; (4) fixed ocean structures; (5) buoy systems; (6) specifications of oceanographic instruments; and (7) ship characteristics.

Section 6 examines the characteristics of phytoplankton in marine waters, concentrating on biomass as related to area, depth, and season, taxon diversity as a function of depth and area, and chemical composition. These microscopic, free-floating plants provide most of the primary production in the world's oceans, serving as the critical base of marine food chains. Encompassing a variety of algal groups, phytoplankton are mainly autotrophic forms consisting of single cells, or relatively simply-organized filamentous or chain-forming types. They have been subdivided on the basis of size into ultraplankton (less than 5 μm in diameter), nanoplankton (5 to 70 μm), microphytoplankton (70 to 100 μm), and macrophytoplankton (greater than 100 μm). Two broad fractions, the net plankton and nanoplankton, often are differentiated by the nominal aperture size of the plankton net employed during field sampling. All phytoplankton retained by the net (approximately 64 μm apertures) comprise the net plankton, and those passing through the net constitute the nanoplankton.

Section 7 details primary productivity of phytoplankton in the oceans. Data are generated on annual primary productivity by regions, and variations in primary productivity by season and depth. Additional information handled in this section includes dissolved organic composition as related to productivity, photosynthetic quotient and gross-net differences, and relationships of primary productivity to productivity at other trophic levels. The term "primary productivity" refers to the rate of production of organic matter by photosynthetic or chemosynthetic organisms. Rates of primary production in the oceans are mainly derived from ^{14}C measurements. The ^{14}C (radiocarbon) method has largely superseded all other techniques of determining primary productivity in the sea.

Section 8 assesses marine zooplankton. These organisms occupy a critical ecological niche in the oceans, with most converting plant to animal matter and serving as the principal herbivorous component of the marine ecosystem. Classification schemes for zooplankton principally rely on two criteria, that is, the size of the organism or its length of planktonic life. Micro-, meso-, and macrozooplankton are size groups used to categorize zooplankton populations. Zooplankton passing through a plankton net with a mesh size of 202 μm comprise the microzooplankton; those individuals retained by this net represent the mesozooplankton. Still larger forms collected with plankton nets having a mesh size of 505 μm are defined as macrozooplankton. Marine ecologists discriminate between meroplankton, holoplankton, and tychoplankton when classifying zooplankton on the basis of duration of planktonic life. Meroplanktonic species live only a part of their life cycle in the plankton, the remainder of the cycle occurring on the seafloor. In contrast, the holoplankton spend their entire lives in the open waters although some of them have resting stages that reside on the bottom.[62] Tychoplankton primarily consist of small benthic fauna temporarily translocated into the water column via wave activity, currents, animal-sediment interactions (i.e., bioturbation), or behavioral activity.

Section 8 addresses seven subject areas. These are: (1) biomass as related to area, depth, and season; (2) taxon diversity as a function of area and depth; (3) chemical compositions; (4) geographical distribution of characteristic faunal groups; (5) indicator species and their associated water masses; (6) length, height, weight, and sample volume relationships in major groups; and (7) sampling equipment and comparative efficiencies. Whenever possible, comparisons are made of zooplankton from different geographical locations.

Organic compounds derived from marine organisms are listed

in Section 9. Both the molecular and structural formulae are shown for each compound. A significant number of hydrocarbons, compounds containing nitrogen, and other novel substances have been recovered from marine organisms. Some of them are especially valuable in microbiological and pharmacological research. For example the compound cycloeudesmol from the red alga *Chondria oppositiclada*, exhibits strong antibiotic activity; chondriol from this red alga shows antiviral activity. Isozonarol and zonarol from the brown alga *Dictyopteris zonarioides* display fungicidal activity. Saxitoxin observed in various molluscs, such as clams and mussels, is the active principle of the paralytic shellfish poison. The abundance of natural compounds derived from marine flora and fauna continues to increase as new biochemical investigations are undertaken on biota from the oceans of the world.

REFERENCES

1. **Press, F. and Siever, R.,** *Earth*, 4th ed., W.H. Freeman, New York, 1986.
2. **Jannasch, H. W. and Mottl, M. J.,** Geomicrobiology of deep-sea hydrothermal vents, *Science*, 229, 717, 1985.
3. **Kenk, V. C. and Wilson, B. R.,** A new mussel (*Bivalvia*, Mytilidae) from hydrothermal vents in the Galapagos Rift zone, *Malacologia*, 26, 253, 1985.
4. **Boss, K. J. and Turner, R. D.,** The giant white clam from the Galapagos Rift, *Calyptogena magnifica* species novum, *Malacologia*, 20, 161, 1980.
5. **McLean, J.,** The Galapagos Rift limpet *Neomphalus*: relevance to understanding the evolution of a major Paleozoic-Mesozoic radiation, *Malacologia*, 21, 291, 1981.
6. **Williams, A. B.,** A new crab family from the vicinity of submarine thermal vents on the Galapagos Rift (Crustacea: Decapoda: Brachyura), *Proc. Biol. Soc. Wash.*, 93, 443, 1980.
7. **Jones, M. L.,** *Riftia pachyptila*, a new genus, new species, the vestimentiferan worm from the Galapagos Rift geothermal vents (Pogonophora), *Proc. Biol. Soc. Wash.*, 93, 1295, 1980.
8. **Galapagos Biology Expedition Participants: Grassle, J. F., Berg, C. J., Childress, J. J., Grassle, J. P., Hessler, R. R., Jannasch, H. W., Karl, D. M., Lutz, R. A., Mickel, T. J., Rhoads, D. C., Sanders, H. L., Smith, K. L., Somero, G. N., Turner, R. D., Tuttle, J. H., Walsh, P. J., and Williams, A. J.,** Galapagos 79: initial findings of a biology quest, *Oceanus*, 22, 2, 1979.
9. **Rona, P. A., Klinkhammer, G., Nelsen, T. A., Trefry, J. H., and Elderfield, H.,** Black smokers, massive sulphides and vent biota at the Mid-Atlantic ridge, *Nature*, 321, 33, 1986.
10. **Tunnicliffe, V., Botros, M., DeBurgh, M. E., Dinet, A., Johnson, H. P., Juniper, S. K., and McDuff, R. E.,** Hydrothermal vents of Explorer Ridge, northeast Pacific, *Deep-Sea Res.*, 33, 401, 1986.
11. **Grassle, J. F.,** The biology of hydrothermal vents: a short summary of recent findings, *Mar. Technol. Soc. J.*, 16, 33, 1982.
12. **Grassle, J. F.,** Introduction to the biology of hydrothermal vents, in *Hydrothermal Processes at Seafloor Spreading Centers*, Rona, P. A., Boström, K., Laubier, L., and Smith, K. L., Jr., Eds., Plenum Press, New York, 1984, 665.
13. **Grassle, J. F.,** Hydrothermal vent animals: distribution and biology, *Science*, 229, 713, 1985.
14. **Grassle, J. F.,** The ecology of deep-sea hydrothermal vent communities, in *Advances in Marine Biology*, Vol. 23, Blaxter, J. H. S. and Southward, A. J., Eds., Academic Press, London, 1986, 301.
15. **Hessler, R. R.,** Oasis under the sea — where sulphur is the staff of life, *New Scientist*, 92, 741, 1981.
16. **Lutz, R. A. and Hessler, R. R.,** Life without sunlight — biological communities of deep-sea hydrothermal vents, *Sci. Teacher*, 50, 22, 1983.
17. **Rona, P. A., Boström, K., Laubier, L., and Smith, K. L., Jr., Eds.,** *Hydrothermal Processes at Seafloor Spreading Centers*, Plenum Press, New York. 1983.
18. **Anderson, R. N.,** *Marine Geology: A Planet Earth Perspective*, John Wiley & Sons, New York, 1986.
19. **Thompson, G.,** Hydrothermal fluxes in the ocean, in *Chemical Oceanography*, Vol. 8, Riley, J. P. and Chester, R., Eds., Academic Press, London, 1983, 271.
20. **Edmond, J. M., Measures, C., MacDuff, R. E., Chan, L. H., Collier, R., Grant, B., Gordon, L. I., and Corliss, J. B.,** Ridge crest hydrothermal activity and the balances of the major and minor elements in the ocean: the Galapagos data, *Earth Planet. Sci. Lett.*, 46, 1, 1979.
21. **Edmond, J. M., Measures, C., Mangum, B., Grant, B., Sclater, F. R., Collier, R., Hudson, A., Gordon, L. I., and Corliss, J. B.,** On the formation of metal-rich deposits at ridge crests, *Earth Planet. Sci. Lett.*, 46, 19, 1979.
22. **Edmond, J. M., Von Damm, K. L., MacDuff, R. E., and Measures, C. I.,** Chemistry of hot springs on the East Pacific Rise and their effluent dispersal, *Nature*, 297, 187, 1982.
23. **Edmond, J. M. and Von Damm, K.,** Hot springs on the ocean floor, *Sci. Am.*, 248, 70, 1983.
24. **Von Damm, K. L., Edmond, J. M., Measures, C. I., and Grant, B.,** Chemistry of submarine hydrothermal solutions at Guaymas Basin, Gulf of California, *Geochim. Cosmochim. Acta*, 49, 2221, 1985.
25. **Von Damm, K. L., Edmond, J. M., Grant, B., Measures, C. I., Walden, B., and Weiss, R. F.,** Chemistry of submarine hydrothermal solutions at 21°N, East Pacific Rise, *Geochim. Cosmochim. Acta*, 49, 2197, 1985.

26. **Comita, P. B., Gagosian, R. B., and Williams, P. M.,** Suspended particulate organic material from hydrothermal vent waters at 21°N, *Nature*, 307, 450, 1984.
27. **Bowman, M. J.,** Introduction and historical perspective, in *Oceanic Fronts in Coastal Processes*, Bowman, M. J. and Esaias, W. E., Eds., Springer-Verlag, New York, 1978, 2.
28. **Le Févre, J.,** Aspects of the biology of frontal systems, in *Advances in Marine Biology*, Vol. 23, Blaxter, J. H. S. and Southward, A. J., Academic Press, London, 1986, 163.
29. **Bowman, M. J. and Esaias, W. E., Eds.,** *Oceanic Fronts in Coastal Processes*, Springer-Verlag, New York, 1978.
30. **Mann, K. H.,** *Ecology of Coastal Waters: A Systems Approach*, University of California Press, Berkeley, 1982.
31. **Joyce, T. and Wiebe, P.,** Warm-core rings of the Gulf Stream, *Oceanus*, 26, 34, 1983.
32. **Sakamoto-Arnold, C. M., Hanson, Jr., A. K., Huizenga, D. L., and Kester, D. R.,** Spatial and temporal variability of cadmium in Gulf Stream warm-core rings and associated waters, *J. Mar. Res.*, 45, 201, 1987.
33. **Schink, D., McCarthy, J. J., Joyce, T., Flierl, G., Wiebe, P., and Kester, D.,** Multidisciplinary program to study warm core rings, *EOS, Trans. Am. Geophys. Union*, 63, 834, 1982.
34. **Davis, C. S. and Wiebe, P. H.,** Evolution of zooplankton size structure and taxonomic composition in a warm-core Gulf Stream ring, *J. Geophys. Res.*, 90, 8871, 1985.
35. **Wiebe, P. H., Barber, V., Boyd, S. H., Davis, C. S., and Flierl, G. R.,** Evolution of zooplankton biomass structure in warm-core rings, *J. Geophys. Res.*, 90, 8885, 1985.
36. **Roman, M. R., Gauzens, A. L., and Cowles, T. J.,** Temporal and spatial changes in epipelagic microzooplankton and mesozooplankton biomass in warm-core Gulf Stream ring 82-B, *Deep-Sea Res.*, 32, 1007, 1985.
37. **Roman, M. R., Yentsch, C. S., Gauzens, A. L., and Phinney, D. A.,** Grazer control of the fine-scale distribution of phytoplankton in warm-core Gulf Stream rings, *J. Mar. Res.*, 44, 795, 1986.
38. **Perry, A. H. and Walker, J. M.,** *The Ocean-Atmosphere System*, Longman, London, 1977.
39. **Gross, M. G.,** *Oceanography: A View of the Earth*, 3rd ed., Prentice-Hall, N. J., 1982.
40. **Etter, P. C., Lamb, P. J., and Portis, D. H.,** Heat and freshwater budgets of the Caribbean Sea with revised estimates for the central American seas, *J. Phys. Oceanogr.*, 17, 1232, 1987.
41. **Gallimore, R. G. and Houghton, D. D.,** Approximation of ocean heat storage by ocean-atmosphere energy exchange: implications for seasonal cycle mixed-layer ocean formulations, *J. Phys. Oceanogr.*, 17, 1214, 1987.
42. **Bowden, K. F.,** *Physical Oceanography of Coastal Waters*, Ellis Horwood, Chichester, U.K., 1983.
43. **Ropelewski, C. F. and Halpert, M. S.,** Global and regional scale precipitation patterns associated with the El Nino/southern oscillation, *Mon. Weather Rev.*, 115, 1606, 1987.
44. **Holthuijsen, L. H., Kuik, A. J., and Mosselman, E.,** The response of wave directions to changing wind directions, *J. Phys. Oceanogr.*, 17, 845, 1987.
45. **Grimshaw, R., Broutman, D., and Sahl, L. E.,** A nondivergent barotropic model for wind-driven circulation in a closed region, *J. Phys. Oceanogr.*, 17, 1114, 1987.
46. **Emanuel, K. A.,** An air-sea interaction model of intraseasonal oscillations in the tropics, *J. Atmos. Sci.*, 44, 2324, 1987.
47. **Dyke, P. P. G., Moscardini, A. O., and Robson, E. H., Eds.,** *Offshore and Coastal Modelling*, Springer-Verlag, New York, 1985.
48. **Robinson, I. S.,** *Satellite Oceanography: An Introduction for Oceanographers and Remote-Sensing Scientists*, Ellis Horwood, Chichester, U. K., 1985.
49. **Mulhearn, P. J.,** The Tasman Front: a study using satellite infrared imagery, *J. Phys. Oceanogr.*, 17, 1148, 1987.
50. **McHugh, J. L.,** *Fishery Management*, Springer-Verlag, New York, 1984.
51. **Schaefer, M. B.,** The potential harvest of the sea, *Trans. Am. Fish. Soc.*, 94, 123, 1965.
52. **Capuzzo, J. M., Burt, W. V., Duedall, I. W., Park, P. K., and Kester, D. R.,** The impact of waste disposal in nearshore environments, in *Wastes in the Ocean*, Vol. 6, Ketchum, B. H., Capuzzo, J. M., Burt, W. V., Duedall, I. W., Park, P. K., and Kester, D. R., Eds., John Wiley & Sons, New York, 1985, 3.
53. **Dawes, C. J.,** *Marine Botany*, John Wiley & Sons, New York, 1981.
54. **Duedall, I., Ketchum, B. H., Park, P. K., and Kester, D. R., Eds.,** *Wastes in the Ocean*, Vol. 1, John Wiley & Sons, New York, 1983.
55. **Kester, D. R., Ketchum, B. H., Duedall, I. W., and Park, P. K., Eds.,** *Wastes in the Ocean*, Vol. 2, John Wiley & Sons, New York, 1983.
56. **Nybakken, J. W.,** *Marine Biology: An Ecological Approach*, Harper & Row, New York, 1982.
57. **Wolfe, D. A.,** Persistent plastics and debris in the ocean: an international problem of ocean disposal, *Mar. Pollut. Bull.*, 18, 303, 1987.
58. **Park, P. K., Kester, D. R., Duedall, I. W., and Bostwick, B. H., Eds.,** *Wastes in the Ocean*, Vol. 3, John Wiley & Sons, New York, 1983.
59. **Thurman, H. V. and Webber, H. H.,** *Marine Biology*, Charles E. Merrill Publishing, Columbus, Ohio, 1984.
60. **Kester, D. R., Burt, W. V., Capuzzo, J. M., Park, P. K., Ketchum, B. H., and Duedall, I. W., Eds.,** *Wastes in the Ocean*, Vol. 5, John Wiley & Sons, New York, 1985.
61. **Longhurst, A. R., Ed.,** *Analysis of Marine Ecosystems*, Academic Press, London, 1981.
62. **Levinton, J. S.,** *Marine Ecology*, Prentice-Hall, Englewood Cliffs, N. J., 1982.
63. **Ketchum, B. H., Ed.,** *Ecosystems of the World*, Vol. 26, Elsevier, New York, 1983.
64. **Rowe, G. T., Ed.,** *Deep-Sea Biology*, John Wiley & Sons, New York, 1983.

65. **Hobbie, J. H. and Williams, P. J. leB., Eds.,** *Heterotrophic Activity in the Sea*, Plenum Press, New York, 1984.
66. **Valiela, I.,** *Marine Ecological Processes*, Springer-Verlag, New York, 1984.
67. **Baker, J. M. and Wolff, W. J., Eds.,** *Biological Surveys of Estuaries and Coasts*, Cambridge University Press, Cambridge, 1987.
68. **Holland, H. D.,** *The Chemistry of the Atmosphere and Oceans*, John Wiley & Sons, New York, 1978.
69. **Liss, P. S. and Slinn, W. G., Eds.,** *Air-Sea Exchange of Gases and Particles*, D. Reidel, London, 1983.
70. **Morel, F. M. M.,** *Principles of Aquatic Chemistry*, John Wiley & Sons, Somerset, N. J., 1983.
71. **Riley, J. P. and Chester, R., Eds.,** *Chemical Oceanography*, Vol. 8, Academic Press, London, 1983.
72. **Rand, G. M. and Petrocelli, S. R.,** *Fundamentals of Aquatic Toxicology*, Hemisphere Publishing, Washington, 1985.
73. **Fanning, K. A., and Manheim, F. T., Eds.,** *The Dynamic Environment of the Ocean Floor*, D. C. Heath, Lexington, Massachusetts, 1982.
74. **Kennett, J.,** *Marine Geology*, Prentice-Hall, Englewood Cliffs, N. J., 1982.
75. **Emery, K. O. and Uchupi, E.,** *The Geology of the Atlantic Ocean*, Springer-Verlag, New York, 1984.
76. **Nairn, A. E. M., Stehli, F. G., and Uyeda, S., Eds.,** *The Ocean Basins and Margins*, Vol. 7A, Plenum Press, New York, 1985.
77. **Beer, T.,** *Environmental Oceanography: An Introduction to the Behavior of Coastal Waters*, Pergamon, Oxford, 1982.
78. **Csanady, G. T.,** *Circulation in the Coastal Ocean*, D. Reidel, Dordrecht, Holland, 1982.
79. **Pickard, G. L. and Emergy, W. J.,** *Descriptive Physical Oceanography: An Introduction*, Pergamon, Oxford, 1982.
80. **Deacon, G.,** *The Antarctic Circumpolar Ocean*, Cambridge University Press, New York, 1985.

Section 1
Air-Sea Interaction

1.1. COMPOSITION OF AIR AND PRECIPITATION

Table 1.1—1
COMPOSITION OF CLEAN DRY AIR NEAR SEA LEVEL

Component	Content, % by vol	Mol wt
Nitrogen	78.084	28.0134
Oxygen	20.9476	31.9988
Argon	0.934	39.948
Carbon dioxide	0.0314	44.00995
Neon	0.001818	20.183
Helium	0.000524	4.0026
Krypton	0.000114	83.80
Xenon	0.0000087	131.30
Hydrogen	0.00005	2.01594
Methane	0.0002	16.04303
Nitrous oxide	0.00005	44.0128
Ozone		47.9982
Summer	0–0.000007	
Winter	0–0.000002	
Sulfur dioxide	0–0.0001	64.0628
Nitrogen dioxide	0–0.000002	46.0055
Ammonia	0–trace	17.03061
Carbon monoxide	0–trace	28.01055
Iodine	0–0.000001	253.8088

(From Wedepohl, K. H., Ed., *Handbook of Geochemistry*, Vol. I, Springer-Verlag, Berlin, 1969. With permission.)

Table 1.1—2
AVERAGE CHEMICAL COMPOSITION OF PRECIPITATION AND RIVER WATER IN JAPAN

	Precipitation, ppm	River water, ppm	River water relative to precipitation normalized to Cl ratio = 1
Na	1.1	5.1	1
K	0.26	1.0	0.8
Mg	0.36	2.4	0.7
Ca	0.94	6.3	1.4
Sr	0.011	0.057	1.1
Cl	1.[illegible]	5.2	1
I	0.0018	0.0022	0.3
I	0.08	0.15	0.4
S	1.5	3.5	0.5
Si	0.83	8.1	0.5
Fe	0.23	0.48	0.4
Al	0.11	0.36	0.7
P	0.014		
Mo	0.00006	0.0006	2.1
V	0.0014	0.0010	0.2
Cu	0.0008	0.0014	0.4
Zn	0.0042	0.0050	0.2
As	0.0016	0.0017	0.2

(From Sugawara, K., personal communication, in Wedepohl, K. H., Ed., *Handbook of Geochemistry*, Vol. I, Springer-Verlag, Berlin, 1969. With permission.)

1.2. WIND AND DYNAMIC TABLES

Table 1.2—1
BEAUFORT WIND SCALE

In 1806 Admiral Sir F. Beaufort devised a scale for recording wind force at sea based on the effect of the wind on a full-rigged man-of-war of that era. In 1838 this scale was adopted by the British Admiralty and with but minor changes had come into general use among mariners for specifying the state of the wind at sea. The International Meteorological Committee (Utrecht, 1874) adopted the Beaufort scale for international use in weather telegraphy, and it now has become the chief scale for specifying the force of the wind and is used in all parts of the world, both on land and on sea.

Since the original Beaufort scale described a state of the atmosphere as manifested by the effects of the wind near the surface, there did not exist originally a set of wind speeds corresponding to the various numbers of the scale. A number of efforts were made to obtain appropriate speed equivalents, but it was found difficult to reach agreement on this matter because the effect of wind variation with height was neglected. The International Meteorological Committee (London, 1921) requested Dr. G. C. Simpson of the British Meteorological Office to investigate the matter, and in 1926 Dr. Simpson proposed a set of speed equivalents that were to apply to anemometers exposed 6 meters above the ground.[1] This scale was adopted by the Committee in Vienna (1926). However, the British Meteorological Office continued to use a scale proposed by Dr. Simpson in 1906[2] and applicable to an anemometer at a height of about 10 meters above the ground, as did the U.S. Weather Bureau. This scale was based on the empirical equation $V=0.836 B^{3/2}$ where V is the wind speed in meters per second and B is the Beaufort force.

In 1946 the International Meteorological Committee meeting in Paris extended the original Beaufort scale to higher values and redefined the speed equivalents to apply to an anemometer at 10 meters above the ground. Up to force 11 these values are consistent with the values for a height of 6 meters adopted in Vienna (1926) and are identical with those proposed by Dr. Simpson in 1906.

Table 1.2—1 gives the speed equivalents of the Paris (1946) resolution and also the "descriptive terms" and "specifications for use on land" from the *Meteorological Observers Handbook* (London, 1939).

[1] Simpson, G.C., The velocity equivalents of the Beaufort scale, Professional Notes No. 44, Air Ministry, Meteorological Office, London, 1926. (See also Anemometers and the Beaufort scale of wind force, *Meteorol. Mag.*, 67, 278, 1933 and Kuhlbrodt, E. *Ann. d. Hydr. & Marit. Meteorol., Zweites Koppen-Heft,* 64, 14, 1936.)

[2] Simpson, G. C., Meteorological Office, Pub. No. 180, London, 1906.

Beaufort Wind Scale

Table A

	Mean wind speeds at 10 m[a]				Limits of wind speed at 10 m[a]			
Force	Knots	m/sec	km/hr	mi/hr	Knots	m/sec	km/hr	mi/hr
0	0	0	0	0	<1	0–0.2	<1	<1
1	2	0.9	3	2	1–3	0.3–1.5	1–5	1–3
2	5	2.4	9	5	4–6	1.6–3.3	6–11	4–7
3	9	4.4	16	10	7–10	3.4–5.4	12–19	8–12
4	13	6.7	24	15	11–16	5.5–7.9	20–28	13–18
5	18	9.3	34	21	17–21	8.0–10.7	29–38	19–24
6	24	12.3	44	28	22–27	10.8–13.8	39–49	25–31
7	30	15.5	55	35	28–33	13.9–17.1	50–61	32–38
8	37	18.9	68	42	34–40	17.2–20.7	62–74	39–46
9	44	22.6	82	50	41–47	20.8–24.4	75–88	47–54
10	52	26.4	96	59	48–55	24.5–28.4	89–102	55–63
11	60	30.5	110	68	56–63	28.5–32.6	103–117	64–72
12	68	34.8	125	78	64–71	32.7–36.9	118–133	73–82
13	76	39.2	141	88	72–80	37.0–41.4	134–149	83–92
14	85	43.8	158	98	81–89	41.5–46.1	150–166	93–103
15	94	48.6	175	109	90–99	46.2–50.9	167–183	104–114
16	104	53.5	193	120	100–108	51.0–56.0	184–201	115–125
17	114	58.6	211	131	109–118	56.1–61.2	202–220	126–136

[a] Resolution 9, International Meteorological Committee, Paris, 1946.

Table 1.2—1 (continued)
BEAUFORT WIND SCALE

Table B

Force	Description of wind[b]	Specifications for use on land[b]
0	Calm	Calm, smoke rises vertically.
1	Light air	Direction of wind shown by smoke drift, but not by wind vanes.
2	Light breeze	Wind felt on face; leaves rustle; ordinary vane moved by wind.
3	Gentle breeze	Leaves and small twigs in constant motion; wind extends light flag.
4	Moderate breeze	Raises dust and loose paper; small branches are moved.
5	Fresh breeze	Small trees in leaf begin to sway; crested wavelets form on inland waters.
6	Strong breeze	Large branches in motion; whistling heard in telegraph wires; umbrellas used with difficulty.
7	Moderate gale	Whole trees in motion; inconvenience felt when walking against wind.
8	Fresh gale	Breaks twigs off trees; generally impedes progress.
9	Strong gale	Slight structural damage occurs (chimney pots and slate removed).
10	Whole gale	Seldom experienced inland; trees uprooted; considerable structural damage occurs.
11	Storm	Very rarely experienced, accompanied by wide-spread damage.
12 or above	Hurricane	

[b] Meteorological Office, *The Meteorological Observers Handbook,* London, 1939.

By permission of the Smithsonian Institution Press. From Smithsonian Meteorological Tables, sixth revised edition, Robert J. List, Ed., Smithsonian Institution, Washington, D. C., 1984.

Figure 1.2—1
WORLD MAP OF WIND REGIMES–FEBRUARY

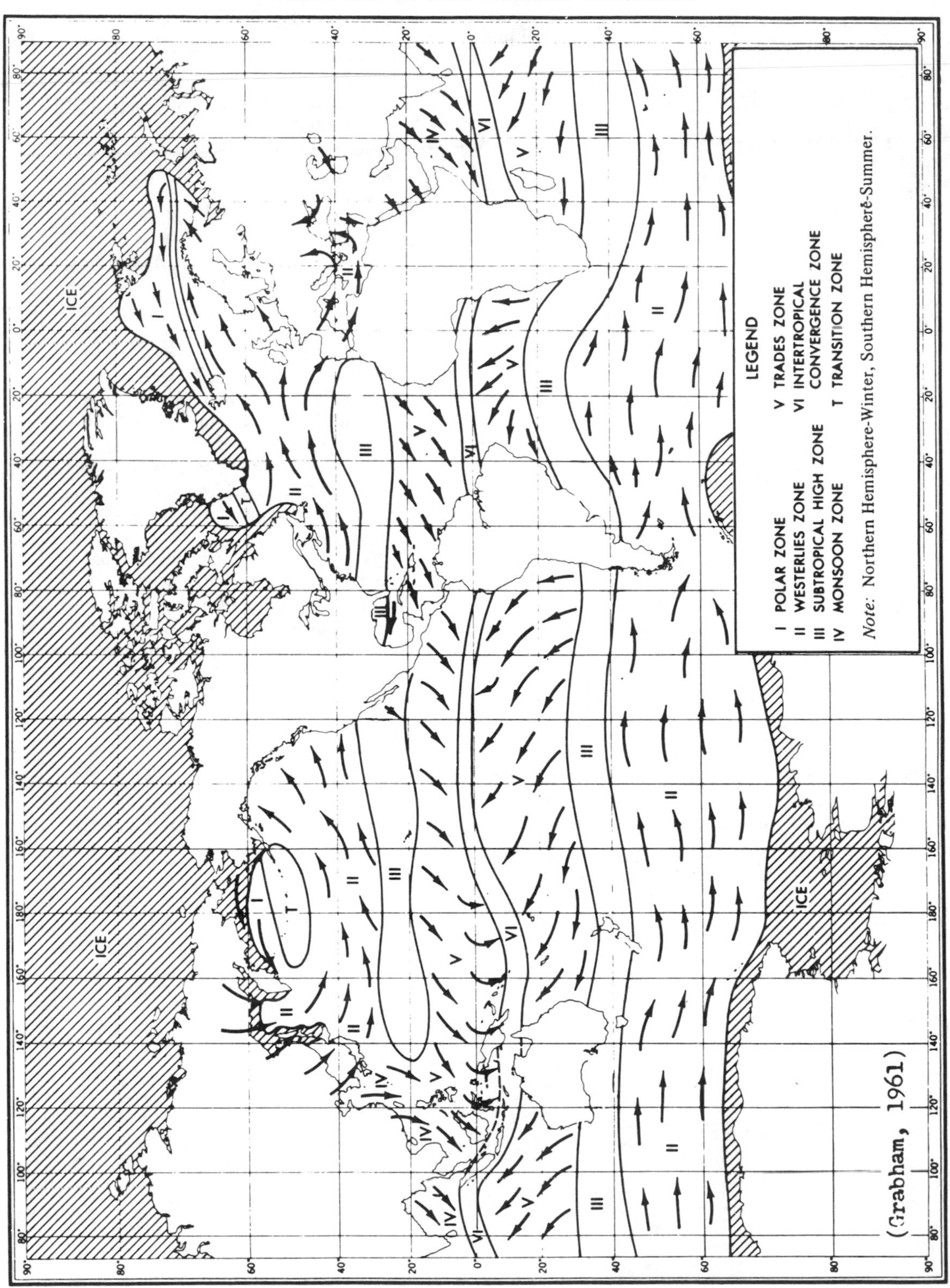

(From Grabham, A.L., *Harbor Analog System*, Part 1, Waves TR-117, U.S. Naval Oceanographic Office, Washington, D.C., 1961.)

Figure 1.2—2
WORLD MAP OF WIND REGIMES–AUGUST

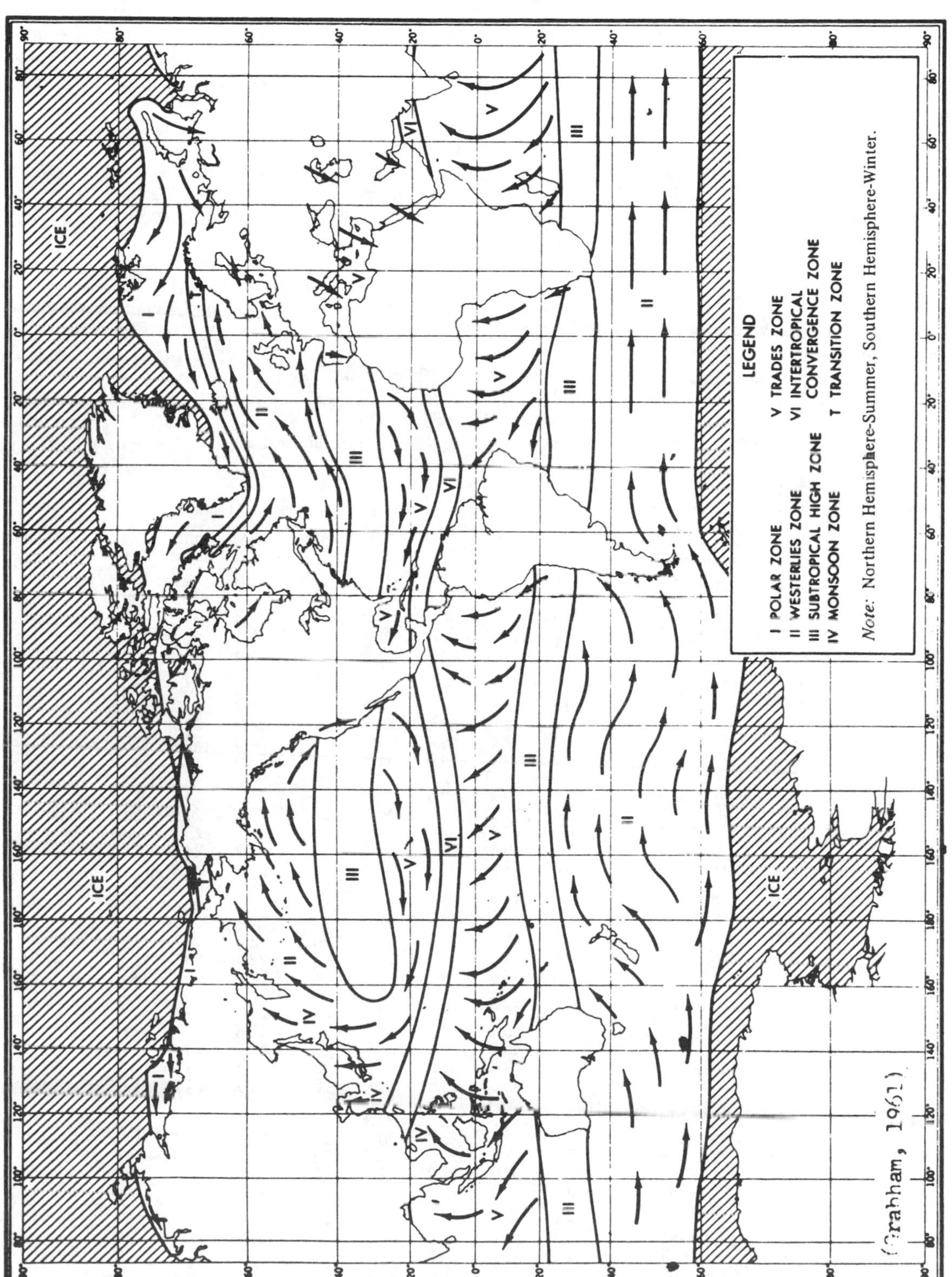

(From Grabham, A.L., *Harbor Analog System*, Part 1, Waves TR-117, U.S. Naval Oceanographic Office, Washington, D.C., 1961.)

Table 1.2—2
GEOSTROPHIC WIND, CONSTANT LEVEL SURFACE

Three Millibar Isobars, Air Density 1 kg m^{-3}

The scalar equation for the geostrophic wind on a constant level surface is

$$V_g = \frac{1}{f\rho}\frac{\partial p}{\partial n}$$

where p is the pressure on a constant level surface, n is distance measured in the surface, f is the Coriolis parameter, ρ is the density of the air, and V_g is the component of the geostrophic wind normal to the direction in which n is measured.

On a constant level surface with a 3 millibar isobaric interval and an air density of 1 kg m^{-3} (0.001 g cm^{-3}) this reduces to

$$V_g(\text{knots}) = \frac{0.0052409}{f\Delta n}$$

where Δn is the isobar spacing measured in degrees of latitude (i.e., one unit of Δn has the length of one degree of latitude at the place for which the isobar spacing is measured). This table gives values of V_g in knots as a function of Δn with auxiliary columns giving equivalents of Δn in kilometers, statute miles, and nautical miles. If the latter are measured by a map scale true at some other latitude, the value should be corrected to the lattitude at which the measurements are taken (see Table 1.4—4).

Since the geostrophic wind is inversely proportional to the isobar spacing and the density ρ, and directly proportional to the isobaric interval (Δpmbar), values of V_g for 1/10 of the indicated spacing may be found by multiplying the tabular values by 10, etc., and for isobaric intervals other than 3 mbar by multiplying the tabular values by $\Delta p/3$. The density ρ_0 of 1 kg m^{-3} (0.001 g cm^{-3}) used in the computations is the average density at about 2 km above sea level; for V_g at other levels multiply the tabular values by ρ_0/ρ (if the density is expressed in kg m^{-3}, simply divide the tabular value by the density).

Isobar spacing				Latitude							
Degrees of Latitude	Kilometers	Statute miles	Nautical miles	10° knots	15° knots	20° knots	25° knots	30° knots	35° knots	40° knots	45° knots
1.0	111	69	60	206.9	138.8	105.1	85.0	71.9	62.7	55.9	50.8
1.1	122	76	66	188.1	126.2	95.5	77.3	65.3	57.0	50.8	46.2
1.2	133	83	72	172.5	115.7	87.6	70.9	59.9	52.2	46.6	42.3
1.3	145	90	78	159.2	106.8	80.8	65.4	55.3	48.2	43.0	39.1
1.4	156	97	84	147.8	99.2	75.0	60.7	51.3	44.8	39.9	36.3
1.5	167	104	90	138.0	92.6	70.0	56.7	47.9	41.8	37.3	33.9
1.6	178	111	96	129.3	86.8	65.7	53.1	44.9	39.2	34.9	31.8
1.7	189	117	102	121.7	81.7	61.8	50.0	42.3	36.9	32.9	29.9
1.8	200	124	108	115.0	77.1	58.4	47.2	39.9	34.8	31.1	28.2
1.9	211	131	114	108.9	73.1	55.3	44.8	37.8	33.0	29.4	26.7
2.0	222	138	120	103.5	69.4	52.5	42.5	35.9	31.3	28.0	25.4
2.1	234	145	126	98.5	66.1	50.0	40.5	34.2	29.8	26.6	24.2
2.2	245	152	132	94.1	63.1	47.8	38.7	32.7	28.5	25.4	23.1
2.3	256	159	138	90.0	60.4	45.7	37.0	31.2	27.2	24.3	22.1
2.4	267	166	144	86.2	57.9	43.8	35.4	29.9	26.1	23.3	21.2
2.5	278	173	150	82.8	55.5	42.0	34.0	28.7	25.1	22.4	20.3
2.6	289	180	156	79.6	53.4	40.4	32.7	27.6	24.1	21.5	19.5
2.7	300	187	162	76.6	51.4	38.9	31.5	26.6	23.2	20.7	18.8
2.8	311	193	168	73.9	49.6	37.5	30.4	25.7	22.4	20.0	18.1
2.9	322	200	174	71.4	47.9	36.2	29.3	24.8	21.6	19.3	17.5
3.0	334	207	180	69.0	46.3	35.0	28.3	24.0	20.9	18.6	16.9
3.2	356	221	192	64.7	43.4	32.8	26.6	22.5	19.6	17.5	15.9
3.4	378	235	204	60.9	40.8	30.9	25.0	21.1	18.4	16.4	14.9
3.6	400	249	216	57.5	38.6	29.2	23.6	20.0	17.4	15.5	14.1
3.8	423	263	228	54.5	36.5	27.6	22.4	18.9	16.5	14.7	13.4
4.0	445	276	240	51.7	34.7	26.3	21.3	18.0	15.7	14.0	12.7
4.2	467	290	252	49.3	33.1	25.0	20.2	17.1	14.9	13.3	12.1
4.4	489	304	264	47.0	31.6	23.9	19.3	16.3	14.2	12.7	11.5
4.6	511	318	276	45.0	30.2	22.8	18.5	15.6	13.6	12.2	11.0
4.8	534	332	288	43.1	28.9	21.9	17.7	15.0	13.1	11.6	10.6
5.0	556	345	300	41.4	27.8	21.0	17.0	14.4	12.5	11.2	10.2
5.5	612	380	330	37.6	25.2	19.1	15.5	13.1	11.4	10.2	9.2
6.0	667	415	360	34.5	23.1	17.5	14.2	12.0	10.4	9.3	8.5
6.5	723	449	390	31.8	21.4	16.2	13.1	11.1	9.6	8.6	7.8
7.0	778	484	420	29.6	19.8	15.0	12.1	10.3	9.0	8.0	7.3
8.0	890	553	480	25.9	17.4	13.1	10.6	9.0	7.8	7.0	6.4
9.0	1001	622	540	23.0	15.4	11.7	9.4	8.0	7.0	6.2	5.6
10.0	1112	691	600	20.7	13.9	10.5	8.5	7.2	6.3	5.6	5.1

Table 1.2—2 (continued)
GEOSTROPHIC WIND, CONSTANT LEVEL SURFACE

Isobar spacing				Latitude							
Degrees of Latitude	Kilometers	Statute miles	Nautical miles	50° knots	55° knots	60° knots	65° knots	70° knots	75° knots	80° knots	85° knots
1.0	111	69	60	46.9	43.9	41.5	39.7	38.2	37.2	36.5	36.1
1.1	122	76	66	42.6	39.9	37.7	36.0	34.8	33.8	33.2	32.8
1.2	133	83	72	39.1	36.6	34.6	33.0	31.9	31.0	30.4	30.1
1.3	145	90	78	36.1	33.7	31.9	30.5	29.4	28.6	28.1	27.7
1.4	156	97	84	33.5	31.3	29.6	28.3	27.3	26.6	26.1	25.8
1.5	167	104	90	31.3	29.2	27.7	26.4	25.5	24.8	24.3	24.0
1.6	178	111	96	29.3	27.4	25.9	24.8	23.9	23.3	22.8	22.5
1.7	189	117	102	27.6	25.8	24.4	23.3	22.5	21.9	21.5	21.2
1.8	200	124	108	26.1	24.4	23.1	22.0	21.2	20.7	20.3	20.0
1.9	211	131	114	24.7	23.1	21.8	20.9	20.1	19.6	19.2	19.0
2.0	222	138	120	23.5	21.9	20.7	19.8	19.1	18.6	18.2	18.0
2.1	234	145	126	22.3	20.9	19.8	18.9	18.2	17.7	17.4	17.2
2.2	245	152	132	21.3	19.9	18.9	18.0	17.4	16.9	16.6	16.4
2.3	256	159	138	20.4	19.1	18.0	17.2	16.6	16.2	15.9	15.7
2.4	267	166	144	19.5	18.3	17.3	16.5	15.9	15.5	15.2	15.0
2.5	278	173	150	18.8	17.5	16.6	15.9	15.3	14.9	14.6	14.4
2.6	289	180	156	18.0	16.9	16.0	15.3	14.7	14.3	14.0	13.9
2.7	300	187	162	17.1	16.2	15.4	14.7	14.2	13.8	13.5	13.4
2.8	311	193	168	16.8	15.7	14.8	14.2	13.7	13.3	13.0	12.9
2.9	322	200	174	16.2	15.1	14.3	13.7	13.2	12.8	12.6	12.4
3.0	334	207	180	15.6	14.6	13.8	13.2	12.7	12.4	12.2	12.0
3.2	356	221	192	14.7	13.7	13.0	12.4	12.0	11.6	11.4	11.3
3.4	378	235	204	13.8	12.9	12.2	11.7	11.2	10.9	10.7	10.6
3.6	400	249	216	13.0	12.2	11.5	11.0	10.6	10.3	10.1	10.0
3.8	423	263	228	12.3	11.5	10.9	10.4	10.1	9.8	9.6	9.5
4.0	445	276	240	11.7	11.0	10.4	9.9	9.6	9.3	9.1	9.0
4.2	467	290	252	11.2	10.4	9.9	9.4	9.1	8.9	8.7	8.6
4.4	489	304	264	10.7	10.0	9.4	9.0	8.7	8.5	8.3	8.2
4.6	511	318	276	10.2	9.5	9.0	8.6	8.3	8.1	7.9	7.8
4.8	534	332	288	9.8	9.1	8.6	8.3	8.0	7.8	7.6	7.5
5.0	556	345	300	9.4	8.8	8.3	7.9	7.6	7.4	7.3	7.2
5.5	612	380	330	8.5	8.0	7.5	7.2	7.0	6.8	6.6	6.6
6.0	667	415	360	7.8	7.3	6.9	6.6	6.4	6.2	6.1	6.0
6.5	723	449	390	7.2	6.7	6.4	6.1	5.9	5.7	5.6	5.5
7.0	778	484	420	6.7	6.3	5.9	5.7	5.5	5.3	5.2	5.2
8.0	890	553	480	5.9	5.5	5.2	5.0	4.8	4.7	4.6	4.5
9.0	1001	622	540	5.2	4.9	4.6	4.4	4.2	4.1	4.1	4.0
10.0	1112	691	600	4.7	4.4	4.1	4.0	3.8	3.7	3.6	3.6

Table 1.2—3
CORIOLIS PARAMETER AND LATITUDINAL VARIATION

Latitude	$2\omega \sin \phi$ $\sec^{-1}$	$\beta =$ $(2\omega \cos \phi)/R$ cm^{-1} $\sec^{-1}$	Latitude	$2\omega \sin \phi$ $\sec^{-1}$	$\beta =$ $(2\omega \cos \phi)/R$ cm^{-1} $\sec^{-1}$
0°	0	2.289×10^{-13}	50°	1.1172×10^{-4}	1471×10^{-13}
5	0.1271×10^{-4}	2.280	55	1.1947	1.313
10	0.2533	2.254	60	1.2630	1.145
15	0.3775	2.211	65	1.3218	0.967
20	0.4988	2.151	70	1.3705	0.783
25	0.6164×10^{-4}	2.075×10^{-13}	75	1.4087×10^{-4}	0.593×10^{-13}
30	0.7292	1.982	80	1.4363	0.398
35	0.8365	1.875	85	1.4529	0.199
40	0.9375	1.754	90	1.4584	0
45	1.0313	1.619			

$2\omega \sin \phi$ = Coriolis parameter.
ϕ = latitude.
ω = angular velocity of the earth = 7.292116×10^{-5} rad $\sec^{-1}$.
R = radius of the earth = 6.371229×10^{6} m (mean radius of the International ellipsoid).
β = $(2\omega \cos \phi)/R$ = rate at which the Coriolis parameter increases northward.

Note: $2\omega \cos \phi = 2\omega \sin (90 - \phi)$.

REFERENCE

1. **Rossby, C.-G.,** *J. Mar. Res.*, 2, 38, 1939.

Table 1.2—4
ROSSBY'S LONG-WAVE FORMULA

Rossby[1] has shown that in the case of sinusoidal perturbations on a zonal current in an ideal, frictionless, homogeneous, and incompressible atmosphere in horizontal motion, the relation between the velocity of the undisturbed zonal current U and the phase velocity of the perturbation c is given by:

$$U - c = \frac{\beta L^2}{4\pi^2}$$

where β is the rate of which the Coriolis parameter increases northward (assumed to be constant with latitude for a given zonal current) and L is the wave length of the perturbation. L is most conveniently measured in terms of degrees of longitude at the latitude in question. Similarly the resulting $\beta L^2/(4\pi^2)$ is measured in terms of degrees of longitude/ 24 hrs; a supplemental column also gives the result in m/sec.

Wave length – degrees of longitude

Latitude	10 °long/24 hr	10 m/sec^{-1}	15 °long/24 hr	15 m/sec^{-1}	20 °long/24 hr	20 m/sec^{-1}	25 °long/24 hr	25 m/sec^{-1}	30 °long/24 hr	30 m/sec^{-1}	35 °long/24 hr	35 m/sec^{-1}
10°	0.5	0.7	1.2	1.5	2.2	2.7	3.4	4.3	4.9	6.2	6.6	8.4
20	0.5	0.6	1.1	1.3	2.0	2.4	3.1	3.7	4.4	5.4	6.0	7.3
30	0.4	0.5	0.9	1.1	1.7	1.9	2.6	2.9	3.8	4.2	5.1	5.7
40	0.3	0.3	0.7	0.7	1.3	1.3	2.0	2.0	3.0	2.9	4.0	4.0
50	0.2	0.2	0.5	0.4	0.9	0.8	1.4	1.2	2.1	1.7	2.8	2.3
60	0.1	0.1	0.3	0.2	0.6	0.4	0.9	0.6	1.3	0.8	1.7	1.1
70	0.1	0.0	0.1	0.1	0.3	0.1	0.4	0.2	0.6	0.3	0.8	0.4
80	0.0	0.0	0.0	0.0	0.1	0.0	0.1	0.0	0.2	0.0	0.2	0.0

Latitude	40 °long/24 hr	40 m/sec^{-1}	45 °long/24 hr	45 m/sec^{-1}	50 °long/24 hr	50 m/sec^{-1}	55 °long/24 hr	55 m/sec^{-1}	60 °long/24 hr	60 m/sec^{-1}	65 °long/24 hr	65 m/sec^{-1}
10°	8.7	11.0	11.0	13.9	13.5	17.2	16.4	20.8	19.5	24.7	22.9	29.0
20	7.9	9.5	10.0	12.1	12.3	14.9	14.9	18.0	17.7	21.5	20.8	25.2
30	6.7	7.5	8.5	9.5	10.5	11.7	12.7	14.1	15.1	16.8	17.7	19.7
40	5.2	5.2	6.6	6.6	8.2	8.1	9.9	9.8	11.8	11.7	13.8	13.7
50	3.7	3.1	4.7	3.9	5.8	4.8	7.0	5.8	8.3	6.9	9.8	8.1
60	2.2	1.4	2.8	1.8	3.5	2.3	4.2	2.7	5.0	3.3	5.9	3.8
70	1.0	0.5	1.3	0.6	1.6	0.7	2.0	0.9	2.4	1.0	2.8	1.2
80	0.3	0.1	0.3	0.1	0.4	0.1	0.5	0.1	0.6	0.1	0.7	0.2

Latitude	70 °long/24 hr	70 m/sec^{-1}	75 °long/24 hr	75 m/sec^{-1}	80 °long/24 hr	80 m/sec^{-1}	85 °long/24 hr	85 m/sec^{-1}	90 °long/24 hr	90 m/sec^{-1}	100 °long/24 hr	100 m/sec^{-1}
10°	26.5	33.6	30.4	38.6	34.6	43.9	39.1	49.6	43.8	55.6	54.1	68.6
20	24.1	29.2	27.7	33.6	31.5	38.2	35.6	43.1	39.9	48.3	49.3	59.7
30	20.5	22.9	23.5	26.3	26.8	39.9	30.2	33.8	33.9	37.9	41.9	46.7
40	16.1	15.9	18.4	18.2	21.0	20.7	23.7	23.4	26.6	26.2	32.8	32.4
50	11.3	9.4	13.0	10.8	14.9	12.3	16.7	13.8	18.7	15.5	23.1	19.2
60	6.9	4.4	7.9	5.1	8.9	5.8	10.1	6.5	11.3	7.3	14.0	9.0
70	3.2	1.4	3.7	1.6	4.2	1.9	4.7	2.1	5.3	2.3	6.6	2.9
80	0.8	0.2	1.0	0.2	1.1	0.2	1.2	0.3	1.4	0.3	1.7	0.4

Table 1.2—4 (continued)
ROSSBY'S LONG-WAVE FORMULA

	110		120		130		140		160		180	
	°long/24 hr	m/sec^{-1}	°long/24 hr	m/sec^{-1}	°long/24 hr	m/sec^{-1}	°long/24 hr	m/sec^{-1}	°long/24 hr	m/sec^{-1}	°long/24 hr	m/sec^{-1}
10°	65.4	83.0	77.9	98.8	91.4	116.0	106.0	134.5	138.5	175.7	175.3	222.4
20	59.6	72.2	70.9	85.9	83.2	100.8	96.5	116.9	126.1	152.7	159.6	193.3
30	50.7	56.6	60.3	67.3	70.7	79.0	82.0	91.6	107.2	119.6	135.6	151.4
40	39.7	39.2	47.2	46.7	55.4	54.8	64.2	63.5	83.9	83.0	106.2	105.0
50	27.9	23.2	33.2	27.6	39.0	32.4	45.2	37.5	59.1	49.0	74.8	62.1
60	16.9	10.9	20.1	13.0	23.6	15.3	27.4	17.7	35.8	23.1	45.3	29.3
70	7.9	3.5	9.4	4.2	11.1	4.9	12.8	5.7	16.8	7.4	21.2	9.4
80	2.0	0.5	2.4	0.5	2.9	0.6	3.3	0.7	4.3	1.0	5.5	1.2

By permission of the Smithsonian Institution Press. From Smithsonian Meteorological Tables, sixth revised edition, Robert J. List, Ed., Smithsonian Institution, Washington, D. C., 1984.

REFERENCE

1. **Rossby, C.-G.,** *J. Mar. Res.*, 2, 38, 1939.

1.3. RADIATION

Table 1.3—1
SPECTRAL DISTRIBUTION OF SOLAR RADIATION AT SEA LEVEL

Using Fowle's[1] data for scattering of solar radiation by water vapor and by air, Moon[2] calculated the spectral transmission factors at sea level with 20 mm of precipitable water vapor in the atmosphere. By comparing these with the mean values of observed transmission, he calculated a spectral transmission for dust.[a] He also evaluated the extraterrestrial solar radiation.

By combining the extraterrestrial solar radiation with the scattering by water vapor, air, and dust, and with the absorptions by water vapor and ozone, Moon calculated the solar radiation that reaches sea level for various optical air masses. The data were computed for a pressure of 1 atmosphere on the basis of arbitrary average values of water vapor, dust, and ozone content. These values are

water vapor, 20 mm precipitable water,
dust, 300 particles cm $^{-3}$ near the ground,
ozone, 2.8 mm path length at N. T. P.

The results given here have been adjusted to a solar constant of 1.94 cal cm $^{-3}$ min. $^{-1}$

Wave-length μ	Optical air mass 1	2	3	4	5
	cal cm $^{-2}$ min $^{-1}$				
0.29–0.40	0.059	0.029	0.015	0.008	0.004
0.40–0.70	.616	.481	.379	.302	.240
0.70–1.1	.454	.393	.343	.301	.266
1.1 –1.5	.140	.103	.084	.071	.060
1.5 –1.9	.075	.066	.060	.056	.052
1.9 –∞	.019	.014	.011	.010	.009
Total	1.363	1.086	0.892	0.748	0.631

[a] This assumes that 20 mm of precipitable water vapor was representative of these observed transmissions, which may not have been the case. Any error so introduced would be small, however.

(From *Smithsonian Meteorological Tables,* 6th rev. ed., List, R. J., Ed., Smithsonian Institution, Washington, D.C., 1971.)

REFERENCES

1. Fowle, F. E., *Smithsonian Misc. Coll.*, 69 (3), 1918.
2. Moon, P., *J. Franklin Inst.*, 5, 583, 1940.

1.3—2
TOTAL SOLAR AND SKY RADIATION

Klein[1] gives equations that permit the evaluation of the total solar and sky radiation Q on a horizontal surface for a cloudless, dust-free atmosphere as a function of the atmospheric transmission obtained from Kimball's chart[2] and the solar zenith distance. Fritz[3] combined Klein's equations and constructed the chart given below. The isopleths give values of Q in cal cm^{-2} min^{-1} as a function of optical air mass m_p (abscissa) and precipitable water vapor w centimeters (ordinate), when the sun is at its mean distance from the earth. To correct for the sun's actual distance from the earth, divide the values by the square of the appropriate radius vector (Table 1.4—7). For elevated stations, multiply the values given in the chart by $p/1013$, where p is the barometric pressure in millibars at the place of observation. Dashed lines indicate extrapolated values.

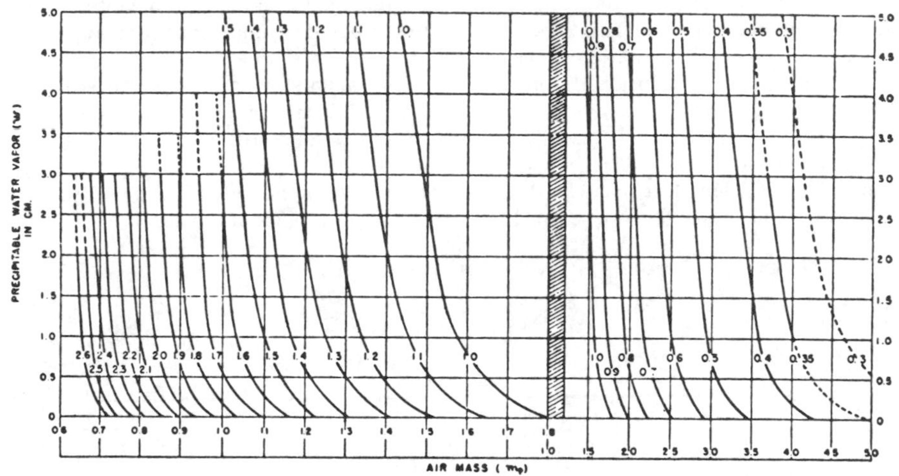

(From *Smithsonian Meteorological Tables,* 6th rev. ed., List, R. J., Ed., Smithsonian Institution, Washington, D.C., 1971.)

REFERENCES

1. **Klein, W. H.,** *J. Meteor.,* 5, 119, 1948.
2. **Kimball, H. H.,** *Month. Weath. Rev.,* 55, 167, 1927; 56, 394, 1928; 58, 43, 1930.
3. **Fritz, S.,** *Heating and Ventilating,* 46, 69, 1949.

Table 1.3—3
SOLAR AND SKY RADIATION

Relation Between the Vertical Component of Direct Solar Radiation and Total Solar and Sky Radiation on a Horizontal Surface

By means of pyrheliometers, Kimball[1] measured the vertical component of direct solar radiation and also the total solar and sky radiation on a horizontal surface for cloudless skies.

	Sun's zenith distance									
	30.0°	48.3°	60.0°	66.5°	70.7°	73.6°	75.7°	77.4°	78.7°	79.8°
Ratio	0.84	0.84	0.80	0.78	0.76	0.72	0.69	0.67	0.65	0.63

(From *Smithsonian Meteorological Tables,* 6th rev. ed., List, R. J., Ed., Smithsonian Institution, Washington, D.C., 1971.)

REFERENCE

1. **Kimball, H. H.,** *Month Weath. Rev.,* 47, 769, 1919. (For variation with height see Klein, W. H., *J. Meteor.,* 5, 119, 1948.)

Table 1.3—4
REFLECTIVITY OF A WATER SURFACE

The reflectivity of a plane water surface for unpolarized light is a function of the angle of incidence of the light and the index of refraction of the water and may be computed from the reflection law of Fresnel

$$R = \frac{1}{2}\left[\frac{\sin^2 (i - r)}{\sin^2 (i + r)} + \frac{\tan^2 (i - r)}{\tan^2 (i + r)}\right]$$

where

R = reflectivity,
i = angle of incidence,
r = angle of refraction,

i and r are related to the index of refraction n of the water by $n = \sin i/\sin r$. Although the values given are valid only for a plane undisturbed water surface, Ångström[1] states: " . . . . it is evident that the observed reflection from disturbed water-surfaces only shows small deviations from the values which are to be expected from the Fresnel formula. . . . . Some investigations which I have carried out on artificially disturbed surfaces seem to indicate that the deviation from the Fresnel formula is positive for slight disturbances of the surface, but negative when the amplitude gets large compared with the wave-length of the water waves. The measurements give strong support to the view that in the average case in geophysical discussions we may base computations of the reflection, absorption, and emission power of water-surfaces on the validity of the Fresnel formula."

Values in this table are computed on the assumption that $n = 1.333$, the value of the index of refraction for pure water. The value for sea water is slightly larger, about 1.3398 for sea water of salinity 35°/oo, but the difference is negligible.

In considering direct solar radiation, the angle of incidence i = sun's zenith distance.

i	0°	10°	20°	30°	40°	50°	60°	70°	80°	85°	90°
R(%)	2.0	2.0	2.1	2.1	2.5	3.4	6.0	13.4	34.8	58.4	100.0

By permission of the Smithsonian Institution Press. From Smithsonian Meteorological Tables, sixth revised edition, Robert J. List, Ed., Smithsonian Institution, Washington, D. C., 1984.

REFERENCE

1. Ångström, A., *Geograf. Ann.*, 7, 323, 1925.

Table 1.3—5
ABSORPTION OF RADIATION BY SEA WATER

Utterback[1] has made observations of the *extinction coefficient* of typical oceanic waters, defined in the same manner as the absorption coefficient *k* of ozone. Sverdrup, Johnson, and Fleming[2] have summarized Utterback's observations as follows:

He has made numerous observations in the shallow waters near islands in the inner part of Juan de Fuca Strait and at four stations in the open oceanic waters off the coast of Washington, and these can be considered typical of coastal and oceanic water, respectively. Table 1.3—5 contains the absorption coefficients of pure water at the wavelengths used by Utterback, the maximum average, and maximum coefficients observed in oceanic wtaer, and the minimum, average, and maximum coefficients observed in coastal water. The minimum and maximum coefficients have all been computed from the four lowest and the four highest values in each group.

	Wavelength[a] – μ						
	0.46	0.48	0.515	0.53	0.565	0.60	0.66
Type water	cm^{-1}	cm^{-1}	cm^{-1}	cm^{-1}	cm^{-1}	cm^{-1}	cm^{-1}
Pure water	.00015	.00015	.00018	.00021	.00033	.00125	.00280
Ocean water							
Lowest	.00038	.00026	.00035	.00038	.00074	.00199	
Avg	.00086	.00076	.00078	.00084	.00108	.00272	
Highest	.00160	.00154	.00143	.00140	.00167	.00333	
Coastal water							
Lowest	.00224	.00230	.00192	.00169		.00375	.00477
Avg	.00362	.00334	.00276	.00269		.00437	.00623
Highest	.00510	.00454	.00398	.00348		.00489	.00760

[a] It should be understood that the wavelength actually stands for a spectral band of finite width.

By permission of the Smithsonian Institution Press. From Smithsonian Meteorological Tables, sixth revised edition, Robert J. List, Ed., Smithsonian Institution, Washington, D. C., 1984.

REFERENCES

1. **Utterback, C. L.,** *Cons. Perm. Int. Explor. Mer,* Rapp. et Proc.-Verb., 101 (4), 15, 1936.
2. **Sverdrup, H. U., Johnson, M. W., and Fleming, R. H.,** *The Oceans,* Prenctice-Hall, Inc., New York, 1942, p. 84.

1.4. GEODETIC AND ASTRONOMICAL TABLES

Table 1.4—1
GEODETIC AND ASTRONOMICAL CONSTANTS

Table A

Dimensions of the Earth	International ellipsoid of reference[1]	Clarke spheroid of 1866[2]
Semimajor axis = a	6,378,388 m	6,378,206.4 m
Semiminor axis = b	6,356,911.946 m	6,356,583.8 m
Mean radius = $\frac{2a + b}{3}$	6,371,299.315 m	
Radius of sphere of same area	6,371,227.709 m	
Radius of sphere of same volume	6,371,221.266 m	
Length of meridian quadrant	10,002,288.299 m	
Length of equatorial quadrant	10,019,148.4 m	
Area of ellipsoid	510,100,934 km^2	
Volume of ellipsoid	1,083,319.78 x 10^6 km^3	
Flattening = f	1/297	

Table B

Mass of the earth[3]	5.975 x 10^{24} kg
Mean distance earth to sun (astronomical unit)[3]	1.4968 x 10^8 km
Mean linear velocity of the earth in its orbit	29.77 km sec^{-1}
Mean linear velocity of the surface of the earth at the equator	465.1 m/sec^{-1}
Obliquity of the ecliptic	23°27′

Note: The Clarke spheroid of 1866 is the reference spheroid for triangulation in the United States, Canada, and Mexico. The International Ellipsoid of Reference is used in South America and in parts of western Europe. It was adopted in 1924 by the International Union of Geodesy and Geophysics, and its use is recommended by that body wherever practicable. (See *Encyclopedia Britannica*, 1947 ed., article on Geodesy, for data concerning other spheroids.)

REFERENCES

1. U.S. Coast and Geodetic Survey, Spec. Publ. No. 200, Washington, D.C., 1935.
2. U.S. Coast and Geodetic Survey, Spec. Publ. No. 5, Washington, D.C., 1946.
3. **Russell, H. N., Dugan, R. S., and Stewart, J. Q.,** *Astronomy*, Ginn & Co., Boston, 1945.

(From *Smithsonian Meteorological Tables*, 6th rev. ed., List, R. J., Ed., Smithsonian Institution, Washington, D.C., 1971.)

Table 1.4—2
CONVERSION FACTORS

Time

1 mean solar second (sec, s)
  = 1.002738 sidereal seconds
1 mean solar minute (min, m)
  = 60 sec (mean solar)
1 mean solar hour (hr, h)
  = 3,600 sec (mean solar)
  = 60 min (mean solar)
1 mean solar day (da., d)
  = 86,400 sec (mean solar)
  = 1,440 min (mean solar)
  = 24 hr (mean solar)
  = 24 hr, 3 min, 56,555 sec of mean sidereal time
1 tropical (mean solar, ordinary) year (yr)
  = 31.5569 × 10^6 sec (mean solar)
  = 525,949 min (mean solar)
  = 8,765.81 hr (mean solar)
  = 365.2422 day (mean solar)
  = 366.2422 sidereal days
1 sidereal second
  = 0.997270 sec (mean solar)
1 sidereal day
  = 86,164.1 sec (mean solar)
  = 23 hr, 56 min, 4.091 sec (mean solar)

By permission of the Smithsonian Institution Press. From Smithsonian Meteorological Tables, sixth revised edition, Robert J. List, Ed., Smithsonian Institution, Washington, D. C., 1984.

Table 1.4—3
DISTRIBUTION OF WATER AND LAND IN VARIOUS LATITUDE BELTS

	Northern hemisphere				Southern hemisphere			
Latitude	Water 10^6 km²	Land 10^6 km²	Water %	Land %	Water 10^6 km²	Land 10^6 km²	Water %	Land %
90–85°	0.979	–	100.0	–	–	0.978	–	100.0
85–80	2.545	0.384	86.9	13.1	–	2.929	–	100.0
80–75	3.742	1.112	77.1	22.9	0.522	4.332	10.7	89.3
75–70	4.414	2.326	65.5	34.5	2.604	4.136	38.6	61.4
70–65	2.456	6.116	28.7	71.3	6.816	1.756	79.5	20.5
65–60	3.123	7.210	31.2	69.8	10.301	0.032	99.7	0.3
60–55	5.399	6.613	45.0	55.0	12.006	0.006	99.9	0.1
55–50	5.529	8.066	40.7	59.3	13.388	0.207	98.5	1.5
50–45	6.612	8.458	43.8	56.2	14.693	0.377	97.5	2.5
45–40	8.411	8.016	51.2	48.8	15.833	0.594	96.4	3.6
40–35	10.029	7.627	56.8	43.2	16.483	1.173	93.4	6.6
35–30	10.806	7.943	57.7	42.3	15.782	2.967	84.2	15.8
30–25	11.747	7.952	59.6	40.4	15.438	4.261	78.4	21.6
25–20	13.354	7.145	65.2	34.8	15.450	5.049	75.4	24.6
20–15	14.981	6.164	70.8	29.2	16.147	4.998	76.4	23.6
15–10	16.553	5.080	76.5	23.5	17.211	4.422	79.6	20.4
10–5	16.628	5.332	75.7	24.3	16.898	5.062	76.9	23.1
5–0	17.387	4.737	78.6	21.4	16.792	5.332	75.9	24.1
90–0°	154.695	100.281	60.7	39.3	206.364	48.611	80.9	19.1

Note: All oceans and seas 361.059 × 10^6 km² 70.8 percent; all land 148.892 × 10^6 km², 29.2 percent.

(From Kossinna, Erwin, Die Tiefen des Weltmeeres, *Berlin Univ., Inst. f. Meereskunde, Veroff.*, N. F., A Geogr.-naturwiss. Reihe, Heft 9, 1921.)

Table 1.4—4
SCALE VARIATION FOR STANDARD MAP PROJECTIONS

Three map projections are widely used in meteorology: the polar stereographic, the Lambert conformal conic, and the Mercator, each of which is conformal. That is, the shape of any small area on the map is the same as the shape of the corresponding small area of the earth, all angles are preserved (except at the pole on the Lambert and Mercator projections), and the scale is the same in all directions at any point, a function only of the latitude of the point for a given assumed figure of the earth.

$$\Delta m = sk\Delta n$$

where:

Δn = (small) distance on the earth,
Δm = corresponding (small) distance on the map,
s = map scale at standard parallel,
k = scale factor for latitude in question.

Values of k are tabulated below assuming the figure of the earth to be spherical and assuming the figure to be that of the International Ellipsoid of Reference.

	Mercator projection Standard parallel 22½°		Lambert conformal conic projection Standard parallels 30° and 60°		Polar stereographic projection Standard parallel 60°	
Latitude	Sphere k	International ellipsoid k	Sphere k	International ellipsoid k	Sphere k	International ellipsoid k
0°	0.924	0.924	1.283	1.281	1.866	1.860
5	0.927	0.928	1.210	1.208	1.716	1.712
10	0.938	0.938	1.149	1.148	1.590	1.586
15	0.956	0.957	1.099	1.098	1.482	1.480
20	0.983	0.983	1.058	1.058	1.390	1.388
25	1.019	1.019	1.025	1.025	1.312	1.310
30	1.067	1.066	1.000	1.000	1.244	1.243
35	1.128	1.127	0.982	0.982	1.186	1.185
40	1.206	1.205	0.970	0.970	1.136	1.136
45	1.307	1.305	0.966	0.966	1.093	1.093
50	1.437	1.435	0.968	0.969	1.057	1.057
55	1.611	1.608	0.979	0.979	1.026	1.026
60	1.848	1.844	1.000	1.000	1.000	1.000
65	2.186	2.181	1.033	1.033	0.979	0.979
70	2.701	2.694	1.084	1.083	0.962	0.962
75	3.570	3.560	1.162	1.162	0.949	0.949
80	5.320	5.306	1.293	1.292	0.940	0.940
85	10.600	10.570	1.566	1.564	0.935	0.936

REFERENCE

Gregg, W. R. and Tannehill, I. R., *Month. Weath. Rev.*, U.S. Dept. of Commerce, Washington, D.C., 65, 415, 1937.

Table 1.4—5
RADIUS OF CURVATURE ON A POLAR STEREOGRAPHIC PROJECTION

In computing gradient wind speeds and in other problems, it is necessary to determine a factor r that depends on curvature of the trajectory. This factor arises in taking account of the horizontal component of the centrifugal force acting on a particle. The problem is twofold: (1) to determine the trajectory of the particle on a map, and (2) to determine the required value of r if the trajectory on the map is known. The first problem is of such nature that it cannot be treated adequately here. (NOTE – In many cases an approximation is made from the curvature of the isobars or streamlines.) The second problem has been solved for the case of a polar stereographic projection, since on this projection a "small circle" on the earth projects as a circle on the map.

Let R be the radius of the earth, r' the true radius of the *small circle* on which the particle is assumed to be traveling at a given instant, and α its angular radius (as seen from the center of the earth). Then $r' = R \sin \alpha$. Since we are concerned with the horizontal component of the centrifugal force, the effective horizontal radius of the curvature required in the gradient wind equation is given by $r = r' \sec \alpha = R \tan \alpha$. If an arc on a map representing the instantaneous trajectory of a particle of air is determined, this arc may be regarded as a portion of a "small circle."

To determine r for a given arc of a trajectory on the map:

1. Complete the circle by extending the arc. (A set of circular templates will prove very useful.)
2. Find the meridian which passes through the center of this circle.
3. Determine the latitudes ϕ_1 and ϕ_2 of the points where this meridian intersects the circle. (Extend the meridian across the pole if necessary.)

4a. If the circle found in Step 1 *does not* contain the pole, find the difference between ϕ_1 and ϕ_2 and enter Part A of the table with this difference as the argument. The corresponding tabular value is the required radius r in statute miles, from the formula $r = R \tan \frac{1}{2}(\phi_1 - \phi_2)$.

4b. If the circle found in Step 1 contains the pole, find the sum $(\phi_1 + \phi_2)$ and enter Part B of the table with this sum as the argument. The corresponding tabular value is the required radius r in statute miles, from the formula $r = R \tan [90° - \frac{1}{2}(\phi_1 + \phi_2)]$.

Table A
Circle not including pole

$\phi_1 - \phi_2$	0 mi	1 mi	2 mi	3 mi	4 mi	5 mi	6 mi	7 mi	8 mi	9 mi
0°	0	35	69	104	138	173	207	242	277	311
10	346	381	416	451	486	521	556	591	627	662
20	698	733	769	805	841	877	914	950	987	1,023
30	1,060	1,097	1,135	1,172	1,210	1,248	1,286	1,324	1,363	1,401
40	1,440	1,479	1,519	1,559	1,599	1,639	1,680	1,721	1,762	1,803
50	1,845	1,887	1,930	1,973	2,016	2,060	2,104	2,148	2,193	2,239
60	2,285	2,331	2,378	2,425	2,473	2,521	2,570	2,619	2,669	2,720
70	2,771	2,822	2,875	2,928	2,982	3,036	3,092	3,148	3,204	3,262
80	3,320	3,380	3,440	3,501	3,563	3,626	3,690	3,755	3,821	3,889
90	3,957									

Table 1.4—5 (continued)
RADIUS OF CURVATURE ON A POLAR STEREOGRAPHIC PROJECTION

Table B

Circle including poles

$\phi_1 + \phi_2$	0 mi	1 mi	2 mi	3 mi	4 mi	5 mi	6 mi	7 mi	8 mi	9 mi
0°		453,433	226,697	151,110	113,313	90,631	75,504	64,697	56,589	50,278
10	45,229	41,093	37,648	34,730	32,227	30,057	28,156	26,477	24,984	23,646
20	22,441	21,350	20,357	19,449	18,616	17,849	17,140	16,482	15,871	15,301
30	14,768	14,269	13,800	13,358	12,943	12,550	12,178	11,826	11,492	11,174
40	10,872	10,583	10,308	10,045	9,794	9,553	9,322	9,100	8,887	8,683
50	8,486	8,296	8,113	7,937	7,766	7,601	7,442	7,288	7,138	6,994
60	6,854	6,718	6,586	6,457	6,332	6,211	6,093	5,978	5,867	5,757
70	5,651	5,547	5,446	5,347	5,251	5,157	5,065	4,975	4,886	4,800
80	4,716	4,633	4,552	4,473	4,395	4,318	4,243	4,170	4,097	4,027
90	3,957	3,889	3,821	3,755	3,690	3,626	3,563	3,501	3,440	3,380
100	3,320	3,262	3,204	3,148	3,092	3,036	2,982	2,928	2,875	2,822
110	2,771	2,720	2,669	2,619	2,570	2,521	2,473	2,425	2,378	2,331
120	2,285	2,239	2,193	2,148	2,104	2,060	2,016	1,973	1,930	1,887
130	1,845	1,803	1,762	1,721	1,680	1,639	1,599	1,559	1,519	1,479
140	1,440	1,401	1,363	1,324	1,286	1,248	1,210	1,172	1,135	1,097
150	1,060	1,023	987	950	914	877	841	805	769	733
160	698	662	627	591	556	521	486	451	416	381
170	346	311	277	242	207	173	138	104	69	35

By permission of the Smithsonian Institution Press. From Smithsonian Meteorological Tables, sixth revised edition, Robert J. List, Ed., Smithsonian Institution, Washington, D. C., 1984.

Table 1.4—6
EPHEMERIS OF THE SUN

All data are for O^h Greenwich Civil Time in the year 1950. Variations of these data from year to year are negligible for most meteorological purposes; the largest variation occurs through the 4 year, leap year cycle. The year 1950 was selected to represent a mean condition in this cycle.

The *declination* of the sun is its angular distance north (+) or south (−) of the celestial equator.

The *longitude* of the sun is the angular distance of the meridian of sun from the vernal equinox (mean equinox of 1950.0) measured eastward along the ecliptic.

The *equation of time* (apparent − mean) is the correction to be applied to mean solar time in order to obtain apparent (true) solar time.

The *radius vector* of the earth is the distance from the center of the earth to the center of the sun expressed in terms of the length of the semimajor axis of the earth's orbit.

Table 1.4—6 (continued)

Date	Declination		Longitude		Equation of time		Radius vector	Date	Declination		Longitude		Equation of time		Radius vector
	°	′	°	′	m	s			°	′	°	′	m	s	
Jan. 1	−23	4	280	1	−3	14	0.98324	Feb. 1	−17	19	311	34	−13	34	0.98533
5	22	42	284	5	5	6	.98324	5	16	10	315	37	14	2	.98593
9	22	13	288	10	6	50	.98333	9	14	55	319	40	14	17	.98662
13	21	37	292	14	8	27	.98352	13	13	37	323	43	14	20	.98738
17	20	54	296	19	9	54	.98378	17	12	15	327	46	14	10	.98819
21	20	5	300	23	11	10	.98410	21	10	50	331	48	13	50	.98903
25	19	9	304	27	12	14	.98448	25	9	23	335	49	13	19	.98991
29	18	8	308	31	13	5	.98493								
Mar. 1	−7	53	339	51	−12	38	0.99084	Apr. 1	+4	14	10	42	−4	12	0.99928
5	6	21	343	51	11	48	.99182	5	5	46	14	39	3	1	1.00043
9	4	48	347	51	10	51	.99287	9	7	17	18	35	1	52	1.00160
13	3	14	351	51	9	49	.99396	13	8	46	22	30	−0	47	1.00276
17	1	39	355	50	8	42	.99508	17	10	12	26	25	+0	13	1.00390
21	−0	5	359	49	7	32	.99619	21	11	35	30	20	1	6	1.00500
25	+1	30	3	47	6	20	.99731	25	12	56	34	14	1	53	1.00606
29	3	4	7	44	5	7	.99843	29	14	13	38	7	2	33	1.00708
May 1	+14	50	40	4	+2	50	1.00759	June 1	+21	57	69	56	+2	27	1.01405
5	16	2	43	56	3	17	1.00859	5	22	28	73	46	1	49	1.01465
9	17	9	47	48	3	35	1.00957	9	22	52	77	36	1	6	1.01518
13	18	11	51	40	3	44	1.01051	13	23	10	81	25	+0	18	1.01564
17	19	9	55	32	3	44	1.01138	17	23	22	85	15	−0	33	1.01602
21	20	2	59	23	3	34	1.01218	21	23	27	89	4	1	25	1.01630
25	20	49	63	14	3	16	1.01291	25	23	25	92	53	2	17	1.01649
29	21	30	67	4	2	51	1.01358	29	23	17	96	41	3	7	1.01662
July 1	+23	10	98	36	−3	31	1.01667	Aug. 1	+18	14	128	11	−6	17	1.01494
5	22	52	102	24	4	16	1.01671	5	17	12	132	0	5	59	1.01442
9	22	28	106	13	4	56	1.01669	9	16	6	135	50	5	33	1.01384
13	21	57	110	2	5	30	1.01659	13	14	55	139	41	4	57	1.01318
17	21	21	113	51	5	57	1.01639	17	13	41	143	31	4	12	1.01244
21	20	38	117	40	6	15	1.01610	21	12	23	147	22	3	19	1.01163
25	19	50	121	29	6	24	1.01573	25	11	2	151	14	2	18	1.01076
29	18	57	125	19	6	23	1.01530	29	9	39	155	5	1	10	1.00986
Sept. 1	+8	35	157	59	−0	15	1.00917	Oct. 1	−2	53	187	14	+10	1	1.00114
5	7	7	161	52	+1	2	1.00822	5	4	26	191	11	11	17	1.00001
9	5	37	165	45	2	22	1.00723	9	5	58	195	7	12	27	0.99888
13	4	6	169	38	3	45	1.00619	13	7	29	199	5	13	30	.99774
17	2	34	173	32	5	10	1.00510	17	8	58	203	3	14	25	.99659
21	+1	1	177	26	6	35	1.00397	21	10	25	207	1	15	10	.99544
25	−0	32	181	21	8	0	1.00283	25	11	50	211	0	15	46	.99433
29	2	6	185	16	9	22	1.00170	29	13	12	214	59	16	10	.99326
Nov. 1	−14	11	217	59	+16	21	0.99249	Dec. 1	−21	41	248	13	+11	16	0.98604
5	15	27	222	0	16	23	.99150	5	22	16	252	16	9	43	.98546
9	16	38	226	1	16	12	.99054	9	22	45	256	20	8	1	.98494
13	17	45	230	2	15	47	.98960	13	23	6	260	24	6	12	.98446
17	18	48	234	4	15	10	.98869	17	23	20	264	28	4	17	.98405
21	19	45	238	6	14	18	.98784	21	23	26	268	32	2	19	.98372
25	20	36	242	8	13	15	.98706	25	23	25	272	37	+0	20	.98348
29	21	21	246	11	11	59	.98636	29	23	17	276	41	−1	39	.98334

(From U.S. Naval Observatory, *The American Ephemeris and Nautical Almanac for the Year 1950,* Washington, D.C., 1948.)

Table 1.4—7
DURATION OF DAYLIGHT

Day of month	Jan.	Feb.	Mar.	Apr.	May	June	July	Aug.	Sept.	Oct.	Nov.	Dec.
	h. m.	h. m.	h. m.	h. m.	h. m.	h. m.	h. m.	h. m.	h. m.	h. m.	h. m.	h. m.
	Latitude 0°											
1	12 07	12 07	12 07	12 06	12 06	12 07	12 07	12 07	12 06	12 06	12 07	12 08
5	12 07	12 07	12 07	12 07	12 07	12 07	12 07	12 06	12 07	12 07	12 07	12 07
9	12 07	12 07	12 07	12 07	12 07	12 08	12 07	12 07	12 07	12 07	12 07	12 07
13	12 07	12 07	12 07	12 07	12 07	12 08	12 07	12 06	12 06	12 07	12 07	12 07
17	12 07	12 07	12 07	12 07	12 07	12 07	12 08	12 06	12 07	12 07	12 08	12 08
21	12 07	12 07	12 07	12 07	12 07	12 07	12 07	12 06	12 06	12 07	12 07	12 07
25	12 07	12 07	12 06	12 06	12 07	12 07	12 07	12 06	12 06	12 07	12 08	12 08
29	12 07	12 07	12 06	12 07	12 07	12 07	12 07	12 06	12 07	12 07	12 07	12 08
	Latitude 5° N.											
1	11 51	11 55	12 01	12 10	12 18	12 24	12 24	12 20	12 12	12 04	11 57	11 52
5	11 51	11 56	12 03	12 11	12 19	12 24	12 25	12 20	12 11	12 03	11 55	11 51
9	11 51	11 57	12 03	12 12	12 19	12 24	12 24	12 19	12 11	12 02	11 55	11 51
13	11 52	11 57	12 05	12 13	12 21	12 24	12 23	12 18	12 10	12 01	11 54	11 50
17	11 52	11 58	12 05	12 14	12 21	12 25	12 23	12 16	12 09	12 01	11 54	11 50
21	11 53	11 59	12 07	12 15	12 22	12 25	12 22	12 16	12 07	11 59	11 52	11 50
25	11 53	12 00	12 08	12 16	12 22	12 25	12 21	12 14	12 06	11 58	11 52	11 50
29	11 54	12 01	12 09	12 17	12 23	12 25	12 21	12 14	12 05	11 57	11 51	11 50
	Latitude 10° N.											
1	11 33	11 42	11 56	12 14	12 29	12 40	12 42	12 33	12 18	12 02	11 47	11 36
5	11 33	11 44	11 58	12 16	12 31	12 41	12 41	12 32	12 16	12 00	11 45	11 35
9	11 35	11 46	12 00	12 18	12 33	12 42	12 40	12 30	12 14	11 58	11 42	11 33
13	11 36	11 48	12 03	12 20	12 34	12 42	12 40	12 28	12 12	11 56	11 41	11 33
17	11 37	11 50	12 05	12 22	12 35	12 42	12 38	12 26	12 09	11 53	11 40	11 32
21	11 39	11 52	12 07	12 24	12 37	12 43	12 37	12 24	12 08	11 51	11 38	11 32
25	11 39	11 54	12 09	12 26	12 38	12 43	12 36	12 22	12 05	11 50	11 36	11 32
29	11 41	11 56	12 12	12 27	12 39	12 43	12 35	12 20	12 03	11 48	11 36	11 33
	Latitude 15° N.											
1	11 15	11 30	11 51	12 16	12 40	12 58	13 00	12 47	12 24	12 00	11 35	11 18
5	11 17	11 32	11 54	12 20	12 43	12 59	12 59	12 44	12 21	11 57	11 33	11 17
9	11 18	11 35	11 57	12 23	12 45	13 00	12 58	12 42	12 18	11 53	11 30	11 16
13	11 19	11 38	12 01	12 27	12 49	13 00	12 57	12 39	12 15	11 50	11 27	11 15
17	11 21	11 41	12 04	12 30	12 51	13 01	12 56	12 36	12 11	11 47	11 25	11 14
21	11 23	11 44	12 07	12 33	12 53	13 01	12 53	12 34	12 08	11 43	11 23	11 14
25	11 25	11 47	12 11	12 36	12 54	13 01	12 51	12 30	12 05	11 40	11 21	11 14
29	11 27	11 51	12 14	12 39	12 57	13 01	12 49	12 27	12 02	11 38	11 19	11 15
	Latitude 20° N.											
1	10 57	11 16	11 45	12 20	12 52	13 16	13 19	13 02	12 32	11 57	11 25	11 00
5	10 59	11 20	11 49	12 25	12 56	13 17	13 19	12 58	12 27	11 53	11 21	10 59
9	11 00	11 23	11 54	12 30	13 00	13 19	13 16	12 55	12 23	11 49	11 17	10 57
13	11 02	11 27	11 59	12 34	13 03	13 20	13 15	12 51	12 18	11 44	11 13	10 56
17	11 05	11 32	12 03	12 38	13 07	13 20	13 12	12 48	12 13	11 40	11 10	10 56
21	11 07	11 36	12 07	12 42	13 09	13 21	13 10	12 42	12 08	11 35	11 07	10 55
25	11 11	11 41	12 12	12 46	13 12	13 21	13 07	12 38	12 05	11 32	11 04	10 56
29	11 13	11 45	12 17	12 51	13 14	13 20	13 05	12 34	12 00	11 27	11 02	10 56
	Latitude 25° N.											
1	10 37	11 02	11 39	12 24	13 06	13 35	13 40	13 18	12 38	11 55	11 13	10 42
5	10 39	11 07	11 45	12 30	13 10	13 37	13 39	13 13	12 33	11 49	11 07	10 39
9	10 41	11 11	11 50	12 36	13 15	13 40	13 36	13 09	12 27	11 44	11 03	10 38
13	10 44	11 17	11 56	12 42	13 19	13 40	13 35	13 04	12 22	11 38	10 58	10 36
17	10 47	11 22	12 02	12 47	13 23	13 42	13 32	12 58	12 15	11 33	10 54	10 36
21	10 51	11 28	12 09	12 52	13 27	13 41	13 28	12 54	12 10	11 27	10 50	10 35
25	10 55	11 33	12 14	12 58	13 30	13 41	13 25	12 48	12 04	11 22	10 46	10 36
29	10 59	11 39	12 20	13 03	13 33	13 41	13 21	12 42	11 58	11 16	10 44	10 36

Table 1.4—7 (continued)
DURATION OF DAYLIGHT

Day of month	Jan.	Feb.	Mar.	Apr.	May	June	July	Aug.	Sept.	Oct.	Nov.	Dec.
	h. m.	h. m.	h. m.	h. m.	h. m.	h. m.	h. m.	h. m.	h. m.	h. m.	h. m.	h. m.
Latitude 30° N.												
1	10 15	10 46	11 33	12 29	13 20	13 57	14 03	13 34	12 46	11 53	10 59	10 22
5	10 17	10 53	11 40	12 36	13 26	13 59	14 01	13 29	12 39	11 46	10 53	10 19
9	10 21	10 59	11 47	12 43	13 31	14 02	13 58	13 23	12 32	11 38	10 48	10 16
13	10 24	11 05	11 54	12 50	13 37	14 04	13 55	13 17	12 25	11 32	10 42	10 14
17	10 27	11 12	12 02	12 57	13 42	14 04	13 52	13 11	12 18	11 25	10 36	10 14
21	10 33	11 18	12 09	13 04	13 47	14 05	13 48	13 04	12 10	11 17	10 32	10 12
25	10 37	11 25	12 16	13 10	13 50	14 05	13 43	12 58	12 03	11 11	10 28	10 13
29	10 43	11 33	12 24	13 17	13 55	14 03	13 39	12 51	11 56	11 05	10 24	10 14
Latitude 35° N.												
1	09 51	10 30	11 26	12 34	13 35	14 21	14 29	13 54	12 55	11 50	10 45	10 00
5	09 53	10 37	11 34	12 42	13 43	14 25	14 27	13 47	12 47	11 41	10 37	09 55
9	09 57	10 45	11 44	12 52	13 50	14 27	14 24	13 40	12 38	11 32	10 31	09 52
13	10 02	10 53	11 52	13 00	13 57	14 30	14 19	13 33	12 29	11 24	10 24	09 50
17	10 06	11 01	12 01	13 08	14 03	14 30	14 16	13 25	12 21	11 16	10 18	09 48
21	10 11	11 08	12 09	13 16	14 09	14 31	14 10	13 18	12 12	11 07	10 11	09 48
25	10 17	11 17	12 19	13 24	14 14	14 31	14 05	13 09	12 03	11 00	10 06	09 48
29	10 25	11 26	12 27	13 32	14 19	14 29	13 59	13 01	11 54	10 51	10 01	09 50
Latitude 40° N.												
1	09 23	10 10	11 18	12 39	13 54	14 49	14 58	14 16	13 05	11 47	10 29	09 33
5	09 27	10 19	11 28	12 50	14 02	14 53	14 55	14 08	12 55	11 36	10 20	09 29
9	09 31	10 28	11 38	13 00	14 11	14 57	14 52	14 00	12 44	11 26	10 11	09 25
13	09 36	10 37	11 50	13 10	14 19	15 00	14 47	13 51	12 34	11 16	10 03	09 22
17	09 42	10 47	12 00	13 20	14 27	15 00	14 42	13 41	12 24	11 06	09 55	09 20
21	09 49	10 58	12 11	13 30	14 34	15 01	14 36	13 32	12 13	10 55	09 48	09 20
25	09 56	11 07	12 21	13 40	14 40	15 01	14 29	13 22	12 03	10 46	09 42	09 20
29	10 03	11 18	12 32	13 49	14 45	14 59	14 22	13 13	11 52	10 37	09 36	09 22
Latitude 42° N.												
1	09 11	10 02	11 14	12 42	14 02	15 02	15 11	14 26	13 09	11 45	10 22	09 22
5	09 15	10 11	11 26	12 53	14 12	15 07	15 09	14 17	12 58	11 34	10 13	09 17
9	09 19	10 21	11 36	13 04	14 21	15 10	15 04	14 08	12 48	11 24	10 03	09 13
13	09 24	10 31	11 48	13 16	14 29	15 12	14 59	13 59	12 36	11 12	09 54	09 10
17	09 31	10 41	12 00	13 26	14 37	15 14	14 54	13 49	12 25	11 02	09 46	09 08
21	09 39	10 52	12 11	13 37	14 45	15 15	14 48	13 38	12 45	10 51	09 38	09 07
25	09 46	11 03	12 23	13 47	14 52	15 15	14 40	13 28	12 03	10 40	09 30	09 08
29	09 55	11 14	12 34	13 57	14 57	15 13	14 32	13 17	11 52	10 29	09 24	09 09
Latitude 44° N.												
1	08 58	09 52	11 10	12 45	14 11	15 16	15 26	14 36	13 14	11 45	10 15	09 09
5	09 01	10 03	11 22	12 57	14 21	15 21	15 23	14 27	13 02	11 32	10 04	09 03
9	09 06	10 13	11 34	13 08	14 31	15 24	15 18	14 17	12 50	11 20	09 54	08 59
13	09 12	10 24	11 46	13 20	14 40	15 28	15 13	14 07	12 39	11 08	09 44	08 56
17	09 19	10 35	11 59	13 32	14 49	15 29	15 07	13 56	12 26	10 57	09 35	08 54
21	09 27	10 47	12 11	13 44	14 47	15 29	15 00	13 45	12 14	10 45	09 27	08 53
25	09 35	10 59	12 23	13 55	15 04	15 29	14 52	13 34	12 03	10 34	09 19	08 54
29	09 45	11 10	12 36	14 06	15 11	15 27	14 44	13 23	11 50	10 23	09 12	08 56
Latitude 46° N.												
1	08 43	09 42	11 06	12 47	14 21	15 30	15 41	14 48	13 19	11 43	10 07	08 56
5	08 47	09 53	11 20	13 00	14 31	15 35	15 38	14 37	13 06	11 30	09 55	08 49
9	08 53	10 05	11 32	13 14	14 42	15 40	15 34	14 27	12 54	11 18	09 44	08 45
13	09 00	10 17	11 46	13 26	14 52	15 42	15 27	14 15	12 41	11 04	09 34	08 42
17	09 07	10 29	11 58	13 38	15 01	15 44	15 21	14 05	12 28	10 52	09 24	08 40
21	09 15	10 42	12 12	13 51	15 10	15 45	15 13	13 53	12 15	10 39	09 15	08 38
25	09 25	10 53	12 25	14 03	15 18	15 45	15 04	13 41	12 02	10 27	09 06	08 39
29	09 35	11 06	12 38	14 14	15 25	15 43	14 56	13 29	11 50	10 15	08 59	08 40

Table 1.4—7 (continued)
DURATION OF DAYLIGHT

Day of month	Jan. h. m.	Feb. h. m.	Mar. h. m.	Apr. h. m.	May h. m.	June h. m.	July h. m.	Aug. h. m.	Sept. h. m.	Oct. h. m.	Nov. h. m.	Dec. h. m.
	Latitude 48° N.											
1	08 27	09 32	11 02	12 51	14 31	15 46	15 59	15 00	13 25	11 41	09 57	08 40
5	08 32	09 43	11 16	13 05	14 42	15 52	15 55	14 49	13 11	11 28	09 45	08 35
9	08 38	09 56	11 30	13 18	14 54	15 57	15 50	14 38	12 57	11 14	09 33	08 29
13	08 44	10 09	11 44	13 32	15 05	16 00	15 43	14 25	12 43	11 00	09 22	08 26
17	08 53	10 21	11 58	13 46	15 15	16 02	15 36	14 13	12 30	10 46	09 12	08 23
21	09 02	10 35	12 12	13 59	15 24	16 03	15 28	14 01	12 16	10 33	09 02	08 22
25	09 12	10 49	12 27	14 11	15 33	16 03	15 18	13 47	12 02	10 20	08 52	08 22
29	09 23	11 02	12 40	14 24	15 41	16 00	15 08	13 35	11 48	10 07	08 44	08 24
	Latitude 50° N.											
1	08 10	09 20	10 58	12 55	14 41	16 04	16 18	15 14	13 31	11 39	09 48	08 24
5	08 15	09 33	11 12	13 09	14 54	16 11	16 13	15 03	13 16	11 24	09 35	08 17
9	08 21	09 46	11 28	13 24	15 07	16 16	16 08	14 50	13 01	11 10	09 22	08 12
13	08 30	10 00	11 42	13 38	15 19	16 20	16 01	14 37	12 47	10 56	09 10	08 08
17	08 38	10 15	11 58	13 53	15 30	16 22	15 53	14 23	12 32	10 40	08 58	08 06
21	08 48	10 28	12 13	14 07	15 40	16 23	15 44	14 09	12 17	10 26	08 47	08 04
25	08 59	10 43	12 28	14 21	15 50	16 21	15 34	13 55	12 02	10 12	08 38	08 05
29	09 11	10 58	12 43	14 34	15 59	16 20	15 22	13 41	11 47	09 59	08 29	08 07
	Latitude 52° N.											
1	07 51	09 08	10 52	12 57	14 53	16 24	16 39	15 29	13 37	11 38	09 38	08 07
5	07 56	09 21	11 09	13 14	15 08	16 31	16 34	15 16	13 21	11 22	09 23	07 59
9	08 03	09 36	11 25	13 30	15 22	16 37	16 28	15 02	13 05	11 06	09 09	07 53
13	08 12	09 51	11 42	13 46	15 34	16 41	16 21	14 48	12 49	10 50	08 56	07 48
17	08 21	10 05	11 58	14 01	15 46	16 43	16 11	14 33	12 34	10 34	08 43	07 46
21	08 33	10 21	12 14	14 17	15 58	16 44	16 02	14 19	12 17	10 19	08 32	07 44
25	08 45	10 37	12 29	14 31	16 09	16 43	15 50	14 03	12 02	10 04	08 21	07 46
29	08 57	10 52	12 45	14 46	16 18	16 41	15 38	13 48	11 46	09 49	08 12	07 48
	Latitude 54° N.											
1	07 29	08 53	10 48	13 02	15 07	16 46	17 03	15 46	13 43	11 36	09 26	07 47
5	07 35	09 09	11 04	13 19	15 23	16 54	16 58	15 31	13 27	11 18	09 11	07 39
9	07 43	09 24	11 22	13 36	15 38	17 01	16 51	15 16	13 10	11 02	08 55	07 32
13	07 52	09 40	11 40	13 53	15 52	17 05	16 43	15 01	12 53	10 44	08 40	07 26
17	08 03	09 57	11 56	14 10	16 05	17 08	16 33	14 45	12 36	10 28	08 27	07 24
21	08 15	10 14	12 14	14 27	16 18	17 09	16 22	14 29	12 19	10 11	08 15	07 22
25	08 28	10 30	12 31	14 43	16 29	17 08	16 10	14 13	12 02	09 55	08 02	07 23
29	08 42	10 48	12 49	15 00	16 39	17 06	15 56	13 56	11 44	09 38	07 52	07 26
	Latitude 56° N.											
1	07 05	08 38	10 42	13 07	15 22	17 12	17 31	16 05	13 51	11 34	09 14	07 24
5	07 11	08 53	11 00	13 25	15 39	17 21	17 25	15 49	13 34	11 16	08 56	07 15
9	07 20	09 11	11 19	13 43	15 56	17 28	17 17	15 32	13 15	10 57	08 40	07 07
13	07 30	09 29	11 38	14 02	16 11	17 33	17 08	15 16	12 57	10 38	08 24	07 02
17	07 43	09 47	11 56	14 20	16 26	17 36	16 57	14 59	12 38	10 20	08 09	06 58
21	07 56	10 05	12 15	14 39	16 40	17 37	16 44	14 41	12 19	10 02	07 55	06 57
25	08 10	10 23	12 34	14 56	16 53	17 37	16 31	14 23	12 01	09 44	07 42	06 58
29	08 25	10 42	12 53	15 14	17 04	17 34	16 16	14 05	11 43	09 27	07 30	07 00
	Latitude 58° N.											
1	06 36	08 20	10 35	13 11	15 39	17 42	18 03	16 27	14 00	11 31	08 59	06 58
5	06 44	08 37	10 55	13 31	15 57	17 52	17 56	16 09	13 40	11 12	08 41	06 49
9	06 53	08 56	11 15	13 51	16 16	18 01	17 47	15 51	13 21	10 52	08 23	06 39
13	07 06	09 16	11 36	14 11	16 34	18 06	17 37	15 32	13 01	10 32	08 04	06 33
17	07 19	09 35	11 56	14 31	16 50	18 10	17 25	15 13	12 41	10 12	07 48	06 29
21	07 33	09 55	12 16	14 51	17 05	18 11	17 10	14 54	12 21	09 52	07 33	06 27
25	07 49	10 15	12 36	15 11	17 19	18 10	16 55	14 35	12 01	09 33	07 18	06 28
29	08 07	10 35	12 57	15 30	17 33	18 06	16 39	14 15	11 41	09 13	07 05	06 32

Table 1.4—7 (continued)
DURATION OF DAYLIGHT

Day of month	Jan.	Feb.	Mar.	Apr.	May	June	July	Aug.	Sept.	Oct.	Nov.	Dec.
	h. m.	h. m.	h. m.	h. m.	h. m.	h. m.	h. m.	h. m.	h. m.	h. m.	h. m.	h. m.
	Latitude 60° N.											
1	06 03	08 00	10 28	13 17	15 58	18 17	18 43	16 51	14 10	11 28	08 43	06 28
5	06 11	08 19	10 49	13 39	16 19	18 30	18 36	16 32	13 48	11 07	08 23	06 17
9	06 23	08 40	11 11	14 01	16 39	18 39	18 25	16 11	13 27	10 46	08 03	06 06
13	06 36	09 01	11 33	14 23	16 59	18 46	18 12	15 50	13 06	10 24	07 44	05 58
17	06 51	09 23	11 55	14 45	17 18	18 50	17 57	15 30	12 44	10 02	07 24	05 54
21	07 08	09 44	12 18	15 05	17 35	18 53	17 41	15 09	12 23	09 41	07 07	05 52
25	07 26	10 06	12 39	15 27	17 52	18 51	17 24	14 48	12 01	09 20	06 50	05 53
29	07 45	10 28	13 01	15 48	18 08	18 46	17 05	14 26	11 39	08 59	06 35	05 57
	Latitude 61° N.											
1	05 43	07 48	10 24	13 20	16 10	18 39	19 07	17 05	14 16	11 27	08 34	06 11
5	05 53	08 09	10 46	13 43	16 31	18 52	18 58	16 44	13 54	11 05	08 12	05 58
9	06 05	08 31	11 09	14 05	16 53	19 03	18 47	16 23	13 31	10 42	07 51	05 47
13	06 20	08 53	11 32	14 29	17 14	19 10	18 32	16 02	13 08	10 20	07 30	05 39
17	06 35	09 15	11 55	14 51	17 34	19 16	18 16	15 39	12 46	09 58	07 11	05 34
21	06 53	09 38	12 18	15 13	17 53	19 17	17 59	15 17	12 23	09 35	06 53	05 32
25	07 12	10 01	12 40	15 36	18 11	19 15	17 41	14 55	12 01	09 13	06 34	05 33
29	07 33	10 24	13 03	15 59	18 28	19 10	17 21	14 32	11 39	08 51	06 18	05 37
	Latitude 62° N.											
1	05 21	07 36	10 19	13 24	16 21	19 03	19 34	17 21	14 22	11 26	08 24	05 52
5	05 32	07 59	10 43	13 47	16 44	19 16	19 24	16 58	13 58	11 02	08 02	05 37
9	05 45	08 22	11 07	14 11	17 07	19 29	19 11	16 35	13 35	10 39	07 39	05 26
13	06 00	08 44	11 31	14 35	17 29	19 38	18 55	16 12	13 11	10 15	07 18	05 16
17	06 19	09 08	11 54	14 59	17 51	19 44	18 38	15 50	12 48	09 52	06 57	05 11
21	06 38	09 32	12 18	15 22	18 11	19 45	18 19	15 26	12 25	09 28	06 37	05 09
25	06 58	09 56	12 42	15 46	18 32	19 43	17 59	15 03	12 00	09 05	06 17	05 10
29	07 19	10 19	13 06	16 09	18 50	19 38	17 37	14 39	11 37	08 42	06 00	05 15
	Latitude 63° N.											
1	04 56	07 22	10 14	13 27	16 34	19 29	20 06	17 37	14 28	11 24	08 14	05 30
5	05 08	07 47	10 39	13 52	16 58	19 46	19 54	17 14	14 03	11 00	07 50	05 15
9	05 23	08 11	11 05	14 17	17 23	20 00	19 39	16 49	13 39	10 35	07 27	05 02
13	05 40	08 36	11 29	14 41	17 47	20 11	19 21	16 25	13 14	10 11	07 04	04 52
17	05 59	09 00	11 54	15 07	18 11	20 16	19 02	16 00	12 50	09 46	06 41	04 45
21	06 20	09 25	12 19	15 32	18 33	20 19	18 40	15 36	12 25	09 22	06 19	04 42
25	06 42	09 50	12 44	15 56	18 54	20 17	18 18	15 11	12 00	08 57	05 58	04 44
29	07 05	10 14	13 08	16 21	19 15	20 10	17 55	14 46	11 37	08 33	05 40	04 49
	Latitude 64° N.											
1	04 28	07 08	10 10	13 31	16 48	20 01	20 46	17 56	14 34	11 22	08 03	05 06
5	04 41	07 33	10 36	13 57	17 14	20 21	20 31	17 30	14 09	10 57	07 38	04 49
9	04 57	08 00	11 01	14 23	17 41	20 38	20 12	17 04	13 43	10 31	07 13	04 34
13	05 17	08 26	11 27	14 49	18 07	20 51	19 51	16 38	13 18	10 05	06 48	04 22
17	05 39	08 51	11 54	15 15	18 33	20 59	19 30	16 12	12 52	09 40	06 23	04 14
21	06 01	09 17	12 20	15 42	18 57	21 01	19 06	15 46	12 26	09 14	05 59	04 12
25	06 24	09 44	12 46	16 08	19 22	20 59	18 41	15 20	12 00	08 48	05 37	04 14
29	06 49	10 10	13 12	16 35	19 45	20 51	18 15	14 54	11 35	08 22	05 16	04 20
	Latitude 65° N.											
1	03 54	06 52	10 04	13 35	17 03	20 40	21 38	18 17	14 42	11 20	07 51	04 38
5	04 09	07 19	10 32	14 02	17 32	21 05	21 18	17 49	14 15	10 54	07 24	04 17
9	04 29	07 47	10 59	14 30	18 00	21 28	20 55	17 21	13 47	10 27	06 57	04 01
13	04 50	08 14	11 26	14 57	18 29	21 46	20 29	16 53	13 20	10 00	06 30	03 47
17	05 15	08 42	11 53	15 25	18 57	21 58	20 02	16 25	12 54	09 33	06 03	03 38
21	05 39	09 09	12 20	15 52	19 26	22 03	19 34	15 56	12 27	09 06	05 37	03 34
25	06 05	09 37	12 48	16 21	19 53	21 58	19 06	15 30	12 00	08 39	05 13	03 37
29	06 32	10 04	13 14	16 49	20 21	21 47	18 38	15 02	11 34	08 11	04 49	03 44

Table 1.4—7 (continued)
DURATION OF DAYLIGHT

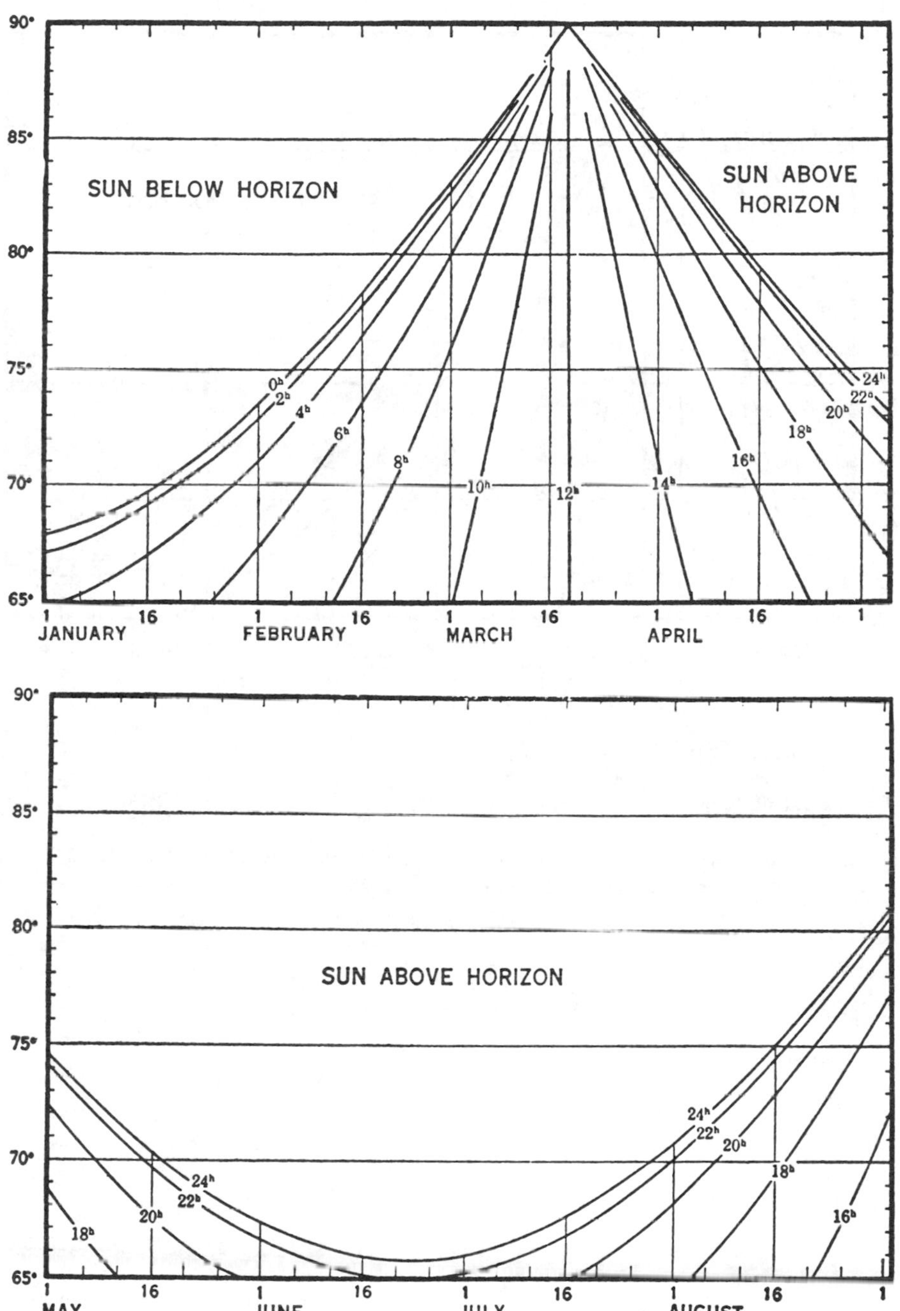

Table 1.4—7 (continued)
DURATION OF DAYLIGHT

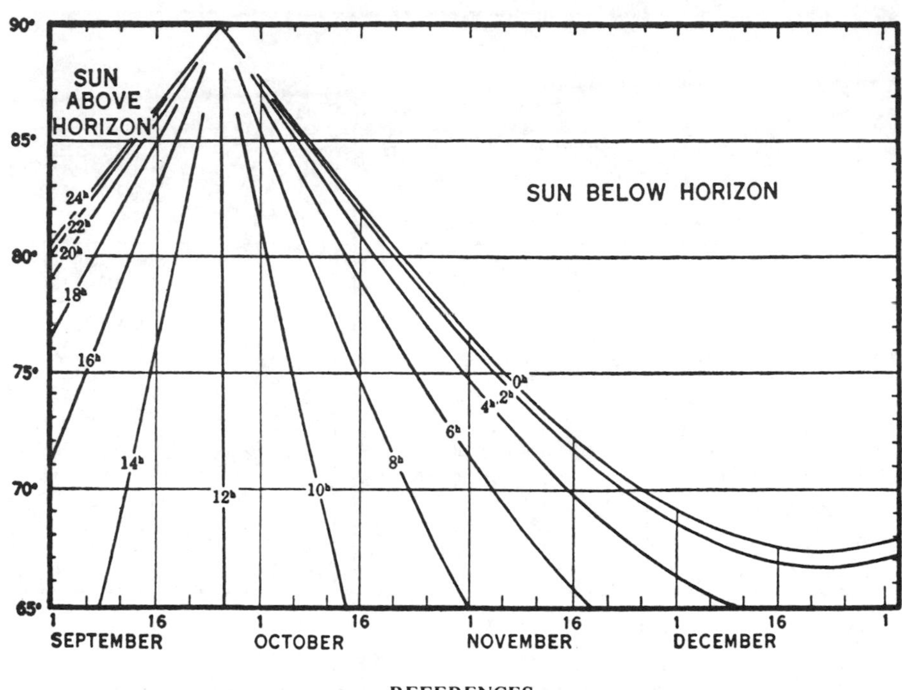

REFERENCES

1. Tables of sunrise and twilight, Supplement to the American ephemeris, 1946, U.S. Naval Observatory, Washington, D.C., 1945.
2. **Kimball, H.,** *Month Weath. Rev.,* 44, 614, 1916.

Table 1.4—8
DAYLIGHT AND TWILIGHT FOR SOUTHERN LATITUDES

Southern latitude date	Corresponding date– northern latitude	Southern latitude date	Corresponding date– northern latitude
Jan. 3	July 5	July 2	Jan. 1
7	9	3	1
11	13	7	5
15	17	11	9
18	21	15	13
22	25	20	17
26	29	24	21
29	Aug. 1	28	25
Feb. 1	5	Aug. 1	29
5	9	5	Feb. 1
9	13	9	5
13	17	13	9
17	21	17	13
20	25	21	17
24	29	26	21
27	Sept. 1	30	25
Mar. 3	5	Sept. 3	Mar. 1
7	9	7	5
11	13	11	9
15	17	15	13
19	21	19	17
23	25	23	21
27	29	27	25
29	Oct. 1		
		Oct. 1	29
Apr. 1	5	2	29
5	9	5	Apr. 1
6	9	9	5
10	13	12	9
14	17	16	13
18	21	20	17
22	25	24	21
26	29	28	25
29	Nov. 1		
		Nov. 1	29
May 3	5	3	May 1
7	9	7	5
11	13	11	9
15	17	15	13
16	17	18	17
20	21	22	21
24	25	26	25
28	29	30	29
30	Dec. 1		
		Dec. 3	June 1
June 3	5	6	5
4	5	10	9
8	9	14	13
12	13	18	17
16	17	25	25
20	21	29	29
25	25	31	July 1
29	29		

By permission of the Smithsonian Institution Press. From Smithsonian Meteorological Tables, sixth revised edition, Robert J. List, Ed., Smithsonian Institution, Washington, D. C., 1984.

1.5. HYGROMETRIC DATA

Table 1.5—1
SATURATION VAPOR PRESSURE OVER WATER OF SALINITY 35°/oo

The saturation vapor pressure over a plane surface of pure water depends only on the temperature of the water. The salinity decreases the saturation vapor pressure slightly, the empirical relation being

$$e_s = e_w (1 - 0.000537S)$$

where

e_s = saturation vapor pressure over sea water at a given temperature,
e_w = saturation vapor pressure over a plane surface of pure water at the same temperature,
S = salinity in parts per thousand (°/oo)

Values of e_s computed for a salinity of 35°/oo are given in the table.

Temp °C	Vapor pressure mbar	Temp °C	Vapor pressure mbar	Temp °C	Vapor pressure mbar	Temp °C	Vapor pressure mbar
−2	5.19	7	9.83	16	17.85	25	31.12
−1	5.57	8	10.52	17	19.02	26	33.01
0	5.99	9	11.26	18	20.26	27	35.02
1	6.44	10	12.05	19	21.57	28	37.13
2	6.92	11	12.88	20	22.96	29	39.33
3	7.43	12	13.76	21	24.42	30	41.68
4	7.98	13	14.70	22	25.96	31	44.13
5	8.56	14	15.69	23	27.59	32	46.71
6	9.17	15	16.74	24	29.30		

(From Sverdrup, H. U., Johnson, M. W., and Fleming, R. H., The Oceans, Their Physics, Chemistry, and General Biology, Prentice-Hall, Inc., Englewood Cliffs, New Jersey, ©1942, renewed 1970, pp 66, 684, 727, 741, 116. Reprinted by permission of Prentice-Hall, Inc., Englewood Cliffs, NJ.)

Figure 1.5—1
SEA WATER VAPOR PRESSURE

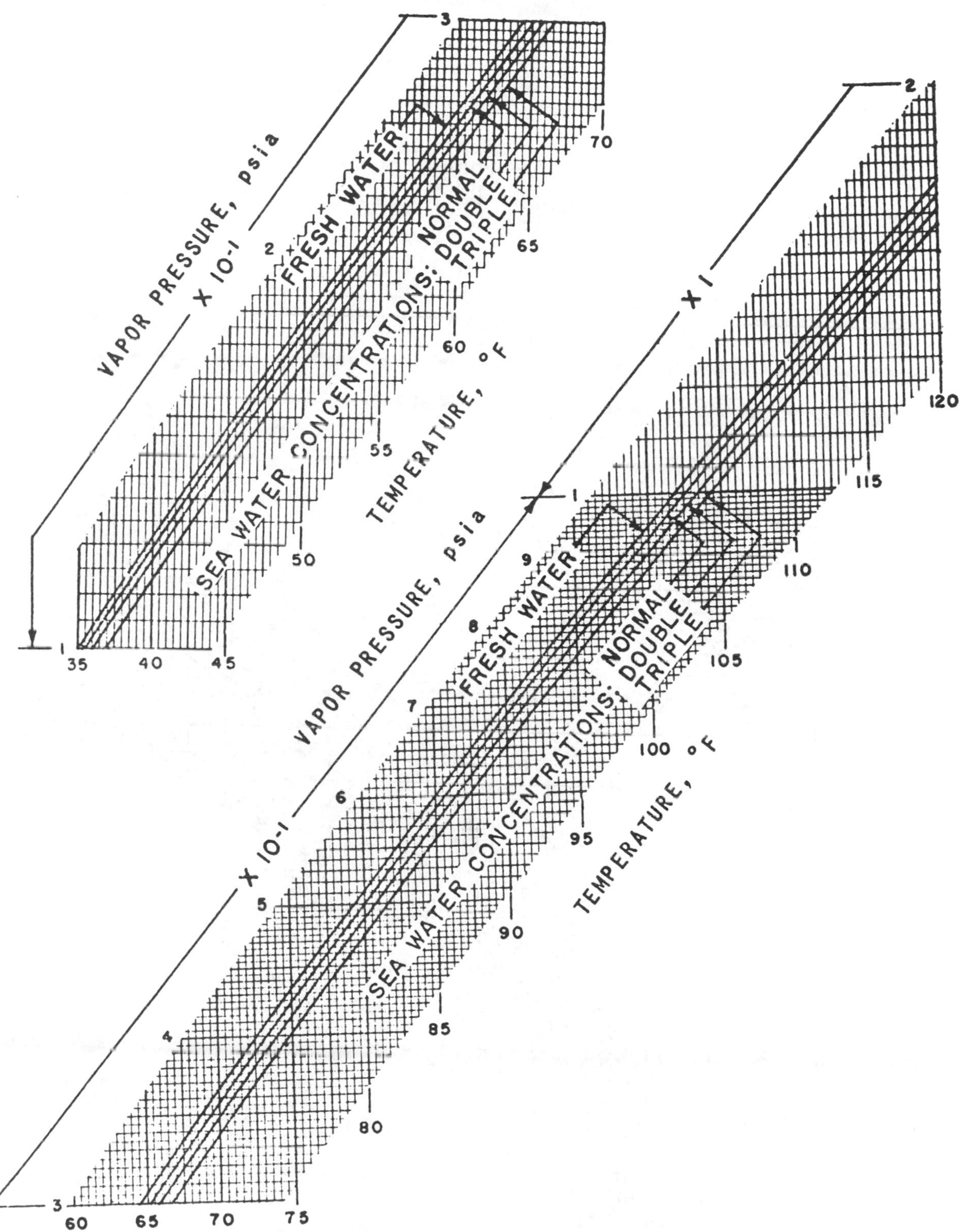

Note: The normal sea water concentration used in this chart has 34.483 g solids/1000 g sea water.

Figure 1.5—1 (continued)
SEA WATER VAPOR PRESSURE

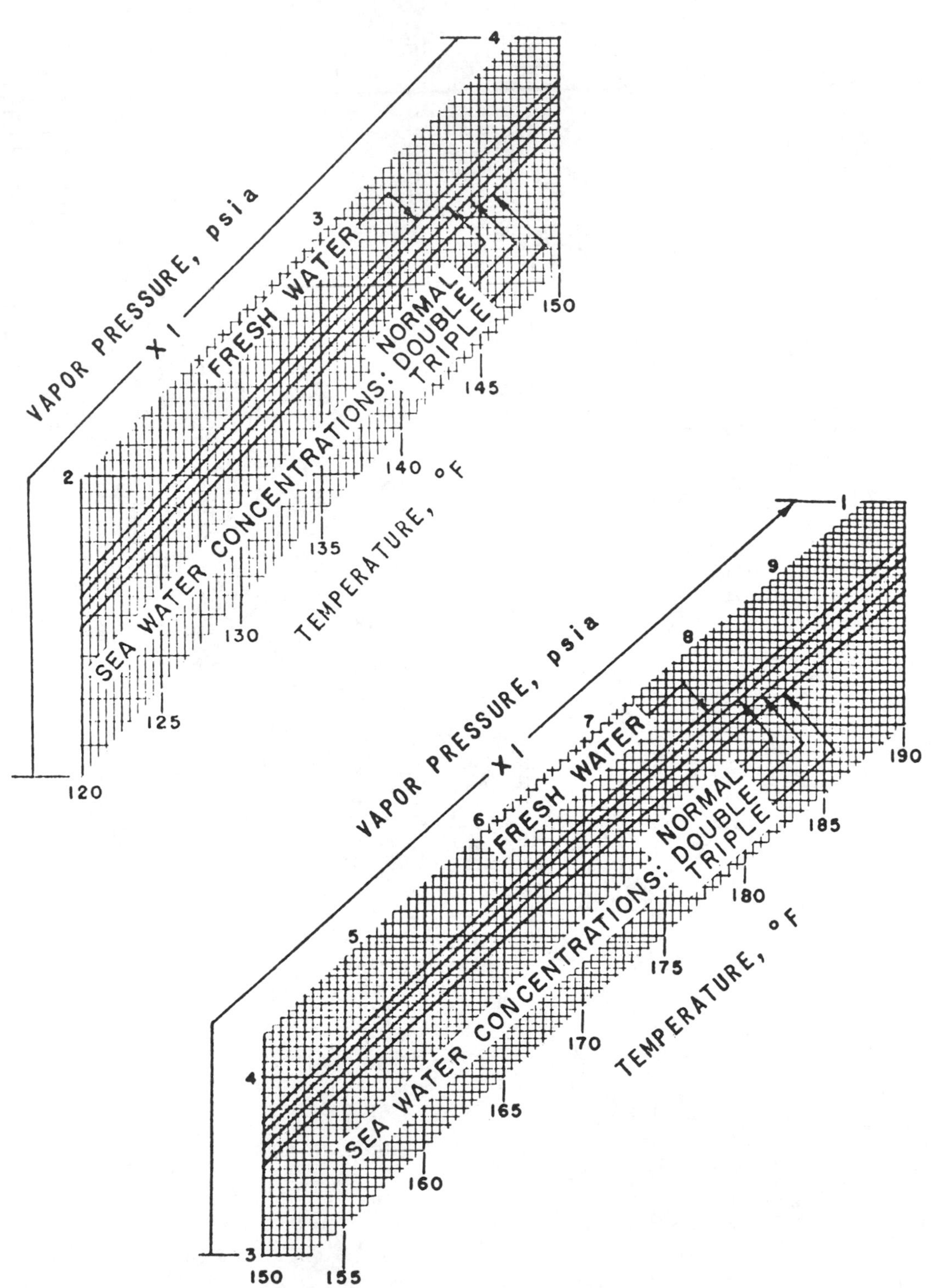

Figure 1.5—1 (continued)
SEA WATER VAPOR PRESSURE

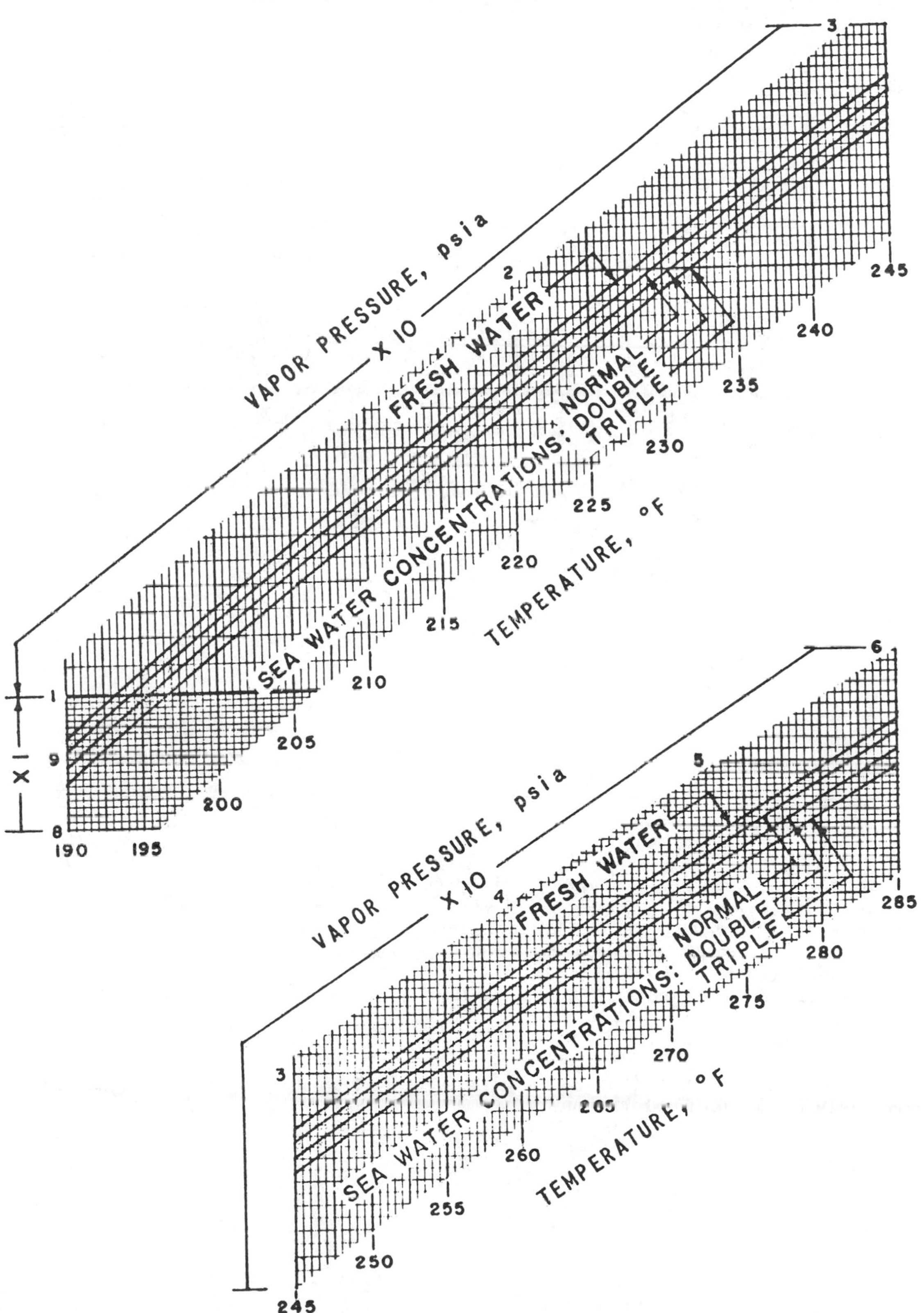

Figure 1.5—1 (continued)
SEA WATER VAPOR PRESSURE

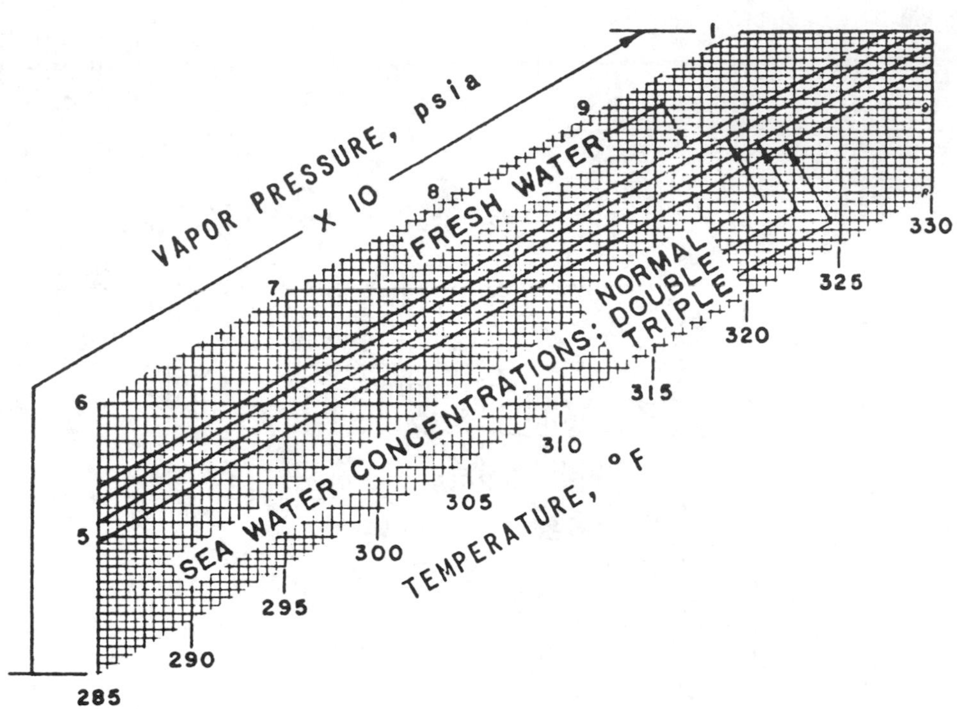

REFERENCES

1. **Frankel, A.,** *Proc. Inst. Mech. Engrs. (London),* 174, 312, 1960.
2. **Keenan, J. H. and Keyes, F. G.,** *Thermodynamic Properties of Steam,* John Wiley & Sons, New York, 1936.
3. **Spiegler, K. S.,** *Salt-Water Purification,* John Wiley & Sons, Inc., New York, 1962, pp. 10, 37.
4. **Sverdrup, H. U., Johnson, M. W., and Fleming, R. H.,** *The Oceans, Their Physics, Chemistry, and General Biology,* Prentice-Hall, Englewood Cliffs, New Jersey, 1942.
5. U.S. Dept. of the Interior, Saline Water Research and Development Progress Report No. 12, November 1956, 70.

(This chart was prepared by the M. W. Kellogg Co. for the Office of Saline Water. Taken from *Saline Water Conversion Engineering Data Book,* 2nd ed., U.S. Government Printing Office, Washington, D.C., November 1971.)

Section 2
Chemical Oceanography

Section 2

CHEMICAL OCEANOGRAPHY

Nearly 80 of the 92 naturally occurring elements have been detected in seawater, most in the form of dissolved ions.[1] The majority of these elements — the minor constituents — frequently behave nonconservatively, their concentrations varying in response to biological processes and chemical reactions. In contrast, the major constituents generally act conservatively; that is, their concentrations are constant except at ocean boundaries where they may be changed by input or output processes, such as dilution by river runoff, anthropogenic activity, or precipitation onto the ocean surface.[2] Most of the major ions have long residence times relative to the mixing time of the ocean. The minor ions, however, often exhibit residence times substantially shorter than the mixing time of the ocean. They typically are more reactive, being removed rather quickly.

Eleven elements comprise more than 99.9% of the total dissolved material in seawater. These include in decreasing order of total mass: chlorine (2.5×10^{16} tonnes), sodium (1.4×10^{16} t), magnesium (1.7×10^{15} t), calcium (5.4×10^{14} t), potassium (5.0×10^{14} t), silicon (2.6×10^{12} t), zinc (6.4×10^{9} t), copper (2.6×10^{9} t), iron (2.6×10^{9} t), manganese (2.6×10^{8} t), and cobalt (6.6×10^{7} t).[2] The amount of dissolved solids in seawater averages 35,000 mg/l, compared to 120 mg/l in river water.[3,4]

In addition to the major and minor dissolved constituents in seawater, trace metals (e.g., antimony, arsenic, cadmium, lead, mercury, nickel, and silver), nutrient elements (e.g., nitrogen, phosphorus, and silicon), dissolved atmospheric gases (e.g., oxygen, nitrogen, and carbon dioxide), and organic matter (dissolved and particulate forms) are important components of seawater. Trace metals, although essential to the growth of many organisms at low concentrations, can be toxic when present at elevated levels.[5] Trace metals enter the ocean via riverine and atmospheric influx, anthropogenic input, and hydrothermal convection of seawater through new oceanic crust at spreading ocean centers. The major nutrients represent the chief limiting elements to phytoplankton production in the sea. Of all atmospheric gases dissolved in seawater, oxygen and carbon dioxide probably are most critical. Oxygen serves as an essential element in the metabolic processes of organisms, as an indicator of water quality of the marine environment, and as a tracer of the motion of deep water masses of the oceans. Carbon dioxide not only buffers natural waters against changes in acidity and alkalinity but also regulates biological processes. A large number of investigations in marine chemistry deal with the sources, forms, and fates of organic matter in seawater.

Chemical oceanography has progressed rapidly since the early 1970s. The review series on chemical oceanography,[6] which deals with a variety of research topics, illustrates some of the significant advances achieved by marine chemists in recent years. These advances range from the chemistry of interstitial waters in sediments[7] to trace elements in seawater[8] and photochemical processes in marine environments.[9]

This section covers the chemical composition of seawater, detailing the major and minor constituents. Solubilities of selected gases and minerals also are treated. In addition, physicochemical properties of seawater are investigated, yielding data on the density, viscosity, and conductivity of seawater, as well as other parameters.

REFERENCES

1. **Smith, G. G., Ed.,** *Cambridge Encyclopedia of Earth Sciences*, Cambridge University Press, Cambridge, 1981.
2. **Kennish, M. J.,** *Ecology of Estuaries*, Vol. 1, CRC Press, Boca Raton, Fla., 1986.
3. **Livingstone, D. A.,** Chemical Composition of Rivers and Lakes, U. S. Geol. Surv. Prof. Pap., 440-G, 1963.
4. **Riley, J. P. and Chester, R.,** *Introduction to Marine Chemistry*, Academic Press, London, 1971.
5. **Spaargaren, D. H. and Ceccaldi, H. J.,** Some relations between the elementary chemical composition of marine organisms and that of seawater, *Oceanologica Acta*, 7, 63, 1984.
6. **Riley, J. P. and Chester, R., Eds.,** *Chemical Oceanography*, Academic Press, London, 1983.
7. **Gieskes, J. M.,** The chemistry of interstitial waters of deep sea sediments: interpretation of deep sea drilling data, in *Chemical Oceanography*, Riley, J. P. and Chester, R., Eds., Academic Press, London, 1983, 221.
8. **Bruland, K. W.,** Trace elements in sea-water, in *Chemical Oceanography*, Riley, J. P. and Chester, R., Eds., Academic Press, London, 1983, 157.
9. **Zafiriou, O. C.,** Natural water photochemistry, in *Chemical Oceanography*, Riley, J. P. and Chester, R., Eds., Academic Press, London 1983, 339.

2.1. PERIODIC TABLE OF THE ELEMENTS

New notation
Previous IUPAC form
CAS version

KEY TO CHART

Atomic Number →	50 +2	← Oxidation States
Symbol →	Sn +4	
1983 Atomic Weight →	118.71	
	18 18 4	← Electron Configuration

1 Group IA	2 IIA	3 IIIA IIIB	4 IVA IVB	5 VA VB	6 VIA VIB	7 VIIA VIIB	8	9 VIIIA VIII	10	11 IB	12 IIB	13 IIIB IIIA	14 IVB IVA	15 VB VA	16 VIB VIA	17 VIIB VIIA	18 VIIIA	Orbit
1 H +1 −1 1.00794 1																	2 He 0 4.00260 2	K
3 Li +1 6.941 2-1	4 Be +2 9.01218 2-2											5 B +3 10.81 2-3	6 C +2 +4 −4 12.011 2-4	7 N +1 +2 +3 +4 +5 −1 −2 −3 14.0067 2-5	8 O −2 15.9994 2-6	9 F −1 18.9984 2-7	10 Ne 0 20.179 2-8	K-L
11 Na +1 22.9898 2-8-1	12 Mg +2 24.305 2-8-2											13 Al +3 26.9815 2-8-3	14 Si +2 +4 −4 28.0855 2-8-4	15 P +3 +5 −3 30.9738 2-8-5	16 S +4 +6 −2 32.06 2-8-6	17 Cl +1 +5 +7 −1 35.453 2-8-7	18 Ar 0 39.948 2-8-8	K-L-M
19 K +1 39.0983 -8-8-1	20 Ca +2 40.08 -8-8-2	21 Sc +3 44.9559 -8-9-2	22 Ti +2 +3 +4 47.88 -8-10-2	23 V +2 +3 +4 +5 50.9415 -8-11-2	24 Cr +2 +3 +6 51.996 -8-13-1	25 Mn +2 +3 +4 +7 54.9380 -8-13-2	26 Fe +2 +3 55.847 -8-14-2	27 Co +2 +3 58.9332 -8-15-2	28 Ni +2 +3 58.69 -8-16-2	29 Cu +1 +2 63.546 -8-18-1	30 Zn +2 65.39 -8-18-2	31 Ga +3 69.72 -8-18-3	32 Ge +2 +4 72.59 -8-18-4	33 As +3 +5 −3 74.9216 -8-18-5	34 Se +4 +6 −2 78.96 -8-18-6	35 Br +1 +5 −1 79.904 -8-18-7	36 Kr 0 83.80 -8-18-8	-L-M-N
37 Rb +1 85.4678 -18-8-1	38 Sr +2 87.62 -18-8-2	39 Y +3 88.9059 -18-9-2	40 Zr +4 91.224 -18-10-2	41 Nb +3 +5 92.9064 -18-12-1	42 Mo +6 95.94 -18-13-1	43 Tc +4 +6 +7 (98) -18-13-2	44 Ru +3 101.07 -18-15-1	45 Rh +3 102.906 -18-16-1	46 Pd +2 +4 106.42 -18-18-0	47 Ag +1 107.868 -18-18-1	48 Cd +2 112.41 -18-18-2	49 In +3 114.82 -18-18-3	50 Sn +2 +4 118.71 -18-18-4	51 Sb +3 +5 −3 121.75 -18-18-5	52 Te +4 +6 −2 127.60 -18-18-6	53 I +1 +5 +7 −1 126.905 -18-18-7	54 Xe 0 131.29 -18-18-8	-M-N-O
55 Cs +1 132.905 -18-8-1	56 Ba +2 137.33 -18-8-2	57* La +3 138.906 -18-9-2	72 Hf +4 178.49 -32-10-2	73 Ta +5 180.948 -32-11-2	74 W +6 183.85 -32-12-2	75 Re +4 +6 +7 186.207 -32-13-2	76 Os +3 +4 190.2 -32-14-2	77 Ir +3 +4 192.22 -32-15-2	78 Pt +2 +4 195.08 -32-16-2	79 Au +1 +3 196.967 -32-18-1	80 Hg +1 +2 200.59 -32-18-2	81 Tl +1 +3 204.383 -32-18-3	82 Pb +2 +4 207.2 -32-18-4	83 Bi +3 +5 208.980 -32-18-5	84 Po +2 +4 (209) -32-18-6	85 At (210) -32-18-7	86 Rn 0 (222) -32-18-8	-N-O-P
87 Fr +1 (223) -18-8-1	88 Ra +2 226.025 -18-8-2	89** Ac +3 227.028 -18-9-2	104 Unq +4 (261) -32-10-2	105 Unp (262) -32-11-2	106 Unh (263) -32-12-2	107 Uns (262) -32-13-2												O P Q

*Lanthanides	58 Ce +3 +4 140.12 -20-8-2	59 Pr +3 140.908 -21-8-2	60 Nd +3 144.24 -22-8-2	61 Pm +3 (145) -23-8-2	62 Sm +2 +3 150.36 -24-8-2	63 Eu +2 +3 151.96 -25-8-2	64 Gd +3 157.25 -25-9-2	65 Tb +3 158.925 -27-8-2	66 Dy +3 162.50 -28-8-2	67 Ho +3 164.930 -29-8-2	68 Er +3 167.26 -30-8-2	69 Tm +3 168.934 -31-8-2	70 Yb +2 +3 173.04 -32-8-2	71 Lu +3 174.967 -32-9-2	N O P
**Actinides	90 Th +4 232.038 -18-10-2	91 Pa +5 +4 231.036 -20-9-2	92 U +3 +4 +5 +6 238.029 -21-9-2	93 Np +3 +4 +5 +6 237.048 -22-9-2	94 Pu +3 +4 +5 +6 (244) -24-8-2	95 Am +3 +4 +5 +6 (243) -25-8-2	96 Cm +3 (247) -25-9-2	97 Bk +3 +4 (247) -27-8-2	98 Cf +3 (251) -28-8-2	99 Es +3 (252) -29-8-2	100 Fm +3 (257) -30-8-2	101 Md +2 +3 (258) -31-8-2	102 No +2 +3 (259) -32-8-2	103 Lr +3 (260) -32-9-2	O P Q

Numbers in parentheses are mass numbers of most stable isotope of that element

From *Chemical and Engineering News*, 63(5), 27, 1985. This format numbers the groups 1 to 18.

2.2. COMPOSITION OF STREAMS

Table 2.2—1
AVERAGE COMPOSITION OF STREAMS

	ppm
HCO_3^-	58.4
$SO_4^=$	11.2
Cl^-	7.8
NO_3^-	1.0
Ca^{++}	15.0
Mg^{++}	4.1
Na^+	6.3
K^+	2.3
(Fe)	(0.67)
SiO_2	13.1
Total	120

(From Livingstone, D., Chemical composition of rivers and lakes, U.S. Geol. Surv. Profess. paper 440-G, 1963.)

Table 2.2—2
THE TRACE ELEMENT COMPOSITION OF STREAMS

	ppb = μg/l	Approximate estimate, ppb	Region	Ref.
Lithium	3.3	3	North America	1
Boron	13	10	USSR	2
Fluorine	88	100	USSR	2
	150		Japan	3
Aluminum	360	400	Japan	3
Phosphorus	19	20	Columbia R.	4
Scandium	0.004	0.004	Columbia R.	4
Titanium	2.7	3	Maine (US) lakes and streams	5
Vanadium	0.9	0.9	Japan	6
Chromium	1.4	1	US streams, Rhone, Amazon	7
	0.3		Maine (US) lakes and streams	5
Manganese	12	7	USSR,	2
	4.0		Maine (US) lakes and streams,	5
	4.8		Columbia R.	4
Cobalt	0.19	0.2	US streams, Rhone, Amazon	7
Nickel	0.3	0.3	Maine (US) lakes and streams	5
Copper	0.9	7	Japan,	3
	10		USSR,	
	12		Maine (US) lakes and streams	5
	4.4		Columbia R.	4
Zinc	5.0	20	Japan,	3
	45		USSR,	
	16		Columbia R.	4

Table 2.2—2 (continued)
THE TRACE ELEMENT COMPOSITION OF STREAMS

	ppb = μg/l	Approximate estimate, ppb	Region	Ref.
Gallium	0.089	0.09	Saale and Elbe (Germany)	8
Germanium				
Arsenic	1.7	2	Japan,	3
	1.6		Columbia R.	4
Selenium	0.20	0.2	US streams, Rhone, Amazon	7
Bromine	19	20	USSR	2
Rubidium	1.1	1	US streams, Rhone, Amazon	7
Strontium	46	50	Eastern US	9
Yttrium		0.7		estimate[a]
Zirconium				
Niobium				
Molybdenum	0.6	1	Japan	3
	1.8		US streams, Amazon	7
Ruthenium				
Rhodium				
Palladium				
Silver	0.39	0.3	US streams, Rhone, Amazon	7
Cadmium				
Indium				
Tin				
Antimony	1.1	1	US streams, Rhone, Amazon	7
Tellerium				
Iodine	7.1	7	USSR	2
Cesium	0.020	0.02	US streams, Rhone, Amazon	7
Barium	11	10	Eastern US	9
Lanthanum	0.2	0.2	Sweden	10
		0.19	Columbia R.	4
Cerium		0.06		estimate[a]
Praseodymium		0.03		estimate[a]
Neodymium		0.2		estimate[a]
Samarium		0.03		estimate[a]
Europium		0.007		estimate[a]
Gadolinium		0.04		estimate[a]
Terbium		0.008		estimate[a]
Dysprosium		0.05		estimate[a]
Holmium		0.01		estimate[a]
Erbium		0.05		estimate[a]
Thulium		0.009		estimate[a]
Ytterbium		0.05		estimate[a]
Lutetium		0.008		estimate[a]
Hafnium				
Tantalum				
Tungsten	0.03	0.03	Sweden	10
Rhenium				
Osmium				
Iridium				
Platinum				

Table 2.2—2 (continued)
THE TRACE ELEMENT COMPOSITION OF STREAMS

	ppb = μg/l	Approximate estimate, ppb	Region	Ref.
Gold	0.002	0.002	Sweden	10
Mercury	0.074	0.07	Saale and Elbe (Germany)	11
Thallium	?			
Lead	3.9	3	Saale and Elbe (Germany)	11
	2.3		Maine (US) lakes and streams	5
Bismuth	?			
Thorium	0.096	0.1	Amazon	12
Uranium	0.06	0.04	Sweden	10
	0.043		Amazon	12
	0.026		North America	13

[a] Estimate based on prorating rare-earth values in streams using La concentration in streams and the relative proportions of rare-earths found in the oceans.

(From Turekian, K. K., in *Handbook of Geochemistry,* Vol. I, Wedepohl, K. H., Ed., Springer-Verlag, Berlin, 1969. With permission.)

REFERENCES

1. **Durum, W. H. and Haffty, J.,** Occurrence of minor elements in water, *U.S. Geol. Surv. Circ.,* 445, 1960.
2. **Konovalov, G. S.,** The transport of microelements by the most important rivers of the U.S.S.R., *Dokl. Akad. Nauk. USSR,* 129, (4) 912, 1959. (Translated into English by M. Fleischer, U.S. Geol. Surv.)
3. **Sugawara, K.,** personal communication, 1967.
4. **Silker, W. B.,** Variations in elemental concentrations in the Columbia River, *Limnol. Oceanogr.,* 9, 540, 1964.
5. **Turekian, K. K. and Kleinkopf, M. D.,** Estimates of the average abundance of Cu, Mn, Pb, Ti, Ni, and Cr in surface waters of Maine, *Bull. Geol. Soc. Am.,* 67, 1129, 1956.
6. **Sugawara, K., Naito, H., and Yamada, S.,** Geochemistry of vanadium in natural waters, *J. Earth Sci. Nagoya Univ.,* 4, 44, 1956.
7. **Kharkar, D. P., Turekian, K. K., and Bertine, K. K.,** Stream supply of dissolved silver, molybdenum, antimony, selenium, chromium, cobalt, rubidium and cesium to the oceans, *Geochim. Cosmochim. Acta,* 32, 285, 1968.
8. **Heide, F. and Kodderitzsch, H.,** Der Galliumgehalt des Saale- und Elbewassers, *Naturwissenschaften,* 51, 104, 1964.
9. **Turekian, K. K.,** Trace elements in sea water and other natural waters, Annual report AEC Contract AT (30-1)-2912, Publ. Yale-2912-12, 1966.
10. **Landström, O. and Wenner, C. G.,** Neutron-activation analysis of natural water applied to hydrogeology, *Aktiebolaget Atomenergi (Sweden),* AE-204, 1965.
11. **Heide, F., Lerz, H., and Bohm, G.,** Gehalt des Saalewassers an Blei und Quecksilber, *Naturwissenschaften,* 44, 441, 1957.
12. **Moore, W. S.,** Amazon and Mississippi River concentrations of uranium, thorium, and radium isotopes, *Earth Planet. Sci. Lett.,* 2, 231, 1967.
13. **Rona, E. L. and Urry, W. D.,** Radioactivity of ocean sediments. VIII. Radium and uranium content of ocean and river waters, *Am. J. Sci.,* 250, 241, 1952.

2.3. COMPOSITION AND RELATED PROPERTIES OF SEA WATER

Table 2.3—1

CONCENTRATION OF THE MAJOR IONS IN SEA WATER (g/kg SEA WATER) NORMALIZED TO 35°/₀₀ SALINITY

Ion	Average value	Range	Reference
Chloride	19.353		
Sodium	10.76	10.72 – 10.80	Culkin and Cox (1966)
	10.79	10.76 – 10.80	Riley and Tongudai (1967)
Magnesium	1.297	1.292 – 1.301	Culkin and Cox (1966)
	1.292	1.296 – 1.287	Riley and Tongudai (1967)
Sulfate	2.712	2.701 – 2.724	Morris and Riley (1966)
Calcium	0.4119	0.4098 – 0.4134	Culkin and Cox (1966)
	0.4123	0.4088 – 0.4165	Riley and Tongudai (1967)
Potassium	0.399	0.393 – 0.405	Culkin and Cox (1966) Riley and Tongudai (1967)
Bicarbonate*	0.145	0.137 – 0.153	Koczy (1956); Postma (1964) Park (1966)
Bromide	0.0673	0.0666 – 0.0680	Morris and Riley (1966)
Boron	0.0046	0.0043 – 0.0051	Culkin (1965)
Strontium	0.0078	0.0074 – 0.0079	Culkin and Cox (1966)
	0.0081	0.0078 – 0.0085	Riley and Tongudai (1967)
Fluoride	0.0013	0.0012 – 0.0017	Greenhalgh and Riley (1963) Riley (1965)

* The values reported for bicarbonate are actually titration alkalinities.

(Compiled by D. R. Kester, Graduate School of Oceanography, University of Rhode Island.)

REFERENCES

1. Culkin, F., The major constituents of sea water, in *Chemical Oceanography,* Vol. 1, Riley, J. P. and Skirrow, G., Eds., Academic Press, London, 1965, 121-161.
2. Culkin, F. and Cox, R. A., Sodium, potassium, magnesium, calcium, and strontium in seawater, *Deep-Sea Res. Oceanogr. Abstr.,* 13, 789, 1966.
3. Greenhalgh, R. and Riley, J. P., Occurrence of abnormally high fluoride concentrations at depth in the oceans, *Nature,* 197, 371, 1963.
4. Koczy, F. F., The specific alkalinity, *Deep-Sea Res. Oceanogr. Abstr.,* 3, 279, 1956.
5. Morris, A. W. and Riley, J. P., The bromide/chlorinity and sulphate/chlorinity ratio in sea water, *Deep-Sea Res. Oceanogr. Abstr.,* 3, 699, 1966.
6. Park, Kilho, Deep-sea pH, *Science,* 154, 1540, 1966.
7. Postma, H., The exchange of oxygen and carbon dioxide between the ocean and the atmosphere, *Neth. J. Sea Res.,* 2, 258, 1964.
8. Riley, J. P., The occurrence of anomalously high fluoride concentrations in the North Atlantic, *Deep-Sea Res. Oceanogr. Abstr.,* 12, 219, 1965.
9. Riley, J. P. and Tongudai, M., The major cation/chlorinity ratios in sea water, *Chem. Geol.,* 2, 263, 1967.

Table 2.3—2

MOLALITY OF THE MAJOR CONSTITUENTS IN SEA WATER AT VARIOUS SALINITIES

Constituent	Salinity		
	30.0°/oo	34.8°/oo	40.0°/oo
Cl^-	0.48243	0.56241	0.64997
Na^+	0.41417	0.48284	0.55801
Mg^{2+}	0.04666	0.05440	0.06287
SO_4^{2-}	0.02495	0.02909	0.03362
Ca^{2+}	0.00909	0.01059	0.01224
K^+	0.00902	0.01052	0.01215
HCO_3^-	0.00211	0.00245	0.00284
Br^-	0.00074	0.00087	0.00100
$B(OH)_3$	0.00038	0.00044	0.00051
Sr^{2+}	0.00008	0.00009	0.00011

Note: These molalities are based on the concentrations in Table 2.3—1. The average of the two sets of analyses was used for Mg, Ca, K, and Sr. The Na^+ was calculated to preserve electroneutrality; the resulting value is within 0.1% of the analytical values of Na^+ reported in the previous table.

(Compiled by D. R. Kester, Graduate School of Oceanography, University of Rhode Island.)

Table 2.3—3

MAJOR CONSTITUENT CONCENTRATION-TO-CHLORINITY RATIOS FOR VARIOUS OCEANS AND SEAS

Ocean or Sea	Na °/₀₀ Cl	Mg °/₀₀ Cl	K °/₀₀ Cl	Ca °/₀₀ Cl	Sr °/₀₀ Cl	SO_4 °/₀₀ Cl	Br °/₀₀ Cl
N. Atlantic	–	–	0.02026	–	–	–	0.00337 – 0.00341
Atlantic	0.5544 – 0.5567	0.0667	0.01953 – 0.0263	0.02122 – 0.02126	0.000420	0.1393	0.00325 – 0.0038
N. Pacific	0.5553	0.06632 – 0.06695	0.02096	0.02154	–	0.1396 – 0.1397	0.00348
W. Pacific	0.5497 – 0.5561	0.06627 – 0.0676	0.02125	0.02058 – 0.02128	0.000413 – 0.000420	0.1399	0.0033
Indian	–	–	–	0.02099	0.000445	0.1399	0.0038
Mediterranean	0.5310 – 0.5528	0.06785	0.02008	–	–	0.1396	0.0034 – 0.0038
Baltic	0.5536	0.06693	–	0.02156	–	0.1414	0.00316 – 0.00344
Black	0.55184	–	0.0210	–	–	–	–
Irish	0.5573	–	–	–	–	0.1397	0.0033
Puget Sound	0.5495 – 0.5562	–	0.0191	–	–	–	–
Siberian	0.5484	–	0.0211	–	–	–	–
Antarctic	–	–	–	0.02120	0.000467	–	0.00347
Tokyo Bay	–	0.0676	–	0.02130	–	0.1394	–
Barents	–	0.06742	–	0.02085	–	–	–
Arctic	–	–	–	–	0.000424	–	–
Red	–	–	–	–	–	0.1395	0.0043
Japan	–	–	–	–	–	–	0.00327 – 0.00347
Bering	–	–	–	–	–	–	0.00341
Adriatic	–	–	–	–	–	–	0.00341

(From Culkin, F. and Cox, R. A., *Deep-Sea Research*, 13, 789, 1966. With permission.)

Table 2.3—4

THE MAJOR CHEMICAL SPECIES IN SEA WATER

Constituent	Percentage of the constituent present as each species at 25°C, 19.375°/₀₀ chlorinity, 1 atm, and pH 8.0
Chloride	–
Sodium	Na^+ (97.7%); $NaSO_4^-$ (2.2%); $NaHCO_3^\circ$ (0.03%)
Magnesium	Mg^{2+} (89%); $MgSO_4^\circ$ (10%); $MgHCO_3^+$ (0.6%); $MgCO_3^\circ$ (0.1%)
Sulfate	SO_4^{2-} (39%); $NaSO_4^-$ (37%); $MgSO_4^\circ$ (19%); $CaSO_4^\circ$ (4%)
Calcium	Ca^{2+} (88%); $CaSO_4^\circ$ (11%); $CaHCO_3^+$ (0.6%); $CaCO_3^\circ$ (0.1%)
Potassium	K^+ (98.8%); KSO_4^- (1.2%)
Bicarbonate	HCO_3^- (64%) $MgHCO_3^+$ (16%); $NaHCO_3^\circ$ (8%); $CaHCO_3^+$ (3%); CO_3^{2-} (0.8%); $MgCO_3^\circ$ (6%); $NaCO_3^-$ (1%); $CaCO_3^\circ$ (0.5%)
Bromide	–
Boron	$B(OH)_3$ (84%); $B(OH)_4^-$ (16%)
Strontium	–
Fluoride	F^- (50-80%); MgF^+ (20-50%)

(From Pytkowicz, R. M. and Kester, D. R., The physical chemistry of seawater, in *Oceanogr. Mar. Biol. Ann. Rev.*, Barnes, H., Ed., 9, 11, 1971. With permission of George Allen and Unwin, Ltd., London.)

Table 2.3—5

EFFECT OF TEMPERATURE AND PRESSURE ON THE DISTRIBUTION OF SULFATE SPECIES IN SEA WATER

Percentage of Total Sulfate as Each Species

T(°C)	P(atm)	SO_4^{2-}	$NaSO_4^-$	$MgSO_4^\circ$	$CaSO_4^\circ$
25	1	39	38	19	4
2	1	28	47	21	4
2	1000	39	32	24	5

(From Kester, D. R. and Pytkowicz, R. M., *Geochim. Cosmochim. Acta,* 34, 1039, 1970. With permission.)

Table 2.3—6
MOLALITY OF CHEMICAL SPECIES AND IONIC STRENGTH OF SEA WATER AT VARIOUS PRESSURES, TEMPERATURES, SALINITIES, AND pH = 8.0*

Pressure	1 atm				500 atm	1000 atm
Temperature	25°C			2°C	2°C	2°C
Salinity	30.0°/oo	34.8°/oo	40.0°/oo	34.8°/oo	34.8°/oo	34.8°/oo
Cl^-	0.48243	0.56241	0.64997	0.56241	0.56241	0.56241
Na^+	0.40556	0.47178	0.54384	0.46911	0.47098	0.47282
Mg^{2+}	0.04131	0.04831	0.05611	0.04795	0.04792	0.04801
SO_4^{2-}	0.01041	0.01136	0.01216	0.00828	0.01009	0.01201
$NaSO_4^-$	0.00845	0.01085	0.01391	0.01351	0.01165	0.00982
K^+	0.00892	0.01039	0.01200	0.01043	0.01041	0.01039
Ca^{2+}	0.00800	0.00936	0.01088	0.00928	0.00928	0.00930
$MgSO_4^\circ$	0.00498	0.00561	0.00614	0.00599	0.00602	0.00592
HCO_3^-	0.00131	0.00143	0.00154	0.00155	0.00147	0.00138
$CaSO_4^\circ$	0.00102	0.00115	0.00126	0.00123	0.00123	0.00121
Br^-	0.00074	0.00087	0.00100	0.00087	0.00087	0.00087
$MgHCO_3^+$	0.00028	0.00036	0.00045	0.00039	0.00037	0.00035
$B(OH)_3$	0.00032	0.00037	0.00042	0.00040	0.00036	0.00032
$NaHCO_3^\circ$	0.00015	0.00020	0.00024	0.00021	0.00020	0.00019
KSO_4^-	0.00010	0.00012	0.00015	0.00009	0.00011	0.00013
Sr^{2+}	0.00008	0.00009	0.00011	0.00009	0.00009	0.00009
μ	0.5736	0.6675	0.7701	0.6605	0.6640	0.6681

* Ionic strength of sea water is given by the empirical expression: $\mu = 0.0054 + 0.01840\,(S°/oo) + 1.78 \times 10^{-5}\,(S°/oo)^2 - 3.0 \times 10^{-4}\,(25 - t°C) + 7.6 \times 10^{-6}\,(P\,atm - 1)$ for $30°/oo \leqslant S°/oo \leqslant 40°/oo$.

(Based on Kester, D. R. and Pytkowicz, R. M., *Limnol. Oceanogr.*, 14, 686, 1969 and *Geochim. Cosmochim. Acta*, 34, 1039, 1970.)

Table 2.3—7
STOICHIOMETRIC ASSOCIATION CONSTANTS, $K^*_{MA} = \frac{[MA]}{[M]\,[A]}$ †

Pressure	1 atm				500 atm	1000 atm	
Temperature	25°C			2°C	2°C	2°C	
Salinity	30.0°/oo	34.8°/oo	40.0°/oo	34.8°/oo	34.8°/oo	34.8°/oo	Ref.
$NaSO_4^-$	2.00	2.02	2.09	3.45	2.43	1.70	1–4
$MgSO_4^\circ$	11.8	10.2	8.9	14.7	13.5	12.5	1–4
$CaSO_4^\circ$	12.4	10.8	9.5	15.3	14.1	13.1	1–4
KSO_4^-	1.03	1.03	1.03	1.03	1.03	1.03	5
$NaHCO_3^\circ$	0.29	0.29	0.29	0.29	0.29	0.29	6
$MgHCO_3^+$	5.22	5.22	5.22	5.22	5.22	5.22	5
$CaHCO_3^+$	5.10	5.10	5.10	5.10	5.10	5.10	5
$NaCO_3^-$	1.58	1.58	1.58	1.58	1.58	1.58	6
$MgCO_3^\circ$	160	160	160	160	160	160	5
$CaCO_3^\circ$	78	78	78	78	78	78	5

† Used in the calculation of the major chemical species in sea water at various pressures, temperatures, and salinities.

Table 2.3—7 (continued)

REFERENCES

1. **Pytkowicz, R. M. and Kester, D. R.,** *Am. J. Sci.*, 267, 217, 1969.
2. **Kester, D. R. and Pytkowicz, R. M.,** *Limnol. Oceanogr.*, 13, 670, 1968.
3. **Pytkowicz, R. M. and Kester, D. R.,** *Limnol. Oceanogr.*, 14, 686, 1969.
4. **Pytkowicz, R. M. and Kester, D. R.,** *Geochim. Cosmochim. Acta*, 9, 11, 1971.
5. **Garrels, R. M. and Thompson, M. E.,** *Am. J. Sci.*, 260, 57, 1962.
6. **Butler, J. N. and Huston, R.,** *J. Phys. Chem.*, 74, 2976, 1970.

Table 2.3—8
MINOR CONSTITUENTS OF SEA WATER EXCLUDING THE DISSOLVED GASES*

Element	Concentration μg/l		References on the distribution in the oceans
	Average	Range	
Lithium	185 (2, 18, 29, 52, 62)	180–195 (2, 18, 29, 52, 62)	(2, 18, 29, 52, 62)
Beryllium	5.7 x 10^{-4} (48)		
Nitrogen	280 (82)	0–560 (82)	(82)
Aluminum	2 (65, 70, 10, 31)	0–7 (65, 70)	(65)
Silicon	2000 (5)	0–4900 (5)	(5)
Phosphorus	30 (4)	0–90 (4)	(4)
Scandium	0.04 (31)		
	<0.004 (68)	0.1–18 x 10^{-4} (37)	
	9.6 x 10^{-4} (37)		
Titanium	1 (34)		
Vanadium	2.5 (11, 12)	2.0–3.0 (11, 12)	(11)
Chromium	0.3 (30)	0.23–0.43 (30)	
	0.05 (41)	0.04–0.07 (41)	
Manganese	1.5 (64)	0.2–8.6 (64)	(64, 78)
	0.9 (78)	0.7–1.3 (78)	
		3.0–4.4 (28)	
Iron	6.6 (78).	0.1–62 (78)	(3, 7, 22, 78)
	2.6 (70)	8–13 (22)	
	0.2 (7)	0–7 (28, 70)	
		0.03–2.56 (7)	
Cobalt	0.27 (68)	0.035–4.1 (68)	(68, 69, 78)
	0.032 (78)	<0.005–0.092 (78)	
Nickel	5.4 (68)	0.43–43 (22, 68)	(68, 69, 73)
	1.7 (73)	0.8–2.4 (73)	
		0.13–0.37 (28)	
Copper	2 (28, 80)	0.2–4 (28, 73, 78, 80)	(1, 73, 78, 80)
	1.2 (78)	0.5–27 (1, 9, 22, 38)	
	0.7 (73)		

Table 2.3—8 (continued)
MINOR CONSTITUENTS OF SEA WATER EXCLUDING THE DISSOLVED GASES

Element	Concentration μg/l Average	Range	References on the distribution in the oceans
Zinc	12.3 (78)	3.9–48.4 (78)	(64, 73, 78, 80)
	6.5 (64, 80)	2–18 (64, 80)	
	2 (73)	1–8 (73)	
		29–50 (9)	
Gallium	0.03 (23)	0.023–0.037 (23)	
Germanium	0.05 (12, 27)	0.05–0.06 (12)	
Arsenic	4 (39)	3–6 (39)	
	0.46 (41)	2–35 (61)	
Selenium	0.2 (15, 68)	0.34–0.50 (15)	(68)
		0.052–0.12 (68)	
Rubidium	120 (8, 52, 63, 71)	112–134 (8, 52, 63, 71)	(8, 29, 52, 71)
		86–119 (29)	
Yttrium	0.03 (31)	0.0112–0.0163 (37)	
	0.0133 (37)		
Zirconium	2.6 x 10^{-2} (88)		
Niobium	0.01 (13)	0.01–0.02 (13)	
Molybdenum	10 (41)	0.24–12.2 (9, 85)	
	1 (9)		
Technetium			
Ruthenium	0.0007 (88)		
Rhodium			
Palladium			
Silver	0.29 (68)	0.055–1.5 (68)	(68, 69)
	0.04 (31)		
Cadmium	0.113 (53)	0.02–0.25 (53)	
Indium	<20 (31)		
Tin	0.8 (31)		
Antimony	0.33 (68)	0.18–1.1 (68)	(68)
Tellurium			
Iodine	63 (6)	48–80 (6)	(6)
	44 (41)		
Cesium	0.4 (8, 63)	0.27–0.33 (8)	(8)
		0.48–0.58 (63)	
Barium	20 (8, 17, 19, 81)	5–93 (8, 17, 19, 81)	(19, 81)
Lanthanum	3 x 10^{-3} (33, 37)	1–6 x 10^{-3} (37)	(35–37)
Cerium	14 x 10^{-3} (14)	4–850 x 10^{-3} (14)	(35–37)
	1 x 10^{-3} 37)	0.6–2.8 x 10^{-3} (37)	
Praseodymium	6.4 x 10^{-4} (33, 37)	4.1–15.8 x 10^{-4} (37)	(35–37)
Neodymium	23 x 10^{-4} (33)	13–65 x 10^{-4} (37)	(35–37)
	28 x 10^{-4} (37)		
Promethium			
Samarium	4.2 x 10^{-4} (33)	2.6–10 x 10^{-4} (37)	(35–37)
	4.5 x 10^{-4} (37)		
Europium	1.14 x 10^{-4}	0.9–7.9 x 10^{-4} (37)	(35–37)
	1.3 x 10^{-4} (37)		

Table 2.3—8 (continued) MINOR CONSTITUENTS OF SEA WATER EXCLUDING THE DISSOLVED GASES

Element	Concentration μg/l Average	Range	References on the distribution in the oceans
Gadolinium	6.0×10^{-4} (33) 7.0×10^{-4} (37)	$5.2–11.5 \times 10^{-4}$ (37)	(35–37)
Terbium	1.4×10^{-4} (37)	$0.6–3.6 \times 10^{-4}$ (37)	(35–37)
Dysprosium	7.3×10^{-4} (33) 9.1×10^{-4} (37)	$5.2–14.0 \times 10^{-4}$ (37)	(35–37)
Holmium	2.2×10^{-4} (33, 37)	$1.2–7.2 \times 10^{-4}$ (33, 37)	(33, 35–37)
Erbium	6.1×10^{-4} (33) 8.7×10^{-4} (37)	$6.6–12.4 \times 10^{-4}$ (37)	(35–37)
Thulium	1.3×10^{-4} (33) 1.7×10^{-4} (37)	$0.9–3.7 \times 10^{-4}$ (37)	(35–37)
Ytterbium	5.2×10^{-4} (33) 8.2×10^{-4} (37)	$4.8–28 \times 10^{-4}$ (33, 37)	(33, 35–37)
Lutetium	2.0×10^{-4} (33) 1.5×10^{-4} (37)	$1.2–7.5 \times 10^{-4}$ (33, 37)	(33, 35–37)
Hafnium	80×10^{-4} (68)		
Tantalum	25×10^{-4} (68)		
Tungsten	0.1 (41)		
Rhenium	8.4×10^{-3} (66)		
Osmium			
Indium	1×10^{-4} (88)		
Platinum			
Gold	0.068 (86)	0.004–0.027 (68)	
Mercury	0.03 (31)		
Thallium	<0.01 (31)		
Lead	0.05 (19)	0.02–0.4 (19, 76, 77)	(19, 76, 77)
Bismuth	0.02 (56)	0.015–0.033 (56)	
Polonium			
Astatine			
Francium			
Radium	8×10^{-8} (55)	$4–15 \times 10^{-8}$ (45, 49, 75)	(45, 49, 75)
Actinium			
Thorium	0.05 (31) 0.02 (55) 6×10^{-4} (51, 72) $<7 \times 10^{-5}$ (42)	$2–40 \times 10^{-4}$ (51, 72)	
Protactinium	2×10^{-6} (31) 5×10^{-8} (55)		
Uranium	3 (50, 55, 79)	2–4.7 (50, 55, 79)	(50, 55, 79)

* The numbers in parentheses refer to the citations listed after the table. The concentrations represent the dissolved and particulate forms of the elements.

(Based on compilations of Pytkowicz, R. M. and Kester, D. R., The physical chemistry of seawater, in *Oceanogr. Mar. Biol. Ann. Rev.*, Barnes, H., Ed., 9, 11, 1971. With permission of George Allen and Unwin, Ltd., London.)

Table 2.3—8 (continued)

REFERENCES

1. **Alexander, J. E. and Corcoran, E. F.,** *Limnol. Oceanogr.*, 12, 236, 1967.
2. **Angino, E. E. and Billings, G. K.,** *Geochim. Cosmochim. Acta,* 30, 153, 1966.
3. **Armstrong, F. A. J.,** *J. Mar. Biol. Assoc. U.K.*, 36, 509, 1957.
4. **Armstrong, F. A. J.,** in *Chemical Oceanography,* Vol. 1, Riley, J. P. and Skirrow, G., Eds., Academic Press, London, 1965, 323-364.
5. **Armstrong, F. A. J.,** in *Chemical Oceanography,* Vol. 1, Riley, J. P. and Skirrow, G., Eds., Academic Press, London, 1965, 409-432.
6. **Barkley, R. A. and Thompson, T. G.,** *Deep Sea Res.*, 7, 24, 1960.
7. **Betzer, P. and Pilson, M. E. Q.,** *J. Mar. Res.*, 28, 251, 1970.
8. **Bolter, E., Turekian, K. K., and Schutz, D. F.,** *Geochim. Cosmochim. Acta,* 28, 1459, 1964.
9. **Brooks, R. R.,** *Geochim. Cosmochim. Acta,* 29, 1369, 1965.
10. **Burton, J. D.,** *Nature,* 212, 976, 1966.
11. **Burton, J. D. and Krishnamurty, K.,** *Rep. Challenger Soc.*, 3, 24, 1967.
12. **Burton, J. D. and Riley, J. P.,** *Nature,* 181, 179, 1958.
13. **Carlisle, D. B. and Hummerstone, L. G.,** *Nature,* 181, 1002, 1958.
14. **Carpenter, J. H. and Grant, V. E.,** *J. Mar. Res.*, 25, 228, 1967.
15. **Chau, Y. K. and Riley, J. P.,** *Anal. Chim. Acta,* 33, 36, 1965.
16. **Chester, R.,** *Nature,* 206, 884, 1965.
17. **Chow, T. J. and Goldberg, E. D.,** *Geochim. Cosmochim. Acta,* 20, 192, 1960.
18. **Chow, T. J. and Goldberg, E. D.,** *J. Mar. Res.*, 20, 163, 1962.
19. **Chow, T. J. and Patterson, C. C.,** *Earth and Planet. Sci. Lett.*, 1, 397, 1966.
20. **Chow, T. J. and Tatsumoto, M.,** in *Recent Researches in the Fields of Hydrosphere, Atmosphere, and Nuclear Geochemistry,* Miyake, Y. and Koyama, T., Eds., Maruzen Co., Tokyo, 1964, 179-183.
21. **Chuecas, L. and Riley, J. P.,** *Anal. Chim. Acta,* 35, 240, 1966.
22. **Corcoran, E. F. and Alexander, J. E.,** *Bull. Mar. Sci. Gulf Caribbean,* 14, 594, 1964.
23. **Culkin, F. and Riley, J. P.,** *Nature,* 181, 180, 1958.
24. **Curl, H., Cutshall, N., and Osterberg, C.,** *Nature,* 205, 275, 1965.
25. **Cutshall, N., Johnson, V., and Osterberg, C.,** *Science,* 152, 202, 1966.
26. **Duursma, E. K. and Sevenhuysen, W.,** *Neth. J. Sea Res.*, 3, 95, 1966.
27. **El Wardani, S. A.,** *Geochim. Cosmochim. Acta,* 15, 237, 1958.
28. **Fabricand, B. P., Sawyer, R. R., Ungar, S. G., and Adler, S.,** *Geochim. Cosmochim. Acta,* 26, 1023, 1962.
29. **Fabricand, B. P., Imbimbo, E. S., Brey, M. E., and Weston, J. A.,** *J. Geophys. Res.*, 71, 3917, 1966.
30. **Fukai, R.,** *Nature,* 213, 901, 1967.
31. **Goldberg, E. D.,** in *Chemical Oceanography,* Vol. 1, Riley, J. P. and Skirrow, G., Eds., Academic Press, London, 1965, 163-196.
32. **Goldberg, E. D. and Arrhenius, G. S.,** *Geochim. Cosmochim. Acta,* 13, 153, 1958.
33. **Goldberg, E. D., Koide, M., Schmitt, R. A., and Smith, R. H.,** *J. Geophys. Res.*, 68, 4209, 1963.
34. **Griel, J. V. and Robinson, R. J.,** *J. Mar. Res.*, 11, 173, 1952.
35. **Høgdahl, O.,** Semi Annual Progress Report No. 5, NATO Scientific Affairs Div., Brussels, 1967.
36. **Høgdahl, O.,** Semi Annual Progress Report No. 6, NATO Scientific Affairs Div., Brussels, 1968.
37. **Høgdahl, O., Melsom, S., and Bowen, V. T.,** Trace inorganics in water, in *Advances in Chemistry Series,* No. 73, American Chemical Society, Washington, D.C., 1968, 308-325.
38. **Hood, D. W.,** in *Oceanogr. Mar. Biol. Annu. Rev.*, Vol. 1, Barnes, H., Ed., George Allen and Unwin, Ltd., London, 1963, 129-155.
39. **Ishibashi, M.,** *Rec. Oceanogr. Works Jap.*, 1, 88, 1953.
40. **Johnson, V., Cutshall, N., and Osterberg, C.,** *Water Resour. Res.*, 3, 99, 1967.
41. **Kappanna, A. N., Gadre, G. T., Bhavnagary, H. M., and Joshi, J. M.,** *Curr. Sci. (India),* 31, 273, 1962.
42. **Kaufman, A.,** *Geochim. Cosmochim. Acta,* 33, 717, 1969.
43. **Kester, D. R. and Pytkowicz, R. M.,** *Limnol. Oceanogr.*, 12, 243, 1967.
44. **Kharkar, D. P., Turekian, K. K., and Bertine, K. K.,** *Geochim. Cosmochim. Acta,* 32, 285, 1968.
45. **Koczy, F. F.,** *Proc. Second U.N. Internat. Conf. Peaceful Uses Atomic Energy,* 18, 351, 1958.

46. **Krauskopf, K. B.,** *Geochim. Cosmochim. Acta,* 9, 1, 1956.
47. **Menzel, D. W. and Ryther, J. H.,** *Deep Sea Res.,* 7, 276, 1961.
48. **Merrill, J. R., Lyden, E. F. X., Honda, M., and Arnold, J.,** *Geochim. Cosmochim. Acta,* 18, 108, 1960.
49. **Miyake, Y. and Sugimura, Y.,** in *Studies on Oceanography,* Yoshida, K., Ed., Univ. of Washington Press, Seattle, 1964, 274.
50. **Miyake, Y., Sugimura, Y., and Uchida, T.,** *J. Geophys. Res.,* 71, 3083, 1966.
51. **Moore, W. S. and Sackett, W. M.,** *J. Geophys. Res.,* 69, 5401, 1964.
52. **Morozov, N. P.,** *Oceanology,* 8, 169, 1968.
53. **Mullin, J. B. and Riley, J. P.,** *J. Mar. Res.,* 15, 103, 1956.
54. **Peshchevitskiy, B. I., Anoshin, G. N., and Yereburg, A. M.,** *Dokl. Earth Sci. Sect.,* 162, 205, 1965.
55. **Picciotto, E. E.,** in *Oceanography,* Sears, M., Ed., Amer. Assoc. Adv. Sci., Washington, D.C., 1961, 367.
56. **Portmann, J. E. and Riley, J. P.,** *Anal. Chim. Acta,* 34, 201, 1966.
57. **Putnam, G. L.,** *J. Chem. Educ.,* 30, 576, 1953.
58. **Pytkowicz, R. M.,** *J. Oceanogr. Soc. Jap.,* 24, 21, 1968.
59. **Pytkowicz, R. M. and Kester, D. R.,** *Deep Sea Res.,* 13, 373, 1966.
60. **Pytkowicz, R. M. and Kester, D. R.,** *Limnol. Oceanogr.,* 12, 714, 1967.
61. **Richards, F. A.,** in *Physics and Chemistry of the Earth,* Vol. 2, Ahrens, L. H., Press, F., Rankama, K., and Runcorn, S. K., Eds., Pergamon Press, New York, 1957, 77-128.
62. **Riley, J. P. and Tongudai, M.,** *Deep Sea Res.,* 11, 563, 1964.
63. **Riley, J. P. and Tongudai, M.,** *Chem. Geol.,* 1, 291, 1966.
64. **Rona, E., Hood, D. W., Muse, L., and Buglio, B.,** *Limnol. Oceanogr.,* 7, 201, 1962.
65. **Sackett, W. and Arrhenius, G.,** *Geochim. Cosmochim. Acta,* 26, 955, 1962.
66. **Scadden, E. M.,** *Geochim. Cosmochim. Acta,* 33, 633, 1969.
67. **Schink, D. R.,** *Geochim. Cosmochim. Acta,* 31, 987, 1967.
68. **Schutz, D. F. and Turekian, K. K.,** *Geochim. Cosmochim. Acta,* 29, 259, 1965.
69. **Schutz, D. F. and Turekian, K. K.,** *J. Geophys. Res.,* 70, 5519, 1965.
70. **Simmons, L. H., Monaghan, P. H., and Taggart, M. S.,** *Anal. Chem.,* 25, 989, 1953.
71. **Smith, R. C., Pillai, K. C., Chow, T. J., and Folson, T. R.,** *Limnol. Oceanogr.,* 10, 226, 1965.
72. **Somayajulu, B. L. K. and Goldberg, E. D.,** *Earth Planet. Sci. Lett.,* 1, 102, 1966.
73. **Spencer, D. W. and Brewer, P. G.,** *Geochim. Cosmochim. Acta,* 33, 325, 1969.
74. **Sugawara, K. and Terada, K.,** *Nature,* 182, 250, 1958.
75. **Szabo, B. J.,** *Geochim. Cosmochim. Acta,* 31, 1321, 1967.
76. **Tatsumoto, M. and Patterson, C. C.,** *Nature,* 199, 350, 1963.
77. **Tatsumoto, M. and Patterson, C. C.,** in *Earth Sciences and Meteoritics,* Geiss, J. and Goldberg, E. D., Compilers, North Holland Publ. Co., Amsterdam, 1963, 74-89.
78. **Topping, G.,** *J. Mar. Res.,* 27, 318, 1969.
79. **Torii, T. and Murata, S.,** in *Recent Researches in the Fields of Hydrosphere, Atmosphere, and Nuclear Geochemistry,* Miyake, Y. and Koyama, T., Eds., Maruzen Co., Tokyo, 1964.
80. **Torii, T. and Murata, S.,** *J. Oceanogr. Soc. Jap.,* 22, 56, 1966.
81. **Turekian, K. K. and Johnson, D. G.,** *Geochim. Cosmochim. Acta,* 30, 1153, 1966.
82. **Vaccaro, R. F.,** in *Chemical Oceanography,* Vol. 1, Riley, J. P. and Skirrow, G., Eds., Academic Press, London, 1965, 365-408.
83. **Veeh, H. H.,** *Earth and Planet. Sci. Lett.,* 3, 145, 1967.
84. **Wangersky, P. J. and Gordon, D. C., Jr.,** *Limnol. Oceanogr.,* 10, 544, 1965.
85. **Weiss, H. V. and Lai, M. G.,** *Talanta,* 8, 72, 1961.
86. **Weiss, H. V. and Lai, M. G.,** *Anal. Chim. Acta,* 28, 242, 1963.
87. **Williams, P. M.,** *Limnol. Oceanogr.,* 14, 156, 1969.
88. **Riley, J. P. and Chester, R.,** *Introduction to Marine Chemistry,* Academic Press, London, 1971.

Table 2.3—9
PERCENTAGE DISTRIBUTION OF CARBONIC ACID SPECIES IN SEA WATER AS A FUNCTION OF pH, TEMPERATURE, AND SALINITY

pH	S°/oo	0°C			10°C			20°C			30°C		
		CO_2	HCO_3	CO_3	CO_2	HCO_3	CO_3	CO_2	HCO_3	CO_3	CO_2	HCO_3	CO_3
7.5	30.7	4.42	94.5	1.09	3.77	94.8	1.44	3.29	94.9	1.85	2.99	94.5	2.49
	34.3	4.22	94.6	1.19	3.60	94.8	1.58	3.14	94.8	2.03	2.85	94.3	2.85
	37.9	4.04	94.7	1.31	3.44	94.9	1.69	3.00	94.8	2.22	2.71	94.2	3.12
7.6		3.53	95.1	1.38	3.01	95.2	1.82	2.62	95.0	2.34	2.37	94.5	3.13
		3.37	95.1	1.51	2.87	95.1	1.99	2.50	94.9	2.56	2.26	94.2	3.58
		3.22	95.1	1.66	2.74	95.1	2.13	2.38	94.8	2.80	2.15	93.9	3.92
7.7		2.82	95.4	1.73	2.40	95.3	2.28	2.08	95.0	2.93	1.88	94.2	3.92
		2.70	95.4	1.90	2.29	95.2	2.50	1.99	94.8	3.20	1.79	93.7	4.48
		2.57	95.3	2.08	2.18	95.1	2.67	1.89	94.6	3.51	1.70	93.4	4.89
7.8		2.25	95.6	2.18	1.91	95.2	2.87	1.65	94.7	3.67	1.49	93.6	4.90
		2.14	95.5	2.39	1.82	95.0	3.14	1.57	94.4	4.01	1.41	93.0	5.59
		2.05	95.3	2.62	1.73	94.9	3.35	1.50	94.1	4.39	1.34	92.6	6.10
7.9		1.78	95.5	2.75	1.51	94.9	3.61	1.30	94.1	4.61	1.17	92.7	6.12
		1.70	95.3	3.01	1.43	94.6	3.94	1.24	93.7	5.03	1.11	91.9	6.97
		1.62	95.1	3.30	1.37	94.4	4.21	1.18	93.3	5.49	1.05	91.4	7.59
8.0		1.41	95.1	3.45	1.19	94.3	4.52	1.02	93.2	5.75	0.91	91.5	7.61
		1.34	94.9	3.78	1.13	93.9	4.93	0.97	92.8	6.27	0.86	90.5	8.64
		1.28	94.6	4.13	1.08	93.7	5.26	0.92	92.2	6.84	0.82	89.8	9.40
8.1		1.10	94.6	4.34	0.93	93.4	5.66	0.80	92.0	7.19	0.71	89.8	9.46
		1.05	94.2	4.75	0.88	92.9	6.18	0.76	91.4	7.82	0.67	88.6	10.71
		1.00	93.8	5.19	0.84	92.6	6.59	0.72	90.8	8.51	0.63	87.7	11.63
8.2		0.87	93.7	5.40	0.73	92.3	7.01	0.63	90.5	8.86	0.55	87.8	11.60
		0.83	93.3	5.89	0.69	91.7	7.64	0.59	89.8	9.63	0.52	86.4	13.09
		0.79	92.8	6.44	0.66	91.2	8.14	0.56	89.0	10.47	0.49	85.3	14.18
8.3		0.68	92.6	6.72	0.57	90.7	8.69	0.49	88.6	10.93	0.43	85.4	14.21
		0.65	92.0	7.33	0.54	90.0	9.45	0.46	87.7	11.86	0.40	83.6	15.97
		0.62	91.4	7.99	0.51	89.4	10.05	0.43	86.7	12.85	0.38	82.4	17.25
8.4		0.54	91.2	8.28	0.45	88.9	10.65	0.38	86.3	13.31	0.33	82.5	17.16
		0.51	90.5	9.00	0.42	88.0	11.55	0.36	85.2	14.41	0.31	80.5	19.21
		0.48	89.7	9.80	0.40	87.3	12.27	0.34	84.1	15.58	0.29	79.0	20.69

(Calculated from Lyman, J., Buffer Mechanism of Seawater, Ph.D. thesis, University of California, Los Angeles, 1956. With permission.)

Table 2.3—10
CHANGE IN pH OF SEA WATER FOR RISE OF 1°C

pH	Cl °/oo = 10			Cl °/oo = 15		
	0–20°	10–20°	20–30°	0–20°	10–20°	20–30°
7.4	-0.0087	-0.0084	-0.0069	-0.0088	-0.0087	-0.0076
7.6	92	92	79	95	96	83
7.8	100	101	89	103	105	90
8.0	108	109	94	110	112	94
8.2	114	115	98	115	117	96
8.4	117	117	99	118	118	98

pH	Cl °/oo = 19.5			Cl °/oo = 21		
	0–20°	10–20°	20–30°	0–20°	10–20°	20–30°
7.4	-0.0089	-0.0087	-0.0081	-0.0092	-0.0089	-0.0079
7.6	95	95	91	97	98	88
7.8	104	104	98	106	108	93
8.0	110	109	102	112	114	96
8.2	114	112	103	116	116	98
8.4	116	114	104	118	119	100

Note: The table contains some irregular values.

(From Buch, K. and Nynäs, O., Studien über neuere pH-Methodik mit besonderer Berücksichtigung des Meerwassers, *Acta Academiae Aboensis,* Ser. B, 12, 1939. With permission.)

Table 2.3—11
EFFECT OF PRESSURE ON THE pH OF SEA WATER

Calculated Values of ($pH_1 - pH_p$) at 34.8°/oo Salinity

Temp (°C)	Pressure (atm)	pH at atmospheric pressure				
		7.6	7.8	8.0	8.2	8.4
0	250	0.112	0.107	0.103	0.100	0.098
	500	0.222	0.213	0.205	0.200	0.196
	750	0.330	0.318	0.308	0.300	0.294
	1,000	0.437	0.422	0.409	0.399	0.391
5	250	0.107	0.102	0.098	0.096	0.094
	500	0.212	0.203	0.197	0.192	0.189
	750	0.316	0.304	0.294	0.288	0.283
	1,000	0.417	0.402	0.391	0.383	0.376
10	250	0.102	0.098	0.094	0.092	0.091
	500	0.203	0.195	0.189	0.185	0.182
	750	0.302	0.291	0.283	0.277	0.272
	1,000	0.401	0.387	0.376	0.369	0.362

(From Culberson, C. and Pytkowicz, R. M., Effect of pressure on carbonic acid, boric acid, and the pH in seawater, *Limnol. Oceanogr.*, 13, 403, 1968. With permission.)

Table 2.3—12
PHASES FORMED DURING THE PROGRESSIVE EVAPORATION OF SEA WATER

Stage no.	Density of brine	Weight % of liquid remaining	Principal solid phases deposited	% Total dissolved solids
I	1.026	100	Calcium carbonate and dolomite	1
II	1.140	50	Gypsum ($CaSO_4 \cdot 2H_2O$)	3
III	1.214	10	Halite (NaCl)	70
IV	1.236	3.9	Sodium-magnesium-potassium sulfates and chlorides	26

(From Riley, J. P. and Chester, R., *Introduction to Marine Chemistry*, Academic Press, London, 1971. With permission.)

Table 2.3—13
FACTORS FOR CONVERSION OF NUTRIENT AND OXYGEN CONCENTRATIONS IN SEA WATER

To convert μg Si/kg	to μg - at Si/kg	multiply by	0.03560
	ppb SiO_2		2.1392
	μg Si/l		ρ_{sw}*
	μg - at Si/l		0.03560 ρ_{sw}
To convert μg NO_3–N/kg	to μg - at NO_3–N/kg	multiply by	0.07138
	ppb NO_3		4.4261
	μg - NO_3–N/l		ρ_{sw}
	μg - at NO_3–N/l		0.07138 ρ_{sw}
To convert μg PO_4–P/kg	to μg - at PO_4–P/kg	multiply by	0.03229
	ppb PO_4		3.0665
	μg PO_4–P/l		ρ_{sw}
	μg - at PO_4–P/l		0.03229 ρ_{sw}
To convert ml O_2/l	to μmol O_2/kg	multiply by	44.643/ρ_{sw}
	μg - at O/kg		89.286/ρ_{sw}
	ppm O_2		1.4286/ρ_{sw}
To convert mol/kg	to mol/l	multiply by	ρ_{sw}
	molality		1/1 - 0.001 S °/₀₀

* ρ_{sw} is the density of sea water.

(Compiled by C. Culberson, Department of Oceanography, Oregon State University.)

Table 2.3—14
THE DISSOLVED ORGANIC CONSTITUENTS OF SEA WATER

Specific Dissolved Organic Compounds Identified in Sea Water*

Name of compound and chemical formula	Concentration				Author(s)	Locality
I. Carbohydrates						
Pentoses $C_5H_{10}O_5$		0–8	mg/l		Collier et al. (1950, 1956)	Gulf of Mexico
Pentoses $C_5H_{10}O_5$		0.5	μg/l		Degens et al. (1964)	Pacific off California
Hexoses		14–36	μg/l		Degens et al. (1964)	Pacific off California
Rhamnosides $C_6H_{12}O_5$ Rhamnosides		0.1–0.4	mg/l		Lewis and Rakestraw (1955)	Pacific Ocean coast U.S.A.
Dehydroascorbic acid (ring: O) $COCOCOCHCH(OH)CH_2OH$		0.1	mg/l		Wangersky (1952)	Gulf of Mexico inshore water
II. Proteins and Their Derivatives						
Peptides C:N ratio = 13.8:1					Jeffrey and Hood (1958)	Gulf of Mexico
Polypeptides and polycondensates of:	(a) μg/l	(b) μg/l	(c) μg/l	(d) μg/l		
Glutamic acid $COOH(CH_2)_2CH(NH_2)COOH$		8–13	8–13	0.1–1.8	(a) Park et al. (1962) (by ion-exchange)	Gulf of Mexico
Lysine $NH_2(CH_2)_4CH(NH_2)COOH$	<1	?	trace–3	0.1–0.9	(b) Tatsumoto et al. (1961) (by paper chromatography)	Gulf of Mexico
Glycine NH_2CH_2COOH		–	trace–3	1.2–3.7	(c) Tatsumoto et al. (1961) (by ion-exchange)	Gulf of Mexico
Aspartic acid $COOHCH_2CH(NH_2)COOH$		3–8	trace–3	0.1–1.0	(d) Degens et al. (1964)	Pacific off California

* det = detected.
tr. = trace.
– = not detected.
? = possibly present.

(From Duursma, E. K., in *Chemical Oceanography*, Vol. 1, Riley, J. P. and Skirrow, G., Eds., Academic Press, London, 1965, 450. With permission.)

Table 2.3—14 (continued)
Specific Dissolved Organic Compounds Identified in Sea Water*

Name of compound and chemical formula	Concentration				Author(s)	Locality
Serine $CH_2OHCH(NH_2)COOH$		?	trace–3	1.8–5.6		
Alanine $CH_3CH(NH_2)COOH$		3–8	trace–3	0.7–3.1	(a) Park et al. (1962) (by ion-exchange)	Gulf of Mexico
Leucine $(CH_3)_2CHCH_2CH(NH_2)COOH$	0.5–1	8–13	trace–3	0.9–3.8	(b) Tatsumoto et al. (1961) (by paper chromatography)	Gulf of Mexico
Valine $(CH_3)_2CHCH(NH_2)COOH$		trace–3	trace–3	0.1–1.7	(c) Tatsumoto et al. (1961) (by ion-exchange)	Gulf of Mexico
Cystine $[SCH_2CH(NH_2)COOH]_2$		trace–3	–	0.0–3.8	(d) Degens et al. (1964)	Pacific off California
Isoleucine $CH_3CH_2CH(CH_3)CH(NH_2)COOH$		8–13	trace–3	–		
Leucine $(CH_3)_2CHCH_2CH(NH_2)COOH$		–	–	0.9–3.8		
Ornithine $NH_2(CH_2)_3CH(NH_2)COOH$		–	trace–3	0.2–2.4		
Methionine sulphoxide $CH_3S(:O)CH_2CH_2CH(NH_2)COOH$				–		
Threonine $CH_3CHOHCH(NH_2)COOH$		–	3–8	0.3–1.3		
Tyrosine $HOC_6H_4CH_2CH(NH_2)COOH$		–	trace–3	tr.–0.5		
Phenylalanine $C_6H_5CH_2CH(NH_2)COOH$	<0.5	–	–	0.1–0.9		
Histidine $C_3H_3N_2CH_2CH(NH_2)COOH$		?	trace–3	tr.–2.4		
Arginine $NH_2C(:NH)NH(CH_2)_3CH(NH_2)COOH$		?	trace–3	0.1–0.6		
Proline C_4H_8NCOOH		?	–	0.3–1.4		
Methionine $CH_3SCH_2CH_2CH(NH_2)COOH$		–	trace–3	tr.–0.4		
Tryptophan $C_8H_6NCH_2CH(NH_2)COOH$		–	trace–3	–		

* tr. = trace.
– = not detected.
? = possibly present.

Table 2.3—14 (continued)
Specific Dissolved Organic Compounds Identified in Sea Water*

Name of compound and chemical formula	Concentration		Author(s)	Locality
Glucosamine $C_6H_{13}NO_5$	–	trace–3		
Free amino acids	(e)	(f) μg/l		
Cystine $[SCH_2CH(NH_2)COOH]_2$	det	–	(e) Palmork (1963a)	Norwegian coastal water
Lysine $NH_2(CH_2)_4CH(NH_2)COOH$	det.	0.2–3.1	(f) Degens et al. (1964)	Pacific off California
Histidine $C_3H_3N_2CH_2CH(NH_2)COOH$	det.	0.5–1.7		
Arginine $NH_2C(:NH)NH(CH_2)_3CH(NH_2)COOH$	det.	0.0		
Serine $CH_2OHCH(NH_2)COOH$	det.	2.3–28.4		
Aspartic acid $COOHCH_2CH(NH_2)COOH$	det.	tr.–9.6		
Glycine NH_2CH_2COOH	det.	tr.–37.6		
Hydroxyproline $C_4H_7N(OH)COOH$	det.	tr.–2.8		
Glutamic acid $COOH(CH_2)_2CH(NH_2)COOH$	det.	1.4–6.8		
Threonine $CH_3CHOHCH(NH_2)COOH$	det.	2.8–11.8		
α-Alanine $CH_3CH(NH_2)COOH$	det.			
Proline C_4H_8NCOOH	det.	0.0		
Tyrosine $HOC_6H_4CH_2CH(NH_2)COOH$	det.	tr.–5.0		
Tryptophan $C_8H_6NCH_2CH(NH_2)COOH$	det.	–		

* det = detected.
tr. = trace.
– = not detected.

Table 2.3—14 (continued)
Specific Dissolved Organic Compounds Identified in Sea Water*

Name of compound and chemical formula	Concentration		Author(s)	Locality
Methionine $CH_3SCH_2CH_2CH(NH_2)COOH$	det.	–		
Valine $(CH_3)_2CHCH(NH_2)COOH$	det.	0.3–2.7		
Phenylalanine $C_6H_5CH_2(NH_2)COOH$	det.	tr.–2.4		
Isoleucine $CH_3CH_2CH(CH_3)CH(NH_2)COOH$	det.	–		
Leucine $(CH_3)_2CHCH_2CH(NH_2)COOH$	det.	0.5–5.5		
Free compounds				
Uracil NHCONHCOCH:CH (ring)		det.	Belser (1959, 1963)	Pacific coast near La Jolla
Isoleucine $CH_3CH_2CH(CH_3)CH(NH_2)COOH$		det.		
Methionine $CH_3SCH_2CH_2CH(NH_2)COOH$		det.		
Histidine $C_3H_3N_2CH_2CH(NH_2)COOH$		det.		
Adenine $C_5H_3N_4NH_2$		det.		
Peptone		det.		
Threonine $CH_3CHOHCH(NH_2)COOH$		det.		
Tryptophan $C_8H_6NCH_2CH(NH_2)COOH$		det.		
Glycine NH_2CH_2COOH		det.		
Purine $C_5H_4N_4$		det.		
Urea CH_4ON_2		det.	Degens et al. (1964)	Pacific off California

* det. = detected.
tr. = trace.
– = not detected.

Table 2.3—14 (continued)

Specific Dissolved Organic Compounds Identified in Sea Water

Name of compound and chemical formula	Concentration			Author(s)	Locality
III. Aliphatic Carboxylic and Hydroxycarboxylic Acids					
	mg/l (0–200 m)	mg/l (200–600 m)	mg/l (>600 m)		
Lauric acid $CH_3(CH_2)_{10}COOH$	0.01–0.32	0.01–0.28	0–0.28	Slowey et al. (1962)	Coastal waters of Gulf of Mexico
Myristic acid $CH_3(CH_2)_{12}COOH$	0.01–0.10	0.01–0.05	0–0.07		
Myristoleic acid $CH_3(CH_2)_3CH{:}CH(CH_2)_7COOH$	traces–0.02	0.01–0.03	0–0.05		
Palmitic acid $CH_3(CH_2)_{14}COOH$	0.01–0.17	0.03–0.42	0–0.38		
Palmitoleic acid $CH_3(CH_2)_5CH{:}CH(CH_2)_7COOH$	0.02–0.16	0.02–0.16	0–0.21		
Stearic acid $CH_3(CH_2)_{16}COOH$	0.04–0.09	0.02–0.13	0–0.10		
Oleic acid $CH_3(CH_2)_7CH{:}CH(CH_2)_7COOH$	0.01	0.02	0		
Linoleic acid $CH_3(CH_2)_4CH{:}CHCH_2CH{:}CH(CH_2)_7COOH$	0.01	0.01	0		
		mg/l (1000–2500 m)			
Fatty acids with:					
12 C-atoms		0.0003–0.02		Williams (1961)	Pacific Ocean coastal water
14 C-atoms		0.0004–0.043			
16 C-atoms		0.0027–0.0209			
16 C-atoms + 1 double bond		0.0003–0.003			
18 C-atoms		0.0037–0.0222			
18 C-atoms + 1 double bond		0.0083			
18 C-atoms + 2 double bonds		0.0000–0.0029			
20 C-atoms		traces–0.0081			
22 C-atoms		traces–0.0014			

Table 2.3—14 (continued)
Specific Dissolved Organic Compounds Identified in Sea Water

Name of compound and chemical formula	Concentration		Author(s)	Locality
	mg/l			
Acetic acid CH_3COOH	<1.0		Koyama and Thompson (1959)	Pacific Ocean
Lactic acid $CH_3CH(OH)COOH$				
Glycolic acid $HOCH_2COOH$				
Malic acid $HOOCCH(OH)CH_2COOH$	0.28		Creac'h (1955)	Atlantic coastal water
Citric acid $HOOCCH_2C(OH)(COOH)CH_2COOH$	0.14			
Carotenoids and brownish-waxy or fatty matter	2.5		Johnston (1955)	North Sea
			Wilson and Armstrong (1955)	English Channel
IV. Biologically Active Compounds (see also Provasoli, 1963)				
Organic Fe compound(s)	3.4–1.6	mμg/l	Harvey (1925)	Deep sea water
Vitamin B_{12} (Cobalamin) $C_{63}H_{88}O_{14}N_{14}PCo$			Vishniak and Riley (1961)	Long Island Sound
Vitamin B_{12}	0.2	mμg/l (summer)	Cowey (1956)	Oceanic surface water
	2.0	mμg/l (winter)		
Vitamin B_{12}	0.2–5.0	mμg/l	Daisley and Fisher (1958)	
Vitamin B_{12}	0–2.6	mμg/l	Kashiwada et al. (1957)	North Pacific Ocean
Vitamin B_{12}	0–0.03	mμg/l	Menzel and Spaeth (1962)	Sargasso Sea 0–05 m.
Thiamine (Vitamin B_1) $C_{12}H_{17}ON_4SCl_2$	0–20	mμg/l	Cowey (1956)	Surface water, possibly from land drainage
Plant hormones (auxins)	3.41	mμg/l	Bentley (1960)	North Sea near Scotland
V. Humic Acids				
"Gelbstoffe" (Yellow substances) Melanoidin-like			Kalle (1949, 1962)	Coastal waters
			Jerlov (1955)	
			Armstrong and Boalch (1961a,b)	

Table 2.3—14 (continued)
Specific Dissolved Organic Compounds Identified in Sea Water

Name of compound and chemical formula	Concentration	Author(s)	Locality
	VI. Phenolic Compounds		
p-Hydroxybenzoic acid HOC_6H_5COOH	1 – 3 μg/l	Degens et al. (1964)	Pacific off California
Vanillic acid $CH_3(HO)C_6H_3COOH$	1 – 3 μg/l		
Syringic acid $(CH_3O)_2(HO)C_6H_2COOH$	1 – 3 μg/l		
	VII. Hydrocarbons		
Pristane: (2, 6, 10, 14-tetramethylpentadecane)	trace	Blumer et al. (1963)	Cape Cod Bay

Table 2.3—14 (continued)

REFERENCES

1. Armstrong, F. A. J. and Boalch, G. T., *Nature (Lond.)*, 192, 858, 1961a.
2. Armstrong, F. A. J. and Boalch, G. T., *J. Mar. Biol. Assoc. U.K.*, 41, 591, 1961b.
3. Belser, W. L., *Proc. Natl. Acad. Sci., Wash.*, 45, 1533, 1959.
4. Belser, W. L., in *The Sea*, Hill, M. N., Ed., Vol. II, Wiley-Interscience, New York, 1963, 220-231.
5. Bentley, Joyce A., *J. Mar. Biol. Assoc. U.K.*, 39, 433, 1960.
6. Blumer, M., Mullin, M. M., and Thomas, D. W., *Science*, 140, 974, 1963.
7. Collier, A., *Spec. Sci. Rep. U.S. Fish Wildl.*, 178, 7, 1956.
8. Collier, A., Ray, S. M., and Magnitzky, A. W., *Science*, 111, 151, 1950.
9. Cowey, C. B., *J. Mar. Biol. Assoc. U.K.*, 35, 609, 1956.
10. Creac'h, P., *C. R. Acad. Sci., (Paris)*, 240, 2551, 1955.
11. Daisley, K. W. and Fisher, L. R., *J. Mar. Biol. Assoc. U.K.*, 37, 683, 1958.
12. Degens, E. T., Reuter, J. H., and Shaw, K. N. F., *Geochim. Cosmochim. Acta*, 28, 45, 1964.
13. Harvey, H. W., *J. Mar. Biol. Assoc. U.K.*, 13, 953, 1925.
14. Jeffrey, L. M. and Hood, D. W., *J. Mar. Res.*, 17, 247, 1958.
15. Jerlov, N. G., *Göteb. Vetensk Samh. Handl.*, F.6. B.6. (14), 1955.
16. Johnston, R., *J. Mar. Biol. Assoc. U.K.*, 34, 185, 1955.
17. Kalle, K., *Dtsch. Hydrogr. Z.*, 2, 117, 1949.
18. Kalle, K., *Kiel. Meeresforsch.*, 18, 128, 1962.
19. Kashiwada, K., Kakimoto, D., Morita, T., Kanazawa, A., and Kawagoe, K., *Bull. Jap. Soc. Sci. Fish.*, 22, 637,1957.
20. Koyama, T. and Thompson, T. G., *Preprints International Oceanographic Congress, 1959*, American Association for Advancement of Science, Washington, D.C., 1959, 925.
21. Lewis, G. J. and Rakestraw, N. W., *J. Mar. Res.*, 14, 253, 1955.
22. Menzel, D. W. and Spaeth, J. P., *Limnol. Oceanogr.*, 7, 151, 1962.
23. Palmork, K. H., *Acta Chem. Scand.*, 17, 1456, 1963a.
24. Park, K., Williams, W. T., Prescott, J. M., and Hood, D. W., *Science*, 138, 531, 1962.
25. Provasoli, L., in *The Sea*, Hill, M. N., Ed., Vol. II, Wiley-Interscience, New York, 1963, 165-219.
26. Slowey, J. F., Jeffrey, L. M., and Hood, D. W., *Geochim. Cosmochim. Acta*, 26, 607, 1962.
27. Tatsumoto, M., Williams, W. T., Prescott, J. M., and Hood, D. W., *J. Mar. Res.*, 19, 89, 1961.
28. Vishniac, H. S. and Riley, G. A., *Limnol. Oceanogr.*, 6, 36, 1961.
29. Wangersky, P. J., *Science*, 115, 685, 1952.
30. Williams, P. M., *Nature (Lond.)*, 189, 219, 1961.
31. Wilson, D. P. and Armstrong, F. A. J., *J. Mar. Biol. Assoc. U.K.*, 31, 335, 1952.

Table 2.3—15
VALUES OF THE FREE ACTIVITY COEFFICIENTS OF THE MAJOR IONS IN SEA WATER*

Ion	Free activity coefficient
Na^+	0.71
K^+	0.63
Mg^{2+}	0.29
Ca^{2+}	0.26
Sr^{2+}	0.24
Cl^-	0.63
Br^-	0.65
SO_4^{2-}	0.21
F^-	0.68
HCO_3^-	0.68
CO_3^{2-}	0.20

* 34.8°/oo salinity, 25°C; based on the MacInnes assumption ($\gamma_K = \gamma_{Cl}$).

(Compiled by D. R. Kester, Graduate School of Oceanography, University of Rhode Island.)

Table 2.3—16
VALUES OF THE TOTAL ACTIVITY COEFFICIENTS OF THE MAJOR IONS IN SEA WATER*

Ion	Total activity coefficient	Ref.
Na^+	0.67	1
K^+	0.62	3
Mg^{2+}	0.33	4
Ca^{2+}	0.21	2
Sr^{2+}	0.21	3
Cl^-	0.63	3
Br^-	0.65	3
SO_4^{2-}	0.082	3
F^-	0.34	5
HCO_3^-	0.55	2
CO_3^{2-}	0.022	2

* 34.8°/₀₀ salinity, 25°C.

(Compiled by D. R. Kester, Graduate School of Oceanography, University of Rhode Island.)

REFERENCES

1 **Platford, R. F.,** *J. Fish. Res. Board Can.,* 22, 885, 1965.
2 **Berner, R. A.,** *Geochim. Cosmochim. Acta,* 29, 947, 1965.
3 Calculated by D. R. Kester.
4 **Thompson, M. E.,** *Science,* 153, 966, 1966.
5 **Elgquist, B.,** Rep. Chem. Sea Water, No. 7, Univ. Göteborg, Sweden, 1969.

Table 2.3—17
APPARENT DISSOCIATION CONSTANTS

HF and HSO_4^- and the H^+ Species in Sea Water at 25°C and 1 atm

Salinity (°/₀₀)	$K'_{HF} \times 10^3$	$K'_{HSO_4^-} \times 10^2$
26.7	2.06 ± 0.03	6.68 ± 0.01
34.6	2.47 ± 0.05	8.16 ± 0.13

Note: $K'_{HF} = {}^aH \times [F^-]/[HF]$ $K'_{HSO_4} = {}^aH \times [SO_4^{2-}]/[HSO_4^-]$

at 34.6°/₀₀ salinity $[H^+]_{free} = 0.74 [H^+]_{total}$

$[HSO_4^-] = 0.24\ [H^+]_{total}$

$[HF] = 0.02\ [H^+]_{total}$.

(Based on Culberson, C., Pytkowicz, R. M., and Hawley, J. E., *J. Mar. Res.,* 28, 15, 1970. With permission.)

Table 2.3—18
RESIDENCE TIMES OF THE ELEMENTS IN SEA WATER

Element	Years	Element	Years	Element	Years
Li	2.0×10^7	Zn	1.8×10^5	Nd	270
Be	150	Ga	1.4×10^3	Sm	180
Na	2.6×10^8	Ge	7.0×10^3	Eu	300
Mg	4.5×10^7	Rb	2.7×10^5	Gd	260
Al	100	Sr	1.9×10^7	Dy	460
Si	8.0×10^3	Y	7.5×10^3	Ho	530
K	1.1×10^7	Nb	300	Er	690
Ca	8.0×10^6	Mo	5.0×10^5	Tm	1800
Sc	5.6×10^3	Ag	2.1×10^6	Yb	530
Ti	160	Cd	5.0×10^5	Lu	450
V	1.0×10^4	Sm	1.0×10^5	W	1000
Cr	350	Sb	3.5×10^5	Au	5.6×10^5
Mn	1400	Cs	4.0×10^4	Hg	4.2×10^4
Fe	140	Ba	8.4×10^4	Pb	2000
Co	1.8×10^4	La	440	Bi	4.5×10^4
Ni	1.8×10^4	Ce	80	Th	350
Cu	5.0×10^4	Pr	320	U	5.0×10^5

(From Goldberg, E. D, in *Chemical Oceanography*, Vol. 1, Riley, J. P. and Skirrow, G., Eds., Academic Press, London, 1965, 163. With permission.)

Table 2.3—19
FRACTION OF INORGANIC PHOSPHATE

A. Pure Water

pH	$H_2PO_4^-$	HPO_4^{2-}	PO_4^{3-}
7.2	0.503	0.479	0.000
7.6	0.287	0.713	0.000
8.0	0.138	0.862	0.000
8.4	0.060	0.940	0.000

B. 0.68 Molar NaCl

pH	$H_2PO_4^-$	HPO_4^{2-}	PO_4^{3-}
7.2	0.133	0.867	0.000
7.6	0.058	0.942	0.000
8.0	0.024	0.975	0.001
8.4	0.010	0.988	0.002

C. 33°/oo Salinity Sea Water

pH	$H_2PO_4^-$	HPO_4^{2-}	PO_4^{3-}
7.2	0.066	0.915	0.020
7.6	0.026	0.923	0.050
8.0	0.010	0.871	0.119
8.4	0.003	0.741	0.255

Note: As each species in A. the absence of ionic interactions, B. the presence of ionic strength effects, and C. the presence of complexes and ionic strength of sea water.

(From Kester, D. R. and Pytkowicz, R. M., *Limnol. Oceanogr.*, 12, 243, 1967. With permission.)

Table 2.3—20
CHEMICAL SPECIES IN SEA WATER

Element	Chemical form	Reference
Hydrogen	H_2O	8
Helium	He(g)	8
Lithium	Li^+	19
Beryllium		19
Boron	$B(OH)_3$, $B(OH)_4^-$	3, 14
Carbon	HCO_3^-, CO_3^{2-}, CO_2, $MgHCO_3^+$, $NaHCO_3°$, $MgCO_3°$, organic coumpounds	3, 4, 7, 12, 14
Nitrogen	NO_3^-, NO_2^-, NH_3, N_2 (g), organic compounds	20
Oxygen	H_2O, O_2 (g), SO_4^{2-}, organic compounds	8
Fluorine	F^-, MgF^+, CaF^+	6
Neon	Ne(g)	8
Sodium	Na^+, $NaSO_4^-$, $NaHCO_3°$	12, 13, 17
Magnesium	Mg^{2+}, $MgSO_4°$, $MgHCO_3^+$, $MgCO_3°$	7, 16, 18
Aluminum	$Al(OH)_3$	19
Silicon	$Si(OH)_4$, $SiO(OH)_3^-$	1
Phosphorus	$H_2PO_4^-$, HPO_4^{2-}, PO_4^{3-}	11
Sulfur	SO_4^{2-}, $NaSO_4^-$, $MgSO_4°$, $CaSO_4°$	12, 13, 17
Chlorine	Cl^-	12, 13
Argon	Ar(g)	8
Potassium	K^+, KSO_4^-	7, 12, 13
Calcium	Ca^{2+}, $CaSO_4°$, $CaHCO_3^+$	7, 12, 13
Scandium		19
Titanium	$Ti(OH)_4$	19
Vandium	$VO_2(OH)_3^{2-}$	19
Chromium	$Cr(OH)_2^+$, $CrOH^{2+}$, CrO_2^-, CrO_4^{2-}, $HCrO_4^-$, H_2CrO_4	5
Manganese	Mn^{2+}, $MnSO_4°$, $Mn(OH)_{3,4}$	8, 19
Iron	$Fe(OH)_3$, $Fe(OH)_2^+$	10
Cobalt	Co^{2+}, $CoSO_4°$	8
Nickel	Ni^{2+}, $NiSO_4°$	8
Copper	Cu^{2+}, $CuSO_4°$, $CuOH^+$	8, 19
Zinc	Zn^{2+}, $ZnSO_4°$, $ZnOH^+$	2, 8
Gallium	–	
Germanium	$Ge(OH)_4°$, $GeO(OH)_3^-$	8
Arsenic	$H_3AsO_4^-$, $H_2AsO_4^-$, $HAsO_4^{2-}$, AsO_4^{3-}	8
Selenium	SeO_4^{2-}	19
Bromine	Br^-	8
Krypton	Kr(g)	8
Rubidium	Rb^+	19
Strontium	Sr^{2+}, $SrSO_4°$	8
Yttrium	–	
Zirconium	–	
Niobium	–	
Molybdenum	MoO_4^{2-}	19
Technetium	–	
Ruthenium	–	
Rhodium	–	

(Compiled by D. R. Kester, Graduate School of Oceanography, University of Rhode Island.)

Table 2.3—20 (continued)
CHEMICAL SPECIES IN SEA WATER

Element	Chemical form	Reference
Palladium	–	
Silver	$AgCl_2^-$, $AgCl_3^{2-}$	8
Cadmium	$CdCl^+$, Cd^{2+}, $CdSO_4{}^\circ$	2, 9
Indium	–	
Tin	–	
Antimony	–	
Tellurium	–	
Iodine	IO_3^-, I^-	8
Zenon	Xe(g)	8
Cesium	Cs^+	19
Barium	Ba^{2+}, $BaSO_4{}^\circ$	8
Lanthanum	La^{3+}, $La(OH)^{2+}$	19
Cerium	–	
Praseodymium	–	
Neodymium	–	
Promethium	–	
Samarium	–	
Europium	–	
Gadolinium	–	
Terbium	–	
Dysprosium	–	
Holmium	–	
Erbium	–	
Thulium	–	
Ytterbium	–	
Lutetium	–	
Hafnium	–	
Tantalum	–	
Tungsten	WO_4^{2-}	19
Rhenium	–	
Osmium	–	
Iridium	–	
Platinum	–	
Gold	$AuCl_4^-$, $AuCl_2^-$	8, 15, 19
Mercury	$HgCl_3^-$, $HgCl_4^{2-}$	8
Thallium	Tl^+	8
Lead	Pb^{2+}, $PbSO_4{}^\circ$, $PbOH^+$	8, 19
Bismuth	–	
Polonium	–	
Astatine	–	
Radon	Rn(g)	8
Francium	–	
Radium	Ra^{2+}, $RaSO_4{}^\circ$	8
Actinium	–	
Thorium	–	
Protactinium	–	
Uranium	$UO_2(CO_3)_3^{4-}$	8

Table 2.3—20 (continued)

REFERENCES

1. **Armstrong, F. A. J.,** in *Chemical Oceanography,* Vol. 1, Riley, J. P. and Skirrow, G., Eds., Academic Press, London, 1965, 409.
2. **Branica, M., Barić, M., and Jeftić, L.,** *Rapp. P.-V. Reun. Cons. Perm. Int. Explor. Scient. Mer Médit.,* 19, 929, 1969.
3. **Culberson, C. and Pytkowicz, R. M.,** *Limnol. Oceanogr.,* 13, 403, 1968.
4. **Duursma, E. K.,** in *Chemical Oceanography,* Vol. 1, Riley, J. P. and Skirrow, G., Eds., Academic Press, London, 1965, 433-475.
5. **Elderfield, H.,** *Earth Planet. Sci. Lett.,* 9, 10, 1970.
6. **Elgquist, B.,** Rep. Chem. Seawater, No. 7, Univ. Göteborg, Sweden, 1969.
7. **Garrels, R. M. and Thompson, M.,** *Am. J. Sci.,* 260, 57, 1962.
8. **Goldberg, E. D.,** in *The Sea,* Vol. 2, Hill, M. N., Ed., Wiley-Interscience, New York, 1963, 3-25.
9. **Goldberg, E. D.,** in *Chemical Oceanography,* Vol. 1, Riley, J. P. and Skirrow, G., Eds., Academic Press, London, 1965, 163-196.
10. **Horne, R. A.,** *Marine Chemistry,* John Wiley & Sons, New York, 1969.
11. **Kester, D. R. and Pytkowicz, R. M.,** *Limnol. Oceanogr.,* 12, 243, 1967.
12. **Kester, D. R. and Pytkowicz, R. M.,** *Limnol. Oceanogr.,* 14, 686, 1969.
13. **Kester, D. R. and Pytkowicz, R. M.,** *Geochim. Cosmochim. Acta,* 34, 1039, 1970.
14. **Lyman, J.,** Buffer Mechanism of Seawater, Ph.D. Thesis, University of California, Los Angeles, 1956.
15. **Peshchevitskiy, B. I., Anoshin, G. N., and Yereberg, A. M.,** *Dokl. Earth Sci. Sect.,* 162, 205, 1963.
16. **Pytkowicz, R. M. and Gates, R.,** *Science,* 161, 690, 1968.
17. **Pytkowicz, R. M. and Kester, D. R.,** *Am. J. Sci.,* 267, 217, 1969.
18. **Pytkowicz, R. M. and Kester, D. R.,** *Oceanogr. Mar. Biol. Annu. Rev.,* Barnes, H., Ed., George Allen and Unwin, London, 9, 11, 1971.
19. **Sillen, L. G.,** in *Oceanography,* Sears, M., Ed., Assoc. Adv. Sci., Washington, D.C., 1959, 549-581.
20. **Vaccaro, R. F.,** in *Chemical Oceanography,* Vol. 1, Riley J. P. and Skirrow, G., Eds., Academic Press, London, 1965, 365-408.

2.4. SOLUBILITIES OF GASES IN SEA WATER

Table 2.4—1
INTERPOLATED VALUES OF OXYGEN SOLUBILITY (ml/l)*

Temp (°C)	Chlorinity (°/₀₀) 0.0	2.0	4.0	6.0	8.0	10.0	12.0	14.0	16.0	18.0	20.0
0.5	10.10	9.84	9.59	9.35	9.12	8.89	8.67	8.46	8.26	8.06	7.87
1.0	9.96	9.71	9.47	9.24	9.01	8.79	8.57	8.36	8.16	7.96	7.77
2.0	9.68	9.45	9.22	9.00	8.78	8.57	8.37	8.16	7.96	7.77	7.58
3.0	9.41	9.19	8.98	8.77	8.56	8.36	8.16	7.96	7.77	7.58	7.40
4.0	9.16	8.95	8.74	8.54	8.34	8.15	7.95	7.76	7.58	7.40	7.22
5.0	8.91	8.71	8.51	8.32	8.13	7.94	7.75	7.57	7.39	7.22	7.04
6.0	8.68	8.49	8.29	8.11	7.92	7.74	7.56	7.39	7.21	7.04	6.88
7.0	8.46	8.27	8.09	7.90	7.73	7.55	7.38	7.21	7.04	6.88	6.72
8.0	8.26	8.07	7.89	7.71	7.54	7.37	7.20	7.04	6.88	6.72	6.57
9.0	8.06	7.88	7.70	7.53	7.36	7.19	7.03	6.87	6.72	6.57	6.42
10.0	7.88	7.70	7.53	7.36	7.19	7.03	6.87	6.72	6.57	6.42	6.28
11.0	7.71	7.53	7.36	7.19	7.03	6.87	6.72	6.57	6.42	6.28	6.15
12.0	7.54	7.37	7.20	7.03	6.87	6.72	6.57	6.42	6.28	6.15	6.02
13.0	7.39	7.21	7.04	6.88	6.72	6.57	6.43	6.28	6.15	6.02	5.89
14.0	7.23	7.06	6.90	6.74	6.58	6.43	6.29	6.15	6.02	5.89	5.77
15.0	7.07	6.91	6.75	6.60	6.45	6.30	6.16	6.03	5.90	5.78	5.66
16.0	6.92	6.76	6.61	6.46	6.31	6.18	6.04	5.91	5.78	5.66	5.55
17.0	6.77	6.62	6.47	6.33	6.19	6.05	5.92	5.80	5.67	5.55	5.44
18.0	6.62	6.48	6.34	6.20	6.07	5.94	5.81	5.69	5.57	5.45	5.34
19.0	6.49	6.35	6.21	6.08	5.95	5.82	5.70	5.58	5.46	5.35	5.24
20.0	6.36	6.22	6.09	5.96	5.84	5.71	5.59	5.48	5.37	5.26	5.15
21.0	6.23	6.10	5.97	5.85	5.72	5.61	5.49	5.38	5.27	5.16	5.06
22.0	6.11	5.98	5.86	5.74	5.62	5.50	5.39	5.28	5.17	5.07	4.97
23.0	6.00	5.87	5.75	5.63	5.51	5.40	5.29	5.18	5.08	4.98	4.89
24.0	5.88	5.76	5.64	5.52	5.41	5.30	5.19	5.09	4.99	4.90	4.80
25.0	5.77	5.65	5.54	5.42	5.31	5.21	5.10	5.00	4.90	4.81	4.72
26.0	5.67	5.55	5.44	5.33	5.22	5.11	5.01	4.91	4.82	4.73	4.64
27.0	5.57	5.45	5.34	5.23	5.13	5.02	4.92	4.83	4.74	4.65	4.56
28.0	5.47	5.36	5.25	5.14	5.04	4.94	4.84	4.75	4.65	4.57	4.48
29.0	5.37	5.26	5.16	5.05	4.95	4.85	4.76	4.66	4.57	4.49	4.40
30.0	5.28	5.17	5.07	4.97	4.87	4.77	4.68	4.59	4.50	4.41	4.33
31.0	5.19	5.09	4.98	4.88	4.79	4.69	4.60	4.51	4.43	4.34	4.26
32.0	5.11	5.00	4.90	4.80	4.71	4.62	4.53	4.44	4.36	4.27	4.20
33.0	5.02	4.92	4.82	4.73	4.63	4.54	4.45	4.37	4.29	4.21	4.13
34.0	4.94	4.84	4.74	4.65	4.56	4.47	4.38	4.30	4.22	4.14	4.07
35.0	4.85	4.76	4.67	4.57	4.49	4.40	4.32	4.23	4.16	4.08	4.00
36.0	4.77	4.68	4.59	4.50	4.41	4.33	4.25	4.17	4.09	4.02	3.94

* From an atmosphere of 20.94% O_2 and 100% relative humidity.

(From Carpenter, J. H., New measurements of oxygen solubility in pure and natural water, *Limnol. Oceanogr.*, 11, 264, 1966. With permission.)

Table 2.4—2
INTERPOLATED VALUES OF OXYGEN SOLUBILITY (μg-at./l)*

Temp (°C)	Chlorinity (°/₀₀) 0.0	2.0	4.0	6.0	8.0	10.0	12.0	14.0	16.0	18.0	20.0
0.5	902	879	857	835	814	794	774	756	737	720	703
1.0	889	867	846	825	804	785	765	747	729	711	694
2.0	864	844	823	804	784	765	747	729	711	694	677
3.0	840	821	802	783	764	746	728	711	694	677	660
4.0	817	799	780	762	745	727	710	693	677	660	644
5.0	796	778	760	743	725	709	692	676	660	644	629
6.0	775	758	741	724	707	691	675	659	644	629	614
7.0	756	739	722	706	690	674	659	643	629	614	600
8.0	737	721	704	689	673	658	643	628	614	600	586
9.0	720	704	688	672	657	642	628	614	600	586	573
10.0	703	688	672	657	642	627	613	600	586	573	561
11.0	688	672	657	642	627	613	600	586	573	561	549
12.0	674	658	643	628	614	600	586	573	561	549	537
13.0	659	644	629	614	600	587	574	561	549	537	526
14.0	645	630	616	601	588	574	562	549	538	526	515
15.0	631	617	603	589	576	563	550	538	527	516	505
16.0	618	603	590	577	564	551	539	528	516	506	495
17.0	604	591	578	565	553	540	529	517	507	496	486
18.0	591	578	566	554	542	530	519	508	497	487	477
19.0	579	567	555	543	531	520	509	498	488	478	468
20.0	568	556	544	532	521	510	499	489	479	469	460
21.0	556	545	533	522	511	500	490	480	470	461	452
22.0	546	534	523	512	501	491	481	471	462	453	444
23.0	535	524	513	502	492	482	472	463	454	445	436
24.0	525	514	504	493	483	473	464	455	446	437	429
25.0	516	505	494	484	474	465	455	447	438	430	421
26.0	506	496	485	475	466	457	447	439	430	422	414
27.0	497	487	477	467	458	448	440	431	423	415	407
28.0	488	478	468	459	450	441	432	424	415	408	400
29.0	480	470	460	451	442	433	425	416	408	401	393
30.0	471	462	453	443	435	426	418	410	402	394	387
31.0	464	454	445	436	427	419	411	403	395	388	381
32.0	456	447	438	429	420	412	404	396	389	382	375
33.0	448	439	430	422	414	406	398	390	383	376	369
34.0	441	432	423	415	407	399	391	384	377	370	363
35.0	433	425	417	408	400	393	385	378	371	364	358
36.0	426	418	410	402	394	387	379	372	365	359	352

* From an atmosphere of 20.94% O_2 and 100% relative humidity.

(From Carpenter, J. H., New measurements of oxygen solubility in pure and natural water, *Limnol. Oceanogr.*, 11, 264, 1966. With permission.)

Table 2.4—3
INTERPOLATED VALUES OF NITROGEN SOLUBILITY

Temp (°C)	Chlorinity (°/₀₀) 0	2	4	6	8	10	12	14	16	18	20
−2	–	–	–	–	–	–	–	–	–	–	14.74
−1	–	–	–	–	–	–	16.13	15.69	15.26	14.82	14.39
0	18.39	17.89	17.45	17.01	16.58	16.16	15.73	15.31	14.89	14.47	14.05
1	17.92	17.43	17.00	16.58	16.16	15.76	15.35	14.93	14.53	14.13	13.72
2	17.48	16.99	16.57	16.17	15.76	15.37	14.97	14.58	14.18	13.79	13.41
3	17.05	16.57	16.17	15.77	15.38	15.00	14.62	14.24	13.86	13.48	13.11
4	16.64	16.17	15.78	15.39	15.01	14.65	14.28	13.90	13.54	13.18	12.81
5	16.25	15.78	15.41	15.03	14.66	14.31	13.95	13.59	13.24	12.89	12.53
6	15.88	15.42	15.05	14.68	14.33	13.98	13.63	13.29	12.94	12.61	12.27
7	15.51	15.07	14.71	14.35	14.01	13.67	13.34	13.00	12.67	12.34	12.01
8	15.17	14.73	14.37	14.03	13.70	13.37	13.05	12.72	12.40	12.08	11.76
9	14.84	14.41	14.06	13.73	13.40	13.09	12.77	12.45	12.14	11.84	11.53
10	14.52	14.09	13.75	13.44	13.12	12.81	12.50	12.20	11.90	11.59	11.29
11	14.21	13.80	13.47	13.16	12.85	12.55	12.25	11.95	11.66	11.37	11.08
12	13.91	13.51	13.19	12.88	12.58	12.29	12.01	11.71	11.43	11.15	10.87
13	13.63	13.23	12.92	12.62	12.33	12.05	11.77	11.49	11.21	10.94	10.66
14	13.36	12.97	12.67	12.37	12.09	11.81	11.54	11.27	11.00	10.73	10.47
15	13.10	12.71	12.41	12.13	11.85	11.59	11.32	11.06	10.79	10.53	10.27
16	12.84	12.47	12.18	11.90	11.63	11.37	11.11	10.85	10.60	10.35	10.09
17	12.60	12.23	11.95	11.68	11.42	11.16	10.91	10.65	10.41	10.16	9.91
18	12.37	12.00	11.73	11.47	11.21	10.96	10.71	10.46	10.23	9.98	9.75
19	12.14	11.79	11.51	11.26	11.01	10.76	10.53	10.28	10.05	9.82	9.58
20	11.92	11.57	11.31	11.06	10.82	10.58	10.34	10.11	9.88	9.65	9.42
21	11.71	11.37	11.11	10.87	10.63	10.40	10.17	9.94	9.71	9.49	9.26
22	11.51	11.17	10.92	10.69	10.45	10.22	9.99	9.77	9.55	9.33	9.12
23	11.31	10.99	10.74	10.50	10.27	10.05	9.83	9.61	9.40	9.19	8.97
24	11.12	10.80	10.56	10.33	10.10	9.90	9.67	9.46	9.25	9.04	8.83
25	10.94	10.62	10.39	10.16	9.94	9.73	9.52	9.31	9.11	8.90	8.70
26	10.76	10.45	10.22	10.00	9.78	9.58	9.37	9.17	8.96	8.77	8.56
27	10.59	10.28	10.06	9.84	9.64	9.43	9.23	9.03	8.83	8.63	8.44
28	10.41	10.13	9.90	9.69	9.49	9.28	9.09	8.89	8.70	8.50	8.31
29	10.25	9.97	9.75	9.54	9.35	9.14	8.96	8.76	8.57	8.38	8.19
30	10.09	9.82	9.60	9.40	9.20	9.01	8.82	8.63	8.45	8.26	8.07
31	9.94	9.67	9.46	9.26	9.07	8.88	8.69	8.51	8.32	8.14	7.96
32	9.79	9.53	9.32	9.12	8.94	8.75	8.57	8.39	8.21	8.03	7.85
33	9.65	9.39	9.19	8.99	8.81	8.63	8.45	8.27	8.09	7.92	7.74
34	9.51	9.25	9.06	8.87	8.69	8.51	8.33	8.15	7.98	7.81	7.64

* In ml/l from an atmosphere of 78.08% N_2 and 100% relative humidity.

(From Murray, C. N., Riley, J. P., and Wilson, T. R. S., Solubility of gases in distilled water and sea water – I Nitrogen, *Deep Sea Res.*, 16, 297, 1969. With permission.)

Table 2.4—4
SOLUBILITY OF CARBON DIOXIDE IN PURE WATER AND IN SEA WATER*

Temp °C / Cl °/oo	0	2	4	6	8	10	12	14	16	18	20	22	24	26
0	771	713	661	614	572	535	499	467	440	413	389	367	347	328 x 10^{-4}
15	670	620	576	536	500	468	437	409	385	364	344	325	308	291
16	664	615	571	531	496	464	433	406	382	360	341	322	305	289
17	658	609	565	526	491	460	429	402	378	357	338	319	303	287
18	652	603	559	521	486	455	425	398	374	354	335	316	300	284
19	645	597	554	515	481	450	421	394	371	350	332	313	297	281
20	638	591	548	510	476	445	416	390	367	347	328	310	294	278

* In mole/liter/atmosphere.

(From Murray, C. N. and Riley, J. P., *Deep Sea Res.*, 18, 533, 1971. With permission.)

Table 2.4—5
SOLUBILITY OF ARGON (ml AT S.T.P./l) RELATIVE TO AN ATMOSPHERE CONTAINING 0.934% OF ARGON AND HAVING A RELATIVE HUMIDITY OF 100%

Temp (°C)	Chlorinity (°/oo) 0	2	4	6	8	10	12	14	16	18	20
−2	–	–	–	–	–	–	–	–	–	–	0.4082
−1	–	–	–	–	–	–	0.4285	0.4280	0.4176	0.4072	0.3968
0	0.4984	0.4811	0.4698	0.4589	0.4483	0.4377	0.4273	0.4170	0.4067	0.3964	0.3863
1	0.4834	0.4681	0.4571	0.4466	0.4363	0.4261	0.4161	0.4061	0.3962	0.3863	0.3765
2	0.4697	0.4561	0.4455	0.4353	0.4253	0.4154	0.4057	0.3960	0.3863	0.3768	0.3674
3	0.4572	0.4447	0.4343	0.4245	0.4147	0.4052	0.3958	0.3865	0.3771	0.3679	0.3588
4	0.4451	0.4337	0.4236	0.4139	0.4044	0.3954	0.3864	0.3774	0.3684	0.3595	0.3506
5	0.4336	0.4234	0.4137	0.4045	0.3955	0.3864	0.3773	0.3684	0.3598	0.3512	0.3426
6	0.4230	0.4119	0.4025	0.3935	0.3848	0.3762	0.3677	0.3592	0.3509	0.3426	0.3343
7	0.4127	0.4016	0.3924	0.3837	0.3752	0.3669	0.3586	0.3505	0.3424	0.3344	0.3264
8	0.4030	0.3917	0.3827	0.3743	0.3661	0.3580	0.3501	0.3421	0.3344	0.3266	0.3189
9	0.3934	0.3821	0.3734	0.3651	0.3573	0.3494	0.3418	0.3341	0.3265	0.3190	0.3116
10	0.3842	0.3730	0.3645	0.3565	0.3488	0.3411	0.3337	0.3264	0.3191	0.3117	0.3045
11	0.3755	0.3644	0.3561	0.3484	0.3409	0.3336	0.3263	0.3192	0.3121	0.3050	0.2980
12	0.3672	0.3563	0.3483	0.3407	0.3335	0.3263	0.3193	0.3124	0.3054	0.2986	0.2918
13	0.3593	0.3485	0.3406	0.3333	0.3264	0.3194	0.3127	0.3060	0.2993	0.2924	0.2857
14	0.3515	0.3410	0.3333	0.3263	0.3195	0.3128	0.3062	0.2997	0.2931	0.2865	0.2800
15	0.3442	0.3337	0.3264	0.3195	0.3129	0.3064	0.3001	0.2938	0.2873	0.2809	0.2746
16	0.3369	0.3265	0.3194	0.3126	0.3062	0.3000	0.2937	0.2874	0.2812	0.2752	0.2690
17	0.3300	0.3196	0.3126	0.3062	0.3000	0.2939	0.2879	0.2817	0.2758	0.2698	0.2639
18	0.3234	0.3131	0.3062	0.2998	0.2938	0.2879	0.2820	0.2762	0.2705	0.2646	0.2589
19	0.3169	0.3068	0.2999	0.2939	0.2879	0.2822	0.2765	0.2709	0.2654	0.2598	0.2540
20	0.3106	0.3007	0.2938	0.2881	0.2823	0.2768	0.2712	0.2659	0.2605	0.2551	0.2493
21	0.3047	0.2946	0.2882	0.2824	0.2769	0.2714	0.2661	0.2608	0.2557	0.2503	0.2450
22	0.2990	0.2888	0.2826	0.2770	0.2716	0.2663	0.2612	0.2561	0.2512	0.2460	0.2409

Table 2.4—6
SOLUBILITIES OF NOBLE GASES IN SEA WATER*

Temp (°C)	He		Ne		Ar		Kr		Xe	
	Pure	Air	Pure	Air	Pure	Air	Pure	Air	Pure	Air
0 ± 1	–	–	9.37	171×10^{-6}	–	–	71.5	81.5×10^{-6}	136	11.70×10^{-6}
1 ± 1	7.91	41.4×10^{-6}	–	–	38.5	0.359	–	–	–	–
5 ± 1	–	–	9.02	164×10^{-6}	35.3	0.329	63.9	72.8×10^{-6}	115	9.89×10^{-6}
10 ± 1	7.40	38.8×10^{-6}	8.67	158×10^{-6}	32.8	0.306	58.2	66.3×10^{-6}	103	8.86×10^{-6}
15 ± 0.5	6.95	36.4×10^{-6}	8.35	152×10^{-6}	30.2	0.282	51.6	58.5×10^{-6}	90.0	7.74×10^{-6}
17.5 ± 0.5	–	–	–	–	–	–	50.2	57.2×10^{-6}	–	–
20 ± 0.5	7.00	36.7×10^{-6}	8.20	149×10^{-6}	26.3	0.245	44.8	51.1×10^{-6}	80.0	6.88×10^{-6}
22.8 ± 0.5	–	–	–	–	–	–	44.2	50.4×10^{-6}	–	–
24 ± 0.5	–	–	–	–	–	–	42.9	48.9×10^{-6}	–	–
25 ± 0.5	–	–	8.07	147×10^{-6}	–	–	–	–	70.2	6.04×10^{-6}

* Equilibrated with 760 torr of each of the gases and calculated for equilibrium with air containing 5.24×10^{-5} atm of He, 1.82×10^{-5} atm of Ne, 9.32×10^{-3} atm of Ar, 1.14×10^{-6} atm of Kr, and 8.6×10^{-8} atm of Xe, respectively. The units are ml (S.T.P.)/kg water.

(From König, H., Über die Löslichkeit der Edelgase in Meerwasser, *Z. Naturforsch.*, 18A, 363, 1963. With permission.)

2.5. MINERAL SOLUBILITIES IN SEA WATER

Table 2.5—1
SOLUBILITY OF CALCIUM CARBONATE IN SEA WATER AT ATMOSPHERIC PRESSURE

Calcite	$K'_{sp} = [Ca^{2+}][CO_3^{2-}] = (0.69 - 0.0063\ t°C) \times 10^{-6} \times S\ (°/_{oo})/34.3$
Aragonite	$K'_{sp} = [Ca^{2+}][CO_3^{2-}] = 0.90 \times 10^{-6}$ at 19°/oo Cl and t = 25°C $\Delta K'_{sp}/\Delta\ t°C = 0.0078 \times 10^{-6}$/°C at 19°/oo Cl from 0 to 40°C

Note: $K'_{sp} = [Ca^{2+}]\ [CO_3^{2-}]$, the concentration product.

(From MacIntyre, W. G., Fisheries Res. Bd. of Canada, Oceanographic and Limnological Series, Manuscript Rep. Ser. 200, Dartmouth, Nova Scotia, 1965. With permission.)

Table 2.5—2
EFFECT OF PRESSURE ON THE SOLUBILITY OF CALCIUM CARBONATE IN SEA WATER*

Mineral	t°C	$(K'_{sp})_{500}/(K'_{sp})_1$	$(K'_{sp})_{1000}/K'_{sp})_1$
Aragonite	2	2.11 ± 0.06	4.23 ± 0.27
	22	1.80 ± 0.01	3.16 ± 0.02
Calcite	2	2.18	4.79
	22	1.88	3.56

Note: $K'_{sp} = [Ca^{2+}]\ [CO_3^{2-}]$

*The subscripts 1, 500, and 1000 refer to the pressure (atm).

(From Hawley, J. and Pytkowicz, R. M., *Geochim. Cosmochim. Acta,* 33, 1557, 1969. With permission.)

Table 2.5—3
SOLUBILITY OF SOME MINERALS IN SEA WATER*

Mineral	Temperature	Cl (°/oo)	K'sp	Reference	Comments
$Ca_5(PO_4)_3F$	25°C	ca. 19	$0.16-16 \times 10^{-32}$	1	
$Mg(OH)_2$	25°C	19.3	$(2.4\pm0.2 \times 10^{-11}$	2	
SiO_2	0°C	?	58–80 mg/l	3	
SiO_2	22-27°C	?	100–110 mg/l	3	
$MgCO_3$	?	?	3×10^{-4}	4	
$ZnCO_3$	?	?	1.4×10^{-8}	5	Rough estimate**
$CuCO_3$	?	?	1.75×10^{-8}	5	Rough estimate**
$PbCO_3$	?	?	1.05×10^{-11}	5	Rough estimate**
BiOCl	?	?	1.4×10^{-8}	5	Rough estimate**
Cd(OH)Cl	?	?	5.1×10^{-10}	5	Rough estimate**
$Ni(OH)_2$	?	?	2.6×10^{-15}	5	Rough estimate**
$CoCO_3$ } $Co(OH)_3$	?	?	5.6×10^{-11}	5	Rough estimate**
$CaCrO_4$ (?)			5×10^{-2}	5	Rough estimate**
$V_2O_5 \cdot nH_2O$	?	?	10^{-3}	5	Rough estimate**
$MgCO_3 \cdot H_2O$	?	?	7×10^{-4}	5	Rough estimate**
$SrCO_3$	?	?	$2.1-11.2 \times 10^{-8}$	5	Rough estimate**
$BaSO_4$	?	?	7×10^{-9}	5	Rough estimate**

*At atmospheric pressure, expressed as K'_{sp}, the concentration solubility products, except where otherwise stated.

**Obtained by multiplying thermodynamic solubility products by estimated activity coefficients as recommended by Krauskopf. These coefficients only reflect the ionic strength and may not be correct if there is a significant extent of ion pair or complex formation.

REFERENCES

1. **Pytkowicz, R. M. and Kester, D. R.**, *Limnol. Oceanogr.*, 12, 714, 1967.
2. **Pytkowicz, R. M. and Gates, R.**, *Science*, 161, 690, 1968.
3. **Krauskopf, K. B.**, *Geochim. Cosmochim. Acta*, 10, 1, 1956.
4. **Wattenberg, H. and Timmerman, E.**, Ann. Hydrogr., Berl., U.S.W. p. 23.
5. Based upon **Krauskopf, K. B.**, *Geochim. Cosmochim. Acta*, 9, 1, 1956.

Table 2.5—4
MEASURED CONCENTRATIONS OF METALS AT SATURATION IN SEA WATER*

Metal	Saturation concentration (ppm)	Metal	Saturation concentration (ppm)
Zn	1.2–2.5	Cr	high
Cu	0.4–0.8	Mo	25–750
Pb	0.3–0.7	W	2–200
Bi	0.04	V^{IV}	4–150
Cd	4–1000	V^{V}	>400
Ni	20–450	Mg	36,000
Co	25–200	Ca	100–480
Hg	100–1000	Sr	22
Ag	2.0–2.5	Ba	0.11

*t = 18–23°C, pH = 7.8–8.2, S ≅ 30°/oo.

(From Krauskopf, K. B., *Geochim. Cosmochim. Acta*, 9, 1, 1956. With permission.)

2.6. PHYSICOCHEMICAL PROPERTIES OF SEA WATER

Table 2.6—1

THE FREEZING POINT LOWERING OF SEA WATER (°C)

Table A

Cl °/oo	Knudsen (1903)	Miyake (1939)
5	0.483	0.514
10	0.969	1.027
15	1.466	1.541
17	1.668	1.746
19	1.872	1.951
21	2.078	2.157

Table B

Salinity, °/oo	Murray et al.	Salinity, °/oo	Murray et al.
5	0.268	33	1.791
10	0.535	34	1.849
15	0.801	35	1.906
20	1.068	36	1.964
25	1.341	37	2.018
30	1.621	38	2.079
31	1.678	39	2.138
32	1.734	40	2.196

1. **Knudsen, M.,** *J. Cons. Cons. Perm. Int. Explor. Mer.*, 5, 11, 1903.
2. **Miyake, Y.,** *Bull. Chem. Soc. Jap.*, 14b, 58, 1939.
3. **Murray, C. N., Murray, L. A., and Riley, J. P.,** unpublished results.

Table 2.6—2
THE VAPOR PRESSURE LOWERING OF SEA WATER (mm Hg)

	Ref. 1	Ref. 2		
Cl °/oo	25°C	25°C	15°C	5°C
5	–	0.10	0.06	0.03
10	0.225	0.20	0.11	0.06
15	0.340	0.31	0.17	0.09
17	0.387	0.35	0.19	0.10
19	0.435	0.40	0.22	0.11
21	0.484	0.45	0.24	0.13

1. **Robinson, R. A.,** *J. Mar. Biol. Assoc. U.K.*, 33, 449, 1954.
2. **Arons, A. B. and Kientzler, C. F.,** *Trans. Am. Geophys. Union*, 35, 722, 1954.

Table 2.6—3
COMPUTED VALUES FOR SPECIFIC HEAT OF SEA WATER AT CONSTANT PRESSURE*

Salinity (g/kg)	Temperature (°C) -2	-1	0	1	2	5	10	15	20	25	30
0	–	–	4.217	4.214	4.210	4.202	4.192	4.186	4.182	4.179	4.178
5	–	–	4.179	4.176	4.174	4.168	4.161	4.157	4.154	4.153	4.152
10	–	–	4.142	4.140	4.138	4.135	4.130	4.128	4.126	4.126	4.126
15	–	–	4.107	4.106	4.105	4.103	4.100	4.099	4.098	4.099	4.100
20	–	4.075	4.074	4.074	4.073	4.072	4.071	4.071	4.071	4.072	4.074
25	–	4.043	4.043	4.043	4.042	4.042	4.042	4.043	4.045	4.046	4.048
30	–	4.013	4.013	4.013	4.013	4.014	4.015	4.016	4.018	4.020	4.023
32	–	4.001	4.002	4.002	4.002	4.003	4.004	4.006	4.008	4.010	4.013
34	–	3.990	3.990	3.991	3.991	3.992	3.993	3.995	3.998	4.000	4.003
35	3.984	3.984	3.985	3.985	3.985	3.986	3.988	3.990	3.993	3.995	3.999
36	3.979	3.979	3.979	3.980	3.980	3.981	3.983	3.985	3.988	3.991	3.994
38	3.968	3.968	3.968	3.968	3.969	3.970	3.972	3.975	3.978	3.981	3.985
40	3.957	3.957	3.957	3.958	3.958	3.959	3.962	3.965	3.968	3.972	3.976

* In absolute joules per gram.

(From Cox, R. A. and Smith, N. D., The specific heat of sea water, *Proc. R. Soc. Lond. A.*, 252, 51, 1959. With permission of the Royal Society.)

Table 2.6—4
SPECIFIC HEAT OF SEA ICE

The specific heat of pure ice depends upon its temperature and varies within narrow limits but that of sea ice is a much more variable property, depending upon the salt or brine content and the temperature. Changing the temperature of sea ice will generally involve either melting or freezing, and the amount of heat required will depend upon the salinity of the ice, as shown in the table. It should be noted that the specific heat of pure ice is less than half that of pure water. Near the initial freezing point, the extremely high specific heat of ice of high salinity is, of course, due to the formation of ice from the enclosed brine or its melting[1]

Salinity	Temperature (°C)										
°/oo	-2°	-4°	-6°	-8°	-10°	-12°	-14°	-16°	-18°	-20°	-22°
2	2.57	1.00	0.73	0.63	0.57	0.55	0.54	0.53	0.53	0.52	0.52
4	4.63	1.50	0.96	0.76	0.64	0.59	0.57	0.57	0.56	0.55	0.54
6	6.70	1.99	1.20	0.88	0.71	0.64	0.61	0.60	0.58	0.57	0.56
8	8.76	2.49	1.43	1.01	0.78	0.68	0.64	0.64	0.61	0.60	0.58
10	10.83	2.99	1.66	1.14	0.85	0.73	0.68	0.67	0.64	0.62	0.60
15	16.01	4.24	2.24	1.46	1.02	0.85	0.77	0.76	0.71	0.68	0.65

(From *Smithsonian Meteorological Tables*, Smithsonian Miscellaneous Collection, V.114, 6th ed., 1966.)

REFERENCE

1. Sverdrup, H. U., Johnson, M. W., and Fleming, R. H., *The Oceans, Their Physics, Chemistry, and General Biology*, Prentice-Hall, Inc., Englewood Cliffs, N.J., © 1942, renewed 1970.

Table 2.6—5
LATENT HEAT OF MELTING OF SEA ICE

Let S be the salinity of the ice and t_g the freezing point of a sea water with the salinity S. If the temperature t lies in the neighborhood of 0 °C, the heat of fusion of pure ice between t and t_g can be considered constant and equal to 80 gram calories. The amount of heat required to melt the sea ice is then the sum of the heat required to melt all the pure ice in one gram of sea ice ($80[1-S(1-A_t)]$) and the amount of heat required to increase the temperature of the pure ice and brine from t to t_g (approximately=0.5 (t_g-t), where A_t is the weight of all the pure ice in 1 gram of sea ice with salinity 1°/oo and temperature t and is equal to $1-1/S_t$ where S_t is the salinity of the brine at t. Thus

$$U = 80(1 - \frac{S}{S_t}) + 0.5(t_g - t;)$$

where U = number of calories required to melt 1 g sea ice of the temperature t and the salinity S.

Temperature	Salinity (°/oo)						
	0	2	4	6	8	10	15
°C	cal	cal	cal	cal	cal	cal	cal
-1	80	72	63	55	46	37	16
-2	81	77	72	68	63	59	48

(From Malmgren, F., On the properties of sea ice, *The Norwegian North Polar Expedition with the Maud, 1918-1925*, Scientific Results, Vol. I, No. 5, 1927. With permission.)

Figure 2.6—1
SEA WATER DENSITY AT VARIOUS TEMPERATURES

The purpose of this graph is to provide the density of sea water at any temperature apt to be encountered when the density at the standard temperature of 59°F(15°C) is known. To convert a density at 59°F(15°C) to density at another temperature, enter the graph horizontally from the left with the known density and downward from the top or upward from the bottom with the desired temperature; the position of the point of intersection with respect to the curves gives the density at the desired temperature. Interpolate between curves when necessary. For example, by this method, water having a density of 1.0162 at 59°F is found to have a density of 1.0124 at 85°F. The densities are referred to the density of fresh water at 4°C(39.2°F) as unity. (From Surface Water Temperature and Density, *Shore and Beach,* 40, 42, 1972. Original source: National Ocean Survey, National Oceanic and Atmospheric Administration, Rockville, Maryland. With permission.)

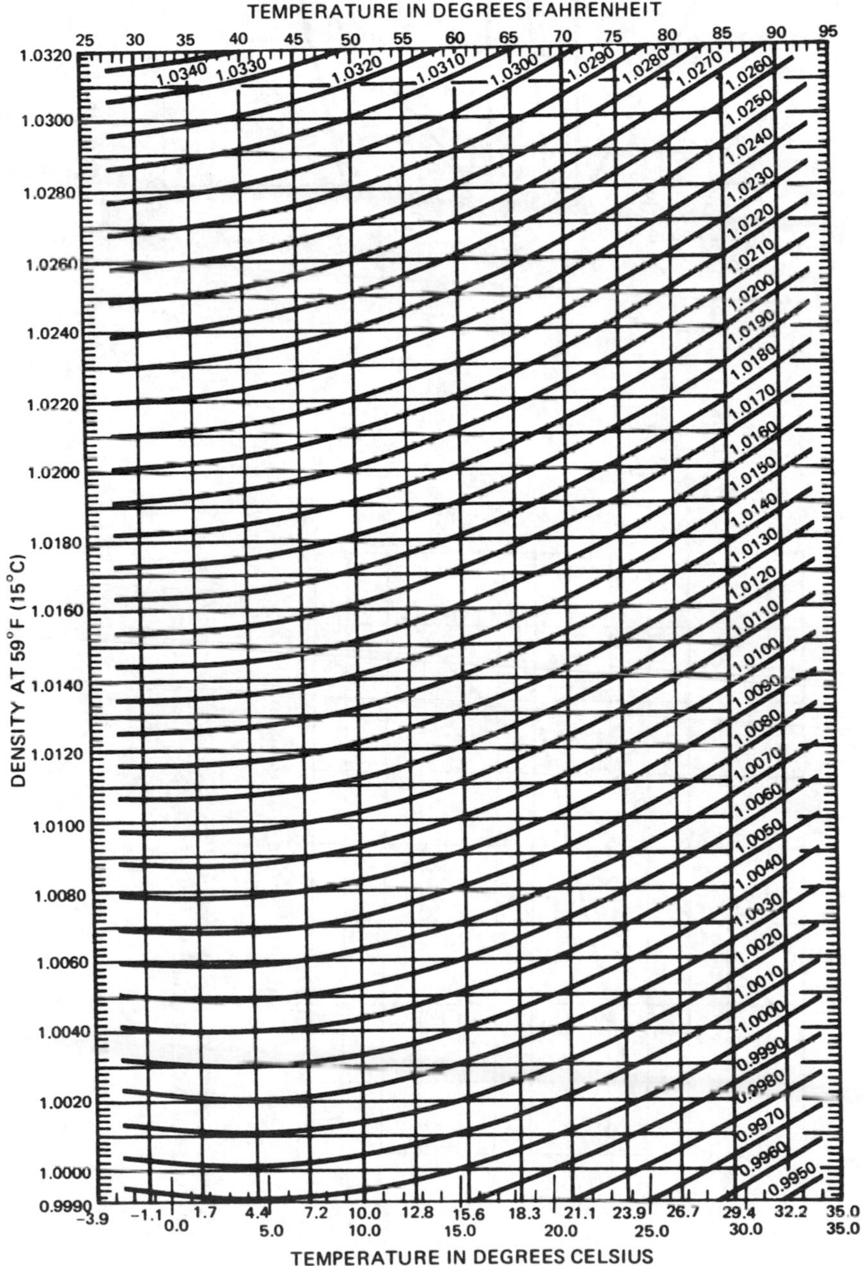

(From National Ocean Survey, National Oceanic and Atmospheric Administration, Rockville, Maryland. With permission.)

Figure 2.6—2
RELATIONSHIP BETWEEN TEMPERATURE OF MAXIMUM DENSITY AND FREEZING POINT FOR WATER OF VARYING SALINITY

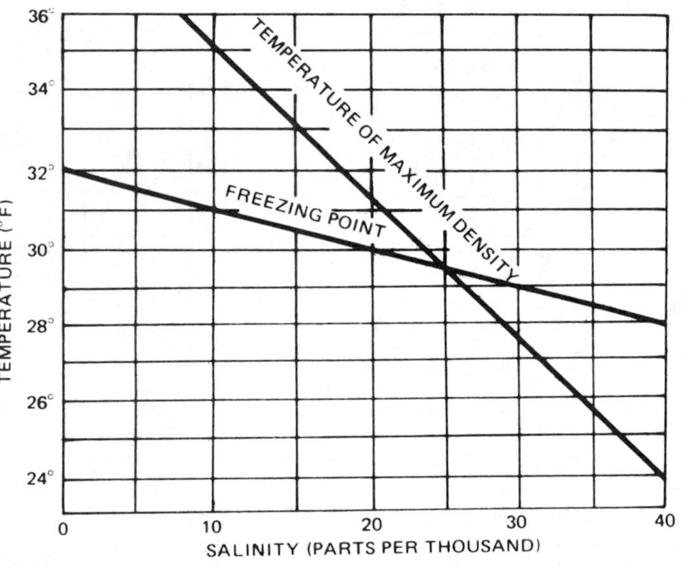

(From *American Practical Navigator,* U.S. Naval Hydrographic Office, Washington, D.C., H.O. Publ. No.9.)

Figure 2.6—3
COLLIGATIVE PROPERTIES OF SEA WATER

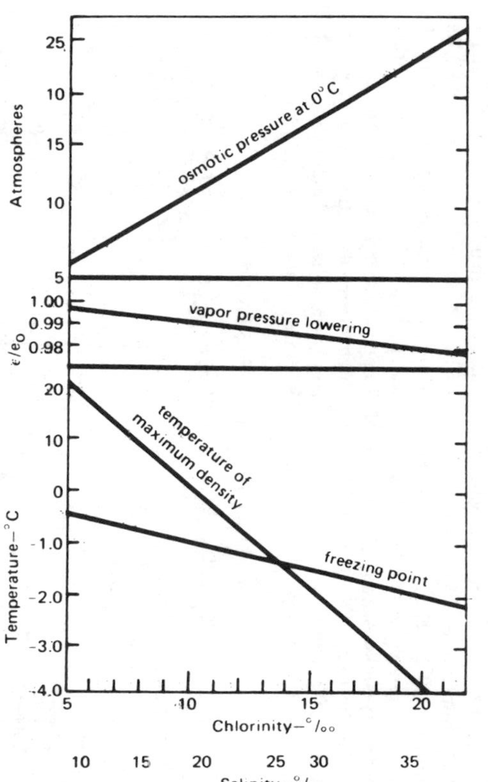

(From Sverdrup, H. U., Johnson, M. W., and Fleming, R. H., *The Oceans, Their Physics, Chemistry, and General Biology,* Prentice-Hall, Inc., Englewood Cliffs, N.J., © 1942, renewed 1970. With permission.)

Table 2.6—6
RELATIVE VISCOSITY OF SEA WATER
At 1 atm[a]

Temp °C	5°/oo	10°/oo	20°/oo	30°/oo	40°/oo
0	1.009	1.017	1.032	1.056	1.054
5	0.855	0.863	0.877	0.891	0.905
10	0.738	0.745	0.785	0.772	0.785
15	0.643	0.649	0.662	0.675	0.688
20	0.568	0.574	0.586	0.599	0.611
25	0.504	0.510	0.521	0.533	0.545
30	0.454	0.460	0.470	0.481	0.491

[a] η/η_O, where η_O is the viscosity of pure water at 0°C (1.787 centipoise).

(From Dorsey, N. E., *Properties of Ordinary Water Substance,* © 1940 by Litton Educational Publishing, Inc. With permission of Van Nostrand Reinhold Co.)

Table 2.6—7
RELATIVE VISCOSITY OF 19.374°/oo CHLORINITY IAPO STANDARD SEA WATER FOR VARIOUS TEMPERATURES AND PRESSURES

Pressure, kg/cm²	η_p/η_1 at -0.024°C	η_p/η_1 at 2.219°C	η_p/η_1 at 6.003°C	η_p/η_1 at 10.013°C	η_p/η_1 at 15.018°C	η_p/η_1 at 20.013°C	η_p/η_1 at 29.953°C
176	0.9828	0.9849	0.9891	0.9908	0.9961	0.9972	1.0010
	0.9828	0.9855	0.9890	0.9929	0.9946	0.9985	0.9983
				0.9911	0.9941	0.9976	0.9999
352	0.9707	0.9742	0.9810	0.9881	0.9924	0.9961	1.0017
	0.9711	0.9747	0.9818	0.9872	0.9917	0.9984	0.9995
		0.9737			0.9938	0.9973	0.9992
527	0.9623	0.9667	0.9762	0.9843	0.9900	0.9969	1.0037
	0.9616	0.9674	0.9770	0.9843	0.9875	0.9992	1.0023
					0.9923	0.9973	1.0033
703	0.9559	0.9622	0.9729	0.9817	0.9916	1.0013	1.0072
	0.9562	0.9629	0.9740	0.9825	0.9909	0.9994	1.0069
					0.9919	0.9987	1.0073
878	0.9528	0.9594	0.9730	0.9834	0.9937	1.0044	1.0143
	0.9538	0.9603	0.9736	0.9837	0.9928	1.0043	1.0114
					0.9933	1.0015	1.0142
1055	0.9519	0.9592	0.9754	0.9875	0.9969	1.0077	1.0174
	0.9536	0.9606	0.9744	0.9872	0.9961	1.0076	1.0183
					0.9964	1.0056	1.0179
1230	0.9527	0.9631	0.9781	0.9904	1.0025	1.0116	1.0253
	0.9540	0.9643	0.9752	0.9899	1.0009	1.0099	1.0235
					1.0009	1.0106	
1406	0.9553	0.9670	0.9825	0.9956	1.0099	1.0177	1.0323
	0.9565	0.9676	0.9817	0.9965	1.0058	1.0168	1.0303
					1.0064	1.0154	

(From Stanley, E. M. and Batten, R. C., Viscosity of sea water at moderate temperatures and pressures, *J. Geophys. Res.,* 74, 3415, 1969. With permission of American Geophysical Union.)

Figure 2.6—4
PRESSURE OF MINIMUM RELATIVE VISCOSITY P_m, FOR STANDARD SEA WATER AND PURE WATER AS A FUNCTION OF TEMPERATURE

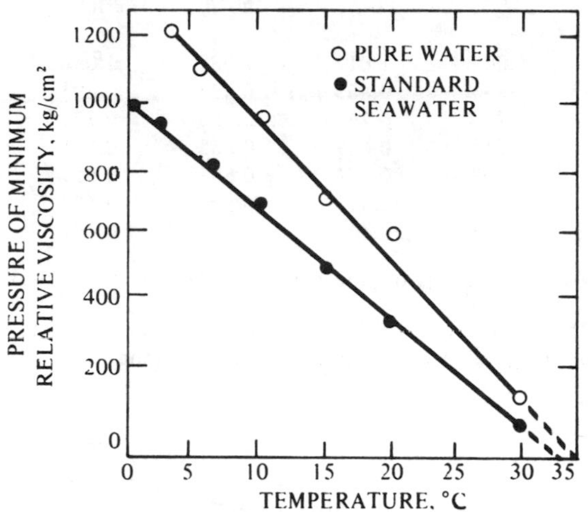

Note: Dashed lines indicate extrapolated values.

(From Stanley, E. M. and Batten, R. C., Viscosity of sea water at moderate temperatures and pressures, *J. Geophys. Res.*, 74, 3415, 1969. With permission of American Geophysical Union.)

Figure 2.6—5
DIFFERENTIAL RELATIVE VISCOSITY AS A FUNCTION OF TEMPERATURE

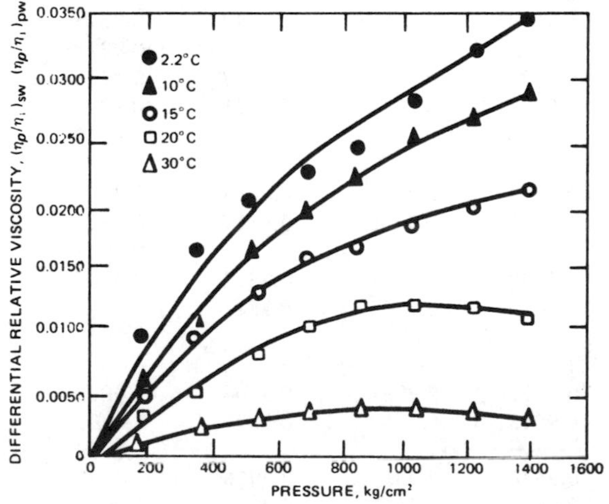

(From Stanley, E. M. and Batten, R. C., Viscosity of sea water at moderate temperatures and pressures, *J. Geophys. Res.*, 74, 3415, 1969. With permission of American Geophysical Union.)

Figure 2.6—6

ISOBARS OF RELATIVE VISCOSITY OF STANDARD SEA WATER AS A FUNCTION OF TEMPERATURE

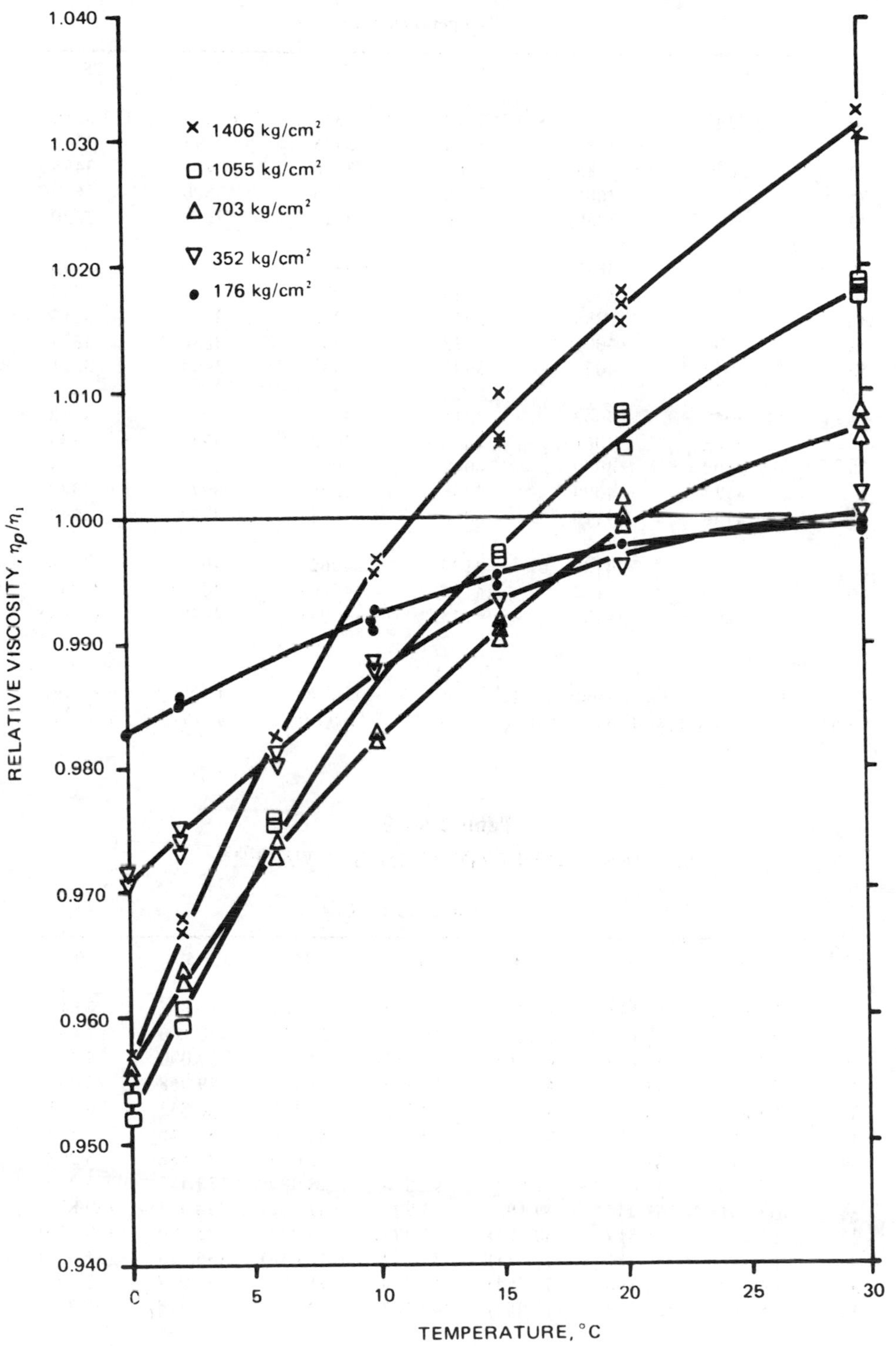

(From Stanley, E. M. and Batten, R. C., Viscosity of sea water at moderate temperatures and pressures, *J. Geophys. Res.*, 74, 3415, 1969. With permission of American Geophysical Union.)

Table 2.6—8
REFRACTIVE INDEX OF SEA WATER*

S°/oo	Temperature (°C) 0	5	10	15	20	25
0	1.3 3395	1.3 3385	1.3 3370	1.3 3340	1.3 3300	1.3 3250
5	3500	3485	3465	3435	3395	3345
10	3600	3585	3565	3530	3485	3435
15	3700	3685	3660	3625	3580	3525
20	3795	3780	3750	3715	3670	3620
25	3895	3875	3845	3805	3760	3710
30	3991	3966	3935	3898	3851	3798
31	4011	3985	3954	3916	3869	3816
32	4030	4004	3973	3934	3886	3834
33	4049	4023	3992	3953	3904	3851
34	4068	4042	4011	3971	3922	3868
35	4088	4061	4030	3990	3940	3886
36	4107	4080	4049	4008	3958	3904
37	4127	4099	4068	4026	3976	3922
38	4146	4118	4086	4044	3994	3940
39	4166	4139	4105	4062	4012	3958
40	(4185)	(4157)	(4124)	(4080)	(4031)	(3976)
41	(4204)	(4176)	(4143)	(4098)	(4049)	(3944)

*For sodium D light.

(From Utterback, C. L., Thompson, T. G., and Thomas, B. D., Refractivity-chlorinity-temperature relationships of ocean waters, *J. Cons. Cons. Perm. Int. Explor. Mer.*, 9, 35, 1934. With permission.)

Table 2.6—9
SPECIFIC CONDUCTIVITY OF SEA WATER*

S°/oo	Temperature (°C) 30	25	20	15	10	5	0
10	(19.127)	17.345	15.628	13.967	12.361	10.816	9.341
20	(35.458)	32.188	29.027	25.967	23.010	20.166	17.456
30	(50.856)	46.213	41.713	37.351	33.137	29.090	25.238
31	(52.360)	47.584	42.954	38.467	34.131	29.968	26.005
32	(53.859)	48.951	44.192	39.579	35.122	30.843	26.771
33	(55.352)	50.314	45.426	40.688	36.110	31.716	27.535
34	(56.840)	51.671	46.656	41.794	37.096	32.588	28.298
35	(58.323)	53.025	47.882	42.896	38.080	33.457	29.060
36	(59.801)	54.374	(49.105)	(43.996)	(39.061)	(34.325)	(29.820)
37	(61.274)	55.719	(50.325)	(45.093)	(40.039)	(35.190)	(30.579)
38	(62.743)	57.061	(51.541)	(46.187)	(41.016)	(36.055)	(31.337)
39	(64.207)	58.398	(52.754)	(47.278)	(41.990)	(36.917)	(32.094)
40	(65.667)	(59.732)	(53.963)	(48.367)	(42.962)	(37.778)	(32.851)

* Conductivity in mmol/cm. Values are calculated from an equation presented by Weyl based on the conductivity data of Thomas, Thompson, and Utterback.

(From Cox, R. A., The physical properties of sea water, in *Chemical Oceanography*, Vol. 1, Riley, J. P. and Skirrow, G., Eds., Academic Press, London, 1965. With permission.)

Table 2.6—9 (continued)

REFERENCES

1. **Weyl, P. K.,** On the change in electrical conductance of seawater with temperature, *Limnol. Oceanogr.*, 9, 75, 1964.
2. **Thomas, B. D., Thompson, T. G., and Utterback, C. L.,** The electrical conductivity of seawater, *J. Cons. Cons. Perm. Int. Explor. Mer.*, 9, 28, 1934.

Table 2.6—10
RELATIVE CONDUCTIVITY OF SEA WATER AT 15° AND 20°C

S %	R_{15}	R_{20}
31	0.89704	0.89732
32	0.92296	0.92327
33	0.94876	0.94890
34	0.97443	0.97450
35	1.00000	1.00000
36	1.02545	1.02537
37	1.05078	1.05062
38	1.07600	1.07577
39	1.10112	1.10080
40	1.12613	1.12572

(From *International Oceanographic Tables,* Vol. 1, National Institute of Oceanography of Great Britain, Wormley, England, and UNESCO, Paris, France, ©1966, reprinted 1971. With permission.)

Table 2.6—11
EFFECT OF TEMPERATURE ON CONDUCTIVITY RATIO EXPRESSED AS CONDUCTIVITY RATIO ANOMALY Δ_t

$R_{15} = R_t + \Delta_t$

R_t	10°	15°	20°	25°
0.875	36	0	−32	−59
0.900	30	0	−26	−49
0.925	23	0	−20	−38
0.950	16	0	−14	−26
0.975	8	0	−7	−13
1.000	0	0	0	0
1.025	−8	0	7	14
1.050	−17	0	15	28
1.075	−27	0	23	43
1.100	−37	0	32	59
1.125	−47	0	41	75

(Values of Δ_t were taken from *International Oceanographic Tables,* Vol. 1, National Institute of Oceanography of Great Britian, Wormley, England and UNESCO, Paris, France, © 1966, reprinted 1971. With permission.)

Table 2.6—12
EFFECT OF PRESSURE ON THE CONDUCTIVITY OF SEA WATER*

	Pressure (db)	S°/oo 31	S°/oo 35	S°/oo 39		S°/oo 31	S°/oo 35	S°/oo 39
	1000	1.599	1.556	1.512		1.032	1.008	0.985
	2000	3.089	3.006	2.922		1.996	1.951	1.906
	3000	4.475	4.354	4.233		2.895	2.830	2.764
	4000	5.759	5.603	5.448		3.731	3.646	3.562
0°C	5000	6.944	6.757	6.569	15°C	4.506	4.403	4.301
	6000	8.034	7.817	7.599		5.221	5.102	4.984
	7000	9.031	8.787	8.543		5.879	5.745	5.612
	8000	9.939	9.670	9.401		6.481	6.334	6.187
	9000	10.761	10.469	10.178		7.031	6.871	6.711
	10000	11.499	11.188	10.877		7.529	7.358	7.187
	1000	1.368	1.333	1.298		0.907	0.888	0.868
	2000	2.646	2.578	2.510		1.755	1.718	1.680
	3000	3.835	3.737	3.639		2.546	2.492	2.438
	4000	4.939	4.813	4.686		3.282	3.212	3.142
5°C	5000	5.960	5.807	5.655	20°C	3.964	3.879	3.795
	6000	6.901	6.724	6.547		4.594	4.496	4.399
	7000	7.764	7.565	7.366		5.174	5.064	4.954
	8000	8.552	8.333	8.114		5.706	5.585	5.464
	9000	9.269	9.031	8.794		6.192	6.060	5.929
	10000	9.915	9.661	9.408		6.633	6.492	6.351
	1000	1.183	1.154	1.125		0.799	0.783	0.767
	2000	2.287	2.232	2.177		1.547	1.516	1.485
	3000	3.317	3.237	3.157		2.245	2.200	2.156
	4000	4.273	4.170	4.067		2.895	2.837	2.780
10°C	5000	5.159	5.034	4.910	25°C	3.498	3.429	3.359
	6000	5.976	5.832	5.688		4.056	3.976	3.896
	7000	6.728	6.565	6.402		4.571	4.481	4.390
	8000	7.415	7.236	7.057		5.045	4.945	4.845
	9000	8.041	7.847	7.652		5.478	5.369	5.261
	10000	8.608	8.400	8.192		5.872	5.756	5.640

*Percentage increase compared with the conductivity at one atmosphere.

(From Bradshaw, A. and Schleicher, K. E., The effect of pressure on the electrical conductance of seawater, *Deep-Sea Res.*, 12, 151, 1965. With permission.)

Section 3
Physical Oceanography

Section 3

PHYSICAL OCEANOGRAPHY

This branch of marine science is concerned with the motions of ocean waters — currents, tides, and waves — and the forcing mechanisms responsible for these motions. It also deals with physical properties of seawater, especially those parameters that influence its density (i.e., temperature and salinity). Thermohaline circulation of deep ocean currents results from density differences mainly attributable to temperature and salinity. Physical oceanography has practical applications to many problems of contemporary society. Some of these concern technical matters related to beach erosion, fisheries, drilling and mining operations, waste disposal, and transportation. Ocean engineering projects, such as the construction of breakwaters, jetties, and seawalls, as well as the dredging of harbor entrances and estuaries, require comprehension of coastal waves and currents. Local weather conditions and global climate regimes similarly are closely coupled to heat fluxes, temperature distributions, and circulation patterns in the sea.

The ocean is a highly turbulent body of water whose circulation and mixing involve both advective and diffusive processes.[1] Whereas oceanographers believe the ocean to be primarily advective, diffusion plays a role in establishing property fields.[2] Despite the turbulent nature of marine waters, few quantitative measures of turbulence are available.[3] Indeed, various phenomena of physical oceanography remain poorly understood; for instance, the role of eddies in the dynamics of the oceans is unresolved even though they occur nearly everywhere in the ocean.[3-5] Additional data must be collected on eddies, together with other entities, to unravel the complexities of ocean circulation and to generate realistic models.[6]

Although physical oceanography as a science has been practiced for more than a century, the greatest advancements in this discipline stem from research conducted since 1950. The development of new, sophisticated monitoring instruments and computers is partially responsible for these advancements. As an outgrowth of space technology, for example, satellite oceanography has greatly facilitated data acquisition on ocean temperature, color, surface roughness, sea surface height, and other parameters.[7] Remote sensing of the sea surface has aided oceanographers in mapping surface ocean currents and water masses.[8] Hence, in the North Atlantic, investigations of the kinematics of eddies and rings have been performed using satellite-tracked buoys.[3] Richardson[9] produced maps of eddy kinetic energy in the western North Atlantic from satellite-tracked buoys, and Krauss and Käse[10] formulated similar ones for the northeastern part of the North Atlantic. Another valuable remote-sensing tool, acoustic tomography, has been used to detect the thermohaline structure of the oceans or to measure velocities by the doppler shift of backscattered sound pulses.[7,11] Acoustic technology in oceanography, in general, has progressed remarkably in the last 10 years.[12,13]

Two subject areas receiving considerable scrutiny during the 1970s and 1980s are coastal upwelling[14,15] and oceanic fronts.[16] Intense studies of coastal upwelling have been executed in the eastern Pacific Ocean along the coasts of Oregon (U.S.) and Peru. Research on fronts expanded rapidly in the late 1970s. Part of this expansion can be ascribed to the utilization of satellite-borne infrared radiometry for sea-surface monitoring which prompted the discovery of new fronts.[17]

In the shallow water environments of estuaries and continental shelves, substantial strides have been achieved in assessing current profiles and turbulence by deploying several field-measuring techniques. Electromagnetic flow meters, acoustic current meters, heated profiling thermistors and related instruments involving heat transfer, impellors, and laser-doppler current meters have proven to be very useful in coastal oceanographic research.[18] The new Introcean S-4 current meters, high technology sampling devices that record velocity, direction, depth, conductivity, salinity, and temperature, display powerful capabilities; yet, their small size makes them ideally suited for a wide range of field applications.

Ocean circulation can be subdivided into two components: (1) surface circulation and (2) thermohaline deep circulation. The radiation balance on earth, rotation of the earth, and relative motions of the earth, moon, and sun account for the major currents in the ocean. The surface currents are principally driven by the mean atmospheric circulation patterns.[19] It is the long-term winds (e.g., prevailing westerlies and trade winds), along with the Coriolis force, which cause a series of gyres or large circulating current systems to form in all ocean basins, centered at approximately 30°N and 30°S. These gyres circulate clockwise in the north Atlantic and north Pacific Oceans and counterclockwise in the south Atlantic, south Pacific, and Indian oceans. The rotation of the earth displaces the gyres toward the western boundary of the oceans, generating stronger currents along this perimeter. Thus, western boundary currents (Gulf Stream, Kiroshio, Brazil, and East Australian currents) are more intense and narrower than eastern boundary currents (e.g., the California, Humboldt, Canaries, and Benguela currents), and they transport warm waters, thereby transferring heat toward the poles. Surface currents flowing along the western coasts of continents toward the equator are deflected away from the coastline due to the Coriolis force. Consequently, nutrient-rich cold water rises from the bottom to replace surface waters. This upwelling process yields regions of exceptionally high biological productivity.[20]

As the wind blows over the sea surface, its energy passes down through the water column. However, because of the Coriolis force, each successively deeper layer of water is deflected toward the right of the wind direction in the northern hemisphere and toward the left in the southern hemisphere. The subsurface currents likewise decrease in velocity with increasing depth. Ultimately a depth is reached (the Ekman depth) where the water flows in a direction opposite to that of the surface current. When represented by vectors, the resultant change of current direction and velocity with depth follows a spiral pattern known as the Ekman spiral. Effects of the wind can be transmitted to depths of approximately 100 m to several hundred meters.

Other conspicuous features of the ocean current system include the north and south equatorial currents. Positioned between them is the weakly flowing equatorial counter current. Also pronounced is the Antarctic circumpolar current (west wind drift), with a continuously eastward moving flow.

Whereas the planetary wind system drives the surface circulation of the ocean, the circulation in the deep sea depends on density differences largely determined by water temperature and salinity as specified above. In profile, the ocean can be divided into three layers: (1) an upper, wind-mixed zone; (2) a thermocline; and (3) a uniform, deep-water region.[19] The upper, wind-mixed zone extends from the air-seawater interface to the depth of the thermocline beginning at about 100 to 300 m. The main thermocline is that portion of the water column having a well-developed vertical temperature gradient characterized by declining temperature with increasing depth appreciably greater than in overlying and underlying waters. Persisting for several hundred meters, this thermocline acts as a stably stratified layer limiting vertical water transfer, thereby precluding surface-induced mixing at greater depths. Below the main thermocline lies the bulk of ocean waters having uniform temperature and salinity. The dynamics of ocean circulation, incorporating numerical models, are treated comprehensively by Abarbanel and Young.[21]

Studies of thermohaline circulation in the deep ocean concentrate on identifiable water masses classified by their source region and depth of occurrence. The Antarctic bottom water (AABW), for example, originates in the Weddell Sea, flows down the Antarctic continental slope, and moves northward at great depths. Having temperature and salinity characteristics of 0.5°C and 34.7 °/oo, respectively, this water mass can be traced into the Atlantic, Pacific, and Indian oceans. Another source region of deep oceanic waters is the Antarctic convergence, a convergence of surface waters in the Antarctic circumpolar current. This area represents the main supply of Antarctic intermediate water (AAIW) (4°C, 34.4 °/oo) that also descends northward, where it overlies North Atlantic deep water (NADW) and AABW. The Norwegian and Greenland Seas are the major source regions of water below the thermocline in the northern hemisphere.[19] Water from these seas sinks and flows southward, mixing with overlying waters to form the NADW (2°C, 34.95 °/oo). Originating at about 60°N latitude, the NADW spreads southward above the AABW to as far as 50°S latitude. The mixing of NADW, AAIW, and AABW produces the largest water mass in the deep sea, the Indian and Pacific Ocean common water (1.5°C, 34.7 °/oo).

Oceanographers identify and trace a deep oceanic water mass by its characteristic temperature and salinity acquired at the sea surface, together with oxygen measurements.[20] Current velocities and mixing rates in the deep sea can be measured by the decay of the isotope carbon-14 (^{14}C). This technique has revealed time scales of between 200 and 1000 years for thermohaline deep circulation of the oceans.[19]

This section on physical oceanography details temperature, salinity, and density distributions in the oceans. Surface currents are illustrated. In addition, transport charts delineate circulation patterns in oceanic systems.

REFERENCES

1. **Bowden, K. F.,** *Physical Oceanography of Coastal Waters*, Ellis Horwood, Chichester, U.K., 1983.
2. **Hogg, N. G.,** A least-squares fit of the advective-diffusive equations to Levitus Atlas data, *J. Mar. Res.*, 45, 347, 1987.
3. **Krauss, W. and Böning, C. W.,** Lagrangian properties of eddy fields in the northern North Atlantic as deduced from satellite-tracked buoys, *J. Mar. Res.*, 45, 259, 1987.
4. **Holland, W. R., Harrison, D. E., and Semtner, A. J.,** Eddy-resolving numerical models of large-scale ocean circulation, in *Eddies in Marine Science*, Robinson, A. R., Ed., Springer-Verlag, New York, 1983, 329.
5. **Robinson, A. R., Ed.,** Overview and summary of eddy science, in *Eddies in Marine Science*, Springer-Verlag, New York, 1983, 3.
6. **Schmitz, W. J., Jr. and Holland, W. R.,** A preliminary comparison of selected numerical eddy-resolving general circulation experiments with observations, *J. Mar. Res.*, 40, 75, 1982.
7. **Robinson, I. S.,** *Satellite Oceanography: An Introduction for Oceanographers and Remote-Sensing Scientists*, Ellis Horwood, Chichester, U.K., 1985.
8. **Gross, M. G.,** *Oceanography: A View of the Earth*, 3rd ed., Prentice-Hall, Englewood Cliffs, N. J., 1982.
9. **Richardson, P. L.,** Eddy kinetic energy in the North Atlantic from surface drifters, *J. Geophys. Res.*, 88, 4355, 1983.
10. **Krauss, W. and Käse, R. H.,** Mean circulation and eddy kinetic energy in the eastern North Atlantic, *J. Geophys. Res.*, 89, 3407, 1984.
11. **Flatte, S. M., Dashen, R. D., Munk, W. H., Watson, K. M., and Zachariasen, R.,** *Sound Transmission Through a Fluctuating Ocean*, Cambridge University Press, London, 1979.
12. **Akal, T. and Berkson, J. M., Eds.,** *Ocean Seismo-Acoustics*, Vol. 16, NATO Conference Series IV: Marine Science, Plenum Press, New York, 1986.
13. **Merklinger, H. M., Ed.,** *Progress in Underwater Acoustics*, Plenum Press, New York, 1987.
14. **Brink, K. H., Halpern, D., and Smith, R. L.,** Circulation in the Peruvian upwelling system near 15S, *J. Geophys. Res.*, 85, 4036, 1980.
15. **Shaffer, G.,** On the upwelling circulation over the wide shelf off Peru. 2. Vertical velocities, internal mixing, and heat balance, *J. Mar. Res.*, 44, 227, 1986.
16. **Le Févre, J.,** Aspects of the biology of frontal systems, in *Advances in Marine Biology*, Vol. 23, Blaxter, J. H. S. and Southward, A. J., Eds., Academic Press, London, 1986, 163.
17. **Legeckis, R.,** A survey of worldwide sea surface temperature fronts detected by environmental satellites, *J. Geophys. Res.*, 83, 4501, 1978.

18. **Wright, L. D.,** Benthic boundary layers of estuarine and coastal environments, *Rev. Aquat. Sci.,* in press.
19. **Smith, D. G., Ed.,** *Cambridge Encyclopedia of Earth Sciences*, Cambridge University Press, London, 1981.
20. **Levinton, J. S.,** *Marine Ecology*, Prentice-Hall, Englewood Cliffs, N. J., 1982.
21. **Abarbanel, I. and Young, W. R., Eds.,** *General Circulation of the Ocean*, Springer-Verlag, New York, 1987.

3.1. TEMPERATURE, SALINITY, AND DENSITY DISTRIBUTION

Figure 3.1—1
AVERAGE SURFACE TEMPERATURE, SALINITY, AND DENSITY VARIATION WITH LATITUDE FOR ALL OCEANS

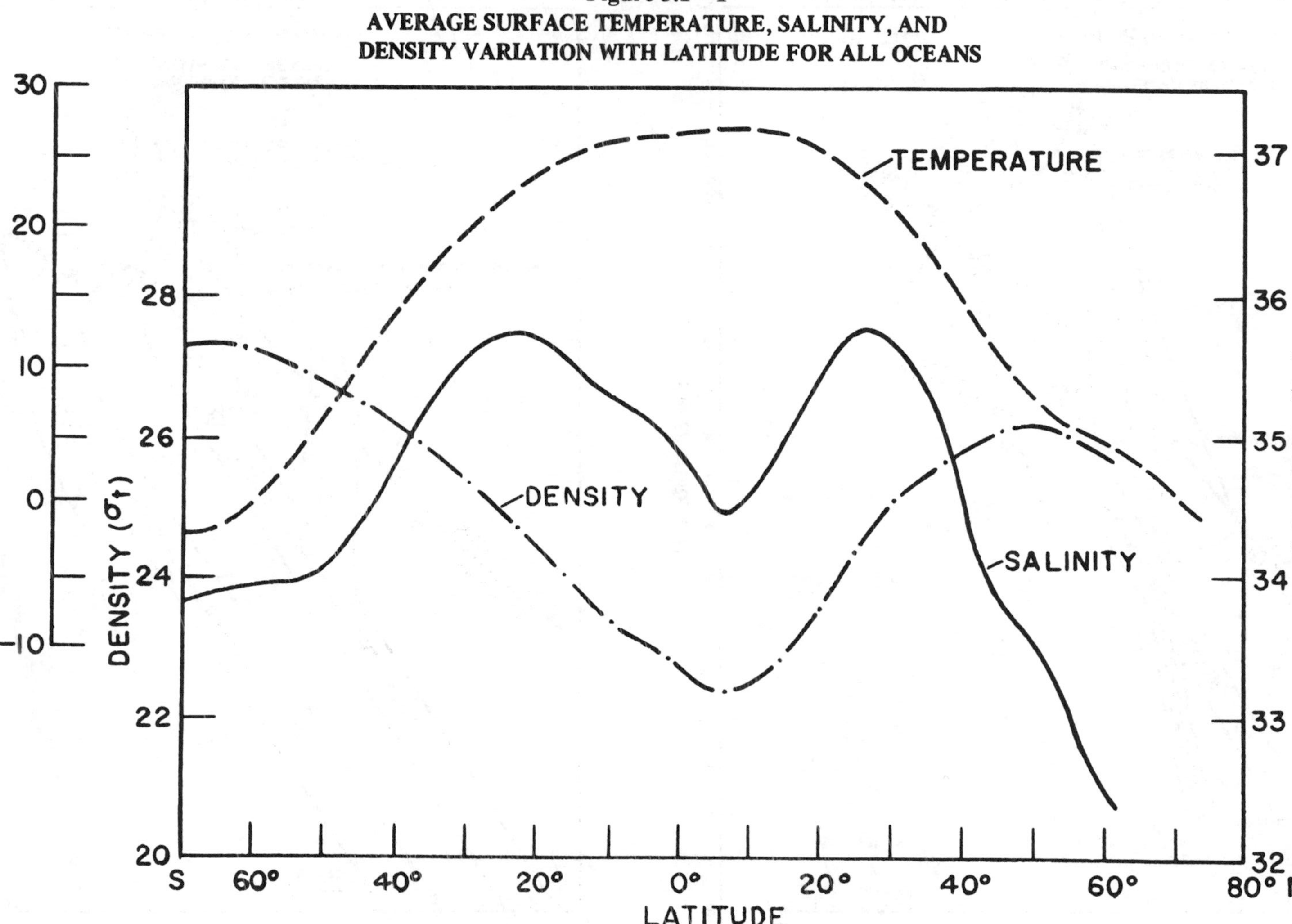

(From Pickard, G.L., *Descriptive Physical Oceanography*, Pergamon Press, New York, 1964. With permission.)

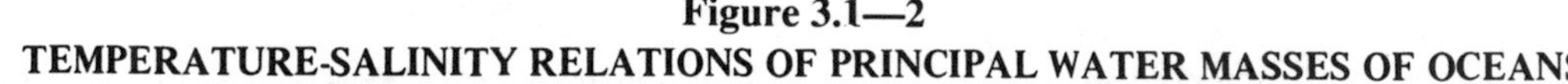

Figure 3.1—2

TEMPERATURE-SALINITY RELATIONS OF PRINCIPAL WATER MASSES OF OCEANS

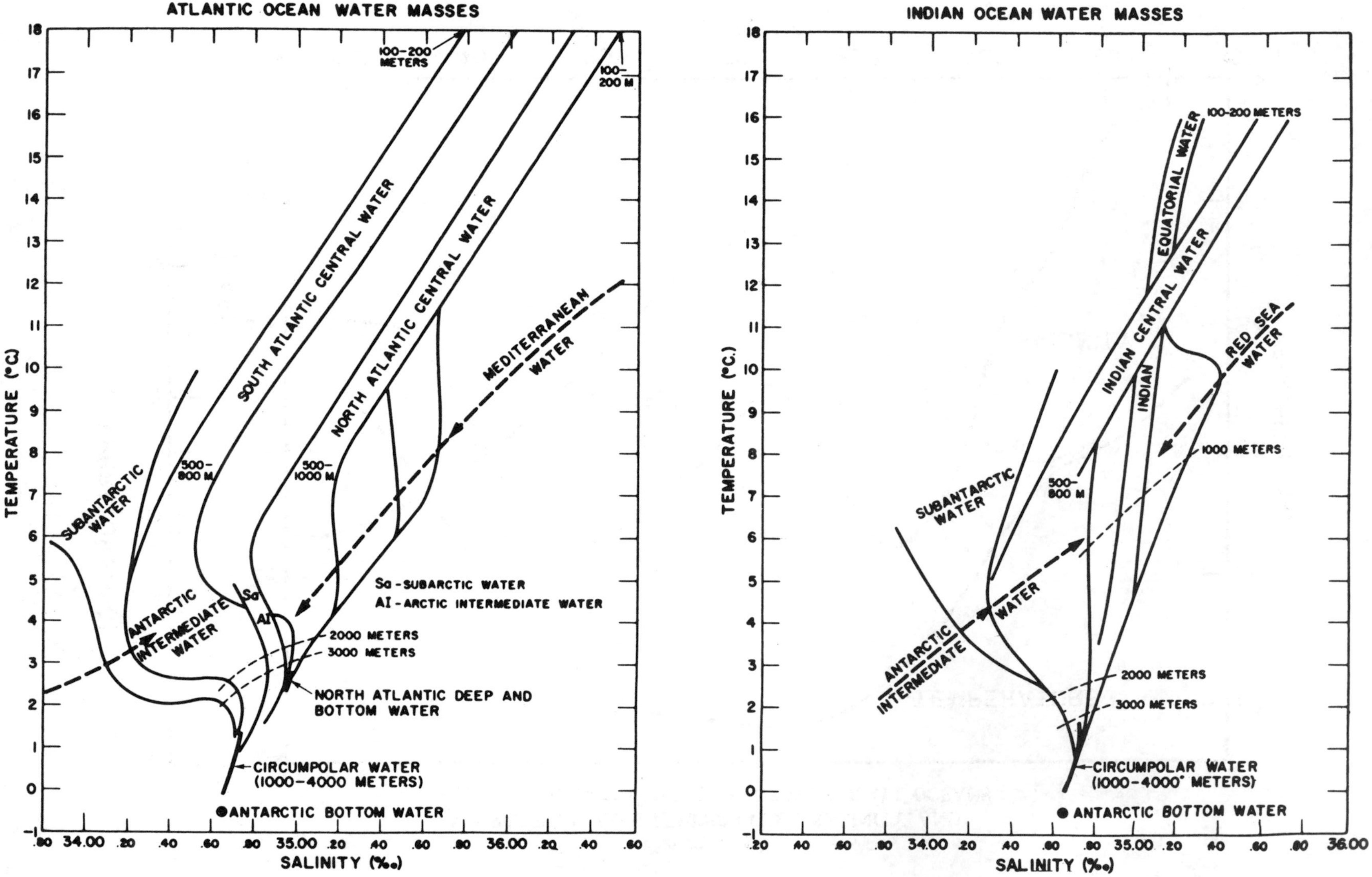

(From Sverdrup, H.U., Johnson, M., and Fleming, R.H., *The Oceans, Their Physics, Chemistry, and General Biology*, Prentice-Hall, Inc., Englewood Cliffs, N.J., © 1942, renewed 1970. With permission.)

Figure 3.1—3
TEMPERATURE-SALINITY RELATIONS OF PRINCIPAL WATER MASSES OF OCEANS

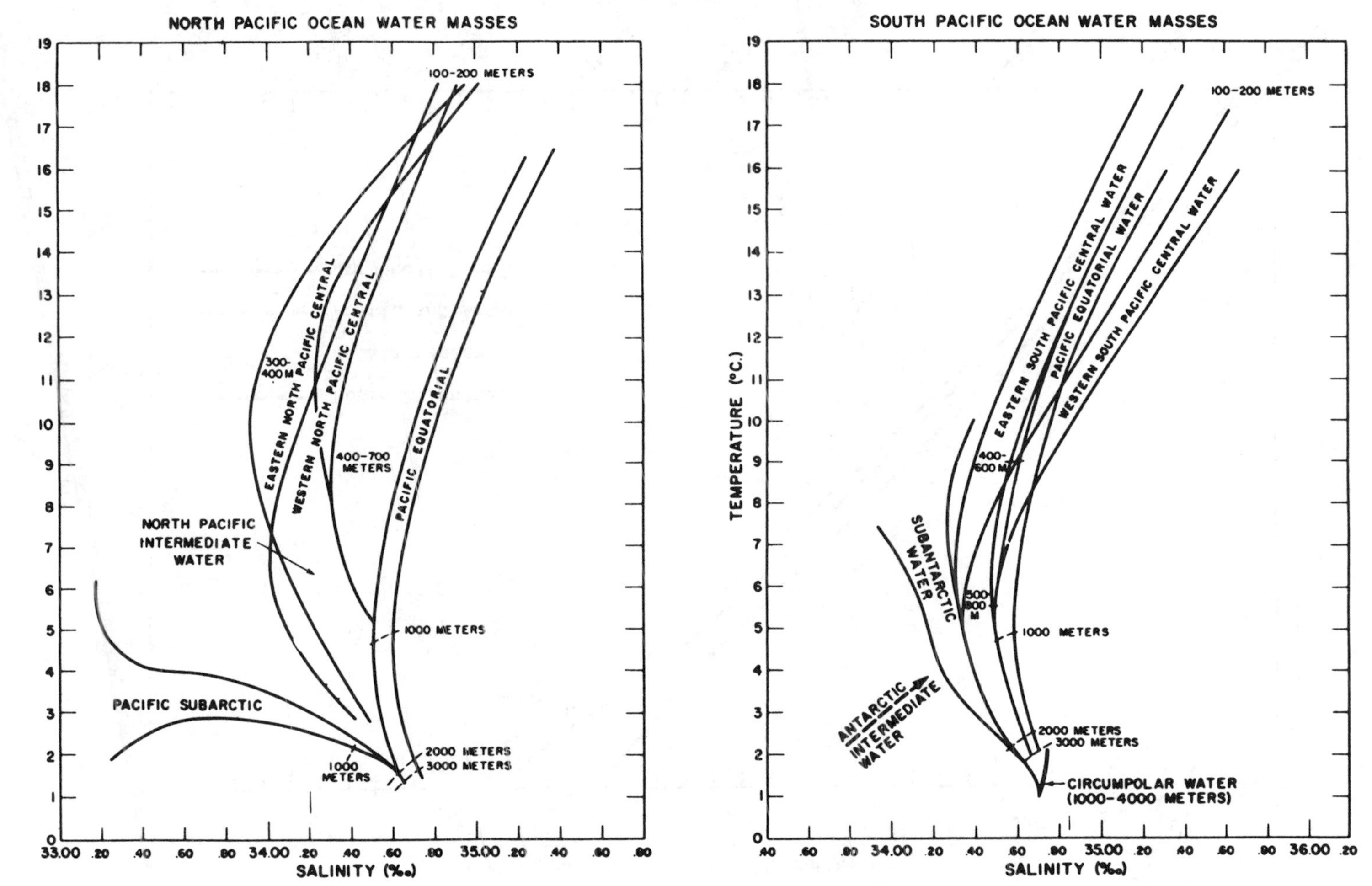

(From Sverdrup, H.U., Johnson, M., and Fleming, R.H., *The Oceans, Their Physics, Chemistry, and General Biology*, Prentice-Hall, Inc., Englewood Cliffs, N.J., © 1942, renewed 1970. With permission.)

Figure 3.1—4
PRESSURE CHANGES WITH DEPTH

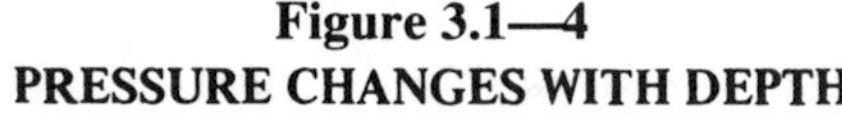

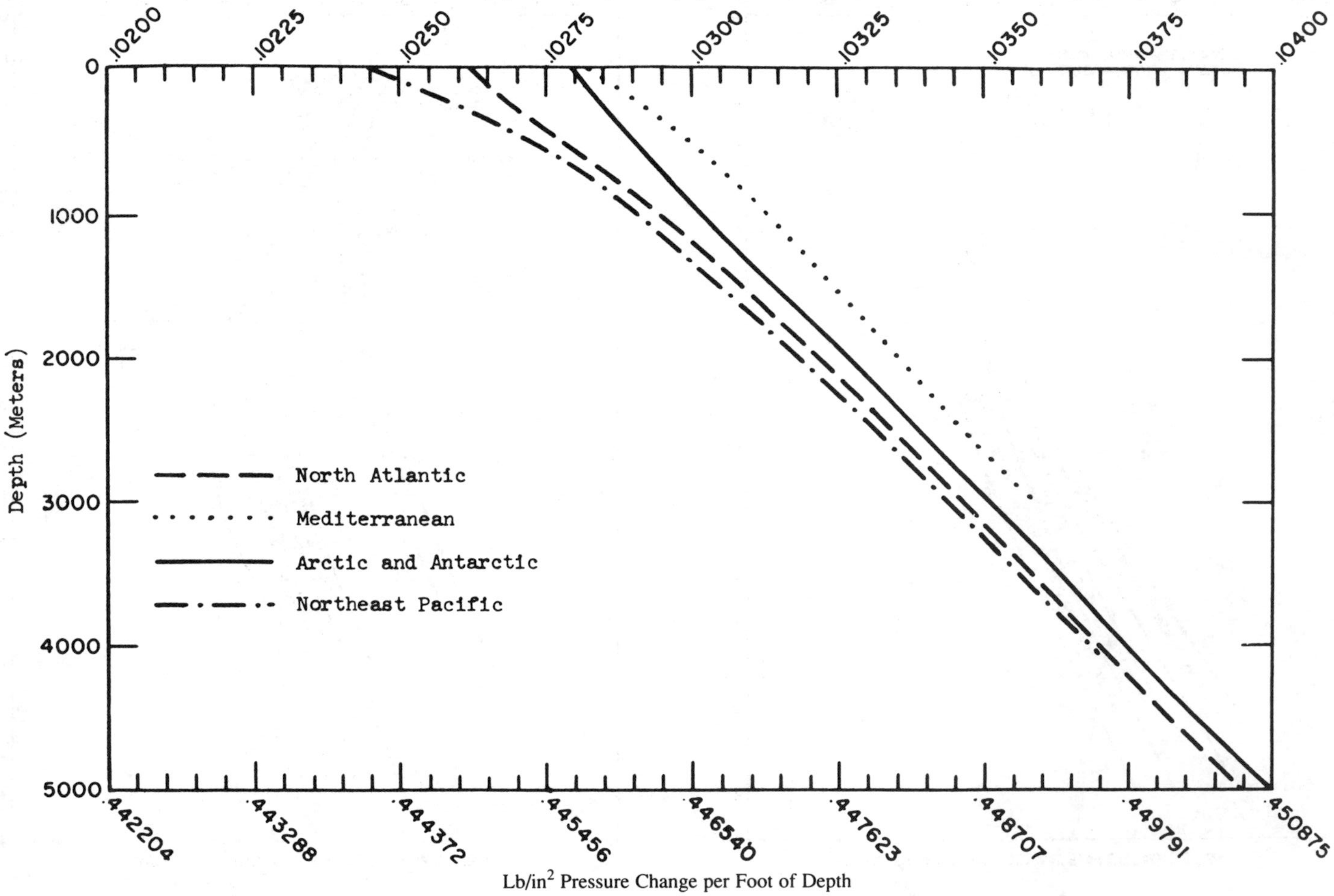

(From Bialek, E. L., Errors in the determination of depth, Int. Hydrograph. Rev., 43, 1966. Tables from *International Oceangraphic Tables*, Vol. I. © Unesco 1966. Reproduced by permission of Unesco.)

Table 3.1—1
MEAN VERTICAL TEMPERATURE (°C) DISTRIBUTION IN THE THREE OCEANS BETWEEN 40°N. AND 40°S.

	Atlantic Ocean		Indian Ocean		Pacific Ocean		Mean	
Depth (m)	°C	Δ°C/ 100 m	°C	Δ°C/ 100 m	°C	Δ°C/ 100 m	°C	Δ°C/ 100 m
0	20.0		22.2		21.8		21.3	
		2.2		3.3		3.1		2.8
100	17.8		18.9		18.7		18.5	
		4.4†		4.7†		4.4†		4.5†
200	13.4		14.3		14.3		14.0	
		1.8		1.6		2.6		2.0
400	9.9		11.0		9.0		10.0	
		1.5		1.2		1.2		1.3
600	7.0		8.7		6.4		7.4	
		0.7		0.9		0.65		0.75
800	5.6		6.9		5.1		5.9	
		0.35		0.7		0.4		0.5
1000	4.9		5.5		4.3		4.9	
		0.20		0.4		0.4		0.35
1200	4.5		4.7		3.5		4.2	
		0.15		0.3		0.2		0.22
1600	3.9		3.4		2.6		3.3	
		0.12		0.15		0.1		0.12
2000	3.4		2.8		2.15		2.8	
		0.08		0.09		0.05		0.07
3000	2.6		1.9		1.7		2.1	
		0.08		0.03		0.03		0.05
4000	1.8		1.6		1.45		1.6	

† Maximum.

(From Defant, A., *Physical Oceanography,* Vol. II, Pergamon Press, New York, 1961. With permission.)

Table 3.1—2
RAPID COMPUTATION OF POTENTIAL TEMPERATURE

Adiabatic cooling in 0.01°C when sea water ($S‰ = 34.85‰$, $\delta_0 = 28.0$), which has a temperature of t_m at the depth of m meters, is raised from that depth to the surface

m/tm	−2°	−1°	0°	1°	2°	3°	4°	5°	6°	7°	8°	9°	10°
	−	−	−	−	−	−	−	−	−	−	−	−	−
1000	2.6	3.5	4.4	5.3	6.2	7.0	7.8	8.6	9.5	10.2	11.0	11.7	12.4
2000	7.2	8.9	10.7	12.4	14.1	15.7	17.2	18.8	20.4	21.9	23.3	24.8	26.2
3000	13.6	16.1	18.7	21.2	23.6	25.9	28.2	30.5	32.7	34.9	37.1	39.2	41.2
4000	21.7	25.0	28.4	31.6	34.7	37.7	40.6	43.5	46.3	49.1	51.9	54.6	57.2
5000	31.5	35.5	39.6	43.4	47.2	50.9	54.4						
6000	42.8	47.5	52.2	56.7	61.1	65.3	69.4						
7000			66.2	71.3	76.2	80.9	85.5						
8000			81.5	87.1	92.5	97.7	102.7						
9000			98.1	104.1	109.9	115.6	121.0						
10000			115.7	122.1	128.3	134.4	140.2						

(From Wüst, G., *Table for Rapid Computation of Potential Temperature,* Technical Report CU-9-61 AT(30-1) 1808 Geol., Lamont Geological Observatory, Palisades, New York, 1961. With permission.)

Table 3.1—3
RAPID COMPUTATION OF POTENTIAL TEMPERATURE

Adiabatic heating in 0.01°C when sea water ($S‰ = 34.85‰$, $\delta_0 = 28.0$), which has a temperature of t_o at the surface, sinks from the surface to a depth of m meters

m/t_o	−2°	−1°	0°	1°	2°	3°	4°	5°	6°	7°	8°	9°	10°
	+	+	+	+	+	+	+	+	+	+	+	+	+
1000	2.6	3.6	4.5	5.4	6.2	7.1	7.9	8.7	9.5	10.3	11.1	11.8	12.5
2000	7.3	9.1	10.9	12.7	14.3	16.0	17.5	19.1	20.7	22.2	23.7	25.1	26.5
3000	13.9	16.6	19.2	21.8	24.2	26.7	28.9	31.2	33.4	35.6	37.8	39.9	41.9
4000	22.4	25.9	29.3	32.6	35.8	39.0	41.9	44.8	47.7	50.5	53.4	56.1	58.7
5000	32.8	37.0	41.2	45.1	49.0	52.8	56.4						
6000	44.9	49.8	54.7	59.3	63.8	68.1	72.3						
7000		64.3	69.8	75.0	80.0	84.8	89.5						
8000		80.4	86.4	92.1	97.6	102.9							
9000		97.9	104.4	110.5	116.5	122.2							
10000		116.7	123.7	130.2	136.6	142.7							

(From Wüst, G., *Tables for Rapid Computation of Potential Temperature,* Technical Report CU-9-61 AT(30-1) 1808 Geol., Lamont Geological Observatory, Palisades, New York, 1961. With permission.)

Table 3.1—4
RAPID COMPUTATION OF POTENTIAL TEMPERATURE

Adiabatic variations of temperature in 0.01°C for the upper 1000 m of sea water at different salinities

S°/oo	0°C	2°	4°	6°	8°	10°	12°	14°	16°	18°	20°	22°
30.0	3.5	5.3	7.0	8.7	10.3	11.8	13.2	14.7	16.1	17.6	18.9	20.3
32.0	3.9	5.7	7.3	9.0	10.6	12.1	13.5	15.0	16.4	17.8	19.1	20.5
34.0	4.3	6.0	7.7	9.4	10.9	12.4	13.8	15.3	16.6	18.0	19.3	20.7
36.0	4.7	6.4	8.1	9.7	11.2	12.7	14.1	15.5	16.9	18.3	19.6	20.9
38.0	5.1	6.8	8.4	10.0	11.6	13.0	14.4	15.8	17.2	18.5	19.8	21.1

(From Wüst, G., *Tables for Rapid Computation of Potential Temperature,* Technical Report CU-9-61 AT(30-1) 1808 Geol., Lamont Geological Observatory, Palisades, New York, 1961. With permission.)

Table 3.1—5

RAPID COMPUTATION OF POTENTIAL TEMPERATURE

Adiabatic variations of temperature in 0.01°C in Mediterranean sea water of (S°/oo = 38.57°/oo, δ_0 31.0)

	t_m (raising)			t_o (sinking)		
m	12°	13°	14°	12°	13°	14°
	–	–	–	+	+	+
1000	14.4	15.1	15.8	14.5	15.3	16.0
2000	30.0	31.4	32.7	30.4	31.8	33.1
3000	46.6	48.6	50.6	47.4	49.4	51.4
4000	64.2	66.7	69.2	65.7	68.3	70.8

(From Wust, G., *Tables for Rapid Computation of Potential Temperature*, Technical Report, CU-9-61 AT(30-1) 1808 Geol., Lamont Geological Observatory, Palisades, New York, 1961. With permission.)

3.2 WAVES

Table 3.2—1
RELATIVE FREQUENCY OF WAVES OF DIFFERENT HEIGHTS IN DIFFERENT REGIONS

Ocean region	Height of waves, ft					
	0–3	3–4	4–7	7–12	12–20	$\geqq$20
North Atlantic, between Newfoundland and England	% 20	% 20	% 20	% 15	% 10	% 15
Mid-equatorial Atlantic	20	30	25	15	5	5
South Atlantic, latitude of Argentina	10	20	20	20	15	10
North Pacific, latitude of Oregon and south of Alaskan Peninsula	25	20	20	15	10	10
East equatorial Pacific	25	35	25	10	5	5
West wind belt of South Pacific, latitude of southern Chile	5	20	20	20	15	15
North Indian Ocean, Northeast monsoon season	55	25	10	5	0	0
North Indian Ocean, Southwest monsoon season	15	15	25	20	15	10
Southern Indian Ocean, between Madagascar and northern Australia	35	25	20	15	5	5
West wind belt of southern Indian Ocean, on route between Cape of Good Hope and southern Australia	10	20	20	20	15	15

(From Bigelow, H. B. and Edmondson, W. T., *Wind Waves at Sea, Breakers, and Surf,* U.S. Navy Hydrographic Office, Washington, D.C., Pub. 602, 1962.)

Table 3.2—2
LENGTH OF STORM WAVES OBSERVED IN DIFFERENT OCEANS

Ocean area	Wave length, ft			No. cases
	Max	Min	Avg	
North Atlantic	559	115	303	15
South Atlantic	701	82	226	32
Pacific	765	80	242	14
Southern Indian	1121	108	360	23
China Sea	261	160	197	3

(From Bigelow, H. B. and Edmondson, W. T., *Wind Waves at Sea, Breakers, and Surf,* U.S. Navy Hydrographic Office, Washington, D.C., Pub. 602, 1962.)

Figure 3.2—1
ISOLINES OF RATIO OF CREST DISPLACEMENT TO WAVE HEIGHT, η_c/H

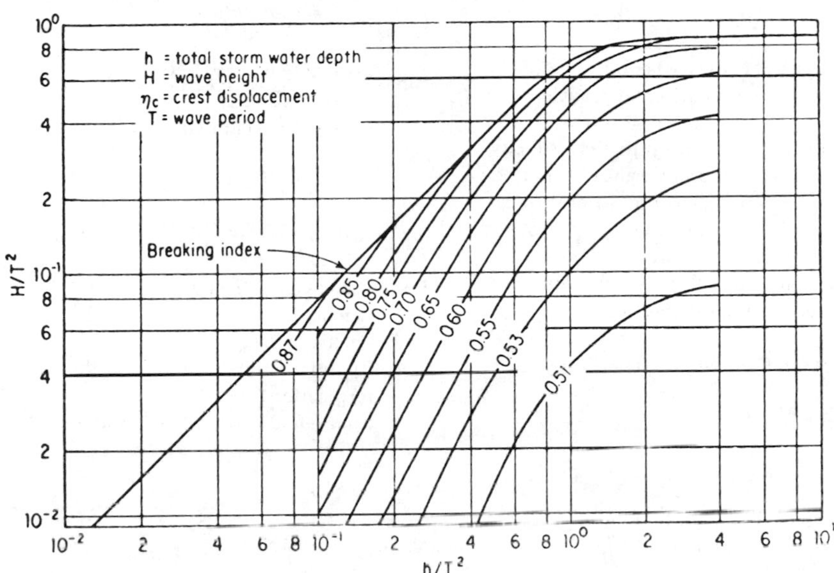

(From Dean, R. G. in *Handbook of Ocean and Underwater Engineering,* Myers, J. J., Ed., McGraw-Hill, New York, 1969. With permission.)

Table 3.2—3
AIRY WAVE

The amplitude A is found to be half the wave height, $H/2$; the coefficients a_i and b_i are unity. Hence

$$\phi_1 = -\frac{H}{2}C\frac{\cosh k(d+z)}{\sinh kd}\sin\theta$$

$$\eta = \frac{H}{2}\cos\theta$$

$$u = \frac{\pi H}{T}\frac{\cosh k(d+z)}{\sinh kd}\cos\theta$$

$$w = \frac{\pi H}{T}\frac{\sinh k(d+z)}{\sinh kd}\sin\theta$$

$$\frac{\partial u}{\partial t} = \frac{2\pi^2 H}{T^2}\frac{\cosh k(d+z)}{\sinh kd}\sin\theta$$

$$\frac{\partial w}{\partial t} = -\frac{2\pi^2 H}{T^2}\frac{\sinh k(d+z)}{\sinh kd}\cos\theta$$

$$p = \rho g\frac{H}{2}\frac{\cosh k(d+z)}{\cosh kd}\cos\theta - \rho gz$$

$$C^2 = \frac{g}{k}\tanh kd$$

where

$$\theta = kx - \omega t$$

$$k = \frac{2\pi}{L}$$

$$\omega = \frac{2\pi}{T}$$

The group velocity (coinciding with the rate of transmission of energy) is given by

$$C_g = nC \quad \text{where}$$

$$n = \frac{1}{2}\left[1 + \frac{2kd}{\sinh 2kd}\right]$$

Because of the conservation of energy, a wave traveling from infinitely deep water into water of finite depth will change height from H_0 to H, where

$$H = K_S H_0$$

and

$$K_S = \tanh kd\left(1 + \frac{2kd}{\sinh 2kd}\right)^{-1/2}$$

The relationships of wavelength, wave period, and water depth are illustrated in figure 3.2—2. A brief table of pertinent hyperbolic functions and a summary of the Airy wave theory is presented in table 3.2—4.

(From Bretschneider, C. L., in *Handbook of Ocean and Underwater Engineering,* Myers, J. J., Ed., McGraw-Hill, New York, 1969. With permission.)

Table 3.2—4
HYPERBOLIC FUNCTIONS AND SUMMARY OF AIRY THEORY

a = horizontal amplitude
b = vertical amplitude
c = wave velocity
c_g = wave group velocity
d = water depth
$k = 2\pi/L$
g = acceleration due to gravity

H = wave height
L = wavelength
p = pressure
T = wave period
U = horizontal particle velocity
V = vertical particle velocity
ρ = unit weight of water
η = height of water surface above mean

Values for deep water are given by the subscript zero. Amplitudes are measured from mean position.

Basic Formulas

$$a_{max}(z) = \frac{H \cosh k(d+z)}{2 \sinh kd}$$

$$b_{max}(z) = \frac{H \sinh k(d+z)}{2 \sinh kd}$$

$$U_{max}(z) = \frac{\pi H \sinh k(d+z)}{T \sinh kd}$$

$$V_{max}(z) = \frac{\pi H \sinh k(d+z)}{T \sinh kd}$$

$$c = \sqrt{\frac{gL}{2\pi} \tanh kd}$$

$$c_g = \frac{1}{2} C \left(1 + \frac{kd}{\sinh kd}\right)$$

$$p_{max}(z) = \rho \frac{H}{2} \frac{\cosh k(d+z)}{2 \cosh kd} + \rho g(d+z)$$

For given values of T and d
1. Calculate $L_0 = g/2\pi T^2 = 5.12T^2$ ft.
2. Calculate d/L_0.
3. Find the desired values from the tables (by linear interpolation).
4. Find L and C from $L = L_0 \tanh kd$ $0 = L/T$.

$\frac{d}{L_0}$	tanh kd	$\frac{d}{L}$	kd	sinh kd	cosh kd	$\frac{2kd}{\sinh 2kd}$	$\frac{H}{H_0}$
0.00	0.000	0.0000	0.000	0.000	1.00	1.000	
.01	.248	.0403	.253	.256	.03	0.958	1.44
.02	.347	.0576	.362	.370	.07	.918	.23
.03	.420	.0714	.448	.463	.10	.877	.12
.04	.480	.0833	.523	.548	.14	.839	.06
0.05	0.531	0.0942	0.592	0.627	1.18	0.800	1.02
.06	.575	.104	.655	.703	.22	.763	0.993
.07	.614	.114	.716	.778	.27	.725	.971
.08	.649	.123	.774	.854	.31	.690	.955
.09	.681	.132	.831	0.930	.36	.654	.942
0.10	0.709	0.141	0.886	1.01	1.42	0.621	0.933
.11	.735	.150	.940	.08	.48	.587	.926
.12	.759	.158	0.994	.16	.54	.555	.920
.13	.780	.166	1.05	.25	.60	.524	.917
.14	.800	.175	.10	.33	.67	.494	.915
0.15	0.818	0.183	1.15	1.42	1.74	0.465	0.913
.16	.835	.192	.20	.52	.82	.437	.913
.17	.850	.200	.26	.61	.90	.410	.913
.18	.864	.208	.31	.72	1.99	.384	.914
.19	.877	.217	.36	.82	2.08	.359	.916
0.20	0.888	0.225	1.41	1.94	2.18	0.335	0.918
.21	.899	.234	.47	2.06	.28	.313	.920
.22	.909	.242	.52	.18	.40	.291	.923
.23	.918	.251	.58	.31	.52	.270	.926
.24	.926	.259	.63	.45	.65	.251	.929
0.25	0.933	0.268	1.68	2.60	2.78	0.233	0.932
.26	.940	.277	.74	.76	2.93	.215	.936
.27	.946	.285	.79	2.92	3.09	.199	.939
.28	.952	.294	.85	3.10	.25	.183	.942
.29	.957	.303	.90	.28	.43	.169	.946

Table 3.2—4 (continued)
HYPERBOLIC FUNCTIONS AND SUMMARY OF AIRY THEORY

$\frac{d}{L_0}$	tanh *kd*	$\frac{d}{L}$	*kd*	sinh *kd*	cosh *kd*	$\frac{2kd}{\sinh 2kd}$	$\frac{H}{H_0}$
0.30	0.961	0.312	1.96	3.48	3.62	0.155	0.949
.31	.965	.321	2.02	.69	3.83	.143	.952
.32	.969	.330	.08	3.92	4.04	.131	.955
.33	.972	.339	.13	4.16	.28	.120	.958
.34	.975	.349	.19	.41	.52	.110	.961
0.35	0.978	0.358	2.25	4.68	4.79	0.100	0.964
.36	.980	.367	.31	4.97	5.07	.091	.967
.37	.982	.377	.37	5.28	.37	.083	.969
.38	.984	.386	.42	.61	5.70	.076	.972
.39	.986	.396	.48	5.96	6.04	.069	.974
0.40	0.988	0.405	2.54	6.33	6.41	0.063	0.976
.41	.989	.414	.60	6.72	6.80	.057	.978
.42	.990	.424	.66	7.15	7.22	.052	.980
.43	.991	.434	.72	7.60	7.66	.047	.982
.44	.992	.443	.79	8.08	8.14	.042	.983
0.45	0.993	0.453	2.85	8.58	8.64	0.038	0.985
.46	.994	.463	.91	9.13	9.19	.035	.986
.47	.995	.472	2.97	9.71	9.76	.031	.987
.48	.995	.482	3.03	10.3	10.4	.028	.988
.49	.996	.492	.09	11.0	11.0	.026	.990
0.50	0.996	0.502	3.15	11.7	11.7	0.023	0.990

(From Lundgren, H., Copenhagen, Denmark, personal communication in *Handbook of Ocean and Underwater Engineering,* Myers, J. J., Ed., McGraw-Hill, New York, 1969. With permission.)

Figure 3.2—2
RELATIONSHIP OF WAVE PERIOD, LENGTH, AND DEPTH

Airy wave theory

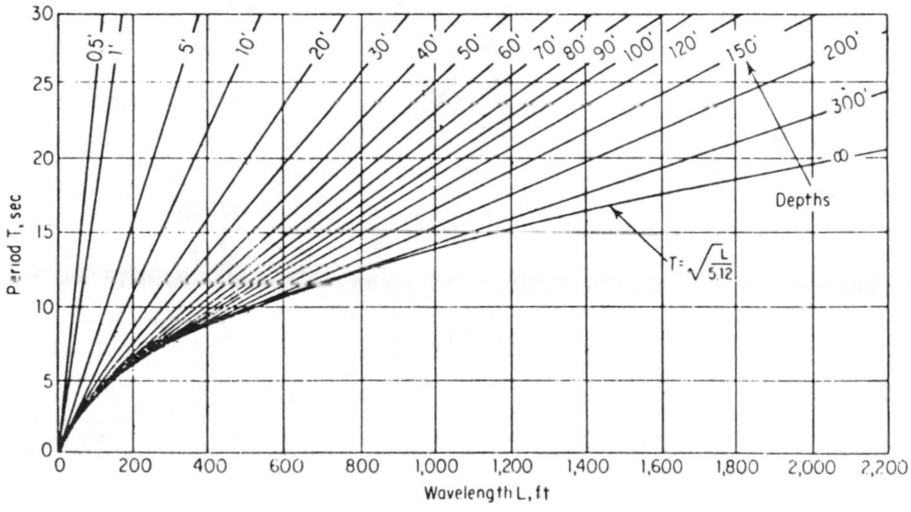

(From Bretschneider, C. L. in *Handbook of Ocean and Underwater Engineering,* Myers, J. J., Ed., McGraw-Hill, New York, 1969. With permission.)

3.3 CURRENTS AND TRANSPORT CHARTS

Figure 3.3—1
TRANSPORT CHART OF THE NORTH PACIFIC

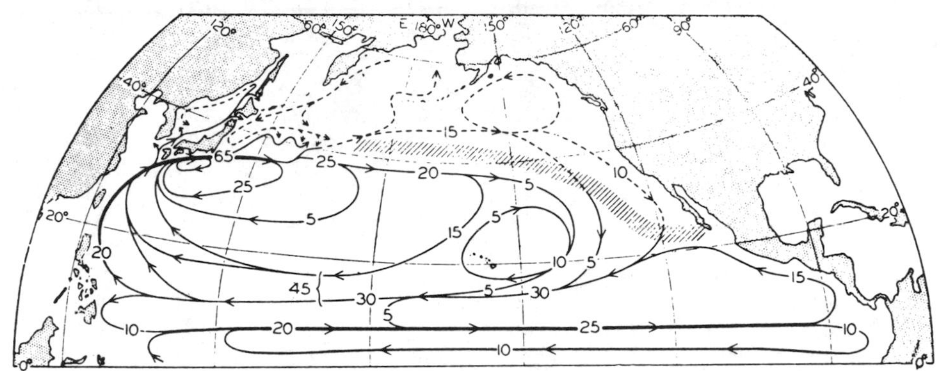

Note: The lines with arrows indicate the approximate direction of the transport above 1500 m, and the inserted numbers indicate the transported volumes in mil m³/sec. Dashed lines show cold currents; full-drawn lines show warm currents.

(From Sverdrup, H. U., Johnson, M. W., and Fleming, R. H., *The Oceans, Their Physics, Chemistry, and General Biology*. © 1942, renewed 1970, pp. 66, 615, 684, 727, 744, 116. Reprinted by permission of Prentice-Hall, Inc., Englewood Cliffs, N. J.)

Figure 3.3—2
TRANSPORT LINES AROUND THE ANTARCTIC CONTINENT

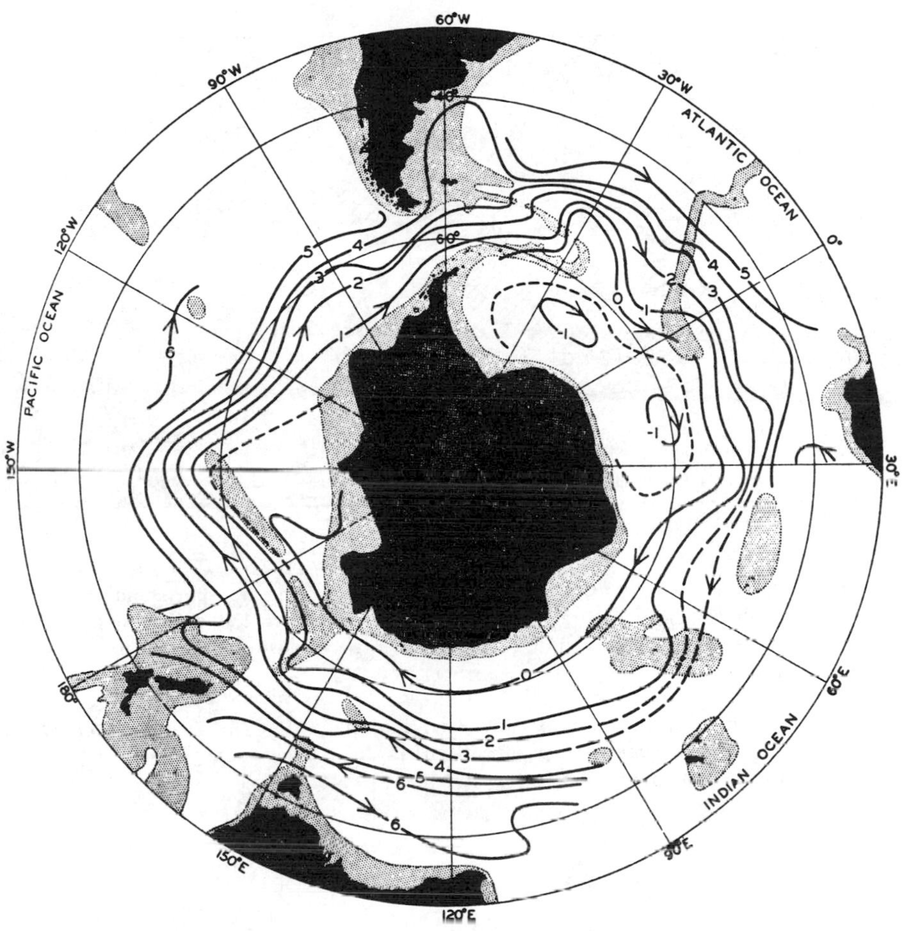

Note: Between two lines the transport relative to the 3000-decibar surface is about 20 mil m^3/sec.

(From Sverdrup, H.U., Johnson, M.W., and Fleming, R.H., *The Oceans, Their Physics, Chemistry, and General Biology,* Prentice-Hall, Inc., Englewood Cliffs, N.J., © 1942, renewed 1970. With permission.)

Figure 3.3—3
TRANSPORT OF CENTRAL WATER AND SUBARCTIC WATER IN THE ATLANTIC OCEAN

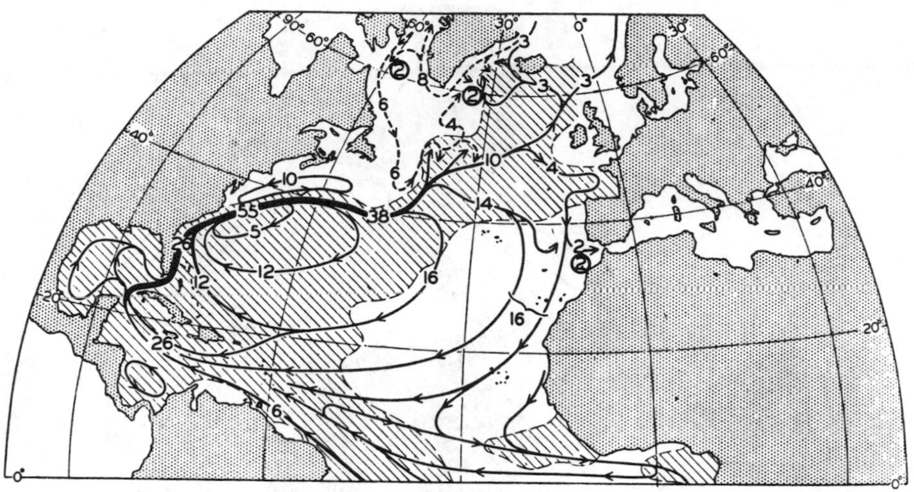

Note: The lines with arrows indicate the direction of the transport, and the inserted numbers indicate the transported volumes in mil m^3/sec. Full-drawn lines show warm currents; dashed lines show cold currents. Areas of positive temperature anomaly are shaded.

(From Sverdrup, H.U., Johnson, M.W., and Fleming, R.H., *The Oceans, Their Physics, Chemistry, and General Biology,* Prentice-Hall, Inc., Englewood Cliffs, N.J., © 1942, renewed 1970. With permission.)

Section 4
Marine Geology

Section 4

MARINE GEOLOGY

The study of the geology of the seafloor, commonly termed marine geology, has been revolutionized in many respects by the theory of plate tectonics. Supported by the concepts of continental drift proposed by Wegener,[1] thermal convection in the mantle invoked by Holmes,[2] and seafloor spreading advocated by Hess (see Reference 3), the theory of plate tectonics was embraced by most earth scientists as a unifying theme in geology during the late 1960s and early 1970s. Data derived from worldwide magnetic anomaly surveys, paleontological investigations, earthquakes, heat flow measurements, determinations of the age and thickness of sedimentary layers of the seafloor, and other areas of research confirmed the existence of continental drift and seafloor spreading and contributed to the final resolution of plate tectonics.

While once viewed as rigid with fixed continents and ocean basins, the earth is now envisaged as a mosaic of a dozen or so dynamic lithospheric plates which move on a partially molten, plastic asthenosphere. Francheteau[4] lists the lithospheric plates as follows: the Eurasian, Anatolian, Arabian, African, Somali, Antarctic, Caribbean, South American, North American, Gorda, Cocos, Nazca, Pacific, Philippine, and Indian Australian plates. The thickness of the plates amounts to approximately 70 km under the oceans and 100 to 150 km under the continents. Individual plates behave as rigid bodies, but at their boundaries, the plates interact in three ways: (1) by spreading or diverging, as at mid-ocean ridges; (2) by converging, as at deep-sea trenches; and (3) by sliding past each other, as at transform faults or fracture zones. Forces operating at the plate boundaries include tension, at areas of divergence; compression, at zones of convergence; and shearing, at transform faults, the latter resulting in shallow-focus earthquakes. The relative motions of the plates account for tectonic activity only at their margins, concentrating most earthquakes and volcanoes on the earth's surface at these perimeters.

Mid-ocean ridges represent regions where new lithosphere is created, as partially molten mantle material upwells at the ridge axis. These spreading centers are sites of active basaltic volcanism, shallow-focus earthquakes, and high rates of heat flow. The rate of formation of new seafloor varies at fast-, intermediate-, and slow-spreading ocean centers. At fast-spreading centers, seafloor genesis generally exceeds 9 cm/year. The East Pacific Rise at 13°N latitude (11 to 12 cm/year) and 20°S latitude (16 to 18 cm/year) provides examples.[5] Intermediate rates of seafloor spreading range from 5 to 9 cm/year;[6] for example, the Galapagos Rift yields values of 6 to 7 cm/year.[7] Low rates of seafloor spreading (less than 2 cm/year) have been discerned on the mid-Atlantic ridge near 26°N latitude.[8] Central rift valleys of 50 to 200 m depth develop at slow and intermediate rates of seafloor spreading, but not at fast-spreading rates.[6]

The lithosphere, while being generated at mid-ocean ridges, is resorbed along subduction zones. At boundaries of convergence, lithospheric plates collide, with one plate overriding another. The overriden plate sinks into the mantle and remelts. Deep-sea trenches mark areas of subducted oceanic crust beneath less dense continental lithosphere. By consumption of lithosphere at convergence boundaries, the earth maintains constancy in size. Characteristic features of convergence boundaries are deep-sea trenches, volcanic island arcs, shallow- and deep-focus earthquakes, adjacent mountain ranges of folded and faulted rocks, and basaltic and andesitic volcanism.[9]

The topography of the oceans can now be explained in light of the mechanisms of plate tectonics. The 75,000-km global length of mid-ocean ridges encircling the earth arises from the injection of mantle material at diverging plate boundaries. These volcanic ridges comprise a mountain system comparable in physical dimensions to those on the continents. As the lithosphere cools and subsides on either side of the mid-ocean ridge, the elevations of the submarine volcanic mountains decline, and the topography becomes less rugged. These abyssal hills typically are thinly veneered with marine sediment. Further removed from the mid-ocean ridge system toward the continents, broad abyssal plains are found. These level segments of the seafloor form by the slow deposition of sand, silt, and clay (turbidites) transported via turbidity currents off of the outer continental shelves and continental slopes, ultimately burying hilly terrane. Also protruding from the seafloor are isolated volcanic mountains (seamounts) rising 1 km or more above their surroundings. Occasionally, the seamounts merge into a chain of aseismic ridges. Flat-topped seamounts, referred to as guyots, develop as the volcanic peaks are eroded during emergence. Thick turbidite deposits underlie the continental rise, a gently sloping topographic feature that extends from the abyssal plain to the continental slope. The original basalt topography, therefore, is buried under a thick apron of sedimentary layers as the lithosphere ages and gradually moves away from the mid-ocean ridge. Proceeding toward a mainland, the relatively steeply inclined continental slope grades into the continental shelf, a gently sloping submerged edge of a continent. The continental shelves characteristically exhibit uneven topography consisting of small hills, basin-like depressions, troughs, and steep-walled submarine canyons cut by turbidity currents. Immediately seaward of submarine canyons lie thick sediments constituting submarine fans. Conspicuous features at the site of a subducting plate are deep-sea trenches, the deepest parts of the oceans reaching 10 km or more below sea level. These long, narrow, and deep troughs caused by the downward thrusting of a lithospheric plate are bordered by volcanic island arcs or a continental margin magmatic belt. A volcanic island arc (e.g., Japan) occurs when the overriding lithospheric plate is oceanic. A continental-margin magmatic belt (e.g., the Andes mountains) exists when the overriding plate is a continent.[9]

In this section, hypsometric data are provided on the ocean basins. Information is given on the location, size, and depths of deep-sea trenches, submarine canyons, and other oceanic areas. Conductivity and heat flow measurements for deep-sea sediments and the Atlantic Ocean, respectively, also are presented.

REFERENCES

1. **Wegener, A.,** *The Origin of Continents and Oceans*, Dover, New York, 1924.
2. **Holmes, A.,** *Principles of Physical Geology*, John Wiley & Sons, New York, 1978.
3. **Anderson, R. N.,** *Marine Geology: A Planet Earth Perspective*, John Wiley & Sons, New York, 1986.
4. **Francheteau, J.,** The oceanic crust, *Sci. Am.*, 249, 114, 1983.
5. **Francheteau, J. and Ballard, R. D.,** The East Pacific Rise near 21°N, 13°N and 20°S: inferences for along-strike variability of axial processes of the mid-ocean ridge, *Earth Planet. Sci. Lett.*, 64, 93, 1983.
6. **Macdonald, K. C.,** Mid-ocean ridges: fine-scale tectonic, volcanic and hydrothermal processes within the plate boundary zone, *Ann. Rev. Earth Planet. Sci.*, 10, 155, 1982.
7. **Grassle, J. F.,** The ecology of deep-sea hydrothermal vent communities, in *Advances in Marine Biology*, Vol. 23, Blaxter, J. H. S. and Southward, A. J., Eds., Academic Press, London, 1986, 301.
8. **Rona, P. A., Klinkhammer, G., Nelsen, T. A., Trefry, J. H., and Elderfield, H.,** Black smokers, massive sulphides and vent biota at the mid-Atlantic ridge, *Nature*, 321, 33, 1986.
9. **Press, F. and Siever, R.,** *Earth*, 4th ed., W. H. Freeman, New York, 1986.

4.1. GENERAL TABLES

Table 4.1–1
EARTH'S DIMENSIONS

	I.E.R.[a]	I.A.U.[b]
Equatorial radius a_e	6,378.388 km	6,378.160 km
Polar radius a_p	6,356.912 km	6,356.775 km
Flattening factor f	1/297	1/298.25

Radius of sphere of equal volume a_0	6,371 km
Area of surface	5.101×10^8 km^2
Volume	1.083×10^{12} km^3
Mass	5.976×10^{27} g
Mean density	5.517 g cm^{-3}
Gravitational constant G	6.670×10^{-8} dynes cm^{-2} g^{-2}
Normal acceleration of gravity at equator g_e (based on Potsdam standard)	978.0436 cm sec^{-2}
Mean solar day d	86,400 sec = 24^{hr}
Sidereal day S	86,164.09 sec = 23^{hr} 56^{m} 4.09^{sec}
Velocity of rotation at equator	465.12 m sec^{-1}
Mean moment of inertia C_0	8.02×10^{44} g cm^2

[a] I.E.R. International Ellipsoid of Reference, 1924.
[b] I.A.U. International Astronomical Union, 1966.

(From Wedepohl, K.H., *Handbook of Geochemistry*, Vol. I, Springer-Verlag, Berlin, 1969. With permission.)

REFERENCES

1. *Astronomer's Handbook*, Transactions of the International Astronomical Union, Vol. XIIC, Academic Press, London, 1966.
2. **Gondolatsch, F.**, Mechanical data of planets and satellites, *Landolt-Bornstein*, New series, Group VI: Astronomy, astrophysics and space research, Vol. I, Springer-Verlag, Berlin, 1965.
3. **MacDonald, G. J. F.**, in Geodetic data, *Handbook of Physical Constants*, Clark, S. P., Jr., Ed., Geological Society of America, Memoir 97, 1966.

Table 4.1–2
EARTH'S INTERIOR, MASSES, AND DIMENSIONS OF THE PRINCIPAL SUBDIVISIONS

	Mass,	Mean density,	Surface area,	Radius or thick-ness,	Volume,	Mean moment of inertia (spherical symmetry),
	10^{25} g	g/cm^3	10^6 km^2	km	10^9 km^3	10^{42} g cm^2
Core	192	11.0	151	3471	1175	90
Mantle	403	4.5			898	705
Below 1000 km	240	5.1	362	1900	474	333
Above 1000 km	163	3.9	505	970–990	424	372
Crust	2.5	2.8	510		8.9	(7)
Continental	2.0	2.75	242[a]	30	7.3	
Oceanic	0.5	2.9	268[a]	6	1.6	
Oceans and marginal seas	0.14	1.03	361[a]	3.8	1.4[a]	(<1)
Whole earth	597.6	5.52	510	6371	1083	802
Atmosphere	0.00051	0.0013[b]	–	8[c]	–	–

[a] See Reference 2.
[b] Surface value.
[c] Scale height of the "homogeneous" atmosphere.

(From Wedepohl, K. H., *Handbook of Geochemistry,* Vol. I, Springer-Verlag, Berlin, 1969. With permission.)

REFERENCES

1. **MacDonald, G. J. F.,** in Geodetic data, *Handbook of Physical Constants,* Clark, S. P., Jr., Ed., Geological Society of America, Memoir 97, 1966.
2. **Poldervaart, A.,** Chemistry of the earth crust, *Geol. Soc. Am. Spec. Pap.,* No. 62, p. 119, 1955.
3. **Schmucker, U.,** in *Handbook of Geochemistry,* Vol. I, Wedepohl, K. H., Ed., Springer-Verlag, Berlin, 1969, chap. 6.

Table 4.1—3
THE SURFACE AREAS OF THE EARTH

	10^6 km^2		10^6 km^2
Continental shield region	105	Land about	
Region of young folded belts	42	29.2% of total	149
Volcanic islands in deep oceanic and suboceanic region	2		
Shelves and continental slopes region	93	Ocean about	
Deep oceanic region	268	70.8% of total	361
		Total surface	510

(From Poldervaart, A., Chemistry of the earth crust, *Geol. Soc. Am. Spec. Pap.*, 62, 119, 1955.)

Table 4.1–4
GEOLOGICAL TIME-SCALES

Era	Period	Epoch	Ref 4[a]	Ref. 1[a]	Ref. 2[a]	Ref. 3,7	Ref. 5
			Time since beginning in mil yr				
Cenozoic	Quaternary	Pleistocene[b]	1	1.5–2	1.5–2	2	
	Tertiary	Pliocence	12	12 ± 1	7	10	
		Miocene	23	26 ± 1	26	27	
		Oligocene	35	37 ± 2	37–38	38	
		Eocene	55	60 ± 2	53–54	55	
		Paleocene	70	67 ± 3	65	65–70	
Mesozoic	Cretaceous		135	137 ± 5	136	130	
	Jurassic		180	195 ± 5	190–195	180	
	Triassic		220	240 ± 10	225	225	
Paleozoic	Permian		270	285 ± 10	280	260	
	Carboniferous		350	340–360	345	340	
	Devonian		400	410 ± 10	395	405	
	Silurian		430	440 ± 15	430–440	435	
	Ordovician		490	500 ± 20	500	480	
	Cambrian		600	570	570	550–570	570
Pre-cambrian	Upper precambrian						1900
	Middle Precambrian						2700
	Lower Precambrian						3500

[a] Used by the I.U.G.S. Commission on Geochronology as base for discussion of a revised geological time-scale.
[b] Subdivisions[6]

Wurmian glaciation 10–70 thousand years
Warthian glaciation 100–120 thousand years
Rissian glaciation 175–210 thousand years
Mindelian glaciation 370–600 thousand years
Gunzian glaciation 750–1,000 thousand years
Begin Calabrian age 1,800 ± 200 thousand years

(From Wedepohl, K. H., *Handbook of Geochemistry,* Vol. I, Springer-Verlag, Berlin, 1969. With permission.)

REFERENCES

1. **Afanassyev, G. D. et al.,** The project of a revised geological time-scale in absolute chronology, Contr. Geol. Sovietiques Congr. geol. inter. 22[e] Sess., India, 287-324, 1964.
2. Holmes' Symposium, Geological Society phanerozoic time-scale, *Q. J. Geol. Soc. Lond.,* 120, 260, 1964.
3. **Krauskopf, K.,** *Introduction to Geochemistry,* McGraw-Hill Book Co., New York, 1967.
4. **Kulp, J. L.,** The geological time-scale, Report of Int. geol. Congr. 21st Sess., Norden, Part III, 18–27, 1960.
5. **Vinogradov, A. P. and Tugarinov, A. I.,** Geochronological scale of the Precambrian, Report Int. geol. Congr. 23rd Sess., Czechoslovakia, 6, 205, 1968.
6. **Zubakov, V. A.,** Geochronology of the continental Pleistocene deposits (based on radiometric data), *Geochem. Int.,* 4, 97, 1967.
7. **Knopf, A.,** private communication.

Table 4.1–5
MEASURES, UNITS, AND CONVERSION FACTORS
Metric and U.S. System

Prefix	Symbol	Meaning	Units
Tera	T	1,000,000,000,000	10^{12}
Giga	G	1,000,000,000	10^{9}
Mega	M	1,000,000	10^{6}
Kilo	k	1,000	10^{3}
Hecto	h	100	10^{2}
Deka	dk	10	10^{1}
Deci	d	0.1	10^{-1}
Centi	c	0.01	10^{-2}
Milli	m	0.001	10^{-3}
Micro	μ	0.000001	10^{-6}
Nano	n	0.000000001	10^{-9}
Pico	p	0.000000000001	10^{-12}

Lengths

Metric system		U.S. system
10^{-8} cm	1 Å	$3.937 \cdot 10^{-9}$ in.
10^{-4} cm	1 μ	$3.937 \cdot 10^{-5}$ in.
	1 cm	0.3937 in.
	2.540 cm	1 in.
	0.3048 m	1 ft
	0.9144 m	1 yd
10^{2} cm	1 m	1.09361 yd
	1.8288 m	1 fath.
10^{5} cm	1 km	0.62137 mi
	1.60935 km	1 mi
	1.852 km	1 int. nautical mi

1 A.U. (astronomical unit) = $1.49598 \cdot 10^{8}$ km

Area

Metric system	U.S. system
1 mm^2	0.00155 $in.^2$ (sq. in.)
1 cm^2	0.155 $in.^2$
6.45163 cm^2	1 $in.^2$
0.0929 m^2	1 ft^2
0.83613 m^2	1 yd^2
1 m^2	10.7639 ft^2
1 km^2	0.3861 mi^2
2.58998 km^2	1 mi^2

Volume

Metric system	U.S. system
1 mm^3	$0.6102 \cdot 10^{-4}$ $in.^3$ (cu. in.)
1 cm^3	0.06102 $in.^3$
16.3872 cm^3	1 $in.^3$
0.02831 m^3	1 ft^3
0.76456 m^3	1 yd^3
1 m^3	1.30794 yd^3

Liquid measures

Metric system	U.S. system	
1 ml	0.0610 $in.^3$	
0.473 L	28.875 $in.^3$	1 pt
0.946 L	57.749 $in.^3$	1 qt
1 L	61.0 $in.^3$	1.0567 qt
3.7853 L	231 $in.^3$	1 gal

Mass

Metric system	U.S. system
1 g	0.035 oz av (ounce av)
28.349 g	1 oz av
453.59 g	1 lb av (lb av.[a])
1 kg	2.20462 lb av
907.1848 kg	1 ton sh (short ton)
1 t	1.1023 ton sh
1016.047 kg	1 ton 1 (long ton)

Density

Metric system	U.S. system
1 g/cm^3	0.036127 $lb/in.^3$
27.68 g/cm^3	1 $lb/in.^3$
0.0160 g/cm^3	1 lb/ft^3

[a] 1 lb av = 1 pound avoirdupois is the mass of 27.692 $in.^3$ of water weighed in air at 4°C, 760 mm pressure.

Table 4.1–5 (Continued)

Energy

	erg	$Joule_{mt}$	$k W_{int}h$	$kcal_l$	Liter-atmos.	BTU
erg	1	0.9997×10^{-7}	2.7769×10^{-14}	2.389×10^{-11}	9.8692×10^{-10}	9.4805×10^{-11}
$Joule_{int}$	1.0002×10^{7}	1	2.7778×10^{-7}	2.390×10^{-4}	9.8722×10^{-3}	9.480×10^{-4}
$kW_{int}h$	3.6011×10^{13}	3.6000×10^{6}	1	8.6041×10^{2}	3.5540×10^{4}	3.413×10^{3}
$kcal_{l\,s}$	4.1853×10^{10}	4.186×10^{3}	1.1622×10^{-3}	1	4.1306×10^{1}	3.9685
Liter-atmos.	1.0133×10^{9}	1.0133×10^{2}	2.8137×10^{-5}	2.421×10^{-2}	1	9.607×10^{-2}
BTU	1.0548×10^{10}	1.0548×10^{3}	2.930×10^{-4}	2.5198×10^{-1}	1.0409×10^{1}	1

Pressure

	bar	Torr	atm.	at	lb/in.^{2}
1 bar (10^6 dynes/cm^2)	1	750	0.98692	1.0197	14.504
1 Torr	0.00133	1	0.00131	0.001359	0.01934
1 atm	1.0133	760	1	1.033	14.696
1 at (1 kg/cm^2)	0.98067	735.56	0.96784	1	14.223
1 lb/in.^{2}	0.06895	51.7144	0.068046	0.07031	1

Temperature

Absolute Centigrade or Kelvin (K)	x°K = T°C + 273.18
Degrees Centigrade (°C)	x°C = 5/9 (T°F−32)
	x°C = 5/4 T°R
Degrees Fahrenheit (°F)	x°F = 9/5 T°C + 32
	x°F = 9/4 T°R + 32
Degrees Réaumur (°R)	x°R = 4/9 (T°F−32)
	x°R = 4/5 T°C

Centigrade to Fahrenheit

C	°F	°C	°F	°C	F°
−200	−328	60	140	200	392
−150	−238	70	158	250	482
−100	−148	80	176	300	572
− 50	− 58	90	194	400	752
0	+ 32	100	212	500	932
10	50	110	230	600	1112
20	68	120	248	700	1292
30	86	130	266	800	1472
40	104	140	284	900	1652
50	122	150	302	1000	1832

Time

1 sidereal second	= 0.99727 mean solar second
1 sidereal day	= 86,164 mean solar seconds
1 solar day	= 86,400 mean solar seconds
1 mean solar year	= 365.242 mean solar days = 3.1557×10^7 mean solar seconds
1 sidereal year	= 365.256 mean solar days = 3.15581×10^7 mean solar seconds

Table 4.1–5 (Continued)

(From Heydemann, A., in *Handbook of Geochemistry,* Vol. I, Wedepohl, K. H., Ed., Springer-Verlag, Berlin, 1969. With permission.)

REFERENCE

1. **Weast, Robert C., Ed.,** *CRC Handbook of Chemistry and Physics,* 69th ed., CRC Press, Boca Raton, Fla., 1988.

Table 4.1–6
ASTRONOMICAL CONSTANTS

(Reference List of Recommended Constants)

Defining constants	
No. ephemeris sec in 1 tropical yr (1900)	s = 31,556,925.9747
Gaussian gravitational constant, defining the A.U.	k = 0.01720209895
Primary constants	
Measure of the A.U. in meters	A = 149,600 x 10^6
Velocity of light in m/sec	c = 299,792.5 x 10^3
Equatorial radius for Earth in meters	a_e = 6,378,160
Dynamical form-factor for Earth	J_2 = 0.0010827
Geocentric gravitational constant (units: m^3 s^{-2})	GE = 398,603 x 10^9
Ratio of the masses of the Moon and Earth	μ = 1/81.30
Sidereal mean motion of Moon in radians/sec (1900)	n^* = 2.661699489 x 10^{-6}
General precession in longitude/tropical century (1900)	p = 5,025.″644
Obliquity of the ecliptic (1900)	ϵ = 23°27′08.″26
Constant of nutation (1900)	N = 9.″210
Derived constants	
Heliocentric gravitational constant (units:m^3 s^{-2})	GS = 132,718 x 10^{15}
Ratio of masses of Sun and Earth	S/E = 322,958
Ratio of masses of Sun and Earth + Moon	$S/E(1 + \mu)$ = 328,912
Perturbed mean distance of Moon, in meters	$a_{☽}$ = 384,400 x 10^3

(From Wedepohl, K. H., *Handbook of Geochemistry,* Vol. I, Springer-Verlag, Berlin, 1969. With permission.)

REFERENCE

1. *Astronomer's Handbook,* Transactions of the International Astronomical Union, Vol. XIIC, Academic Press, London, 1966.

Table 4.1—7
SOLAR DIMENSIONS

Radius	6.960×10^{10} cm
Surface area	6.087×10^{22} cm^2
Volume	1.412×10^{33} cm^3
Mass	1.989×10^{33} g
Mean density	1.409 g cm^{-3}
Density at the center	98 g cm^{-3}
Gravitational acceleration at the solar surface	2.740×10^4 cm sec^{-2}
Escape velocity at the surface	6.177×10^7 cm sec^{-1}
Effective temperature	5,785° K
Temperature at the center	13.6×10^6 ° K
Radiation	3.9×10^{33} erg sec^{-1}
Specific surface emission	6.41×10^{10} erg cm^{-2} sec^{-1}
Specific mean energy production	1.96 erg g^{-1} sec^{-1}
Solar constant = extraterrestrial energy flux at the mean distance between earth and sun	1.39×10^6 erg cm^{-2} sec^{-1} 2.00 cal cm^{-2} min^{-1}

(From Wedepohl, K. H., *Handbook of Geochemistry*, Vol. I, Springer-Verlag, Berlin, 1969. With permission.)

REFERENCE

1. **Waldmeier, M.**, The quiet sun, *Landolt-Bornstein*, New series, Group VI: Astronomy, astrophysics and space research, Vol. I, Springer-Verlag, Berlin, 1965.

Table 4.1–8
DIMENSIONS OF THE PLANETS AND THE MOON

Symbols:

- a = semi-major axis of the orbit.
- P = sidereal period = true period of the planet's revolution around the Sun (with respect to the fixed star field).
- g_{Eq} = total acceleration, including centrifugal acceleration, at equator.
- v_e = velocity of escape at equator.
- A = Albedo = total reflectivity, wavelength λ_{eff} = 5,500 Å.
- T_{max} = max temp for the subsolar point of a slowly rotating planet or satellite (computed from the visual albedo).
- T_{av} = avg temp of a rapidly rotating sphere.

Atm. constituents = main atmospheric constituents.

Name	Symbol	a, 10^6 km	P a	Diameter, km	Mass,[a] 10^{26} g	Volume, 10^{10} km^3
Mercury	☿	57.9	0.24085	4,840	3.333	5.958
Venus	♀	108.2	0.61521	12,228	48.70	95.765
Earth	♁	149.6	1.00004	12,742.06	59.76	108.332
Moon	☽	0.384[b]	0.07480[c]	3,476	0.735	2.192
Mars	♂	227.9	1.88089	6,770	6.443	16.250
Jupiter	♃	778	11.86223	140,720	18,993	145,923.204
Saturn	♄	1,427	29.4577	116,820	5,684	83,469.806
Uranus	♅	2,870	84.0153	47,100	867.6	5,481.599
Neptune	♆	4,496	164.7883	44,600	1,029	4,636.610
Pluto	♇	5,881.9 to 5,946.5	247.7	6,000	55.3	10.833

[a] Mass without moons.
[b] Mean distance from Earth.
[c] True period of the Moon's revolution around Earth (with respect to the fixed star field).

Table 4.1–8 (Continued)

Name	ρ g/cm³	g_{Eq}, cm/s²	v_e, km/s	A	T_{max}, °K	$T_{av.}$, °K	Atm. constituents
Mercury	5.62	380	4.29	0.056	625	–	((^{40}Ar))
Venus	5.09	869	10.3	0.76	324	229	CO_2, H_2O
Earth	5.517	978	11.2	0.39	349	246	N_2, O_2
Moon	3.35	162	2.37	0.067	387	274	–
Mars	3.97	372	5.03	0.16	306	216	N_2, CO_2, H_2O
Jupiter	1.30	2301	57.5	0.67	131	93	H_2, CH_4, NH_4
Saturn	0.68	906	33.1	0.69	95	68	H_2, CH_4, NH_3
Uranus	1.58	972	21.6	0.93[d]	67[d]	47[d]	He, H_2, CH_4
Neptune	2.22	1347	24.6	0.84[d]	53[d]	38[d]	He, H_2, CH_4
Pluto	–	–	–	0.14	60	43	Uncertain

[d] Since the albedos of Uranus and Neptune are very low in the red and infrared an effective value A = 0.7 has been adopted for calculating the temperatures.

(From Wedepohl, K. H., *Handbook of Geochemistry*, Vol. I, Springer-Verlag, Berlin, 1969. With permission.)

REFERENCES

1. **Gondolatsch, F.,** Mechanical data of planets and satellites, *Landolt-Bornstein,* New series, Group, VI, Astronomy, astrophysics and space research, Vol. I, Springer, Berlin, 1965.
2. **Kuiper, G. P.,** Physics of planets and satellites, *Landolt-Bornstein,* New series, Group VI, Astronomy, astrophysics, and space research, Vol. I, Springer-Verlag Berlin, 1965.

Table 4.1–9

Conversion Tables

Time

1 sidereal sec	= 0.99727 mean solar sec
1 sidereal day	= 86,164 mean solar sec
1 solar day	= 86,400 mean solar sec
1 mean solar year	= 365.242 mean solar days = 3,1557 x 10^7 mean solar sec
1 sidereal year	= 365.256 mean solar days = 3.15581 x 10^7 mean solar sec

(From Wedepohl, K. H., *Handbook of Geochemistry*, Vol. I, Springer Verlag, Berlin, 1969. With permission.)

4.2. TOPOGRAPHIC DATA

Table 4.2—1
DEPTH DISTRIBUTION OF WORLD OCEAN

	Area				Cumulative area (Shallower than deeper limit of depth interval)		
	10^6 km^2		Percent		10^6 km^2	Percent	
Depth, km	**Menard and Smith**	**Kossinna**	**Menard and Smith**	**Kossinna**	**Menard and Smith**	**Menard and Smith**	**Kossinna**
0–0.2	27.123	27.491	7.49	7.6	27.123	7.49	7.6
0.2–1	16.012	15.437	4.42	4.3	43.135	11.91	11.9
1–2	15.844	15.184	4.38	4.2	58.978	16.29	16.1
2–3	30.762	24.347	8.50	6.8	89.740	24.79	22.9
3–4	75.824	70.800	20.94	19.6	165.565	45.73	42.5
4–5	114.725	119.092	31.69	33.0	280.289	77.42	75.5
5–6	76.753	84.317	21.20	23.3	357.042	98.62	98.8
6–7	4.461	3.919	1.23	1.1	361.503	99.85	99.9
7–8	0.380	0.328	0.10		361.883	99.96	
8–9	0.115	0.126	0.03		361.998	99.99	
9–10	0.032	0.018	0.01	0.1	362.031	100.00	100.0
10–11	0.002		0.00		362.033	100.00	

(From Menard, H. W. and Smith, S. M., Hypsometry of ocean basin provinces, *J. Geophys. Res.*, 71, 4305, 1966. With permission of American Geophysical Union.)

REFERENCE

Kossinna, E., Die Tiefen des Weltmeeres, *Inst. Meereskunde, Veroff., Georg.-naturwiss.*, 9, 70, 1921.

Table 4.2—2
AREA, VOLUME, AND MEAN DEPTH OF THE OCEANS

Oceans and adjacent seas	Area, 10^6 km^2	Volume, 10^6 km^3	Mean depth, m — Menard and Smith	Mean depth, m — Kossinna (our boundaries)
Pacific	166.241	696.189	4188	4282
Asiatic Mediterranean	9.082	11.366	1252	1182
Bering Sea	2.261	3.373	1492	1437
Sea of Okhotsk	1.392	1.354	973	838
Yellow and East China seas	1.202	0.327	272	188
Sea of Japan	1.013	1.690	1667	1350
Gulf of California	0.153	0.111	724	813
Pacific and adjacent seas, total	181.344	714.410	3940	4013
Atlantic	86.557	323.369	3736	3805
American Mediterranean	4.357	9.427	2164	2216
Mediterranean	2.510	3.771	1502	1487
Black Sea	0.508	0.605	1191	1115
Baltic Sea	0.382	0.038	101	55
Atlantic and adjacent seas, total	94.314	337.210	3575	3641
Indian	73.427	284.340	3872	3963
Red Sea	0.453	0.244	538	491
Persian Gulf	0.238	0.024	100	25
Indian and adjacent seas, total	74.118	284.608	3840	3929
Arctic	9.485	12.615	1330	1240
Arctic Mediterranean	2.772	1.087	392	277
Arctic and adjacent seas, total	12.257	13.702	1117	1020
Totals and mean depths	362.033	1349.929	3729	3814

(From Menard, H. W. and Smith, S. M., Hypsometry of ocean basin provinces, *J. Geophys. Res.*, 71, 4305, 1966. With permission of American Geophysical Union.)

REFERENCE

Kossinna, E., Die Tiefen des Weltmeeres, *Inst. Meereskunde, Veroff., Georg.-naturwiss.*, 9, 70, 1921.

Table 4.2—3
DEPTH ZONES IN THE OCEANS*

Table A

Ocean	Depth interval in kilometers												Total area (ocean)
	0–0.2	0.2–1	1–2	2–3	3–4	4–5	5–6	6–7	7–8	8–9	9–10	10–11	
Pacific Ocean	2.712	4.294	5.403	11.397	36.233	58.162	44.691	2.896	0.313	0.105	0.032	0.002	166.241
Asiatic Mediterranean	4.715	0.841	0.948	1.104	0.608	0.707	0.149	0.007	0.005	0	0	0	9.082
Bering Sea	1.050	0.135	0.172	0.234	0.670	0	0	0	0	0	0	0	2.261
Sea of Okhotsk	0.368	0.549	0.311	0.047	0.115	0	0	0	0	0	0	0	1.392
Yellow and East China seas	0.977	0.137	0.072	0.015	0.001	0	0	0	0	0	0	0	1.202
Sea of Japan	0.238	0.154	0.199	0.204	0.218	0	0	0	0	0	0	0	1.013
Gulf of California	0.071	0.032	0.040	0.010	0	0	0	0	0	0	0	0	0.153
Atlantic Ocean	6.080	4.474	3.718	7.436	16.729	28.090	19.324	0.639	0.058	0.010	0	0	86.557
American Mediterranean	1.021	0.465	0.589	0.667	0.906	0.586	0.112	0.008	0.002	0	0	0	4.357
Mediterranean	0.513	0.564	0.437	0.766	0.224	0.006	0	0	0	0	0	0	2.510
Black Sea	0.177	0.064	0.117	0.149	0	0	0	0	0	0	0	0	0.508
Baltic Sea	0.381	0.001	0	0	0	0	0	0	0	0	0	0	0.382
Indian Ocean	2.622	1.971	2.628	7.364	18.547	26.906	12.476	0.911	0.001	0	0	0	73.427
Red Sea	0.188	0.195	0.068	0.003	0	0	0	0	0	0	0	0	0.453
Persian Gulf	0.238	0	0	0	0	0	0	0	0	0	0	0	0.238
Arctic Ocean	3.858	1.569	0.968	1.249	1.573	0.269	0	0	0	0	0	0	9.485
Arctic Mediterranean	1.913	0.567	0.174	0.118	0	0	0	0	0	0	0	0	2.772
Total each depth	27.123	16.012	15.844	30.762	75.824	114.725	76.753	4.461	0.380	0.115	0.032	0.002	362.033

*Area in millions of square kilometers.

(From Menard, H. W. and Smith, S. M., Hypsometry of ocean basin provinces, *J. Geophys. Res.*, 71, 4305, 1966. With permission of American Geophysical Union.)

Table 4.2—3
DEPTH ZONES IN THE OCEANS*

Table B

Ocean	Depth interval in kilometers												Percent of world ocean in each ocean
	0–0.2	0.2–1	1–2	2–3	3–4	4–5	5–6	6–7	7–8	8–9	9–10	10–11	
Pacific Ocean	1.631	2.583	3.250	6.856	21.796	34.987	26.884	1.742	0.188	0.063	0.019	0.001	45.919
Asiatic Mediterranean	51.913	9.255	10.433	12.151	6.698	7.780	1.636	0.076	0.058	0	0	0	2.509
Bering Sea	46.443	5.975	7.623	10.330	29.629	0	0	0	0	0	0	0	0.625
Sea of Okhotsk	26.475	39.479	22.383	3.403	8.260	0	0	0	0	0	0	0	0.384
Yellow and East China seas	81.305	11.427	5.974	1.239	0.055	0	0	0	0	0	0	0	0.332
Sea of Japan	23.498	15.176	19.646	20.096	21.551	0.033	0	0	0	0	0	0	0.280
Gulf of California	46.705	20.848	25.891	6.556	0	0	0	0	0	0	0	0	0.042
Atlantic Ocean	7.025	5.169	4.295	8.590	19.327	32.452	22.326	0.738	0.067	0.012	0	0	23.909
American Mediterranean	23.443	10.674	13.518	15.313	20.796	13.440	2.572	0.193	0.051	0	0	0	1.203
Mediterranean	20.436	22.475	17.413	30.515	8.940	0.221	0	0	0	0	0	0	0.693
Black Sea	34.965	12.587	23.077	29.371	0	0	0	0	0	0	0	0	0.140
Baltic Sea	99.832	0.168	0	0	0	0	0	0	0	0	0	0	0.105
Indian Ocean	3.570	2.685	3.580	10.029	25.259	36.643	16.991	1.241	0.001	0	0	0	20.282
Red Sea	41.454	43.058	14.920	0.568	0	0	0	0	0	0	0	0	0.125
Persian Gulf	100.000	0	0	0	0	0	0	0	0	0	0	0	0.066
Arctic Ocean	40.673	16.539	10.209	13.167	16.580	2.834	0	0	0	0	0	0	2.620
Arctic Mediterranean	69.013	20.454	6.274	4.260	0	0	0	0	0	0	0	0	0.766
Percent of world ocean in each depth interval	7.492	4.423	4.376	8.497	20.944	31.689	21.201	1.232	0.105	0.032	0.009	0.001	

*Area in percent of each ocean.

(From Menard, H. W. and Smith, S. M., Hypsometry of ocean basin provinces, *J. Geophys. Res.*, 71, 4305, 1966. With permission of American Geophysical Union.)

HYPSOMETRY OF OCEAN BASIN PROVINCES

Physiographic provinces – These 'provinces' are regions or groups of features that have distinctive topography and usually characteristic structures and relations to other provinces. Province boundaries are based on detailed physiographic diagrams where available [Heezen and Tharp, 1961, 1964; Heezen et al., 1959; Menard, 1964] supplemented by more generalized physiographic and bathymetric charts. Provinces do not overlap nor are they superimposed in this study. Thus the area of a volcano rising from an ocean basin is included only in province VOLCANO and excluded from province OCEAN BASIN. The provinces identified in this study, the capitalized province names used in the text, the abbreviations used in data processing, and the corresponding numbers appearing in illustrations are

1. Continental SHELF AND SLOPE (CONS), the whole region from the shoreline to the base of the steep continental slope. Shelf and slope are grouped because they are merely the top and front of the margins of continental blocks.

2. CONTINENTAL RISE and partially filled sedimentary basins (CNRI). Gently sloping or almost flat, they appear to have characteristic features resulting from the accumulation of a thick fill of sediment eroded from an adjacent continent and overlying an otherwise relatively normal oceanic crust. In this respect, the Gulf of Mexico and the western basin of the Mediterranean differ from the continental rise off the eastern United States only because they are relatively enclosed.

3. OCEAN BASIN (OCBN), the remainder after removing all other provinces. Abyssal plains and abyssal hills and archipelagic aprons are common features of low relief.

4. Oceanic RISE AND RIDGE (RISE), commonly called "mid-ocean ridges" despite the fact that they continue across ocean margins. They form one worldwide system with many branches. Boundaries are taken in most places as outer limit of essentially continuous slopes from crest.

5. RIDGE NOT KNOWN TO BE VOLCANIC (RIDG), relatively long and narrow and with steep sides. Most have unknown structure and some or most may be volcanic.

6. Individual VOLCANO (VOLC), with a boundary defined as the base of steep side slopes.

7. Island ARC AND TRENCH (TNCH), includes whole system of low swells and swales subparallel to trenches. Continental equivalents or extensions of island arcs, such as Japan, are excluded.

8. Composite VOLCANIC RIDGE (VRCM), formed by overlapping volcanoes and with a boundary at the base of steep side slopes.

9. POORLY DEFINED ELEVATION (BLOB), with nondescript side slopes and length no more than about twice width. Crustal structure unknown; may be thin continental type.

Tabulation of data and measuring procedure — Data were tabulated by 10° squares of latitude and longitude. Squares containing more than one ocean were split, and each ocean was treated separately. Within a square, the areas between the depth intervals 0—200 m, 200—1,000 m, and between 1-km contours down to 11-km were compiled for each physiographic province.

The polar planimeters (Keuffel and Esser models 4236 and 4242) used for measuring areas were read to the nearest unit on the vernier scale, and measurements were tabulated directly for card punching. These values were converted to square kilometers during computer processing by a scale factor derived from a measurement of the total number of units in the square. The area of a square was calculated assuming a spherical earth with a radius of 6,371.22 km.

(From Menard, H. W. and Smith, S. M., Hypsometry of ocean basin provinces, *J. Geophys. Res.*, 71, 4305, 1966. With permission of American Geophysical Union.)

REFERENCES

Heezen, B. C. and Tharp, M., Physiographic diagram of the South Atlantic, Geological Society of America, N.Y., 1961.

Heezen, B. C. and Tharp, M., Physiographic diagram of the Indian Ocean, Geological Society of America, N.Y., 1964.

Heezen, B. C., Tharp, M., and Ewing, M., The floors of the oceans, 1, the North Atlantic, Geological Society of America, Special paper 65, 1959.

Menard, H. W., *Marine Geology of the Pacific,* McGraw-Hill, New York, 1964.

Table 4.2—4
PROVINCE AREAS IN EACH OCEAN AND TOTAL AREAS OF PROVINCES AND OCEANS (10^6 km^2)*

Oceans and adjacent seas	RISE	OCBN	VOLC	CONS	TNCH	CNRI	VRCM	RIDG	BLOB	Total area of each ocean
Pacific Ocean	65.109	77.951	2.127	11.299	4.757	2.690	1.589	0.494	0.227	166.241
Asiatic Mediterranean	0	0	0.003	7.824	0.023	1.233	0	0	0	9.082
Bering Sea	0	0	0	1.286	0.281	0.694	0	0	0	2.261
Sea of Okhotsk	0	0	0	1.254	0.023	0.115	0	0	0	1.392
Yellow and East China seas	0	0	0	1.119	0.082	0	0	0	0	1.202
Sea of Japan	0	0	0.005	0.798	0	0.210	0	0	0	1.013
Gulf of California	0.042	0	0	0.111	0	0	0	0	0	0.153
Atlantic Ocean	30.519	35.728	0.882	12.658	0.447	5.381	0	0.412	0.530	86.557
American Mediterranean	0	1.346	0.060	1.889	0.201	0.861	0	0	0	4.357
Mediterranean	0	0	0	1.465	0	1.046	0	0	0	2.510
Black Sea	0	0	0	0.263	0	0.245	0	0	0	0.508
Baltic Sea	0	0	0	0.382	0	0	0	0	0	0.382
Indian Ocean	22.426	36.426	0.358	6.097	0.256	4.212	0.407	2.567	0.679	73.427
Red Sea	0	0.070	0	0	0.383	0	0	0	0	0.453
Persian Gulf	0	0	0	0.238	0	0	0	0	0	0.238
Arctic Ocean	0.513	0	0	5.874	0	2.267	0.302	0	0.528	9.485
Arctic Mediterranean	0	0	0	2.483	0	0.289	0	0	0	2.772
Total area each province	118.607	151.522	3.435	55.421	6.070	19.242	2.298	3.473	1.965	362.033

* Abbreviations are defined in text preceding this table.

(From Menard, H. W. and Smith, S. M., Hypsometry of ocean basin provinces, *J. Geophys. Res.*, 71, 4305, 1966. With permission of American Geophysical Union.)

Table 4.2—5
PERCENT OF PROVINCES IN OCEANS AND ADJACENT SEAS*

Oceans and adjacent seas	RISE	OCBN	VOLC VRCM RIDG BLOB	CONS	TNCH	CNRI	Percent of world ocean in each ocean group
Pacific and adjacent seas	35.9	43.0	2.5	13.1	2.9	2.7	50.1
Atlantic and adjacent seas	32.3	39.3	2.0	17.7	0.7	8.0	26.0
Indian and adjacent seas	30.2	49.2	5.4	9.1	0.3	5.7	20.5
Arctic and adjacent seas	4.2	0	6.8	68.2	0	20.8	3.4
Percent of world ocean in each province	32.7	41.8	3.1	15.3	1.7	5.3	

* Abbreviations are defined in text preceding Table 2.1–4.

(From Menard, H. W. and Smith, S. M., Hypsometry of ocean basin provinces, *J. Geophys. Res.*, 71, 4305, 1966. With permission of American Geophysical Union.)

Figure 4.2—1
INDEX MAP OF SOURCE CHARTS AND OCEAN BASIN BOUNDARIES

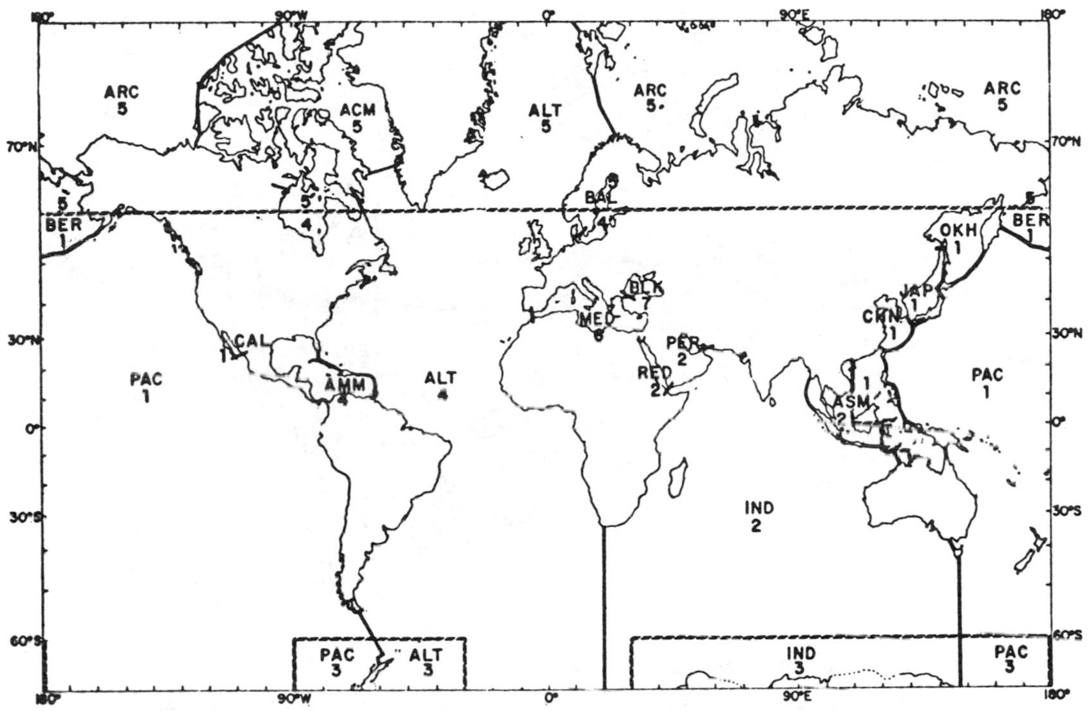

FIGURE 4.2—1. Source charts are bounded by dashed lines. Ocean abbreviations are explained in the text, and their boundaries (solid lines) are shown in more detail in Figures 4.2—4 to 4.2—7.

(From Menard, H. W. and Smith, S. M., Hypsometry of ocean basin provinces, *J. Geophys. Res.*, 71, 4305, 1966. With permission of American Geophysical Union.)

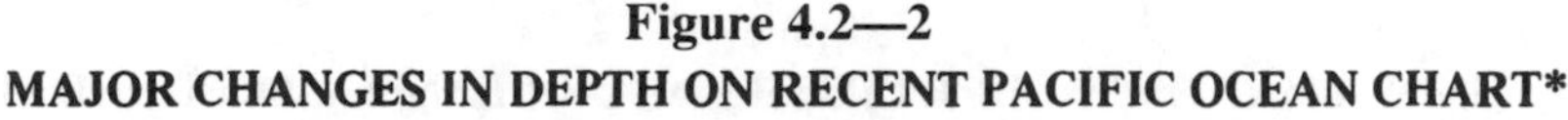

Figure 4.2—2
MAJOR CHANGES IN DEPTH ON RECENT PACIFIC OCEAN CHART*

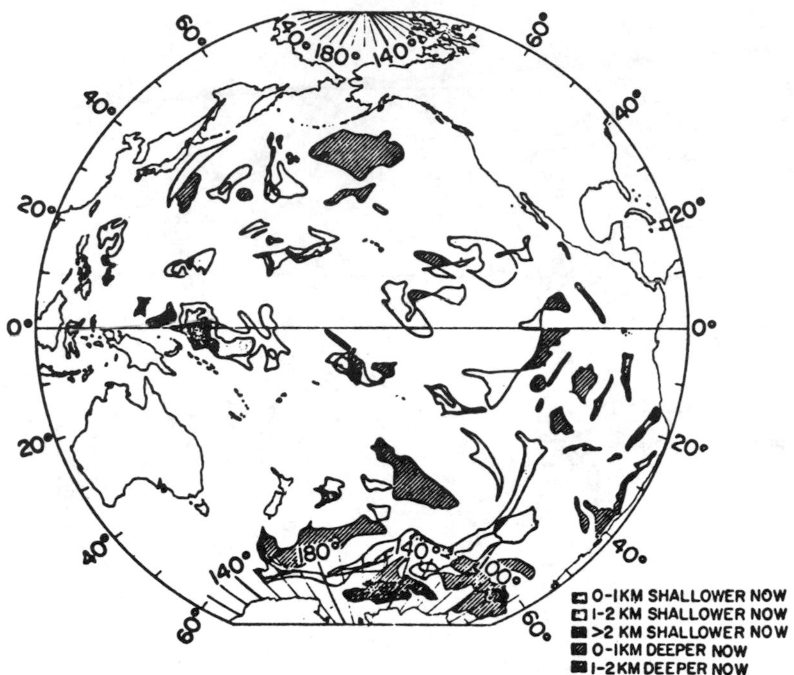

* Compared with chart used by Kossinna (1921).

(From Menard, H. W. and Smith, S. M., Hypsometry of ocean basin provinces, *J. Geophys. Res.*, 71, 4305, 1966. With permission of American Geophysical Union.)

REFERENCE

Kossinna, E., Die Tiefen des Weltmeeres, *Inst. Meereskunde, Veroff., Georg.-naturwiss.*, 9, 70, 1921.

Figure 4.2—3
MAJOR CHANGES IN DEPTH ON RECENT ATLANTIC AND INDIAN OCEAN CHARTS*

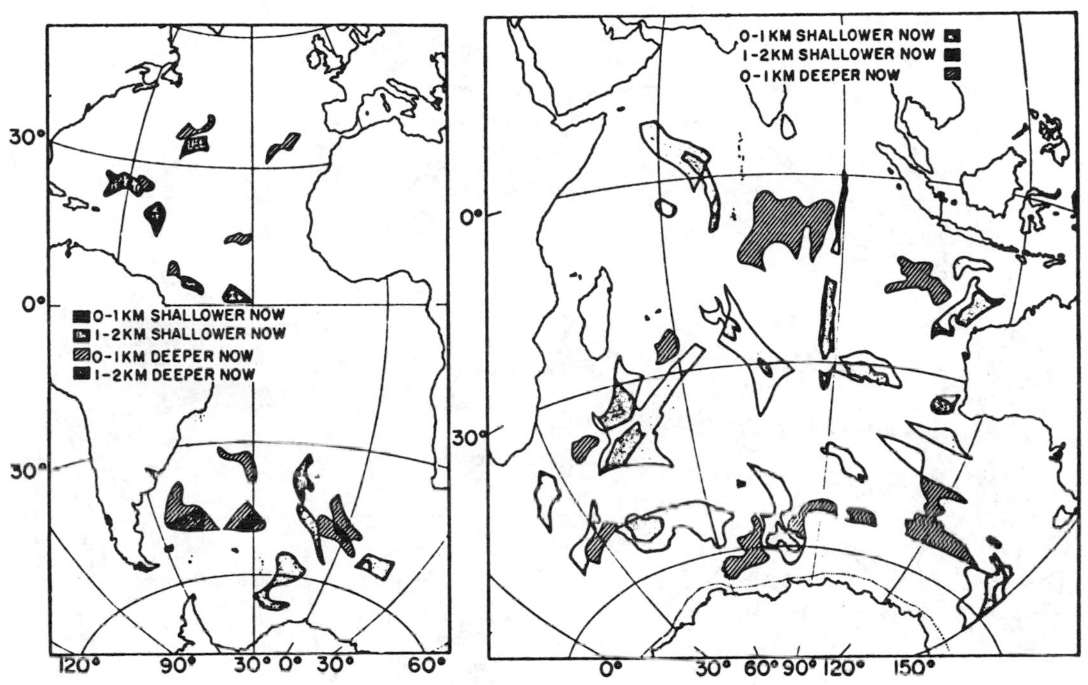

* Compared with charts used by Kossinna (1921).

(From Menard, H. W. and Smith, S. M., Hypsometry of ocean basin provinces, *J. Geophys. Res.*, 71, 4305, 1966. With permission of American Geophysical Union.)

REFERENCE

Kossinna, E., Die Tiefen des Weltmeeres, *Inst. Meereskunde, Veroff., Georg.-naturwiss.*, 9, 70, 1921.

Figure 4.2—4
PACIFIC OCEAN – PHYSIOGRAPHIC PROVINCES

FIGURE 4.2—4. Text contains key to province numbers. Individual volcanoes (VOLC) in black.

(From Menard, H. W. and Smith, S. M., Hypsometry of ocean basin provinces, *J. Geophys. Res.*, 71, 4305, 1966. With permission of American Geophysical Union.)

Figure 4.2—5
ATLANTIC OCEAN – PHYSIOGRAPHIC PROVINCES

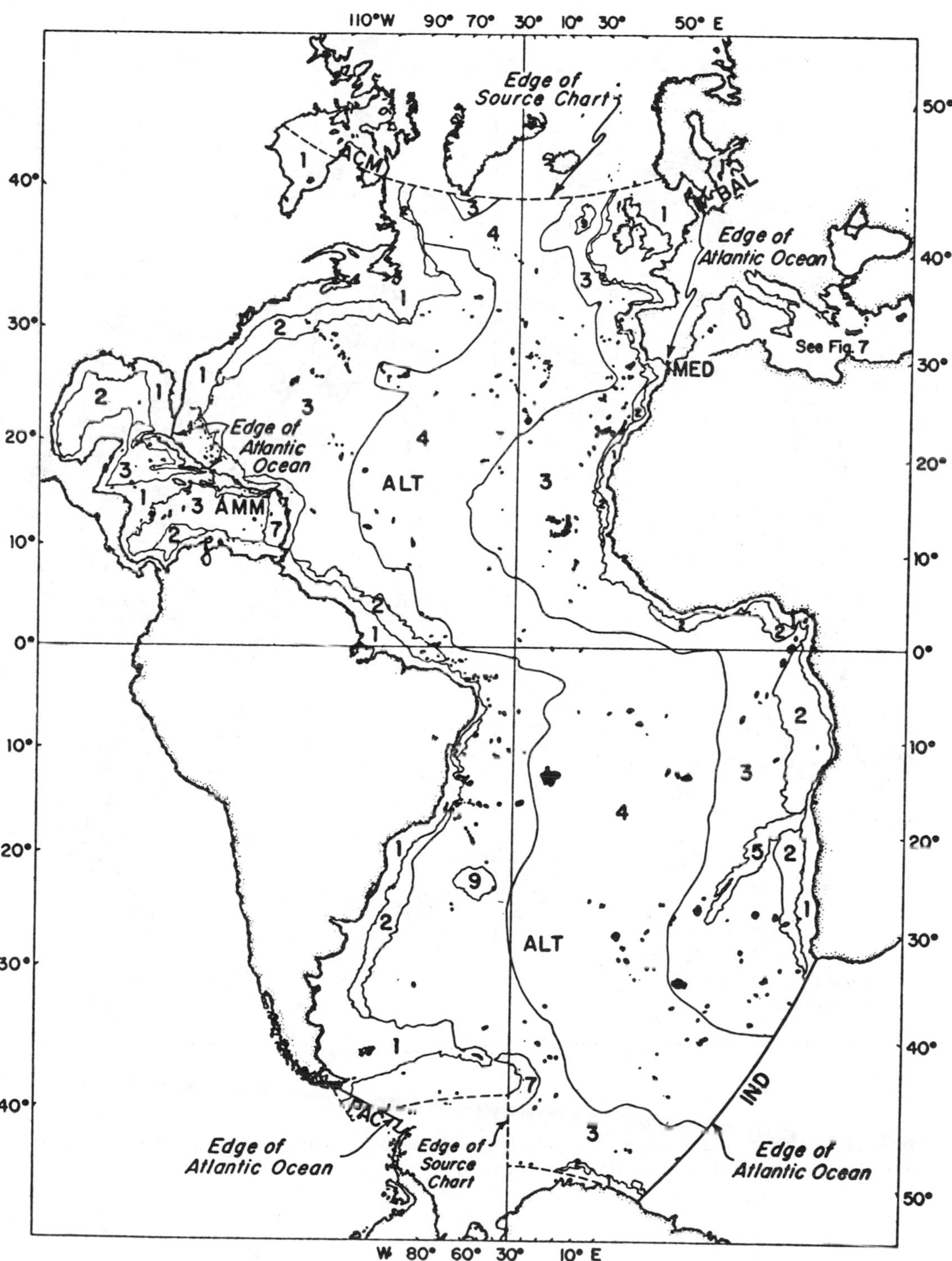

(From Menard, H. W. and Smith, S. M., Hypsometry of ocean basin provinces, *J. Geophys. Res.*, 71, 4305, 1966. With permission of American Geophysical Union.)

Figure 4.2—6
INDIAN OCEAN – PHYSIOGRAPHIC PROVINCES

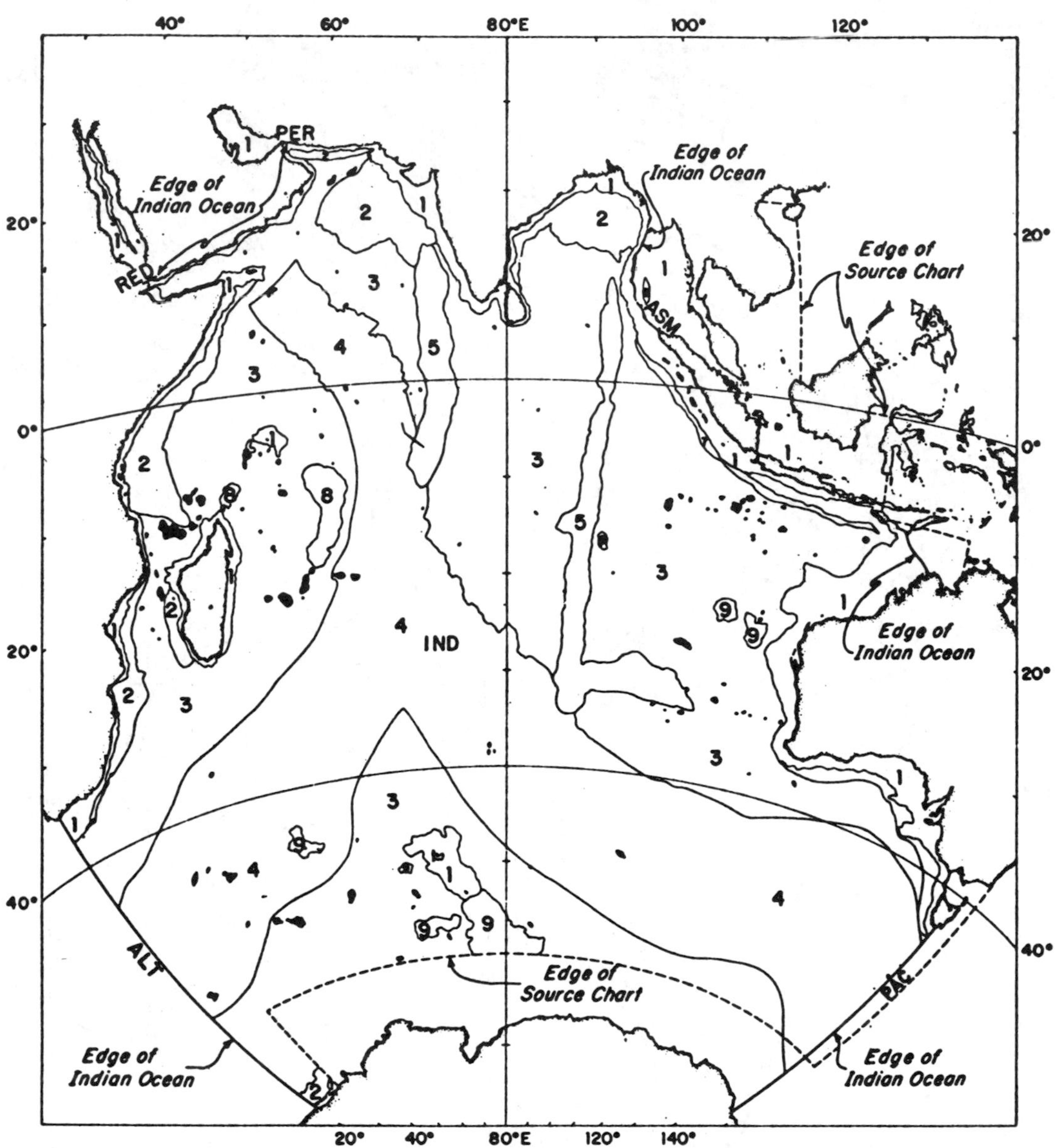

(From Menard, H. W. and Smith, S. M., Hypsometry of ocean basin provinces, *J. Geophys. Res.*, 71, 4305, 1966. With permission of American Geophysical Union.)

Figure 4.2—7

SMALLER OCEANS AND SEAS – PHYSIOGRAPHIC PROVINCES

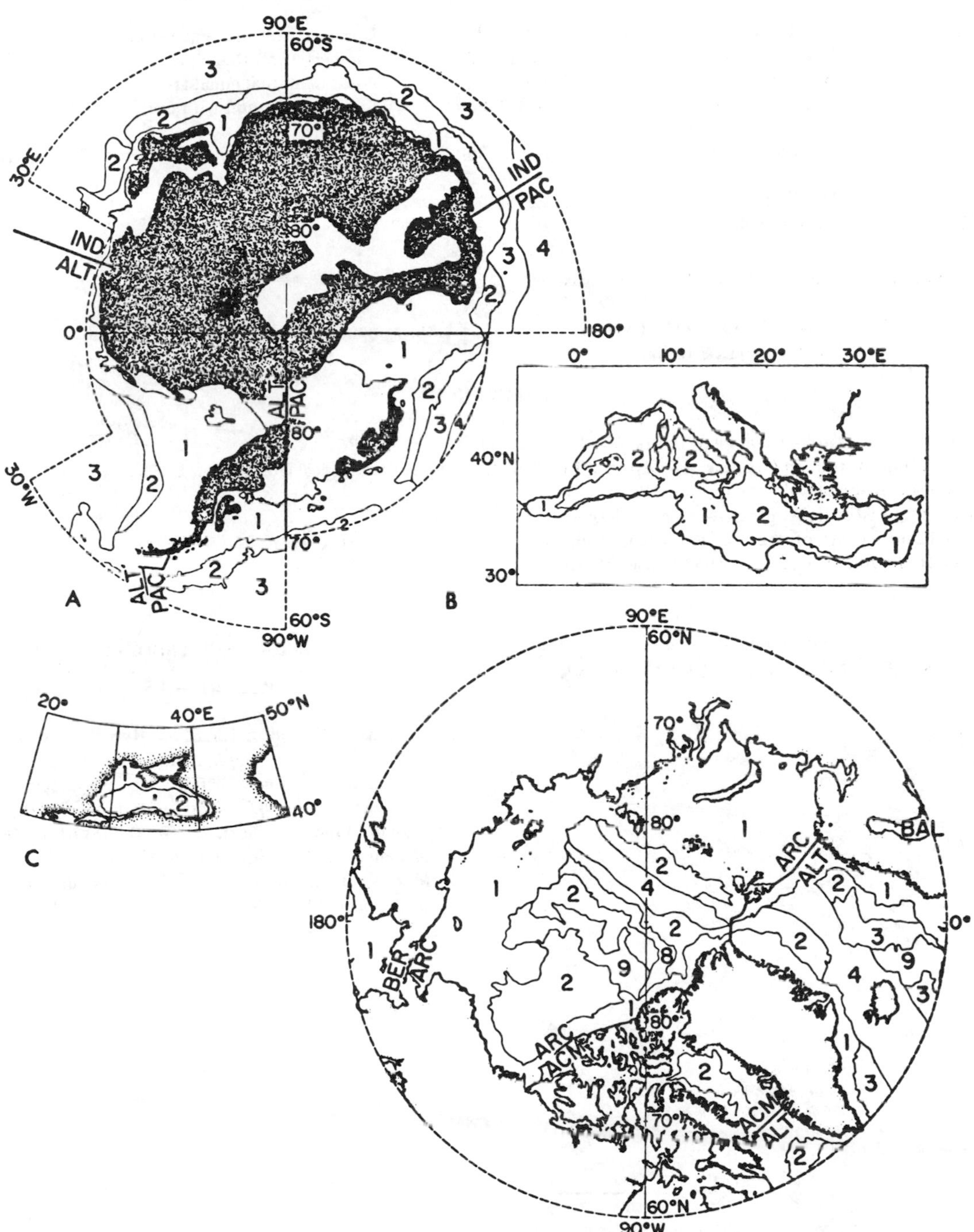

FIGURE 4.2—7. Antarctic sub-ice in white is below sea level.

(From Menard, H. W. and Smith, S. M., Hypsometry of ocean basin provinces, *J. Geophys. Res.*, 71, 4305, 1966. With permission of American Geophysical Union.)

Table 4.2—6
BATHYMETRIC CHARTS USED FOR HYPSOMETRIC CALCULATIONS

Source No.	Title	Scale	Projection	Reference
1	Pacific Ocean	1:7,270,000*	Lambert azimuthal equal-area	A
2	Indian Ocean	1:7,510,000*	Lambert azimuthal equal-area	B
3	Antarctica	1:9,667,000	Polar azimuthal equal-area	C
4	Atlantic Ocean	1:10,150,000*	Lateral projection with oval isoclines	B
5	Tectonic Chart of the Arctic	1:10,000,000	Polar azimuthal equal-area	D
6	Mediterranean Sea	1:2,259,000	Mercator	E
7	Northern Hemisphere	1:25,000,000	Polar azimuthal equal-area	B

* Scale of photographic enlargement used for measuring.

(From Menard, H. W. and Smith, S. M., Hypsometry of ocean basin provinces, *J. Geophys. Res.*, 71, 4305, 1966. With permission of American Geophysical Union.)

REFERENCES

A. **Menard, H. W.,** *Marine Geology of the Pacific,* McGraw-Hill, New York, 1964.
B. Main Administration in Geodesy and Cartography of the Government Geological Committee, USSR.
C. American Geographical Society, New York.
D. Geological Institute, Academy of Science, Moscow.
E. Unpublished chart of the Mediterranean, modified from contours compiled by R. Nason from various sources. U.S. Navy Hydrographic Office chart 4300 used as base.

Figure 4.2—8
HYPSOMETRY OF ALL OCEAN BASINS*

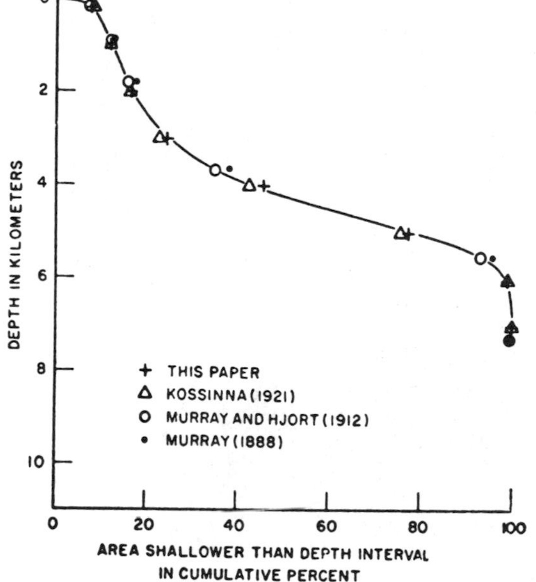

* According to various studies.

(From Menard, H. W. and Smith, S. M., Hypsometry of ocean basin provinces, *J. Geophys. Res.*, 71, 4305, 1966. With permission of American Geophysical Union.)

Figure 4.2—8 (continued)

REFERENCES

1. **Menard, H. W. and Smith, S. M.,** Hypsometry of ocean basin provinces, *J. Geophys. Res.*, 71, 4305, 1966.
2. **Kossinna, E.,** Die Tiefen des Weltmeeres, *Inst. Meereskunde, Veroff., Geogr.-naturwiss.*, 9, 70, 1921.
3. **Murray, John and Hjort, J.,** *The Depths of the Ocean,* Macmillan and Co., London, 1912.
4. **Murray, John,** On the height of the land and the depth of the ocean, *Scot. Geogr. Mag.*, 4, S. 1, 1888.

Figure 4.2—9
HYPSOMETRY OF INDIVIDUAL MAJOR OCEAN BASINS

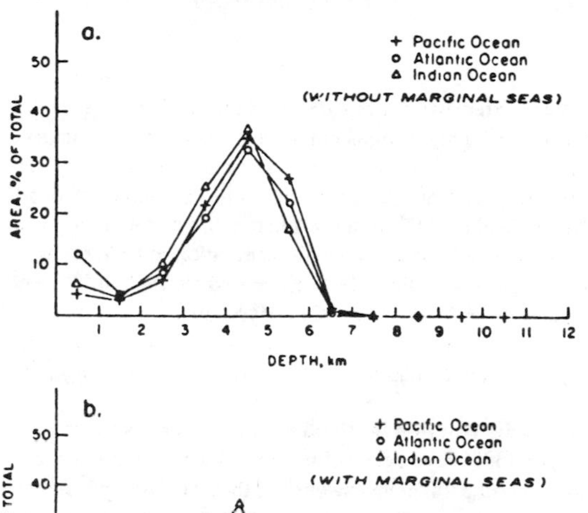

(From Menard, H. W. and Smith, S. M., Hypsometry of ocean basin provinces, *J. Geophys. Res.*, 71, 4305, 1966. With permission of American Geophysical Union.)

Figure 4.2—10
HYPSOMETRY OF ALL OCEAN BASINS*

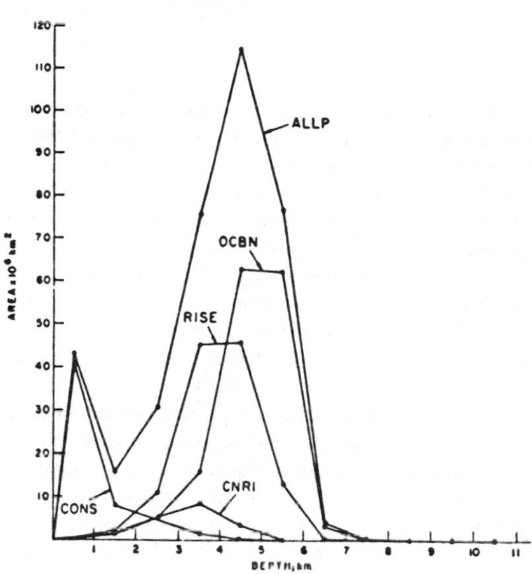

FIGURE 4.2—10. This diagram is for all privinces combined (ALLP) and for individual major provinces.

* Abbreviations are identified in the text preceding Table 4.2–4.

(From Menard, H. W. and Smith, S. M., Hypsometry of ocean basin provinces, *J. Geophys. Res.*, 71, 4305, 1966. With permission of American Geophysical Union.)

Figure 4.2—11
HYPSOMETRY OF OCEAN BASINS PLOTTED CUMULATIVELY BY PROVINCES*

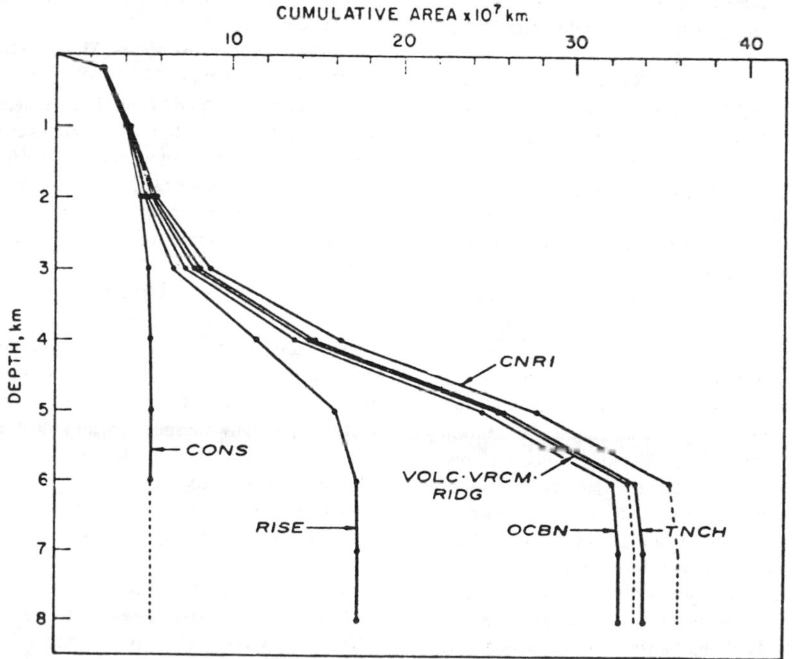

* Abbreviations are identified in the text preceding Table 4.2—4.

(From Meynard, H. W. and Smith, S. M., Hypsometry of ocean basin provinces, *J. Geophys. Res.*, 71, 4305, 1966. With permission of American Geophysical Union.)

HYPSOMETRY OF OCEAN BASIN PROVINCES (continued)

Depth distribution in different oceans as a function of provinces — The depth distribution in provinces in different ocean basins, as seen in Figure 4.2—14, closely resembles the composite distribution in the world ocean (Figure 4.2—10). The sum of the depth distribution of all provinces in an ocean basin is double peaked, but for the individual provinces it is single peaked and relatively symmetrical. However, the depth distributions are sufficiently different to warrant some discussion. The mean depths of all provinces in the three major ocean basins, including marginal seas, range from 3575 m for the Atlantic Ocean to 3940 m for the Pacific (Table 4.2—7). The range of mean depths of the OCEAN BASIN province in each of these ocean basins is similar. The smallest mean OCEAN BASIN depth of 4530 m in the Indian Ocean may be the result of epirogenic movement of the oceanic crust, but it is also partially attributable to sedimentaion. The eastern and southwestern parts of the Indian Ocean are deeper than 5000 m and are thus below the mean depth of the world ocean. The northwestern and southeastern parts, however, are exceptionally shallow. Seismic stations and topography show that the northwestern region has been shoaled by deposition of turbidities spreading from the mouths of the great Indian and east African rivers [Menard, 1961; Heezen and Tharp, 1964].

The mean depths of RISE and RIDGE have a limited range — from 3945 m in the Indian Ocean to 4008 m in the Atlantic Ocean (Table 4.2—7). This uniformity seems remarkable considering the widespread and diverse evidence that oceanic rises and ridges are tectonically among the more unstable features of the surface of the earth. It is all the more remarkable because the local, relief, or elevation above the adjacent OCEAN BASIN, differ substantially in different oceans. The relief of RISE AND RIDGE in an ocean basin can be estimated by subtracting the mean depth from that of OCEAN BASIN or by determining the deepening required to give a best fit of individual hypsometric curves for each province. Comparing the means gives the relief of RISE AND RIDGE ABOVE OCEAN BASIN as 585 m in the Indian Ocean, 662 m in the Atlantic, and 928 m in the Pacific. The reliefs from matching curves are 800, 900, and 1200 m, respectively. The greater relief obtained by the curve-matching method results from ignoring the shallow tails of the depth distribution. Thus the range in relief is about 6 times as great as the range in mean depths of RISE AND RIDGE, which may be explained if the sea floor is not only elevated by epirogeny but is also depressed. It seems reasonable to assume that the depth intervals in OCEAN BASIN with the largest areas (4 to 5 and 5 to 6 km) are those underlain by normal crust and mantle. A uniform oceanic process in the mantle acting on a uniform oceanic crust at a uniform depth may produce oceanic rises and ridges of uniform depth. Many current hypotheses for the origin of rises and ridges suggest just such an elevation. However, it is at least implicit that the mantle under the ocean basins cannot become denser and thus epirogenically depress the crust. Moreover, it is assumed by advocates of convection that if the curst is dragged down dynamically it forms a long narrow oceanic trench. The symmetrical distribution curves for OCEAN BASIN indicate that a considerable area is below the most common depth interval. Very extensive regions deeper than 6000 m exist in the northwestern Pacific and eastern Indian oceans, and there may be places where the normal oceanic curst is epirogenically depressed by more than a kilometer below the 4753-m mean depth of the OCEAN BASIN for the world ocean. Formation of broad depressions would alter the depth distribution in OCEAN BASIN and thereby vary the relief of RISE AND RIDGE in different oceans. If these broad epirogenic depressions exist, they may have a significant effect on the possible range of sea level changes relative to continents. This will be considered under "Discussion".

Depth distribution in island arc and trench provinces — These provinces have been defined to include not only trenches but also the subparallel low swells and the island arcs which rise above some of them. The justification for this definition is that these features probaby are caused by the same process; one question that can be answered by this type of study is whether the process elevates or depresses the sea floor. The volcanoes, some capped with limestone, which form most islands in this province, have a rather minor volume and have hardly any effect on the hypsometry.

The median depth for island ARC AND TRENCH is somewhat less than 4 km, which is less than the median depth for all ocean basins and considerably less than for the OCEAN BASIN province. The average depth would be much shallower if it were possible insome simple way to include the elevations above normal continents of the mountain ranges parallel to the Peru-Chile, Central America, Japan, and Java trenches. This would require some elaborate assumptions, but it is clear that the process which forms trenches and related features generally elevates the crust.

Volume of the ocean — Murray [1888] calculated the volume of the ocean at 323,722,150 cubic miles, which equals about 1.325×10^9 km^3. Kossina [1921] obtained 1.370×10^9 km^3, and we obtain 1.350×10^9 km^3. It appears unlikely that this value is in error by more than a few percent. Our method of calculation is essentially the same as that of Murray and Kossinna. The midpoint value of a depth interval is multiplied by the area of that interval, and the volumes of the intervals are summed.

Discussion

Sea floor epirogeny and sea level changes — Sea floor epirogeny is only one of a multitude of causes of sea level change of which the wax and wane of glaciers is probably the most intense. Epirogeny is especially important because it may have occurred at any time in the history of the earth in contrast to relatively brief periods of glaciation. That eustatic changes in sea level have occurred during geological time is suggested by widespread epicontinental seas alternating with apparently high continents.

The hypothesis that oceanic rises are ephemeral [Menard, 1958] provides a basis for quantitative estimates of epirogenic effects on sea level. If the approximate volume of existing rises and ridges is compared with the area of the oceans, it appears that uplift of the existing rises has elevated the sea level 300 m.

Likewise, subsidence of the ancient Darwin rise has lowered it by 100 m [Menard, 1964].

The present study suggests that the sea floor may be depressed epirogenically in places where this movement does not merely restore the equilibrium disturbed by a previous uplift. The argument derives from the fact that the mean depth in the OCEAN BASIN province is about 4700 m. Considering that the crust has about the same thickness everywhere in the province, variations from this depth generally are caused by differences in density in the upper mantle. (We assume that where the mean depth of the crust is "normal" it is underlain by a "normal" mantle.) Thus the deeper regions, which are roughly 70 million km^2 in area, have been depressed by a density increase in the mantle. If large areas of the sea floor can be depressed as well as elevated, the resulting changes in sea level would be highly complex.

Only the most general conclusions can be drawn from this analysis, but they may be significant. First, a plausible mechanism is available to explain the eustatic changes in sea level observed in the geological record. At present, the mechanism places no constraints on the sign of a change but appears to limit the amount to a few hundred meters. Second, in large regions the upper mantle may possibly become denser than normal. Substantial evidence exists that it is less dense than normal under rises and ridges [Le Pichon et al., 1965]. If it can also be more dense than normal in large regions, these facts can provide very useful clues regarding the composition of the upper mantle and processes acting below the crust. The implications of possible densification of normal mantle can be avoided by defining the "normal" depth as the deepest that is at all widespread. If this definition is accepted as reasonable (it does not appear so to us), small decreases in density of the upper mantle occur under most of the world ocean. The volume of ocean basin elevated above normal is consequently large, and the possible range of sea level changes is thus at least 1 km.

Sea floor spreading and continental drift — Several aspects of our data appear to have some bearing on modern hypotheses of global tectonics. The relationships are not definitive, however, and at this time we prefer merely to indicate some of the equations which have arisen.

1. The proportion of RISE AND RIDGE to OCEAN BASIN in a basin could range from zero to infinity, but it is 0.84 for the Pacific, 0.82 for the Atlantic, and 0.61 for the Indian Ocean. The sample is very small, and consequently the similarity of the proportions may be coincidental. However, it suggests that the area of RISE AND RIDGE is proportional to the whole area of an ocean basin. This in turn suggests that the size of the basin is related to the existence of rises and ridges.

2. The proportion of SHELF AND SLOPE to OCEAN BASIN plus RISE AND RIDGE is relatively constant for large ocean basins and quite different from the proportion for small ocean basins. This relationship may require modification of at least many of the details of the hypothesis that the Atlantic Ocean basin was formed when an ancient continent split. When the supposed splitting began, the whole basin was SHELF AND SLOPE. Consequently, the proportion of SHELF AND SLOPE has since decreased. In the Pacific basin, on the other hand, the proportion of SHELF AND SLOPE to OCEAN BASIN plus RISE AND RIDGE was smaller than now and has since increased. If the Atlantic split apart at a constant rate and is still splitting as the Pacific contracts, the present equality of the proportions of SHELF AND SLOPE in the two ocean basins requires a striking coincidence. No coincidence is necessary if the splitting occurred relatively rapidly until it reached some dynamic equilibrium state, perhaps when the proportion of RISE AND RIDGE to OCEAN BASIN in each ocean basin reached about 0.8 to 0.9.

Acknowledgments — Some of these data were compiled by Isabel Taylor, Surendra Mathur, and Sarah Buffington. We wish especially to thank Mrs. Taylor for her careful rechecking of the measurements. We are indebted to Dr. G. B. Udintsev for providing Russian bathymetric charts from the first press runs.

This research was supported by Office of Naval Research Long Range Research Contract 2216(12) and National Science Foundation Grant NSF gp-4235.

(From Menard, H. W. and Smith, S. M., Hypsometry of ocean basin provinces, *J. Geophys. Res.*, 71, 4305, 1966. With permission of the American Geophysical Union.)

REFERENCES

Heezen, B. C. and Tharp, M., Physiographic diagram of the Indian Ocean, Geological Society of America, N.Y., 1964.

Heezen, B. C., Tharp, M., and Ewing, M., The floors of the oceans, 1, the North Atlantic, Geological Society of America, Special paper 65, 1959.

Kossinna, E., Die Tiefen des Weltmeeres, *Inst. Meereskunde, Veroff., Georg.-naturwiss.,* 9, 70, 1921.

Le Pichon, X., Houtz, R. E., Drake, C. L., and Nafe, J. E., Crustal structure of the mid-ocean ridges, 1, Seismic refraction measurements, *J. Geophys. Res.,* 70(2), 319, 1965.

Menard, H. W., Development of median elevations in ocean basins, *Bull. Geol. Soc. Am.,* 69(9), 1179, 1958.

Menard, H. W., *Marine Geology of the Pacific,* McGraw-Hill, New York, 1964.

Menard, H. W., Some rates of regional erosion, *J. Geol.,* 69(2), 154, 1961.

Murray, John, On the height of the land and the depth of the ocean, *Scot. Geogr. Mag.,* 4, S. 1, 1888.

Figure 4.2—12
HYPSOMETRY OF ALL ARC AND TRENCH PROVINCES AND OF SOME GROUPS OF ARCS AND TRENCHES

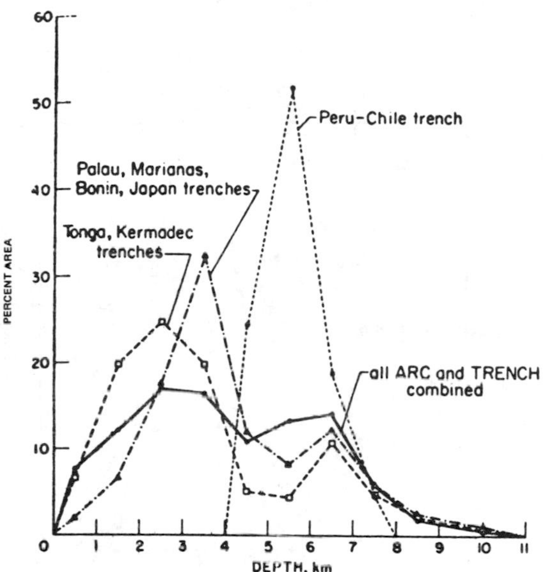

(From Menard, H. W. and Smith, S. M., Hypsometry of ocean basin provinces, *J. Geophys. Res.*, 71, 4305, 1966. With permission of American Geophysical Union.)

Figure 4.2—13
HYPSOMETRIC CURVE OF OCEAN BASINS

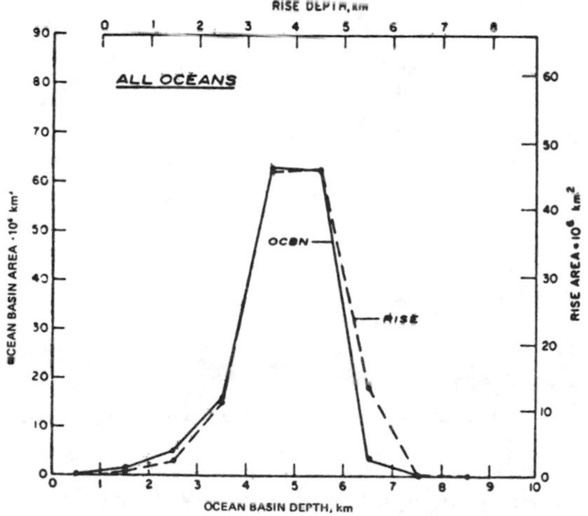

FIGURE 4.2—13. Hypsometric curve of all ocean basins for RISE and RIDGE province normalized to curve for OCEAN BASIN province to show close similarity.

(From Menard, H. W. and Smith, S. M., Hypsometry of ocean basin provinces, *J. Geophys. Res.*, 71, 4305, 1966. With permission of American Geophysical Union.)

Figure 4.2—14
HYPSOMETRY OF ALL PROVINCES (ALLP) AND INDIVIDUAL PROVINCES IN MAJOR BASINS

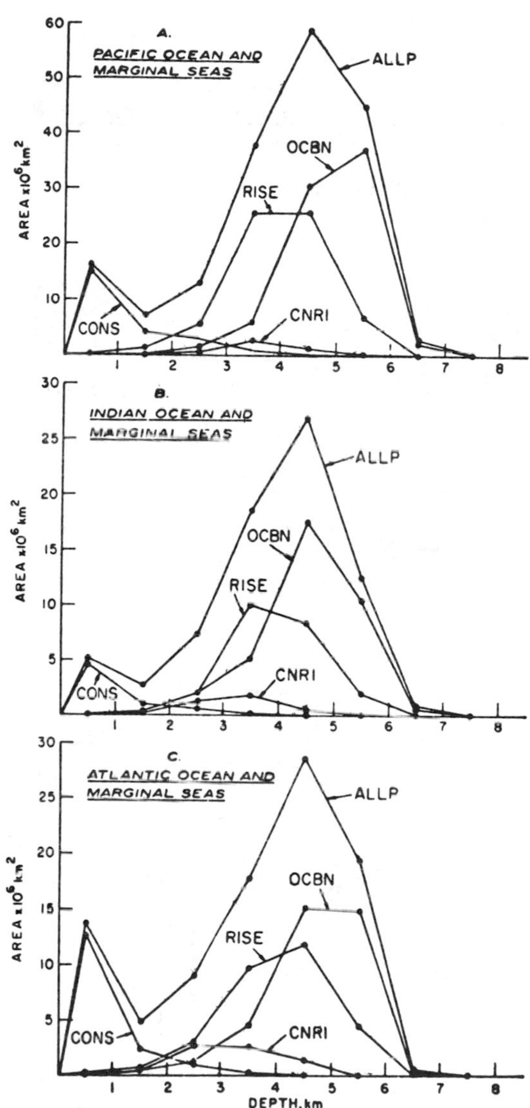

(From Menard, H. W. and Smith, S. M., Hypsometry of ocean basin provinces, *J. Geophys. Res.*, 71, 4305, 1966. With permission of American Geophysical Union.)

Figure 4.2—15

HYPSOMETRIC CURVES FOR RISE AND RIDGE PROVINCES

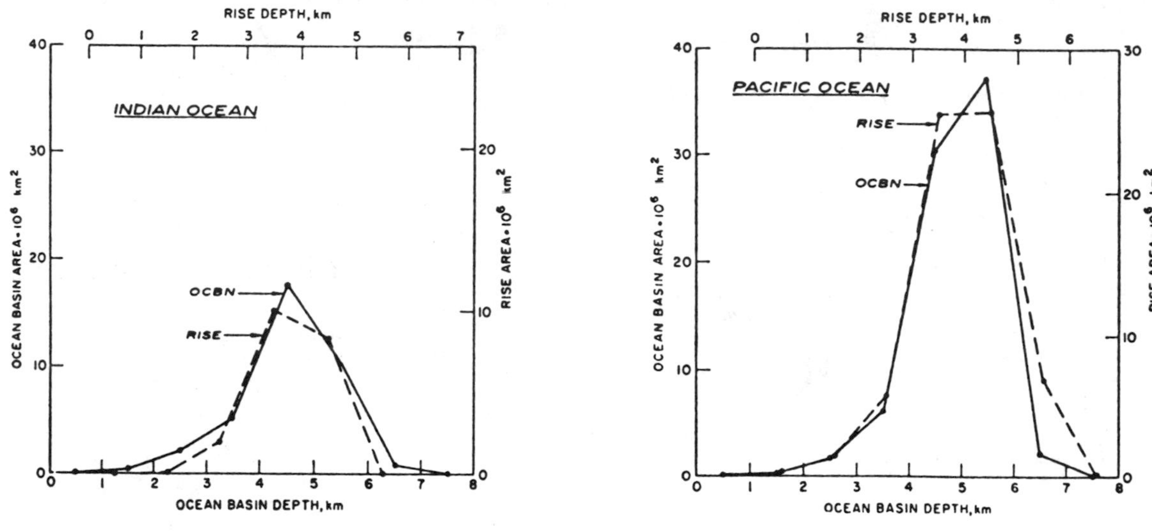

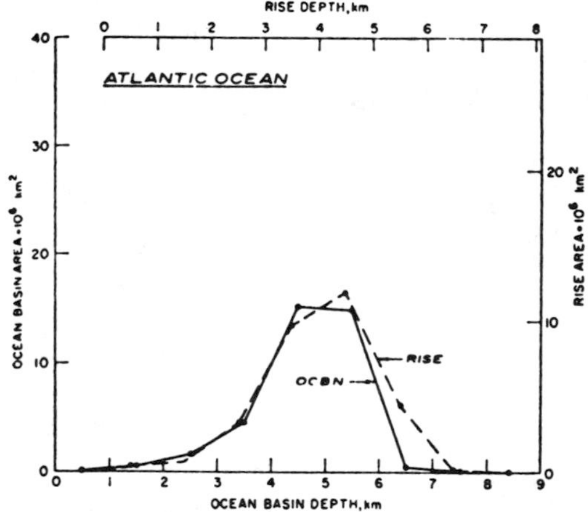

FIGURE 4.2—15. Hypsometric curves for RISE and RIDGE provinces in major ocean basins normalized to curves for OCEAN BASIN provinces.

(From Menard, H. W. and Smith, S. M., Hypsometry of ocean basin provinces, *J. Geophys. Res.*, 71, 4305, 1966. With permission of the American Geophysical Union.)

Table 4.2—7
CHARACTERISTICS OF OCEANIC RISES

	Mean depth, m			Relief, m	
	All provinces	OCBN (1)	RISE (2)	Mean (1) –Mean (2)	Shift of distribution curves*
World ocean	3729	4753	3970	783	1000
Pacific Ocean and marginal seas	3940	4896	3968	928	1100
Atlantic Ocean and marginal seas	3575	4670	4008	662	900
Indian Ocean and marginal seas	3840	4530	3945	585	800

* See Figure 4.2—13.

(From Menard, H. W. and Smith, S. M., Hypsometry of ocean basin provinces, *J. Geophys. Res.*, 71, 4305, 1966. With permission of American Geophysical Union.)

Table 4.2—8
MAJOR DEEPS AND THEIR LOCATION, SIZE, AND DEPTHS

Trench	Max depths, m	Ref.
Marianas Trench	11,034 ± 50	5
(specifically Challenger Deep)	10,915 ± 20[a]	
	10,915 ±	8
	10,863 ± 35	1
	10,850 ± 20[a]	
Tonga	10,882 ± 50	5
	10,800 ± 100	4
Kuril–Kamchatka	10,542 ± 100	13
	9,750 ± 100[b]	2
Philippine		
(vicinity of Cape Johnson Deep)	10,497 ± 100	6
	10,265 ± 45	16
	10,030 ± 10[a]	
Kermadec	10,047 ±	14
Idzu–Bonin		
(includes "Ramapo Deep" of the Japan Trench)	9,810	13
(vicinity of Ramapo Depth)	9,695	11
Puerto Rico	9,200 ± 20	9
New Hebrides (North)	9,165 ± 20[a]	
North Solomons (Bougainville)	9,103 ±	14
	8,940 ± 20[a]	
Yap (West Caroline)	8,527 ±	7
New Britain	8,320 ±	14
	8,245 ± 20[a]	
South Solomons	8,310 ± 20[a]	
South Sandwich	8,264	10
Peru–Chile	8,055 ± 10	3
Palau	8,054 ±	14
	8,050 ± 10[a]	

[a] These soundings were taken during Proa Expedition, April–June, 1962, aboard R.V. *Spencer F. Baird.* A Precision Depth Recorder was employed, and the ship's track crossed over (within the limits of celestial navigation) points from which maximum depths had been reported.

[b] This is the maximum sounding obtained in the vicinity of the Vitiaz Depth (Udintsev, 1959) by French and Japanese vessels in connection with dives of the bathyscaph *Archimède,* July, 1962.

Table 4.2—8 (continued)

Trench	Max depths, m	Ref.
Aleutian (uncorrected, taken with nominal sounding velocity of 1,500 m/sec)	7,679	
Nansei Shoto (Ryuku)	7,507	
Java	7,450	15
New Hebrides (South)	7,070 ± 20[a]	
Middle America	6,662 ± 10	

[a] These soundings were taken during Proa Expedition, April–June, 1962, aboard R.V. *Spencer F. Baird.* A Precision Depth Recorder was employed, and the ship's track crossed over (within the limits of celestial navigation) points from which maximum depths had been reported.

REFERENCES

1. **Carruthers, J.N. and Lawford, A.L.,** The deepest oceanic sounding, *Nature,* 169, 601, 1952.
2. **Delauze,** personal communication, 1962.
3. **Fisher, R.L.,** in *Preliminary Report on Expedition Downwind,* I.G.Y. General Report Ser., 2, I.G.Y. World Data Center A, Washington, 1958.
4. **Fisher, R.L. and Revelle, R.,** A deep sounding from the southern hemisphere, *Nature,* 174, 469, 1954.
5. **Hanson, P.P., Zenkevich, N.L., Sergeev, U.V., and Udintsev, G.B.,** Maximum depths of the Pacific Ocean, *Priroda (Mosk),* 6, 84, 1959 (in Russian).
6. **Hess, H.H. and Buell, M.W.,** The greatest depth in the oceans, *Trans. Am. Geophys. Un.,* 31, 401, 1950.
7. **Kanaev, V.F.,** New data on the bottom relief of the western part of the Pacific Ocean, *Oceanological Researches,* 2, 33, 1960 (in Russian).
8. **Lyman, J.,** personal communication, 1960.
9. **Lyman, J.,** The deepest sounding in the North Atlantic, *Proc. R. Soc. Lond.,* A222, 334, 1954.
10. **Maurer, H. and Stocks, T.,** *Die Echolötungen des Meteor. Wiss. Ergebn. Deut. Atlant. Exped. 'Meteor,' 1925–27,* 2, 1, 1933.
11. **Nasu, N., Iijima, A., and Kagami, H.,** Geological results in the Japanese Deep Sea Expedition in 1959, *Oceanog. Mag.,* 11, 201, 1960.
12. **Udintsev., G.B.,** Discovery of a deep-sea trough in the western part of the Pacific Ocean, *Priroda (Mosk),* 7, 85, 1958 (in Russian).
13. **Udintsev, G.B.,** Relief of abyssal trenches in the Pacific Ocean (abst.), *Intern. Oceanog. Cong. Preprints,* Am.Assoc. Adv. Sci., Washington, 1959.
14. **Udintsev, G.B.,** Bottom relief of the western part of the Pacific Ocean, *Oceanological Researches,* 2, 5, 1960 (in Russian).
15. **van Riel, P.M.,** The bottom configuration in relation to the flow of bottom water, *The 'Snellius' Exped.,* E.J. Brill, Leiden Neth., 2(2), Ch. 2, 1933.
16. **Wiseman, J.D.H. and Ovey, C.D.,** Proposed names of features on the deep-sea floor, *Deep-Sea Res.,* 2, 93, 1955.

(From Hill, M. N., Ed., *The Sea*, Vol. III, Wiley-Interscience, New York, 1963. With permission. In part after the compilation of Wiseman and Ovey,[16] 1955.)

Figure 4.2—16
BATHYMETRIC EXPLORATION OF THE CHALLENGER DEEP, MARIANAS TRENCH

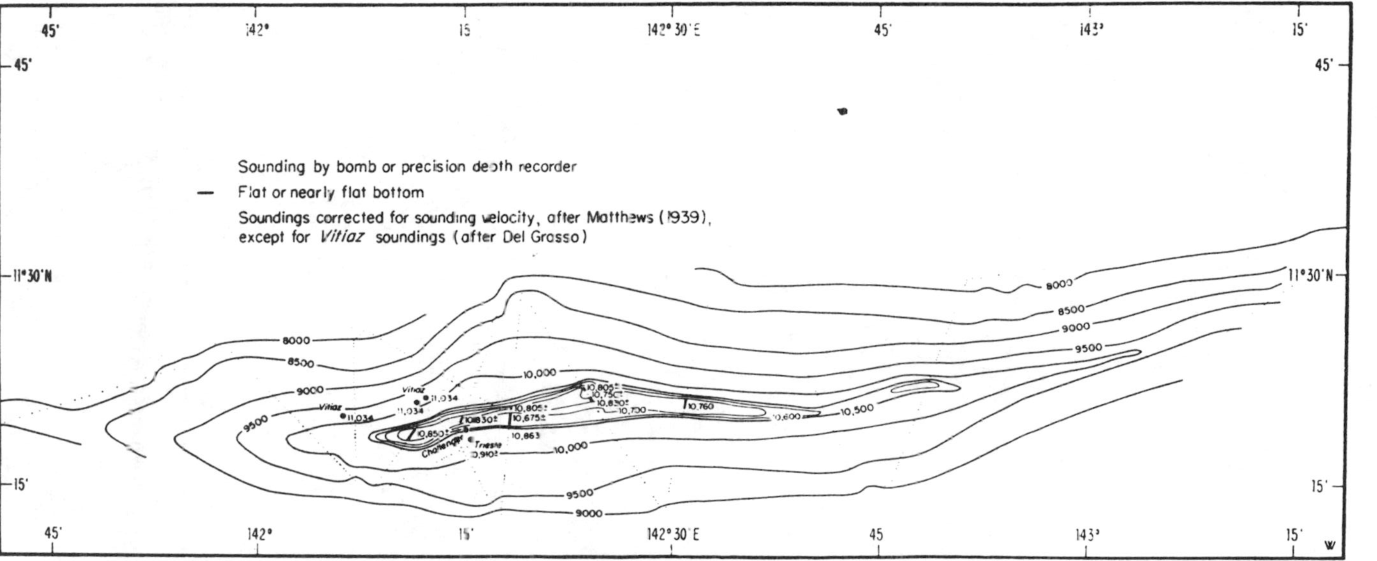

Note: By S.I.O. R.V. *Stranger,* Naga Expedition. July, 1959.

(From Hill, M. N., Ed., The *Sea*, Vol. III, Wiley-Interscience, New York, 1963. © by John Wiley & Sons. With permission.)

Figure 4.2—17
CROSS SECTIONS OF THE PHILIPPINE TRENCH

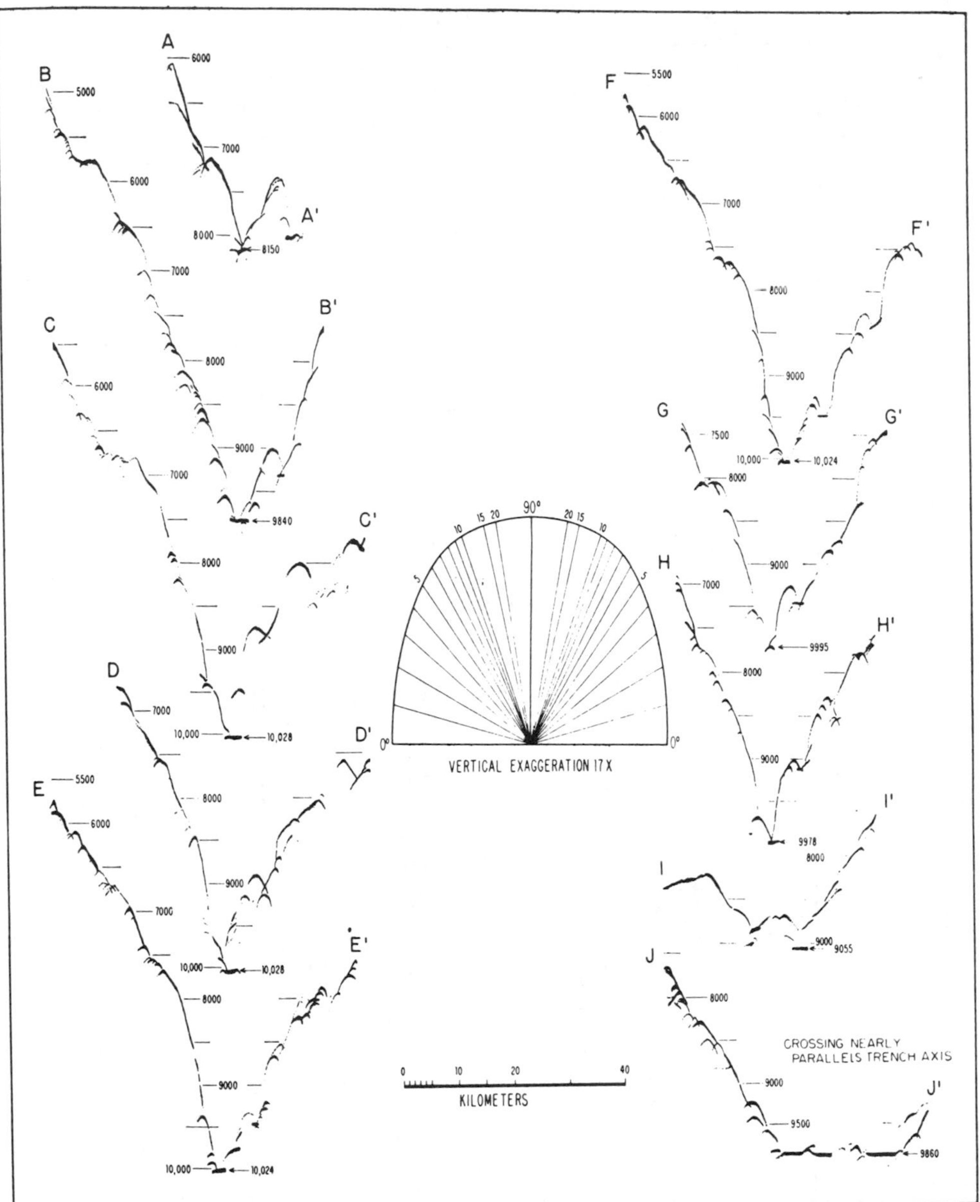

FIGURE 4.2—17. Traced from R. V. *Stranger* Precision Depth Recorder tapes. Sounding scales corrected after Matthews (1939).

Note: Since a wide-beam sounder was employed, many side echoes appear as early, late, or faint returns. The upper surface, ordinarily accepted as the sounding trace, is actually the envelope of minimum reflection times.

(From Hill, M. N., Ed., The *Sea*, Vol. III, Wiley-Interscience, New York, 1963. © by John Wiley & Sons. With permission.)

REFERENCE

Matthews, D. J., Tables of the velocity of sound in pure water and sea water for use in echo-sounding and sound-ranging, *Brit. Admiralty Hydrog. Dep. Pub.*, H.D. 282, 2nd ed.

Figure 4.2—18
CROSS SECTIONS OF THE MIDDLE AMERICA TRENCH

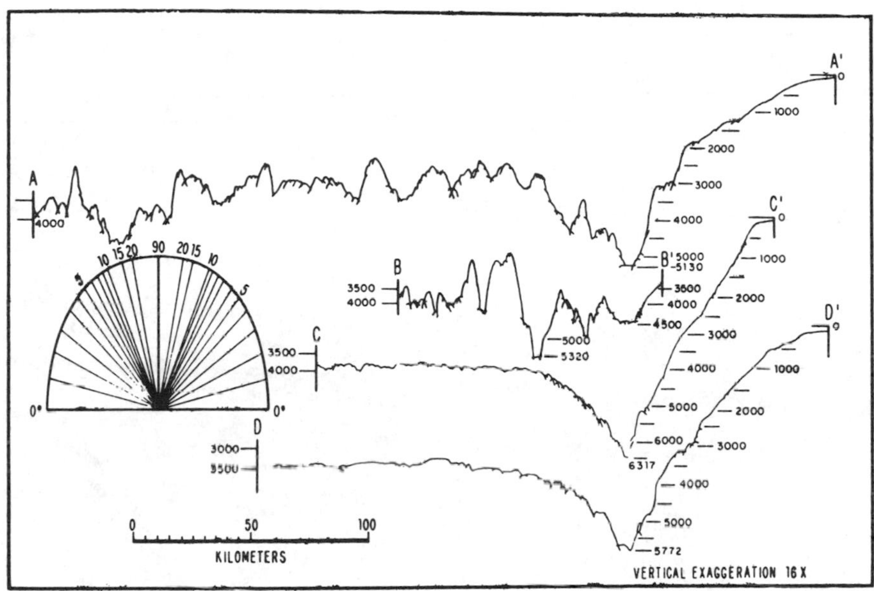

FIGURE 4.2—18. Traced from contiguous-sounding records. Sounding scales corrected after Matthews (1939).

(From Hill, M. N., Ed., *The Sea*, Vol. III, Wiley-Interscience, New York, 1963. © by John Wiley & Sons. With permission.)

REFERENCE

Matthews, D. J., Tables of the velocity of sound in pure water and sea water for use in echo-sounding and sound-ranging, *Brit. Admiralty Hydrog. Dep. Pub.*, H.D. 282, 2nd ed.

Figure 4.2—19
SUBMARINE TOPOGRAPHY OF THE NEW BRITAIN-NEW HEBRIDES REGION, SOUTHWEST PACIFIC*

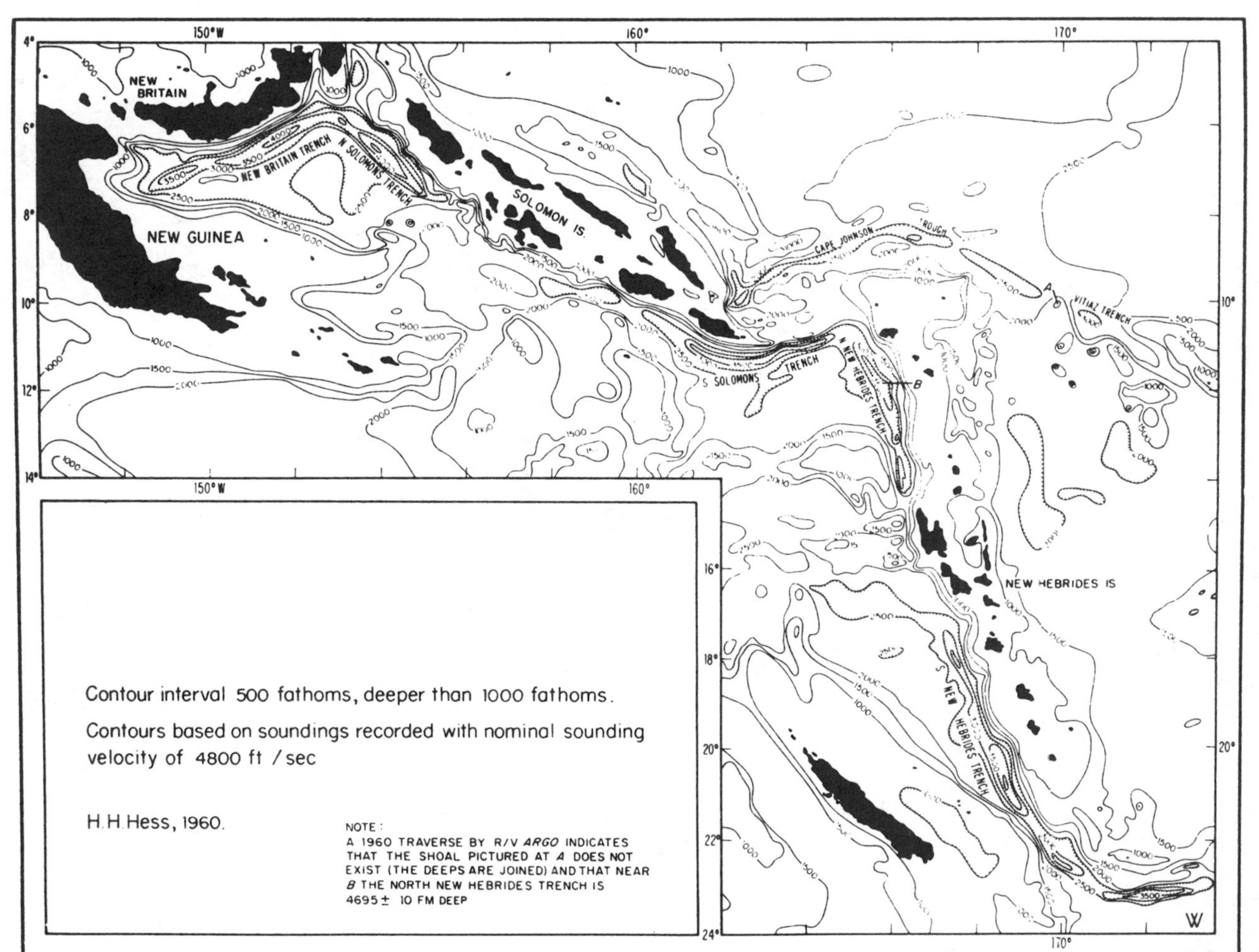

*** Near 10°S, 170°E; *B* in North New Hebrides Trench.**

(From Hill, M. N., Ed., The *Sea*, Vol. III, Wiley-Interscience, New York, 1963. © by John Wiley & Sons. With permission.)

Figure 4.2—20

ISLAND ARC AND CONTINENTAL MARGIN STRUCTURE SECTIONS DEDUCED FROM SEISMIC-REFRACTION DATA

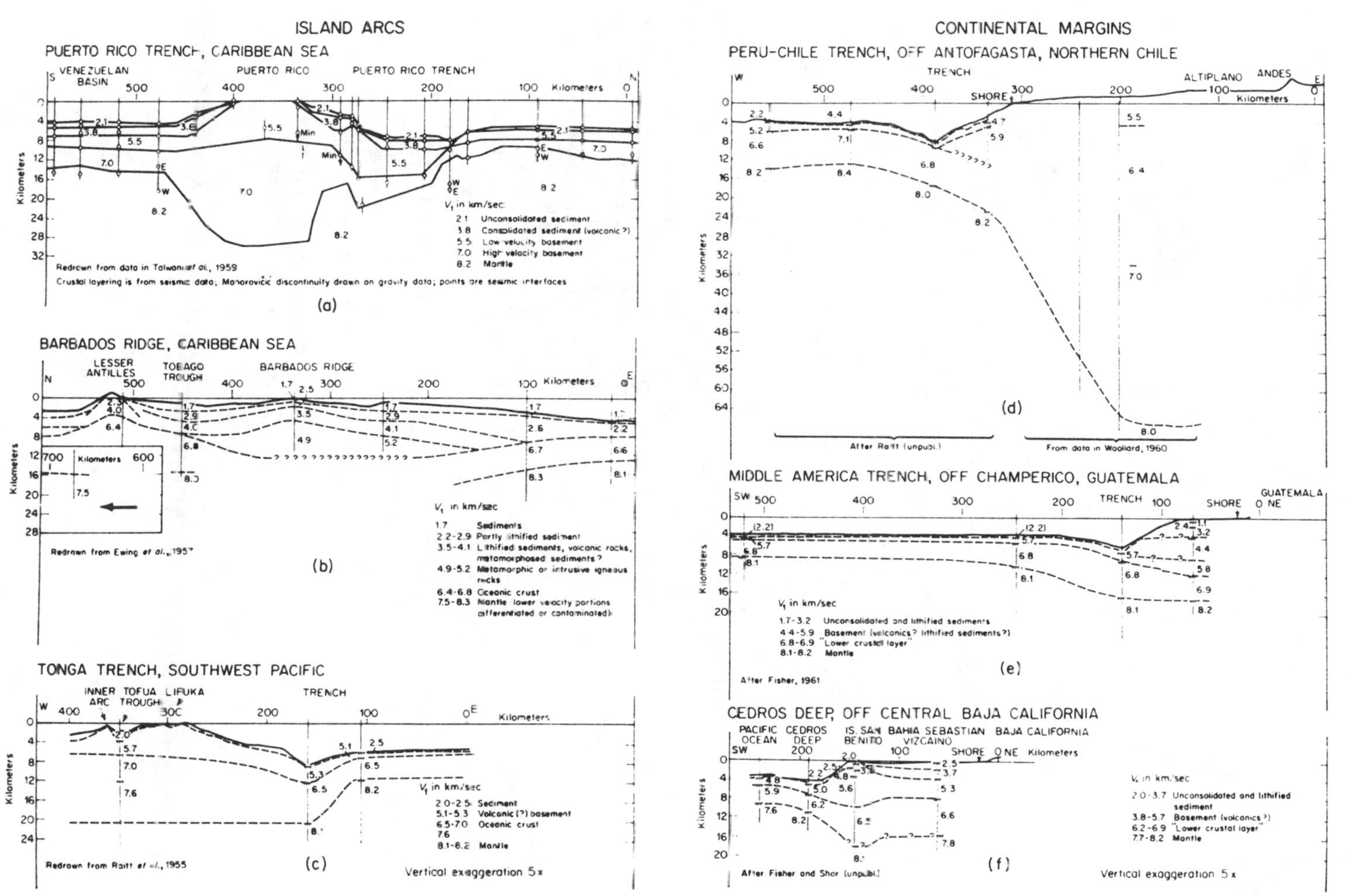

(From Hill, M. N., Ed., The *Sea*, Vol. III, Wiley-Interscience, New York, 1963. © by John Wiley & Sons. With permission.)

Figure 4.2—21
SUPPOSED STRUCTURE FOR A HYPOTHETICAL TRENCH-ISLAND ARC ASSOCIATION*

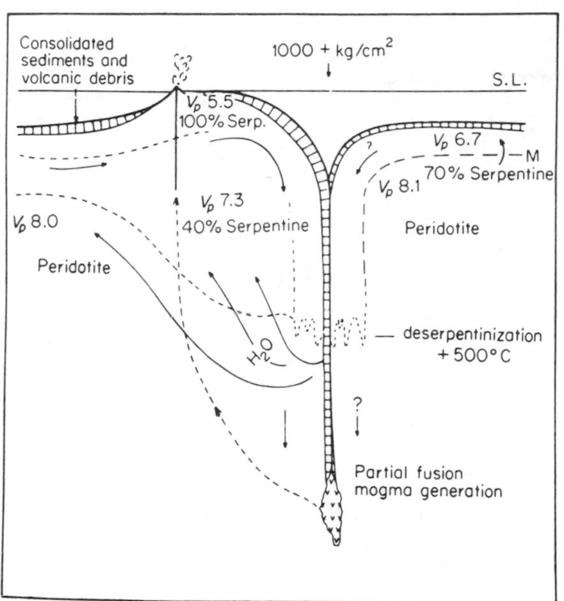

* With typical seismic velocities (V_p in km/sec).

(From Hill, M. N., Ed., The *Sea*, Vol. III, Wiley-Interscience, New York, 1963. © by John Wiley & Sons. With permission.)

Figure 4.2—22
POSTULATED MECHANICAL SITUATION WHERE THE CRUST MOVES INTO A CURVED TRENCH

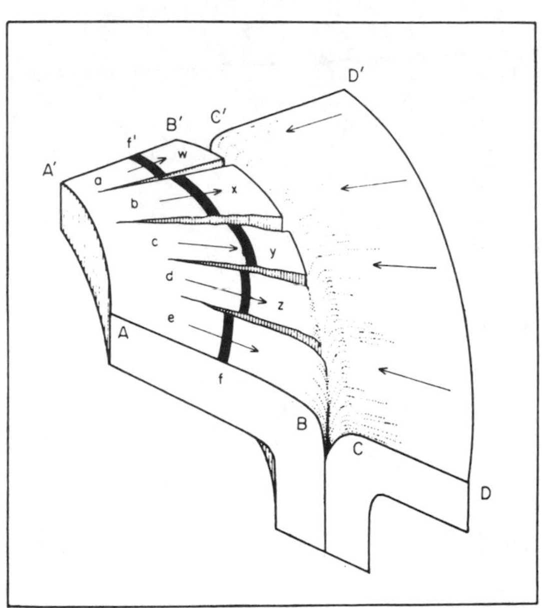

FIGURE 4.2—22. Movement of A′A to B′B would result in extension in the direction B′B, providing potential fissures to permit excape of magma to volcanoes at w, x, y, and z. These volcanoes are in all cases on the concave side of the trench and about 180 km from the trench axis. Immediately beneath the trench, deformation is severe, and plastic flow might be expected to close the fissures. Movement on the convex side from D′D to C′C would cause compression along C′C so that open fissures would not form and hence volcanoes are absent. Individual sectors a, b, c, d, and e might move differentially as portrayed by the displaced bed f′f. Strike-slip faults would bound the sectors and would tend to be sinistral toward A′B′ and dextoral toward AB.

Note: This figure represents the deformation of a layer of arbitrary thickness, not necessarily the curst.

(From Hill, M. N., Ed., *The Sea,* Vol. III, Wiley-Interscience, New York, 1963. With permission.)

Table 4.2—9

SUBMARINE CANYONS: LOCATION, SIZE, DEPTH

Symbols

1. Length of canyon measured along axis (nautical miles)
2. Depth at canyon head (feet)
3. Depth at canyon terminus (feet)
4. Character of coast inside canyon head
   - A. Heads in estuary
   - B. Heads off embayment
   - C. Heads off straight beach or barrier
   - D. Heads off relatively straight cliff
   - E. Uncertain
5. Relation of canyon head to points of land
   - A. On upcurrent side of point
   - B. Relatively near upcurrent side of point
   - C. No relation to point
6. Relation of canyon head to river valleys
   - A. Probable connection
   - B. No connection
   - C. Uncertain
7. Source of sediments to canyon head
   - A. Receives good supply
   - B. Supply restricted now, greater during lowered sea level stages
   - C. Little known supply of sediment because of depth
8. Gradient of axis in meters per kilometer
9. Nature of longitudinal profile
   - A. Generally concave upward
   - B. Generally convex upward
   - C. Relatively even slope
   - D. Local step-like steepening along axis
10. Maximum height of walls in feet
11. Channel curvature
    - A. Straight
    - B Slightly curving
    - C. Twisting or winding
    - D. Meandering
    - E. One meandering bend
    - F. Right-angled bends
12. Abundance of tributaries
    - A. As common as typical land valleys
    - B. Less common than typical land valleys
    - C. Confined to canyon head
    - D. No known tributaries
13. Character of transverse profile
    - A. Predominantly V-shaped
    - B. V-shaped inner canyon, trough-shaped outer canyon
    - C. Predominantly trough-shaped
    - D. Uncertain
14. Nature of canyon wall material
    - A. Crystalline rock dredged
    - B. Rock dredged, but all sedimentary
    - C. Mud only dredged on wall
    - D. Unknown
15. Nature of core sediment from axis
    - A. Includes sand layers
    - B. Includes sand and gravel layers
    - C. Mud cores only
    - D. Unknown

Table 4.2—9 (continued)
SUBMARINE CANYONS: LOCATION, SIZE, DEPTH

	1	2	3	4	5	6	7	8
Canyon name and location	Canyon length	Depth at canyon head	Depth at canyon terminus	Coast character	Relation to points	Relation to river valleys	Sediment sources at head	Gradient in m/km
California								
Coronado	8.0	240	5,580	C	C	C	B	58
La Jolla	7.3	50	1,800	C	B	A	A	40
Scripps (tributary)	1.45	60	900	D	C	A	A	97
Redondo	8.0	30	1,920	C	B	B	A	39
Dume	3.0	120	1,860	C	A	A	A	97
Mugu	8.0	40	2,400	C	B	C	A	49
Sur Partington	49.0	300	10,200	D	C	C	A	34
Carmel (tributary)	15.0	30	6,600	A	A	A	A	73
Monterey	60.0	50?	9,600?	C	C	B	A	26.5
Delgada	55.0	90	8,400	D	C	B	A	25
Mattole	16.0	60	5,720	B	B	A	A	59
Eel	27.0	250	8,500	C	C	A	B	51
Total or Average (12)	21.5	110	5,290	1A, 1B, 7C, 3D	2A, 4B, 6C	6A, 3B, 3C	10A, 2B	54
Oregon-Washington								
Columbia	37.0?	360	6,130?	B	B	A	C	26
Willapa	60.0	500+	7,000	B	C	A	C	24
Gray	30.0	500	6,440	B	C	A	C	33
Quinault	25.0	500±	5,750	C	C	A	C	35
Juan de Fuca	31.0+	800±	4,520+	B	B	C	C	20
Total or Average (5)	36.6	532	5,968	4B, 1C	2B, 3C	4A, 1C	5C	27.6
Bering Sea								
Umnak	160.0	900	10,850	B	C	C	C	10.4
Bering	220.0	600±	11,160	B	C	C	C	8
Pribilof	86.0	500±	10,700	C?	C	C	C	20
Total or Average (3)	155.3	667	10,903	2B, 1C	3C	3C	3C	12.8

Table 4.2—9 (continued)
SUBMARINE CANYONS: LOCATION, SIZE, DEPTH

	1	2	3	4	5	6	7	8
Canyon name and location	Canyon length	Depth at canyon head	Depth at canyon terminus	Coast character	Relation to points	Relation to river valleys	Sediment sources at head	Gradient in m/km
U.S. East Coast								
Corsair	14+	360	5,400	E	C	C	B	23
Lydonia	16+	370	4,400+	E	C	C	B	42
Gilbert	20+	480	7,680+	E	C	C	C	60
Oceanographer	17+	600+	7,230+	E	C	C	C	65
Welker	27+	400	6,450+	E	C	C	B	38
Hydrographer	27+	450+	6,600+	E	C	C	B	37
Hudson	50	300	7,000	B	C	A	B	25
Wilmington	23+	320	6,940+	E	C	C	B	48
Baltimore	28+	400	6,110+	B?	C	C	B	34
Washington	28+	360	6,740+	E	C	C	B	38
Norfolk	38	320	8,300	E	C	C	B	35
Total or Average (11)	26.2	395	6,623	2B, 9E	11C	1A, 10C	9B, 2C	40
Hawaiian-Molokai								
Halawai	6.0+	300±	3,540+	B	C	A	B	90
Naiwa	7.5	380	4,880	D	C	A	B	100
Waikolu	9.0	<600	6,540	B	C	A	B	110
Pelekunu	10.0	<320	6,320	B	C	A	A	100
Hawaiian-Kauai								
Hanakapiai	6.0+	280	7,480	D	C	A	B	200
Hanakoa	3.7	600±	4,820	D	C	C	C	190
Hanopu	3.6	300	5,100	D	C	C	B	220
Total or Average (7)	6.5	397	5,526	3B, 4D	7C	5A, 2C	1A, 5B, 1C	144
Western Europe								
Shamrock	30+	1,200±	14,400	B?	C	C	C	28
Black Mud	30+	900±	12,200	E	C	C	B	57
Audierne	27	600±	10,500	B	C	C	C	60
Cap Ferret	50+	800	11,647	C	C	C	C	31
Cap Breton	135 or 70	400±	13,100	D	C	C	B	58
Aviles	65	60±	8,000	C	B	A (old river)	A	20 or 16

Table 4.2—9 (continued)
SUBMARINE CANYONS: LOCATION, SIZE, DEPTH

	1	2	3	4	5	6	7	8
Canyon name and location	Canyon length	Depth at canyon head	Depth at canyon terminus	Coast character	Relation to points	Relation to river valleys	Sediment sources at head	Gradient in m/km
Llanes	38	450	13,300	D	B	C	B	45
Nazare	93	200±	14,764	C	A	A	A	36
Lisbon	21	400	6,450	B	B	A?	B	48
Setubal	33+	350	6,880	C	C	A	B	33
Total or Average (10)	52.2 or 45.7	536	11,224	3B, 4C, 2D	1A 3B, 5C	4A, 5C	2A, 4B, 3C	41.6 or 41.2
Mediterranean mainland								
Grand Rhone	15+	600±	5,550	C	C	A	C	55
Marseille	20+	600±	6,840±	B	C	C	C	52
Canon de la Cassidaigne	19+	360	6,630+	B	B	A	B	55
Toulon	12+	260	6,600	A	C	C	B	110
Stoechades	17+	300	4,380+	borderline A	C	C	A	40
St. Topez	25+	60?	5,750	A	C	A	A	38
Cannes	17+	100?	6,600?	borderline C or A	C	B	A	65
Var	15	160	6,550	D	A	A	A	71
Nice	12	150	5,840	delta D	C	B	A	79
Cap d'Ail	14	320	6,870	B	C	C	B	78
Nervia	16	330	6,280	C	C	A	B	62
Taggia	12	300±	7,500	C	C	A	B	100
Mele	31	200	6,150	C	C	B	B	32
Noli	14	120	4,990	curving D	C	A?	A	58
Polcevera	49	300	8,830	curving C	C	A?	B	29
Genoa	20	260	6,260	B	C	A	B	50
Total or Average (16)	17.4	276.5	6,351	3A, 4B, 5C, 3D	1A, 1B, 14C	9A, 3B, 4C	7A, 7B, 2C	60.9
Mediterranean Islands								
Crete	4	<300	3,300	B?	C	A	A	200

Table 4.2—9 (continued)
SUBMARINE CANYONS: LOCATION, SIZE, DEPTH

	1	2	3	4	5	6	7	8
Canyon name and location	Canyon length	Depth at canyon head	Depth at canyon terminus	Coast character	Relation to points	Relation to river valleys	Sediment sources at head	Gradient in m/km
West Corsica								
St. Florent	25	150±	7,850	A	A	A	A	51
Calvi	13	200±	7,800	B	C	A	B	97
Porto	20	150±	8,200	A	C	A	A	67
Sagone	29	150±	6,200	A	C	A	A	35
Ajaccio	34	150±	8,200	A	C	A	A	39
Valinco	35	150±	8,000	A	C	A	A	37
Total or Average (7)	22.8	178	7,078	5A, 2B	1A, 6C	7A	6A, 1B	75
Baja, California								
San Pablo	20+	<400	8,400+	D	A	B	A	67
Cardonal	16+	<450	7,500+	C	B	B	B	73
Vigia	10	?	7,200	C + D	C	A or B	A	115
San Lucas –	19	30	6,900?	A-bay	C	A	A	70
Santa Maria	(24)		8,000					56
San Jose	32	50	7,200?	C	C	A?	A	41
Vinorama – Salado	9	200	6,300	C	A	A	B	113
Los Frailes	9.5	10	5,200	A-bay	A (S. wind)	A	A	91
Saltito	6	1,200	5,100	B?	C	B	C	108
Palmas – Pescadero	13 9.3	100?	5,300	C	A (S. wind)	C	A	65 or 91
Total or Average (9)	15.2	305	6,710	2A, 1B, 4½C, 1½D	4A, 1B, 4C	4½A, 3½B, 1C	6A, 2B, 1C	81
East Honshu								
Ninomiya	4.8	400	2,600	C	C	C	B	77
Sagami	5.0	310	3,300	C	C	C	B	100
Enoshima	6.7	450	3,250	E	C	C	B	70
Hayama	13	300		E	C	C	B	54
			4,600					
Miura	15	173		E	C	C	B	48
Misaki	14.5	330	4,600	E	C	C	B	49

Table 4.2—9 (continued)
SUBMARINE CANYONS: LOCATION, SIZE, DEPTH

	1	2	3	4	5	6	7	8
Canyon name and location	Canyon length	Depth at canyon head	Depth at canyon terminus	Coast character	Relation to points	Relation to river valleys	Sediment sources at head	Gradient in m/km
Jogashima	10	330		E	C	C	B	71
Tokyo	30	300	4,900	A	C	C	B	26
Mera	20	190	5,500	B (bay)	B	C	B	44
Kamogawa	25	200	9,100	B	C	A	B	59
Total or Average (10)	29.25	298	4,731	1A, 2B, 2C, 5E	1B, 9C	1A, 9C	10B	60
Miscellaneous								
Great Bahama	125	4,800	14,060	B	C	C	C	13
Congo	120	80	7,000	A	C	A	A	96
Ceylon Trincomalee	20+	30+	9,500+	A	A?	A	A	79
Manila	31+	300	7,800+	B	C	A	B	40
Bacarra NW Luzon	15	300	6,000+	C + D	C	A	B	63
San Antonio, Chile	20+	<150	2,700+	C?	C	A?	A	32
Total or Average (6)	55	943	7,843	2A, 2B, 1½C, ½D	1A, 5C	5A, 1C	3A, 2B, 1C	54

(From Shepard, F. P. and Dill, R. F., *Submarine Canyons and Other Sea Valleys,* Rand McNally & Co., Chicago, 1966. With permission.)

Table 4.2—9 (continued)
SUBMARINE CANYONS: LOCATION, SIZE, DEPTH

	9	10	11	12	13	14	15	16
Canyon name and location	Nature of long profile	Max wall heights to nearest 1000 ft	Channel curvature	Abundance of tributaries	Transverse profile character	Name of canyon wall material	Sediment found in axial cores	Relation to fan-valleys
California								
Coronado	A	1,000	C	B	A	B	A	A
La Jolla	A	1,000	C	A	A	B	B	A
Scripps (tributary)	A	<1,000	B	C	A	B	B	B
Redondo	A	1,000	C	B	C	B	A	A
Dume	A	1,000	C	B	A	A	D	C
Mugu	A	<1,000	D	A	A	B	A	C
Sur Partington	A	2,000	C	A	A	B	D	C
Carmel (tributary)	A	2,000	C	A	A	A	A	B
Monterey	A + D	6,000	C or D	A	B	A	B	A
Delgada	A	2,000	D	B	A	D	D	A
Mattole	A	3,000	C	B	A	B	D	C
Eel	B	4,000	E	B	A	D	D	C
Total or Average (12)	10A, 1B, 1D	2,083	½B, 8½C, 2½D, 1E	5A, 6B 1C	10A, 1B, 1C	3A, 7B, 2D	4A, 3B, 5D	5A, 2B, 5C
Orgeon-Washington								
Columbia	A	2,000	C	B	D	D	D	C
Willapa	A	2,000	C	A	D	B	A	C
Gray	A	1,000	C	A	D	D	D	C
Quinault	A	3,000	C	A	D	D	D	C
Juan de Fuca	A	2,000	C	D	D	D	D	C
Total or Average (5)	5A	2,400	5C	3A, 1B, 1D	5D	4D, 1B	4D, 1A	5C
Bering Sea								
Umnak	A	4,000	C	A	D	D	D	C
Bering	A + D	6,000	C	A	A	B?	D	C
Pribilof	A	7,000	C	A	D	D	D	C
Total or Average (3)	2½A ½D	5,667	3C	3A	1A, 2D	1B, 2D	3D	3C

Table 4.2—9 (continued)
SUBMARINE CANYONS: LOCATION, SIZE, DEPTH

Canyon name and location	9 Nature of long profile	10 Max wall heights to nearest 1000 ft	11 Channel curvature	12 Abundance of tributaries	13 Transverse profile character	14 Nature of canyon wall material	15 Sediment found in axial cores	16 Relation to fan-valleys
U.S. East Coast								
Corsair	A	2,000	B	B	A	B	D	C
Lydonia	A	3,000	B	B	A	B	C	C
Gilbert	A + D	3,000	B	A	A	B	D	C
Oceanographer	A	2,000	B	B	A	D	D	C
Welker	A + D	4,000	B	B	A	B	A	C
Hydrographer	A + D	3,000	B	B	A	C	A	A
Hudson	A + D	4,000	B	A	A	B	A	A
Wilmington	A	3,000	C	A	A	C	C	C
Baltimore	A + D	3,000	B	B	A	C	C	C
Washington	A + D	2,000	B	A	A	C	C	C
Norfolk	A + D	3,000	B	B	A	B	A	C
Total or Average (11)	7½A, 3½D	2,900	10B, 1C	4A, 7B	11A	6B, 4C, 1D	4A, 4C, 3D	2A, 9C
Hawaiian-Molokai								
Halawai	C	1,000	C	D	A	D	D	C
Naiwa	C	1,000	B	D	A	A	A	B
Waikolu	A	2,000	B	B	A	A	B	A
Pelekunu	A	1,000	C	B	A	D	D	B
Hawaiian-Kauai								
Hanakapiai	B + D	1,000	B	B?	A	A	D	C
Hanakoa	B	1,000	B	B?	A	A	A	C
Hanopu	B + D	2,000	B	B?	A	B	A	A
Total or Average (7)	2A, 2B 2C, 1D	1,286	5B, 2C	5B, 2D	7A	4A, 1B, 2D	3A, 1B, 3D	2A, 2B, 3C
Western Europe								
Shamrock	C + D	3,000	B	B?	D	D	D	C
Black Mud	D	3,000	B	B	B	B	A	A
Audierne	D	4,000	B	D	D	D	D	C
Cap Ferret	A	3,000	C	A	D	D	D	A

Table 4.2—9 (continued)
SUBMARINE CANYONS: LOCATION, SIZE, DEPTH

	9	10	11	12	13	14	15	16
Canyon name and location	Nature of long profile	Max wall heights to nearest 1000 ft	Channel curvature	Abundance of tributaries	Transverse profile character	Nature of canyon wall material	Sediment found in axial cores	Relation to fan-valleys
Cap Breton	A	6,000	B	B	D	D	D	C
Aviles	C + D	5,000 or 6,000	B	A?	B	B	D	A?
Llanes	C	5,000	C	A	D	D	D	C
Nazare	D	5,000	F	B	D	A?	D	C
Lisbon	B + D	4,000	B	D	D	B	D	C
Setubal	C?	2,000	C?	B	D	B?	D	C
Total or Average (10)	2A, ½B, 3C, 3½D	4,000	5B, 3C, 1F	3A, 4B, 2D	1B, 8D	1A, 3B, 5D	9D	2A, 7C
Mediterranean mainland								
Grand Rhone	C?	2,000	B	C	A	D	D	C
Marseille	A + D	2,000	C	A	A	D	D	C
Canon de la Cassidaigne	D	3,000	C	B	A	D	D	C
Toulon	D?	4,000	B	A	A	D	D	C
Stoechades	A	4,000	B	A	A	B?	D	C
St. Topez	C?	3,000	C	A	A	D	D	C
Cannes	A	3,000	C	A	A	D	D	C
Var	A	3,000	C?	A	B	D	B	A
Nice	A + D	2,000	C?	A	B	B?	B	A
Cap d'Ail	D	1,000	B	B	B	D	D	A
Nervia	D	2,000	B	B	B	D	B	A
Taggia	A	2,000	C	B	A	D	D	A
Mele	A + D	1,000	C	B	B	D	D	B
Noli	A	2.000	C	B	B	D	D	B
Polcevera	A + D	3,000	B	B	B	D	D	C
Genoa	D	2,000	B	B	C?	D	D	B
Total or Average (16)	7A, 2C, 7D	2,400	7B, 9C	7A, 8B, 1C	8A, 7B, 1C	2B, 15D	3B, 13D	4A, 3B, 8C
Mediterranean Islands								
Crete	D	1,000	C	B	A	D	B	A

Table 4.2—9 (continued)
SUBMARINE CANYONS: LOCATION, SIZE, DEPTH

	9	10	11	12	13	14	15	16
Canyon name and location	Nature of long profile	Max wall heights to nearest 1000 ft	Channel curvature	Abundance of tributaries	Transverse profile character	Nature of canyon wall material	Sediment found in axial cores	Relation to fan-valleys
West Corsica								
St. Florent	A	3,000	C	A	A	D	A	A?
Calvi	A + D	3,000	C	Bor A	A	D	D	A
Porto	A	4,000	C	A	A	A?	D	C
Sagone	A	3,000	C	A	A	D	D	A?
Ajaccio	A	4,000	B	A	A	D	D	B?
Valinco	A + D	4,000	B	A	A	D	D	C
Total or Average (7)	6A, 1D	3,100	2B, 5C	5½A, 1½B	7A	1A, 6D	1A, 1B, 5D	4A, 1B, 2C
Baja California								
San Pablo	B + D	3,000	C	B	A	B	D?	C
Cardonal	A + D	3,000	C	A	A	B	D	C
Vigia	A	3,000	B	C	B	A	B	A
San Lucas – Santa Maria	A + D	3,000	C	A	A	A	B	A
San Jose	A + D	3,000	C	A	A	A (sed.)	B	A
Vinorama – Salado	A	1,000	C	A	A	A	B	A
Los Frailes	A + D	2,000	C	C	A	A	B	B
Saltito	A	1,000	C?	A	A	A	D	A
Palmas – Pescadero	A	2,000	C	A	A	A	A	A
Total or Average (9)	6A, ½B, 2½D	2,333	1B, 8C	6A, 1B, 2C	8A, 1B	7A, 2B	1A, 5B, 3D	6A, 1B, 2C
East Honshu								
Ninomiya	A	<1,000	B	C	A	B	D	C
Sagami	A	<1,000	C	A	B	B	D	C
Enoshima	A	<1,000	B	C	A	B	B	C
Hayama	A	2,000	C	B	A	A	D	C
Miura	A	2,000	C	B	A	B	B	C
Misaki	A	2,000	C	A	A	B	A	C
Jogashima	D	1,000	C	A	A	A	D	C
Tokyo	A + D	3,000	C	A	B	A	B	A

Table 4.2—9 (continued)
SUBMARINE CANYONS: LOCATION, SIZE, DEPTH

	9	10	11	12	13	14	15	16
Canyon name and location	**Nature of long profile**	**Max wall heights to nearest 1000 ft**	**Channel curvature**	**Abundance of tributaries**	**Transverse profile character**	**Nature of canyon wall material**	**Sediment found in axial cores**	**Relation to fan-valleys**
Mera	C + D	2,000	C	A	A	B	B	C
Kamogawa	A + D	5,000	B	B	B	B	A	C
Total or Average (10)	7A, ½C, 2½D	2,000	3B, 7C	5A, 3B, 2C	7A, 3B	3A, 7B	2A, 4B, 4D	1A, 9C
Miscellaneous								
Great Bahama	A	14,000	C	A	A	C	B	A
Congo	A	4,000	B	C	A?	D	A sand	A
Ceylon Trincomalee	A	4,000	F	A	A	A?	D	C
Manila	A + D	6,000	C	A	D	D	D	C
Bacarra NW Luzon	C	3,000	C	A	B?	D	D	C
San Antonio, Chile	A	3,000	B + C	B	D	D	D	C
Total or Average (6)	4½A, 1C, ½D	5,666	1½B, 3½C, 1F	4A, 1B, 1C	3A, 1B, 2D	1A, 1C, 4D	1A, 1B, 4D	4½A, 1C, ½D

(From Shepard, F. P. and Dill, R. F., *Submarine Canyons and Other Sea Valleys*, Rand McNally & Co., Chicago, 1966. With permission.)

4.3. CONDUCTIVITY AND HEART FLOW DATA

Table 4.3—1
CONDUCTIVITY OF DEEP-SEA SEDIMENTS

Sediment type	Water content, % wet wt	Density, gm/cm³	Conductivity, 10^{-3} cal/cm sec °C	Ref.
Red clay	52	1.43	1.93	3
	54	1.39	1.93	
	56.5	1.38	1.93	
	50	1.47	2.17	
	50	1.47	2.20	
	42.5	1.58	2.37	
	43.5	1.57	2.43	
	52.5	1.41	1.91	
	52	1.40	1.96	
	69.5	1.20	1.68	
	61.8	1.27	1.73	
Mud	55	1.32	1.91	
	52.5	1.36	1.90	
	56.5	1.31	1.88	
	51.5	1.37	1.94	
	46	1.47	2.06	
Globigerina ooze and glacial clay	41.3	1.58	2.31	1
	39.8	1.62	2.40	
	44.7	1.52	2.24	
	43.8	1.56	2.23	
	40.5	1.55	2.52	
	37.5	1.61	2.60	
	31.5	1.83	2.72	
	50.0	1.44	2.04	
	47.0	1.50	2.19	
	43.1	1.55	2.27	
	20.2	2.14	3.24	
	38.2	1.59	2.54	
	43.7	1.46	2.27	
	40.3	1.56	2.44	
	32.2	1.72	2.68	
Globigerina ooze	37.8	1.54	2.33	2
	43.8	1.47	2.07	
	43.4	1.47	2.22	
	36.9	1.55	2.55	
	38.5	1.54	2.52	
Dark mud	46.8	1.43	2.08	
	45.7	1.45	2.17	
	44.4	1.47	2.24	
	44.6	1.47	2.24	
	42.6	1.49	2.24	
	38.0	1.57	2.39	
	37.8	1.57	2.30	
	38.9	1.56	2.44	

Note: These measurements show a close correspondence between conductivity and water content and little dependence on type of sediment.[4]

(From Clark, S. P., Jr., *Handbook of Physical Constants*, Geological Society of America, Memoir 97, 1966. With permission.)

REFERENCES

1. **Bullard, E.,** *Proc. R. Soc. (Lond.)*, A 222, 408, 1954.
2. **Bullard, E.,** unpublished.
3. **Butler, D. W.,** unpublished.
4. **Ratcliffe, E. H.,** *J. Geophys. Res.*, 65, 1535, 1960.

Table 4.3—2
HEAT FLOW IN THE ATLANTIC OCEAN

Including Black Sea, Caribbean Sea, and Mediterranean Sea

SYMBOLS

Lat = Station latitude in degrees and minutes
Long = Station longitude in degrees and minutes
Elev. = Station elevation on land in meters
Depth = Station depth at sea in meters
∇T = Temperature gradient in 10^{-3} °C/cm
K = Thermal conductivity in 10^{-3} cal/cm sec °C
Q = Heat flow in 10^{-6} cal/cm² sec

No. = Number of heat-flow values averaged together
Ref. = Reference number
Yr = Year of publication
() = Heat-flow value derived from estimated conductivity
* = Heat-flow value obtained when penetration of the temperature gradient probe is partial
? = Heat-flow value questionable

Station	Lat	Long	Depth	∇T	K	Q	No.	Ref.	Yr
Black Sea	–	–	2269	.48	4.0	1.9?	7	1	61
CH21-1	29°51′N	54°36′W	5610	.50	2.08	1.04	1	2	64
CH21-4	28°56′N	46°44′W	4370	.30	2.24	.67	1	2	64
CH21-5	28°47′N	44°55′W	3940	.51	2.22	1.13	1	2	64
CH21-10	29°04′N	43°12′W	3080	.4	1.96	<.8	1	2	64
CH21-12	28°51′N	42°49′W	3520	.38	2.11	.81	1	2	64
CH21-13	29°02′N	41°10′W	4060	.2	1.94	.4	1	2	64
CH19-C	20°13′N	66°35′W	5810	.56	2.27	1.28	1	2	64
CH19-7-1	20°14′N	66°35′W	5770	.75	2.05	1.54	1	2	64
A-282-3	23°20′N	70°02′W	5480	.54	2.09	1.12	1	3	63
A-282-5	23°28′N	72°18′W	5300	.66	1.77	1.17	1	3	63
A-282-6	25°14′N	73°16′W	5310	.53	2.03	1.08	1	3	63
A-282-7	26°59′N	72°13′W	5150	.58	1.86	1.09	1	3	63
A-282-9	25°18′N	69°01′W	5580	.55	2.11	1.17	1	3	63
A-282-10	23°37′N	67°54′W	5650	.53	2.00	1.06	1	3	63
A-282-11	21°47′N	68°51′W	5560	.61	2.10	1.27	1	3	63
A-282-12	20°22′N	67°23′W	5410	.87	2.01	1.76	1	3	63
A-282-13	21°54′N	66°37′W	5640	.61	1.94	1.19	1	3	63
A-282-14	23°40′N	65°37′W	5800	.59	1.92	1.13	1	3	63
A-282-15	25°29′N	64°34′W	5680	.57	1.92	1.09	1	3	63
A-282-17	25°26′N	66°40′W	5580	.64	1.90	1.22	1	3	63
A-282-18	27°05′N	67°56′W	5200	.57	1.88	1.07	1	3	63
A-282-20	28°44′N	69°05′W	5330	.58	2.06	1.18	1	3	63
A-282-21	28°51′N	66°50′W	5240	.62	1.93	1.19	1	3	63
A-282-22	28°54′N	64°39′W	4900	.61	1.80	1.11	1	3	63
A-282-23	30°27′N	67°58′W	5230	.55	1.91	1.05	1	3	63
AII-1-1	32°02′N	74°09′W	4870	.40	2.05	.81	1	3	63
AII-1-3	30°56′N	74°36′W	3430	.47	1.99	.94	1	3	63
AII-1-5	29°10′N	76°22′W	4990	.46	2.51	1.17	1	3	63
C-36-1	21°08′N	65°02′W	5696	.53	1.82	.96	1	4	64
C-36-3	19°24′N	61°30′W	5468	.73	1.89	~1.37	1	4	64
C-36-5	16°45′N	57°38′W	5853	.12	2.28	>.27	1	4	64
C-36-6	16°47′N	57°49′W	5853	.15	2.0	>.3	1	4	64
C-36-7	16°34′N	57°52′W	4330	.54	1.96	1.06	1	4	64
C-36-8	16°35′N	57°54′W	4330	.54	1.93	1.05	1	4	64

Table 4.3—2 (continued)

Station	Lat	Long	Depth	∇*T*	*K*	*Q*	No.	Ref.	Yr
C-36-9	16°57′N	58°24′W	5890	.22	2.01	>.44	1	4	64
C-36-10	16°18′N	58°37′W	5599	.60	1.86	1.11	1	4	64
ATS296-4	39°32′N	65°50′W	4330	.47	2.29	>1.08	1	4	64
ATS296-6	39°33′N	66°17′W	4325	.56	2.37	>1.33	1	4	64
ATS296-7	39°47′N	65°16′W	4467	.48	2.22	1.07	1	4	64
ATS296-8	39°26′N	65°09′W	4757	.54	2.10	1.14	1	4	64
ATS296-9	39°46′N	66°28′W	3922	.56	2.11	<1.18	1	4	64
C-39-1	20°00′N	59°11′W	5811	.47	1.96	.92	1	4	64
C-39-2	25°18′N	55°44′W	5932	.72	1.93	~1.39	1	4	64
C-39-3	24°04′N	55°14′W	5984	.33	1.82	.60	1	4	64
C-39-5	28°30′N	57°59′W	5800	.48	1.98	.95	1	4	64
C-39-6	29°56′N	60°33′W	5715	.72	1.84	1.33	1	4	64
C-39-7	29°47′N	62°12′W	4865	.66	1.81	1.19	1	4	64
B-D-6	39°36′N	12°13′W	3020	.46	2.30	1.06*	1	5	61
B-D-7	35°59′N	9°59′W	4534	.37	2.31	.87	1	5	61
B-D-8	35°58′N	4°34′W	1251	.57	2.13	1.22	1	5	61
B-D-9	45°28′N	5°47′W	4592	.33	2.26	.75	1	5	61
B-D-10	46°32′N	13°04′W	4413	.50	2.17	1.09	1	5	61
B-D-11	46°30′N	22°58′W	4084	.57	2.25	1.29	1	5	61
B-D-12	46°37′N	27°18′W	4109	3.15	2.07	6.52*	1	5	61
B-D-13	36°20′N	21°00′W	4844	.54	2.12	1.14	1	5	61
B-D-14	35°36′N	19°02′W	5375	.67	2.01	1.34*	1	5	61
B-D-15	35°34′N	18°56′W	5380	.46	2.01	.93*	1	5	61
B-D-16	36°39′N	17°21′W	5146	.53	2.13	1.14	1	5	61
B-D-17	44°55′N	10°45′W	4844	.64	2.18	1.39	1	5	61
B-D-18	40°59′N	15°09′W	5305	.49	2.32	1.14*	1	5	61
B-D-19	42°18′N	11°53′W	3063	.36	2.18	.78	1	5	61
B-D-20	41°27′N	14°40′W	5260	.55	2.18	1.21*	1	5	61
B-D-21	43°42′N	12°39′W	5030	.51	2.29	1.16	1	5	61
CHAIN-1	35°35′N	61°08′W	4590	.62	1.92	1.20	1	6	61
CHAIN-2	35°35′N	61°15′W	4680	.68	1.92	1.31	1	6	61
CHAIN-3	51°18′N	29°35′W	3260	3.7	1.7	>6.2	1	6	61
CHAIN-4	53°53′N	24°05′W	3350	.73	2.10	1.54	1	6	61
V-15-3	00°59′S	38°10′W	4137	.66	2.31	1.52	1	7	62
V-15-4	00°12′N	39°54′W	4111	.48	2.23	1.07	1	7	62
V-15-5	02°30′N	40°55′W	4285	.63	2.19	1.38	1	7	62
V-15-6	05°04′N	41°01′W	4544	.83	2.23	1.85	1	7	62
V-15-7	06°59′N	41°04′W	4636	.90	2.25	2.03	1	7	62
V-15-8	10°45′N	41°21′W	5002	1.51	2.23	3.37	1	7	62
V-15-10	14°14′N	57°06′W	5002	.73	2.19	1.60	1	7	62
V-15-12	17°21′N	65°11′W	4169	.52	2.23	1.16	1	7	62
V-15-13	20°49′N	66°25′W	5227	.68	2.23	1.52	1	7	62
V-15-14	23°14′N	66°36′W	5605	.61	2.23	1.36	1	7	62
V-15-16	21°34′N	67°06′W	5115	.75	2.23	1.67	1	7	62
V-15-19	19°50′N	65°53′W	7934	.52	2.23	1.16	1	7	62
V-15-23	32°35′N	74°24′W	4521	.46	2.23	1.03	1	7	62
V-15-24	32°47′N	74°49′W	4462	.47	2.22	1.04	1	7	62
LSDA-55	33°45′S	15°00′E	4170	.77	2.45	1.88	1	8	64
LSDA-56	33°15′S	11°59′E	4630	.43	2.37	(1.01)	1	8	64
LSDA-57	32°30′S	09°01′E	5040	.40	2.01	.8*	1	8	64

Table 4.3—2 (continued)

Station	Lat	Long	Depth	∇T	K	Q	No.	Ref.	Yr
LSDA-58B	32°00′S	06°06′E	5210	.55	2.01	(1.1)*	1	8	64
LSDA-59	31°37′S	02°47′E	4215	.04	2.18	(.09)	1	8	64
LSDA-60	31°21′S	01°58′E	4190	1.00	2.18	2.17	1	8	64
LSDA-61	30°52′S	00°56′W	3810	.41	2.18	(.90)	1	8	64
LSDA-63	30°16′S	04°21′W	4890	.46	2.15	.99	1	8	64
LSDA-64	30°06′S	05°45′W	4340	.34	2.19	(.74)	1	8	64
LSDA-65	29°43′S	07°16′W	4150	.22	2.23	.48	1	8	64
LSDA-66	29°48′S	08°24′W	4155	.12	2.23	(.27)	1	8	64
LSDA-67	29°51′S	09°25′W	3940	.21	2.32	.48	1	8	64
LSDA-68	29°49′S	10°18′W	3735	.51	2.28	(1.16)	1	8	64
LSDA-69	29°51′S	11°07′W	3690	.50	2.28	(1.15)	1	8	64
LSDA-70	29°55′S	11°54′W	3400	.18	2.28	(.41)	1	8	64
LSDA-71	29°51′S	12°46′W	3200	.50	2.24	1.12	1	8	64
LSDA-72B	29°45′S	14°11′W	3385	.48	2.24	(1.08)	1	8	64
LSDA-73	29°50′S	14°51′W	3735	.15	2.24	(.34)	1	8	64
LSDA-74	29°50′S	15°33′W	3405	.32	2.24	(.72)	1	8	64
LSDA-75	27°22′S	12°34′W	3520	.99	2.27	(2.24)	1	8	64
LSDA-76	27°27′S	10°56′W	3580	.59	2.27	1.34	1	8	64
LSDA-77	26°47′S	13°54′W	2480	.78	2.27	(1.7)*	1	8	64
LSDA-78	25°58′S	14°51′W	3785	.44	2.27	(1.0)*	1	8	64
LSDA-79	24°03′S	15°32′W	4100	.05	2.18	.10	1	8	64
LSDA-80	23°47′S	14°27′W	4000	.41	2.18	(.9)	1	8	64
LSDA-81	23°42′S	12°12′W	3580	.51	2.18	(1.12)	1	8	64
LSDA-82	22°43′S	13°07′W	3605	3.44	2.27	(7.8)*	1	8	64
LSDA-83	21°21′S	11°35′W	2515	3.58	2.27	8.14	1	8	64
LSDA-85	21°15′S	10°39′W	3535	.45	2.18	(.97)	1	8	64
LSDA-86	20°10′S	11°30′W	2925	3.35	2.18	(7.3)*	1	8	64
LSDA-87	19°53′S	12°26′W	2710	1.73	2.18	(3.78)	1	8	64
LSDA-88	19°44′S	12°55′W	3500	.48	2.18	1.04	1	8	64
LSDA-89	18°58′S	12°49′W	3125	.51	2.18	(1.11)	1	8	64
LSDA-90	18°58′S	12°00′W	2510	2.14	2.27	(4.85)	1	8	64
LSDA-91	18°32′S	10°15′W	3395	.21	2.15	.45	1	8	64
LSDA-92	18°08′S	11°15′W	3305	.34	2.18	(.75)	1	8	64
LSDA-93	17°39′S	12°22′W	3440	.74	2.18	(1.61)	1	8	64
LSDA-94	17°15′S	13°20′W	3340	.22	2.18	(.47)	1	8	64
LSDA-95	16°46′S	14°30′W	3455	.62	2.18	(1.35)	1	8	64
LSDA-96	16°15′S	15°45′W	3435	.20	2.18	(.43)	1	8	64
LSDA-97	15°48′S	16°50′W	3820	1.07	2.18	(2.33)	1	8	64
LSDA-98	15°23′S	17°54′W	4390	.23	2.18	(.51)	1	8	64
LSDA-99	14°55′S	19°22′W	4230	.19	2.24	.43	1	8	64
LSDA-100	10°00′S	15°26′W	3595	.13	2.23	.29	1	8	64
LSDA-101	09°11′S	13°20′W	2690	.04	2.16	.08	1	8	64
LSDA-102	09°03′S	10°29′W	3550	.18	2.23	(.40)	1	8	64
LSDA-103	06°43′S	13°27′W	3245	.12	2.18	(.26)	1	8	64
LSDA-104	05°41′S	11°12′W	2905	1.18	2.18	2.58	1	8	64
LSDA-105	04°57′S	09°28′W	3500	.53	2.18	(1.15)	1	8	64
LSDA-106	00°56′S	10°37′W	4040	.50	2.12	(1.07)	1	8	64
LSDA-107	00°28′S	10°51′W	4350	.42	2.12	(.89)	1	8	64
LSDA-108	00°03′N	11°02′W	4125	.68	2.12	1.45	1	8	64
LSDA-109	00°26′N	11°14′W	4215	.85	2.12	(1.80)	1	8	64

Table 4.3—2 (continued)

Station	Lat	Long	Depth	∇*T*	*K*	*Q*	No.	Ref.	Yr
LSDA-110	00°52′N	11°28′W	4950	.07	2.12	(.15)	1	8	64
LSDA-111	02°38′N	12°12′W	4735	.76	1.81	1.37	1	8	64
LSDA-112	05°01′N	12°45′W	4390	.82	1.91	1.56	1	8	64
LSDA-113	07°24′N	17°08′W	4800	.71	1.95	1.39	1	8	64
LSDA-114	06°47′N	19°18′W	4360	.46	2.09	.96	1	8	64
LSDA-115	06°21′N	20°49′W	3590	.58	2.12	(1.22)	1	8	64
LSDA-116	05°07′N	25°15′W	4360	.92	2.16	1.99	1	8	64
LSDA-117	03°21′N	30°52′W	2590	.16	2.34	.37	1	8	64
LSDA-118	03°18′N	31°00′W	2820	1.16	2.31	(2.68)	1	8	64
LSDA-119	03°15′N	31°35′W	2415	1.00	2.31	(2.3)*	1	8	64
LSDA-120	03°57′N	34°04′W	3340	.82	2.28	1.87	1	8	64
LSDA-121	05°42′N	32°51′W	2955	2.23	2.28	(5.08)	1	8	64
LSDA-122	05°59′N	32°28′W	3300	2.26	2.31	(5.22)	1	8	64
LSDA-124	08°26′N	34°23′W	4790	.76	2.04	1.56	1	8	64
LSDA-125	09°39′N	37°40′W	4045	.11	2.16	(.23)	1	8	64
LSDA-126	09°34′N	39°32′W	3340	.57	2.28	(1.31)	1	8	64
LSDA-127	09°41′N	40°49′W	2315	.74	2.28	(1.7)*	1	8	64
LSDA-128	09°45′N	41°18′W	3295	.74	2.28	(1.70)	1	8	64
LSDA-130	11°35′N	44°03′W	2755	1.05	2.28	(2.4)*	1	8	64
LSDA-131	11°34′N	44°48′W	3830	.40	2.12	.84	1	8	64
LSDA-132	11°34′N	45°33′W	4105	1.11	2.08	(2.30)	1	8	64
LSDA-133	12°17′N	46°13′W	4515	.22	2.08	(.46)	1	8	64
LSDA-134	14°59′N	58°19′W	3535	.32	2.22	.72	1	8	64
LSDA-135	15°04′N	59°58′W	4480	.32	2.20	.71	1	8	64
LSDA-136	15°04′N	60°30′W	2335	.93	2.15	2.0*	1	8	64
LSDA-137	15°02′N	62°15′W	2720	.93	2.15	(2.0)	1	8	64
LSDA-139	15°00′N	63°50′W	2082	.66	2.06	1.36	2	8	64
ZEP-4	13°36′N	71°59′W	4232	.72	2.0	1.4	1	9	64
ZEP-5	13°43′N	68°38′W	5042	.58	1.9	1.1	1	9	64
ZEP-8	14°22′N	62°19′W	2877	.70	1.9	1.3*	1	9	64
ZEP-9	16°24′N	57°39′W	4647	.39	1.8	.7?	1	9	64
ZEP-11	19°10′N	52°03′W	5344	.81	1.7	1.4	1	9	64
ZEP-12	20°12′N	49°01′W	4632	.30	1.5	.5	1	9	64
ZEP-13	21°06′N	46°30′W	3912	.16	1.9	.3	1	9	64
ZEP-14	21°04′N	44°57′W	3255	.84	2.1	1.8	1	9	64
ZEP-15	21°56′N	45°46′W	3372	3.24	2.0	6.5	1	9	64
ZEP-16	23°06′N	45°39′W	3983	1.48	2.0	3.0	1	9	64
ZEP-17	23°34′N	44°14′W	4960	.81	2.0	1.6	1	9	64
ZEP-18	23°57′N	44°59′W	3493	1.34	2.1	2.8*	1	9	64
ZEP-19	23°36′N	42°28′W	4113	.23	2.1	.5*	1	9	64
ZEP-20	24°16′N	39°06′W	5439	.19	1.9	.4	1	9	64
ZEP-22	25°05′N	34°13′W	5602	.36	1.9	.7	1	9	64
ZEP-23	26°14′N	26°27′W	5210	.59	2.0	1.2	1	9	64
ZEP-25	26°57′N	19°58′W	4298	.46	2.1	1.0	1	9	64
ZEP-26	31°12′N	11°50′W	3210	.50	2.2	1.1*	1	9	64
ZEP-27	33°35′N	9°43′W	4340	.45	2.2	1.0	1	9	64
ZEP-32	40°37′N	5°50′E	2720	.56	2.2	1.2?	1	9	64
D 4775	29°02′N	25°27′W	5342	–	–	1.39	1	10	63
D 4777	28°60′N	25°26′W	5344	–	–	1.20	1	10	63
D 4778	29°03′N	25°33′W	5342	–	–	1.13	1	10	63

Table 4.3—2 (continued)

Station	Lat	Long	Depth	∇T	K	Q	No.	Ref.	Yr
D 4784	29°04′N	25°27′W	5339	–	–	1.21	1	10	63
D 4788	29°05′N	25°15′W	5299	–	–	1.29	1	10	63
D 4809	28°51′N	25°27′W	4871	–	–	1.11	1	10	63
D 4813	28°50′N	25°24′W	4862	–	–	1.05	1	10	63
D 4817	29°34′N	25°18′W	5400	–	–	1.03	1	10	63
D 4821	29°35′N	25°23′W	5297	–	–	1.23	1	10	63
D 4822	29°08′N	24°19′W	5281	–	–	1.33	1	10	63
D 4528	45°19′N	11°27′W	4143	–	–	1.13	1	11	63
D 4531	45°19′N	11°28′W	4125	–	–	1.00	1	11	63
C19-6-17	31°54′N	64°44′W	4262	–	–	.97	1	11	63
CH21-8	29°04′N	44°11′W	–	–	–	+?	1	11	63
CH21-14	34°00′N	15°51′W	3810	–	–	.57	1	11	63
CH21-16	34°06′N	14°24′W	4315	–	–	.94	1	11	63
CH21-18	39°31′N	05°26′E	2826	–	–	>.87	1	11	63
CH21-19	42°14′N	07°09′E	2731	–	–	2.5	1	11	63
D 4790	27°10′N	21°06′W	4702	–	–	1.06	1	11	63
D 4794	27°10′N	21°00′W	4682	–	–	~1.2	1	11	63
D 4795	27°13′N	21°05′W	4707	–	–	.92	1	11	63
D 4805	29°35′N	23°52′W	5240	–	–	1.13	1	11	63
D 4824	43°06′N	19°50′W	5959	–	–	1.30	1	11	63
V18-151	19°51′N	84°56′W	4564	.7	2.0	(1.4)	1	12	64
V18-153	26°35′N	88°49′W	2582	.22	2.3	.5	1	12	64
V18-155	26°28′N	68°25′W	5284	.55	2.0	1.1	1	12	64
V18-158	38°45′N	67°33′W	4184	.55	1.8	1.0	1	12	64
V18-159	39°11′N	65°26′W	4730	.55	2.0	(1.1)	1	12	64
V19-1	34°50′N	70°15′W	4716	.42	1.9	(.8)	1	12	64
V19-2	32°36′N	71°19′W	5392	.63	1.9	(1.2)	1	12	64
V19-3	28°20′N	68°06′W	5261	.68	1.9	1.3	1	12	64
V19-4	27°28′N	68°27′W	2858	.47	1.9	.9	1	12	64
V19-5	24°16′N	67°11′W	5562	.63	1.9	(1.2)	1	12	64
V19-6	16°06′N	66°29′W	4520	.6	2.0	1.2	1	12	64
C7-2	13°06′N	63°09′W	1060	.55	2.0	(1.1)	1	12	64
C7-3	12°34′N	66°18′W	4529	.4	2.0	(.8)	1	12	64
C7-4	13°59′N	71°43′W	3948	.75	2.0	(1.5)	1	12	64
C7-5	12°04′N	74°54′W	3611	.5	2.0	(1.0)	1	12	64
C7-6	14°11′N	76°32′W	4087	.6	2.0	(1.2)	1	12	64
C7-9	14°50′N	73°50′W	3460	.5	2.0	(1.0)	1	12	64
C7-10	15°23′N	73°17′W	3324	.75	2.0	(1.5)	1	12	64
C7-11	16°08′N	72°48′W	2893	.55	2.0	(1.1)	1	12	64
C7-12	14°36′N	70°57′W	3525	.5	2.0	(1.0)	1	12	64
Bullard 1	49°46′N	12°30′W	2032	.426	2.59	1.10	1	13	54
Bullard 2	49°58′N	18°33′W	4017	.548	2.58	1.42	1	13	54
Bullard 3	49°09′N	17°38′W	4532	.237	2.43	.58	1	13	54
Bullard 4	48°14′N	16°58′W	4670	.254	2.28	.58	1	13	54
Bullard 5	48°52′N	15°00′W	4710	.455	2.64	1.20	1	13	54

(From Clark, S. P., Jr., *Handbook of Physical Constants,* Geological Society of America, Memoir 97, 1966. With permission.)

Table 4.3—2 (continued)

REFERENCES

1. **Sisoev,** *Okeanologiya,* 1, 886, 1961.
2. **Lister, C. R. B. and Reitzel, J. S.,** *J. Geophys. Res.,* 69, 2151, 1964.
3. **Reitzel, J.,** *J. Geophys. Res.,* 68, 5191, 1963.
4. **Birch, F. S.,** M.Sc. thesis, University of Wisconsin, Madison, 1964.
5. **Bullard, E.,** *Geophys. J.,* 4, 282, 1961.
6. **Reitzel, J.,** *J. Geophys. Res.,* 66, 2267, 1961.
7. **Gerard, R. et al.,** *J. Geophys. Res.,* 67, 785, 1962.
8. **Vacquier, V. and Von Herzen, R. P.,** *J. Geophys. Res.,* 69, 1093, 1964.
9. **Nason, R. D. and Lee, W. H. K.,** *J. Geophys. Res.,* 69, 4875, 1964.
10. **Lister, C. R. B.,** *J. Geophys. Res.,* 68, 5569, 1963.
11. **Lister, C. R. B.,** *Geophys. J.,* 7, 571, 1963.
12. **Langseth, M. G. and Grim, P. J.,** *J. Geophys. Res.,* 69, 4916, 1964.
13. **Bullard, E.,** *Proc. R. Soc. Lond.,* 222A, 408, 1954.

Section 5
Ocean Engineering

Section 5

OCEAN ENGINEERING

The field of marine engineering has experienced substantial expansion in response to man's increased use of oceanic systems. The application of modern engineering practices to the marine environment is most pervasive in the coastal zone where beach and inlet stabilization, dredging and dredge-spoil disposal, land reclamation and development, and other major habitat alterations have been imposed by man.[1] The sea serves as a repository for many substances such as sewage sludge, acid wastes, industrial and municipal effluents, and dredged material, thereby fostering the implementation of a wide range of waste management plans and engineering controls to rectify their impact.[2] The occurrence of heavy metals and toxic chemicals in dredged materials, for example, has prompted marine engineers to devise sub-bottom containment of contaminated spoils by covering or capping the disposed sediments with clean sediments. This technique minimizes reentry of the toxic substances into the water column.[1,3,4] For highly contaminated, dredged sediments, the formation of artificial islands has isolated potentially dangerous materials to confined areas.[5]

Another major concern of marine engineering involves the effect of micro- and macrofouling organisms on the structural integrity and utility of submerged wooden structures and on corrosion processes of metals. The longevity of man-made materials in the marine environment generally is compromised by the proliferation of epiflora and epifauna. The settlement and growth of biofouling organisms in intake structures, condensers, valves, and pipe lines of electric generating stations often reduce the pumping efficiency of cooling water systems and the capability to transfer heat.[6] The marine biofouling community is diverse, consisting of at least 1975 plant and animal species,[7] and probably double this number.[8] Workers in marine engineering and materials science have developed numerous fouling-control technology measures to deal with this complex community. These measures include scrubbing techniques, chemical controls (i.e., antifouling agents), sonics, electric currents and fields, magnetic fields, optical methods, nuclear methods, thermal controls, osmotic controls, surface modification, explosive removal, and velocity controls.[9]

The infestation of untreated wood by marine borers — primarily crustaceans and molluscs — creates countless problems that account for millions of dollars worth of damage each year to waterfront structures. Estimated annual losses to the U.S. Navy alone due to the destruction of wharf structures along the U.S. coastline amount to $200 million.[9,10] Proper design, selection of treated materials, preparation of materials, and construction techniques must be adhered to in order to mitigate marine engineering problems related to waterfront structures.[11] Destruction of wooden piles, docks, bulkheads, boats, and barges by teredinids (shipworms) and wood-boring isopods (gribbles) has resulted in widespread use of creosote-treated materials and broad-spectrum toxins to control the pests.[12]

The scope of marine engineering extends to many other areas as well. For instance, the construction of offshore oil platforms, the development of military hardware systems for deployment at sea, and the utilization of submersible vehicles for oceanographic research all incorporate various aspects of marine engineering. Indeed, most oceanographic studies on fixed ocean structures, mechanical properties of materials, and instrumentation require interaction with this discipline.

This section on ocean engineering is subdivided into seven subject areas. These include: (1) materials for marine applications; (2) ropes, chains, and shackles; (3) marine power sources; (4) fixed ocean structures; (5) buoy systems; (6) specifications of oceanographic instruments; and (7) ship characteristics. Among the tabular and illustrative data on ocean engineering are useful discussions on pilings, buoys, and corrosion behavior of metals in seawater.

REFERENCES

1. **Barnes, R. S. K.,** *The Coastline*, John Wiley & Sons, Chichester, U.K., 1977.
2. **Capuzzo, J. M., Burt, W. V., Duedall, I. W., Park, P. K., and Kester, D. R.,** The impact of waste disposal in nearshore environments, in *Wastes in the Ocean*, Vol. 6, Ketchum, B. H., Capuzzo, J. M., Burt, W. V., Duedall, I. W., Park, P. K., and Kester, D. R., Eds., John Wiley & Sons, New York, 1985, 3.
3. **Bokuniewicz, H. J.,** Submarine borrow pits as containment sites for dredged sediment, in *Wastes in the Ocean*, Vol. 2, Kester, D. R., Ketchum, B. H., Duedall, I. W., and Park, P. K., Eds., John Wiley & Sons, New York, 1983, 215.
4. **Morton, R. W.,** Precision bathymetric study of dredged-material capping experiment in Long Island Sound, in *Wastes in the Ocean*, Vol. 2, Kester, D. R., Ketchum, B. H., Duedall, I. W., and Park, P. K., Eds., John Wiley & Sons, New York, 1983, 99.
5. **Gordon, R. B., Bohlen, W. F., Bokuniewicz, H. J., dePicciotto, M., Johnson, J., Kamlet, K. S., McKinney, T. F., Schubel, J. R., Suszkowski, D. J., and Wright, T. D.,** Management of dredged material, in *Ecological Stress and the New York Bight: Science and Management*, Mayer, G. F., Ed., Estuarine Research Federation, Columbia, S.C., 1982, 113.
6. **Loveland, R. E. and Shafto, S. S.,** Fouling organisms, in *Ecology of Barnegat Bay, New Jersey*, Kennish, M. J. and Lutz, R. A., Eds., Springer-Verlag, New York, 1984, 226.
7. **Woods Hole Oceanographic Institution,** *Marine Fouling and Its Prevention*, Naval Institute Press, Annapolis, Md., 1952.

8. **Crisp, D. J.,** The role of the biologist in anti-fouling research, in *Proc. 3rd Int. Congr. Marine Corrosion and Fouling*, Northwestern University Press, Evanston, Ill., 1973, 88.
9. **Fischer, E. C., Castelli, V. J., Rodgers, S. D., and Bleile, H. R.,** Technology for control of marine biofouling — a review, in *Marine Biodeterioration: An Interdisciplinary Study*, Costlow, J. D. and Tipper, R. C., Eds., Naval Institute Press, Annapolis, Md., 1984, 261.
10. **Fischer, E. C., Birnbaum, L. S., DePalma, J., Muraoka, J. S., Dear, H., and Wood, F. G.,** Survey Report: Navy Biological Fouling and Biodeterioration, U.S. Off. Nav. Res. Tech. Rept., NUC TP-456, 1975.
11. **Roe, T., Jr.,** Maintenance of waterfront structures, in *Marine Biodeterioration: An Interdisciplinary Study*, Costlow, J. D. and Tipper, R. C., Eds., Naval Institute Press, Annapolis, Md., 1984, 38.
12. **Boyle, P. J. and Mitchell, R.,** The microbial ecology of crustacean wood borers, in *Marine Biodeterioration: An Interdisciplinary Study*, Costlow, J. D. and Tipper, R. C., Eds., Naval Institute Press, Annapolis, Md., 1984, 17.

5.1. MATERIALS FOR MARINE APPLICATION

Table 5.1–1
MECHANICAL PROPERTIES OF ALUMINUM ALLOYS

Material	Yield strength, psi X 10^3	Ultimate strength, psi X 10^3	Young's modulus, psi X 10^6	Elongation, %	Density, lb/in.^{3}
5052 H-38	33	39	10.1	4	0.097
2024-T4	8–46	60–65	10.6	6–17	0.100
2014-T6	57–60	63–70	10.5	2–8	0.100
7075-T6	60–72	75–80	10.5	1–5	0.100
7079-T6	57–69	72–77	10.5	2–8	0.100
6061-T6	35	42	10.1	8–10	0.096

(From Myers, J. J., Ed., *Handbook of Ocean and Underwater Engineering*, McGraw-Hill, New York, 1969. With permission.)

Table 5.1—2
MECHANICAL PROPERTIES OF STEELS

Material	Yield strength, psi X 10^3	Ultimate strength, psi X 10^3	Elongation, %	Modulus of elasticity psi X 10^6	Density, lb/in.^{3}
HY-80	80	100	14	29–30	0.28
HY-100	100	120	14	29–30	0.28
HY-140	140–160	–	16	29–30	0.28
T-1	90–110	110–130	14	29–30	0.28
A302 Grade B	70	85	27	29–30	0.28
2.25% Cr-1% Moly	100	120	19	29–30	0.28
H-11	200–220	240–260	5–9	29–30	0.28
4340	210–230	260–280	5–9	29–30	0.28
18 Ni 200	190–225	195–230	6–12	26	0.29
18 Ni 250	240–265	245–270	6–10	27	0.29
HP 9-4-25	220	250	12	29–30	0.29
HP 9-4-45	250	290	7	29–30	0.29
Stainless 301, annealed	40	105	60	28–29	0.28
Stainless 301, half-hard	95	150	54	28–29	0.28
Stainless 316, annealed	38	76	60	28–29	0.28
Stainless 410, annealed	87	110	21	28–29	0.28
Stainless 440 A	240–250	260–270	5	28–29	0.28

(From Myers, J. J., Ed., *Handbook of Ocean and Underwater Engineering*, McGraw-Hill, New York, 1969. With permission.)

Corrosion Behavior

Aluminum – Selected aluminum alloys are used successfully for long-lived marine structures, provided the proper precautions are taken. These alloys include members of the 5000 and 6000 series, but not unalloyed aluminum (1100) or the 2000 series. The very high strength 7000-series alloys (particularly 7075 and 7079) and being used successfully for intermittent or short-term marine exposures with extremely careful attention to paint coatings supplemented with cathodic protection, such as the use of anode-grade zinc, when practicable. Without such protective measures, the 7000-series alloys are susceptible to layer corrosion and to stress-corrosion cracking.

The aluminum alloy 5086 is virtually inert to clean sea water, but it usually presents a major procurement problem. The alloy 6061 will in most cases probably be selected because of its good fabricability, good balance of strength and toughness, reasonably good resistance to marine corrosion, and ready availability in a wide variety of forms. This alloy pits in sea water at the rate of about 0.01 in./yr, but the pitting can be mitigated, and perhaps even prevented altogether, by painting or providing cathodic protection (using anode-grade zinc or a reliable commercial aluminum anode having about the same potential as zinc), or both.

There are several cautions to be observed in using aluminum alloys successfully in sea water. All the aluminum alloys exhibit such active potentials in sea water that dissimilar-metal corrosion cells can be a serious problem when the aluminum alloys are connected with many of the standard marine alloys. The more common 5000- and 6000-series alloys are compatible, however, with galvanized steel as long as the galvanizing lasts. It is desirable to avoid having corrodible copper alloys in the vicinity of aluminum alloys, even where the two alloy classes are not connected by a metallic conductor; the reason is that copper-corrosion products are carried to the surface of the aluminum by convection, and the copper is reduced and remains on the surface of the aluminum, setting up galvanic cells which greatly accelerate corrosion of the aluminum. Antifouling paints that corrosion copper or mercury compounds are to be avoided for the same reason, even though a barrier coat of vinyl is interposed between the aluminum and the antifouling coat (if an antifouling paint is required, organic tin formulations are preferable to the copper or mercury formulations).

Mud can present special corrosion problems, perhaps because of bacterial action. The firm ocean floor of the Atlantic and the calcareous bottom of the Tongue of the Ocean have not presented such problems, but in an area of the Pacific just off the California coast, where the bottom is described as "green ooze," variable performance has been observed in the normally resistant aluminum alloys. Finally, if it is decided to use cathodic protection, only zinc anodes or reliable aluminum anodes with about the same potential as zinc should be used. One alloy, 5257-H25, has been observed to be more electronegative than zinc in sea water. If there is any doubt that the zinc will actually be negative (anodic) to an unfamiliar alloy, there is no safe substitute for making the determination experimentally in sea water.

Brasses and Bronzes – If a copper alloy contains more than about 15% zinc and does not contain an inhibitor additive such as arsenic, it is susceptible to dezincification in sea water and in the marine atmosphere as well. It is generally considered that the alloys of moderate zinc content, such as admiralty brass (29 percent Zn), can be made resistant to dezincification by the addition of small amounts of arsenic. The higher-zinc alloys, such as Muntz metal, Naval brass, and many manganese bronzes, cannot be satisfactorily inhibited against dezincification. There are unresolved contradictions as to whether cathodic-protection measures are effective in preventing dezincification. For these reasons it is mandatory to exclude the high-zinc alloys (those containing more than 30% zinc) from critical components intended for prolonged service in the sea.

The venerable marine alloys, such as G bronze, the phosphor bronzes, and inhibited admiralty brass (inhibited with arsenic against dezincing), have low susceptibility to crevice corrosion, pitting, and general corrosion (of the order of a 0.001 in./year) at slow relative water velocities. The more recent aluminum bronzes, containing nickel as an inhibitor against dealuminification, have comparable resistance to corrosion.

An argument of some merit for total exclusion of all brasses and bronzes where feasible is the quality-control problem in the fabricator's shops posed by the substitution of uninhibited admiralty for the required inhibited grade, the substitution of a high-zinc bronze for the specified bronze, etc. These substitutions present a problem of detection, and for this reason one would do well in many instances to consider the merits of employing an iron-bearing cupronickel instead of brass or bronze.

Cadmium – Cadmium is often used as a plating to protect steel articles against atmospheric corrosion. It is generally considered to be more compatible with aluminum than is steel or stainless steel (because of its position in the galvanic series), and for this reason it is sometimes specified as a plating for steel or stainless steel fasteners for aluminum structures. Such coatings are invariably thin and must therefore be regarded as a highly temporary measure for sea water service. There is a possibility that because of its solution potential in sea water, cadmium may prove to be a useful anode to provide a limited degree of cathodic protection to high-strength steels for critical structures.

Copper – In clean, quiet sea water, copper is reasonably resistant to corrosion, exhibiting a general surface-recession rate of the order of 0.001 in./year. For sea water moving faster than about 5 ft/sec, copper begins to corrode so rapidly that the cupronickel alloys described below are preferable.

Cupronickels – This family of alloys has two members in particular whose remarkably good corrosion resistance would repay in many cases the extra effort that may be required to procure them in the desired form. These alloys, nominally 70% copper – 30% nickel and 90% copper – 10% nickel, each with appreciable iron added, are becoming increasingly available in various forms. They are very resistant to crevice corrosion, and the pitting rate is low (of the order of 0.001 in./year). The 70–30 alloy has the added advantage of being readily distinguished by its color from the high-zinc brasses and bronzes, which as a class are unsuitable for lengthy service in the sea. Thus, part of the premium for the 70–30 alloy might be justified on the basis of quality control.

Lead – Lead is resistant to attack by sea water and corrodes uniformly at a rate of about 0.0005 in./year or less. It can act to stimulate serious dissimilar-metal-corrosion cells when coupled to structural steel or aluminum alloys. It can be corroded by excessive degrees of cathodic "protection."

Magnesium – Magnesium is useful as a galvanic-anode material. Magnesium alloys corrode so rapidly in sea water that they are not considered suitable for marine structural use. The alloy designated AZ31X, for example, was observed to undergo pitting attack at a rate of about 0.2 in./year. Alloying and surface-treatment attempts have not been found adequate to make magnesium acceptably corrosion resistant for use in the sea.

Monel® – This well-known and widely used alloy, more specifically designated as Monel alloy 400, has good corrosion resistance (general corrosion of the order of 0.001 in./year) in moving sea water. Under stagnant conditions, however, it may pit seriously (0.05 in./year or more), and it is susceptible to crevice corrosion.

Nickel and Nickel-base Alloys – Nickel in quiet sea water pits rapidly (of the order of 0.05 in./year). Two nickel-base alloys, however, are reported to be essentially inert to all forms of attack in sea water.[1] One of these is the relatively new alloy designated Inconel® alloy 625, available in various wrought forms, including sheet and tubing. The second apparently inert alloy is Hastelloy C®. The high molybdenum content of these alloys is believed responsible for their resistance to pitting and crevice corrosion. Other roughly comparable alloys are in the advanced developmental stage and may well become of interest for marine use.

Stainless Steels – The stainless steels as a group resist corrosion as long as the thin oxide film on the surface can be kept intact. In this condition they are described as being *passive*. No special acid or other chemical treatment is required to confer this passivity, and even special treatment for passivity does not confer permanent passivity in sea water. Passive films tend to break down in the pressure of chloride ions, and if there is not adequate oxygen to continuously repair the oxide film at a given point, the steel becomes active (corrodes) at that point. Crevices are especially vulnerable points for initiation of this form of attack, but particularly in the more susceptible steels, the localized attack may start on surfaces free from visible attached particles or organisms and progress as a pit.

All the stainless steels are susceptible to crevice corrosion and pitting in quiet sea water, but they differ greatly in degree of susceptibility. Generally speaking, those lower in chromium and nickel, such as AISI types 205, 304, and 410, are among the most susceptible. Type 316 is the least susceptible of the more commonly available medium-alloy stainless steels, but pits have been observed to grow even in this alloy at rates greater than 0.25 in./year. The premium low-carbon grades are no less susceptible to crevice corrosion and pitting than the standard grades.

Crevice corrosion and pitting of stainless steels in sea water can be prevented by coupling to ordinary unalloyed steel (if the unalloyed steel is not too well painted) or to anode-grade zinc. As a consequence, the unalloyed steel or zinc corrodes at an accelerated rate, but the attack on the stainless steel can thereby be prevented (when geometry and cathode-anode ratios are favorable).

In summary, all the stainless steel alloys now available commercially are susceptible to pitting and crevice corrosion to varying degrees in quiet sea water, and none should be considered for critical items for prolonged immersion unless they are cathodically protected, as by coupling to structural steel.

The successful application of stainless steel for marine propellers is useful to illustrate several corrosion-engineering points. The propellers of most ships are electrically grounded to the hull when the ships are anchored. Under these conditions the unalloyed-steel hull galvanically protects the stainless steel propeller against pitting attack, and the high conductivity of sea water and large ratio of wetted hull area to propeller area prevent disastrous attack on the hull (the principle of relative anode/cathode area). When the ship is under way, particularly in turbine-driven ships, the lubrication films insulate the propeller from the ship electrically, but when stainless steels are exposed to rapidly moving sea water the oxide film remains in repair and the steels do not pit.

The high-strength hardenable stainless steels are susceptible to stress-corrosion cracking in sea water (in addition to being susceptible to crevice corrosion and pitting). At the highest strength levels this may be so serious as to prohibit the use of these alloys. There appears to be a correlation between susceptibility to stress-corrosion cracking, susceptibility to hydrogen embrittlement, and susceptibility to fast "brittle fracture" in all these steels (though the rigorous evidence for this correlation is very scant), and the changes in heat treatment which improve fracture toughness appear to improve resistance to cracking.

There is no clear-cut dividing line according to strength levels between steels which are immune to stress-corrosion cracking and those which are susceptible, but as a very general rule, all hardenable steels at and above yield strengths of about 175,000 psi should be examined carefully for cracking. For these materials at high strength levels, cathodic-protection methods would have to be custom designed, requiring research data not now available, since coupling to a galvanic anode more electronegative than cadmium would possibly introduce hydrogen-embrittlement cracking. The technology for the use of high-strength hardenable steels (including the hardenable stainless steels) with yield strengths of about 150,000 psi or more in massive forms in sea water must be regarded as an area under development. However, steels both stainless and otherwise have long been used successfully in thin sections, as in wire rope, at even higher strengths, where this strength has been conferred by cold-drawing rather than by heat treatment, except that, again, in sea water the stainless steel is susceptible to crevice corrosion, pitting, and their consequences.

Nonstainless Steels – The corrosion rates of unalloyed steels, low-alloy (nonstainless) steels, and wrought iron in sea water are comparable and appear to be about the same regardless of geographical location. The proprietary copper-bearing grades, which are superior to the non-copper-bearing steels in resisting atmospheric corrosion, do not appear to exhibit any superiority in corrosion-resistance behavior in the sea. The *average* penetration rate is of the order of 0.005 in./year, but the rate of pit growth may be up to ten times this much. For some, applications of corrosion can be tolerated and the wastage accepted, whereas for other applications, such as in thin walled containers, pitting might produce a premature failure by perforation, as it sometimes does in ship plate. Pitting of components cyclically stressed may greatly accelerate the initiation of a corrosion-fatigue crack. Both pitting and general corrosion of steel can be readily controlled by cathodic-protection measures if geometry considerations permit.

The high-strength low-alloy steels that owe their strength to heat treatment are susceptible to stress-corrosion and hydrogen-embrittlement cracking under tensile stress. As with the hardenable stainless steels, there is no sharply defined strength level separating immune steels from susceptible ones, but careful attention is again recommended at yield strengths of 175,000 psi or greater, and at even lower yield strengths if the stresses are high. As with the hardenable stainless steels, cathodic-protection methods are not presently a practical solution to stress-corrosion cracking because of the possibility of hydrogen-embrittlement cracking. It does, however, appear practicable and safe to cathodically protect low-alloy and unalloyed steels in which the strength is conferred by cold-drawing, as, for example, in the case of wire rope.

The higher the alloy content in steels, the greater is the tendency for pitting instead of general corrosion. The maraging steels, having high nickel content, follow this rule. The main reason for interest in maraging steels, however, is their high strength and attractive toughness. At the strength levels where these steels might be reasonably justifiable, however, the corrosion phenomenon of prinicpal concern is stress-corrosion cracking. Alloy technology is changing so rapidly that, again, the answer to questions on this point will more likely have to be sought in the laboratory than in the library for a long time to come.

Titanium – Titanium, like the two nickel-base alloys cited above, appears to be essentially inert to all forms of corrosion attack in sea water at ambient temperature. Several titanium alloys, however, have recently been found to be susceptible to cracking and rapid corrosion fatigue when stressed in sea water. This subject is in a state of rapid development, and the designer would be well advised to check the current status before selecting an alloy for an application involving appreciable tensile stresses. Unalloyed titanium and the 6% aluminum – 4% vanadium alloy with low oxygen content appear to be reasonably resistant. Several other alloys which are acutely susceptible in some conditions of heat treatment can be made essentially immune by selected heat treatment, but this immunity can be removed by subsequent welding. The phenomenon apparently does not occur if the section is sufficiently thin. The critical thickness varies, depending upon composition and heat treatment, but 0.05 in. is an example for the 8% aluminum – 1% molybdenum – 1% vanadium alloy. Cathodic protection does not now appear to provide a practical safeguard against stress-corrosion cracking in susceptible titanium systems.

Zinc – Zinc not coupled to other metals corrodes in sea water at an average rate of 0.0005 to 0.001 in./year, with pitting or crevice attack up to ten times this rate. When coupled to large areas of steel or other more cathodic metals, it can corrode at a rate faster by several orders of magnitude. For successful use of zinc as a reliable galvanic anode, MIL SPEC 18001 grade zinc should be used. High purity alone is not adequate.

(From Brown, B. F., in *Handbook of Ocean and Underwater Engineering,* Myers, J. J., Ed., McGraw-Hill, New York, 1969. With permission.)

Table 5.1—3
EXAMPLES OF GALVANIC COUPLES IN SEA WATER

Metal A	Metal B	Comments
Couples that usually give rise to undesirable results on one or both metals		
Magnesium	Low-alloy steel	Accelerated attack on **A**, danger of hydrogen damage on **B**
Aluminum	Copper	Accelerated pitting on **A**; ions from **B** attack **A**. Reduced corrosion on **B** may result in biofouling on **B**
Bronze	Stainless steel	Increased pitting on **A**
Borderline, may work, but uncertain		
Copper	Solder	Soldered joint may be attacked but may have useful life
Graphite	Titanium or Hastelloy C	
Monel-400	Type 316 SS	Both metals may pit
Generally compatible		
Titanium	Inconel 625	
Lead	Cupronickel	

(From Fink, F. W. and Boyd, W. K., The corrosion of metals in a marine environment, Defense Metals Information Center Report No. 245, Battelle Memorial Institute, Columbus, Ohio, 1970. With permission.)

Table 5.1—4
FIVE-YEAR WEIGHT LOSS AS DETERMINED GRAPHICALLY FOR STEEL

With Copper, Nickel, or Chromium Additions

	Copper			Nickel				Chromium			
Amount added, wt %	0	0.2	1	0	0.2	1	2	0	0.2	1	2
Weight loss/ paneling	>50	34	28	48	42	31	25	32	30	22	17

(From Fink, F. W. and Boyd, W. K., The corrosion of metals in a marine environment, Defense Metals Information Center Report No. 245, Battelle Memorial Institute, Columbus, Ohio, 1970. With permission.)

Table 5.1—5
CORROSION PENETRATION OF ALLOY STEELS

Immersed in the Pacific Ocean Near the Panama Canal Zone after 8 Years

		Penetration, mils							
		Mean tide[b]				14 ft below surface[b]			
Steel	Type	1	2	3	Ratio[a]	1	2	3	Ratio[a]
A	Low carbon	23.2	40	65	1.7	25.5	66	86	2.6
D	Copper bearing	24.2	45	63	1.9	27.7	63	108	2.3
E	Ni (2%)	22.9	39	50	1.7	31.7	94	179	3.0
F	Ni (5%)	20.0	39	75	2.0	32.0	117	214	3.7
G	Cr (3%)	25.7	82	93	3.2	40.5	65	78	1.6
H	Cr (5%)	24.5	88	99	3.6	32.0	63	90	2.0
I	Low alloy (Cu-Ni)	39.7	70	134	1.8	26.4	82	152	3.2
J	Low alloy (Cu-Cr-Si)	21.1	47	54	2.2	43.2	80	175	1.8
K	Low alloy (Cu-Ni-Mn-Mo)	24.8	40	94	1.6	25.5	56	139	2.2
L	Low alloy (Cr-Ni-Mn)	20.5	39	50	1.9	43.9	97	259(p)[c]	2.2

[a] Ratio of average of 20 deepest pits of weight-loss penetration. The higher the number, the greater is the pitting tendency in relation to the corrosion rate.
[b] 1 = calculated from weight loss.
2 = average of 20 deepest pits.
3 = deepest pit.
[c] p = completely perforated.

(From Southwell, C. R. and Alexander, A. L., Corrosion of structural ferrous metals in tropical environments – Sixteen years' exposure to sea and fresh water, Paper No. 14, Preprint, 1968 NACE Conference, Cleveland, Ohio.)

Table 5.1—6
COMPREHENSIVE EVALUATION OF CORROSION DAMAGE OF STAINLESS STEELS

Exposed to Marine Environments in the Panama Canal Zone

Metal	Stainless steel	Type of exposure	Corrosion rate, mpy			Avg of 20 deepest pits[a], mil			Deepest pit[a], mil			Loss in tensile strength[b], %	Type corrosion attack[c]
			1 yr	4 yr	8 yr	1 yr	4 yr	8 yr	1 yr	4 yr	8 yr	4 yr	4 yr
A	Type 410 (13 Cr)	Sea water immersion	2.98	1.97	1.75	61(11)	148	161	260(p)	260(p)	259(p)	d	KQH
		Sea water mean tide	0.50	0.41	0.42	46	67	67	66	173	152	0	JKQ
		Seashore	0.040	0.013	0.005	0(0)	3	0(0)	0(0)	5	0(0)	0	KR
B	Type 430 (17 Cr)	Seashore	0.025	0.008	0.004	0(0)	0(0)	0(0)	0(0)	0(0)	0(0)	0	KR
C	Type 301 (17 Cr, 7 Ni)	Seashore	0.00	0.001	0.001	0(0)	0(0)	0(0)	0(0)	0(0)	0(0)	0	K
D	Type 302 (18 Cr, 8 Ni)	Sea water immersion	1.46	0.88	0.69	70(12)	107	140	261(p)	286(p)	236	d	KQ
		Sea water mean tide	0.18	0.12	0.11	7(13)	26	57	16	82	110	0	JK
E	Type 316 (18-13 and Mo)	Sea water immersion	0.59	0.07	0.25	44(7)	48	154	245(p)	93	245(p)	0	KQ
		Sea water mean tide	0.06	0.03	0.02	5(9)	7(12)	16	23	22	30	1	JK
		Seashore	0.00	0.00	0.03	0(0)	0(0)	0(0)	0(0)	0(0)	0(0)	1	K
F	Type 321 (17-10 and Ti)	Sea water immersion	1.16	0.81	0.62	64(8)	175	193	270(p)	273(p)	272(p)	d	KQ
		Sea water mean tide	0.13	0.08	0.08	8(11)	37	56	26	60	93	d	JKQ
		Seashore	0.005	0.001	0.001	0(0)	0(0)	0(0)	0(0)	0(0)	0(0)	1	K

[a] Pit depths referred to the original surface of the metal either by measurement from an uncorroded surface or by calculation using the original and final average measured thickness of the sample. Average of 20 deepest pits represents average of the 5 deepest pits measured on each side of duplicate specimens. (Area, 2.25 ft^2 on immersed specimens, and 0.89 ft^2 on atmospheric specimens); values in parentheses indicate total number averaged when less than 20 measurable pits. Perforation of plate by deepest pit is indicated by (p).

[b] Changes in tensile strength calculated on basis of ¼ in. thick metal (average of 4 tests for immersed specimens, average of 3 tests on atmospheric specimens).

[c] H – concentration cell, J – marine fouling contact, K – no visible attack, Q – pitting attack (random), R – localized attack (random).

[d] Intensity and distribution of pitting prevented satisfactory tensile testing.

(From Alexander, A. L. et al., *Corrosion*, 17(7), 345t, 1961. Reproduced by kind permission of the International Council for the Exploration of the Sea.)

Table 5.1—7
SUMMARY OF CORROSION OF TITANIUM AND TITANIUM ALLOYS IN SEA WATER

Alloy	No. exposures	Exposure time, days	Exposure depth, ft	Type of corrosion observed
Unalloyed Ti (grade unknown)	10	123–1064	2350–6780	None visible (<0.1 mpy)[a]
Unalloyed Ti (Grade RC 55)	4	90–199	4250–4500	None visible (<0.1 mpy)[a]
Unalloyed Ti (Grade 75 A)	8	123–751	5–6780	None visible (0.0 mpy)
Ti-5Al-2.5Sn	12	123–751	5–6780	None visible (<0.1 mpy)
Ti-7Al-12Zr	1	123	5640	None visible (0.0 mpy)
Ti-7Al-2Cb-1Ta	2	181	5	None visible (Fouling stains)
Ti-8Mn	1	402	2,370	
Ti-4Al-3Mo-1V	3	402–1064	2370–330	None visible (0.0 mpy)
Ti-6Al-4V	20	123–1064	5–6780	None visible (<0.1 mpy, mostly 0.0)[a]
Ti-13V-11Cr-3Al	12	123–751	5–6780	None visible (<0.1 mpy, mostly 0.0)

[a] One panel was reported as 0.19 mpy.

(From Reinhart, F. M., Corrosion of materials in hydrospace, Part III, Titanium and titanium alloys, U.S. Naval Civil Eng. Lab., Port Hueneme, Calif., Technical Note N-921, September, 1967.)

Table 5.1—8
CORROSION OF ALUMINUM ALLOYS

Exposed 16 Years in Three Tropical Environments in the Panama Canal Zone

	Average penetration[a], mils			Depth of pitting,[b] mils: Avg of 20 deepest pits			Depth of pitting,[b] mils: Deepest pits			Tensile strength loss, %[c]	Type of corrosion attack,[d]
	1 yr	8 yr	16 yr	1 yr	8 yr	16 yr	1 yr	8 yr	16 yr	8 yr	16 yr
Alloy 1100											
Immersion											
Sea water	0.28	0.61	0.97	9(13)	11	17	15	19	33	2	J
Mean tide	0.06	0.31	0.53	11(9)	14	39	29	37	67	1	JQ
Atmospheric											
Marine	0.01	0.02	0.11	N	N	N	N	N	N	0	A
Alloy 6061											
Immersion											
Sea water	0.28	0.73	0.91	N	23	14	N	49	79	0	J
Mean tide	0.04	0.13	0.29	N	N	17	N	N	41	0	J
Atmospheric											
Marine	0.03	0.03	0.11	N	N	N	N	N	N	1	A

[a] Calculated from weight loss and specific gravity.
[b] Represents depth of penetration from original surface; N – measurable pits; number in parentheses gives number of measurable pits when less than 20.
[c] Percent change in tensile strength calculated on basis of 1/4 in. thick metal and average of four tests for underwater specimens, and 1/16 in. thick metal and average of three tests for atmospheric specimens.
[d] A – uniform attack; J – marine fouling contact; Q – pitting attack (random).

(From Southwell, C. R., Alexander, A. L., and Hummer, C. W., Jr., *Materials Protection,* 4(12), 30, 1965.)

Table 5.1—9
CORROSION OF LEAD, SOLDER, TIN, AND ZINC IN SEA WATER

Metal	Days	Depth, ft	Weight loss, mpy	Pit depth, mils	Remarks
Chemical lead[a]	181	5	1.2	–	Uniform attack
	197	2340	0.3	–	Uniform attack
	123	5640	0.8	–	Uniform attack
Tellurium lead[b]	181	5	1.0	–	Uniform attack
	197	2340	0.3	–	Uniform attack
	123	5640	1.1	–	Uniform attack
Antimonial lead[c]	181	5	1.2	–	Uniform attack
	197	2340	0.3	–	Uniform attack
	123	5640	0.8	–	Uniform attack
Solder[d]	181	5	3.7	–	Uniform attack
	197	2340	0.5	–	Uniform attack
	123	5640	0.5	–	Uniform attack
Tin[e]	181	5	8.3	30	Perforated
	197	2340	1.8	2	Crevice attack
	123	5640	0.5	=	General attack
Zinc[f]	181	5	4.5	5	Pitting
	197	2340	2.3	2	Pitting
	123	5640	6.7	13	Pitting

[a] 99.9 Pb.
[b] 99+ Pb, 0.04 Te.
[c] 94.0 Pb, 6.0 Sb.
[d] 67 Pb-33 Sn.
[e] 99.9 Sn.
[f] 0.01 Fe, 0.09 Pb.

(From Reinhart, F. M., Corrosion of materials in surface sea water after 6 months exposure, Naval Civil Eng. Lab., Port Hueneme, Calif., Technical Note N-1023, March, 1969.)

5.2. ROPES, CHAINS, AND SHACKLES

Figure 5.2–1
WIRE-ROPE CONSTRUCTION CLASSES

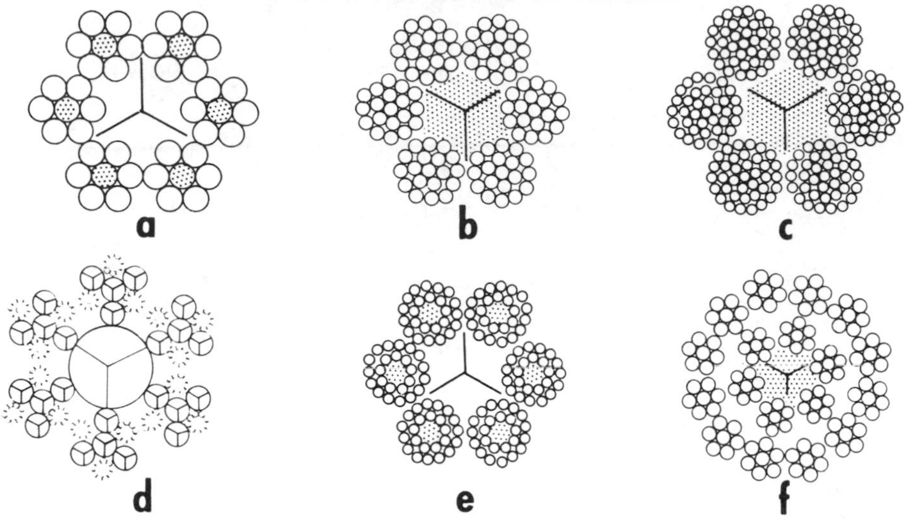

a – 6 × 7; b – 6 × 19; c – 6 × 37; d – 18 × 7; nonrotating; e – mooring lines, 6 × 12 or 6 × 24; f – spring-lay, 6 × 3 × 19.

(From Meals, W. D. in *Handbook of Ocean and Underwater Engineering,* Myers, J. J., Ed., McGraw-Hill, New York, 1969. With permission.)

Table 5.2–1
NOMINAL WIRE-ROPE BREAKING STRENGTH

in 2,000 lb Tons, for Bright Improved-plow-grade Steel

Rope diam, in.	6 × 7	6 × 19 class		6 × 37 class		Non-rotating	Mooring line	Spring lay
	FC	IWRC	FC	IWRC	FC	18 × 7	6 × 24[a]	6 × 3 × 19[a]
3/8	5.86	6.56	6.10	6.20	5.77	5.59	4.77	
1/2	10.3	11.5	10.7	11.0	10.2	9.85	8.40	
5/8	15.9	18.0	16.7	17.0	15.8	15.3	13.0	
3/4	22.7	25.6	23.8	24.3	22.6	21.8	18.6	
7/8	30.7	34.6	32.2	32.9	30.6	29.5	25.3	13.5
1	39.7	45.0	41.8	42.8	39.8	38.3	32.8	17.5
1 1/8	49.8	56.6	52.6	53.9	50.1	48.2	41.2	22.1
1 1/4	61.0	65.0	64.6	66.1	61.5	59.2	50.7	27.2
1 3/8	73.1	83.5	77.7	79.7	74.1	71.8	61.0	32.8
1 1/2	86.2	98.9	92.0	94.5	87.9	84.4	72.3	38.9
1 5/8	–	115.0	107.0	111.0	103.0	98.4	84.5	45.6
1 3/4	–	134.0	124.0	128.0	119.0	114.0	97.5	52.7
2	–	172.0	160.0	166.0	154.0	–	126.0	68.5

[a] Strengths shown are for galvanized ropes. (a) For all galvanized ropes other than 6 × 24 and 6 × 3 × 19, subtract 10 percent from listed strengths unless rope is made from redrawn galvanized wire. (b) For extra-improved-plow in 6 × 19 and 6 × 37 class with IWRC add 15 percent to tabulated strength. (c) Minimum breaking strength to allow for testing variations is 97.5 percent of nominal breaking strength.

(Source: Federal Specification RR 410. From Meals W. D. in *Handbook of Ocean and Underwater Engineering,* Myers, J. J., Ed., McGraw-Hill, New York, 1969. With permission.)

Table 5.2—2
CHEMICAL PROPERTIES OF FIBER ROPES

Property	Manila	Nylon	Dacron	Polypro-pylene
Specific gravity	1.38	1.14	1.38	0.90
Resistance to mildew	Poor	Excellent	Excellent	Excellent
Resistance to acid	Poor	Fair	Good	Excellent
Resistance to alkali	Poor	Excellent	Fair	Excellent
Resistance to sunlight	Fair	Good	Good	Good
Resistance to organic solvent	Good	Good	Good	Fair
Critical temp	300°F	350°F	350°F	250°F
Melting point	–	480°F	482°F	330°F

(From Haas, F. J. in *Handbook of Ocean and Underwater Engineering,* Myers, J. J., Ed., McGraw-Hill, New York, 1969. With permission.)

Table 5.2—3
STRENGTH AND RESISTANCE PROPERTIES OF FIBER ROPES

Property	Manila	Nylon	Dacron	Polypro-pylene
Strength/wt ratio, avg	1.0	2.84	2.03	2.32
Relative resistance to impact load	1.0	8.6	4.0	5.2
Resistance to abrasion	Good	Excellent	Superior	Good
Water absorption, %	25.0	4.5	1.5	0

(From Haas, F. J. in *Handbook of Ocean and Underwater Engineering,* Myers, J. J., Ed., McGraw-Hill, New York, 1969. With permission.)

Figure 5.2–2

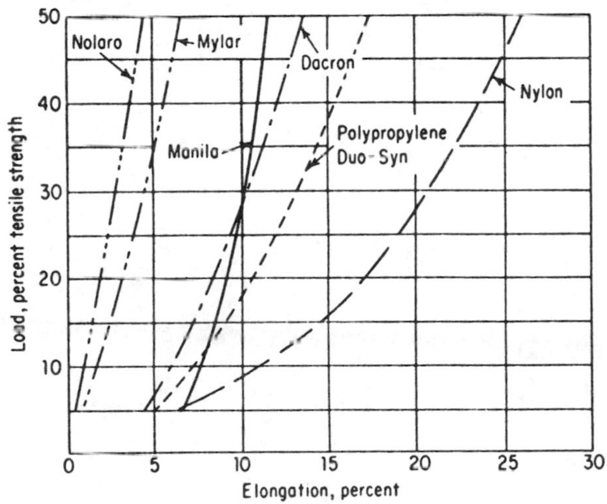

(From Haas, F. J. in *Handbook of Ocean and Underwater Engineering,* Myers, J. J., Ed., McGraw-Hill, New York, 1969. With permission.)

Table 5.2—4
BUOY CHAIN

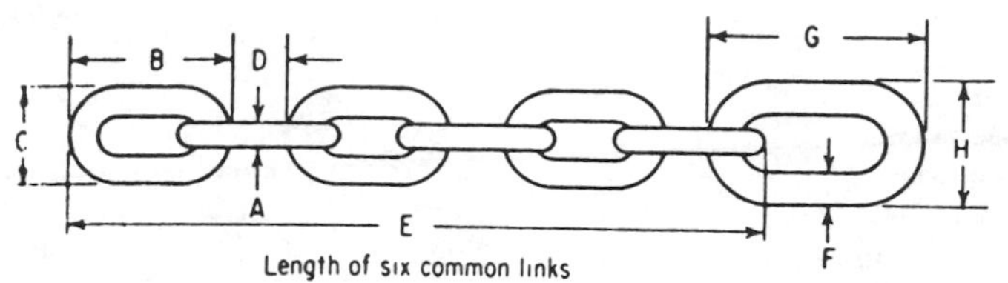

Common links					End links						
Wire diam, *A*	Length, *B*	Width, *C*	Space be-tween ends of links, *D*	Length of 6 links, *E*	Wire diam, *F*	Length, *G*	Width *H*	Link toler-ance, length and width	Proof load, lb	Break load, lb	Wt / 15 fath-oms, lb
1/2	3	1 7/8	1	13	3/4	4 1/2	2 5/8	1/32	7500	15,000	210
5/8	3 3/4	2 1/4	1 1/4	16 1/4	3/4	4 1/2	2 5/8	1/32	11,500	23,000	323
3/4	4 1/2	2 5/8	1 1/2	19 1/2	7/8	5 1/4	3 1/8	1/16	16,000	32,000	442
7/8	5 1/4	3 1/8	1 3/4	22 3/4	1 1/8	6 3/4	3 7/8	1/16	22,000	44,000	608
1	6	3 1/2	2	26	1 1/4	7 1/2	4 3/8	3/32	29,000	58,000	780
1 1/8	6 3/4	3 7/8	2 1/4	29 1/4	1 1/4	7 1/2	4 3/8	3/32	38,500	77,000	990
1 1/4	7 1/2	4 3/8	2 1/2	32 1/2	1 1/2	9	5 1/4	1/8	45,500	91,000	1245
1 1/2	9	5 1/4	3	39	1 7/8	11 1/4	6 1/2	5/32	65,500	131,000	1762
1 5/8	9 3/4	5 11/16	3 1/4	42 1/4	1 7/8	11 1/4	6 1/2	5/32	76,500	153,000	2040
1 3/4	10 1/2	6 1/16	3 1/2	45 1/2	2 1/8	12	7 3/16	3/16	86,500	173,000	2370
1 7/8	11 1/4	6 1/2	3 3/4	48 3/4	2 1/8	12	7 3/16	3/16	100,000	200,000	2640

(From Linnenbank, C. D. in *Handbook of Ocean and Underwater Engineering,* Myers, J. J., Ed., McGraw-Hill, New York, 1969. With permission.)

Table 5.2—5
HIGH-TEST CHAIN

Trade size, in.	Material size, in.	Inside length of link, in.	Inside width of link, in.	Outside length of link, in.	Outside width of link, in.	Links/ ft	Max wt / 100 ft, lb	Working-load limit, lb	Trade size, in.
1/4	9/32	0.82	0.39	1.38	0.95	14 1/2	80	2500	1/4
5/16	11/32	1.01	0.48	1.70	1.17	12	123	4000	5/16
3/8	13/32	1.15	0.56	1.96	1.37	10 1/2	175	5100	3/8
7/16	15/32	1.29	0.65	2.23	1.59	9 1/4	235	6600	7/16
1/2	17/32	1.43	0.75	2.49	1.81	8 1/2	300	8200	1/2
5/8	21/32	1.79	0.90	3.10	2.21	6 3/4	450	11,500	5/8
3/4	25/32	1.96	1.06	3.52	2.62	6	655	16,200	3/4

Note: Chain is heat-treated and is of higher carbon content than proof coil or BBB chain. High strength-weight ratio permits the use of smaller and lighter chain for a given purpose. Available in continuous lengths in fiber drums or in short lengths. Joints are electric-welded.

(From Linnenbank, C. D. in *Handbook of Ocean and Underwater Engineering*, Myers, J. J., Ed., McGraw-Hill, New York, 1969. With permission.)

Table 5.2—6
WROUGHT-IRON CRANE CHAIN

Trade size, in.	Material size, in.	Inside length of link, in.	Inside width of link, in.	Outside length of link, in.	Outside width of link, in.	Links/ ft	Wt / 100 ft, lb	Working-load limit, lb	Trade size, in.
3/8	13/32	1 3/32	5/8	1 29/32	1 7/16	11	166	2385	3/8
1/2	17/32	1 11/32	3/4	2 13/32	1 13/16	9	275	4240	1/2
5/8	21/32	1 11/16	7/8	3	2 3/16	7 1/8	430	6630	5/8
3/4	25/32	1 7/8	1	3 7/16	2 9/16	6 3/8	615	9540	3/4
7/8	29/32	2 1/4	1 1/4	4 1/16	3 1/16	5 3/8	820	12,960	7/8
1	1 1/32	2 9/16	1 3/8	4 5/8	3 7/16	4 3/4	1045	16,950	1
1 1/8	1 5/32	2 7/8	1 5/8	5 3/16	3 15/16	4 1/4	1310	20,040	1 1/8
1 1/4	1 9/32	3 1/16	1 3/4	5 5/8	4 5/16	4	1600	24,750	1 1/4
1 3/8	1 13/32	3 5/8	1 7/8	6 7/16	4 11/16	3 3/8	1930	29,910	1 3/8
1 1/2	1 17/32	3 7/8	2	6 15/16	5 1/16	3 1/8	2335	35,600	1 1/2

Note: The elastic quality of the iron permits recovery from strain produced by shock. When overloaded, the links will stretch before breaking, providing a valuable and important safety feature. Joints are fire-welded, and standard packaging is in bulk.

(From Linnenbank, C. D. in *Handbook of Ocean and Underwater Engineering*, Myers, J. J., Ed., McGraw-Hill, New York, 1969. With permission.)

Table 5.2—7
ALLOY CHAIN

Trade size, in.	Material size, in.	Inside length of link, in.	Inside width of link, in.	Outside length of link, in.	Outside width of link, in.	Links/ ft	Wt/ 100 ft, lb	Working-load limit, lb	Trade size, in.
1/4	9/32	0.85	0.39	1.41	.95	14	73	3250	1/4
5/16	11/32	0.98	0.44	1.67	1.13	12 1/4	110	4250	5/16
3/8	13/32	1.10	0.52	1.91	1.33	11	163	6600	3/8
1/2	17/32	1.49	0.72	2.55	1.78	8	270	11,250	1/2
5/8	21/32	1.74	0.91	3.05	2.22	7	422	16,500	5/8
3/4	25/32	2.05	0.97	3.61	2.53	5 3/4	590	23,000	3/4
7/8	7/8	2.25	1.09	4.00	2.84	5 1/3	730	28,750	7/8
1	1	2.62	1.25	4.62	3.25	4 1/2	965	38,750	1
1 1/8	1 1/8	3.00	1.37	5.25	3.62	4	1200	44,500	1 1/8
1 1/4	1 1/4	3.25	1.65	5.75	4.15	3 3/4	1525	57,500	1 1/4

Note: This chain, heat-treated to develop maximum strength consistent with proper ductility, has more than twice the strength of low-carbon chain of the same size. It will elongate well in excess of the minimum requirements. This safety feature provides a warning when the chain is overloaded or when it is used in applications where links are subject to extreme bending stresses. Heat treating develops a Brinell hardness of 240 to 270. This chain does not work-harden under normal conditions, making stress relief or re-heat treatment unnecessary. Min. elongation is 15%. Joints are electric-welded. Alloy chain is available in continuous lengths in fiber drums or in short lengths. Test certificates are generally available on request.

Alloys: Low-alloy heat-treatable steel[a] has enjoyed a rapid rise in popularity for chain in recent years. Such chain is sold under a variety of trade names. After heat treatment typical chain has the properties of minimum tensile strength of 125,000 psi, min. elongation of 15%, and hardness of 25-28R_c. Softer, even more ductile chain may be produced for specific application by control of the heating cycle.

[a] ASTM Specification A391-65.

(From Linnenbank, C. D. in *Handbook of Ocean and Underwater Engineering*, Myers, J. J., Ed., McGraw-Hill, New York, 1969. With permission.)

Table 5.2—8
SWIVELS

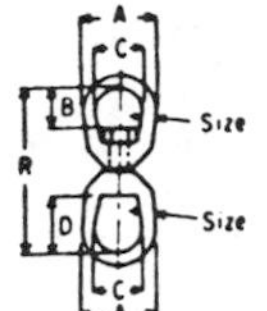

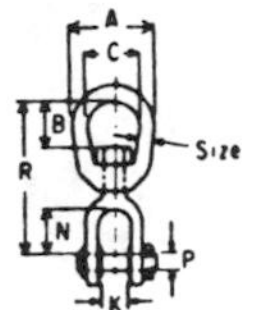

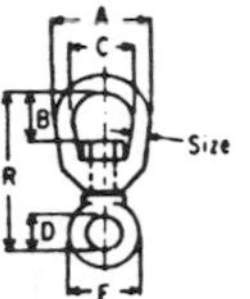

Regular

Size, in.	Est. UTS, tons	Swivel dimensions, in.					Wt, lb
		A	B	C	D	R	
1/4	1.8	1 1/4	11/16	3/4	1 1/16	2 15/16	0.19
5/16	2.6	1 5/8	7/8	1	1 1/4	3 9/16	0.31
3/8	4.9	2	1	1 1/4	1 1/2	4 5/16	0.68
1/2	7.9	2 1/2	1 3/8	1 1/2	2	5 7/16	1.25
5/8	11.7	3	1 21/32	1 3/4	2 3/8	6 9/16	2.25
3/4	16.2	3 1/2	1 13/16	2	2 5/8	7 3/16	3.5
7/8	21.2	4	2 1/8	2 1/4	3 1/16	8 3/8	5.4
1	26.7	4 1/2	2 7/16	2 1/2	3 1/2	9 5/8	8.8
1 1/8	33.9	5	2 7/16	2 3/4	3 3/4	10 3/8	12
1 1/4	40.4	5 5/8	2 3/4	3 1/8	2 3/4	11 1/8	16
1 1/2	113.0	7	4 1/4	4	4 1/4	17 1/8	49

Jaw End

Size, in.	Est. UTS, tons	Swivel dimensions, in.							Wt, lb
		A	B	C	K	N	P	R	
1/4	1.8	1 1/4	11/16	3/4	15/32	7/8	1/4	2 5/8	0.22
5/16	2.6	1 5/8	7/8	1	1/2	7/8	5/16	2 15/16	0.31
3/8	4.9	2	1	1 1/4	5/8	1 1/16	3/8	3 5/8	0.56
1/2	7.9	2 1/2	1 3/8	1 1/2	3/4	1 5/16	1/2	4 1/2	1.25
5/8	11.7	3	1 21/32	1 3/4	15/16	1 1/2	5/8	5 5/16	2.13
3/4	16.2	3 1/2	1 13/16	2	1 1/8	1 3/4	3/4	6 1/16	3.5
7/8	21.2	4	2 1/8	2 1/4	1 3/16	2 1/16	7/8	7	5.3
1	26.7	4 1/2	2 7/16	2 1/2	1 3/4	2 13/16	1 1/8	8 9/16	9.8
1 1/8	33.9	5	2 7/16	2 3/4	1 3/4	2 13/16	1 1/8	8 15/16	14
1 1/4	40.4	5 5/8	2 3/4	3 1/8	2 1/16	2 13/16	1 3/8	9 7/16	17
1 1/2	113.0	7	4 1/4	4	2 7/8	4 7/16	2 1/4	14 3/4	49

Chain

Size, in.	Est. UTS, tons	Swivel dimensons, in.						Wt, lb
		A	B	C	D	E	R	
1/4	1.8	1 1/4	11/16	3/4	7/16	15/16	2 1/4	0.13
5/16	2.6	1 5/8	7/8	1	1/2	1 1/8	2 23/32	0.25
3/8	4.9	2	1	1 1/4	3/4	1 1/2	3 7/16	0.5
1/2	7.9	2 1/2	1 3/8	1 1/2	7/8	1 7/8	4 1/4	1
5/8	11.7	3	1 21/32	1 3/4	1 1/16	2 3/16	5 1/8	1.75
3/4	16.2	3 1/2	1 13/16	2	1 1/4	2 5/8	5 25/32	2.88
7/8	21.2	4	2 1/8	2 1/4	1 7/16	2 15/16	6 5/8	4.3
1	26.7	4 1/2	2 7/16	2 1/2	2	4	8 1/16	6.8

(Source: Federal Specification RRc 272a. From Linnenbank, C. D. in *Handbook of Ocean and Underwater Engineering*, Myers, J. J., Ed., McGraw-Hill, New York, 1969. With permission.)

Table 5.2—9
SCREW-PIN SHACKLES

Anchor shackles: round pin, screw pin

Safe working load, tons	Size, in.	Inside length, in.	Inside width, in.		Diam, in.		Tolerance, plus or minus		Wt, lb
			At pin	At bow	Pin	Outside of eye	Length	Width	
1/3[a]	3/16	7/8	3/8	19/32	1/4	9/16	1/16	1/16	0.05
1/2	1/4	1 1/8	15/32	25/32	5/16	11/16	1/16	1/16	0.12
3/4	5/16	1 7/32	17/32	27/32	3/8	13/16	1/16	1/16	0.18
1	3/8	1 7/16	21/32	1 1/32	7/16	31/32	1/8	1/16	0.3
1 1/2	7/16	1 11/16	23/32	1 5/32	1/2	1 1/16	1/8	1/16	0.49
2	1/2	1 7/8	13/16	1 5/16	5/8	1 3/16	1/8	1/16	0.74
3 1/4	5/8	2 3/8	1 1/16	1 11/16	3/4	1 9/16	1/8	1/16	1.44
4 3/4	3/4	2 13/16	1 1/4	2	7/8	1 7/8	1/4	1/16	2.16
6 1/2	7/8	3 5/16	1 7/16	2 9/32	1	2 1/8	1/4	1/16	3.37
8 1/2	1	3 3/4	1 11/16	2 11/16	1 1/8	2 3/8	1/4	1/16	5.3
9 1/2	1 1/8	4 1/4	1 13/16	2 29/32	1 1/4	2 5/8	1/4	1/16	7
12	1 1/4	4 11/16	2 1/32	3 1/4	1 3/8	3	1/4	1/16	9.6
13 1/2	1 3/8	5 1/4	2 1/4	3 5/8	1 1/2	3 5/16	1/4	1/8	12.6
17	1 1/2	5 3/4	2 3/8	3 7/8	1 5/8	3 5/8	1/4	1/8	17.3
25	1 3/4	7	2 7/8	5	2	4 5/16	1/4	1/8	27.8
35	2	7 3/4	3 1/4	5 3/4	2 1/4	5	1/4	1/8	41.1
50	2 1/2	10 1/2	4 1/8	7 1/4	2 3/4	6	3/4	1/8	83.5
75[b]	3	13	5	7 7/8	3 1/4	6 1/2	3/4	1/8	119

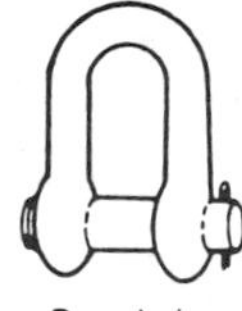

Chain shackles: round pin, screw pin

Safe working load, tons,	Size, in.	Inside length, in.	Inside width, in.	Diam, in.		Tolerance, plus or minus		Wt, lb
				Pin	Outside of eye	Length	Width	
1/2	1/4	7/8	15/32	5/16	11/16	1/16	1/16	0.11
3/4	5/16	1 1/32	17/32	3/8	13/16	1/16	1/16	0.17
1	3/8	1 1/4	21/32	7/16	31/32	1/8	1/16	0.29
1 1/2	7/16	1 7/16	23/32	1/2	1 1/16	1/8	1/16	0.42
2	1/2	1 5/8	13/16	5/8	1 5/16	1/8	1/16	0.68
3 1/4	5/8	2	1 1/16	3/4	1 9/16	1/8	1/16	1.21

[a] Furnished in screw pin only.
[b] Furnished in round pin only.

Table 5.2—9 (continued)
SCREW-PIN SHACKLES

Safe working load, tons,	Size, in.	Inside length, in.	Inside width, in.	Diam, in.		Tolerance, plus or minus		Wt , lb
				Pin	Outside of eye	Length	Width	
4 3/4	3/4	2 3/8	1 1/4	7/8	1 7/8	1/4	1/16	2.14
6 1/2	7/8	2 13/16	1 7/16	1	2 1/8	1/4	1/16	3.1
8 1/2	1	3 3/16	1 11/16	1 1/8	2 3/8	1/4	1/16	4.5
9 1/2	1 1/8	3 9/16	1 13/16	1 1/4	2 5/8	1/4	1/16	6.6
12	1 1/4	3 15/16	2 1/32	1 3/8	3	1/4	1/8	8.9
13 1/2	1 3/8	4 7/16	2 1/4	1 1/2	3 5/16	1/4	1/8	12
17	1 1/2	4 7/8	2 3/8	1 5/8	3 5/8	1/4	1/8	16.2
25	1 3/4	5 3/4	2 7/8	2	4 5/16	1/4	1/8	25
35	2	6 3/4	3 1/4	2 1/4	5	1/4	1/8	36
50	2 1/2	8	4 1/8	2 3/4	6	3/4	1/8	74
75[b]	3	8 1/2	5	3 1/4	6 1/2	3/4	1/8	10 1/8

(Source: Federal Specification RRc 271a. From Linnenbank, C. D. in *Handbook of Ocean and Underwater Engineering,* Myers, J. J., Ed., McGraw-Hill, New York, 1969. With permission.)

Table 5.2—10
SAFETY AND TRAWLING SHACKLES

Safety-type anchor shackles with thin head bolt-nut with cotter pin

Safety-type chain shackles thin hex head bolt-nut with cotter pin

Trawling shackle with thin square head with screw pin

Safety Anchor and Chain

Safe working load, tons	Size, in.	Inside length, in.		Inside width at pin, in.	Diam, in.		Tolerance, plus or minus		Wt , lb	
		2130	2150		Pin	Out-side of eye	Length	Width	2130	2150
2	1/2	1 7/8	1 5/8	13/16	5/8	1 5/16	1/8	1/16	0.82	0.76
3 1/4	5/8	2 3/8	2	1 1/16	3/4	1 9/16	1/8	1/16	1.58	1.56
4 3/4	3/4	2 13/16	2 3/8	1 1/4	7/8	1 7/8	1/4	1/16	2.82	2.62
6 1/2	7/8	3 5/16	2 13/16	1 7/16	1	2 1/8	1/4	1/16	3.95	3.65
8 1/2	1	3 3/4	3 3/16	1 11/16	1 1/8	2 3/8	1/4	1/16	5.6	5.35
9 1/2	1 1/8	4 1/4	3 9/16	1 13/16	1 1/4	2 5/8	1/4	1/16	7.85	7.27
12	1 1/4	4 11/16	3 15/16	2 1/32	1 3/8	3	1/4	1/16	11.2	10.2
13 1/2	1 3/8	5 1/4	4 7/16	2 1/4	1 1/2	3 5/16	1/4	1/8	15.2	13.35
17	1 1/2	5 3/4	4 7/8	2 3/8	1 5/8	3 5/8	1/4	1/8	19.5	18.5
25	1 3/4	7	5 3/4	2 7/8	2	4 5/16	1/4	1/8	31.3	28.5
35	2	7 3/4	6 3/4	3 1/4	2 1/4	5	1/4	1/8	46.3	41.1
50	2 1/2	10 1/2	8	4 1/8	2 3/4	6	3/4	1/8	94	84.5
75	3	13	9	5	3 1/4	6 1/2	3/4	1/8	145	123
100	3 1/2	15	10 1/2	5 3/4	3 3/4	8	1	1/4	250	218
130	4	17	12 1/2	6 1/2	4 1/4	9	1	1/4	358	310

Table 5.2—10 (continued)
SAFETY AND TRAWLING SHACKLES

Trawling Shackles

Safe working load, tons	Size, in.	Inside length, in.	Inside width at pin, in.	Diam, in.		Tolerance, plus or minus		Wt, lb
				Pin	Outside of eye	Length	Width	
2	1/2	1 5/8	13/16	5/8	1 5/16	1/8	1/16	0.68
3 1/4	5/8	2	1 1/16	3/4	1 9/16	1/8	1/16	1.21
4 3/4	3/4	2 3/8	1 1/4	7/8	1 7/8	1/4	1/16	2.14
6 1/2	7/8	2 13/16	1 7/16	1	2 1/8	1/4	1/16	3.07
8 1/2	1	3 3/16	1 11/16	1 1/8	2 3/8	1/4	1/16	4.53
12	1 1/4	3 15/16	2 1/32	1 3/8	3	1/4	1/8	8.87

(Source: Federal Specification RRc 271a. From Linnenbank, C. D. in *Handbook of Ocean and Underwater Engineering*, Myers, J. J., Ed., McGraw-Hill, New York, 1969. With permission.)

5.3. MARINE POWER SOURCES

Table 5.3—1
CELL CHARACTERISTICS: FIVE COMMON BATTERY TYPES

	Composition, charged state			Cell potential, V		Time to discharge				Shelf life in charged condition: Without maintenance	Shelf life in charged condition: With maintenance			Life in operation	
Battery type	Pos.	Neg.	Electrolyte	Open circ.	Discharging	Fastest, min	Avg, hr	Slowest, days	Shelf life if discharged, wet	Charge loss, %	Shelf life	If charged each	Shelf life	Cycles	Float
Lead-acid	PbO_2	Pb	H_2SO_4	2.14	2.1–1.46	3–5	8	>3	Not permitted	High-rate 50%/10 days Low rate 15–20%/yr	Days Months	30–45 days	Years	To 500	To 14 yr
Nickel-iron	NiO_2	Fe	KOH	1.34	1.3–0.75	10	5	>3	Decades	15–25%/mo	Weeks	30–45 days	Years	100–3000	To 30 yr
Nickel-cadmium	NiO_2	Cd	KOH	1.34	1.3–0.75	5	5	>3	Years	Pocket 20–40%/yr Sintered 10–15%/mo	Months Weeks	30–45 days	Years	100–2000 25–500	8–14 yr 4–8 yr
Silver-zinc	AgO	Zn	KOH	1.86	1.55–1.1	<0.5	5	>90	Years	15–20%/yr	3–12 mo	6 mo	1–2 yr	100–300 low dis. 5–100 high dis.	1–2 yr
Silver-cadmium	AgO	Cd	KOH	1.34	1.3–0.8	5	5	>90	Years	50%/2 yr	1–2 yr	6 mo	2–3 yr	500–1000	2–3 yr

(From Yeaple, F. D., *Prod. Eng.*, 36, 100, 1965.)

Table 5.3—2
RELATIONSHIP OF TEMPERATURE, DISCHARGE RATE, AND PERFORMANCE

Battery type	Cell rated capacity, A-hr	Discharge rate, A	Potential at midpoint, V			Capacity, A-hr, % rated			Energy density W-hr/lb			Energy density W-hr/in.^{3}		
			80°F	0°F	-40°F	80°F	0°F	-40°F	80°F	0°F	-40°F	80°F	0°F	-40°F
Lead-acid	5	0.5	1.95	1.89	1.85	100	54	30	10.8	5.6	3.1	0.76	0.40	0.22
		1.0	1.92	1.84	1.80	88	50	21	9.3	5.1	2.1	0.66	0.36	0.15
		10.0	1.81	1.60	1.40	46	16	3	4.7	1.4	0.24	0.33	0.10	0.02
	60	10.0	1.92	1.89	1.82	100	54	26	12.2	6.5	3.1	0.91	0.48	0.23
		25.0	1.90	1.80	1.65	87	31	10	10.5	3.5	1.0	0.78	0.26	0.07
		50.0	1.87	1.70	–	63	18	–	7.5	2.0	–	0.56	0.14	–
		100.0	1.70	–	–	39	–	–	4.1	–	–	0.31	–	–
Nickel-iron	10	2.0	1.20	0.98	–	100	67	–	10.6	7.1	–	0.91	0.61	–
	75	15.0	1.20	0.98	–	100	67	–	10.3	7.2	–	0.73	0.49	–
Nickel-cadmium	5	1.0	1.22	–	–	100	–	–	10.6	–	–	0.76	–	–
		10.0	1.11	1.05	1.05	94	67	21	9.1	6.2	2.0	0.65	0.45	0.14
	75	10.0	1.23	1.16	1.14	100	86	64	11.4	9.2	6.7	0.87	0.70	0.51
		25.0	1.20	1.14	1.06	97	82	48	10.8	8.7	4.7	0.83	0.66	0.36
		50.0	1.18	1.07	1.00	94	72	34	10.3	7.0	3.1	0.78	0.54	0.24
		100.0	1.17	–	–	82	–	–	8.4	–	–	0.64	–	–
Silver-zinc	5	0.5	1.52	1.45	–	100	75	–	43.0	31.0	–	2.2	1.6	–
		1.0	1.50	1.42	–	96	70	–	40.7	30.0	–	2.1	1.6	–
		10.0	1.40	1.26	–	85	63	–	33.7	27.0	–	1.7	1.4	–
	60	10.0	1.52	1.46	–	100	92	–	46.5	39.0	–	3.15	2.6	–
		25.0	1.49	1.42	–	97	79	–	44.7	39.0	–	3.0	2.6	–
		50.0	1.48	1.42	–	92	75	–	42.2	39.0	–	2.8	2.6	–
		100.0	1.42	1.30	–	84	69	–	36.9	35.0	–	2.5	2.4	–
Silver-cadmium	5	0.5	1.08	1.03	0.9	100	85	44	26.8	21.5	10	1.78	1.4	0.67
		5.0	1.05	0.99	0.7	95	80	40	21.0	16.0	5.6	1.4	1.05	0.37
	60	6.0	1.10	1.08	–	100	96	–	31.7	30.6	–	2.9	2.8	–
		60.0	1.00	0.94	0.9	95	80	40	24.5	22.6	13	2.2	2.0	1.2

Note: These performance data are taken at random from a variety of cells. They show pronounced effects of temperature and discharge rate on performance. Dashes indicate that heaters are needed to warm cell in cold ambients. Nickel-iron cells, however, are not usually used at temperatures below 0°F.

(From Yeaple, F. D., *Prod. Eng.*, 36, 100, 1965. With permission.)

5.4. FIXED OCEAN STRUCTURES

PILINGS – DESIGN CONSIDERATIONS

For structures composed of a single vertical piling it is possible that the natural period of the structure and the vortex shedding period will be sufficiently close to cause coupling between the vortices and the piling, resulting in a self-excited resonant oscillation; that is, the lateral oscillation of the piling enhances the vortex shedding, which in turn causes larger vortex forces, and so on. This effect has been reasonably well investigated for steady flows such as wind blowing on smokestacks, but little relevant information is available for cases of unsteady flows such as water waves.

The wave-force treatment described below pertains to the deterministic forces, that is, those forces directly related to the water particle velocities and accelerations.

For a vertical circular piling of diameter D, in a wave with horizontal components of water-particle velocity u and acceleration $\dot{u}$, the elemental horizontal force ΔF on an incremental piling length ΔS is

$$\Delta F = C_D \rho D \frac{u|u|}{2} \Delta S + C_M \frac{\rho \pi D^2}{4} \dot{u} \Delta S$$

where C_D and C_M are drag and inertia coefficients, respectively, and ρ is the mass density of the water (see figure 5.4—1). It should be emphasized that Eq. (1) and the following force and moment treatment include only the effects of wave motion; tidal, wind-driven, or other currents are not included but should be accounted for in design if they are deemed important in a specific location.

The primary problems in the calculation of wave forces are (1) the accurate description of the kinematic flow field for nonlinear waves and (2) correct values of the C_D and C_M coefficients. Wilson and Reid[1] have tabulated available published values of drag and inertia coefficients obtained from laboratory and field investigations. There is considerable discrepancy in the reported results, and it is well known that the steady flow circular-cylinder-drag coefficient depends on Reynolds number and cylinder surface roughness. The coefficient averages of the summary are probably the most representative published values available; the averages are $\bar{C}_D = 1.05$ and $\bar{C}_M = 1.40$, with ranges $0.40 < C_D < 1.60$ and $0.93 < C_M < 2.30$.

These ranges indicate the difficulties in selecting valid design coefficients from the available published literature. It should be mentioned that most of the tests included in the summary were based on reasonably smooth cylinders and were conducted at Reynolds number ranges below those that would be appropriate for design. On the basis of steady-state drag-coefficient variations with Reynolds number, the larger-design Reynolds numbers would be expected to result in a design drag coefficient less than the average C_D value. Tests[2] indicate that the effect of the rougher design piling surface would result in an increased drag coefficient; these two effects would therefore tend to cancel.

The total force at one time on a single vertical piling is determined by integrating Eq. (1) over the submerged portion of the piling. The total forces and moments will vary with time as the wave passes the piling; however, it is the maximum forces and moments that are of most interest. Because of the many parameters involved in the wave-force problem, it is difficult to develop a compact graphical or tabular representation of forces and moments on a piling for all wave conditions (including nonlinearities) and piling diameters of interest. It can be shown, however, that if the drag and inertia coefficients are considered constants, then the maximum force F_m on a single vertical piling can be expressed in dimensionless form as

$$\frac{F_m}{\gamma C_D H^2 D} = \phi_m \quad \frac{h}{T^2}, \frac{H}{T^2}, W$$

The maximum total moment M_m on a single vertical piling can be expressed in a corresponding dimensionless form as

$$\frac{M_m}{\gamma C_D H^2 D h} = \alpha_m \quad \frac{h}{T^2}, \frac{H}{T^2}, W$$

where γ is the specific weight of water in pounds per cubic foot and W is defined as

$$W = \frac{C_M}{C_D} \frac{D}{H}$$

To illustrate, suppose we wish to calculate the crest elevation η_c, the maximum total force F_m, and maximum total moment M_m on a single vertical piling for the following wave conditions:

Wave height H	=	40 ft
Water depth h	=	80 ft
Wave period T	=	12 sec
Piling diameter D	=	5 ft

For purposes of this example, drag and inertia coefficients will be taken as the average values ($C_D = 1.05$ and $C_M = 1.40$). The use of these average coefficients here, however, should not be interpreted as endorsement for design.

The following parameters are calculated:

$$\frac{h}{T^2} = \frac{80}{12^2} = 0.556$$

$$\frac{H}{T^2} = \frac{40}{12^2} = 0.278$$

PILING – DESIGN CONSIDERATIONS (*Continued*)

$$W = \frac{C_M D}{C_D H} = \frac{1.4 \times 5}{1.05 \times 40} = 0.164$$

(From Dean, R. G. in *Handbook of Ocean and Underwater Engineering,* Myers, J. J., Ed., McGraw-Hill, New York, 1969. With permission.)

REFERENCES

1. **Wilson, B. W. and Reid, R. O.,** *J. Waterways Harbors Div. Am. Soc. Civil Eng.,* 89(WW1), 61, 1963.
2. **Blumberg, R. and Rigg, A. M.,** Petrol. Session, Am. Soc. Met. Eng. meeting, Los Angeles, June 14, 1961.

Figure 5.4—1
WAVE AND PILING SYSTEM DEFINITION SKETCH

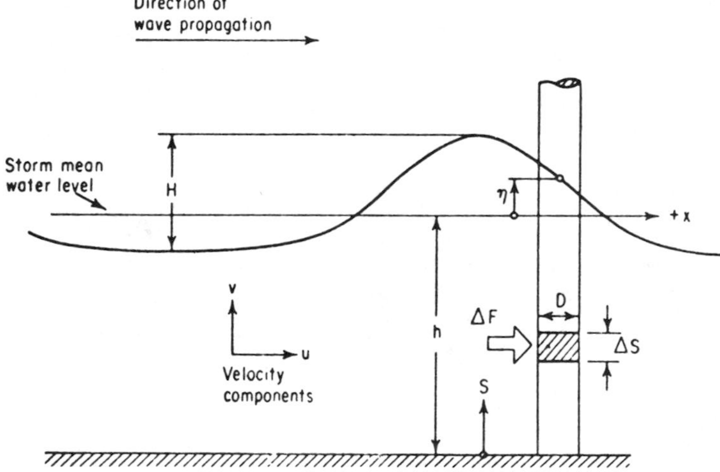

(From Dean, R. G. in *Handbook of Ocean and Underwater Engineering,* Myers, J. J., Ed., McGraw-Hill, New York, 1969. With permission.)

Table 5.4—1
ESSENTIAL PROPERTIES OF SOME AMERICAN SHEET PILES

Section No.[a]							
U.S. Steel	Bethlehem Steel	Area, in.^{2}	Width, in.	Web or flange thickness, min in.	Wt, lb/ft^2 of wall	Section modulus, in.^{3}/linear ft of wall	Interlock strength, lb/in.[a]
MZ 38	ZP-38	16.77	18	3/8	38	46.8	8000
MZ 32	ZP-32	16.47	21	3/8	32	38.3	8000
MZ 27	ZP-27	11.91	18	3/8	27	30.2	8000
MP 110	DP-1	12.56	16	3/8	32	15.3	8000
MP 116	DP-2	10.59	16	3/8	27	10.7	8000
MP 115	AP-3	10.59	19 5/8	3/8	22	5.4	8000
MP 112	SP-4	8.99	16	3/8	23	2.4	12,000
MP 113	SP-5	10.98	16	1/2	28	2.5	12,000
MP 101	SP-6a	10.29	15	3/8	28	2.4	16,000
MP 102	SP-7a	11.76	15	1/2	32	2.4	16,000
MP 117	AP-8	11.41	15	–	31	7.1	10,000
	SP-9	4.38	8 1/2	13/64	21	1.4	8000

[a] See catalogs of respective companies for additional information.

(From Chellis, R. D. in *Handbook of Ocean and Underwater Engineering,* Myers, J. J., Ed., McGraw-Hill, New York, 1969. With permission.)

Table 5.4—2
H-PILE DIMENSIONS AND PROPERTIES FOR DESIGNING

Section No.[a]			Area of section,	Depth of section,	Flange		Web thickness,	Axis XX			Axis YY		
U.S. Steel	Bethlehem Steel	Wt/ft, lb	in.^{2}	in.	Width, in.	Thickness, in.	in.	I, in.^{4}	S, in.^{3}	r, in.	I', in.^{4}	S', in.^{3}	r', in.
CBP 146	BP 14	117	34.44	14.23	14.885	0.805	0.805	1228.5	172.6	5.97	443.1	59.5	3.59
CBP 146	BP 14	102	30.01	14.03	14.784	0.704	0.704	1055.1	150.4	5.93	379.6	51.3	3.56
CBP 146	BP 14	89	26.19	13.86	14.696	0.616	0.616	909.1	131.2	5.89	326.2	44.4	3.53
CBP 146	BP 14	73	21.46	13.64	14.586	0.506	0.506	733.1	107.5	5.85	261.9	35.9	3.49
CBP 124	BP 12	74	21.76	12.12	12.217	0.607	0.607	566.5	93.5	5.10	184.7	30.2	2.91
CBP 124	BP 12	53	15.58	11.78	12.046	0.436	0.436	394.8	67.0	5.03	127.3	21.2	2.86
CBP 103	BP 10	57	16.76	10.01	10.224	0.564	0.564	294.7	58.9	4.19	100.6	19.7	2.45
CBP 103	BP 10	42	12.35	9.72	10.078	0.418	0.418	210.8	43.4	4.13	71.4	14.2	2.40
CBP 83	BP 8	36	10.60	8.03	8.158	0.446	0.446	119.8	29.9	3.36	40.4	9.9	1.95

[a] See catalogs of respective companies for additional information.

(From Chellis, R. D. in *Handbook of Ocean and Underwater Engineering*, Myers, J. J., Ed., McGraw-Hill, New York, 1969. With permission.)

Table 5.4—3
PROPERTIES OF SOME AMERICAN STEEL-PIPE-PILE SECTIONS

Nominal size, in.	Diam, in.[a]		Wall thickness, in.		Wt/ft, lb	Gross area of metal, in.^{2}	Moment of inertia I, in.^{4}	Section modulus I/c, in.^{3}	Radius of gyration r	Area of concrete		Effective area of metal, in.^{2} [b]
	OD	ID	Dec.	Frac.						in.^{2}	ft^2	
10	10.00	**9.782**[c]	**0.109**[c]	7/64	11.52	3.39	41.37	8.47	3.50	75.15	0.5209	1.52
		9.656	**0.172**	11/64	18.05	5.31	64.13	12.82	3.48	73.29	0.5085	3.36
10	10.75	**10.532**[c]	**0.109**[c]	7/64	12.39	3.64	51.58	9.59	3.77	87.12	0.6050	1.54
		10.312[c]	**0.219**[c]	7/32	24.60	7.25	100.4	18.68	3.72	83.52	0.5800	5.15
		10.250	0.250	1/4	28.04	8.24	113.7	21.16	3.71	82.52	0.5730	6.15
		10.189	0.281	–	31.44	9.24	126.8	23.60	3.70	81.52	0.5665	7.14
		10.136	*0.307*	–	34.24	10.07	137.4	25.57	3.69	80.68	0.5603	7.97
		9.750	*0.500*	1/2	54.73	16.10	212.0	39.43	3.63	74.66	0.5184	14.02
12	12.00	**11.782**[c]	**0.109**[c]	7/64	13.84	4.07	71.98	12.00	4.21	109.03	0.7571	1.73
		11.656	**0.172**	11/64	21.73	6.39	111.8	18.64	4.19	106.70	0.7410	4.05
12	12.75	**12.532**[c]	**0.109**[c]	7/64	14.71	4.33	86.49	13.57	4.48	123.35	0.8566	1.84
		12.126[c]	**0.3125**[c]	5/16	41.51	12.19	236.3	37.06	4.40	115.49	0.8018	9.70
		12.090	*0.330*	–	43.77	12.88	248.5	38.97	4.39	114.80	0.7972	10.38
		11.750	*0.500*	1/2	65.41	19.24	361.5	56.71	4.34	108.43	0.7530	16.75
14	14.00	**13.782**[c]	**0.109**[c]	7/64	16.17	4.75	114.7	16.39	4.92	149.18	1.0360	2.01
		13.438[c]	**0.281**[c]	9/32	41.21	12.11	283.1	40.44	4.85	141.83	0.9849	9.37
		13.375[c]	0.3125[c]	5/16	45.68	13.42	314.9	44.98	4.84	140.48	0.9756	10.68
		13.250	*0.375*	3/8	54.56	16.05	372.8	53.25	4.82	137.89	0.9575	13.31
		13.000	*0.500*	1/2	72.09	21.21	483.8	69.11	4.78	132.73	0.9217	18.47

Note: Sizes in boldface type furnished, spiral-welded pipe, by the American Rolling Mill Company. Sizes in italics furnished, seamless and welded pipe, by the National Tube Company. Other sizes furnished by both companies. Pipe furnished by the American Rolling Mill Company conforms to ASTM Standard Specifications for Electric-fusion (Arc) Welded Steel Pipe (sizes 8 in. to but not including 30 in.) A139, Grade B, except that hydrostatic testing will not be required, or ASTM Standard Specifications for Welded and Seamless Steel Pipe Piles A252, for all sizes except 6, 8, 10, and 12 in. OD, which conform to ASTM Standard Specifications for Spiral-welded Steel or Iron Pipes A211 and ASTM Standard Specifications for Low Tensile Strength Carbon-steel Plates of Structural Quality for Welding A78, Grade B.

Pipe furnished by the National Tube Company conforms to ASTM Standard Specifications for Welded and Seamless Steel Pipe Piles A252.

a Lightest and heaviest sections of each group are shown. Full ranges of values for intermediate items appear in pipe manufacturer's catalogs.
b After deducting the outer 1/14 in. of pipe wall.
c Not carried in stock but available from the American Rolling Mill Company in minimum quantities of 10,000 lb.

Table 5.4—3 (continued)
PROPERTIES OF SOME AMERICAN STEEL-PIPE-PILE SECTIONS

Nominal size, in.	Diam, in.[a]		Wall thickness, in.		Wt/ft, lb	Gross area of metal, in.^{2}	Moment of inertia I, in.^{4}	Section modulus I/c, in.^{3}	Radius of gyration r	Area of concrete		Effective area of metal, in.^{2} [b]
	OD	ID	Dec.	Frac.						in.^{2}	ft^2	
16	16.00	**15.782**[c]	**0.109**[c]	7/64	18.50	5.44	171.8	21.47	5.64	195.62	1.3585	2.31
		15.438[c]	**0.281**[c]	9/32	47.22	13.88	429.1	53.64	5.56	187.19	1.2999	10.75
		15.375[c]	0.3125[c]	5/16	52.36	15.38	473.9	59.24	5.55	185.69	1.2893	12.25
		15.125	*0.4375*	7/16	72.71	21.39	648.1	81.01	5.52	179.67	1.2477	18.26
		15.000	*0.500*	1/2	82.77	24.35	731.9	91.49	5.50	176.72	1.2271	21.22
18	18.00	**17.782**[c]	**0.109**[c]	7/64	20.83	6.13	245.1	27.24	6.35	248.34	1.7246	2.61
		17.438[c]	**0.281**[c]	9/32	53.22	15.64	614.5	68.28	6.27	238.83	1.6585	12.12
		17.375[c]	0.3125[c]	5/16	59.03	17.34	679.2	75.47	6.25	237.13	1.6466	13.81
		17.250	0.375	3/8	70.59	20.76	806.6	89.62	6.23	233.71	1.6229	17.24
		17.125	*0.4375*	7/16	82.06	24.14	931.3	103.5	6.21	230.33	1.5995	20.62
		16.750	*0.625*	5/8	115.97	34.12	1289.0	143.2	6.15	220.35	1.5302	30.59
20	20.00	**19.782**[c]	**0.109**[c]	7/64	23.15	6.81	336.9	33.69	7.02	307.35	2.1343	2.84
		19.438[c]	**0.281**[c]	9/32	59.23	17.41	847.1	84.71	6.97	296.75	2.0608	13.49
		19.375[c]	0.3125[c]	5/16	65.71	19.30	936.7	93.67	6.96	294.80	2.0474	15.38
		19.250	0.375	3/8	78.60	23.12	1113.5	111.3	6.94	291.04	2.0210	19.20
		19.125	*0.4375*	7/16	91.40	26.89	1287.0	128.7	6.92	287.27	1.9949	22.92
		18.750	*0.625*	5/8	129.33	38.04	1787.0	178.7	6.85	276.12	1.9174	34.13
22	22.00	**21.718**[c]	**0.141**[c]	9/64	32.92	9.68	578.4	52.58	7.75	370.45	2.5726	5.36
		21.438[c]	**0.281**[c]	9/32	65.18	19.17	1130.2	102.7	7.69	360.96	2.5067	14.85
		21.375[c]	0.3125[c]	5/16	72.38	21.29	1252.0	113.8	7.67	358.83	2.4919	16.97
		21.250[c]	0.375[c]	3/8	86.60	25.48	1490.0	135.4	7.65	354.66	2.4628	21.16
		21.125	*0.3475*	7/16	100.75	29.64	1723.0	156.6	7.62	350.50	2.4339	25.32
		20.750	*0.625*	5/8	142.68	41.97	2399.0	218.1	7.56	338.16	2.3483	37.65
24	24.00	**23.718**[c]	**0.141**[c]	9/64	35.83	10.57	752.1	62.67	8.45	441.82	3.0682	5.87
		23.438[c]	**0.281**[c]	9/32	71.25	20.94	1472.7	122.8	8.39	431.45	2.9962	16.24
		23.375[c]	0.3125[c]	5/16	79.06	23.22	1631.7	135.9	8.38	429.17	2.9800	18.52
		23.250	0.375	3/8	94.62	27.83	1942.0	161.9	8.35	424.56	2.9482	23.13
		23.125	*0.3475*	7/16	110.09	32.39	2248.0	187.4	8.33	420.00	2.9166	27.69
		22.750	*0.625*	5/8	156.03	45.90	3137.0	261.4	8.27	406.49	2.8228	41.20

[a] Lightest and heaviest sections of each group are shown. Full ranges of values for intermediate items appear in pipe manufacturer's catalogs.
[b] After deducting the outer 1/14 in. of pipe wall.
[c] Not carried in stock but available from the American Rolling Mill Company in minimum quanitites of 10,000 lb.

Table 5.4—3 (continued)
PROPERTIES OF SOME AMERICAN STEEL-PIPE-PILE SECTIONS

Nominal size, in.	Diam, in.[a]		Wall thickness, in.		Wt/ft, lb	Gross area of metal, $in.^2$	Moment of inertia I, $in.^4$	Section modulus I/c, $in.^3$	Radius of gyration r	Area of concrete		Effective area of metal, $in.^2$ [b]
	OD	ID	Dec.	Frac.						$in.^2$	ft^2	
26	26.00	25.500[c]	0.250[c]	1/4	68.76	20.22	1676.4	129.0	9.12	510.70	3.5466	15.12
		25.000[c]	0.500[c]	1/2	136.19	40.06	3257.0	250.5	9.04	490.86	3.4088	34.96
28	28.00	27.500[c]	0.250[c]	1/4	74.09	21.79	2098.1	149.9	9.83	593.96	4.1247	17.60
		27.000[c]	0.500[c]	1/2	146.88	43.20	4084.8	291.8	9.75	572.56	3.9761	37.71
30	30.00	29.500[c]	0.250[c]	1/4	79.44	23.36	2585.1	172.3	10.52	683.49	4.7465	17.49
		29.000[c]	0.500[c]	1/2	157.53	46.34	5043.0	336.1	10.43	660.52	4.5870	40.46
36	36.00	35.500[c]	0.250[c]	1/4	95.39	28.06	4485.8	249.2	12.66	989.80	6.8736	21.00
		35.000[c]	0.500[c]	1/2	189.57	55.76	8786.0	488.1	12.55	962.12	6.6813	48.70

[a] Lightest and heaviest sections of each group are shown. Full ranges of values for intermediate items appear in pipe manufacturer's catalogs.
[b] After deducting the outer 1/14 in. of pipe wall.
[c] Not carried in stock but available from the American Rolling Mill Company in minimum quanitites of 10,000 lb.

(From Chellis, R. D. in *Handbook of Ocean and Underwater Engineering*, Myers, J. J., Ed., McGraw-Hill, New York, 1969. With permission.)

Table 5.4—4
PROPERTIES OF PRETENSIONED-CONCRETE BEARING PILES

Diam, in.	Shape	Solid or hollow core	No. strands, strand diam, in.	Effective prestress in concrete, psi[a]	A_c, in.^{2} [b]	I, in.^{4}	I/c, in.^{3}	Perim., in.	Wt / lin. ft, lb	Allowable design load[c] Kips	Allowable design load[c] Tons	Allowable moment, kip-in.[d]	Allowable moment for earthquakes, kip-in.[e]	Allowable unsupported length, ft[f]	Normal max length, ft
10	Octag.	S	6, 3/8	839	83	547	109	33	86	91	46	124	157	32	75
12	Octag.	S	8, 3/8	780	119	1135	189	40	124	131	65	204	261	38	100
14	Octag.	S	10, 3/8	716	162	2103	300	46	169	178	89	305	395	45	130
16	Octag.	S	13, 3/8	711	212	3587	448	53	221	233	117	453	587	51	140
18	Octag.	S	16, 3/8	701	268	5746	638	60	279	295	147	637	828	58	150
20	Octag.	S	20, 3/8	701	331	8758	876	66	345	364	182	877	1140	64	160
20	Octag.	11 in. H	16, 3/8	786	236	8039	804	66	246	260	130	873	1114	73	160
20	Square	S	19, 7/16	749	398	13,146	1315	78	415	438	219	1379	1774	72	160
20	Square	11 in. H	20, 3/8	766	308	12,427	1243	78	316	333	167	1325	1698	80	160
26	Square	16 in. H	23, 7/16	760	475	32,272	2482	97	495	523	261	2631	3376	103	160
30	Round	20 in. H	19, 7/16	759	393	31,907	2127	94	409	431	216	2252	2891	113	180
36	Round	26 in. H	24, 7/16	774	487	60,016	3334	113	507	536	268	3581	4581	138	200
48	Round	38 in. H	32, 7/16	744	675	158,222	6593	151	703	743	371	6883	8861	191	225
54	Round	44 in. H	36, 7/16	734	770	233,409	8645	170	802	847	424	8939	11,532	217	250

a Effective prestress in based on a final effective force of 11,600 lb for 3/8-in.-diam strand and 15,700 lb for 7/16-in.-diam strand, or 145,000 psi.

b All holes are circular; 1-in. chamber on 20-in.-square pile corners and 3-in. chamfer on 26-in.-square pile corners.

c Allowable design load is based on 1100 psi on the concrete section. Where driving and soil conditions are favorable, this may be raised accordingly.

d Allowable moment is based on a tension of 300 psi with an effective prestress as given in the table. Where bending resistance is critical, the allowable moment may be increased by using more strands to raise the effective prestress to about 1200 psi maximum.

e Allowable moment for earthquake or similar loads is based on a tension of 600 psi with an effective prestress as given in the table.

f Allowable unsupported length is computed for E_c = 5 million psi, with a factor of safety of 2 on the allowable direct load, assuming pin ending at both ends. If the external direct load is smaller, the length can be increased. Note that this length is for transient loads; for sustained loads the value must be revised for a modulus E_c of 2 million psi. If eccentricity is expected, allowable length should be reduced.

(From Ben C. Gerwick, Inc., in Chellis, R. D. in *Handbook of Ocean and Underwater Engineering*, Myers, J. J., Ed., McGraw-Hill, New York, 1969. With permission.)

5.5. BUOY SYSTEMS

Buoy-system Scope

Calculation of the equilibrium position of a buoy in relation to its anchor under specific conditions is generally a relatively complex calculation, particularly if the system is subsurface supported and the scope is critical. If the system is a simple surface-supported buoy using synthetic line, a simple geometric estimate is probably as meaningful as any.

Realistic evaluation of the scope of single-point moorings by the conventional catenary equations is not a straightforward process. The addition of drag forces on the mooring line introduces even more complications. Various tables are available to cover the multitude of cases possible, but their use in design problems is made unwieldy by the need for repeated interpolation. It is convenient to be able to evaluate the scope of buoy systems in terms of numerous design variables, such as cable size, buoy drag, ocean-current profiles, and water depth. This evaluation may be done with the use of a formula[1] that gives the mooring's scope directly for the range of cases where straight-line approximations are too inaccurate and catenary methods are too cumbersome. It expresses the shape of a mooring line as an exponential curve similar to the steep limb of a catenary.

The formula has a wide range of application to cases where the scope (horizontal distance from the buoy to a point vertically above the anchor) of a single mooring is of the order of one-fifth or less of the total depth of water. This proves to be a realistic situation for many deep-water oceanographic buoy systems as well as for nearly all single-line taut moorings.[a] The derivation of the formula follows, with a discussion of its applicability and some numerical examples.

The balance of horizontal forces on a vertically hanging uniform string subject to a transverse load q is given, with reference to figure 5.5—1, by

$$T_{k+1}\frac{y_{k+1}-y_k}{h}-T_k\frac{y_k-y_{k-1}}{h}+q_k h=0$$

where $T_k + 1 = T_k + wh$, with w being the wt/unit length of cable. Sorting terms and passing to infinitesimals, the following differential equation is obtained:

$$(T_1+wx)\frac{d^2y}{dx^2}+w\frac{dy}{dx}+q=0$$

which has the solution

$$y=C_1-\frac{qx}{w}+C_2\ln\frac{T_1+wx}{T_1} \qquad (1)$$

in which the constants C_1 and C_2 can be solved for appropriate end conditions. For a line secured at the lower end (a mooring) the general expression for the shape of the mooring line curve at any point becomes

$$y=\frac{H}{w}\ln\frac{T_1+wx}{T_1}-\frac{qx}{w}+\frac{qT_2}{w^2}\ln\frac{T_1+wx}{T_1}$$

and that for the scope of a buoy at $x = L$ becomes

$$y_L=\frac{H}{w}\ln\frac{T_2}{T_1}-\frac{qL}{w}+\frac{qT_2}{w^2}\ln\frac{T_2}{T_1}$$

$$=x_1-x_3+x_2 \qquad (2)$$

where x may be computed with the aid of the nomograms.

These equations rest on the assumption that drag forces are normal to the cable and that the depth of water L and cable length are approximately equal. The first assumption is conservative, but it implies that the validity is restricted to cables with moderate inclinations from the vertical. The second assumption means that for any given depth of water, a buoy would reach the scope calculated at a depth slightly less than that used in the calculation.

Equation (2) is, by suitable manipulation of the end conditions, also applicable to a variety of towing problems for heavy cables or low towing speeds. By changing the sign of w, the scope of moorings using positively buoyant lines, such as polypropylene, may also be calculated.

Equation (2) can readily be used for quick design estimates. For current profiles with several current shears, or for moorings using various types of cable, it is applied to each subsection and the displacements are added vectorially in the horizontal plane.

As a sample solution, let us calculate scope for the buoy system of fig. 5.6–2 using Eq. (2) and the following parameters:

Buoy:
- Diam = 6 ft
- Wt = 1700 lb
- Net buoyancy = 5500 lb $\cong T_2$

Cable:
- w = 0.15 lb/ft
- L = 15,000 ft
- Diam = 0.25 in.

Current:
- Velocity at buoy depth = 1.5 knots
- Velocity over cable length = 0.4 knot

Given $C_D = 1.2$, the total drag force on the cable is given by

$$qL=C_DAp\frac{v^2}{2}=180\text{ lb}$$

from which

$$q=1.2\times10^{-2}\text{ lb/ft.}$$

[a] A taut mooring may be defined as one that will have an upward tension at the anchor with no impressed currents.

Given C_D - 0.8, the total drag force on the buoy is given by

$$H = C_D A\rho \frac{v^2}{2} = 142 \text{ lb}$$

The vertical force at the sea floor is

$$T_1 \cong T_2 - wL = 3200 \text{ lb}$$

For the mooring scope, Equation (2) yields:

$$y = \frac{142}{0.15} \ln 1.72 - \frac{180}{0.15} + \frac{5500}{0.15^2} \times 1.2 \times 10^{-2} \ln 1.72$$

$$= 513 - 1200 + 1590$$

$$= 903 \text{ ft}$$

REFERENCES

1. **Lampietti, F. J.,** *Am. Soc. Mech. Engrs. Paper,* 63-WA-101, 1963.

(From Lampietti, F. J. and Snyder, R. M., *Geo-Marine Technol.,* 1(6), 29, 1965. With permission.)

Figure 5.5—1
SCHEMATIC OF SINGLE-POINT MOORING

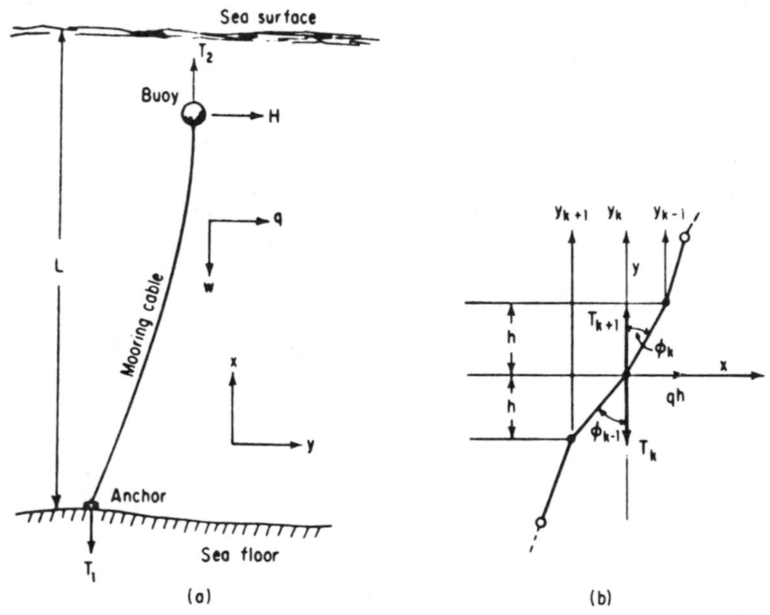

Note: (a) System; (b) force diagram in infinitesimals. x = vertical direction, y = horizontal direction, L = depth of water below buoy, ft, w = weight of cable in water, lb/ft, q = drag force on cable, lb/ft, T_1 = vertical force at sea floor, lb, T_2 = vertical force at buoy, lb, and H = horizontal drag force on buoy and all components above it.

(From Lampietti, F. J. and Synder, R. M., *Geo-Marine Technol.,* 1(6), 29, 1965. With permission.)

Buoy-system Motions

Because ocean currents are not steady but continuously shift direction and change speed, a mooring cannot remain in static equilibrium with the current and must continuously seek a new equilibrium configuration. This continual readjustment is termed *mooring motion*. Its magnitude determines the usefulness of the mooring for current measurements and position keeping.

One of the major incentives for studying mooring motion in detail is that the motion should be predictable for the measured relative flows past the mooring. Hence, in principle, the measured flows can be corrected for the motion. Also, if such corrections can be made with precision, it is not necessary to attempt to reduce the motion to a minimum by designing large rigid systems or using multiple anchoring. A relatively *soft* (compliant) mooring may be capable of giving a performance equivalent to that of a much more rigid platform for current measurements if numerical corrections can be applied precisely.

In deep water, because of the decrease of velocity with depth, the major contribution to horizontal drag is made by the upper 10 to 20 percent of the mooring. As far as displacement of the mooring is concerned, the action of the horizontal drag can be approximated by a horizontal force of the same magnitude acting on the mooring float (buoy). The restoring force for small displacements is proportional to the displacement of the float from its vertical equilibrium position.

Because both the horizontal drag and the restoring force are very large in comparison to accelerations, the inertia of the mooring can be neglected and the balance equations must be simple: the horizontal drag force is equal to the horizontal restoring force or

$$\tfrac{1}{2}\rho C_D A V^2 = F_B \frac{r}{L}$$

where ρ = water density
C_D = drag coefficient
A = effective mooring cross section
V = water speed
F_B = float buoyancy
r = horizontal displacement
L = mooring length

(see fig. 5.5—3). The displacement is in the same direction as the velocity vector.

The force equation can be rewritten in the form

$$r = KV^2 \qquad (3)$$

where

$$K = \tfrac{1}{2}\rho C_D A \frac{L}{F_B} \qquad (4)$$

is a measure of the softness (compliance), of a mooring. The larger the value of K, the greater the horizontal displacement of the buoy at a given current speed. In deep water a single-line mooring with good performance characteristics will have a K value of 5 to 20 sec²/cm. The complicance constant K has a usefulness beyond specifying the softness of the mooring. It can be related to the time constant or recovery time of a mooring.

Consider a mooring that is displaced a horizontal distance r_0 in still water and released at time $t = 0$. The speed of flow past the buoy is dr/dt. Hence Eq. (3) becomes

$$r = K\left(\frac{dr}{dt}\right)^2$$

or, solving for dr/dt,

$$\frac{dr}{dt} = -K^{-1/2} r^{1/2}$$

The negative sign is chosen because the motion is such as to decrease r. The equation for r is nonlinear but is easily solved to give

$$r = \frac{1}{4K}(t_0 - t)^2$$

where t_0 is a constant of integration.

The mooring reaches equilibrium in a finite time t_0 that is given by

$$r_0 = \frac{t_0^{\,2}}{4K}$$

The initial speed is

$$V_0 = \left|\frac{dr}{dt}\right|_{t=0} = \frac{t_0}{2K}$$

If the time constant τ of the mooring is defined to be $1/2t_0$, these equations can be expressed in the form

$$r_0 = Kv_0^{\,2} = V_0 r$$
$$r = Kv_0$$

Thus the compliance constant Kv specifies not only the magnitude of the displacement for a given velocity v_0, but also a typical time required to produce the displacement.

In a current of 100 cm/sec (about 2 knots) a mooring with a compliance constant of 10 sec²/cm would have a time constant of 1000 sec, or about 17 min. The horizontal displacement would be 10^5 cm, or 1 km.

The value of the single-level model as given by Eq. (3) is that explicit solutions are possible. In general, the displacement r is in the direction of the relative current v. For a changing current the equation can be written in component form as follows:

Displacement along the x axis:

$$x = r\frac{u_r}{\sqrt{u_r^{\,2} + v_r^{\,2}}} = r\frac{u - dx/dt}{(r/K)^{1/2}} = (Kr)^{1/2}\left(u - \frac{dx}{dt}\right)$$

Displacement along the y axis:

$$y = r\frac{v_r}{\sqrt{u_r^{\,2} + v_r^{\,2}}} = (Kr)^{1/2}\left(v - \frac{dy}{dt}\right)$$

where u_r and v_r are components of the relative current and u and v are components of the true current referred to the horizontal axes x and y.

Solving for the derivatives yields the mooring motion equations

$$\frac{dx}{dt} = u - \frac{x}{\sqrt{Kr}}$$

$$\frac{dy}{dt} = v - \frac{y}{\sqrt{Kr}} \qquad (5)$$

where $r = (x^2 + y^2)^{1/2}$ and u and v are assumed to be arbitrary specified functions of time t. These equations constitute the single-level approximation to mooring motion. Solutions can be obtained explicitly for a step increase or decrease in speed and for a rotating current of constant speed. Numerical solutions can be readily obtained for more complex currents.

The case of a uniformly rotating current is instructive because the mooring motion is most pronounced under these circumstances. Substituting

$$u = V_0 \cos wt, \qquad v = V_0 \sin wt$$
$$x = r\cos\theta, \qquad y = r\sin\theta$$

into Eq. (5) and equating coefficients yields the equation for mooring speed:

$$r\omega = \omega K (V_0{}^2 - r^2\omega^2)$$

At low rotation rates ($t \ll 1$) the ratio of mooring speed to water speed is

$$\frac{r\omega}{V_0} \simeq \omega K V_0 = \omega r$$

$$\simeq \frac{2\pi\tau}{T} \qquad T \ll \tau$$

where T is the period of rotation of the current.

Thus, in the example given earlier, if a current of 100 cm/sec is rotary with an inertial period of 17.5 hr, the float speed would be a steady 10 cm/sec. The speed of the mooring cable would decrease approximately linearly with depth. At mid-depth (2000 to 3000 m) the speed would still be about 5 cm/sec, which is comparable to the speeds expected in deep water. Hence the degree of mooring motion must be determined before fluctuations of currents at depth can be accepted with confidence.

The compliance constant can be estimated by computing the cross section of the upper third of a moor and substituting into Eq. (4). Such a calculation yielded a value of 12 sec²/cm for a subsurface mooring launched in a depth of 2000 m. Another method is to measure the pressure (depth) overshoot during the launching transient. If the currents are weak, the change of pressure (depth) with time after the anchor reaches the sea floor can be approximated by

$$\Delta p \simeq \rho g\,\Delta Z = \frac{\frac{1}{2}\rho g r^2}{L} = \frac{1}{2}\,\frac{\rho g}{L}\,\frac{1}{(4K)^2}\,(t_0 - t)^4$$

$$\simeq \frac{\rho g r_0{}^2}{2L}\left(1 - \frac{t}{2\tau}\right)^4 = \Delta p_{max}\left(1 - \frac{t}{2\tau}\right)^4$$

as the mooring moves toward the vertical equilibrium. Therefore $(\Delta P/\Delta P_{max})^{1/4}$ is a linear function of time with a slope equal to $\frac{1}{2}\tau$. An estimate of the slope together with the maximum pressure overshoot ΔP_{max} is sufficient to determine K. For the mooring mentioned above the measured overshoot was 15.6 m and yielded an estimate of 14 sec²/cm for K. It was considered to be in good agreement with the calculated value of 12 sec²/cm. Similarly, an estimate can be made from the measured speed of the mooring during the launching transient.

If the compliance constant is known, the measured relative velocity u_r and v_r can be corrected for mooring motion with Eqs. (3) and (5) in the form

$$x = KV_r u_r$$
$$y = KV_r v_r$$
$$V_r = \sqrt{u_r{}^2 + v_r{}^2}$$

to calculate the mooring motion dx/dt and dy/dt. The measured relative velocities must be filtered to remove high frequencies before differentiating; otherwise the noise level will be too high.

Multilevel extensions of this model are straightforward to derive, but the constants involved cannot be as easily estimated from measurements. However, it is fairly simple to calculate the relative magnitudes of the compliance constants for each level from the known cross section and mass distribution of a mooring. The number of levels chosen can be made to correspond with the number of current meters on the mooring.

The single-level model is a necessary first step in understanding and developing the more complex, and hopefully more precise, multilevel model. It provides a simple, concrete dynamical model of a mooring in a fluctuating current and can be used to define a single parameter to compare the motion characteristics of a wide variety of surface and subsurface moorings.

(From Fofonoff, N. P., *Geo-Marine Technol.*, 1(7), 10, 1965. With permission.)

Nonlinear Response of Buoys

Advances in the art of naval architecture and structural engineering, although of some value to the buoy system engineer, are not really in the area of his greatest concern. Essentially all practical work in fluid dynamics has been concerned with objects which are purposefully moved through the fluid. The fluid field is generally considered to be stationary even though the flow may be turbulent. Ship and airplane hulls are designed to receive flow from a single direction, with allowance for minor variations. Because the magnitude of the purposeful flow is much greater than the natural motions in the fluid, these natural motions can generally be ignored. The bodies may be streamlined or at least made to operate beyond the critical Reynolds number regime.

None of this is true for a moored buoy system. The moored buoy system must operate in the natural field, whatever it might be. Only in the major ocean-current systems can the steady component of flow be considered large in comparison with the unsteady flow. Even here it cannot be relied upon for long periods of time.

Good definitions of energy spectra for surface waves have been developed. These imput spectra have been used in analyzing the motions of ships at sea. However, the approach has generally been statistical, and with good reason. The overall length of ships is greater than their beam and they must be capable of moving in any direction. They must be considered large in relation to a wavelength. Their motions are extremely complex, with coupling between heave, roll, pitch, yaw, surge, and sway. In addition, there are torsional, lateral, and longitudinal oscillations set up in a ship's hull in response to engine torque, slamming, and cantilevering across and between waves.

With a problem this complex, it is necessary to linearize the equations of motion to keep the problem tractable. Essentially all theoretical work with ships and buoys floating on the sea surface have been pursued with the linearized equations. This work has led to an increased understanding of the problems of ship designs, and those who have attempted to analyze buoy motion have naturally borrowed on the work of the ship-motion analyst. In the limit, however, the buoy-system designer's problems are quite different. For example, with few exceptions, buoys are symmetrical about the vertical axis. Most buoys that are not symmetrical have length/beam ratios less than 2. Buoys are not required to travel through the seas in any direction; they are essentially stationary with respect to the passing seas. They can generally be considered small in comparison to a wavelength. Their motions are much less complex than the motions of a ship. This simpler nature of buoys, comforming with the flexibility of modern analog computers, opens up the possibility of performing nonlinear analyses of buoy motions. The problem remains complex, but if care is taken to proceed in a logical stepwise manner, certain tendencies can perhaps be detected and obviated by careful design.

Figure 5.5—2
BUOY SYSTEM

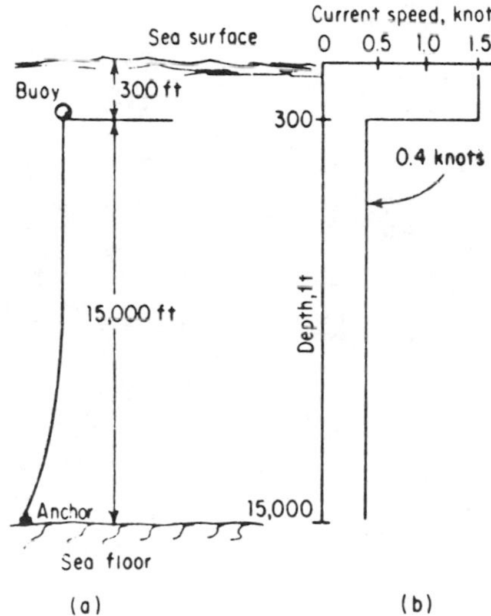

Note: For example: (a) system; (b) current velocity profile (see text for system parameters).

Figure 5.5—3
GEOMETRY FOR COMPUTING
BUOY-SYSTEM MOTION

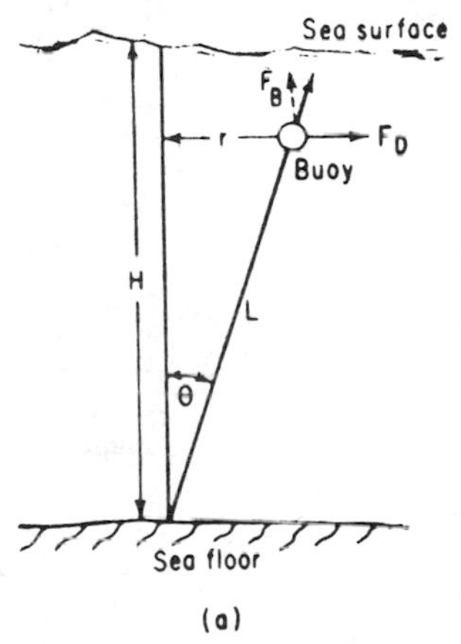

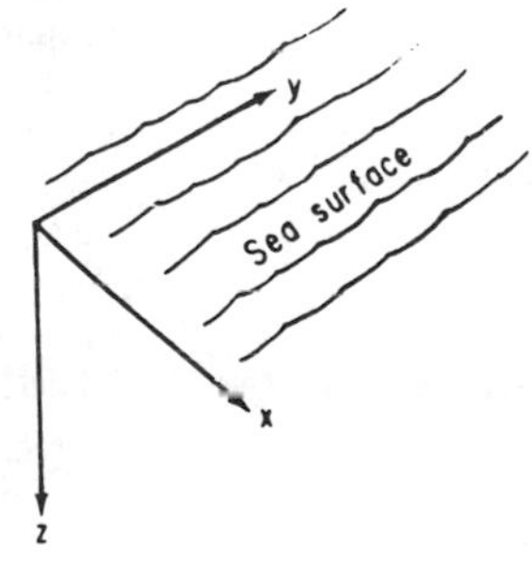

(b)

Note: (a) Buoy system; (b) coordinates.

5.6. SPECIFICATIONS OF OCEANOGRAPHIC INSTRUMENTS

Table 5.6—1
BOTTOM SAMPLERS, DREDGES

Bottom Grab

Manufacturer	Type	Size	Total wt, kg
Hagen	Van Veen	–	–
Hydro-Bios	Van Veen	0.1/0.022 m^2	25/12
Hydrow	Van Veen	0.1 m^2	40/60
Kelvin	Van Veen	14.2 liters	–
Lab. Oceanogr	Van Veen	0.1/0.2 m^2	32/70
GM	Emery dredge (lever type)	2 liters	5
Bergen Nautik	Petersen	0.1/0.2 m^2	41/53
Lab. Oceanogr	Petersen	0.1/0.2 m^2	46/118
GM	Modified Petersen	0.023 m^2	4
Kahl	Modified Petersen	0.023 m^2	4
Mashpribor	Modified Petersen	0.025 m^2	18
Rigosha	Modified Petersen	0.5 liters	5.2
Hydro-Bios	Ekman-Birge	25 × 25 × 30 cm	3.1/5.6
GM	Ekman-Birge	15 × 15 × 15 cm	5.4
Kahl	Ekman-Birge	15 × 15 × 15 cm	5.4
Rigosha	Ekman-Birge	15 × 15/20 × 20 cm	5.2/9.2
Hydro-Bios	Ekman-Birge-Lenz	25 × 25 × 40 cm	5.6
Rigosha	Edman-Birge-Lenz	15 × 15 × 22 cm	7.8
GM	Rectangular box sediment B.S.	0.025/0.1 m^2	40/70
Kahl	Rectangular box sediment B.S.	0.025/0.1 m^2	40/70
Ballant	Mud snapper		
GM	Mud snapper	–	1.5
Kahl	Mud snapper	–	1.5
Rigosha	Snapper (Marukawa)	300 cm^3	9
T.S.K.	Snapper (Marukawa)	300 cm^3	16
	Snapper	–	9
Hydro-Products	Shipek	20 × 20 cm	70
Lab. Oceanogr	Knudsen	0.1 m^2	140
Sepine	Heavy duty	–	27
GM	Heavy duty (Dietz-Lafond)	–	27
Kahl	Dietz-Lafond Heavy Duty	–	27
GM	Orange Peel Dredge	1.65/4.9 liters	20/-
Hayward	Dwarf Orange Peel Bucket	1.6/3.6/4.9/24.5/28.3 liters	
Hytech	Dwarf Orange Peel Bucket	1.6/3.6/4.9/24.5/28.3 liters	16/18/20/86/95
Mécabolier	Benne preneuse à griffe	0.1 m^2	–
GM	Bacterial bottom sampler (Emery)	14 cm diam	–

Table 5.6-1 (continued)
BOTTOM SAMPLERS, DREDGES

Bottom Grab

Manufacturer	Type	Size	Total wt, kg
Kahl	Bacterial bottom sampler (Emery)	14 cm diam	11
EG & G	Photo-grab system (combined with underwater camera)	60 × 100 cm	340

Plummet Bottom Samplers

Manufacturer	Type	Size	Total wt, kg
Rigosha	Single-bowl plummet	60 m liters	5.3
T.S.K.	Single-bowl plummet		

Dredge

Manufacturer	Type	Size: Mouth	Size: Length	Total wt, kg
GM	Pipe dredge	15 cm diam	45 cm	25
Kahl	Pipe dredge	15 cm diam	45 cm	25
GM	Rectangular with net and canvas	25	–	–
Hydro-Bios	Rectangular Lubek (silk net)	30 × 40	25	1.3
Kahl	Rectangular with net and canvas	28	46	11

(From Takenouti, A. Y., reprinted from *International Marine Science,* a quarterly newsletter, Vol 4, 3, 1966, by permission of Unesco and FAO; also in *Handbook of Ocean and Underwater Engineering,* Myers, J. J., Ed., McGraw-Hill, New York, 1969.)

Table 5.6—2
CURRENT METERS

Indicator or Recorder on Instrument

Manufacturer	Type/name	Mode of use	Max depth	Max duration	Method of recording		Remarks
					Speed	Direction	
Bergen Nautik	Ekman C.M.	From ship	No limit	No limit	Dials	Balls and cabinet	
Rigosha	Ekman-Merz C.M.	From ship	No limit	No limit	Dials	Balls and cabinet	
T.S.K.	Ekman-Merz C.M.	From ship	No limit	No limit	Dials	Balls and cabinet	
GM	Gemware C.M.	From ship	No limit	No limit	Dials	Balls and cabinet	Modified Ekman type
Kahl	Gemware C.M.	From ship	No limit	No limit	Dials	Balls and cabinet	Modified Ekman type
Mashpribor	BM-M Modernized C.M.	From ship	No limit	No limit	Dials	Balls and cabinet	Modified Ekman type
T.S.K.	T.S. Multiple C.M.	From ship	No limit	No limit	Dials	Balls and cabinet	For serial observation
Bergen Nautik	Fjeldstad C.M.	From ship	No limit	No limit	Printed on tinfoil		Messengers excluded
GM	Fjeldstad C.M.	From ship	No limit	No limit	Printed on tinfoil		
Kahl	Fjeldstad C.M.	From ship	No limit	No limit	Printed on tinfoil		
Askania	Böhnecke C.M.	From ship	No limit	25 hr	Printed on tinfoil		
Kyowa Shoko	Ono C.M.	From buoy	300 m	15 days	Printed on paper		
Rigosha	Ono C.M.	From buoy	50 m	3 days	Printed on paper		
Mécabolier	Carrentograph, Type E (No. 1031)	From buoy	1000 m	8 days	Printed on paper		
	Carrentograph, Type B (No. 1203)	From buoy	1000 m	20 days	Photographed on 2 × 8-mm film		
T.S.K.	T.S. Self-Direction-Recording C.M.	From ship/ buoy	200 m	30 hr	Printed on paper		

Table 5.6—2 (continued)
CURRENT METERS

Indicator or Recorder on Instrument

Manufacturer	Type/name	Mode of use	Max depth	Max duration	Method of Recording: Speed	Method of Recording: Direction	Remarks
Geodyne	Woods Hole C.M.	From buoy	6000 m	1 yr	Photographed on 16-mm film		
	A850	From buoy	6000 m	No limit	Digital on magnetic tape		Data-processing service available
Plessey	Recording C.M. MO21	From buoy	2000 m	80 days	Binary signal on magnetic tape		Acoustic link permits monitoring
Marine Adviser	Data Acquisition System Q-122	From buoy	6100 m	1 yr	Binary signal on magnetic tape		Temperature also
Mashpribor	Alexejev C.M. Type 2-r	From ship/buoy	250 m	1 mo	Printed on paper		
	Alexejev C.M. Type 2	From ship/buoy	2000 m	1 mo	Printed on paper		
Braincon	Recording C.M. Type 188	From ship/buoy	6000 m	90 days	Photographed on 16-mm film		
T.S.K.	Nanniti C.M. Type 2	From ship	–	–	Indication aboard	Balls and cabinet at the equipment	
	T.S.-N1 Type C.M.	From ship	1000/3000	–	On smoked glass		

(From Takenouti, A. Y., reprinted from *International Marine Science*, a quarterly newsletter, Vol 4, 3, 1966, by permission of Unesco and FAO; also in *Handbook of Ocean and Underwater Engineering*, Myers, J. J., Ed., McGraw-Hill, New York, 1969.)

Table 5.6—3
THERMOGRAPHS, SALINOMETERS

In Situ Salinity, Temperature, and Depth-measuring Instruments

Manufacturer	Salinity, %	Temp, °C	Depth, m	Indicator or recorder	Cable length, m	Remarks
Beckman RS 5-3	0–40 ± 0.3	0–40 ± 0.5	None	Digital readout	400	
Beckman RS 6	0–40 ± 0.2	0–30 ± 0.2	0–130 ± 2.5	Digital readout	130	
Beckman RS 3[a]	32–39.4 ± 0.05	None	None	Recording	30 ea. sensor	Temp corrected 12-points electrode-less salinity recorder
Geodyne	30–40 ± 0.02	−2–35 ± 0.05	0–9000 ± 0.25%	Digital tape, *XY* plot		
GM	0–40 ± 0.3	0–40 ± 0.5	None	Dial	15	
Hydro-Bios	0–40 ± 0.3	−1–30 ± 0.03	–	Indicator	200	
Hytech (model 9006)[b]	30–40 ± 0.01	−2–35 ± 0.05	–	X_1X_2Y recorder		Sound velocity optional
Kahl	0–40 ± 0.3	0–40 ± 0.5	None	Dial	15	
Kjeler	30–40 ± 0.3	−2–35 ± 0.02	0–2500	Meter readout,		
Howaldswerke		–	0–2500 0–2500 0–200 ± 1.5%	*X*-2*Y* recorder, magnetic tape recorder, digital readout, digital printer	Up to 2000	
O.S.K.	27.2–34.3 ± 0.1	0–40 ± 0.2	None	Indicator	20	
Plessey	–	–	–	Digital, magnetic tape, punched card		Sound velocity also
T.S.K.	29–36 ± 0.03	−2–32 ± 0.2	0–100 ± 3%	*XY* recorder	–	
	0–40 ± 0.3	0–40 ± 0.5	None	Indicator	30	
Whitney	–	–	–	Meter readout	–	

[a] Metal-coated staballoy BT slides are supplied by Hytech at $12.50 per box of 50 slides.
[b] Including accessories (grids, viewer, etc.) and 200 smoked-glass slides.

Table 5.6—3 (continued)
THERMOGRAPHS, SALINOMETERS

Bathythermograph[a] (Standard Model)

Manufacturer	Temp range, °C	Depth, m	Remarks
Belfort	−1−30	60/135/250	Fahrenheit-feet scale is available
	−2−32	55/137/274	Fahrenheit-feet scale is available
GM	−2−32	60/137/274	Fahrenheit-feet scale is available
Jules Richard	−2−30	50/150/300	
Kahl	−2−30	60/137/274	Fahrenheit-feet scale is available
Mashprib	−2−30	200	
Mécabolièr	–	1000	
Wallace and Tieman	−1−30	60/135/270	
T.S.K.	−2−32	75/150/270	

Temperature-Depth Recorder (Excluding STD)

Manu-facturer	Model	Temp, °C Range	Temp, °C Accuracy, %	Depth, m Range	Depth, m Accuracy, %	Recording system	Remarks
Askania	6481	0−30	0.2	0−100	1.5	Portable temp-indicating instrument	Mechanical
	6433	0−30	0.15	0−300	5	Dots on chart	Battery 6 V, 6.5 amp-hr
Francis	Expendable BT	−2−30	0.2	0−460	2/5	Analog/digital	Ship speed 0−30 knots; price is for single probe; launcher and recorder on-board cost approx. $5000

[a] Metal-coated staballoy BT slides are supplied by Hytech at $12.50 per box of 50 slides.
[b] Including accessories (grids, viewer, etc.) and 200 smoked-glass slides.

(From Takenouti, A. Y., reprinted from *International Marine Science*, a quarterly newsletter, Vol 4, 3, 1966, by permission of Unesco and FAO; also in *Handbook of Ocean and Underwater Engineering*, Myers, J. J., Ed., McGraw-Hill, New York, 1969.)

Table 5.6—4
WATER-SAMPLING BOTTLES

Closing Water Bottles (No Thermometer Frames Attached)

Manufacturer	Name/type	Capacity, liters	Wt, kg	Material of tube (inside)	Remarks
For serial sampling:					
Hytech	Frautschy	0,5/1	–	Polyvinyl chloride	With valves
Rigosha	Doty's transparent	0,5	2,5	Plastic	
Hydro Products	Van Dorn	1/2/4	1,6/1,8/2,6	Plexiglass	
	Van Dorn	6/8	3,1/3,4	Plastic	
GM	PVC Water Bottle	2/3/6	2,7/3,1/4	Plastic transp.	
Kahl	PVC Water Bottle	2/3/6	2,7/3,1/4	Plastic transp.	
For single sampling:					
Hydro-Bios	Ruttner	0,5/1/2	–	Plexiglass	
Valco	Van Dorn	5	–	Plastic	
T.S.K.	Van Dorn	50	70	Plastic	
Int. Ag. 14	Insulating W.B.	0,375	10	Plexyglass/Sanyl	Modified Petterson
	Insulating W.B.	1	14,5	Plexyglass/Sanyl	Nansen bottles
Kahl	Closing W.B.	0,35	2,3	Tin-plated	
G.M.	Closing W.B.	0,35	2,3	Tin-plated	
Mécabolier	Closing W.B.	4	8	Plastic/chromed	
Rigosha	C.W.B.	10/20	15/18	Brass	
T.S.K.	C.W.B.	5/10/20	–	Plastic	
Lab. Oceanogr	Plankton Sampler	8	18	Nickel-plated	
	Steeman-Nielsen	100	–	Nylon and PVC	
Valco	Jitts Twin Sampler	0,35 × 2	–	Plastic	Transparent and brass-sheathed
T.S.K.	T.S. "TOMEI" T.R.W.B.	0,5/2	3,5/6,7	Plastic	
	T.S. "TOMEI" Kitahara W.B.	1	–	Plastic	

Table 5.6—4 (continued)
WATER-SAMPLING BOTTLES

Manufacturer	Name/type	Capacity, liters	Wt, kg	Material of tube (inside)	Remarks
Horizontal closing water bottle:					
Hydro-Bios	Horizontal C.W.E.	–	16,8	–	With frame for reversing therm
Mécabolier	Horizontal C.W.B.	1/2/4	12/24/48	–	
Rigosha	Horizontal W.B.	0,5	4	Brass	Open and close with 2 mess. up to 50 m
T.S.K.	Horizontal W.B.	0,7/1,2	7,3/10	–	Open and close with piston by 1 mess.

Sterile Water Sampler

Manufacturer	Name/type	Capacity, ml	Material of ampule	Remarks
GM	Zo Bell sampler	230/250	Rubber/glass	
Kahl	Zo Bell sampler	230/250	Rubber/glass	
T.S.K.	Zo Bell type	200	Rubber	
Mécabolier	Reversing	250	–	
Hytech	ABC sampler	100	Pyrex	
Hydro Products	Cobet sampler	110/230	Neoprene	For Kit, including 1 mechanism and clamp assembly, 12 flint-sealed capillary tubes, one 4 oz bulb and inlet tube, one 8 oz bulb and inlet tube

(From Takenouti, A. Y., reprinted from *International Marine Science,* a quarterly newsletter, Vol 4, 3, 1966, by permission of Unesco and FAO; also in *Handbook of Ocean and Underwater Engineering,* Myers, J. J., Ed., McGraw-Hill, New York, 1969.)

5.7 SHIP CHARACTERISTICS

Table 5.7-1
OCEANOGRAPHIC-RESEARCH VESSELS

Characteristics	*H. V. Sverdrup*	AGOR-3 class	*Atlantis II*	*Catamaran* (proposed)
Length overall	127′-7″	208′-0″	209′-9″	141′-0″
Length at waterline	111′-6″	196′-0″	195′-0″	130′-0″
Beam	24′-11″	37′-0″	44′-0″	52′7″ max 17′-6″ ea. hull
Draft	–	14′-3″	16′-0″	8′-0″
Full-load displacement, tons	400	1373	2110	640
Gross tons	295	–	1100	
Coefficients				
Block	–	0.424	0.537	
Longitudinal	–	0.530	0.614	
Midship	–	0.800	0.875	
Engine	Diesel[a]	Diesel-elect.	Uniflow-steam	Diesel
Shp	600	1000	1400	950
Trial speed, knots	11.5	13	13	13.5
Endurance speed, knots	10	12	12	12
Endurance				
Nautical mi.	5000	12,000	8000	5000
Days	28	45		
Ship-service-generator capacity, kW	–	600	300	
Accommodations				
Officer	4	8	9	
Crew	5–7	14	19	16
Scientists	10–8	15	25	15
Laboratories, ft²	Oceanographic 65 Sound recording 160 General electronic 375	Wet 250 Dry 800	Below main deck 458 On main deck 1010 Above main deck, 1010	Main deck 1160 Below main deck 120

[a] Controllable reversible-pitch propeller.

Table 5.7-1 (continued)
OCEANOGRAPHIC-RESEARCH VESSELS

Characteristics	*H. V. Sverdrup*	AGOR-3 class	*Atlantis II*	*Catamaran* (proposed)
Winches, wire or line-haul capacity	Deep-sea anchoring 5000 m × 12 mm Hydrographic (2) 5500 m × 4 mm Cable 7000 m × 6 mm Cargo 3-ton Crane 1.5-ton Towing 4-ton Boat 1.5-ton	Deep sea 6800 lb @ 600 ft/min, 30,000 lb @ 133 ft/min, 60,000 lb static Hydrographic 2000 lb @ 350 ft/min, 4000 lb static (2)	Deep sea 30,000 ft × ½ in. Towed instrument 1800-ft chain Hydrographic 30,000 ft × 3/16 in. (2)	Trawl Hydrographic (2)
Special features	All winches hydraulically operated; 360 ft^3 explosive storage	300 kW gas-turbine–electric generator for quiet operation; 10 ton crane @ 30 ft outreach, 3 ton crane @ 50 ft outreach; passive antirolling tanks; bow thruster	5 ton crane; 1 ton cranes (2); passive antirolling tanks; bow thruster; internal well; bow observation chamber	5 ton cranes (2); centerwells (3)

(From Miller, R. T., in *Handbook of Ocean and Underwater Engineering*, Myers, J. J., Ed., McGraw-Hill, New York, 1969. With permission.)

Table 5.7—2
TYPICAL SHIP CONVERSIONS FOR OCEANOGRAPHIC-RESEARCH VESSELS

Characteristics	*Paolina T.*, converted purse seiner	*Crawford*, converted Coast Guard cutter	*Atlantis*, ketch	*Horizon*, converted tug	*Chain*, converted rescue tug	*AMS-AGS*, converted minesweeper
Length overall, ft	80.25	125.0	143.25	143.0	213.5	220.92
Length bet. perp., ft	–	120.0	105.0	–	207.0	214.83
Beam, ft	22.0	24.0	28.0	33.0	41.25	31.92
(Mean) draft, ft	9.75	8.15	19.0 (max)	16.42	15.0	10.08
Full-load displacement, tons	110	304	560	768	2051	1220
Engine	Diesel	Diesel	Diesel	Diesel-elect.	Diesel-elec.	Diesel-elec.
Shp	250	800	400	1500	3000	3532
Speed, knots	9	12	8	11.5	12	16
Endurance						
Nautical mi.	2450	3000	2700	7000	10,500	7650
Days	30	90	120	45	75	30
Accommodations						
Officers	3			7	10	9
Crew	6	15	20	11	19	96
Scientists	5	7	9	18	26	4
Laboratory area, ft^2	90	300	430	600	1637	309
Special features	Hydrographic winch used for cable laying		Auxiliary ketch	Dredge winch, 22,000-ft cable	Thermistor chain winch; 5 ton crane: dredge winch; silent ship capability	Oceanographic winch

(From Miller, R. T., in *Handbook of Ocean and Underwater Engineering,* Myers, J. J., Ed., McGraw-Hill, New York, 1969. With permission.)

Section 6
Phytoplankton

Section 6

PHYTOPLANKTON

GENERAL CHARACTERISTICS

Plankton of the ocean may be defined as floating or drifting organisms with limited powers of locomotion that are transported primarily by prevailing water movements. Subdivisions of the plankton include bacterioplankton (bacteria), phytoplankton (plants), and zooplankton (animals). Phytoplankton are free-floating, microscopic plants (unicellular, filamentous, or chain-forming species) inhabiting surface waters (photic zone) of open oceanic and coastal environments. Although unicellular forms comprise the bulk of phytoplankton, some green and blue-green algae are filamentous (i.e., they develop thread-like cell systems). Colonial diatoms and the blue-green alga, *Trichodesmium thiebautii* provide examples.[1] In addition, a number of diatoms and dinoflagellates produce chains of loosely associated cells.[2] Not all pelagic photosynthetic organisms are microscopic. For example, macroscopic multicellular brown algae, *Sargassum* spp., yield substantial biomass in the Sargasso Sea.

Despite being composed of single cells or of relatively simply organized, small colonies, phytoplankton encompass a rather wide diversity of algal groups.[3] These diminutive autotrophs, which are largely holoplanktonic, serve a major function in the world's oceans, being responsible for at least 90% of the photosynthesis, with the remaining 10% attributable mainly to benthic macroalgae and vascular plants (e.g., seagrasses, salt marsh grasses, and mangroves) in intertidal and shallow subtidal environments.[4] Because the ocean covers approximately 72% of the surface of the earth, phytoplankton as a group are the most important primary producers on the planet. They play a vital role in initiating the flow of energy in a useful form through oceanic ecosystems.

Among classification schemes, phytoplankton are commonly categorized on the basis of size into four classes: (1) ultraplankton (less than 5 μm in diameter); (2) nanoplankton (5 to 70 μm); (3) microphytoplankton (70 to 100 μm); and (4) macrophytoplankton (greater than 100 μm). More than half of all phytoplankton belong to the ultraplankton and nanoplankton. For practical purposes, phycologists often sort the microscopic plants into net plankton and nanoplankton, determined by the nominal aperture size of the plankton net deployed in the field. Net plankton samples embody all phytoplankton retained by the finest nets (about 64 μm apertures) that can be conveniently towed, and in coastal waters, tend to be dominated by diatoms and dinoflagellates. The nanoplankton pass through the fine-mesh nets; large numbers of coccolithophores, other flagellates such as those of the Chrysophyceae and Cryptophyceae, and small species of diatoms are the most abundant components.[5] Because nets are now available with mesh dimensions of 20 to 30 μm, some workers recommend limiting the term nanoplankton to all planktonic algae less than 30 μm in size.

TAXONOMY

Diatoms (class Bacillariophyceae), dinoflagellates (class Dinophyceae), coccolithophores (class Prymnesiophyceae), silicoflagellates (class Chrysophyceae), and blue-green algae (class Cyanophyceae) constitute the principal taxa of planktonic producers in the ocean.[6] In estuarine or enclosed bodies of water, other taxonomic assemblages may also be locally important, for example, the green-colored algae (class Chlorophyceae), brown-colored phytoflagellates (class Haptophyceae), and euglenoid flagellates (class Euglenophyceae). The classification of phytoplankton relies on studies of fine cell structure, life-history investigations, and biochemical research.

Algalogists utilize certain cell characteristics to identify phytoplankton. Chief among these are the cell shape, cell dimensions, cell wall, mucilage layers, chloroplasts, flagella, reserve substances (e.g., starch, oil, and leucosin), and other cell features (e.g., cell vacuoles and trichocysts).[2,7] Advances in electron microscopy have greatly aided investigators in elucidating the fine structure of cell constituents, which has facilitated taxonomic work. Nevertheless, numerous species, particularly among the nanoplankton, are poorly understood; consequently, revisions to the classification of phytoplankton continually take place.

DIATOMS (CLASS BACILLARIOPHYCEAE)

These microalgae dominate phytoplankton communities in high latitudes of the Arctic and Antarctic, in the neritic zone of boreal and temperate waters, and in areas of upwelling (both coastal and equatorial).[8,9] Many specialists regard diatoms as the most important phytoplankton group, contributing substantially to oceanic productivity, especially in coastal waters. Consisting of a single cell or cell chains, diatoms secrete an external rigid silicate skeleton (pectin impregnated with silica) called a frustule, which encases the vegetative protoplast. The frustule or silica cell wall is composed of two valves, the epitheca and hypotheca; the epitheca overlaps the hypotheca in the girdle region, where they link via pectinaceous bands or minute teeth. The valves, therefore, fit together like a petri dish, and they often contain highly ornamented sculptured markings. The complex ornamentation may be differentiated into four types, namely puncta, areolae or alveoli, canaliculi, and costae.[3] Cavities or thin areas of the frustule expedite the exchange of nutrients, gases, and metabolic products across the cell wall.[4]

The shape and symmetry of the frustule assist taxonomists in the classification of diatoms. Based on these features, two orders are recognized, that is, centric diatoms (Centrales or Centricae), having circular or dome-shaped valves and a predominantly planktonic existence, and pennate diatoms, possessing oblong or "boat-shaped" valves and a mainly benthic habit. The frustule of centric diatoms has radial symmetry about the pervalar axis, that of pennate diatoms, bilateral symmetry with reference to the apical plane.[8,9] In pennate diatoms, a long slit, the raphe, may extend to both valves and connect to thickenings known as nodules.[10] By protruding their cytoplasm through the raphe and making contact with the substrate, benthic diatoms creep along the seafloor.[4] Most marine diatoms are planktonic, and of the 10,000 species of diatoms that have been identified, about half live in marine environments.

Diatoms range in size from less than 10 μm to approximately 200 μm.[6,10] Bearing no flagella, cilia, or other organs of locomotion, the planktonic species are nonmotile and sink in nonturbulent waters. According to Smayda,[11] sinking rates of diatoms and other phytoplankton depend on cell size and shape, colony dimensions, physiological condition, and age. Live diatom cells descend at a rate of 0 to 30 m/day through the water column, but dead cells fall more quickly, exceeding 60 m/day in some cases.[3] Buoyancy declines with age, resulting in greater sinking rates for senescent populations. As the cell or colony size increases, so does the sinking rate, with the ratio of surface area to volume being a key factor in the sinking process.

Smayda[11] postulated several morphological, physiological, and physical adaptations which foster the suspension of diatoms. Flotation of nonmotile forms may be enhanced by modification of the frustule. For instance, needle-like extensions, spines, and siliceous projections increase the frustule surface area, thereby mitigating sinking rates. Among physiological mechanisms, ion regulation in the cytoplasm and vacuole, gas production via photosynthesis, and secretion of mucilage promote diatom flotation.[1] Water viscosity and vertical circulation in the ocean are two physical factors considered to be critical to diatom suspension.

Cell constituents also have been implicated in the sinking of phytoplankton populations. Diatoms with large concentrations of lipids sink faster than those with small concentrations. Since aged individuals accumulate higher lipid concentrations, they generally sink more rapidly. Higher silica content of the cell walls raises density levels, inducing settlement. The amount of silicon in diatoms varies markedly with species, with estimates ranging from 4 to 50% of the dry weight.[3]

The storage products of diatom photosynthesis are chrysolaminarin, a polysaccharide, and lipids. The lipids, which appear as tiny oil droplets dispersed in the cytoplasm, reportedly generate oily patches in some marine waters during final stages of a diatom bloom. The lipid level in the cell is a function of environmental conditions and the species composition. Because of these storage compounds, diatoms furnish zooplankton with high-energy ration.

Most diatoms reproduce by vegetative cell division, leaving two daughter cells of smaller size than the parent cell. The vegetative stage is the only one commonly observed in many species. Vegetative cells may exist independently or develop into distinct colonies.[9] During cell fission, the two valves of the frustule separate slightly, and each retains the products of a mitotic division. Subsequently, each daughter cell secretes a new hypotheca. Due to this type of asexual reproduction executed over several cell generations, the mean size of a population of diatoms gradually diminishes. When a species attains a lower limit in size, restoration of maximum cell size occurs through sexual reproduction and the formation of an auxospore, which then secretes a large-sized frustule. A few diatom species do not experience a progressive decline in cell size, but seemingly maintain constancy of dimensions. This method of asexual reproduction accounts for the large variability in the size of a population of diatoms of a given species, the smallest being up to 30 times more diminutive than the largest.[4]

Eppley[12] mentioned optimal cell doubling rates in diatoms equal to 0.5 to 6 doublings per day, which allow blooms to develop. Numerous centric diatoms produce resting spores during unfavorable conditions.[13] Resting cells, similar to vegetative cells, also are found in some species, particularly those in temperate waters.[14] As Steidinger and Walker[14] assert, asexual resting spores and resting cells (dormant stages) may be important in species occurrence, species succession, and distribution patterns, as well as in population survival.

DINOFLAGELLATES (CLASS DINOPHYCEAE)

Dinoflagellates typically are unicellular, biflagellated, autotrophic forms, and, like diatoms, supply a major fraction of primary production in many regions. Individual dinoflagellate cells range from about 5 μm to greater than 200 μm in size, but some species (e.g., *Polykrikos* spp.) often grow in larger chains or pseudocolonies.[15] Widely distributed in marine and estuarine environments, dinoflagellates often dominate phytoplankton communities in subtropical and tropical waters. In addition, they are usually abundant in temperate and boreal autumn assemblages.[8] Some 1000 to 1500 species of dinoflagellates occur in marine and freshwater habitats;[15] however, most of them (more than 90%) are marine.[4]

Lacking an external, siliceous skeleton, many dinoflagellates bear an armor of cellulose thecal plates. This group, the Peridinales, can be readily distinguished with a light microscope from the unarmored or naked forms, the Gymnodiniales, which only have a firm periplast. Representative genera of the Peridinales include *Ceratium, Gonyaulax*, and *Peridinium*; representative genera of the Gymnodiniales are *Amphidinium, Ptychodiscus (Gymnodinium)*, and *Gyrodinium*.

Two unequal flagella (longitudinal flagellum and transverse flagellum) lie in separate grooves on the body surface of most dinoflagellates, propelling and stabilizing the organisms. The two grooves are: (1) the girdle or annulus, a transverse groove that surrounds the cell and divides it into two subequal parts (epicone and hypocone) and; (2) the sulcus, a longitudinal furrow passing along the posterior end of the cell. Thecate forms may have a cingulum or perforated plate covering the annulus.

Asexual reproduction via binary cell division is the normal means of reproduction in dinoflagellates, although some individuals reproduce sexually as well (e.g., *Ceratium* and *Glenodinium*). Simple binary fission in dinoflagellates does not translate into a progressive reduction in size of successive generations as it does in diatoms. The rates of cell division of these plants are comparable to those of diatoms, but can be highly variable contingent upon environmental conditions. Hence, *Peridinium* has a doubling rate of 10 to 50 h; *Exuviaella*, a doubling rate of 15 to 90 h; and *Prorocentrum*, a doubling rate of 12 to 127 h.[3]

The majority of dinoflagellates contain chloroplasts with chlorophyll *a* and *c* and other pigment cells used in photosynthesis. Starch and lipids provide food reserves. Heterotrophy (e.g., *Kofoidinium* and *Polykrikos*) is not rare, and various species feed on phytoplankton and zooplankton. A number of others consume decaying organic matter (holozoic) or obtain nutrition through parasitic activity (e.g., *Blastodinium* and *Oodinium*). The zooxanthellae enter into a symbiotic relationship with the giant tropical clam, *Tridacna*, and with corals. Various species also are symbiotic with some sea anemones, radiolarians, and fish.

Certain dinoflagellate species generate toxins. When blooms of these microflora reach high densities from 5×10^5 to 2×10^6 cells/l,[1] the cumulative effect of the toxins liberated periodically induces mass mortality of fish, shellfish, and other organisms. Blooms of dinoflagellates occasionally impart a red or brown color to the water causing the so-called "red tide". The genera *Gonyaulax* and *Ptychodiscus* are responsible for occurrences of toxic red tides in estuaries, two noteworthy species promoting this phenomenon being *Gonyaulax polyhedra* and *Ptychodiscus brevis* (= *Gymnodinium breve*). Neurotoxin released by dinoflagellate cells of a red tide can kill fish directly as the cells pass through the gills. In addition, shellfish populations, which accumulate the toxic agent saxitoxin in their siphons and hepatopancreas by filtering the dinoflagellates from the water, cause a neurological disorder, paralytic shellfish poison, in humans who consume them.[16,17] Conditions favoring the success of dinoflagellates and the development of red-tide blooms include the sudden input of nutrients by upwelling, washout of nutrients from land, influx of vitamins from shore, and tidal turbulence.[8,18,19]

Some dinoflagellates are bioluminescent, lighting surface waters at night. The genera *Gymnodidium, Noctiluca,* and *Pyrocystis* provide excellent examples. A few, such as the genus *Noctiluca*, while experiencing bioluminesence, are not photosynthetic.[4,20] Spector[21] treats the subject of dinoflagellates comprehensively, and additional information on these microflora can be obtained from his volume.

COCCOLITHOPHORES AND OTHER BROWN-COLORED ALGAE (CLASS HAPTOPHYCEAE)

The coccolithophores are biflagellate, unicellular algae covered by calcareous plates, called coccoliths, embedded in a gelatinous sheath surrounding the cell.[22] These small phytoplankters (5 to 50 µm in size) reach peak abundance in tropical and subtropical, open-ocean waters, but sometimes also proliferate in coastal environments. Although most species live in warmer seas, a few (e.g., *Pontosphaera huxleyi* and *Syracosphaera* spp.) attain maximum abundance in colder regions. The major fraction of photosynthesis in certain areas is attributable to coccolithophores, which periodically comprise a significant portion of calcareous sediment (i.e., *Globigerina* ooze) on the seafloor. Despite a predominantly autotrophic existence, a few taxa obtain energy heterotrophically below the photic zone.

Other brown-colored algae of this class show a range of body form from unicellular motile species, palmelloid colonial structure, or filamentous microscopic colonies.[3] The brown-colored algae typically possess body scales; however, genera such as *Diacrateria* and *Isochrysis* do not. When present, the scales are unmineralized. Body scales can be used to distinguish members of the Haptophyceae from those of the Chrysophyceae. Most individuals fall in the nanoplanktonic size range and have paired, smooth flagella generally of equal length.[2] They occur in inshore as well as offshore waters.[8]

SILICOFLAGELLATES (CLASS CHRYSOPHYCEAE)

These planktonic microflora are single-celled, uniflagellated or biflagellated organisms, usually less than 30 µm in diameter. They secrete an internal skeleton composed of siliceous spicules. Whereas the majority of this class consist of photosynthetic plants, some may be heterotrophic. Silicoflagellates can outnumber all other phytoplankton in temperate marine waters, but are most abundant in cold nutrient-rich environments, where they reproduce by simple cell division.

BLUE-GREEN ALGAE (CLASS CYANOPHYCEAE)

Representatives of this class differ from all other algae in being prokaryotic. The organisms have a relatively simple construction, lacking organized nuclei, nuclear membranes, and chromosomes. They are photosynthetic with chitinous walls. In addition to chlorophyll *a*, blue-green algae contain phycobilins and carotenoids, responsible for the varied color in different species. The pigment phycocyanin causes the blue-green color in many individuals of the group. The coloration of the Red Sea has been ascribed, in part, to the presence of *Trichodesmium erythraeum*.

Blue-green algae are frequently encountered in shallow, nearshore tropical seas, but appear in low densities in nearly all regions. Occasionally, they build blooms, especially red tides, in brackish or nearshore habitats. A favored location for blue-green algae is the salt-marsh biotope.

The size of blue-green algae ranges from less than 1 µm for single-celled forms to more than 100 µm for filamentous types. Pelagic Cyanophyceae encompasses species of *Haliarachne, Katagnymene, Oscillatoria,* and *Trichodesmium*.[10] Benthic species often construct mats on mud flats which break free from the substratum and float on the surface of advancing tidal waters. Blue-green algae reproduce by simple cell division, by fragmentation, and by nonmotile spores.[3]

GREEN-COLORED ALGAE (CLASS CHLOROPHYCEAE)

The true green algae exist as unicellular, filamentous, or colonial forms. Both flagellated and nonflagellated cells have been documented. Generally of ultraplanktonic or nanoplanktonic size, these autotrophs can be quite important in estuaries and enclosed seas, producing blooms in both ecosystems from time to time, particularly in late summer and fall.[8]

EUGLENOID FLAGELLATES (CLASS EUGLENOPHYCEAE)

This class characteristically consists of unicellular, biflagellated forms, with a fusiform or cylindrical shape. The flagella emerge from an anterior invagination. Members may be pigmented or colorless, and many species obtain their nutrition saprophytically. Those that photosynthesize normally store paramylum (paramylon), a starch-like carbohydrate.[3] Binary fission is the usual mode of reproduction. Euglenoids, although largely freshwater, also are common in estuaries, especially in somewhat polluted systems.[2,3]

SPECIES DIVERSITY

GENERAL PATTERNS

When transgressing from high to low latitudes or from neritic to oceanic waters, the species diversity of phytoplankton changes. Fewer species occupy high latitudes, but here, a tendency exists for greater numbers of each species and higher total biomass. Similarly, nearshore, the absolute abundance of a species and the total biomass often surpass those of oceanic waters. However, fewer species are observed in the neritic environment.

A trend in species diversity of phytoplankton also can be seen in the vertical plane of the photic zone of open, oceanic waters. A lower number of species and higher abundance of individuals are evident near the surface. With increasing depth in the photic zone, species diversity rises, but fewer numbers of each species can be found.[4]

SAMPLING METHODS

A number of publications deal with methods of phytoplankton sampling.[1,23-25] Plankton nets, plankton pumps, and water bottles represent the three most frequently used types of gear for collecting phytoplankton field samples. Plankton nets, typically with mesh openings ranging from 20 to 64 μm, selectively capture nanoplankton and larger phytoplankters. Fragile forms often suffer cell damage during sampling. Plankton pumps permit a continuous stream of water to be pumped from a desired depth to the surface, where phytoplankton can be concentrated by filtration. Therefore, this method allows phytoplankton samples to be taken at any depth through the water column. Water bottles or closing samplers capture phytoplankton by closing automatically at a selected depth when activated by a solenoid hooked to a pressure-sensing device or by a messenger released from the surface to the sampler.[1] The nansen bottle, niskin sampler, and van dorn sampler are three commonly used closing samplers.

Phytoplankton samples may be stored live for several hours in a refrigerator or ice chest. The samples usually are preserved in a buffered fixative, such as Lugol's solution (10 g of iodine and 20 g of potassium iodide dissolved in 200 ml of distilled water and 20 g of glacial acetic acid), which is added in a ratio of 1 part to 100 parts of seawater sample. Another fixative incorporates a solution of formalin neutralized with hexamine (200 g of hexamine [hexamethylenetetramine] added to 1l of commercial 40% [37% formaldehyde] formalin). This solution is filtered after about 1 week.[1]

PLAN OF THIS SECTION

The major objective of this section is to provide useful data on oceanic phytoplankton, particularly in respect to the biomass, taxon diversity, and chemical composition of these important primary producers. Information contained in tables and figures assesses phytoplankton biomass as related to area, depth, and season. Taxon diversity, as a function of depth and area, also is covered. Physical, chemical, and biological factors affecting phytoplankton populations in the ocean can be found in Section 7, together with data on primary productivity of the microflora.

REFERENCES

1. **Dawes, C. J.**, *Marine Botany,* John Wiley & Sons, New York, 1981.
2. **Boney, A. D.**, *Phytoplankton*, Edward Arnold, London, 1975.
3. **Raymont, J. E. G.**, *Plankton and Productivity in the Oceans*, 2nd ed., Vol. 1, Pergamon Press, Oxford, 1980.
4. **Thurman, H. V. and Webber, H. H.**, *Marine Biology,* Charles E. Merrill, Columbus, Ohio, 1984.
5. **Mann, K. H.**, *Ecology of Coastal Waters: A Systems Approach*, University of California Press, Berkeley, 1982.
6. **Valiela, I.**, *Marine Ecological Processes,* Springer-Verlag, New York, 1984.
7. **Dodge, J. D.**, *The Fine Structure of Algal Cells*, Academic Press, New York, 1973.
8. **Levinton, J. S.**, *Marine Ecology*, Prentice-Hall, Englewood Cliffs, N. J., 1982.
9. **Garrison, D. L.**, Planktonic diatoms, in *Marine Plankton Life Cycle Strategies*, Steidinger, K. A. and Walker, L. M., Eds., CRC Press, Boca Raton, Fla., 1984, 1.
10. **McConnaughey, B. H., and Zottoli, R.**, *Introduction to Marine Biology,* C. V. Mosby, St. Louis, 1983.
11. **Smayda, T. J.**, The suspension and sinking of phytoplankton in the sea, *Oceanogr. Mar. Biol. Ann. Rev.*, 8, 853, 1970.
12. **Eppley, R. W.**, The growth and culture of diatoms, in *The Biology of Diatoms*, Werner, D., Ed., Blackwell Scientific, Oxford, 1977, 24.
13. **Hargraves, P. E.**, Studies on marine planktonic diatoms. II. Resting spore morphology, *J. Phycol.*, 12, 118, 1976.
14. **Steidinger, K. A. and Walker, L. M.**, Introduction, in *Marine Plankton Life Cycle Strategies*, Steidinger, K. A. and Walker, L. M., Eds., CRC Press, Boca Raton, Fla., 1984.
15. **Walker, L. M.**, Life histories, dispersal, and survival in marine, planktonic dinoflagellates, in *Marine Plankton Life Cycle Strategies*, Steidinger, K. A. and Walker, L. M., Eds., CRC Press, Boca Raton, Fla., 1984, 19.
16. **Steidinger, K. A.**, Phytoplankton ecology: a conceptual review based on eastern Gulf of Mexico research, *CRC Crit. Rev. Microbiol.*, 3, 49, 1973.
17. **Steidinger, K. A.**, A re-evaluation of toxic dinoflagellate biology and ecology, in *Progress in Phycological Research*, Vol. II, Round, F. and Chapman, D., Eds., Elsevier North-Holland, New York, 1983, 147.
18. **Hutner, S. H. and McLaughlin, J. J. A.**, Poisonous tides, *Sci. Am.*, 199, 92, 1958.
19. **Steidinger, K. A. and Ingle, R. M.**, Observations of the 1971 summer red tide in Tampa Bay, Florida, *Environ. Lett.*, 3, 271, 1972.
20. **Nybakken, J. W.**, *Marine Biology: An Ecological Approach*, Harper & Row, New York, 1982.
21. **Spector, D. L., Ed.**, *Dinoflagellates*, Academic Press, Orlando, Fla., 1984.
22. **Gross, M. G.**, *Oceanography: A View of the Earth*, 3rd ed., Prentice-Hall, Englewood Cliffs, N. J., 1982.
23. **Parsons, T. R. and Takahashi, M.**, *Biological Oceanographic Processes*, Pergamon, New York, 1973.
24. **Stein, J. R., Ed.**, *Handbook of Phycological Methods: Culture Methods and Growth Measurements*, Cambridge University Press, London, 1973.
25. **UNESCO**, A Review of Methods Used for Quantitative Phytoplankton Studies, *UNESCO Tech. Pap. Mar. Sci.*, 18, 1974.
26. **Hellebust, J. A. and Craigie, J. S.**, *Handbook on Phycological Methods: Physiological and Biochemical Methods*, Cambridge University Press, London, 1978.

6.1. BIOMASS AS RELATED TO AREA, DEPTH AND SEASON

Table 6.1—1
VERTICAL AND SEASONAL DISTRIBUTIONS OF CHLOROPHYLL *a* AND PRIMARY PRODUCTIVITY AT SELECTED STATIONS OFF THE OREGON COAST — 1961

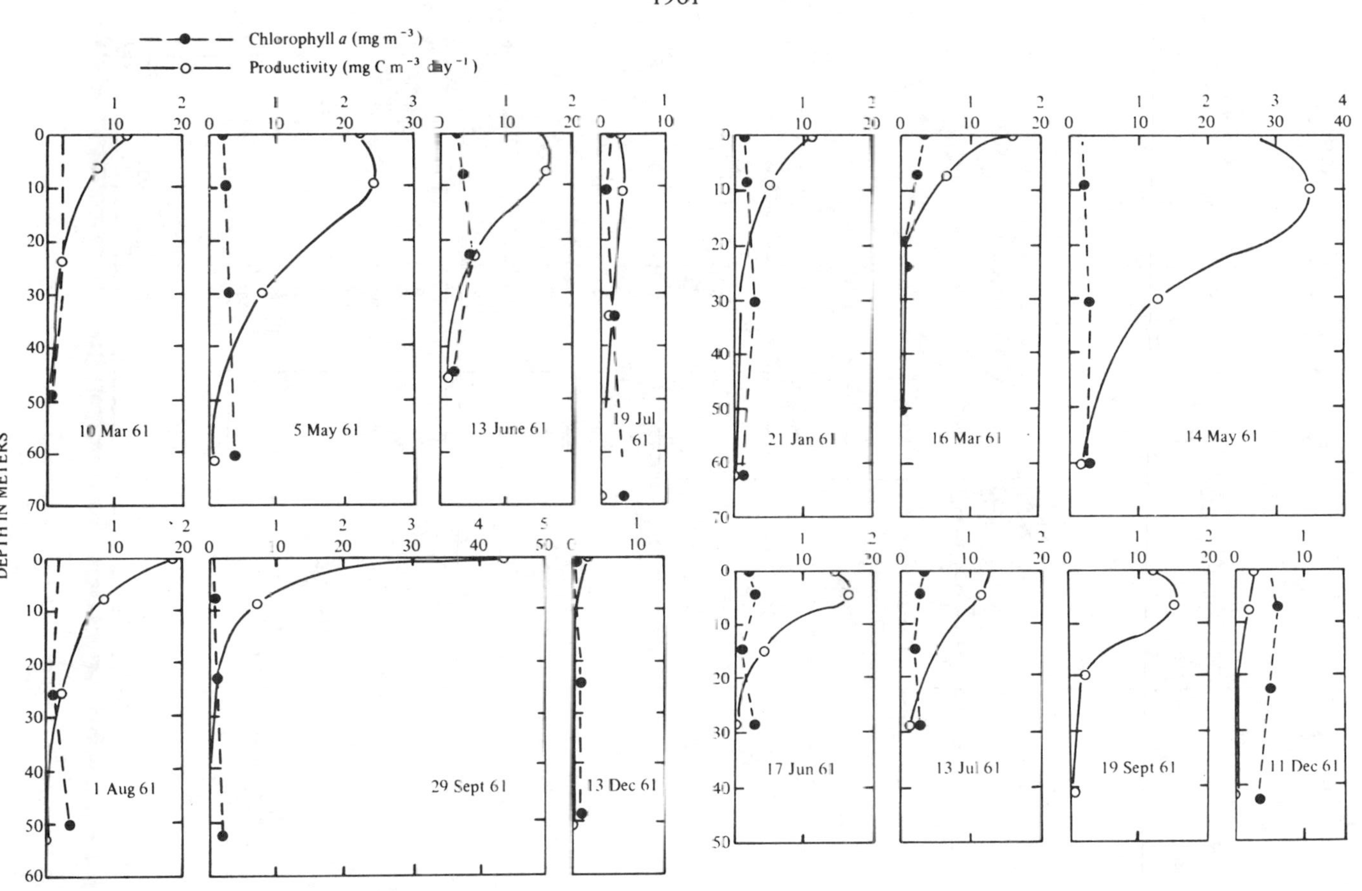

(From Anderson, G. C., The seasonal and geographic distribution of primary productivity off the Washington and Oregon coasts, *Limnol. Oceanogr.*, 9, 294, 1964. With permission.)

Figure 6.1—1

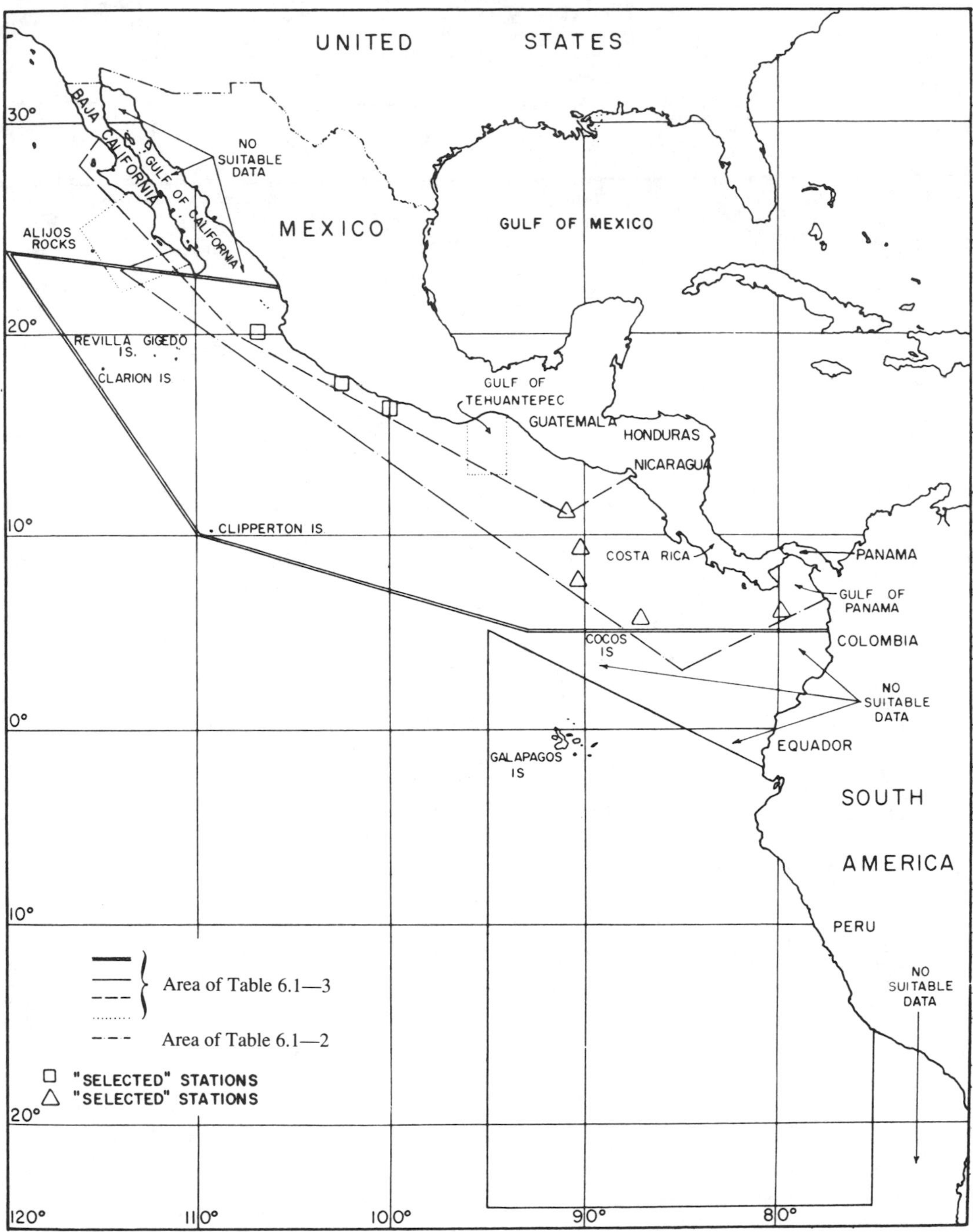

Figure 6.1—2. Sampling stations for Tables 6.1—2 and 6.1—3.

(From Blackburn, M., Relationships between standing crops at three successive trophic levels in the eastern tropical Pacific, *Pac. Sci.*, 20, 41, 1966.)

Table 6.1—1
MEASUREMENTS OF STANDING CROPS OF CHLOROPHYLL a[1] AND ZOOPLANKTON[2] AT OCEANOGRAPHIC STATIONS IN THE EASTERN TROPICAL PACIFIC[3]

Date	Chlorophyll	Zooplankton
November, 1955	24	193
	22	164
	25	143
Costa Rica Dome, November–December, 1959	20	84
	22	250
	8.8	120
	9.6	43
	12	100
	10	190
	9.4	140
	20	70
	8.7	120
	11	51
	14	56
	12	59
November–December, 1956	24	49
	25	37
	43	32
	44	54
	51	85
	74	314
	60	125
	24	95
	27	250
	36	135
	45	166
	70	104
	32	95
	33	139
	32	114
	27	96
	59	233
	30	33
	33	77
	49	47

[1] Mg/m^2.
[2] $Ml/10^3\ m^3$.
[3] In water columns or layers to about 100 m and 300 m, respectively.

(From Blackburn, M., Relationships between standing crops at three successive trophic levels in the eastern tropical Pacific, *Pac. Sci.*, 20, 40, 1966. With permission.)

Table 6.1—2
MEASUREMENTS OF STANDING CROPS OF CHLOROPHYLL a[1], ZOOPLANKTON[2], CARNIVOROUS MICRONEKTON[2] AND COPEPODS[2] AT OCEANOGRAPHIC STATIONS IN THE EASTERN TROPICAL PACIFIC[3]

Noon chlorophyll	Noon zooplankton	Night zooplankton	G.M. zooplankton	Night micronekton	Noon copepods	Night copepods	G.M. copepods
April 23–June 20, 1958							
9.7	23	24	23	2.6	0.36	1.51	0.73
6.4	18	21	19	1.0	0.77	0.96	0.86
14	10	20	14	3.3	0.51	0.89	0.67
10	13	31	20	2.8	0.64	1.50	0.98
16	30	50	39	6.7	2.46	1.42	1.87
13	79	57	67	1.0	1.89	4.52	2.92
12	49	67	57	4.4	1.67	1.94	1.80
15	72	100	85	8.5	3.22	5.45	4.19
12	85	84	84	4.0	7.00	6.75	6.87
14	86	56	69	10.0	1.74	7.30	3.56
22	63	87	74	9.5	3.87	4.25	4.06
46	100	163	131	13.9	7.85	15.06	10.87
39	270	299	284	19.1	30.96	11.90	19.20
36	85	101	93	6.5	3.63	1.82	2.57
81	74	79	76	4.1	4.57	3.97	4.26
41	113	148	129	7.5	5.28	3.05	4.01
29	30	104	91	5.9	2.28	6.16	3.74
45	90	160	120	10.0	2.23	7.05	3.96
39	130	150	140	7.9	6.63	5.49	6.03
50	114	182	144	16.7	3.66	6.88	5.02
120	117	206	155	15.5	4.16	10.30	6.55
49	74	111	91	24.5	0.85	6.12	2.28
27	108	84	95	15.4	3.74	5.81	4.66
26	127	143	135	12.7	5.77	4.26	4.96
11	48	48	48	16.7	2.51	4.72	3.43
23	107	122	114	13.8	6.31	7.38	6.82
37	44	52	48	14.5	2.61	4.68	3.49
April 30–May 27, 1960							
11	37	77	53	1.1	1.08	5.63	2.47
7.4	22	32	27	2.2	1.28	1.32	1.30
6.0	56	50	53	5.5	2.72	3.70	3.17
7.4	38	59	47	5.9	2.50	1.65	2.03
4.8	61	63	62	4.0	1.45	2.82	2.02
6.0	42	41	41	2.8	2.94	0.61	1.34
9.3	68	57	62	6.8	1.42	2.68	1.95
19	53	95	71	1.4	1.01	7.85	2.82
19	53	115	78	5.5	1.01	1.44	1.21

1 Mg/m^2.
2 $Ml/10^3\ m^3$.
3 In water columns or layers to about 100 m, 300 m, 90 m, and 300 m, respectively, with exceptions noted. G. M. is geometric mean.

Table 6.1—2 (continued)

MEASUREMENTS OF STANDING CROPS OF CHLOROPHYLL *a*,[1] ZOOPLANKTON,[2] CARNIVOROUS MICRONEKTON[2] AND COPEPODS[2] AT OCEANOGRAPHIC STATIONS IN THE EASTERN TROPICAL PACIFIC[3]

September 15–December 14, 1960				
12		150		3.2
18		170		8.0
11		150		10.8
15		120		11.1
8.8		80		6.6
3.0		20		2.3
12		60		4.4
16		80		7.4
34		200		1.4
5.7		50		0.6
3.0		50		1.2
9.2		140		2.0
7.7		40		2.7
2.4		40		1.4
2.8		10		1.6
5.0		30		4.8
4.2		50		6.1
1.8		80		5.7
January 15–February 25, 1959				
24	24	47	34	12.9
18	9	18	13	0.0
41	18	10	13	0.0
22	37	88	57	12.1
33	111	95	103	6.6
29	98	43	65	8.1
21	197	396	279	6.6
24	166	239	199	11.9
105	196	175	185	12.1
25	91	71	80	5.5
30	158	77	110	9.7
107	164	130	146	5.9
36	60	121	85	6.2
31	54	33	42	1.4

1 Mg/m^2.
2 $Ml/10^3\ m^3$.
3 In water columns or layers to about 100 m, 300 m, 90 m, and 300 m, respectively, with exceptions noted. G. M. is geometric mean.

Table 6.1—2 (continued)

MEASUREMENTS OF STANDING CROPS OF CHLOROPHYLL *a*,[1] ZOOPLANKTON,[2] CARNIVOROUS MICRONEKTON[2] AND COPEPODS[2] AT OCEANOGRAPHIC STATIONS IN THE EASTERN TROPICAL PACIFIC[3]

August 13–September 22, 1959	
6.6	4.3
7.5	4.6
16	1.4
12	3.2
23	31.4
9.7	8.0
7.6	5.1
6.8	6.1
12	12.7
4.4	0.5
11	8.0
12	9.6
10	8.0

[1] Mg/m^2.
[2] $Ml/10^3\ m^3$.
[3] In water columns or layers to about 100 m, 300 m, 90 m, and 300 m, respectively, with exceptions noted. G. M. is geometric mean.

(From Blackburn, M., Relationships between standing crops at three successive trophic levels in the eastern tropical Pacific, *Pac. Sci.*, 20, 38, 1966. With permission.)

6.2. TAXON DIVERSITY AS A FUNCTION OF DEPTH AND AREA

Table 6.2—1
DISTRIBUTION OF CELLULAR CARBON BETWEEN PHYTOPLANKTON GROUPS IN WATER SAMPLES FROM VARIOUS DEPTHS OFF THE COAST OF SOUTHERN CALIFORNIA

	Depth, m				
	0	50	100	200	600
Organic C (μg/liter) in:					
Diatoms	2.1	5.5	0.2	0	0
Dinoflagellates	7.1	6.5	1.6	0.2	0
Coccolithophorids	1.7	9.2	0.6	0	0.1
Monads	1.2	3.2	0.9	0.5	0.4
Phytoplankton (total)	12.1	24.4	3.3	0.7	0.5
Biomass (μg C/liter) based on:					
Direct examination	13	25	3.8	1.1	0.5
ATP	24	22	5.6	1.9	0.5
Chlorophyll	14	22	5.0	0.2	–
DNA	250	200	95	50	25
Total particulate organic carbon (μg/liter)	45	30	9.0	9.0	13

Note: Also shown for comparative purposes are the total particulate organic carbon value and the biomass estimates based on ATP, DNA, and chlorophyll. ATP, DNA, and chlorophyll values (in μg) have been converted to organic carbon (in μg) by multiplying by the factors 250, 50, and 100, respectively. These factors are based on laboratory studies of phytoplankton cultures.

Table 6.2—2
OCCURRENCE OF BACTERIA IN VARIOUS ESTUARINE ENVIRONMENTS

(Percentage of Species)

Species	Water			Sea grass community
	Bottom	1 m from bottom	Surface	
Bacillus subtilis	45	39.5	22	10–25
B. megaterium	18	7	5.5	0
B. sphaericus	0	7.5	0	0
Corynebacterium globiforme	0	7.0	0	0
C. flavum	0	0	0	10
C. miltinum	0	0	0	5
Actinomyces spp.	18	0	5.5	10–25
Staphylococcus candidus	0	8	8	0
S. roseus	0	8	0	0
Mycoplana dimorpha	19	23	54	40
M. citrea	0	0	0	5
Sarcina lutea	0	0	5.5	0
Pigmented strains	20	38	27.5	50
Ratio g pos./g neg. strains	1.9	2.0	0.7	0.9

Table 6.2—3
FLORA OF DIATOMS FOUND IN THE PLANKTON-COLORED LAYER AT THE BOTTOM OF ARCTIC SEA ICE OFF BARROW, ALASKA

Amphiprora kryophila
Gomphonema exiguum v. arctica
Navicula algida
N. crucigeroides
N. directa

N. gracilis v. inaequalis
N. kjellmanii
N. obtusa
N. transitans

N. transitans v. derasa
N. transitans v. erosa
N. trigoncephla
N. valida
Nitzschia lavuensis

Pinnularia quadratarea
P. quadratarea v. biconstracta
P. quadratarea v. capitata
P. quadratarea v. constricta
P. quadratarea v. stuxbergii

P. semiinflata
P. semiinflata v. decipiens
Pleurosigma stuxbergii
P. stuxbergii v. rhomboides
Stenoneis inconspiqua

Table 6.2—4

COUNTS OF PHYTOPLANKTON AND PROTISTANS (AS INFUSORIA) FROM THE INDICATED VOLUMES OF SEA WATER TAKEN FROM VERTICAL SERIAL SAMPLES IN SEVERAL NORWEGIAN FJORDS

Drøbak (Group I Samples)

Depth, m	0	5	10	20	30	10	50
Temperature, °C	0.0	1.46	1.99	2.19	4.41	5.24	5.56
Salinity, °/oo	30.08	31.25	31.58	32.27	33.31	33.89	34.08
Density, σ_t	24.17	25.03	25.26	25.74	26.43	26.78	26.91
O_2, cc/l	–	7.46	7.59	7.24	6.72	6.61	6.53
O_2, °/o	–	93.0	96.5	94.5	92.0	92.5	92.0
No. cc centrifuged	100	160	100	100	100	100	100
Diatoms							
I, sum of all species	2725	920	860	470	100	–	130
Cerataulina bergonii	–	38	30	–	–	–	–
Chaetoceras constrictum	–	88	40	–	–	–	–
debile	–	–	80	–	–	–	–
decipiens	170	–	–	–	–	–	–
laciniosum	–	181	160	–	–	–	–
Gyrosigma fasciola	–	–	–	–	10	–	–
Lauderia glacialis	960	150	130	110	70	–	–
Navicula sp.	–	6	–	–	10	–	–
Rhizosolenia setigera	5	–	–	–	–	–	
Skeletonema costatum	50	–	–		–	–	130
Thalassiosira decipiens	20	6	70	–	–	–	–
gravida	780	438	350	360	10	–	–
Thalassiothrix nitzschioides	10	–	–	–	–	–	–
Tropidoneis lepidoptera	–	13	–	–	–	–	–
Flagellata							
Eutreptia lanowii	740	63	30	–	–	20	10
Peridiniales							
Ceratium longipes	–	6	–	–	–	–	–
tripos	–	6	–	–	–	–	–
Diplopsalis lenticula	–	–	10	–	–	–	–
Gymnodinium lohmanni	30	56	–	10	–	10	–
sp.	–	25	–	–	–	–	–
Peridinium achromaticum	–	13	20	–	–	–	–
conicum	–	6	–	–	–	–	–
parallelum	–	6	–	–	–	–	–
pellucidum	–	6	–	–	–	–	–
roseum	–	–	10	–	–	–	–
sp.	–	19	–	–	–	–	10
Pouchetia sp.	–	–	–	–	–	–	–
Torodinium robustum	80	19	20	–	–	–	–

Table 6.2—4 (continued)

COUNTS OF PHYTOPLANKTON AND PROTISTANS (AS INFUSORIA) FROM THE INDICATED VOLUMES OF SEA WATER TAKEN FROM VERTICAL SERIAL SAMPLES IN SEVERAL NORWEGIAN FJORDS

Infusoria							
Laboea conica	10	19	–	–	–	–	–
crassula	10	231	90	20	10	–	–
strobila	–	31	–	–	–	–	–
vestita	–	77	–	–	–	–	–
sp.	–	–	10	–	–	–	–
Lohmaniella oviformis	170	231	–	–	–	–	–
sp.	–	–	30	10	–	–	–
Mesodinium, small	–	75	260	–	–	–	–
bigger	20	513	60	20	30	–	10
Ptychocylis urnula	–	6	–	–	–	–	–
Tintinnopsis sp.	–	13	–	10	–	–	–
Infusoria indeterminata	–	6	10	–	–	–	–

Drøbak (Group II Samples)

Depth, m	0	2	5	10	30
Temperature, °C	1.0	5.57	5.69	6.09	6.25
Salinity, °/₀₀	31.33	34.15	34.21	34.56	34.80
Density, σ_t	25.11	26.95	26.98	27.21	27.39
O_2, cc/l	7.44	5.91	5.97	5.70	5.62
O_2, °/₀	92.5	83.0	84.5	81.5	81.0
No. cc centrifuged	100	100	100	100	100
Diatoms					
All species, sum	4030	1140	480	10	100
Biddulphia aurita	10	–	–	–	–
Chaetoceras decipiens	50	20	–	–	70[1]
diademia	–	260	–	–	–
Coscinodiscus radiatus	–	10	–	–	–
Lauderia glacialis	1150	10	130	10[1]	10
Licmophora sp.	20	10	–	–	–
Navicula sp.	20	20	20	–	10
Nitzschia sp.	–	–	–	–	10
Rhabdonema arcuatum	10	–	–	–	–
Skeletonema costatum	–	240	–	–	–
Thalassiosira decipiens	250	–	10	–	–
gravida	2200	470	300	–	–
nordenskioldii	320	–	–	–	–
Thalassiothrix nitzschioides	–	–	20	–	–
Flagellata					
Eutreptia lanowii	200	10	–	–	–

[1] Dead cells.

Table 6.2—4 (continued)

COUNTS OF PHYTOPLANKTON AND PROTISTANS (AS INFUSORIA) FROM THE INDICATED VOLUMES OF SEA WATER TAKEN FROM VERTICAL SERIAL SAMPLES IN SEVERAL NORWEGIAN FJORDS

Cilioflagellata					
Ceratium furca	10	20	–	–	–
Dinophysis acuta	–	10	–	–	–
Gymnodinium lohmanni	30	20	10	20	–
sp.	–	10	–	10	–
Peridinium pallidum	–	–	10	–	–
sp.	–	–	10	–	–
Pouchetia sp.	–	10	–	–	–
Torodinium robustum	60	10	10	–	–
Infusoria					
Cyttarocylis denticulata	–	–	10	–	–
Luboea crassula	10	10	–	–	–
sp.	–		–	–	20
Lohmanniella oviformis	40	90	70	30	–
sp.					
Mesodinium, small	–	10	40	–	–
bigger	50	150	30	10	–
Strombidium spinosum	–	10	70	–	–
Tintinnus acuminatus	–	–	–	–	10
Infusoria indeterminata	140	40	–	–	–

Drøbak (Group III Samples)

Depth, m	0	2	5	10	20	30
Temperature, °C	1.0	1.02	4.68	5.79	6.14	6.11
Salinity, °/₀₀	30.69	30.91	33.41	34.22	34.71	34.73
Density, σ_t	24.60	24.79	26.48	26.98	27.32	27.35
O_2, cc/l	7.95	7.77	6.34	5.87	5.73	5.90
O_2, °/₀	98.0	95.5	87.0	83.0	82.0	84.5
P_H	7.92	7.95	7.96	7.98	7.98	8.02
No. cc centrifuged	25	25	25	100	50	100
Diatoms						
All species, sum	26,040	34,480	16,680	6060	60	240
Biddulphia aurita	120	680	–	90	–	–
sinensis	–	–	–	–	20	–
Chaetoceras decipiens	1000	640	–	–	–	–
debile	280	1640	–	–	–	40
teres	120	–	–		–	–
Lauderia glacialis	9520	12,840	6720	1960	40	30
Licmophora sp.	10	40	–	20	–	–
Melosira borreri	–	–	–	380	–	80
Navicula sp.	–	40	80	20	–	–
Nitzschia closterium	–	–	–	–	–	10
sp.	–	40	–	–	–	–
Pleurosigma sp.	–	40	–	–	–	–

Table 6.2—4 (continued)
COUNTS OF PHYTOPLANKTON AND PROTISTANS (AS INFUSORIA) FROM THE INDICATED VOLUMES OF SEA WATER TAKEN FROM VERTICAL SERIAL SAMPLES IN SEVERAL NORWEGIAN FJORDS

Rhizosolenia semispina	–	–	40	–	–	–
Skeletonema costatum	–	–	–	200	–	–
Thalassiosira decipiens	–	–	–	20	–	–
gravida	12,720	15,960	6520	1590	–	80
Nordenskioldii	2240	2320	3320	1780	–	–
Thalassiothrix nitzschioides	–	240	–	–	–	–
Flagellata						
Eutreptia lanowii	1760	1600	40	10	–	–
Peridiniales						
Ceratium furca	–	–	–	20	–	–
fusus	–	–	–	10	–	–
Dinophysis norvegica	20	–	–	10	–	–
Gonyaulax sp.	40	–	40	–	–	–
Gymnodinium lohmanni	120	–	40	50	20	–
sp.	–	40	–	–	–	–
Peridinium achromaticum	–	40	–	–	–	–
ovatum	–	40	–	10	–	–
Pouchetia sp.	80	40	40	20	–	–
Prorocentrum micans	–	–	–	10	–	–
Torodinium robustum	40	80	–	–	–	–
Infusoria						
Laboea crassula	80	240	–	70	–	–
strobila	40	40	–	20	–	–
vestita	80	80	280	–	–	–
Lohmanniella oviformis	80	–	–	–	20	10
Mesodinium	–	240	280	100	20	–
Strombidium reflexum	80	160	40	10	–	–
Infusoria indeterminata	–	–	–	10	–	10

Drøbak (Group IV Samples)

Depth, m	0	2	5	10	20	30	40	50
Temperature, °C	1.7	1.43	1.38	1.38	1.62	3.04	5.38	6.04
Salinity, °/oo	30.06	30.10	30.12	30.23	30.45	31.93	34.12	34.71
Density, σ_t	24.06	24.11	24.13	24.21	24.38	25.47	26.95	27.34
O_2, cc/l	8.36	8.31	8.35	8.26	7.98	6.92	6.02	5.88
O_2, °/o	104.0	103.0	103.5	102.5	99.5	90.5	84.5	84.0
No. cc centrifuged	10	20	10	10	25	25	50	50
Diatoms								
All species, sum	347,200	371,450	476,700	393,600	272,180	94,820	75,600	200
Asterionella japonica	200	–	–	–	–	–	–	–
Biddulphia aurita	37,600	43,200	41,000	58,700	33,040	12,520	900	–
Cerataulina bergonii	–	–	–	–	80	–	–	–

Table 6.2—4 (continued)

COUNTS OF PHYTOPLANKTON AND PROTISTANS (AS INFUSORIA) FROM THE INDICATED VOLUMES OF SEA WATER TAKEN FROM VERTICAL SERIAL SAMPLES IN SEVERAL NORWEGIAN FJORDS

Chaetoceras boreale	2200	3400	2200	2200	1480	280	–	–
constrictum	9700	9400	14,000	14,600	3120	1760	–	–
compressum	7000	3800	13,600	22,500	6520	440	–	–
curvisetum	2800	3500	7400	3600	100	600	100	–
debile	45,800	43,400	58,300	41,200	22,920	8920	300	–
decipiens	1000	1600	2700	900	720	680	–	–
diadema	7000	11,500	18,900	11,700	4760	720	–	–
breve	200	200	200	–	–	–	–	–
laciniosum	9800	6800	10,300	9300	3960	1760	–	–
scolopendra	3000	9500	7500	5300	600	320	–	–
simile	700	–	200	–	–	–	–	–
teres	800	600	600	2500	880	–	–	–
Detonula cystifera	2000	400	500	600	–	–	–	–
Lauderia glacialis	6000	6800	8500	9800	8160	6040	1260	80
Leptocylindrus danicus	800	500	1400	500	1200	–	20	–
minimus	600	200	500	1500	720	840	–	–
Navicula sp.	1800	900	1600	2600	680	–	140	40
Nitzschia seriata	–	300	200	200	280	–	–	–
delicatissima	–	–	400	–	–	–	–	–
Rhizosolenia semispina	2700	900	200	1900	1460	620	–	–
setigera	100	300	100	300	160	160	–	–
Thalassiosira decipiens]	–	200	–	600	–	–	40	–
gravida	26,900	20,100	19,500	22,700	20,080	11,480	1300	80
nordenskioldii	94,100	112,300	135,500	165,100	137,840	39,040	3420	–
sp.	–	400	200	–	–	–	–	–
Thalassiothrix longissima	–	50	100	–	–	–	–	–
nitzschioides	400	400	–	500	80	160	–	–
Coscinodiscus radiatus	–	–	–	–	–	40	–	–
Skeletonema costatum	86,000	90,800	129,300	24,800	12,440	8440	80	–
Flagellata								
Eutreptia lanowii	3400	3700	4100	5100	3760	1040	40	–
Peridiniales								
Dinophysis acuminata	–	–	–	100	–	–	40	20
Gymnodinium lohmanni	200	50	400	500	320	–	–	–
Peridinium achromaticum	–	–	–	–	–	40	–	–
ovatum	–	–	–	–	–	–	20	–
Prorocentrum micans	–	–	–	–	–	40	–	–
Torodinium robustum	100	–	100	–	–	–	–	–
Infusoria								
Laboea conica	100	700	600	–	–	–	–	–
crassula	–	200	–	–	–	–	–	–
strobila	–	50	–	100	120	40	–	–
vestita	3000	5400	6500	3500	720	80	20	–
Lohmanniella oviformis	100	200	200	240	40	–	–	–
Mesodinium	900	1900	1800	1100	480	480	20	80
Strombidium sp.	100	200	100	600	40	–	–	–

Table 6.2—4 (continued)
COUNTS OF PHYTOPLANKTON AND PROTISTANS (AS INFUSORIA) FROM THE INDICATED VOLUMES OF SEA WATER TAKEN FROM VERTICAL SERIAL SAMPLES IN SEVERAL NORWEGIAN FJORDS

Drøbak (Group V Samples)

Depth, m	0	5	10	20	30	50
Temperature, °C	1.4	0.52	0.55	0.55	4.50	5.98
Salinity, °/oo	24.90	25.05	25.12	25.42	33.27	34.85
Density, σ_t	19.95	20.10	20.16	20.40	26.38	27.45
O_2, cc/l	8.84	8.84	8.87	8.80	6.33	6.03
O_2, °/o	105.5	103	103.5	103	86.5	86
No. cc centrifuged	10	10	10	10	25	50
Diatoms						
All species, sum	142,750	342,000	431,750	258,750	46,640	1290
Biddulphia aurita	2900	18,700	32,700	25,200	3340	–
sinensis	–	–	–	–	–	20
Chaetoceras boreale	700	1100	1100	2300	–	–
breve	500	500	200	400	–	–
constrictum	700	3000	2600	1300	320	–
compressum	500	1800	1400	4300	680	–
curvisetum	1300	5100	1600	4300	1160	–
debile	17,400	35,400	29,000	32,400	4280	–
decipiens	1000	600	800	–	–	–
diadema	4000	4900	6100	2600	480	–
laciniosum	1100	2700	3500	2100	320	–
scolopendra	2000	4400	1500	5400	–	–
simile	800	200	800	–	–	–
teres	700	800	1800	1400	80	–
Detonula cystifera	600	700	100	400	–	–
Eucampia groenlandica	–	–	–	100	–	–
Lauderia glacialis	–	100	200	900	7120	340
Leptocylindrus danicus	–	–	–	400	–	–
Navicula sp.	–	900	500	2000	40	20
Nitzschia seriata	800	–	–	200	–	–
Rhizosolinia semispina	1550	3050	3250	3550	280	10
setigera	100	100	700	500	80	–
Skeletonema costatum	99,200	188,800	231,000	94,100	4360	80
Thalassiosira decipiens	–	–	–	–	1600	–
gravida	300	2100	1800	5900	12,840	580
nordenskioldii	4800	64,200	108,700	67,700	9520	200
sp.	200	700	600	–	–	–
Thalassiothrix longissima	–	–	100	–	40	–
nitzschioides	1600	2200	800	1300	–	40
Flagellata						
Eutreptia lanowii	–	1500	500	500	200	40

Table 6.2—4 (continued)

COUNTS OF PHYTOPLANKTON AND PROTISTANS (AS INFUSORIA) FROM THE INDICATED VOLUMES OF SEA WATER TAKEN FROM VERTICAL SERIAL SAMPLES IN SEVERAL NORWEGIAN FJORDS

Peridiniales						
Dinophysis norvegica	–	–	100	–	–	–
Diplopsalis lenticula	–	–	100	–	–	–
Glenodinium bipes	–	200	100	–	–	–
Gonyaulax triacantha	–	–	100	–	–	–
Gymnodinium lohmanni	200	–	100	–	–	40
Peridinium pellucidum	–	200	–	300	–	–
steinii	–	–	–	100	–	–
sp.	100	200	700	–	–	–
Prorocentrum micans	–	–	–	–	–	40
Protoceratium reticulatum	–	–	100	–	–	–
Infusoria						
Laboea conica	100	1200	200	–	–	–
crassula	–	400	300	–	–	–
strobila	–	–	100	200	–	–
vestita	100	6000	3600	500	40	–
Lohmanniella oviformis	–	–	100	–	–	–
Mesodinium	–	600	–	100	–	–
Strombidium sp.	–	–	–	200	–	–

Nesodden (Group VI Samples)

Depth about 50 m

Depth, m	1	5	10	20	30	40	50
Temperature, °C	–	–	–	15.13	5.41	–	–
Salinity, °/₀₀	20.43	20.46	20.49	30.56	32.04	32.22	33.15
Density, σ_t	–	–	–	22.55	25.20	–	–
O_2, cc/l	6.47	6.49	6.37	4.57	4.09	4.01	3.18
O_2, °/₀	–	–	–	76	57		–
No. cc centrifuged	100(10)	100(10)	100	100	50	50	50
Diatoms							
Cerataulina bergonii	20	60	180	–	–	–	–
Chaetoceras curvisetum	320	260	300	–	–	–	100
Coscinodiscus radiatus	40	140	90	–	–	–	–
Lauderia glacialis	–	–	–	–	20	–	–
Skeletonema costatum	6000	2580	17,380	60	1340	–	–
Rhizosolenia fragilissima	8240	7660	10,460	30	–	–	–
Peridiniales							
Ceratium furca	2620	2030	1340	–			
fusus	3160	2130	1870	–	–	–	–
macroceros	10	–	–	–	–	–	–
tripos	2450	1960	1050	–	–	–	–
Dinophysis acuminata	300	160	180	10	60	20	–
acuta	380	180	120	–	20	–	–
rotundata	180	20	100	–	–	–	–
Gonyaulax polyhedra	17,280	6680	3340	10	–	–	–
spinifera	60	140	60	–	–	–	–

Table 6.2—4 (continued)
COUNTS OF PHYTOPLANKTON AND PROTISTANS (AS INFUSORIA) FROM THE INDICATED VOLUMES OF SEA WATER TAKEN FROM VERTICAL SERIAL SAMPLES IN SEVERAL NORWEGIAN FJORDS

Peridinium achromaticum	–	20	–	–	–	–	–
conicum	–	–	30	–	–	–	–
depressum	–	20	–	30	–	–	–
divergens	140	240	120	40	60	–	–
pallidum	40	40	40	–	–	–	–
pellucidum	20	–	–	–	–	–	–
pyriforme	–	20	–	–	–	–	–
steinii	880	420	340	30	–	–	–
Prorocentrum micans	35,100	26,500	24,480	220	280	180	–
Torodinium robustum	–	–	–	70	20	–	–
Infusoria							
Amphorella subulata	180	80	300	–	–	–	–
Cyttarocylis denticulata	–	40	–	–	–	–	–
Laboea conica	–	80	20	–	–	–	–
delicatissima	–	–	40	10	–	–	–
emergens	120	240	340	10	–	20	–
strobila	60	40	40	–	–	–	–
vestita	60	380	40	–	–	–	–
Lohmanniella oviformis	100	80	160	10	20	40	–
Mesodinium	35,500	47,500	13,320	320	580	–	–
Tintennopsis sp.	120	80	20	–	–	–	–
Infusoria indeterminata	460	440	140	–	–	–	–
Nauplii of *Copepoda*	100	210	200	30	–	–	–

Steilene (Group VII Samples)

Depth 53 m

Depth, m	1	5	10	20	30	40	50
Temperature, °C	17.59	17.58	17.71	8.19	5.51	5.42	–
Salinity, °/₀₀	20.24	20.23	20.58	29.59	32.08	32.76	33.05
Density, σ_t	14.14	14.14	14.39	23.03	25.32	25.87	–
O_2, cc/l	6.39	6.36	6.30	5.04	4.82	4.31	3.92
O_2,°/₀	104	104	103	73	67	60	–
No. cc centrifuged	100	50	50	50	50	50	50
Diatoms							
Chaetoceras curvisetum	2920	160	–	–	–	–	–
Coscinodiscus radiatus	180	170	120	40	40	40	20
Rhizosolenia fragilissima	100	–	–	–	–	–	–
Peridiniales							
Ceratium bucephalum	20	20	–	–	–	–	–
furca	70	60	–	–	–	–	–
fusus	480	460	20	–	–	–	–
macroceras	30	10	–	–	–	–	–
tripos	800	730	200	–	–	–	–

Table 6.2—4 (continued)

COUNTS OF PHYTOPLANKTON AND PROTISTANS (AS INFUSORIA) FROM THE INDICATED VOLUMES OF SEA WATER TAKEN FROM VERTICAL SERIAL SAMPLES IN SEVERAL NORWEGIAN FJORDS

Dinophysis acuminata	80	100	–	20	–	20	–
acuta	80	180	–	–	–	20	–
norvegica	20	–	–	–	–	–	–
rotundata	40	20	–	–	–	–	–
Gonyaulax polyhedra	60	100	20	–	–	–	20
Peridinium conicum	–	40	60	–	–	–	–
depressum	20	–	–	20	–	–	–
divergens	210	200	40	–	–	–	–
steinii	–	60	20	–	–	20	–
Prorocentrum micans	4540	4300	2180	60	20	20	40
Torodinium robustum	–	–	–	80	–	–	–
Infusoria							
Laboea conica	940	720	140	–	–	–	–
delicatissima	280	60	–	–	–	–	–
emergens	260	240	520	–	–	–	–
strobila	80	80	20	–	–	–	–
Lohmanniella oviformis	80	160	–	–	–	20	–
Infusoria indeterminata	500	380	240	–	–	–	–
Mesodinium	32,000	25,580	5020	380	40	40	20
Nauplii of *Copepoda*	80	80	40	20	–	–	–

Drøbak (Group VIII Samples)

	St. 4				St. 3			
Depth, m	1	5	10	20	20	30	40	50
Temperature, °C	17.0	16.91	16.92	14.07	12.99	12.03	–	6.54
Salinity, °/₀₀	20.16	20.35	20.57	24.38	28.24	28.80	30.12	32.16
Density, σ_t	14.21	14.38	14.54	18.02	21.20	21.82	–	25.27
O_2, cc/l	6.27	6.22	6.18	5.78	5.57	5.60	5.48	6.01
O_2, °/₀	101	100	100	91	88	87	–	85
No. cc centrifuged	100	100(10)	50	50	50	50	50	50
Halosphaera viridis	–	20	–	–	–	40	20	–
Diatoms								
Chaetoceras curvisetum	3780	520	2340	1600	–	–	120	–
Coscinodiscus radiatus	190	240	160	80	60	–	–	–
Thalassiothrix nitzschioides	–	–	–	–	–	–	–	220
Cerataulina bergonii	–	–	–	–	–	–	–	60
Peridiniales								
Ceratium bucephalum	60	–	10	10	–	20	–	–
furca	140	90	200	30	–	–	–	–
fusus	1350	810	580	200	–	20	–	–
longipes	10	–	–	–	–	20	–	–
macroceros	30	30	70	20	20	–	–	–
tripos	1320	1160	870	340	–	20	20	–

Table 6.2—4 (continued)
COUNTS OF PHYTOPLANKTON AND PROTISTANS (AS INFUSORIA) FROM THE INDICATED VOLUMES OF SEA WATER TAKEN FROM VERTICAL SERIAL SAMPLES IN SEVERAL NORWEGIAN FJORDS

	St. 4				St. 3			
Dinophysis acuminata	80	220	120	–	–	–	–	–
acuta	200	280	220	100	–	–	–	–
rotundata	20	–	–	20	–	–	–	–
Gonyaulax polyedra	140	180	220	80	–	–	–	–
spinifera	–	20	–	–	–	–	–	–
Peridinium conicum	–	–	–	20	–	–	–	–
depressum	–	–	30	–	–	–	–	–
divergens	640	740	590	100	–	–	–	–
pallidum	–	20	20	–	–	–	–	–
steinii	–	–	–	–	–	–	–	40
Prorocentrum micans	9240	6600	6200	3600	260	360	40	80
Protoceratium reticulatum	20	–	20	–	–	–	–	–
Torodinium robustum	–	–	–	–	80	–	–	–
Infusoria								
Laboea conica	660	560	1120	240	60	20	–	–
delicatissima	–	–	100	20	–	–	–	–
emergens	100	260	360	60	40	40	40	100
strobila	50	180	100	–	–	–	–	–
Lohmanniella oviformis	180	40	20	20	60	20	–	60
Mesodinium	72,500	77,100	52,200	7420	260	140	–	–
Tintinnopsis campanula	20	–	–	–	–	–	–	–
Infusoria indeterminata	460	360	340	240	20	20	–	–
Nauplii of *Copepoda*	80	100	140	60	100	80	–	–

Oslofjord, W. of Mölen (Group IX Samples)

Depth about 200 m

Depth, m	1	5	10	20	30	40	50	75
Temperature, °C	16.95	17.09	17.24	15.92	14.56	13.26	12.88	5.82
Salinity, °/₀₀	14.43	20.20	23.87	29.84	31.15	31.87	32.64	33.08
Density, σ_t	9.87	14.22	16.98	21.82	23.11	23.96	24.61	26.08
O_2, cc/l	6.11	6.10	5.36	5.54	5.63	5.72	5.70	6.24
O_2, °/₀	95	99	89	93	93	93	93	88
P_H	8.12	8.16	8.08	8.08	8.08	8.06	8.06	8.04
Oxidizability:								
No. cc $\frac{n}{100}$ $KMnO_4$ used/l	81.0	53.6	36.5	25.7	13.6	24.2	–	17.8
No. cc O_2 used/l	4.52	2.99	2.04	1.44	0.76	1.35	–	0.99
No. cc centrifuged	100	50	100	50	50	50	50	50
Halosphaera viridis	–	–	30	40	20	20	–	–
Diatoms								
Cerataulina bergonii	20	–	–	–	–	–	–	–
Chaetoceras curvisetum	200	220	–	–	–	–	–	–
Coscinodiscus radiatus	140	320	20	20	–	20	20	20
Leptocylindrus danicus	60	100	–	–	–	–	–	–
Melosira borreri	320	–	–	–	–	–	–	–
Rhizosolenia fragilissima	140	–	–	–	–	–	–	–

Table 6.2—4 (continued)
COUNTS OF PHYTOPLANKTON AND PROTISTANS (AS INFUSORIA) FROM THE INDICATED VOLUMES OF SEA WATER TAKEN FROM VERTICAL SERIAL SAMPLES IN SEVERAL NORWEGIAN FJORDS

Peridiniales								
Ceratium bucephalum	–	80	50	–	–	–	–	–
furca	70	–	–	–	–	–	–	–
fusus	570	80	50	–	–	–	–	–
longipes	–	–	10	–	–	–	–	–
macroceros	10	–	10	–	–	–	–	–
tripos	210	20	20	–	–	–	–	–
Dinophysis acuminata	40	–	20	–	–	40	–	–
acuta	410	–	–	–	–	20	–	–
norvegica	20	–	–	–	–	20	–	–
rotundata	–	–	–	–	–	20	–	–
Gonyaulax polyedra	60	–	–	–	–	–	–	–
Gymnodinium sp.	40	20	20	–	–	–	–	–
Peridinium conicum	20	–	–	–	–	–	–	–
divergens	330	20	–	–	–	20	20	20
pallidum	20	–	–	–		–	–	–
steinii	20	20	20	–	–	20	–	20
Prorocentrum micans	2560	840	840	40	–	40	60	100
Infusoria								
Laboea conica	160	180	–	–		–	–	–
emergens	–	220	80	–	–	20	40	–
strobila	20	–	–	–	–	–	–	–
Lohmanniella oviformis	200	20	80	–	–	20	20	40
Leprotintinnus sp.	20	–	–	–	–		–	–
Mesodinium	120	500	240	20	60	140	40	140
Tintinnopsis campanula	20	40	–	–	–	–	–	–
sp.	–	80	40	20	20	80	40	–
Infusoria of *Copepoda*	200	140	60	–	–	–	–	–
Nauplii of *Copepoda*	90	60	20	40	–	–	20	–

Svelvikfjord, by Knivsfjeld (Group X Samples)

Depth 16 m

Depth, m	1	5	10	15
Temperature, °C	16.05	17.56	18.68	16.05
Salinity, °/₀₀	7.16	19.77	24.14	27.10
Density, σ_t	4.49	13.80	16.87	19.70
O_2, cc/l	6.02	5.84	5.23	4.91
O_2, °/₀	88.5	95	89	81
P_H	7.68	8.08	8.06	8.01
Oxidizability:				
No. cc $\frac{n}{100}$ $KMnO_4$ used/l	100.0	65.0	36.2	35.0
No. cc O_2 used/l	5.59	3.63	2.02	1.96
No. cc centrifuged	100	50	50	50
Halosphaera viridus	–	–	20	–

Table 6.2—4 (continued)
COUNTS OF PHYTOPLANKTON AND PROTISTANS (AS INFUSORIA) FROM THE INDICATED VOLUMES OF SEA WATER TAKEN FROM VERTICAL SERIAL SAMPLES IN SEVERAL NORWEGIAN FJORDS

Diatoms				
Chaetoceras curvisetum	–	600	–	–
Coscinodiscus radiatus	40	320	40	20
Melosira distans	–	20	20	–
Nitzschia sigma	–	–	–	80
Pleurosigma sp.	–	–	20	–
Tabellaria flocculosa	–	40	–	–
Peridiniales				
Ceratium bucephalum	–	40	20	–
fusus	20	100	–	20
birundinella	–	20	–	–
longipes	20	–	–	–
Dinophysis acuminata	–	20	–	–
rotundata	–	60	–	–
Gymnodinium sp.	–	220	–	–
Peridinium conicum	–	20	–	–
divergens	–	60	–	20
pallidum	–	40	–	–
steinii	–	–	20	–
sp. (brown)	20	–	–	–
Prorocentrum micans	40	20	–	–
Infusoria				
Laboea emergens	–	80	–	–
sp.	–	80	–	–
Lohmanniella oviformis	40	160	20	40
Tintinnopsis beroidea	–	140	20	–
campanula	–	80	–	–
sp.	–	460	140	–
Infusoria indeterminata	–	20	20	–
Nauplii of *Copepoda*	–	60	60	–

Drammensfjord, by Hernaetstangen (Group XI Samples)

Depth, m	1	5	10	15	20	30
Temperature, °C	15.10	15.02	16.10	12.36	7.18	4.65
Salinity, °/₀₀	0.21	0.25	11.40	20.32	26.06	29.25
Density, σ_t	0.55	0.53	7.72	15.22	20.40	23.19
O_2, cc/l	7.00	6.94	4.68	4.51	3.11	0.77
O_2, °/₀	97	96	71	67	43	10
P_H	7.58	7.58	7.49	7.59	7.40	7.40
Oxidizability:						
No. cc $\frac{n}{100}$ $KMnO_4$ used/l	100.0	118.0	78.8	48.0	40.0	33.2
No. cc O_2 used/l	5.60	6.60	4.40	2.68	2.23	1.85
No. cc centrifuged	50	50	50	50	50	50

Table 6.2—4 (continued)

COUNTS OF PHYTOPLANKTON AND PROTISTANS (AS INFUSORIA) FROM THE INDICATED VOLUMES OF SEA WATER TAKEN FROM VERTICAL SERIAL SAMPLES IN SEVERAL NORWEGIAN FJORDS

Diatoms						
Coscinodiscus radiatus	20	–	20	–	–	–
Cyclotella sp.	–	20	–	–	–	–
Navicula sp.	–	60	–	–	–	–
Tabellaria flocculosa	–	200	–	–	–	–
Peridiniales						
Ceratium fusus	–	–	–	–	40	–
hirundinella	60	40	–	–	–	–
Dinophysis acuminata	–	–	60	–	–	–
acuta	–	–	40	–	–	–
Gymnodinium sp.	–	–	80	20	20	–
Peridinium depressum	–	–	–	–	20	–
sp. (brown)	–	40	100	–	–	–
Infusoria						
Laboea emergens	220	–	–	–	–	–
Lohmanniella oviformis	810	280	120	20	–	60
Tintinnopsis sp.	40	–	240	–	–	–
Infusoria indeterminata	20		40	–	–	–
Nauplii of *Copepoda*	–	–	–	20	–	–

Drøbak (Group XII Samples)

Depth, m	1	5	10	20	30	40
Temperature, °C	16.75	16.36	15.30	9.88	8.06	7.56
Salinity, °/₀₀	19.64	20.21	26.59	29.14	30.63	31.26
Density, σ_t	13.87	14.39	19.49	22.43	23.86	24.42
O_2, cc/l	6.22	5.96	5.27	4.94	4.87	5.04
O_2, °/₀	100	95	86	74	71	73
P_H	8.19	8.13	7.99	7.87	7.79	7.79
Oxidizability:						
No. cc $\frac{n}{100}$ $KMnO_4$ used/l	58.0	48.4	29.4	24.7	21.2	23.2
No. cc O_2 used/l	3.23	2.70	1.64	1.38	1.18	1.29
No. cc centrifuged	100	100	50	50	50	50
Halosphaera viridus	–	10	60	–	–	–
Distephanus speculum	20	–	–	–	–	–
Diatoms						
Chaetoceras curvisetum	100	40	40	180	–	–
Cerataulina bergonii	–	–	–	–	20	–
Coscinodiscus radiatus	1040	920	20	40	40	20
Rhizosolenia alata	–	60	–	–	–	–

Table 6.2—4 (continued)
COUNTS OF PHYTOPLANKTON AND PROTISTANS (AS INFUSORIA) FROM THE INDICATED VOLUMES OF SEA WATER TAKEN FROM VERTICAL SERIAL SAMPLES IN SEVERAL NORWEGIAN FJORDS

Peridiniales						
Ceratium furca	270	70	–	–	–	20
fusus	1230	880	60	–	–	–
longipes	–	–	20	–	–	–
macroceros	–	–	20	–	–	–
tripos	980	420	–	–	–	–
Dinophysis acuminata	1180	20	–	–	–	–
acuta	1020	380	100	–	–	–
rotundata	100	40	–	–	–	–
Gonyaulax polyedra	380	80	–	–	–	–
spinifera	30	–	–	–	–	–
Gymnodinium sp.	–	–	40	–	–	–
Peridinium depressum	–	–	–	–	–	20
divergens	1250	220	–	–	–	20
pallidum	130	40	20	–	–	–
pellucidum	60	–	–	–	–	–
pyriforme	–	20	–	–	–	–
steinii	120	70	40	–	–	–
Prorocentrum micans	5540	2340	600	–	–	–
Torodinium robustum	–	–	80	40	40	20
Infusoria						
Amphorella subulata	70	–	–	–	–	–
Cyttarocylis annulata	40	–	–	–	–	–
claparedei	20	–	–	–	–	–
Laboea conica	400	1700	60	–	–	–
delicatissima	–	80	–	–	–	–
emergens	320	1220	40	40	–	–
strobila	50	40	20	–	–	–
Lohmanniella oviformis	300	300	40	20	120	100
Mesodinium	6880	4760	80	100	200	940
Infusoria indeterminata	1080	620	60	–	–	–
Nauplii of *Copepoda*	70	60	20	80	20	–

Drøbak (Group XIII Samples

Depth, m	1	5	10	20	30	40	50	70
Temperature, °C	14.2	14.2	14.0	12.9	12.22	11.96	9.78	8.11
Salinity, °/°°	20.97	21.41	26.21	29.73	31.22	31.75	31.825	32.12
Density, σ_t	15.38	15.73	19.44	22.36	23.64	24.09	24.54	25.02
O_2, cc/l	6.16	6.12	5.70	5.13	5.20	5.50	5.45	5.64
O_2, °/°	95	95	91	82	82	87	82	82
P_H	8.13	8.13	8.10	8.01	7.98	7.98	7.94	7.92
Oxidizability:								
No. cc $\frac{n}{100}$ $KMnO_4$ used/l	36.5	34.1	17.1	10.5	10.7	10.2	8.2	10.8
No. cc O_2 used/l	2.04	1.91	0.96	0.59	0.60	0.59	0.45	0.61
No. cc centrifuged	50	100	50	50	50	50	50	50
Distephanus speculum	140	100	–	–	20	–	–	–

Table 6.2—4 (continued)

COUNTS OF PHYTOPLANKTON AND PROTISTANS (AS INFUSORIA) FROM THE INDICATED VOLUMES OF SEA WATER TAKEN FROM VERTICAL SERIAL SAMPLES IN SEVERAL NORWEGIAN FJORDS

Diatoms								
Chaetoceras curvisetum	1060	560	100	–	–	–	–	–
Coscinodiscus radiatus	3920	2910	1590	760	260	100	40	40
Guinardia flaccida	–	–	–	20	–	–	–	–
Leptocylindrus danicus	40	360	240	–	–	–	–	–
Rhizosolenia fragilissima	800	340	790	–	–	–	–	–
Skeletonema costatum	–		60	–	–	–	–	–
Peridiniales								
Ceratium furca	4340	1880	410	–	–	–	–	–
fusus	5620	3970	370	60	20	–	–	–
longipes	–	10	10	–	–	–	–	–
tripos	960	2260	120	–	–	–	–	–
Dinophysis acuminata	80	20	–	–	–	–	–	–
acuta	1660	1760	140	–	20	–	–	–
rotundata	20	–	–	–	–	–	–	–
Gonyaulax polyedra	320	100	20	–	–	–	–	–
spinifera	40	20	–	–	–	–	–	–
Gymnodinium lohmanii	–	80	60	20	20	20	–	–
Peridinium achromaticum	–	–	20	–	–	–	–	–
divergens	280	100	140	–	–	–	–	–
pallidum	–	–	–	20	–	–	–	–
steinii	80	180	40	20	–	–	–	–
Prorocentrum micans	16,600	11,180	3360	480	20	–	20	–
Torodinium robustum	–	–	60	–	–	–	–	–
Infusoria								
Laboea conica	580	440	80	–	–	–	–	–
emergens	–	80	160	80	20	–	–	–
strobila	100	40	–		–	–	–	–
vestita	1340	80	20	–	–	–	–	–
Liohmanniella oviformis	80	100	40	–	–	–	–	–
spiralis	–	–	140	–	–	–	–	–
Amphorella subulata	–	60	–	–	–	–	–	–
Infusoria indeterminata	60	160	–	–	–	–	–	–
Mesodinium	1660	220	60	60	20	–	–	–
Nauplii of *Copepoda*	40	20	20	–	–	–	–	–

6.3. CHEMICAL COMPOSITION

Table 6.3—1
ELEMENTARY COMPOSITION OF SOME AQUATIC PLANTS IN PERCENT ASH-FREE DRY WEIGHT ± STANDARD DEVIATION

Organism and reference	No. analyses	C	H	O	N	P
Diatoms	18	50.54	10.21	28.83	7.0	1.55
		±3.90	±1.90	±7.10	±3.7	±0.88
Peridineans	7	48.12	7.50	33.85	10.40	0.80
		±1.90	±1.70	±3.90	±1.40	±.36
Chlorophyceae	18	54.55	7.54	31.21	7.7	2.94
		±1.30	±0.50	±4.35	±0.56	±0.83

(From Ryther, J. H., The measurement of primary production, *Limnol. Oceanogr.*, 1(2), 73, 1956. With permission.)

Table 6.3—2
RELATIVE AMOUNTS OF AMINO ACIDS IN MEMBERS OF THE OCEANIC FOOD CHAIN

	G amino acid/16 g amino acid N		
	Phytoplankton	*Calanus*	Cod[1]
Glutamic acid	13.7	14.5	16.6
Aspartic acid	11.7	11.5	10.6
Lysine	10.4	8.9	10.3
Glycine	9.4	8.6	5.1
Leucine	9.3	9.3	9.3
Alanine	9.1	9.5	7.2
Valine	7.9	7.3	5.8
Serine	6.8	5.2	5.4
Arginine	6.8	7.8	6.7
Threonine	6.5	5.7	5.3
Phenylalanine	5.9	4.3	4.7
Isoleucine	5.7	4.9	4.9
Proline	5.3	4.5	4.2
Tyrosine	3.7	4.7	4.0
Histidine	2.1	1.9	3.5

Table 6.3—3
AVERAGE CONCENTRATIONS OF CARBON IN PHYTOPLANKTON, HERBIVORES, AND CARNIVORES

On the Continental Shelf of the North Atlantic
1956–1958

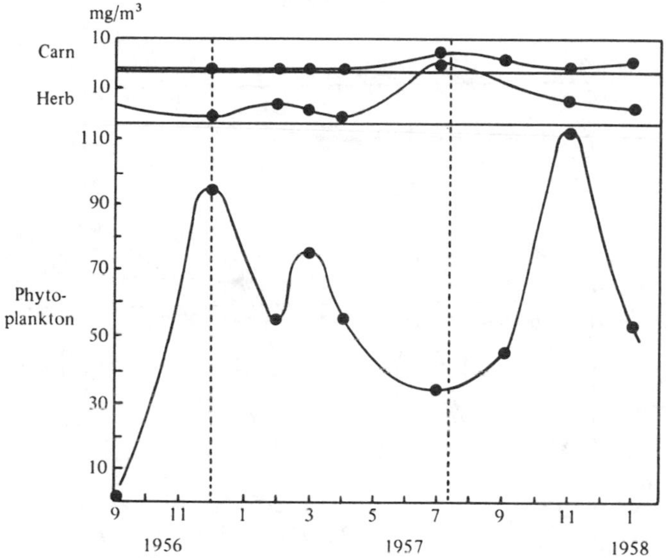

Table 6.3—4
AVERAGE CONCENTRATIONS OF PHOSPHORUS IN THREE TROPHIC LEVELS AND DISSOLVED PHOSPHORUS IN SEA WATER

On the Continental Shelf of the North Atlantic
1956–1958

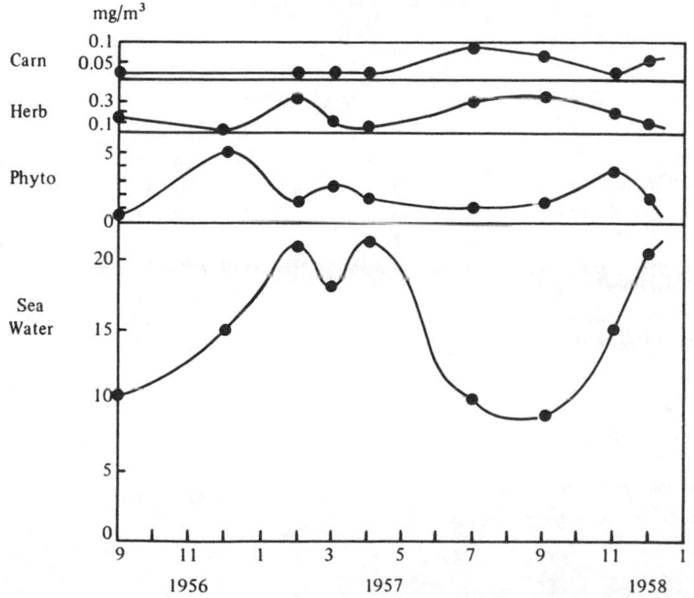

Table 6.3—5
AVERAGE CONCENTRATIONS OF NITROGEN IN THREE TROPHIC LEVELS AND AVAILABLE NITROGEN IN SEA WATER*

On the Continental Shelf of the North Atlantic
1956–1958

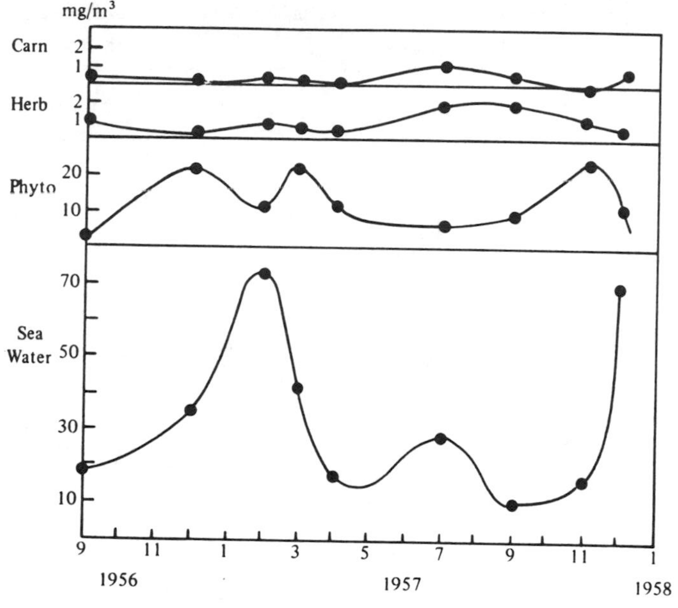

*As ammonia, nitrite, and nitrate totaled.

Table 6.3—6
THE RELATIVE COMPOSITION BY ATOMS OF C,[1] H, AND O, AND THE CALCULATED PHOTOSYNTHETIC QUOTIENT[2] IN VARIOUS PHYTOPLANKTERS

Organism	No. analyses	C-H-O	PQ
Diatoms	8	100–251–48	1.38
Peridians	7	100–184–54	1.16
Stichococcus bacillaria	6	100–160–46	1.17
Chlorella pyrenoidosa	8	100–168–41	1.21
Chlorella vulgaris	1	100–168–44	1.20
Scenedesmus obliquus 1	1	100–180–41	1.25
Scenedesmus obliquus 2	1	100–180–39	1.25
Scenedesmus basilensis	1	100–184–39	1.27
Nitzschia closterium	1	100–164–42	1.20

[1] C = 100.
[2] $O_2/-CO_2$.

(From Ryther, J. H., The measurement of primary production, *Limnol. Oceanogr.*, 1 (2), 75, 1956. With permission.)

Table 6.3—7
TOTAL CARBON, CARBONATE CARBON, AND ORGANIC CONCENTRATION IN PHYTOPLANKTONIC MARINE ORGANISMS

Phytoplankton, net collections		Total carbon, % dry wt	Carbonate carbon, % dry wt	Organic carbon	
				% dry wt	% a-f dry wt
12a	Phytoplankton (diatoms)	12.0	–	12.0	40.0
33a	Phytoplankton (diatoms)	7.1	–	7.1	35.3
36b	*Sargassum* sp.	31.5	–	31.5	39.3
	Average	16.8	–	38.2	

Note: All figures given are % of dry wt, excepting last column, which is ash-free dry wt. Where no results are shown, inorganic carbonate was not detectable.

Table 6.3—8
VITAMIN CONTENT OF ALGAE

Species	Vitamin	Quantities/100g dry matter	Ref.
Chlorella pyrenoidosa	thiamine	1–4.1 mg	1
	riboflavin	3.6–8 mg	
	nicotinic ac.	12–24 mg	
	pyridoxine	2.3 mg	
	pantothenic ac.	0.8–2.0 mg	
	biotin	14.8 μg	
	choline	300 mg	
	B_{12} (*E.g.*)*	2.2–10 μg	
Chlorella vulgaris	B_{12} (*E.c*)*	6.3 μg	2
Chlorella ellipsoidea	B_{12} (*E.g.*)*	4.2–8.9 μg	3

Note: See also Kanazawa, A., *Mem. Fac. Fish. Kagoshima Univ.*, 10, 38, 1962.

*B_{12} as assayed with microorganism: *E.g.* = *Euglena gracilis, E.c.* = *E. coli.*

(From Provasoli, L., Organic regulation of phytoplankton fertility, *The Sea*, Vol. 2, Hill, M. N., Ed., Interscience, New York, 1963, 182. © by John Wiley & Sons. With permission.)

REFERENCES

1. **Combs, C. F.,** *Science,* 116, 453, 1952.
2. **Brown, F., Cuthbertson, W. F. J., and Fogg, G. E.,** *Nature,* 177, 188, 1955.
3. **Hashimoto, Y.,** *J. Vitaminol.,* 1, 49, 1954.

Table 6.3—9
PRODUCTION OF CARBOHYDRATES IN MARINE ALGAE

Species	Carbohydrate produced (mg/l)	
	Before maximal growth	Highest value (stationary phase)
CHLOROPHYTES		
Dunaliella euchlora	3.1	9.0
Chlorella sp. (No. 580, Indiana Univ. Cult. Coll.)	–	9.0
Chlamydomonas sp. ("Y" R. Lewin)	2.1	10.6
Chlorococcum sp.	–	27.0
Pyramimonas inconstans	2.8	5.4
DIATOMS		
Cyclotella sp.	–	1.5
Nitzschia brevirostris	–	25.6
Melosira sp.	–	60.0
CHRYSOMONADS and **CRYPTOMONADS**		
Isochrysis galbana	–	25.0
Monochrysis lutheri	1.7	15.7
Prymnesium parvum	5.8–15.9	123.0
Rhodomonas sp.	1.9	8.8
DINOFLAGELLATES		
Amphidinium carteri	–	>5.0
Prorocentrum sp.[1]	–	~20.0
Katodinium dorsalisulcum[2]	–	0.6–2.4 g/l

(From Guillard, R. R. L. and Wangersky, P. J., the production of extracellular carbohydrates by some marine flagellates, *Limnol. Oceanogr.*, 3, 449, 1958. With permission.)

REFERENCES

1. **Collier, A.,** *Limnol. Oceanogr.*, 3, 33, 1958.
2. **McLaughlin, J. J. A., Zahl, P. A., Novak, Z., Marchisotto, J., and Prager, J.,** *Ann. N. Y. Acad. Sci.*, 90, 856, 1960.

Section 7
Primary Productivity

Section 7

PRIMARY PRODUCTIVITY

MEASUREMENTS

Odum (p. 43)[1] defines primary productivity of an ecological system, community, or any part thereof, "as the rate at which radiant energy is stored by photosynthetic and chemosynthetic activity of producer organisms (chiefly green plants) in the form of organic substances which can be used as food materials." It is generally expressed as grams of carbon produced in a column of water intersecting one square meter of sea surface per day (g C/m^2/d) or as grams of carbon produced in a given cubic meter per day (g C/m^3/d).[2] By the process of photosynthesis, green plants utilize light energy to synthesize energy-rich organic molecules (carbohydrate) from carbon dioxide and water in the presence of chlorophyll (contained in special organelles, the chloroplasts), with oxygen being released as a by-product. Each mole of carbohydrate manufactured requires the absorption of 120 kcal of radiant energy.[3] The basic photosynthetic equation can be written as

$$\underset{\text{carbon dioxide}}{6\,CO_2} + \underset{\text{water}}{6\,H_2O} + \underset{\text{energy}}{120\text{ kcal}} \;\underset{\text{Respiration}}{\overset{\text{Photosynthesis (chlorophyll)}}{\rightleftarrows}}\; \underset{\text{carbohydrate}}{C_6H_{12}O_6} + \underset{\text{oxygen}}{O_2}$$

This equation does not reflect the complicated details of the multistep process of photosynthesis which consists of two independent series of reactions.[4] When plants or animals respire, the photosynthetic process reverses; oxygen is taken up and chemical bonds are broken. In respiration, release of energy occurs together with the oxidation of carbohydrate.

Estimates of primary productivity in the ocean involve two distinct approaches: (1) measuring the exchange of individual chemicals between marine plants and their environment; and (2) calculating the increase in biomass of a plant population.[5] In the first approach, measurements of oxygen or carbon provide values of *in situ* photosynthesis. Somewhat less preferable are delineations of biomass (second approach) which, at best, yield estimates of *in situ* growth. Productivity, unlike biomass, is a rate phenomenon. Biomass or standing crop accounts for the amount, by weight, of plant (or animal) matter per volume of seawater (g/m^3, mg/l) or in a whole column of seawater over a square meter of the seafloor (g/m^2).[2,3] The rate of appearance of new algal biomass over time ascertains net primary production. The chlorophyll content per volume of seawater gives a direct reading of the total biomass of plants present.[6] The determination of chlorophyll *a*, a pigment comprised of complex organic molecules with a chemical composition of $C_{55}H_{72}O_5N_4Mg$, represents the most common method of assessing the standing crop of phytoplankton in the sea.

Primary productivity is difficult to measure accurately in the ocean. The light- and dark-bottle oxygen technique has been employed for many years in investigations of primary productivity; however, it has been superseded by the ^{14}C (radiocarbon) method as the preferred way to determine productivity of autotrophs in marine waters. Today, rates of primary production in coastal and oceanic waters are basically derived from ^{14}C measurements.[4] A third method, particle counting, has gained popularity with the advent of electronic particle counters, but difficulties in its application require considerable diligence, making it a less favorable choice. Comprehensive descriptions of the aforementioned techniques can be found elsewhere.[1,3-9]

Although the ^{14}C method has been widely adapted by marine scientists[10,11] since being introduced by Steemann Nielsen,[12] it continues to be scrutinized extensively. Frequently encountered questions relate to the reliability and validity of the radiocarbon technique. The method is subject to potential sources of error that may arise at four levels: (1) methodological mistakes; (2) physiological problems; (3) containment deficiencies; and (4) sampling and incubation strategies.[11,13-15] Oviatt et al.,[16] Gieskes et al.,[17] and Gieskes and Kraay[18] discuss the drawbacks of this procedure. One major controversy centers on whether the ^{14}C method measures gross photosynthesis, net photosynthesis, or some intermediate value.[5,19,20] Raymont (p. 392)[9] states that "the method is usually supposed to estimate net rather than gross primary production or some intermediate value." Most actual measurements, according to Valiela,[4] yield some number closer to net than gross production. Despite possible weaknesses, the ^{14}C method has garnered widespread appeal and acceptance in research on marine phytoplankton and primary productivity.

Over the past 35 years, elaborate illustrations have been published on the productivity of phytoplankton in the world's oceans, largely derived from ^{14}C measurements. Primary productivity, although varying widely from place to place, displays an overall pattern of higher values in upwelling zones and in shallow temperate waters and lower values in tropical regions. The centers of ocean gyres with depleted nutrient concentrations characteristically have low productivity. A stably stratified thermocline precludes the vertical exchange of water in the ocean gyres, thereby preventing the replenishment of nutrients for phytoplankton in surface waters. Here, primary productivity averages about 50 g C/m^2/year compared to 100 g C/m^2/year in coastal zones and 300 g C/m^2/year in upwelling areas. Coastal and upwelling regions benefit from greater nutrient supplies, resulting in elevated productivity levels. Coastal upwelling fosters the highest fertility of all, with coastal waters devoid of upwelling processes displaying intermediate productivity.

Upwelling takes place along much of the eastern margin of the Pacific Ocean (e.g., California and the Peruvian-Chilean coast) in addition to areas off India, Kamchatka, and Japan. In the Atlantic Ocean, similar processes operate off the southeast coast of South America, northeast Brazil, and west Africa. Divergences in the open ocean along the equator experience upwelling conditions as do the borders between the Antarctic and mid-oceanic current gyres.[3] As noted by Gross,[3] the Indian Ocean harbors the largest proportion of productive areas, especially in temperate and equatorial latitudes. Relatively broad, unproductive waters can be found in the Arctic and Pacific Oceans and in the tropics.

Productivity attributable to upwelling near the equator in tropical oceanic waters of the Atlantic and Pacific amounts to approximately 0.3 to 0.4 g C/m^2/d. Between latitudes 10 and 40° and within tropical gyres (e.g., Sargasso Sea and North Pacific gyre), productivity drops to levels approaching 0.1 g C/m^2/d.[2] Higher figures than this exist near Bermuda. Nevertheless, the Atlantic Ocean generally has lower primary productivity than the Pacific Ocean, due in part, to upwelling effects. For example, upwelling associated with the Peru Current alone contributes 10 g C/m^2/d to the Pacific. Productivity in the eastern part of the Indian Ocean averages about 0.19 g C/m^2/d; it equals about 0.24 g C/m^2/d in the western part of this ocean. The Arctic and Antarctic produce 1 g C/m^2/d, and production under sea ice can be considerable.[2]

In spite of the substantial production ascribable to coastal upwelling, these zones cover only a small fraction of the total world ocean; consequently, they supply only about 0.1 × 10^9 tons of organic carbon to the annual total of 20 × 10^9 tons for the hydrosphere.[3] Upwelling waters remain important to man. In particular, they support many major fisheries that feed humanity worldwide.

Coastal waters and estuaries typically are much more productive than the open ocean because of the occurrence of multiple plant subsystems, specifically benthic macrophytes (e.g., seagrasses, seaweeds, salt marsh grasses, mangroves) and phytoplankton.[21,22] The shallowness of these habitats along with rich sources of nutrients, creates an excellent environment for plant growth. A comparison of the productivity of marine autotrophs indicates that attached macroalgae and vascular plants are more productive than phytoplankton. Oceanic phytoplankton, especially, have low production.[4] It is not surprising, therefore, to discover the average annual (net) production rates of estuaries exceeding those of the continental shelf by four times, those of upwelling water by five times, and those of the open ocean by ten times.[4]

FACTORS AFFECTING PRIMARY PRODUCTIVITY

PHYSICAL AND CHEMICAL FACTORS

Light

Light and nutrient concentration comprise the most important physical and chemical factors respectively, affecting primary productivity in the ocean.[4,6,23] Hydrographic components, such as currents, diffusion, and upwelling, interact with these two factors to limit or enhance phytoplankton productivity. Thus, they exert a major influence on primary productivity in the ocean as well. Of all biological factors, grazing by herbivorous zooplankton most significantly limits phytoplankton production.[8,23,24]

Primary productivity by phytoplankton, as alluded to above, is closely coupled to sunlight as a source of radiant energy for photosynthesis.[25,26] Four aspects of light have been considered in studies of phytoplankton primary production: (1) the intensity of incident light; (2) changes in light on passing from air into water; (3) changes in light with increasing water depth; and (4) the utilization of radiant energy by phytoplankton cells.[25] A portion of light incident upon the ocean is lost by scattering and reflection at the water surface. The angle of the sun, the degree of cloud cover, and the roughness of the water surface regulate the amount of solar radiation reflected at the surface.[27] Further attenuation of light happens in the water column by absorption and scattering due to water molecules, suspended particles, and dissolved matter.[2,4] Beer's law can be used to calculate the total amount of light penetrating to any depth of water.[4] This law is given by

$$I_z = I_0 e^{-kz}$$

where I_z is the intensity of light at depth z, I_0 is the intensity at the surface, and k is the extinction coefficient of water, which varies from location to location and is wavelength specific.

Many publications deal with the relationship between light intensity and photosynthetic rate. Laboratory investigations reveal that photosynthesis by phytoplankton increases logarithmically with increasing light intensity until a maximum value, known as the light saturation value, is reached. Inhibition of photosynthesis is evident near the sea surface, where strong light intensities suppress phytoplankton growth. However, as light intensity decreases exponentially with increasing depth, the photosynthetic rate gradually increases to a peak level and then progressively diminishes to the compensation depth. At this depth, light intensity falls to about 1% of the surface radiation, and the oxygen liberated in photosynthesis balances that consumed in respiration.[21] Net primary production at the compensation depth, in other words, equals zero. Phytoplankton production is confined to the photic zone of the ocean.

The diminution of light intensity with increasing depth of water is greater in estuaries and coastal waters than in the open ocean, principally because of the higher concentrations of suspended particulate matter and dissolved organic substances nearshore which strongly absorb light. Selective absorption of wavelengths of light generates different spectral distributions for coastal and oceanic waters. Hence, the maximum transmittance in turbid coastal waters occurs at approximately 575 nm, and in the open ocean, at about 465 nm. Phycologists are mainly concerned with photosynthetically active radiation (PAR), approximately 400 to 700 nm, the wavelengths at which photosynthesis takes place in ordinary photoautotrophic plants.[28] Within this range of usable wavelengths, phytoplankton primarily utilize chlorophyll pigments in their chloroplasts to absorb light of greater than 600 nm, and accessory pigments (e.g., fucoxanthin and peridinin) to absorb light of less than 600 nm. Photosynthetic-irradiance curves have proven to be most helpful in studies of the physiological adjustments and primary production of phytoplankton to changing light intensity and quality.[24,29,30]

Light, interacting with temperature, triggers spring blooms of phytoplankton in temperate waters.[23] In mid and high latitudes, the photoperiod follows conspicuous seasonal cycles that are related to plant growth and production. Insufficient light during winter inhibits photosynthesis. In spring, increased light and water temperature initiate blooms in temperate latitudes as a thermocline forms, isolating the mixed layer in the photic zone. Once this condition develops, the rate of photosynthesis of the phytoplankton will surpass that of respiration and spring growth can proceed.[22,23]

Nutrients

In addition to light, marine plants need a number of nutrients for adequate growth and reproduction, the most critical being nitrogen, phosphorus, and silicon. Among these three elements, nitrogen (as nitrate, NO_3^-) and phosphorus (as phosphate, PO_4^{3-}) impact primary productivity most greatly; both are necessary for survival of autotrophs, yet exist in very small concentrations in seawater. For example, nitrate levels in seawater equal about 1 µg-atom/l or less and rarely exceed 25 µg-atom/l, whereas phosphate values generally range from 0 to 3 µg-atom/l.[4] Silicon, when present in very low amounts, represses metabolic activity of the cell[31] and can limit phytoplankton production. It represents an essential element for the skeletal growth of diatoms, as well as radiolarians and certain sponges. Elements other than nitrogen and phosphorus also are required by autotrophs, but their availability usually does not limit growth. These encompass the major elements (e.g., calcium, carbon, magnesium, oxygen, and potassium), minor and trace elements (e.g., cobalt, copper, iron, molybdenum, vanadium, and zinc), and organic nutrients (e.g., biotin, thiamine, and vitamin B_{12}).[2] Phytoplankton utilize essential elements in both particulate and dissolved forms.[2] Fogg[32] lists at least 18 minerals and various growth factors necessary for the growth of planktonic microalgae, any of which may mitigate the growth process.[21]

Nitrogen

Nitrogen is the chief limiting element to primary production in estuarine and oceanic waters.[33-37] Ammonia (NH_3), nitrite (NO_2), and nitrate (NO_3^-) compose the three principal dissolved inorganic forms of nitrogen in the ocean, with nitrate found in highest concentrations. Dissolved organic forms, such as urea, amino acids, and peptides, may also be valuable nutrient sources for autotrophic growth in the sea. Phytoplankton assimilate ammonia, nitrite, and nitrate. Whereas most phytoplankton exhibit good growth on nitrate, some do not.[9] A number of euglenids, cryptomonads, and green algae, for instance, require ammonia and amino acids (reduced nitrogen) for growth. Ammonia and organic nitrogenous substances are the preferred forms of nitrogen for phytoplankton.[2,38] The uptake of ammonia serves a significant advantage for these microscopic plants since it can be used directly in the synthesis of amino acids. Wheeler[39] explores the aspects of phytoplankton nitrogen metabolism that control, in part, nutrient uptake and growth rates.

Biogeochemical cycling of nitrogen in shallow water systems (e.g., estuaries) incorporates several significant processes in the water column and bottom sediments. In the water column, these processes include uptake, remineralization, and oxidation, and in bottom sediments, they involve burial, remineralization, biological uptake, oxidation, reduction, nitrous oxide production, and denitrification.[40] Benthic-pelagic coupling of nutrients in nearshore environments may play a major role in phytoplankton dynamics.

A distinction can be drawn between new and regenerated forms of nitrogen for supporting phytoplankton production in the sea. Dugdale and Goering[41] differentiated new and regenerated nitrogen: new nitrogen pertains to nitrate allochthonously entering the photic zone from coastal and upwelling sources; and regenerated nitrogen refers to ammonia and urea autochthonously regenerated by zooplankton in the upper part of the water column. Kemp et al.[42] extended this concept to estuarine waters, considering new nitrogen to be that entering allochthonously from the watershed or from offshore waters and regenerated nitrogen to be that recycled either within the water column or across the sediment-water interface.

Nitrogen and phosphorus undergo marked seasonal cycles in temperate oceanic waters, peaking in the winter, subsiding rapidly in the spring, remaining low in the summer, and rising in the late fall.[43] This seasonal cycle has been related to that of phytoplankton abundance. As phytoplankton populations increase in the spring, they assimilate nutrients which become depleted. In the winter, phytoplankton populations decline, and the nutrients attain maximum levels. This seasonal pattern is less conspicuous in tropical regions characterized by more rapid nutrient cycling attributable to higher light energies and temperatures that allow continuous phytoplankton production.[23] Nutrient replenishment appears to control phytoplankton growth in the tropics.[44]

Phosphorus

This nutrient element enters rivers, estuaries, and embayments in dissolved inorganic form largely as the phosphate anion (PO_4^{3-}). In marine waters, phosphorus is present not only as dissolved inorganic phosphorus but also as dissolved organic phosphorus and particulate phosphorus. Orthophosphates — phosphoric acid (H_3PO_4) and its dissociation products ($H_2PO_4^-$, HPO_4^{2-}, and PO_4^{3-}) — constitute the major fraction of dissolved inorganic phosphorus. The main ion in seawater is HPO_4^{2-} which exceeds PO_4^{3-} by about one order of magnitude.[4] Orthophosphate is the preferred form for phytoplankton;[2] however, these autotrophs probably assimilate some dissolved organic phosphorus as well, particularly during periods of deficiency.[25]

Zooplankton grazing and excretion account for rapid regeneration of phosphorus in pelagic waters.[45] In estuaries, bottom sediments provide a sink for this nutrient element.[22] The flux of phosphorus across the sediment-water interface in shallow water systems supplies a continuum of nutrients for plant growth.

The ratio of nitrogen to phosphorus (N:P) in phytoplankton is about 16:1, which approximates the ratio of N:P in seawater. Wide variations can take place in the N:P ratios; over a 12-month period, for instance, Harris and Riley[46] discerned ratios of 12:1 to 19.8:1 in phytoplankton of Long Island Sound. Smith (p. 1149)[35] remarks that "phosphorus vs. nitrogen limitation is a function of the relative rates of water exchange and internal biochemical processes acting to adjust the ratio of ecosystem N:P availability." As noted by Levinton,[2] the absolute amount of phosphate in seawater probably limits the phytoplankton standing crop, but the minimum concentration limiting the growth rate has not been established. Nitrogen shortages rather than phosphorus limitation are deemed to be responsible for halting the growth of phytoplankton populations in marine ecosystems.

BIOLOGICAL FACTORS

Zooplankton Grazing

Estimates of the biomass or standing crop of a phytoplankton community may not reflect actual production because of grazing by herbivorous organisms, especially zooplankton.[23] Whereas some phytoplankton losses in the ocean arise via sinking below the photic zone, the vast majority of cells disappear by zooplankton grazing. Grazing intensity by these herbivores varies both in space and time. In some areas, a pronounced alternation in the abundance of phytoplankton and zooplankton has been observed, with the decline in phytoplankton numbers being ascribed to zooplankton grazing. In other cases, zooplankton grazing has been discounted as the major factor in the decrease of phytoplankton populations. A lack of control by zooplankton grazing on phytoplankton is most often perceived in inshore waters experiencing greater environmental variability than in offshore waters.[2] Zooplankton grazing exerts a more consistent controlling influence on phytoplankton in the open ocean.

The potential effectiveness of zooplankton grazing becomes apparent when examining the daily demands of zooplankton populations for maintenance and growth. Daily requirements of zooplankton populations generally approach 30 to 50% of their weight each day. The daily demand in exceptional cases (e.g., *Calanus*) may surpass 300%.[9] In Long Island Sound, tintinnids (the dominant component of the microzooplankton community) consume approximately 27% of the annual primary production, and copepods remove about 44%.[47]

Selective feeding by zooplankton potentially governs the composition of the phytoplankton community.[2] Copepods prefer to crop larger-sized phytoplankton cells. In addition, they apparently have an ability to graze upon the most abundant size classes, thereby favoring the persistence of the less common size classes and less abundant species. When grazing is intense and phytoplankton abundance decreases below a critical level, zooplankton abundance likewise wanes after a lag period. The short generation time of phytoplankton relative to that of zooplankton enhances rapid recovery of the plants.

In seasonal plankton cycles of estuaries, phytoplankton blooms are often superseded by a peak in zooplankton abundance, although strong oscillations in both phytoplankton and zooplankton abundance tend to be suppressed.[2] The time lag between a phytoplankton bloom and a subsequent rise in zooplankton abundance in these systems usually varies, leading to differences in phytoplankton concentrations. In open oceanic environments, the time lag may be slight, especially in the tropics. Zooplankton grazing in open oceanic, tropical waters acts as a stabilizing force on the phytoplankton standing crop year-round.[48]

Both phytoplankton and zooplankton exhibit a patchy distribution in marine waters,[49,50] with the patches ranging in size from several meters to many kilometers.[9] The patchiness of phytoplankton is often noticeable in estuaries and coastal waters where environmental conditions (e.g., currents, wind-induced turbulence, and river runoff) change abruptly. Parsons and Takahashi[51] recount six processes which promote patchiness: (1) grazing; (2) advective effects of water movement; (3) physical-chemical boundary conditions; (4) reproductive rates within a population; (5) social behavior in populations of the same species; and (6) intraspecific interactions causing attraction or repulsion between species. The tendency for denser patches of phytoplankton to be spatially separated or to alternate with those of zooplankton has been attributed to zooplankton grazing.[21]

Zooplankton grazing accelerates at night and subsides during the day. Stimulated by changes in the level of ambient illumination, zooplankton undergo vertical migration, descending to greater depths of the water column in the day and ascending toward the surface at night. This diel vertical migration confers advantages to the zooplankton populations, enabling them to avoid predators in well lit surface waters and to utilize cooler, deeper waters to lower metabolic rates. Additional advantages include the maximization of food intake and utilization, maximization of fecundity, and maximization of various strategies associated with horizontal dispersion and transport.[21,52]

PLAN OF THIS SECTION

Primary productivity in this volume covers various facets, initially concentrating on annual productivity by geographic region and subsequently treating variations of marine productivity by season and depth. Data also are presented on dissolved organic composition related to productivity as well as photosynthetic quotients and gross-net differences. Finally, this section addresses the relationship of primary productivity to productivity at other trophic levels.

REFERENCES

1. **Odum, E. P.,** *Fundamentals of Ecology*, 3rd ed., W. B. Saunders, Philadelphia, 1971.
2. **Levinton, J. S.,** *Marine Ecology*, Prentice-Hall, Englewood Cliffs, N. J., 1982.
3. **Gross, M. G.,** *Oceanography: A View of the Earth*, 3rd ed., Prentice-Hall, Englewood Cliffs, N. J., 1982.
4. **Valiela, I.,** *Marine Ecological Processes*, Springer-Verlag, New York, 1984.
5. **Dring, M. J.,** *The Biology of Marine Plants*, Edward Arnold, London, 1982.
6. **Nybakken, J. W.,** *Marine Biology: An Ecological Approach*, Harper & Row, New York, 1982.
7. **Strickland, J. D. H. and Parsons, T. R.,** A practical handbook of seawater analysis, *J. Fish. Res. Board Can.*, Bull. 167, 1968.
8. **Vollenweider, R. A.,** *A Manual on Methods for Measuring Primary Production in Aquatic Environments*, I. B. P. Handbook 12, 2nd ed., Blackwell Scientific, Oxford, 1974.
9. **Raymont, J. E. G.,** *Plankton and Productivity in the Oceans*, 2nd ed., Vol. 1, Pergamon Press, Oxford, 1980.
10. **Colijn, F. and de Jonge, V. N.,** Primary production of microphytobenthos in the Ems-Dollard estuary, *Mar. Ecol. Prog. Ser.*, 14, 185, 1984.
11. **Davies, J. M. and Williams, P. J. le B.,** Verification of ^{14}C and O_2 derived primary organic production measurements using an enclosed ecosystem, *J. Plankton Res.*, 6, 457, 1984.
12. **Steemann Nielsen, E.,** The use of radioactive carbon (^{14}C) for measuring organic production in the sea, *J. Cons. Perm. Int. Explor. Mer*, 18, 117, 1952.
13. **Sieburth, J. McN.,** International Helgoland Symposium: convener's report on the informal session on biomass and productivity of microorganisms in planktonic ecosystems, *Helgol. Wiss. Meeresunters.*, 30, 697, 1977.
14. **Peterson, B. J.,** Aquatic primary productivity and the ^{14}C-CO_2 method: a history of the productivity problem, *Annu. Rev. Ecol. Syst.*, 11, 359, 1980.
15. **Colijn, F., Gieskes, W. W. C., and Zevenboom, W.,** Problems with the measurement of primary production: conclusions and recommendations, *Hydrobiol. Bull.*, 17, 29, 1983.

16. **Oviatt, C., Buckley, B., and Nixon, S.,** Annual phytoplankton metabolism in Narragansett Bay calculated from survey field measurements and microcosm observations, *Estuaries*, 4, 167, 1981.
17. **Gieskes, W. W. C., Kraay, G. W., and Baars, M. A.,** Current ^{14}C methods for measuring primary production: gross underestimates in oceanic waters, *Neth. J. Sea Res.*, 13, 58, 1979.
18. **Gieskes, W. W. C. and Kraay, G. W.,** State-of-the-art in the measurement of primary production, in *Flows of Energy and Materials in Marine Ecosystems: Theory and Practice*, Fasham, M. J. R., Ed., Plenum Press, New York, 1984, 71.
19. **Dring, M. J. and Jewson, D. H.,** What does ^{14}C uptake by phytoplankton really measure? A theoretical modelling approach, *Proc. R. Soc. London*, B214, 351, 1982.
20. **Holligan, P. M., Williams, P. J. le B., Purdie, D., and Harris, R. P.,** Photosynthesis, respiration and nitrogen supply of plankton populations in stratified, frontal and tidally mixed shelf waters, *Mar. Ecol. Prog. Ser.*, 17, 201, 1984.
21. **Mann, K. H.,** *Ecology of Coastal Waters: A Systems Approach*, University of California Press, Berkeley, 1982.
22. **Kennish, M. J.,** *Ecology of Estuaries*, Vol. 1, CRC Press, Boca Raton, Fla., 1986.
23. **Dawes, C. J.,** *Marine Botany*, John Wiley & Sons, New York, 1981.
24. **Marra, J., Wiebe, P. H., Bishop, J. K. B., and Stepien, J. C.,** Primary production and grazing in the plankton of the Panama Bight, *Bull. Mar. Sci.*, 40, 255, 1987.
25. **Boney, A. D.,** *Phytoplankton*, Edward Arnold, London, 1975.
26. **Marra, J., Heinemann, K., and Landriau, G., Jr.,** Observed and predicted measurements of photosynthesis in a phytoplankton culture exposed to natural irradiance, *Mar. Ecol. Prog. Ser.*, 24, 43, 1985.
27. **Peterson, D. H. and Festa, J. F.,** Numerical simulation of phytoplankton productivity in partially mixed estuaries, *Est. Coastal Shelf Sci.*, 19, 563, 1984.
28. **Steemann Nielsen, E.,** *Marine Photosynthesis: With Emphasis on the Ecological Aspects*, Elsevier Scientific, Amsterdam, 1975.
29. **Lewis, M. R., Warnock, R. E., and Platt, T.,** Absorption and photosynthetic action spectra for natural phytoplankton populations: implications for production in the open ocean, *Limnol. Oceanogr.*, 30, 794, 1985.
30. **Lewis, M. R., Warnock, R. E., Irwin, B., and Platt, T.,** Measuring photosynthetic action spectra of natural phytoplankton populations, *J. Phycol.*, 21, 310, 1985.
31. **Werner, D.,** Silicate metabolism, in *The Biology of Diatoms*, Werner, D., Ed., Blackwell Scientific, Oxford, 1977, 110.
32. **Fogg, G. E.,** Primary productivity, in *Chemical Oceanography*, Vol. 2, 2nd ed., Riley, J. P. and Skirrow, G., Eds., Academic Press, London, 1975, 385.
33. **Carpenter, E. J. and Capone, D. G., Eds.,** *Nitrogen in the Marine Environment*, Academic Press, New York, 1983.
34. **Holm-Hansen, O., Bolis, L., and Gilles, R., Eds.,** *Marine Phytoplankton and Productivity*, Springer-Verlag, Berlin, 1984.
35. **Smith, S. V.,** Phosphorus versus nitrogen limitation in the marine environment, *Limnol. Oceanogr.*, 29, 1149, 1984.
36. **Howarth, R. W. and Cole, J. J.,** Molybdenum availability, nitrogen limitation, and phytoplankton growth in natural waters, *Science*, 229, 653, 1985.
37. **Kanda, J., Saino, T., and Hattori, A.,** Nitrogen uptake by natural populations of phytoplankton and primary production in the Pacific Ocean: regional variability of uptake capacity, *Limnol. Oceanogr.*, 30, 987, 1985.
38. **Lippson, A. J., Haire, M. S., Holland, A. F., Jacobs, F., Jensen, J., Moran-Johnson, R. L., Polgar, T. T., and Richkus, W. A.,** *Environmental Atlas of the Potomac Estuary*, Johns Hopkins University Press, Baltimore, 1981.
39. **Wheeler, P. A.,** Phytoplankton nitrogen metabolism, in *Nitrogen in the Marine Environment*, Carpenter, E. J. and Capone, D. G., Eds., Academic Press, New York, 1983, 309.
40. **Nixon, S. W. and Pilson, M. E. Q.,** Nitrogen in estuarine and coastal marine ecosystems, in *Nitrogen in the Marine Environment*, Carpenter, E. J. and Capone, D. G., Eds., Academic Press, New York, 1983, 565.
41. **Dugdale, R. C. and Goering, J. J.,** Uptake of new and regenerated forms of nitrogen in primary productivity, *Limnol. Oceanogr.*, 12, 196, 1967.
42. **Kemp, W. M., Wetzel, R. L., Boynton, W. R., D'Elia, C. F., and Stevenson, J. C.,** Nitrogen cycling and estuarine interfaces: some current concepts and research directions, in *Estuarine Comparisons*, Kennedy, V. S., Ed., Academic Press, New York, 1982, 209.
43. **Corner, E. D. S. and Davies, A. G.,** Plankton as a factor in the nitrogen and phosphorous cycles in the sea, *Adv. Mar. Biol.*, 9, 102, 1971.
44. **Steeman Nielsen, E.,** Primary production in tropical marine areas, *J. Mar. Biol. Assoc. India*, 1, 7, 1959.
45. **Pomeroy, L. R., Mathews, H. M., and Min, H. S.,** Excretion of phosphate and soluble organic phosphorus compounds by zooplankton, *Limnol. Oceanogr.*, 8, 50, 1963.
46. **Harris, E. and Riley, G. A.,** Oceanography of Long Island Sound, 1952—1954. VIII. Chemical composition of the plankton, *Bull. Bingham Oceanogr. Coll.*, 15, 315, 1956.
47. **Capriulo, G. M. and Carpenter, E. J.,** Grazing by 35 to 202 μm microzooplankton in Long Island Sound, *Mar. Biol.*, 56, 319, 1980.
48. **Steemann Nielsen, E.,** The balance between phytoplankton and zooplankton in the sea, *J. Cons. Perm. Int. Explor. Mer.*, 23, 178, 1958.
49. **Steele, J. H., Ed.,** *Spatial Pattern in Plankton Communities*, Plenum Press, New York, 1978.
50. **Bennett, A. F. and Denman, K. L.,** Phytoplankton patchiness: inferences from particle statistics, *J. Mar. Res.*, 43, 307, 1985.
51. **Parsons, T. R. and Takahashi, M.,** *Biological Oceanographic Processes*, Pergamon Press, Oxford, 1973.
52. **Longhurst, A. R.,** Vertical migration, in *The Ecology of the Seas*, Cushing, D. H. and Walsh, J. J., Eds., Blackwell Scientific, Oxford, 1976, 116.

7.1. ANNUAL, BY REGIONS

Table 7.1—1
ANNUAL RATE OF CARBON FIXATION
g/sq. m of sea surface

	Total depth (m)	$gC/m^2/yr$
Long Island Sound	25	380
Continental Shelf	25–50	160
	50–1,000	135
	1,000–2,000	100
North Central Sargasso Sea	>5,000	78

(From Ryther, J. H. and Yentsch, C. S., Primary production of continental shelf waters off New York, *Limnol.Oceanogr.*, 3(3), 334, 1958. With permission.)

Figure 7.1—1

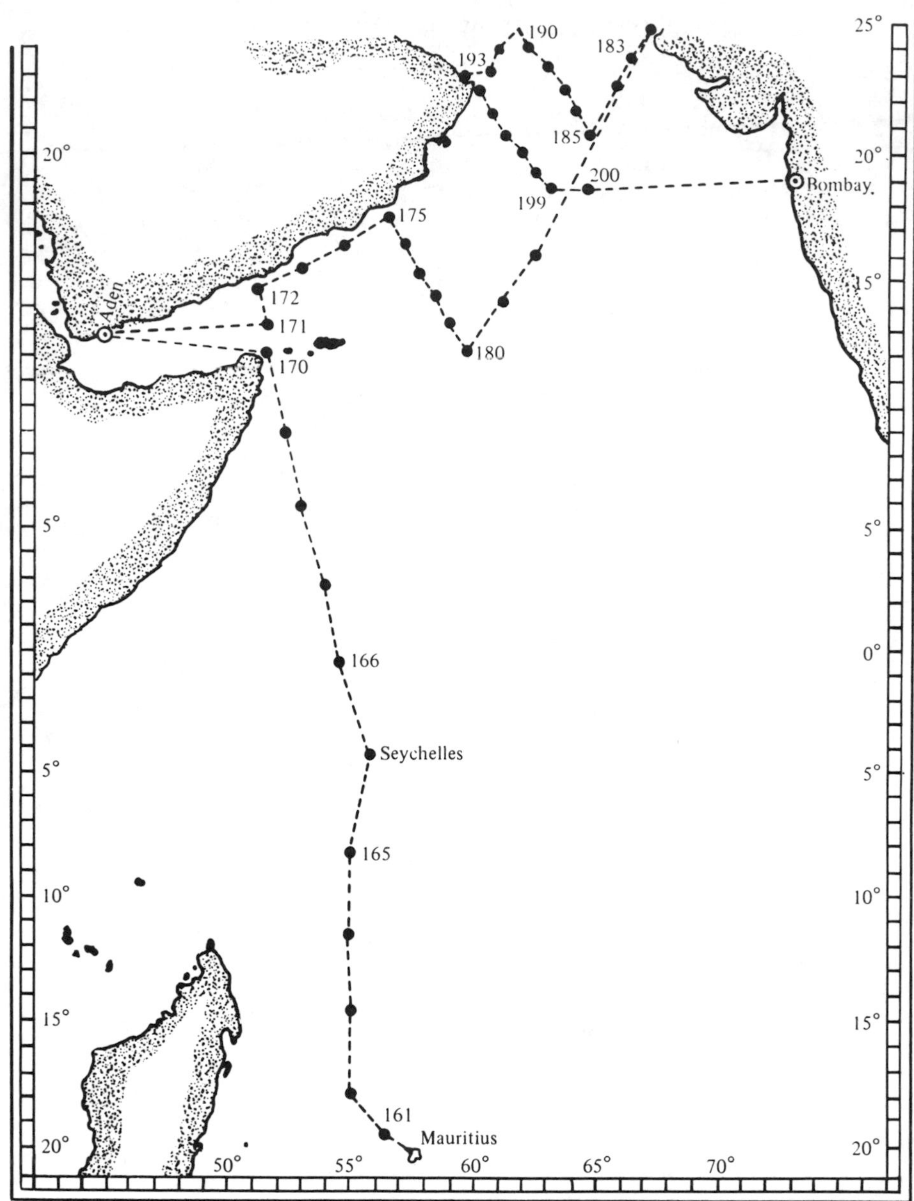

Figure 7.1—1. Cruise course in western Indian Ocean with total living carbon and daily carbon production (A) and vertical distribution of temprature, phosphate, and integrated primary productivity (B) (C) along numbered transects.

(From Ryther, J. H. and Menzel, D. W., on the production, composition, and distribution of organic matter in the western Arabian Sea, *Deep-Sea Res.*, 12, 200, © 1965, Pergamon Press. With permission.)

Figure 7.1—1 (continued)

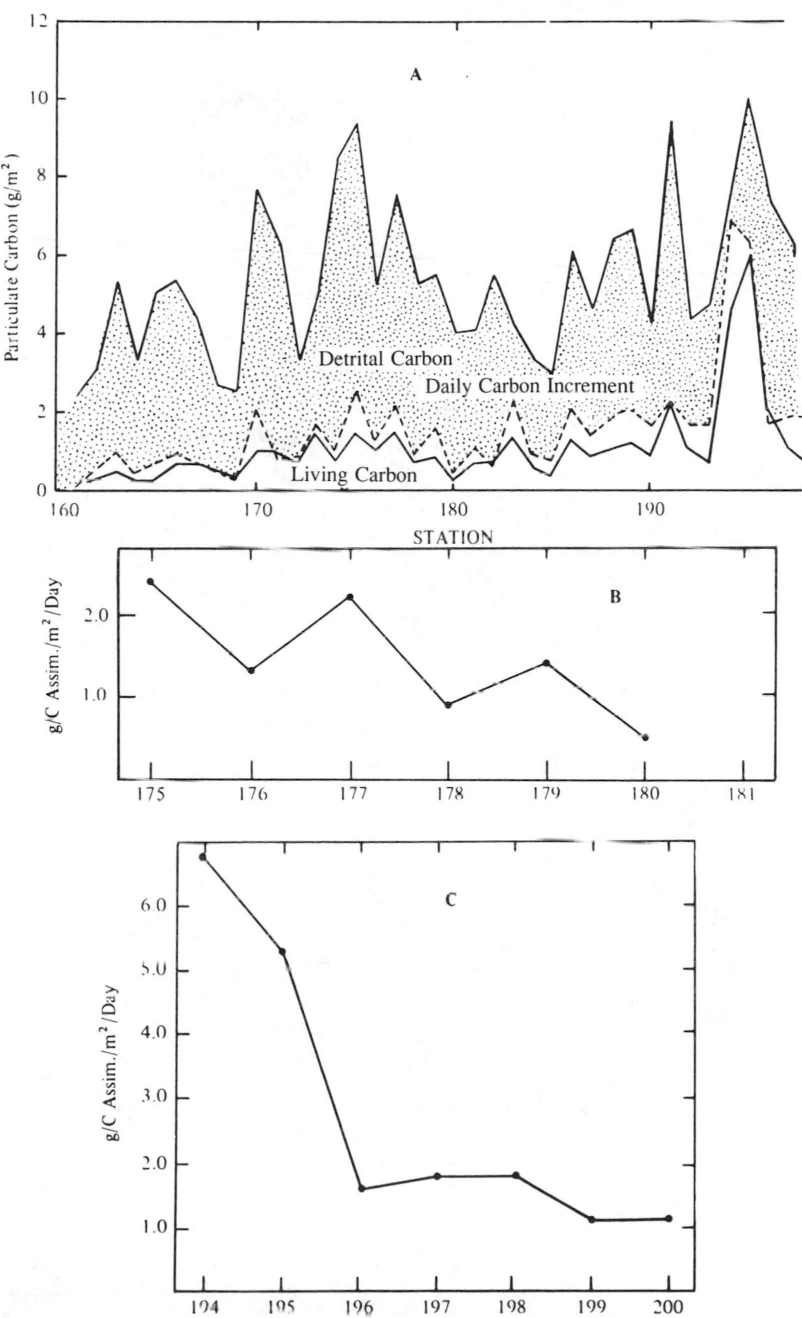

Figure 7.1—2

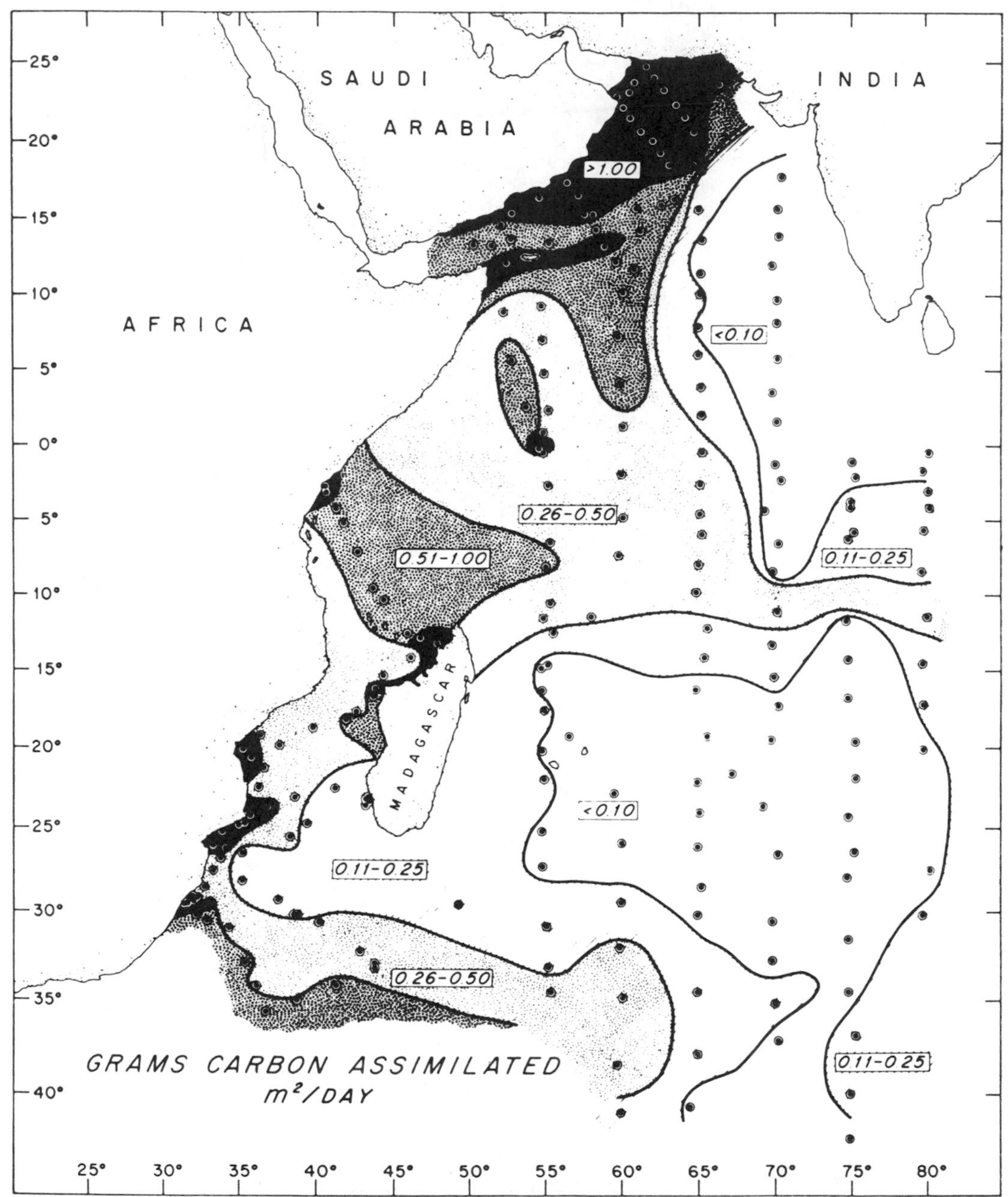

Figure 7.1—2. The general level of primary organic production in the Indian Ocean in grams of carbon assimilated per square meter per day.

(From Ryther, J. H., Hall, J. R., Pease, A. K., Bakun, A., and Jones, M. M., Primary organic production in relation to the chemistry and hydrography of the western Indian Ocean, *Limnol. Oceanogr.*, 11(3), 375, 1966. With permission.)

Figure 7.1—3

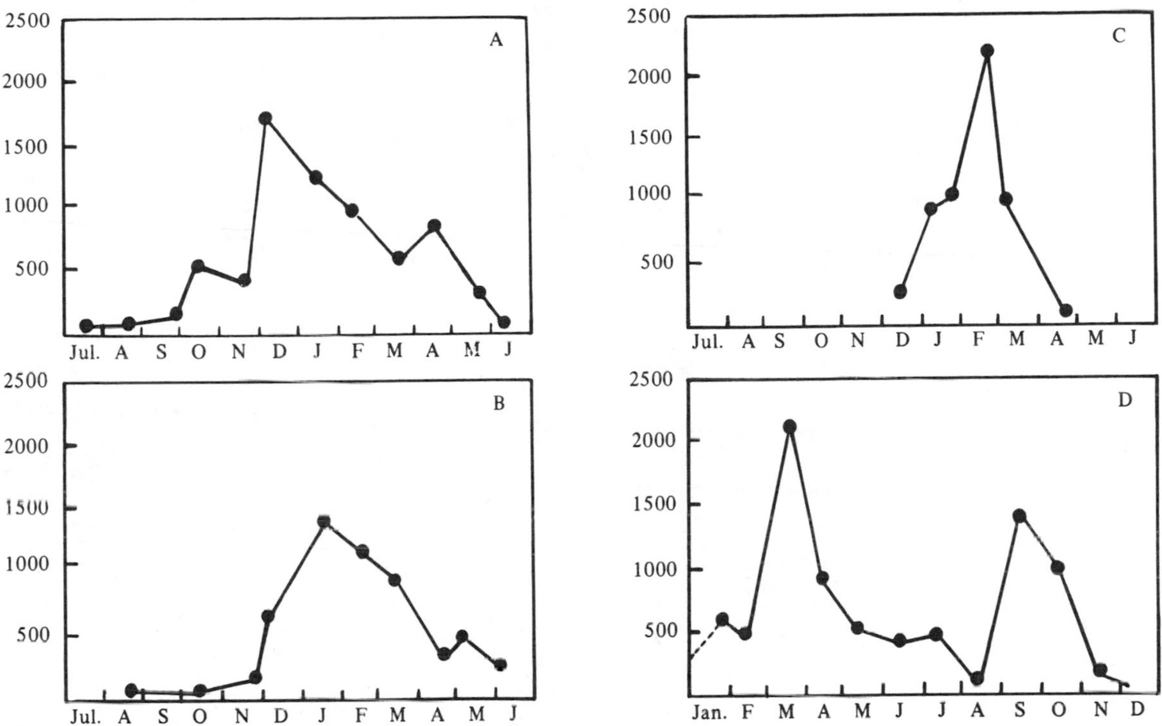

Figure 7.1—3. Plant pigments (Harvey units/m^3) in the northern (A), intermediate (B), and southern (C) regions of the Antarctic Ocean, and in the English Channel (D).

(From Hart, T. J., Phytoplankton periodicity in Antarctic surface waters, *Discovery Rep.*, 21, 261—356, 1942. With permission.)

Table 7.1—2
DIVISION OF THE OCEAN INTO PROVINCES ACCORDING TO THEIR LEVEL OF PRIMARY ORGANIC PRODUCTION

Province	% Ocean	Area (km^2)	Mean productivity (g $C/m^2/yr$)	Total productivity (10^3 tons of C/yr)
Open ocean	90.0	326.0×10^6	50	16.3
Coastal zone*	9.9	36.0×10^6	100	3.6
Upwelling areas	0.1	3.6×10^5	300	0.1
Total				20.0

*Indicates offshore areas of high productivity.

Table 7.1—3
ESTIMATES OF ANNUAL PRIMARY PRODUCTION IN AREAS OFF THE WASHINGTON AND OREGON COASTS

Area	Annual production (g C m^{-2} yr^{-1})	Range (g C m^{-2} yr^{-1})	Mean daily production (g C m^{-2} day^{-1})
Oceanic	61	43–78	0.17
Plume	60	46–73	0.16
River mouth	88		0.24
Upwelling	152		0.42

(From Anderson, G. C., The seasonal and geographic distribution of primary productivity off the Washington and Oregon coasts, *Limnol. Oceanogr.*, 9, 298, 1964. With permission.)

Figure 7.1—4

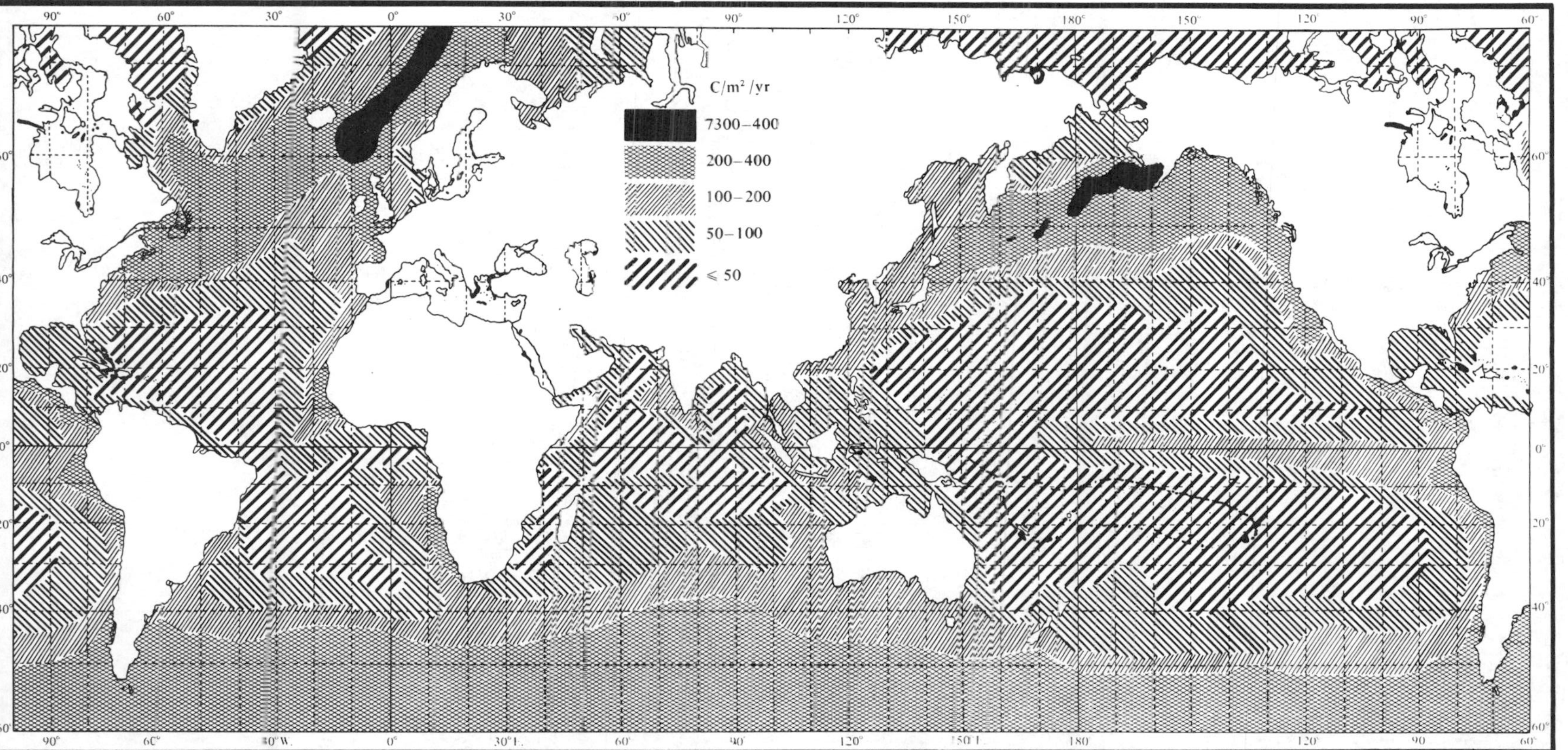

Figure 7.1—4. Estimation of organic production in the oceans. Contours are in grams of carbon biologically fixed under each square meter of sea surface per year.

7.2. VARIATIONS BY SEASON AND DEPTH

Table 7.2—1
DISTRIBUTION OF THE PLANKTONIC BIOMASS AT A 24-HR STATION IN THE SOUTHWESTERN PART OF THE BERING SEA*

Sampled layer, m	$0^{00}-1^{05}$	$3^{55}-4^{30}$	$7^{55}-9^{07}$	$12^{01}-13^{05}$	$16^{00}-17^{00}$	$20^{00}-21^{05}$
0–10	23.9	18.2	10.2	8.4	17.9	14.8
10–25	36.1	23.2	28.8	13.5	11.9	41.2
25–50	22.5	32.6	22.9	31.5	37.8	24.2
50–100	6.1	15.8	26.8	37.5	19.2	5.0
100–200	3.2	3.1	2.9	2.5	4.1	4.7
200–500	8.2	7.1	8.4	6.6	9.5	10.1

* % mg/m^3.

Table 7.2—2
DISTRIBUTION OF THE PLANKTONIC BIOMASS AT A DIURNAL STATION IN THE KURILE-KAMCHATKA REGION[a]

Sampled layer, m	$19^{40}-0^{50}$	$2^{00}-5^{45}$	$9^{00}-13^{40}$ [b]
0–10	8.1	4.8	0.4
10–25	29.3	25.0	11.6
25–50	23.0	17.2	50.2
50–125	12.8	15.8	9.5
125–200	11.3	17.1	9.6
200–400	9.9	11.7	10.3
400–750	2.7	4.8	5.8
750–1250	2.0	2.0	2.1
1250–1500	1.0	1.6	1.3

[a] % mg/m^3.

[b] Sampling started in the lower layers and terminated in the upper layers. Therefore, the data on the amount of plankton in the upper 200 m layer apply to the last 20 min of the indicated period of time.

Figure 7.2—1

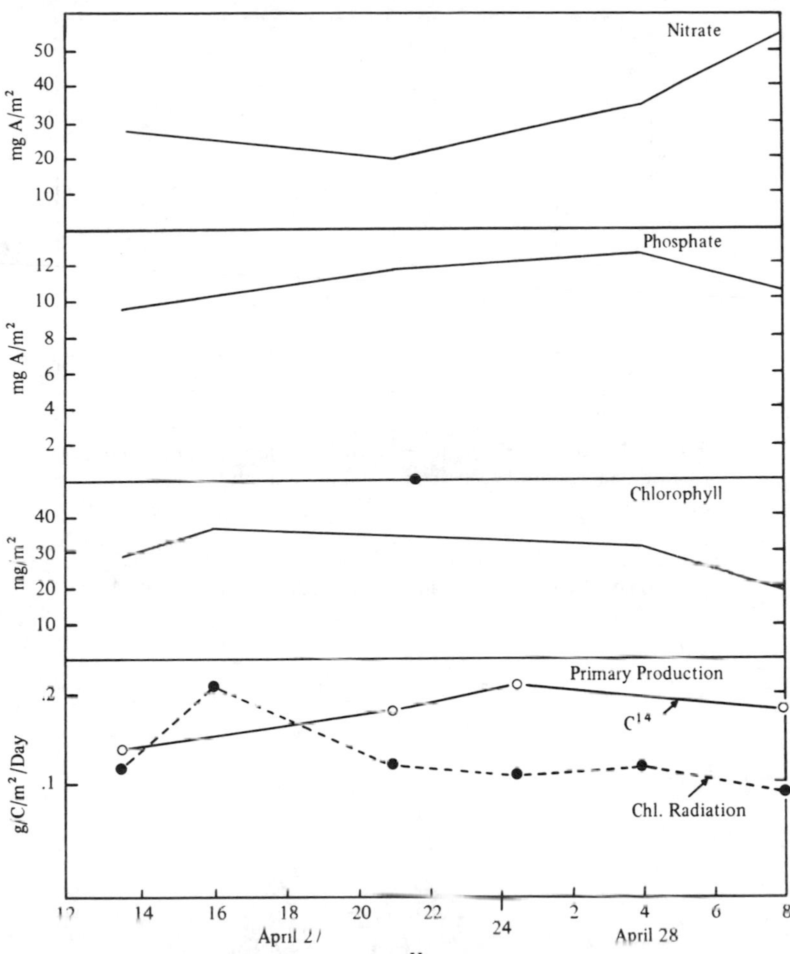

Figure 7.2—1. Diurnal variations in nitrate, phosphate, chlorophyll and primary production in the middle of the Sargasso Sea.

(From Ryther, J. H., Menzel, D. W., and Vaccaro, R. F., Diurnal variations in some chemical and biological properties of the Sargasso Sea, *Limnol. Oceanogr.*, 6(2), 152, 1961. With permission.)

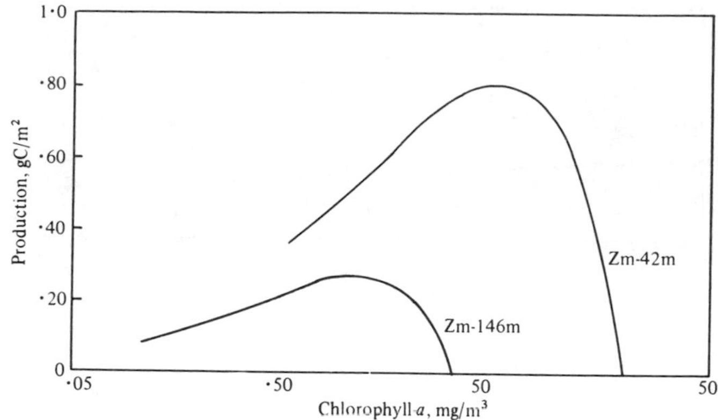

Figure 7.2—2. The relationship between net production and chlorophyll-*a* concentration for two depths of mixed layer (Zm) off Bermuda.

(From Steele, J. H. and Menzel, D. W., Conditions for maximum primary production in the mixed layer, *Deep-Sea Res.*, 9, 45, © 1962 Pergamon Press. With permission.)

Table 7.2—3
PHOTOSYNTHESIS (C^{14} UPTAKE)

For 24 Hr at 1,500 ft. candles of Water from Different Depths off Bermuda

Depth (m)	Mg C assim m^3/24 hr
0	3.05
17	2.10
34	2.20
60	2.90
120	0.47
150	0.00
200	0.00
250	0.00
300	0.00
350	0.00
400	0.00

(From Menzel, D. W. and Ryther, J. H., The annual cycle of primary production in the Sargasso Sea off Bermuda, *Deep-Sea Res.*, 6, 362, © 1960 Pergamon Press. With permission.)

Figure 7.2—3

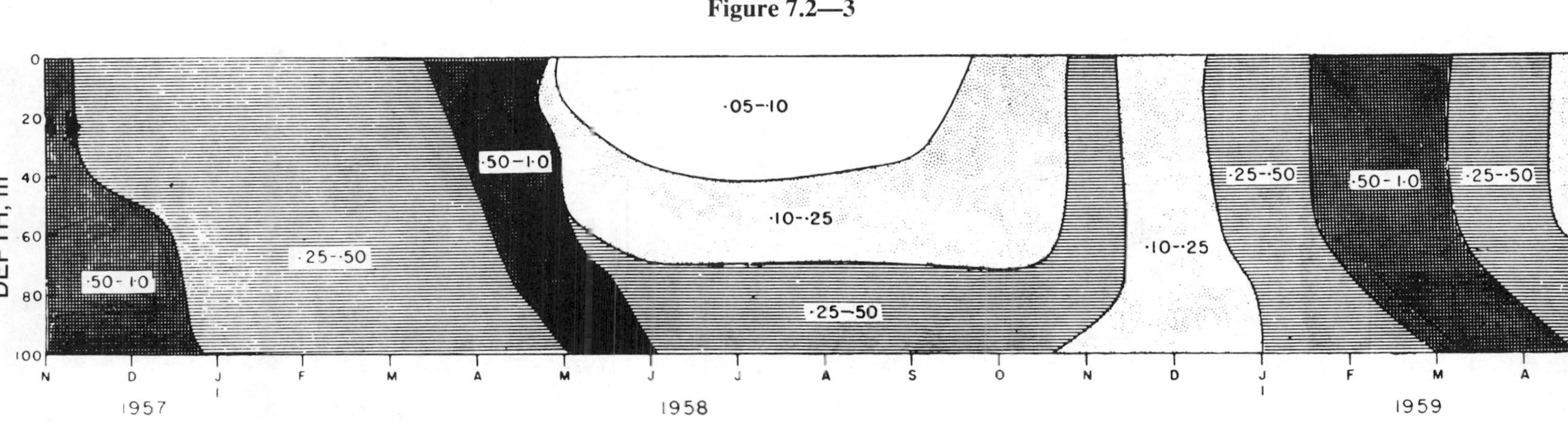

Figure 7.2—3. A three-year seasonal depth profile of chlorophyll-*a* (mg/m^3) off Bermuda.

(From Ryther, J. H. and Menzel, D. W., The annual cycle of primary production in the Sargasso Sea off Bermuda, *Deep-Sea Res.*, 6, 360, © 1960 Pergamon Press. With permission.)

Figure 7.2—4

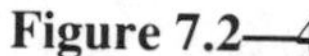

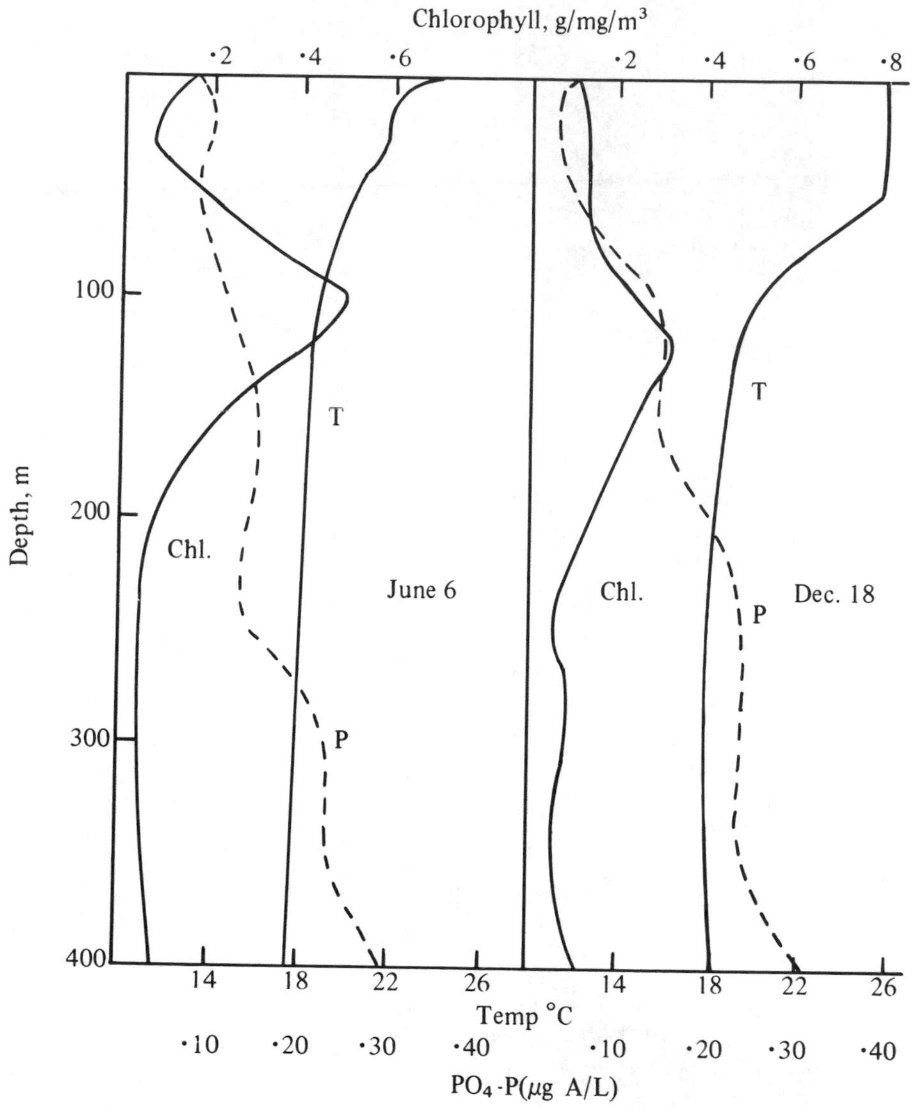

Figure 7.2—4. The vertical distribution of chlorophyll-*a* (Chl), temperature (T), and phosphate-phosphorus (P) off Bermuda, 1958.

(From Menzel, D. W., and Ryther, J. H., The annual cycle of primary production in the Sargasso Sea off Bermuda, *Deep-Sea Res.*, 6, 361, © 1960 Pergamon Press. With permission.)

Table 7.2—4
PRIMARY PRODUCTION RATES FOR WATER COLUMN 0–100 M AT NOON STATIONS IN THE EASTERN TROPICAL PACIFIC

Stn.	Prod.
6	5.2
8	7.1
13	3.2
15	1.8
23	11.5
32	5.0
34	9.5
49	28.3
56	11.0
58	11.8
62	31.0
72	8.0
74	5.0
76	74.0
79	10.8
86	3.2
100	14.8
122	23.4
139*	15.0

* In $mgC/m^2/hr.$; obtained by the C_{14} method, samples incubated at 1000 ft-candles.

Figure 7.2—5

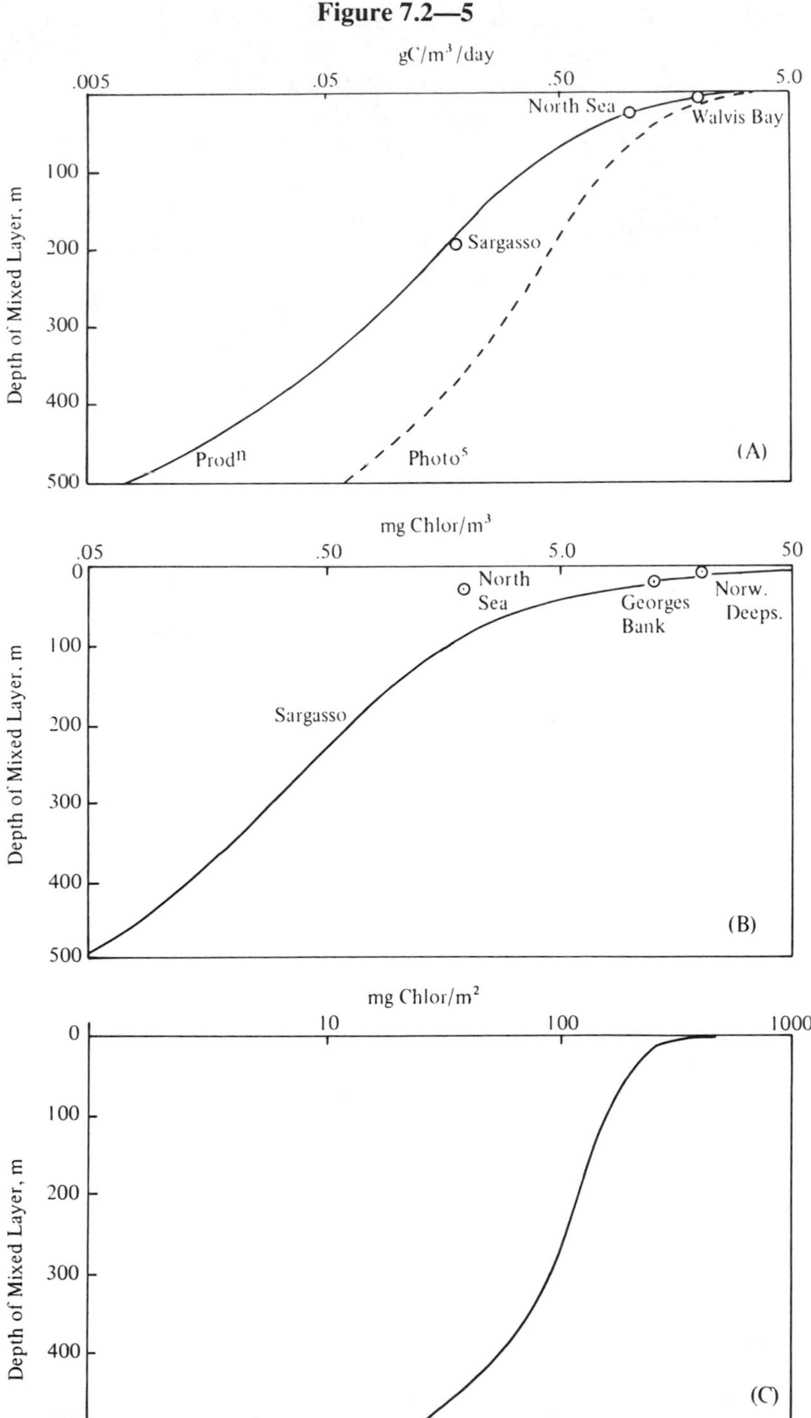

Figure 7.2—5. Relationships between the depth of the mixed layer (A), net production and photosynthetic rates below each square meter of sea surface (B), and chlorophyll-*a* concentration per cubic meter (C), at indicated points in the Atlantic Ocean.

(From Steele, J. H. and Menzel, D. W., Conditions for maximum primary production in the mixed layer, *Deep-Sea Res.*, 9, 44, 1962. With permission.)

Figure 7.2—6

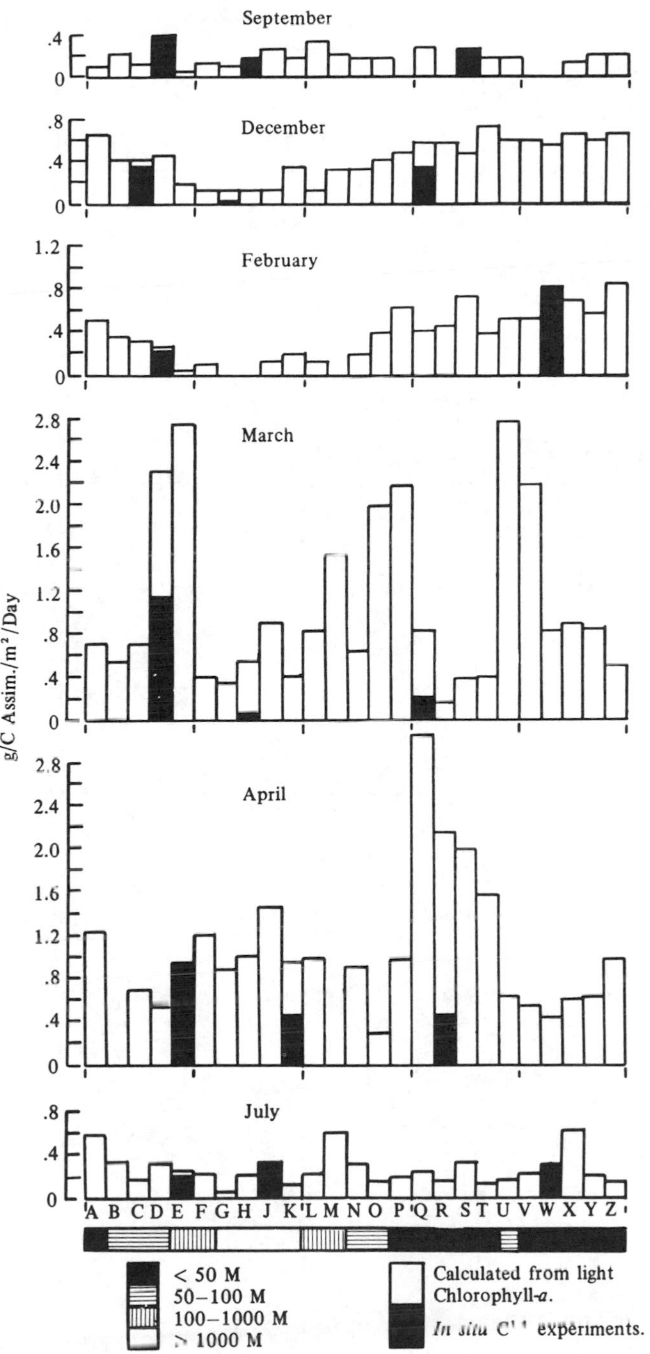

Figure 7.2—6. *In situ* primary production measured by ^{14}C method and/or calculated from chlorophyll, radiation, and light penetration off New Jersey.

(From Ryther, J. H. and Yentsch, C. S., Primary production of continental shelf waters off New York, *Limnol. Oceanogr.*, 3(3), 331, 1958. With permission.)

Figure 7.2—7

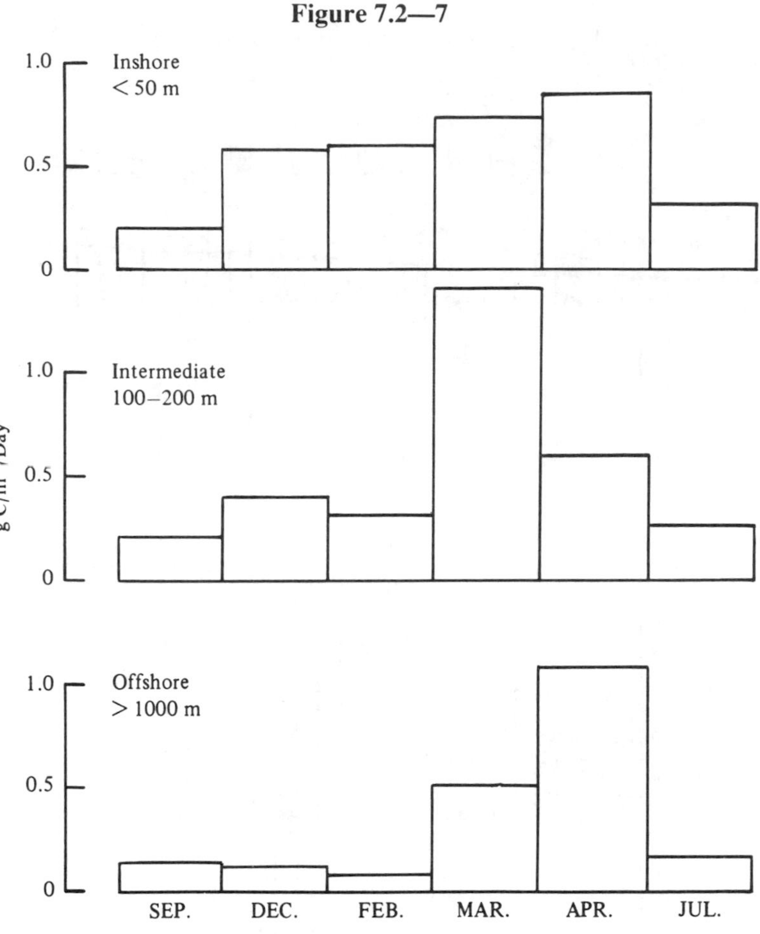

Figure 7.2—7. Mean daily primary production beneath a square meter of sea surface at five shallow, five intermediate, and five deep stations on the continental shelf off New York.

(From Ryther, J. H. and Yentsch, C. S., Primary production of continental shelf waters off New York, *Limnol. Oceanogr.*, 3(3), 333, 1958. With permission.)

Figure 7.2—8

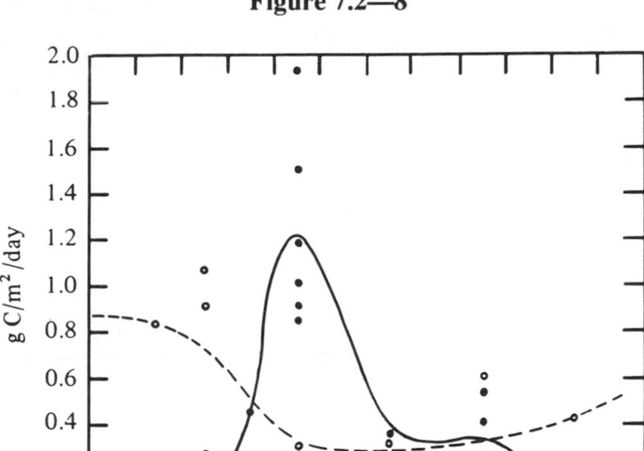

Figure 7.2—8. Primary production as determined by *in situ* ^{14}C measurements at offshore stations (solid line, filled circles) and shallow stations (broken line, open circles) off New York.

(From Ryther, J. H., Geographic variations in productivity, *The Sea*, Vol. 2, Hill, M. N., Ed., Interscience, New York, 1963, 362. © by John Wiley & Sons. With permission.)

Figure 7.2—9

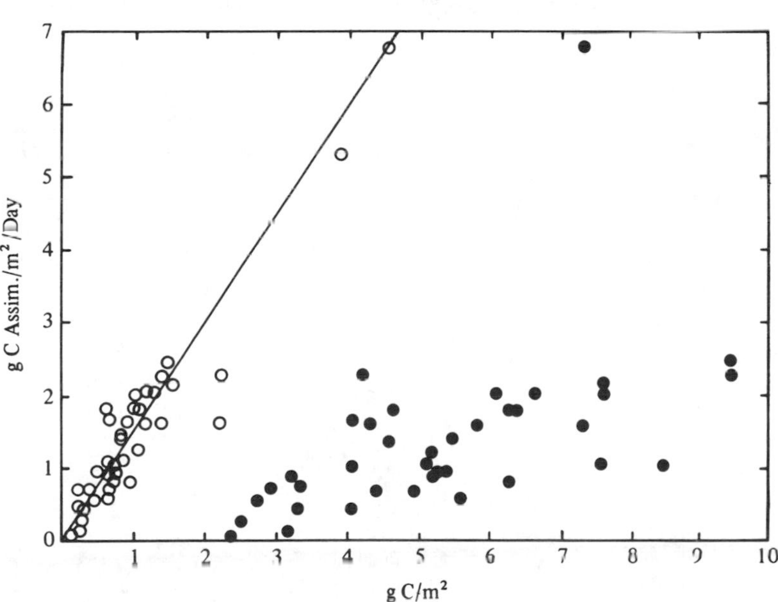

Figure 7.2—9. The regression of carbon assimilation (natural light) on living carbon (open circles) as integrated for the entire euphotic zone in the western Arabian Sea. Broken line is least squares fit for open circles.

(From Ryther, J. H. and Menzel, D. W., On the production, composition, and distribution of organic matter in the western Arabian Sea, *Deep-Sea Res.*, 12, 207, © 1965 Pergamon Press. With permission.)

Figure 7.2—10

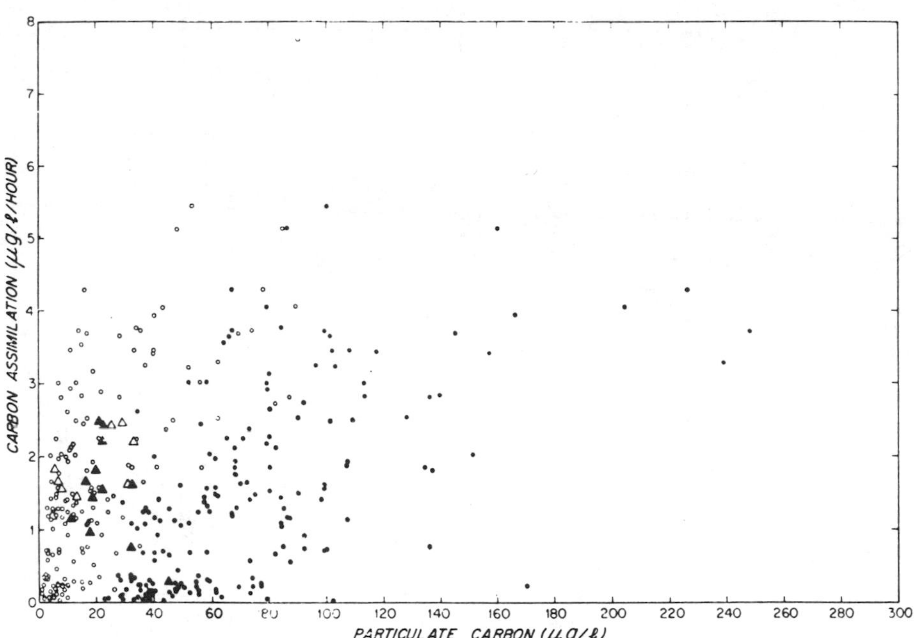

Figure 7.2—10. The regression of carbon assimilation (at 1000 foot candles) on living carbon (open circles and triangles) and on total particulate carbon (filled circles and triangles) in the Indian Ocean.

(From Ryther, J. H. and Menzel, D. W., On the production, composition, and distribution of organic matter in the western Arabian Sea, *Deep-Sea Res.*, 12, 206, © 1965 Pergamon Press. With permission.)

Figure 7.2—11

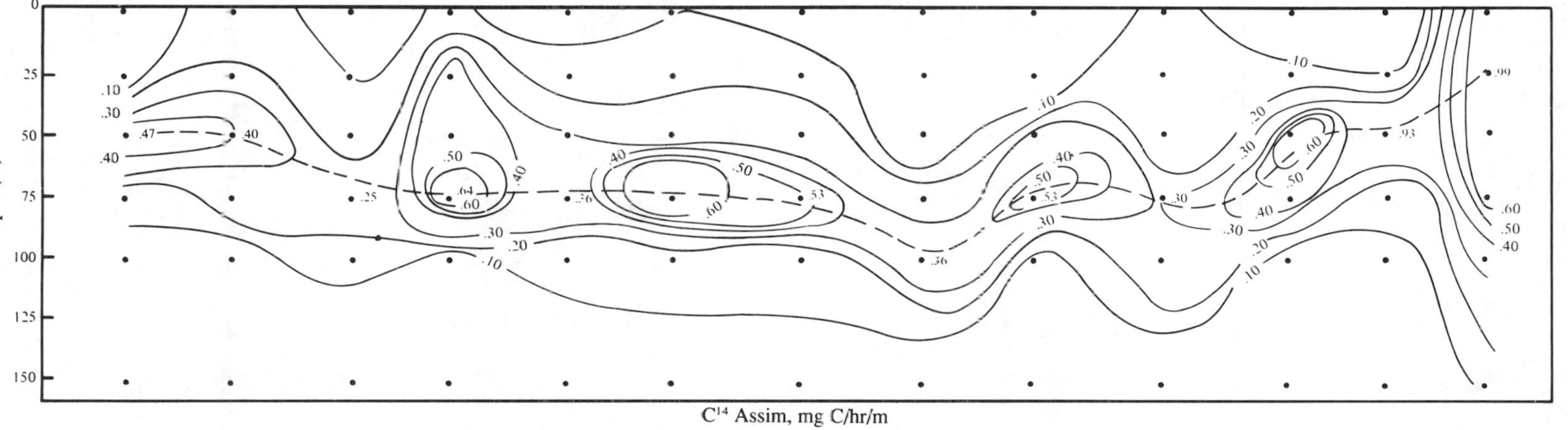

Figure 7.2—11. Profile of ^{14}C assimilation along a section from North Cape, New Zealand, to Cape Howe, Australia, in the East Australian current.

Figure 7.2—12

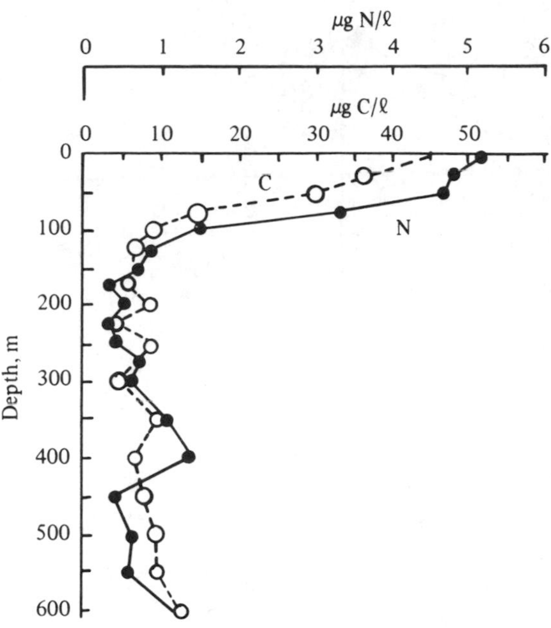

Figure 7.2—12. Profiles of total particulate organic carbon and nitrogen off southern California.

Figure 7.2—13

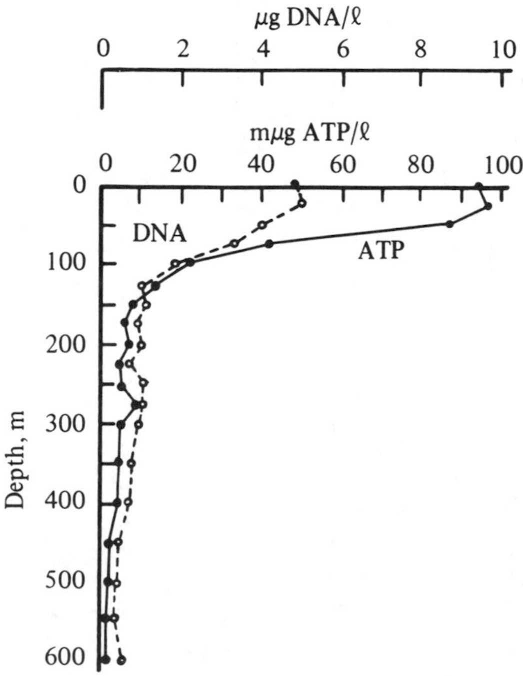

Figure 7.2—13. Profiles of particulate ATP and DNA off southern California.

7.3. DISSOLVED ORGANIC COMPOSITION RELATED TO PRODUCTIVITY

Table 7.3—1
DISSOLVED ORGANIC MATTER IN SEA-WATER

Locality	mg C/l	mg N/l	Ref.
Atlantic Ocean (Bermuda)	2.35 ± 0.07 (along vertical)	0.244 ± 0.08 (same)	1
Black Sea	2.4	–	2
Black Sea	2.83–3.36 (seasonal variations)	–	3
Sea of Azov	4.63–6.02	–	4
Baltic	2.0–4.6 (maximum in euphotic zone)	–	5
Pacific (3 stations)	0.6–2.7	–	6
North Atlantic	1.04–1.97	–	
Atlantic Ocean	2.40–2.48	0.24–0.26	7
Pacific	0.98–2.68	0.07–0.11	
Greenland Sea	2.0–2.1	0.03–0.38	
North Atlantic	0.2–1.3	0.04–0.40	8
Norwegian Sea	0.45–1.38	0.10–0.21	
North Sea	0.5–1.8	0.08–0.54	
Wadden Sea	1.0–8.0	0.10–0.60	

(From Provasoli, L., Organic regulation of phytoplankton fertility, *The Sea,* Vol. 2, Hill, M. N., Ed., Interscience, New York, 1963, 172. With permission.)

REFERENCES

1. **Krogh, A.,** *Ecol. Monogr.,* 4, 421, 1934.
2. **Dazko, V. G.,** *Dokl. Akad. Nauk S.S.S.R.,* 24, 294, 1939.
3. **Dazko, V. G.,** *Dokl. Akad. Nauk S.S.S.R.,* 77, 1059, 1951.
4. **Dazko, V. G.,** *Akad. Nauk S.S.S.R., Hydrotech. Inst. Novocherkask, Hydrochem.,* Mat., 23, 1, 1955.
5. **Kay, H.,** *Kiel. Meeresforsch.,* 10, 26, 1954.
6. **Plunkett, M. A. and Rakestraw, N. W.,** *Deep-Sea Res.,* 3(suppl.), 12, 1955.
7. **Skopintsev, B. A.,** *Preprints Intern. Oceanogr. Cong. A.A.A.S.,* p. 953, 1959.
8. **Duursma, E. K.,** *Neth. J. Mar. Res.,* 1, 1, 1960.

Table 7.3—2
ORGANIC COMPOUNDS IDENTIFIED IN SEA-WATER

Substances	Quantities	Locality	Method	Ref.
Rhamnoside Dehydroascorbic acid	Up to 0.1 g/l present	Inshore waters, Gulf of Mexico	Activated charcoal absorption, ethanol elution	1
Carbohydrates–arabinose equivalents	0.0–20 mg/l	Estuary, Gulf of Mexico	*N*-Ethyl carbazole	2
Carbohydrates–sucrose equivalents	0.14–0.45 mg/l	Pacific Coast, U.S.A.	Anthrone and *N*-ethyl carbazole	3
Carbohydrates–arabinose equivalents	0.0–2.6 mg/l (max. of 12 mg/l at surface, 29° N, 80°, 31′W)	South Atlantic (30° N–25° N)	*N*-Ethyl carbazole	4
	0.0–3.0 mg/l (23% = 0.0; 50% = 0.2–1 mg/l)	Continental Shelf, Gulf of Mexico (50 mg/l in red tides of *G. breve*)	*N*-Ethyl carbazole	5
Citric acid	0.025–0.145 mg/l	Littoral Atlantic French coast		6
Malic acid Acetic and formic acids[a]	0.028–0.277 mg/l <0.1 mg/l	Northeast Pacific, surface and inshore	Chloroform or ether extraction at pH 3; partition chromatography on silica gel column	7
Fatty acids (up to 20 carbons)	0.4–0.5 mg/l (weight of methyl esters)	Gulf of Mexico	Ethyl acetate extraction at pH 2; Gas-liquid chromatography	8
Amino acids – hydrolyzed proteins	Traces to 13 mg/m^3[b]	Gulf of Mexico, Yucatan Strait Reef (British Honduras), Caribbean	Coprecipitation of organic material with $FeCl_3$ + NaOH; acid hydrolysis; paper and ion-exchange chromatography	9

[a] Acetic, formic, lactic, and glycolic (up to 1.4 mg/l) acids are liberated from breakdown of larger organic molecules during the long extraction procedure (4–5 weeks).
[b] 18 amino acids were found in the hydrolysates. The amounts and kind of amino acids vary widely in samples.

Table 7.3—2 (continued)
ORGANIC COMPOUNDS IDENTIFIED IN SEA-WATER

Substances	Quantities	Locality	Method	Ref.
Vitamin B_{12}	Present			10
Plant hormones	Present	North Sea	Chloroform extraction at pH 5; ether extract of residue, measured biologically	

(From Provasoli, L., Organic regulation of phytoplankton fertility, *The Sea*, Vol. 2, Hill, M. N., Ed., Interscience, New York, 1963, 1974. © by John Wiley & Sons. With permission.)

REFERENCES

1. **Wangersky, P. J.,** *Science,* 115, 685, 1960.
2. **Collier, A., Ray, S. M., Magritsky, A. W., and Bell, J. O.,** *U.S. Dept. Int. Fish and Wildlife Service, Fish Bull.,* 84, 167, 1953.
3. **Lewis, G. J. and Rakestraw, N. W.,** *J. Mar. Res.,* 14, 253, 1955.
4. **Anderson, W. W. and Gehringer, J. W.,** *Spec. Sci. Rep.* (Fisheries), 265, 1, 1953; 303, 1, 1958.
5. **Collier, A.,** *Limnol. Oceanogr.,* 3, 33, 1958.
6. **Creac'h, P.,** *C.R. Acad. Sci. Paris,* 240, 2551, 1955.
7. **Koyama, T., and Thompson, T. G.,** *Preprints Intern. Oceanogr. Cong. A.A.A.S.,* p. 925, 1959.
8. **Slowey, J. F., Jeffrey, L. M., and Hood, D. W.,** *Geochim. Cosmochim. Acta,* 26, 607, 1962.
9. **Tatsumoto, M., Williams, W. T., Prescott, J. M., and Hood, D. W.,** *J. Mar. Res.,* 19, 89, 1961.
10. **Bentley, J. A.,** *Preprints Intern. Oceanogr. Cong. A.A.A.S.,* p. 910, 1959.

7.4. PHOTOSYNTHETIC QUOTIENT AND GROSS-NET DIFFERENCES

Table 7.4—1
SOME MEASUREMENTS OF THE PHOTOSYNTHETIC QUOTIENT (O_2/-CO_2) IN AQUATIC PLANTS

Organism	PQ
Chlorella	0.98
Chlorella	ca. 1.00
Chlorella	1.09–1.12
Diatoms	1.12–1.13
Nitzschia closterium	1.05–1.08
Nitzschia palea	1.03–1.05
Peridinium sp.	1.03–1.11
Synechoccus sp.	1.08 ± 0.03
Homidium flaccidum	0.92–1.07
Gelidium	0.94 ± 0.05
Gigartina	0.99–1.07
Raw seawater	1.09 ± 0.23

(From Ryther, J. H., The measurement of primary production, *Limnol. Oceanogr.*, 1(2), 74, 1956. With permission.)

Figure 7.4—1

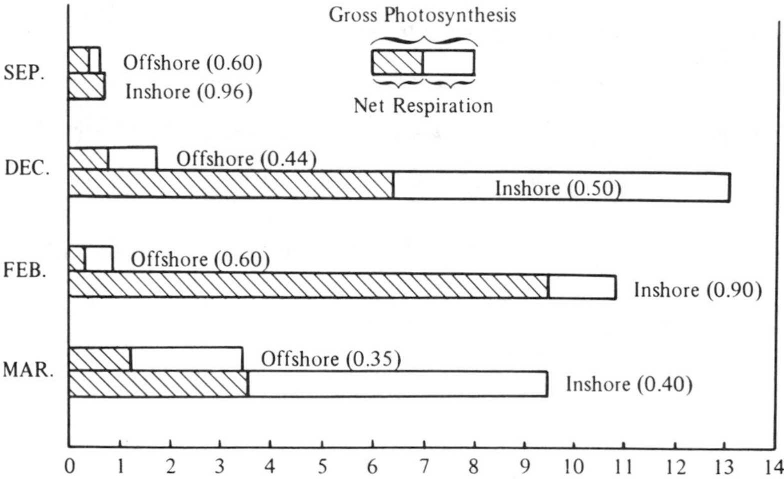

Figure 7.4—1. Comparison of net and gross photosynthesis for inshore and offshore (deeper than 500 fm) locations off Massachusetts. Figures in parentheses are ratios of net:gross.

(From Raymont, J. E. G., Factors affecting primary production – II Light and temperature, *Plankton and Productivity in the Oceans,* Pergamon Press, London, 1963, 227. With permission.)

Figure 7.4—2

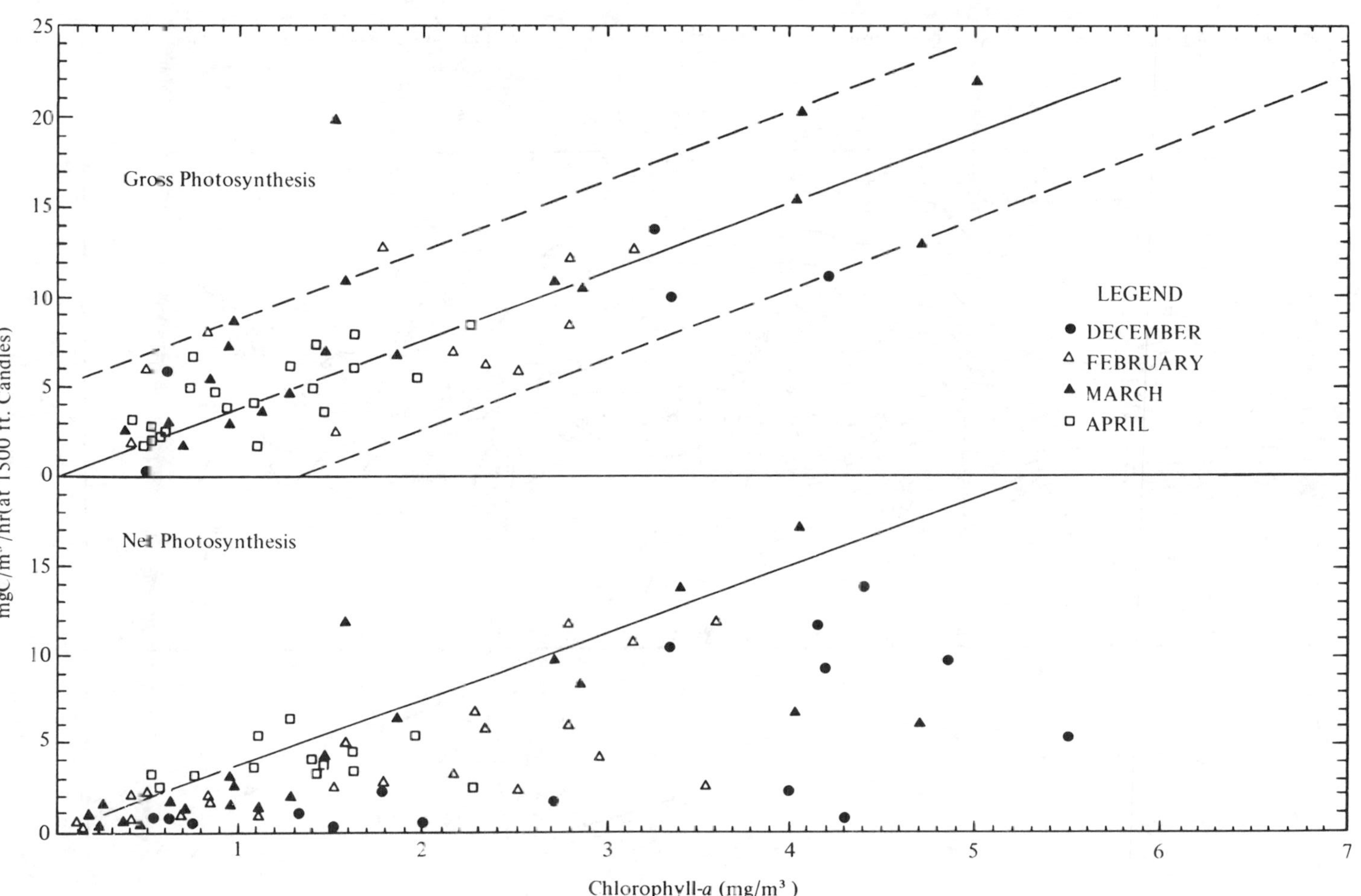

Figure 7.4—2. Gross and net photosynthesis at 1500 foot candles related to the chlorophyll-*a* concentration. Gross photosynthesis was determined by dark and light oxygen bottle experiments. The envelope corresponds to an oxygen titration difference of ± 0.05 ml/l. Net photosynthesis was determined by C^{14} assimilation experiments, uncorrected for respiration. Determinations made off Massachusetts.

Figure 7.4—3

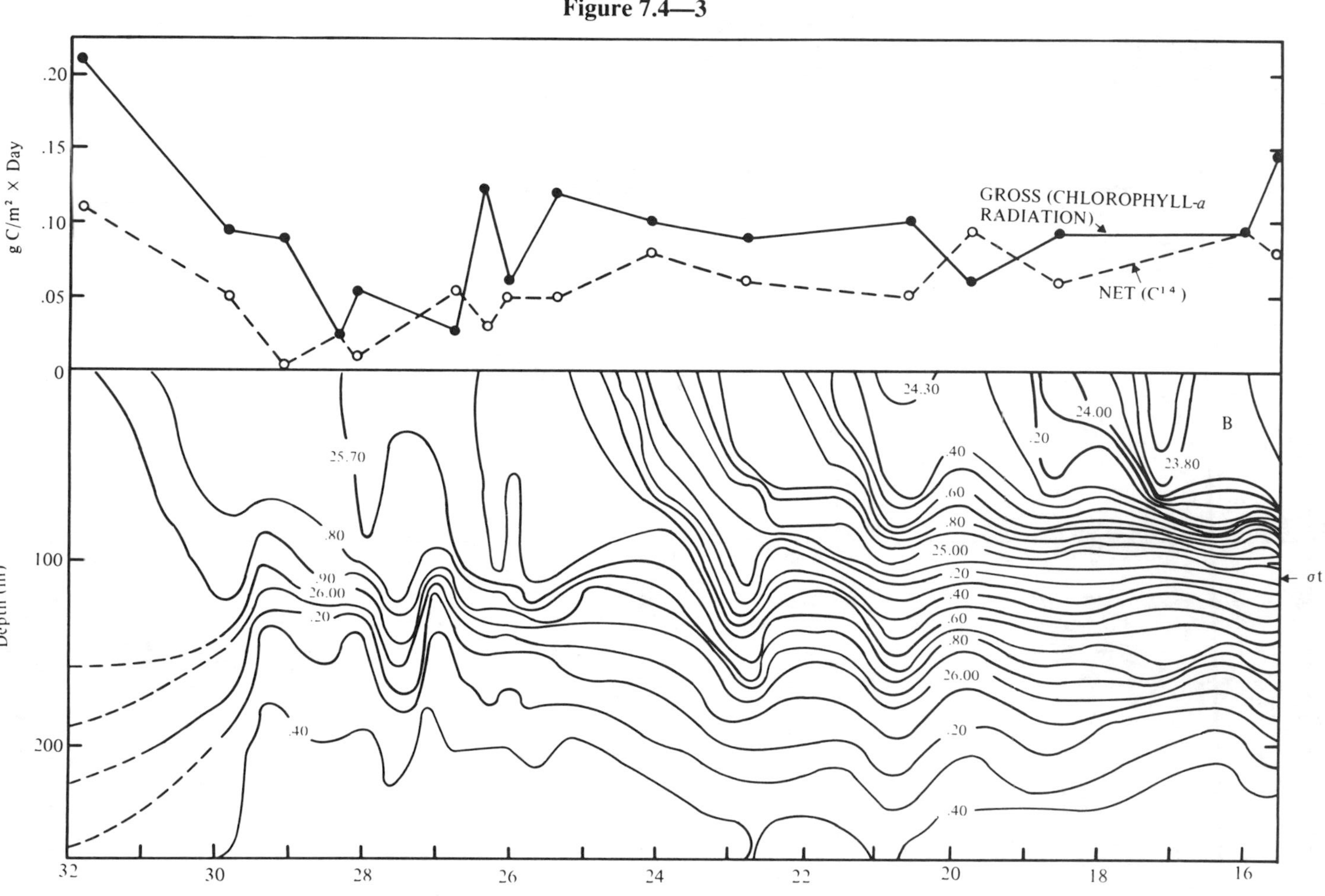

Figure 7.4—3. A. Gross (closed circles, solid line) and net (open circles, broken line) primary production. B. Vertical profile of density (σ) in the surface water as related to latitude off Bermuda.

(From Ryther, J. H. and Menzel, D. W., Primary production in the southwest Sargasso Sea, January-February, 1960, *Bull. Mar. Sci. Gulf Caribb.*, 11, 383, 1961. With permission.)

Table 7.4—1
ULTRAPLANKTON RESPIRATION AT STATIONS IN THE WESTERN NORTH ATLANTIC

Mg atoms O m^{-3} day^{-1}

Sample depth (m)[6]	May 23–26, 1966 33°30′N, 72°00′W Sargasso Sea	Aug. 13–17, 1966 33–32°N, 71°20′W Sargasso Sea	Aug. 18–19, 1966 33°N, 75°W Sargasso Sea	Apr. 4, 1966 35°55′N, 73°32′W Gulf Stream	Apr. 2–3, 1966 36°10′N, 74°38′W Slope water
0.1	**1.3**, 1.9, **1.3**	0.9, 0.8	0.8, **0.7**	2.4	3.8
5	–	–	–	–	15.5
10	–	0.5	–	–	–
15	–	–	2.6	–	5.4
20	–	0.2	–	–	–
25	**0.9**, 0.2	–	–	–	–
30	–	–	0.4	–	–
50	**0.9**, 0.6	0.2	–	3.1	0.5
75	0.6	–	–	–	–
100	0.3	0.3	–	0.7	0.7
150	0.3	–	–	–	–
200	0.3	0.5	–	–	–
250	0.3	0.1	–	0.3	0.9
300	0.5	–	–	–	
400	0.5	0.3	–	–	–
500	0.2	0.3	0.5	0.5	0.8
550	–	0.3	–	–	–
700	–	0.3	–	–	–
750	–	–	–	0.05	–
800	0.1	0.05	–	–	

* Bold numbers are in the scattering layer.

Table 7.4—2
ULTRAPLANKTON RESPIRATION IN THE UPPER 500 METERS, AND P/R RATIOS AT STATIONS IN THE SOUTHEAST PACIFIC OFF PERU

Total depth (m)	Location	Respiration (mg atoms O m^{-2}/day^{-1})	P/R
41	07°58′S 79°21′W	375	1.6
80	08°20′S 79°33′W	400	–
200	08°17′S 79°52′W	520	0.9
520	07°57′S 80°32′W	568	5.3
568	11°51′S 77°33′W	878	1.0
600	08°34′S 80°00′W	216	1.3
1900	08°16′S 80°52′W	469	–
3100	08°22′S 80°45′W	250	1.0
4500	08°24′S 81°40′W	315	1.3
4500	06°22′S 81°47′W	169	1.1
4900	05°49′S 82°05′W	319	0.7

Table 7.4—3
ULTRAPLANKTON RESPIRATION AT STATIONS IN THE SOUTHEAST PACIFIC

Mg atoms O m^{-3}/day^{-1}

Sample depth (m)	Oct. 4, 5 05°50′S 82°04′W	Oct. 7 06°21′S 82°14′W	Oct. 10, 11 06°21′S 81°46′W	Oct. 14 07°57′S 81°40′W	Oct. 15 08°24′S 81°04′W	Oct. 17 08°16′S 80°52′W	Oct. 17 08°22′S 80°43′W	Oct. 18 08°22′S 80°45′W	Oct. 19 08°17′S 79°52′W	Oct. 19 07°58′S 79°21′W
0.1	0.61	—	13.9, 0.89, 0.71, 0.63	9.11	2.80	6.3, 8.6	2.35	2.46	4.00	5.46
10	—	—	—	—	—		3.12	2.17	7.66	13.32
20	—	—	—	—	—	—		—	—	7.05
25	1.31	—	1.80	6.60	0.80	7.4	2.86	0.91	4.93	—
30	—	—	—	—		—	—	—	—	6.50
50	1.04	—	1.96, 0.85	1.04	0.95	3.51	0.28	0.83	3.40	—
60	—	—	—	—	—	—	—			—
75	—	—	—	—	—	0.00	0.00	1.13	2.31	—
100	0.45	0.73	0.05	0.67	0.34	—	—			—
175	—	—	—	—	—	—	—		1.33	—
200	0.21	0.00	0.24	—	0.09	0.68	—	—	—	
250	—	—	—	0.22	—	—	—		—	
300	1.90	0.00	0.00	—	0.99	—		—	—	—
400	—	—	—	—	—	—	—	—	—	—
500	0.66	0.00	0.00	—	—	—	—	—	—	—

Table 7.4—3 (continued)
ULTRAPLANKTON RESPIRATION AT STATIONS IN THE SOUTHEAST PACIFIC

Sample depth (m)	Oct. 20 08°20′S 79°33′W	Oct. 20 08°34′S 80°00′W	Oct. 25 11°51′S 77°33′W	Oct. 25 11°53′S 77°49′W	Oct. 26 12°01′S 78°30′W	Oct. 27 11°58′S 78°35′W	Oct. 28 12°00′S 78°46′W	Oct. 31–Nov. 1 08°24′S 80°25′W	Nov. 2 09°03′S 81°29′W	Nov. 5 09°01′S 80°45′W
0.1	2.24	4.16	4.50	0.00	0.32	0.82	1.60, 0.60	0.41	0.93	0.98
10	6.56	—	—	—	—	—	—	—	—	—
20	—	—	—	—	—	—	—	—	—	—
25	6.44	3.48	3.65	—	2.80	—	1.65	0.53	0.43	2.53
30	—	—	—	—	—	0.10	—	—	—	—
50	2.40	1.33, 1.46	2.66	2.06	0.67	1.10	0.22	0.78	0.20	1.42
60	2.40	—	—	—	—	0.70	—	—	—	—
75	—	0.04	—	—	—	—	—	—	—	—
100	—	—	3.20	—	0.00	—	0.20	0.27	0.39	1.07
175	—	—	—	—	—	—	—	—	—	—
200	—	—	4.00	0.80	0.13	—	0.10	0.20	0.29	0.44
250	—	—	—	—	—	—	—	—	—	—
300	—	0.27	—	—	0.22	—	—	—	0.24	0.20
400	—	—	—	—	0.12	—	—	—	0.76	0.62
500	—	—	—	—	—	—	—	—	—	—

Table 7.4—4

STANDARD RELATIONSHIPS BETWEEN CHLOROPHYLL, STANDING CROP OF ORGANIC MATTER, TRANSPARENCY, ORGANIC PRODUCTION, AND NITROGEN AVAILABILITY AND REQUIREMENT

1	2	3	4	5	6	7	8
Chl. *a* g/l	Standing crop, mg/m^3 dry wt	Extinc. coeff., *k*	Depth euphotic zone, m	Calc. org. prod., g dry wt/m^2/day	N initially available in euphotic zone, mg	N required to produce existing population, mg	N requirement, mg/day
0	0	0.04	120	0	25,300	0	0
0.1	10	0.07	66	0.20	13,800	66	20
0.5	50	0.13	36	0.50	7600	180	50
1.0	100	0.19	24	0.75	5000	240	75
2.0	200	0.29	15	1.00	3100	300	100
5.0	500	0.50	10	1.40	2100	500	140
10.0	1000	0.79	6	1.80	1260	600	180
20.0	2000	1.30	3.5	2.20	735	700	220

(From Ryther, J. H., Geographic variations in productivity, *The Sea*, Vol. 2, Hill, M. N., Ed., Interscience, New York, 1963, 352. © by John Wiley & Sons. With permission.)

Table 7.4—5
GROSS AND NET ORGANIC PRODUCTION OF VARIOUS NATURAL AND CULTIVATED SYSTEMS

g/dry wt/sq. m/day

System	Gross	Net
A. Theoretical potential		
Average radiation (200–400 g cal/cm² day)	23–32	8–19
Maximum radiation (750 g cal/cm² day)	38	27
B. Mass outdoor *Chlorella* culture		
Mean		12.4
Maximum		28.0
C. Land (maximum for entire growing seasons)		
Sugar cane		18.4
Rice		9.1
Wheat		4.6
Spartina marsh		9.0
Pine forest (best growing years)		6.0
Tall prairie		3.0
Short prairie		0.5
Desert		0.2
D. Marine (maxima for single days)		
Coral reef	24	(9.6)
Turtle grass flat	20.5	(11.3)
Polluted estuary	11.0	(8.0)
Grand Banks (April)	10.8	(6.5)
Walvis Bay	7.6	
Continental Shelf (May)	6.1	(3.7)
Sargasso Sea (April)	4.0	(2.8)
E. Marine (annual average)		
Long Island Sound	2.1	0.9
Continental Shelf	0.74	(0.40)
Sargasso Sea	0.88	0.40

(From Ryther, J. H., Potential productivity of the sea, *Science*, 130(3376), 606, 1959. With permission.)

Figure 7.4—4

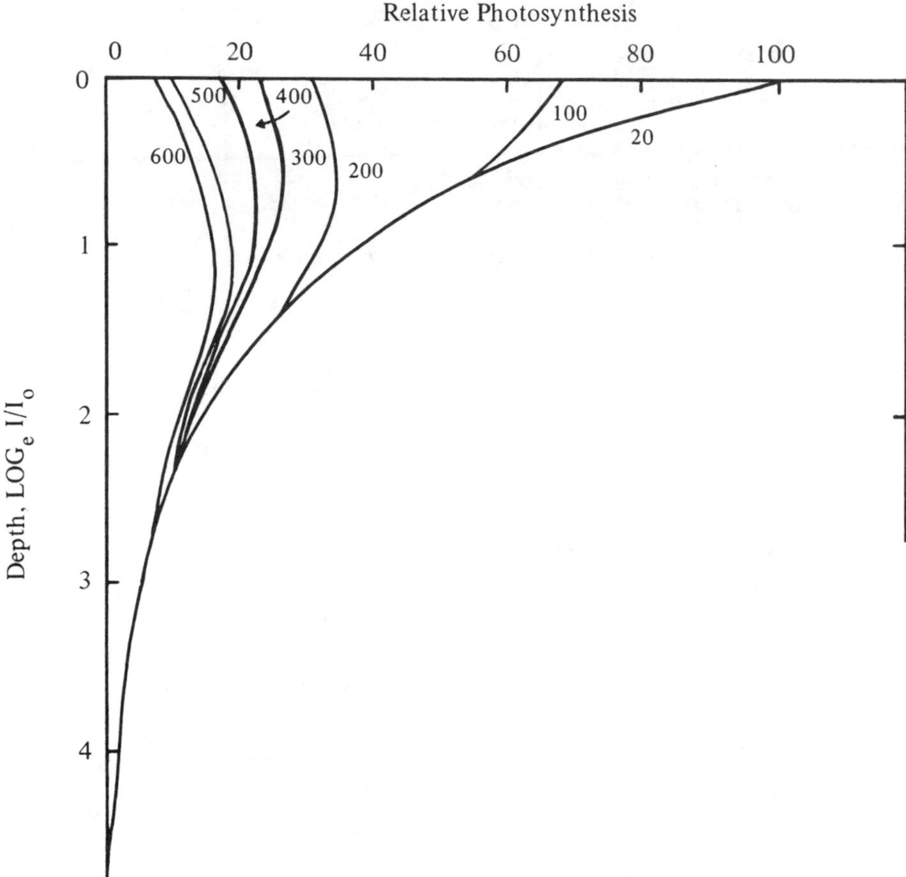

Figure 7.4—4. Relative photosynthesis as a function of water depth for days of different incident radiation. Numbers beside curves shown gram calories per square centimeter per day.

(From Ryther, J. H., Potential productivity of the sea, *Science,* 130(3376), 606, 1959. With permission.)

Figure 7.4—5

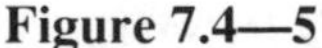

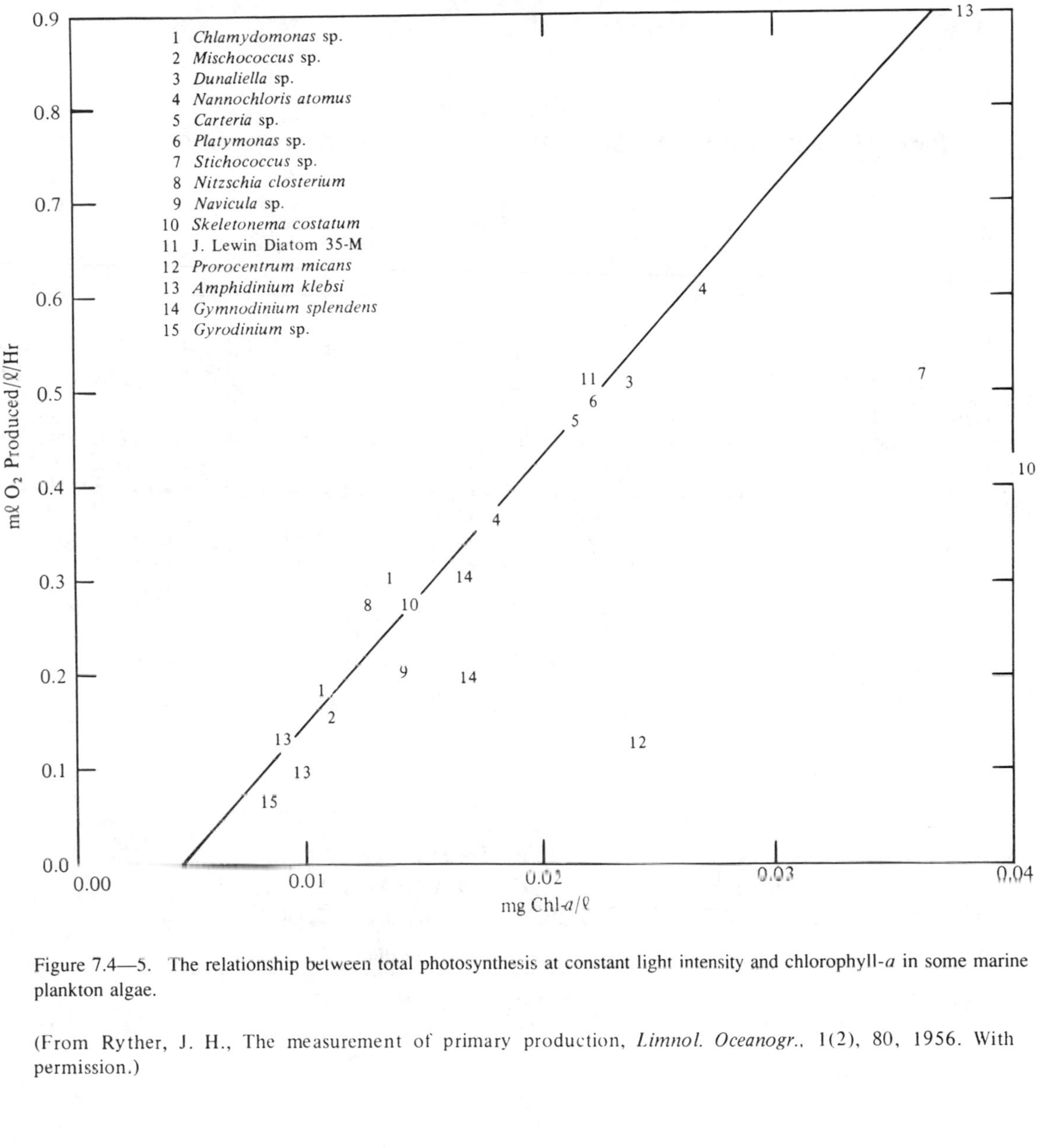
1 Chlamydomonas sp.
2 Mischococcus sp.
3 Dunaliella sp.
4 Nannochloris atomus
5 Carteria sp.
6 Platymonas sp.
7 Stichococcus sp.
8 Nitzschia closterium
9 Navicula sp.
10 Skeletonema costatum
11 J. Lewin Diatom 35-M
12 Prorocentrum micans
13 Amphidinium klebsi
14 Gymnodinium splendens
15 Gyrodinium sp.
mℓ O₂ Produced/ℓ/Hr
mg Chl-a/ℓ
0.0
0.1
0.2
0.3
0.4
0.5
0.6
0.7
0.8
0.9
0.00
0.01
0.02
0.03
0.04

Figure 7.4—5. The relationship between total photosynthesis at constant light intensity and chlorophyll-*a* in some marine plankton algae.

(From Ryther, J. H., The measurement of primary production, *Limnol. Oceanogr.*, 1(2), 80, 1956. With permission.)

7.5. RELATIONSHIPS TO PRODUCTIVITY AT OTHER TROPHIC LEVELS

Table 7.5—1
ESTIMATES OF POTENTIAL YIELDS (PER YEAR) AT VARIOUS TROPHIC LEVELS, IN METRIC TONS

	Ecological efficiency factor					
	10%		15%		20%	
Trophic level	Carbon (tons)	Total weight (tons)	Carbon (tons)	Total weight (tons)	Carbon (tons)	Total weight (tons)
0. Phytoplankton (net particulate production)	1.9×10^{10}	–	1.9×10^{10}	–	1.9×10^{10}	–
1. Herbivores	1.9×10^{9}	1.9×10^{10}	2.8×10^{9}	2.8×10^{10}	3.8×10^{9}	3.8×10^{10}
2. 1st stage carnivores	1.9×10^{8}	1.9×10^{9}	4.2×10^{8}	4.2×10^{9}	7.6×10^{8}	7.6×10^{9}
3. 2nd stage carnivores	1.9×10^{7}	1.9×10^{8}	6.4×10^{7}	6.4×10^{8}	15.2×10^{7}	15.2×10^{8}
4. 3rd stage carnivores	1.9×10^{6}	1.9×10^{7}	9.6×10^{6}	9.6×10^{7}	30.4×10^{6}	30.4×10^{7}

Table 7.5—2
DAILY PRIMARY PRODUCTION AND ZOOPLANKTON BIOMASS AT THREE OCEANIC LOCATIONS

		Location			
Depth		5°N-4°S 15°W	1.5°N-1.5°S 35°W	15°S 17-35°W	
			mgC/m²	mgC/m²	mgC/m²
0–100 m	Primary production	880	480	252	
	Herbivores	397	658	140	
	Carnivores	377	572	148	
	Total zooplankton	774	1,230	288	
0–500 m	Total zooplankton	1,750	1,900	680	

Figure 7.5—1

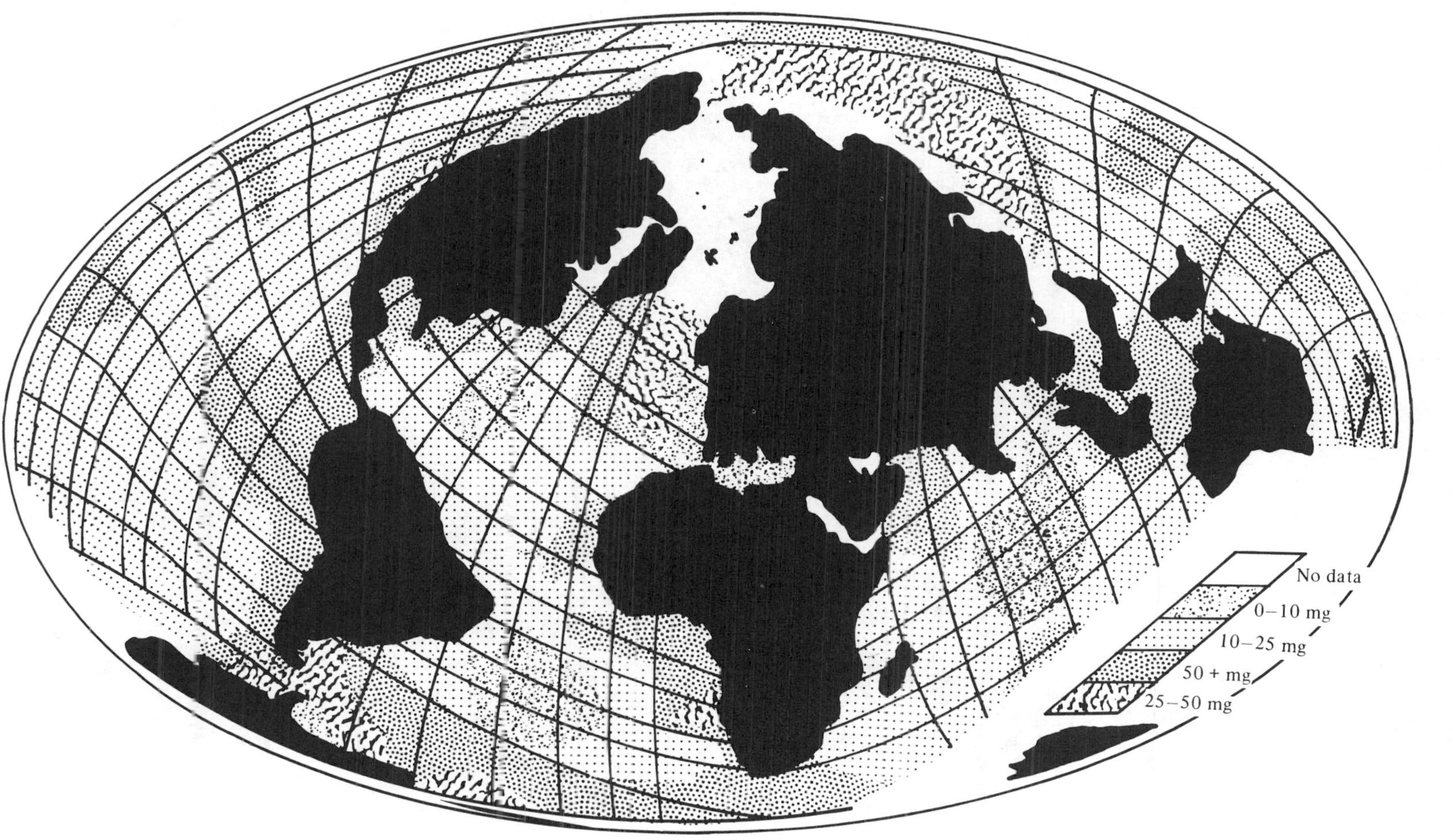

Figure 7.5—1. Biological productivity of the seas. Density of shading is roughly proportional to the degree of biological productivity as measured by the amount of organic matter (in milligrams) produced annually per cubic meter of sea water.

Section 8
Zooplankton

Section 8

ZOOPLANKTON

GENERAL CHARACTERISTICS

The vast majority of the faunal assemblages of the plankton consist of zooplankton — volumetrically abundant animals usually several microns to 2 cm in size — that drift passively in currents due to limited capabilities of locomotion. These lower-trophic-level consumers constitute the principal herbivorous component of oceanic systems. Whereas most zooplankton consume phytoplankton or detritus and serve as an essential link in aquatic food chains by converting plant to animal matter, others are primary carnivores. Some species obtain nutrition by the direct uptake of dissolved organic nutrients. Zooplankton basically gather food via filter feeding or raptorial feeding.[1] Filter feeders employ tiny hairs or mucous surfaces to capture food, mainly phytoplankton.[2] Raptorial feeders seize and eat individual cells, removing a few selected prey; they are principally carnivores. Grazing pressure by herbivorous zooplankton commonly regulates the standing crop of phytoplankton populations.

Both biological and physical-chemical conditions in marine waters affect the species composition, abundance, and distribution of zooplankton. These microfauna must adapt to varying stresses associated with biological (e.g., scarcity of food, predation, and competition) and physical-chemical factors (e.g., temperature, salinity, mass movements of water, and dissolved oxygen levels).[3] General accounts of zooplankton ecology are contained in a number of recent publications.[1-11]

CLASSIFICATIONS

Marine scientists typically classify zooplankton according to their size or length of planktonic life. Three major size categories of zooplankton are recognized, including microzooplankton, mesozooplankton, and macrozooplankton. Microzooplankton comprise those forms which pass through plankton nets with a mesh size of 202 μm, and mesozooplankton, those forms retained by these nets. Larger zooplankton captured by plankton nets with a mesh size of 505 μm embody the macrozooplankton.[12,13]

In regard to the duration of planktonic life, zooplankton may be grouped into three classes: (1) holoplankton; (2) meroplankton; and (3) tychoplankton. Holoplankton are those organisms which spend their entire life in the plankton, in contrast to meroplankton that remain planktonic for only a portion of their life cycle. Tychoplankton refer to small animals, primarily benthic organisms, temporarily translocated into the water column by current action, behavioral activity (e.g., diurnal vertical migration), or other mechanisms.

CLASSIFICATION BY SIZE

Microzooplankton

Zooplankters smaller than approximately 60 μm consist largely of protozoans, especially foraminiferans, radiolarians, and tintinnids. These three groups of protozoans are almost exclusively marine and are found in all oceans.[7] Foraminiferans and radiolarians attain such high abundance in the oceans that their shells or tests produce thick layers of *Globigerina* and radiolarian oozes on widespread areas of the seafloor.[6] Trochophore larvae of polychaetes, veliger larvae of molluscs, copepod nauplii, and arthropod larvae (e.g., barnacle nauplii and crab zoeae) form a significant fraction of the microzooplankton in some waters. Many protozoans graze heavily on bacteria and provide a food source for larger microzooplankton.[13,14] Some microzooplankton (e.g., planktonic ciliates) compete with macrozooplankton for microplanktonic foods.[15] They also act as a trophic link between nanophytoplankton and macrozooplankton.[16,17] Larger planktonic ciliates, although consuming much phytoplankton,[18,19] reportedly ingest smaller ciliates as well.[20] Omnivory and carnivory among these organisms can regulate the trophic dynamics and structure of microplanktonic communities.[21]

Mesozooplankton

Copepods, cladocerans, rotifers, and meroplankton (mostly larval forms of benthic species) dominate the mesozooplankton. Together with the macrozooplankton, the mesozooplankton are considered to be the principal components of the zooplankton.[9] Copepods are the most important mesozooplankton, serving as a vital coupling between phytoplankton and larger carnivores.[6] Free-living planktonic species of the orders Calanoida, Cyclopoida, and Harpacticoida frequently appear in zooplankton samples. Herbivorous copepods, particularly calanoids, tend to have the greatest absolute abundance and biomass.[4,22] Harpacticoids are locally abundant.[5] Omnivorous and carnivorous copepods likewise play a critical role in the flow of energy to higher trophic levels.

In general, herbivorous copepods filter the water to collect phytoplankton during feeding. Carnivorous forms (e.g., some cyclopoids and calanoids) actively prey on other zooplankton by seizing the individual with their appendages.[23] Odhner[24] and Davis[22] present lucid descriptions of the feeding mechanisms of various life-history stages of copepods.

Copepods display pronounced spatial and temporal changes in distribution;[8] biotic as well as abiotic factors may be responsible for these changes. Physical factors, such as fluctuations in temperature or photoperiod and seasonal variations in riverine discharges and coastal hydrography, have been implicated in the temporal changes of zooplankton communities.[25] Other physical factors (e.g., tidal currents, upwelling, stratification, and turbidity) can modulate organismal position. Chief among the biotic factors affecting zooplankton distribution are the environmental preferences of the species, food availability, predator-prey interactions, larval behavior, larval abundance, and vertical migration patterns.

Vertical migration and heterogeneous vertical distribution of copepod populations are well established. For example, the calanoid copepod, *Acartia tonsa*, undergoes a pronounced vertical migration, ascending the water column at night and descending to greater depths during the day.[26,27] By migrating, the copepods avoid predation pressure. Nevertheless, predation occasionally regulates seasonal cycles of abundance of some populations. Hence, the ctenophore *Mnemiopsis leidyi* inflicts heavy mortality upon copepods in Narragansett Bay, Rhode Island, accounting, in part, for seasonal abundance cycles.[28-30]

Conspicuous variation in the concentration of copepods is perhaps demonstrated most clearly along the longitudinal axis of an estuary. Estuarine circulation causes a landward transport of copepods that migrate into the near-bottom layer, thereby concentrating the zooplankton in specific areas.[25] Zooplankton, in general, exhibit a patchy distribution from the head to the mouth of an estuary, attributable to hydrography, grazing, and other factors.[4,5,31]

In marine waters, the spatial heterogeneity of zooplankton communities may be manifested as shifts in the absolute abundance of a single species or as alterations in species composition.[22] At least six scales of spatial patterns in zooplankton communities are observable: (1) micro (1 cm to 1 m); (2) fine (1 m to 1 km); (3) coarse (1 to 100 km); (4) meso (100 to 1000 km); (5) macro (1000 to 3000 km); and (6) mega (greater than 3000 km).[22,32] The processes generating these spatial patterns have been treated in modeling studies.[32] Most models of single-species populations have dealt with smaller scale processes of less than 50 km.[33,34]

Macrozooplankton

Sampling with a 505-µm mesh plankton net generally collects two groups of zooplankton. The first group, jellyfish, consists principally of hydromedusae, comb jellies, and true jellyfishes.[35] This group may dominate the zooplankton in terms of total volume. The second group, crustaceans, incorporates a variety of different forms (e.g., amphipods, isopods, mysid shrimp, and true shrimp). Shrimp-like euphausiids, also known as krill, are most abundant in the Antarctic, but occur worldwide in pelagic waters of high productivity.[2,5] Growing larger than copepods, euphausiids average about 2 to 5 cm in length. They are a main staple of baleen whales in Antarctic waters; many fish also consume them. Both herbivorous and carnivorous euphausiids exist, with carnivorous species being well represented in warm waters.

CLASSIFICATION BY LENGTH OF PLANKTONIC LIFE

Meroplankton

These organisms, as recounted above, spend only a portion of their life in the plankton. They primarily encompass planktonic larvae of the benthic invertebrates, benthic chordates, and nekton (i.e., ichthyoplankton). Sexual jellyfish stages of the hydrozoan and scyphozoan coelenterates (e.g., *Sarsia* spp. and *Cyanea* spp.) comprise an additional meroplanktonic component.[36] Nearly all animal phyla contribute individuals to the meroplankton.

In estuaries and coastal waters, the meroplankton of benthic invertebrates are numerically significant. More than 125,000 species of benthic fauna have been identified, the bulk of them characterized by a free-swimming larval stage persisting for a few weeks.[2] Many species pass through a succession of larval stages prior to becoming adult; therefore, multiple stages of the same species may be represented concurrently in the plankton. Most taxa of the Crustacea, for instance, usually have more than one larval stage, and some decapods exhibit as many as 18 stages.[6]

Benthic ecologists discriminate between two major types of planktonic larvae of marine benthic invertebrates, that is, planktotrophic and lecithotrophic forms. Planktotrophic larvae actively ingest particulate food while in the plankton. Lecithotrophic larvae, however, utilize stored food reserves from the egg for their development.[7] A third type of planktonic larval development, termed facultative planktotrophy, has been enlisted to define a larva capable of feeding while in the plankton but harboring sufficient yolk reserves to develop and metamorphose.[37,38] Emlet[39] alluded to several advantages of facultative planktotrophy, which may be a transition between planktotrophy and lecithotrophy, explaining that the augmentation of nutritive stores by larval feeding could allow an extended competent period or could enhance growth and survivorship subsequent to metamorphosis.

For planktotrophic larvae, development to metamorphosis generally takes place in 2 to 6 weeks. Lecithotrophic larvae usually require less time to reach metamorphosis, typically several days to about 2 weeks.[40] When a larva of a benthic marine invertebrate attains metamorphic competency, it is developmentally capable of settlement to the seafloor and metamorphosis. However, metamorphosis may be delayed during the competency period because of the lack of an appropriate stimulus.[40] A competent larva can delay metamorphosis for weeks while searching the seafloor for a suitable habitat. During this search of the substratum, the larva is subject to predation from a host of sources, such as protozoans, larger omnivorous and carnivorous zooplankton, benthic invertebrates, and fish.[41]

In addition to planktotrophy and lecithotrophy, three alternate developmental patterns of benthic marine invertebrate larvae have been noted. These are: (1) demersal development; (2) direct development; and (3) viviparity. Mileikovsky[42,43] and Chia[38] provide a description of them. An estimated 70% of the species of benthic marine invertebrates experience pelagic planktotrophic development.[40,44]

Occasionally overlooked in meroplanktonic investigations are the ichthyoplankton. Although adult fishes may be benthic, benthopelagic, mesopelagic, or pelagic, most produce planktonic eggs and/or larvae.[45,46] Leiby[45] details the methods by which the eggs and larvae of estuarine and marine fishes enter the planktonic community. Recruitment to the adult population is dependent on the survivorship of these early life-history stages.[46]

The meroplankton of benthic marine invertebrates and the ichthyoplankton are susceptible to the vagaries of environmental conditions that often severely deplete their numbers. Temperature, salinity, turbidity, circulation, and seafloor conditions, as well as other physical factors, influence larval development, distribution, and survivorship. Biological factors, including predation, availability of food, and seasonal abundance of adults and larvae, also affect meroplankton success.[47] Predation alone causes larval mortality to exceed 90%.[2] In spite of immense numbers of eggs and larvae produced by benthic marine invertebrates (e.g., barnacles, bivalves, cyphonautes, gastropods, and polychaetes) in coastal waters, the majority do not survive due to the stresses imposed by the physical and biological conditions of the marine environment.

Holoplankton

Three protozoan phyla from the protistid kingdom and seven phyla from the animal kingdom supply individuals to the holoplankton. Based on absolute numbers, three groups of planktonic protozoans — foraminiferans, radiolarians, and tintinnids (and other ciliates) — often are most important. Approximately 25% of the animal phyla (7 out of 29) contain holoplanktonic species, with the Arthropoda having the greatest significance to the community. Excluding the protozoans, crustaceans (i.e., largely copepods, amphipods, mysids, euphausiids, and decapods) tend to outnumber the other elements of the zooplankton. Copepods, in particular, may account for as much as 95% of a zooplankton sample.[1]

Thurman and Webber[1] and Gross[2] review the taxa encountered among the holoplankton. Tintinnids ("bell animals") and other ciliates are small zooplankton, usually less than 10 μm in size, that periodically compose more than 50% of the number of organisms in a sample. In regions harboring small phytoplankton cells, tintinnids frequently reach high numbers. Less than 2% of foraminiferan taxa are planktonic, and the majority of these belong to the genus *Globigerina*. They secrete calcareous tests in contrast to radiolarians, which build siliceous or strontium sulfate skeletons. Being carnivorous, planktonic foraminifera extend pseudopodia (thin protoplasm extrusions) through openings in their shells to capture small plankton. Radiolarians feed on diatoms, copepods, and other planktonic organisms.

In the animal kingdom, Rotifera, can be abundant in coastal and open oceanic regions. Two, almost exclusively planktonic phyla, the Chaetognatha (with less than 60 species) and Ctenophora (with about 90 species) consume copepods and other small zooplankton. These predators sometimes dominate the zooplankton, inhabiting waters throughout the photic zone. Two groups of planktonic cnidarians, the Scyphozoa (jellyfish) and the Siphonophora, have large representatives, ranking them near the top in size of all zooplankton. The tunicates (phylum Chordata) consist of two assemblages of holoplanktonic forms, the first being salps, doliolids, and the genus *Pyrosoma* and the second, members of the Appendicularia.[1] Another phylum occasionally abundant in the holoplankton is the Mollusca (class Gastropoda). Three groups of gastropods found in some cases in large swarms, especially in tropical and subtropical waters, include the violet snails (*Ianthina* spp.), heteropods, and pteropods. These snails embrace both herbivorous filter feeders and carnivorous types, the carnivores generally ingesting medusae and small fish.

Tychoplankton

Demersal zooplankton may be periodically inoculated into the plankton by bottom currents, wave action, and bioturbation. These organisms do not normally constitute a quantitatively significant fraction of the zooplankton community. However, they are an important food source for planktivorous fishes and carnivorous zooplankton. Tychopelagic species commonly recovered from the stomachs of planktivorous fishes are amphipods, cumaceans, isopods, and mysids.[48,49] The recruitment process of planktonic species, as well as the aperiodic entry of tychoplankton into the water column of shallow regions, can be fostered by bioturbation of benthic fauna, for example, the conveyor-belt-feeding polychaetes *Cistenides (Pectinaria) gouldii* and *Clymenella torquata*.[50,51] Storms and currents which roil bottom sediments also promote the upward translocation and lateral distribution of the tychoplankton.

ZOOPLANKTON DYNAMICS

EFFECTS OF LIGHT

Light is a major environmental factor regulating diel vertical migration of zooplankton.[52] Acting as an environmental cue that triggers faunal migration, changes in illumination at sunrise and sunset elicit vertical movements of the organisms, with downward migration taking place at sunrise and upward migration at sunset. Vertical migration appears to be responsive to the following light cues: (1) a change in depth of a particular light intensity; (2) a change in underwater spectra; (3) a change in the polarized light pattern; (4) an absolute amount of change in light intensity; and (5) a relative rate of intensity change.[27,53] As summarized by Forward et al. (p. 146),[54] "the cue for initiating vertical movements is the rate and direction of change in light intensity from the ambient level (adaptation intensity) which itself can change over a day." Factors other than light (i.e., temperature, salinity, and the organism's age, physiological condition, and reproductive stage) may alter the diel vertical migration pattern, thereby adding complexity to this area of study.[22]

EFFECTS OF TEMPERATURE

Temperature influences zooplankton physiology and ecology.[3,55] Metabolic rates are a function of temperature, with fecundity, duration of life, adult size, and other parameters being modulated by temperature levels. Because higher temperatures accelerate molting, with less growth occurring during intermolt periods, zooplankton living in warmer seas usually grow to a smaller size than do individuals occupying cooler waters.[3] Fecundity is lower in adults of smaller size. Therefore, lower temperatures favor zooplankton with higher fecundity.[56] The vertical migration of zooplankton populations into deeper, colder areas of a thermally stratified system confers an advantage to the organism in regard to fecundity.

Ecological effects ascribable to temperature include seasonal variations in the species composition and abundance of zooplankton in shallow-water systems, which are particularly well documented for mid-latitude estuaries. Seasonal changes in community structure arise from successions of constituent populations, leading to fluctuating dominance patterns during the year. The community of zooplankton in mid-Atlantic Bight estuaries (U.S.), for instance, follows a succession of species populations, with winter-spring dominants (e.g., *Acartia hudsonica*) being replaced by summer-fall dominants (e.g., *A. tonsa*).

Temperature triggers spawning of many benthic invertebrates, leading to pulses of meroplankton during the warm months. These pulses contribute to the high-standing crops of zooplankton delineated in many estuaries and coastal waters during the summer.[57] A number of resident and migratory fish populations spawn seasonally in estuaries, adding icthyoplankton to the total concentration of zooplankters.[58]

Whereas the holoplankton of the open ocean generally peak in abundance in spring and fall, those in estuaries often experience a different seasonal pattern of events.[36] Holoplanktonic species subject to variable seasonal densities may survive through the off seasons by having a few individuals persist to proliferate as environmental conditions improve. Alternatively, the propagation of populations takes place via individuals developing from resting eggs generated the previous year. The seasonal success of copepods in terms of growth and development hinges largely on temperature and food supply.[59]

EFFECTS OF SALINITY

Salinity effects are of little consequence in the open ocean, but are of major significance in estuaries and coastal waters. As mentioned previously, zooplankton are responsive to salinity levels encountered along the longitudinal axis of an estuarine system. They respond to salinity changes in the vertical plane as well. The salinity tolerances of holoplankton and meroplankton differ among species and may vary among ontogenetic stages of a single species. Thus, the patterns of species succession and dominance along the longitudinal axis of an estuary can be affected by salinity concentrations. In the vertical plane, salinity limitations restrain the vertical migration of zooplankton populations. Salinity extremes often arrest growth and increase mortality of susceptible forms.

Effects of salinity on zooplankton populations are evident in mid-Atlantic estuaries. Differences in the tolerances of *A. tonsa* and *A. hudsonica* to reduced salinity, for example, play a role in the seasonal dominance patterns of the copepods in the estuaries.[60] Although temperature is the main factor causing the seasonal exchanges of dominance in the lower reaches of these systems, the sensitivity of the copepods to salinities upestuary promotes succession of the species. *Acartia hudsonica*, being more sensitive to brackish water conditions than *A. tonsa*, is replaced by *A. tonsa* far upestuary early in the spring, with replacement gradually spreading seaward and taking one to several months to progress to the estuarine mouth.

COPEPOD FEEDING MECHANISMS

Observations on copepods indicate that they feed by means of passive filtration of small particles or by active capture of large particles.[8,61-64] In the active capture of prey, a copepod initially detects a food particle by chemo- or mechanoreception, and subsequently seizes it. The prey can be detected up to about 1 mm away from the animal via mechanoreception.[64] Within this distance, the copepod has the capability of sensing the shape and dimensions of its prey. Once a particle is captured, the copepod orients the particle and determines its suitability for consumption. Some particles are ingested and others rejected. While the mechanical properties of the feeding apparatus may be partially responsible for some of the particle selectivity, decision making and chemosensation probably play a paramount role.[65] Indeed, distance and contact chemoreception, along with mechanoreception, possibly are all involved in the detection and selection process.[64,66] Particle size has been invoked as a primary factor influencing the selectivity of copepod feeding.[67] However, other factors might also affect the selectivity process, including particle shape or smell.[62]

Diel vertical migration of zooplankton with greater nocturnal feeding rates has been scrutinized with respect to grazing impacts on phytoplankton and bacteria.[68,69] With increased densities of zooplankton in surface waters at night, a decline in the concentration of food particles purportedly occurs, which is inversely proportional to the length of time the zooplankton remain in the upper layers.[70,71] Conclusions from other investigations negate this view, suggesting that diel vertical migration behavior and feeding activity may not be strongly linked, as is illustrated by the copepods *Calanus* and *Pseudocalanus*.[72] On theoretical grounds, an energetic advantage has been postulated in which nocturnal grazing could impart a greater energy gain for the zooplankton than continuous feeding.[73]

The rate of ingestion of zooplankton increases with increasing (phytoplankton) prey density up to a limit above which the ingestion rate is nearly constant. Two models can be employed to describe the relationship of food intake vs food concentration. The first is the Ivlev equation expressed as

$$I = I_m[1 - e^{-\delta(p - p')}]$$

where I is the rate of ingestion, I_m is the maximum rate of ingestion, δ is a constant, p is the phytoplankton concentration, and p′, is the phytoplankton concentration at which no ingestion results. The second is the Michaelis-Menten equation, given as

$$I = \frac{I_m(p - p')}{K + (p - p')}$$

where K is a constant, and the other parameters are the same as above.

Parsons et al.[74,75] and McAllister[76] advocate that zooplankton do not feed below a threshold concentration of food particles. Such feeding thresholds are of major ecological significance because they preclude total exploitation and extermination of phytoplankton populations by the grazers, thereby providing a more stable system.[77] The phytoplankton, not being completely depleted in numbers, regenerate as the zooplankton grazing pressure diminishes. Feeding thresholds incorporated into the aforementioned models consequently furnish a means for recovery of the phytoplankton that has been cropped.

In the open ocean, grazing zooplankton typically remove 50% or more of the net production of phytoplankton. The intensity of grazing pressure of zooplankton in many shallow coastal and estuarine areas is in contrast much lower. In these systems, consumption by zooplankton frequently amounts to less than 30% of the net production of phytoplankton throughout the year.

PLAN OF THIS SECTION

Tabular and illustrative data presented on zooplankton in Section 8 are subdivided into seven major subject areas. These include: (1) biomass as related to area, depth, and season; (2) taxon diversity as a function of area and depth; (3) chemical compositions; (4) geographical distribution of characteristic faunal groups; (5) indicator species and their associated water masses; (6) length, height, weight, and sample volume relationships in major groups; and (7) sampling equipment and comparative efficiencies. As in previous biological sections, data have been organized according to geography, ocean depth, chemical parameters, and taxonomic category.

REFERENCES

1. **Thurman, H. V. and Webber, H. H.,** *Marine Biology*, Charles E. Merrill Publishing, Columbus, OH, 1984.
2. **Gross, M. G.,** *Oceanography: A View of the Earth*, 3rd ed., Prentice-Hall, Englewood Cliffs, N.J., 1982.
3. **Heinle, D. R.,** Zooplankton, in *Functional Adaptations of Marine Organisms*, Vernberg, F. J. and Vernberg, W. B., Eds., Academic Press, New York, 1981, 85.
4. **Raymont, J. E. G.,** *Plankton and Productivity in the Oceans*, 2nd ed., Vol. 1, Pergamon Press, Oxford, 1980.
5. **Levinton, J. S.,** *Marine Ecology*, Prentice-Hall, Englewood Cliffs, N.J., 1982.
6. **Nybakken, J. W.,** *Marine Biology: An Ecological Approach*, Harper & Row, New York, 1982.
7. **McConnaughey, B. H., and Zottoli, R.,** *Introduction to Marine Biology*, C. V. Mosby, St. Louis, 1983.
8. **Miller, C. B.,** The zooplankton of estuaries, in *Estuaries and Enclosed Seas*, Ketchum, B. H., Ed., Elsevier Scientific, Amsterdam, 1983, 103.
9. **Omori, M. and Ikeda, T.,** *Methods in Marine Zooplankton Ecology*, John Wiley & Sons, New York, 1984.
10. **Steidinger, K. A. and Walker, L. M., Eds.,** *Marine Plankton Life Cycle Strategies*, CRC Press, Boca Raton, Fla., 1984.
11. **Valiela, I.,** *Marine Ecological Processes*, Springer-Verlag, New York, 1984.
12. *Biological Methods Panel Committee on Oceanography*, Recommended Procedures for Measuring the Productivity of Plankton Standing Stock and Related Oceanic Properties, National Academy of Sciences, Washington, D.C., 1969.
13. **Lippson, A. J., Haire, M. S., Holland, A. F., Jacobs, F., Jensen, J., Moran-Johnson, R. L., Polgar, T. T., and Richkus, W. A.,** *Environmental Atlas of the Potomac Estuary*, Johns Hopkins University Press, Baltimore, 1981.
14. **Roman, M. R.,** Ingestion of detritus and microheterotrophs by pelagic marine zooplankton, *Bull. Mar. Sci.*, 35, 477, 1984.
15. **Smetacek, V.,** The annual cycle of protozooplankton in the Kiel Bight, *Mar. Biol.*, 63, 1, 1981.
16. **Stoecker, D. K. and Sanders, N. K.,** Differential grazing by *Acartia tonsa* on a dinoflagellate and a tintinnid, *J. Plankton Res.*, 7, 85, 1985.
17. **Verity, P. G.,** Growth rates of natural tintinnid populations in Narragansett Bay, *Mar. Ecol. Prog. Ser.*, 29, 117, 1986.
18. **Capriulo, G. M. and Carpenter, E. J.,** Abundance, species composition, and feeding impact of tintinnid micro-zooplankton in central Long Island Sound, *Mar. Ecol. Prog. Ser.*, 10, 277, 1983.
19. **Verity, P. G.,** The Physiology and Ecology of Tintinnids in Narragansett Bay, Rhode Island, Ph.D. Thesis, University of Rhode Island, Kingston, 1984.
20. **Robertson, J. R.,** Predation by estuarine zooplankton on tintinnid ciliates, *Estuarine Coastal Shelf Sci.*, 16, 27, 1983.
21. **Verity, P. G.,** Grazing of phototrophic nanoplankton by microzooplankton in Narragansett Bay, *Mar. Ecol. Prog. Ser.*, 29, 105, 1986.
22. **Davis, C. C.,** Planktonic Copepoda (including Monstrilloida), in *Marine Plankton Life Cycle Strategies*, Steidinger, K. A. and Walker, L. M., Eds., CRC Press, Boca Raton, Fla., 1984, 67.
23. **Mann, K. H.,** *Ecology of Coastal Waters: A Systems Approach*, University of California Press, Berkeley, 1982.
24. **Odhner, M. X.,** Life Histories and Nutrition of Copepods at Calvert Cliffs, Technical Report for Baltimore Gas and Electric Company, Academy of Natural Sciences, Philadelphia, 1977.
25. **Ambler, J. W., Cloern, J. E., and Hutchinson, A.,** Seasonal cycles of zooplankton from San Francisco Bay, *Hydrobiologia*, 129, 177, 1985.
26. **Stearns, D. E. and Forward, R. B., Jr.,** Photosensitivity of the calanoid copepod *Acartia tonsa*, *Mar. Biol.*, 82, 85, 1984.
27. **Stearns, D. E. and Forward, R. B., Jr.,** Copepod photobehavior in a simulated natural light environment and its relation to nocturnal vertical migration, *Mar. Biol.*, 82, 91, 1984.
28. **Kremer, P.,** Predation by the ctenophore *Mnemiopsis leidyi* in Narragansett Bay, Rhode Island, *Estuaries*, 2, 97, 1979.
29. **Durbin, A. G. and Durbin, E. G.,** Standing stock and estimated production rates of phytoplankton and zooplankton in Narragansett Bay, R.I., *Estuaries*, 4, 24, 1981.
30. **Deason, E. E.,** *Mnemiopsis leidyi* (Ctenophora) in Narragansett Bay, 1975—79: abundance, size composition and estimation of grazing, *Estuarine Coastal Shelf Sci.*, 15, 121, 1982.
31. **Omori, M. and Hamner, W. M.,** Patchy distribution of zooplankton: behavior, population assessment and sampling problems, *Mar. Biol.*, 72, 193, 1982.
32. **Haury, L. R., McGowan, J. A., and Wiebe, P. H.,** Patterns and processes in the time-space scales of plankton distributions, in *Spatial Pattern in Plankton Communities*, Steele, J. H., Ed., Plenum Press, New York, 1978, 277.
33. **Wroblewski, J. S.,** A simulation of the distribution of *Acartia clausi* during the Oregon USA upwelling August, 1973, *J. Plankton Res.*, 2, 46, 1980.
34. **Wroblewski, J. S.,** Interaction of currents and vertical migration in maintaining *Calanus marshallae* in the Oregon upwelling zone — a simulation, *Deep-Sea Res.*, 29, 665, 1982.
35. **Lippson, A. J. and Lippson, R. L.,** *Life in the Chesapeake Bay*, Johns Hopkins University Press, Baltimore, 1984.
36. **Perkins, E. J.,** *Biology of Estuaries and Coastal Waters*, Academic Press, London, 1974.
37. **Vance, R. R.,** On reproductive strategies in marine benthic invertebrates, *Am. Nat.*, 107, 339, 1973.
38. **Chia, F. S.,** Classification and adaptive significance of developmental patterns in marine invertebrates, *Thalassia Jugosl.*, 10, 121, 1974.
39. **Emlet, R. B.,** Facultative planktotrophy in the tropical echinoid *Clypeaster rosaceus* (Linnaeus) and a comparison with obligate planktotrophy in *Clypeaster subdepressus* (Gray) (Clypeasteroida: Echinoidea), *J. Exp. Mar. Biol. Ecol.*, 95, 183, 1986.

40. **Day, R. and McEdward, L.,** Aspects of the physiology and ecology of pelagic larvae of marine benthic invertebrates, in *Marine Plankton Life Cycle Strategies*, Steidinger, K. A. and Walker, L. M., Eds., CRC Press, Boca Raton, Fla., 1984, 93.
41. **Rumrill, S. S., Pennington, J. T., and Chia, F. S.,** Differential susceptibility of marine invertebrate larvae: laboratory predation of sand dollar, *Dendraster excentricus* (Eschscholtz), embryos and larvae by zoeae of the red crab, *Cancer productus* Randall, *J. Exp. Mar. Biol. Ecol.*, 90, 193, 1985.
42. **Mileikovsky, S. A.,** Types of larval development in marine bottom invertebrates, their distribution and ecological significance: a reevaluation, *Mar. Biol.*, 10, 193, 1971.
43. **Mileikovsky, S. A.,** Types of larval development in marine bottom invertebrates: an integrated ecological scheme, *Thalassia Jugosl.*, 10, 171, 1974.
44. **Thorson, G.,** Reproduction and larval ecology of marine bottom invertebrates, *Biol. Bull.*, 25, 1, 1950.
45. **Leiby, M. M.,** Life history and ecology of pelagic fish eggs and larvae, in *Marine Plankton Life Cycle Strategies*, Steidinger, K. A., and Walker, L. M., Eds., CRC Press, Boca Raton, Fla., 1984, 121.
46. **Fortier, L. and Leggett, W. C.,** A drift study of larval fish survival, *Mar. Ecol. Prog. Ser.*, 25, 245, 1985.
47. **Norcross, B. L. and Shaw, R. F.,** Oceanic and estuarine transport of fish eggs and larvae: a review, *Trans. Am. Fish. Soc.*, 113, 153, 1984.
48. **Robertson, A. I. and Howard, R. K.,** Diel trophic interactions between vertically-migrating zooplankton and their fish predators in an eelgrass community, *Mar. Biol.*, 48, 207, 1978.
49. **Alldredge, A. L. and King, J. M.,** The distance demersal zooplankton migrate above the benthos: implications for predation, *Mar. Biol.*, 84, 253, 1985.
50. **Marcus, N. H.,** Recruitment of copepod nauplii into the plankton: importance of diapause eggs and benthic processes, *Mar. Ecol. Prog. Ser.*, 15, 47, 1984.
51. **Marcus, N. H. and Schmidt-Gengenbach, J.,** Recruitment of individuals into the plankton: the importance of bioturbation, *Limnol. Oceanogr.*, 31, 206, 1986.
52. **Forward, R. B., Jr.,** Light and diurnal vertical migration: photobehavior and photophysiology of plankton, *Photochem. Photobiol. Rev.*, 1, 157, 1976.
53. **Forward, R. B., Jr.,** Behavioral responses of larvae of the crab *Rhithropanopeus harrisii* (Brachyura: Xanthidae) during diel vertical migration, *Mar. Biol.*, 90, 9, 1985.
54. **Forward, R. B., Jr., Cronin, T. W., and Stearns, D. E.,** Control of diel vertical migration: photoresponses of a larval crustacean, *Limnol. Oceanogr.*, 29, 146, 1984.
55. **Heinle, D. R.,** Temperature and zooplankton, *Chesapeake Sci.*, 10, 186, 1969.
56. **McLaren, I. A.,** Demographic strategy of vertical migration by a marine copepod, *Am. Nat.*, 108, 91, 1974.
57. **Stickney, R. R.,** *Estuarine Ecology of the Southeastern United States and Gulf of Mexico*, Texas A&M University Press, College Station, Texas, 1984.
58. **Haedrich, R. L.,** Estuarine fishes, in *Estuaries and Enclosed Seas*, Ketchum, B. H., Ed., Elsevier Scientific, Amsterdam, 1983, 183.
59. **Diel, S. and Klein Breteher, W. C. M.,** Growth and development of *Calanus* spp. (Copepoda) during spring phytoplankton succession in the North Sea, *Mar. Biol.*, 91, 85, 1986.
60. **Jeffries, H. P. and Johnson, W. C.,** Distribution and abundance of zooplankton, in *Coastal and Offshore Environmental Inventory: Cape Hatteras to Nantucket Shoals*, Marine Publication Series, No. 2, University of Rhode Island, Kingston, 1973, 4-1.
61. **Price, H. J., Paffenhöfer, G.-A., and Strickler, J. R.,** Modes of cell capture in calanoid copepods, *Limnol. Oceanogr.*, 28, 116, 1983.
62. **Price, H. J. and Paffenhöfer, G.-A.,** Effects of feeding experience in the copepod *Eucalanus pileatus:* a cinematographic study, *Mar. Biol.*, 84, 35, 1984.
63. **Koehl, M. A. R.,** The morphology and performance of suspension-feeding appendages, *J. Theoret. Biol.*, 105, 1, 1983.
64. **Legier-Visser, M. F., Mitchell, J. G., Okubo, A., and Fuhrman, J. A.,** Mechanoreception in calanoid copepods: a mechanism for prey detection, *Mar. Biol.*, 90, 529, 1986.
65. **Lehman, J. T.,** Grazing, nutrient release, and their impacts on the structure of phytoplankton communities, in *Trophic Interactions Within Aquatic Ecosystems*, Meyers, D. G. and Strickler, J. R., Eds., Am. Assoc. Adv. Sci., Selected Symp. 85, Westview Press, Boulder, Colorado, 1984, 49.
66. **Strickler, J. R.,** Sticky water: a selective force in copepod evolution, in *Trophic Interactions Within Aquatic Ecosystems*, Meyers, D. G. and Strickler, J. R., Eds., Am. Assoc. Adv. Sci., Selected Symp. 85, Westview Press, Boulder, Colorado, 1984, 187.
67. **Harris, R. P.,** Comparison of the feeding behavior of *Calanus* and *Pseudocalanus* in two experimentally manipulated enclosed ecosystems, *J. Mar. Biol. Assoc. U.K.*, 62, 71, 1982.
68. **Daro, M. H.,** Feeding rhythms and vertical distribution of marine copepods, *Bull. Mar. Sci.*, 37, 487, 1985.
69. **Lampert, W., and Taylor, B. E.,** Zooplankton grazing in a eutrophic lake: implications of diel vertical migration, *Ecology*, 66, 68, 1985.
70. **Mackas, D. L. and Bohrer, R.,** Fluorescence analysis of zooplankton gut contents and an investigation of diel feeding patterns, *J. Exp. Mar. Biol. Ecol.*, 25, 77, 1976.
71. **Arashkevich, Ye C.,** Relationship between the feeding rhythm and the vertical migrations of *Cypridina sinuosa* (Ostracoda, Crustacea) in the western part of the equatorial Pacific, *Oceanology*, 17, 466, 1977.

72. **Bohrer, R. N.,** Experimental studies on diel vertical migration in evolution and ecology of zooplankton communities, in *Evolution and Ecology of Zooplankton Communities*, Kerfoot, W. C., Ed., University Press, London, 1980, 111.
73. **Emright, J. T.,** Diurnal vertical migration: adaptive significance and timing. I. Selective advantage: a metabolic model, *Limnol. Oceanogr.*, 22, 856, 1977.
74. **Parsons, T. R., LeBrasseur, R. T., and Fulton, J. D.,** Some observations on the dependence of zooplankton grazing on the cell size and concentration of phytoplankton blooms, *J. Oceanogr. Soc. Jpn.*, 23, 10, 1967.
75. **Parsons, T. R., LeBrasseur, R. T., Fulton, J. D., and Kennedy, O. D.,** Production studies in the Strait of Georgia. II. Secondary production under the Fraser River plume, February to May, 1967, *J. Exp. Mar. Biol. Ecol.*, 3, 39, 1969.
76. **McAllister, C. D.,** Zooplankton rations, phytoplankton mortality, and the estimation of marine production, in *Marine Food Chains*, Steele, J. H., Ed., University of California Press, Berkeley, 1970, 419.
77. **Steele, J. H.,** *Structure of Marine Ecosystems,* Harvard University Press, Cambridge, 1974.

8.1. BIOMASS AS RELATED TO AREA, DEPTH, AND SEASON

Table 8.1—1
RATIOS OF THE AVERAGE MONTHLY VALUES OF THE MAXIMAL AND MINIMAL AMOUNTS (WEIGHT) OF PLANKTON DURING A YEAR IN DIFFERENT REGIONS OF THE OCEAN

Region	0–100 m	0–1,000 m	Kind of data
North Atlantic-Norwegian Sea	1:406	1:2.36	Average of several samples per month
Northwestern Pacific and Bering Sea	1:12	1:3*	Average of trip stations
Northeastern Pacific	1:10	1:2	Average of single sample per month
Subantarctic	1:5.1	1:1.8	Average of many trip stations
Antarctic	1:3.4	1:1.9	Average of many trip stations

* Collections made at a depth of 0–500 m.

Figure 8.1—1

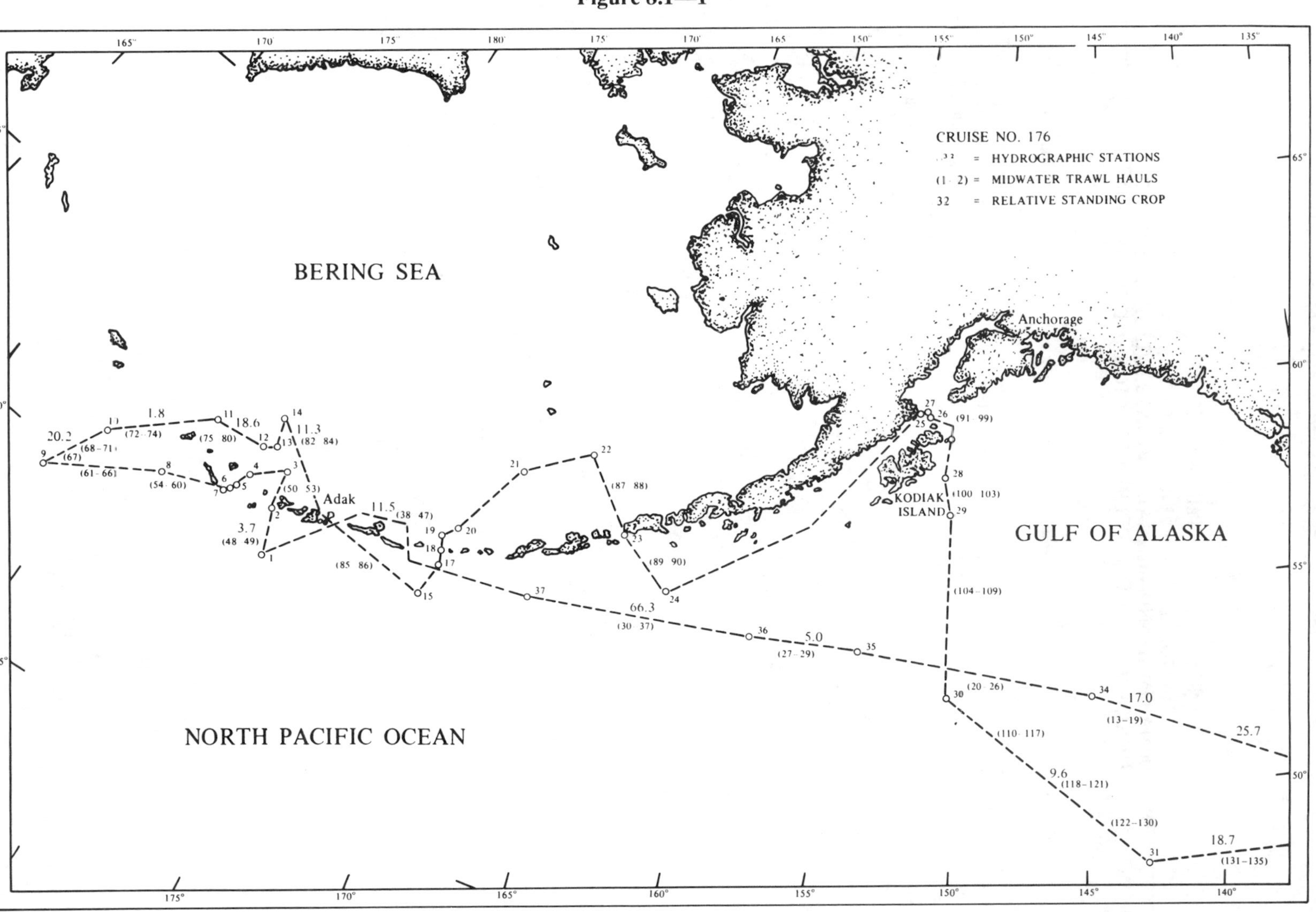

Figure 8.1—1. Cruise tracks showing position of hydrographic stations, position of hauls, and relative standing crop of plankton in the eastern north Pacific.

Figure 8.1—1 (continued)

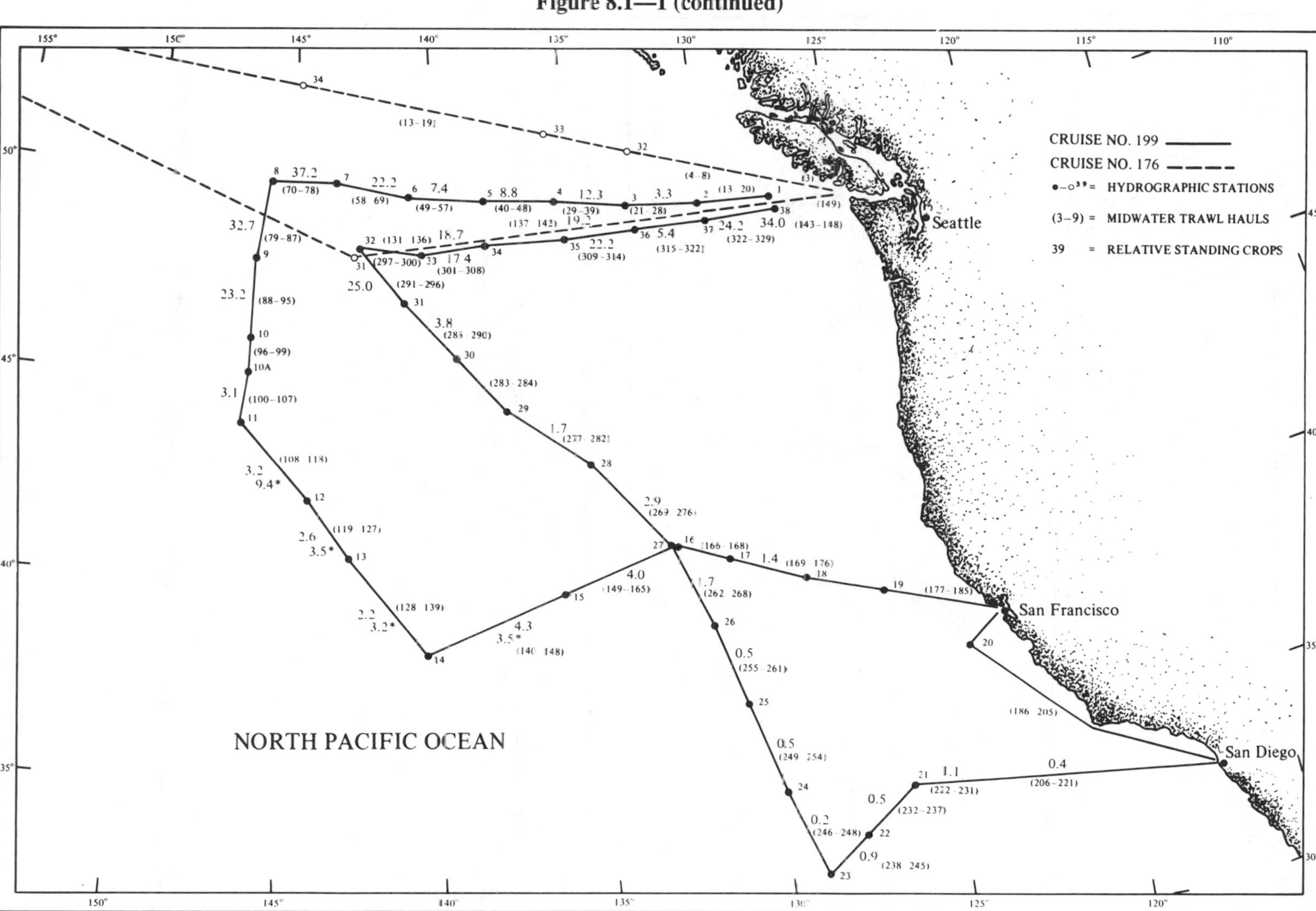

(From Aron, W., The distribution of animals in the eastern north Pacific and its relationship to physical and chemical conditions, *J. Fish. Res. Bd. Canada*, 19(2), 290, 1962. With permission.)

Figure 8.1—2

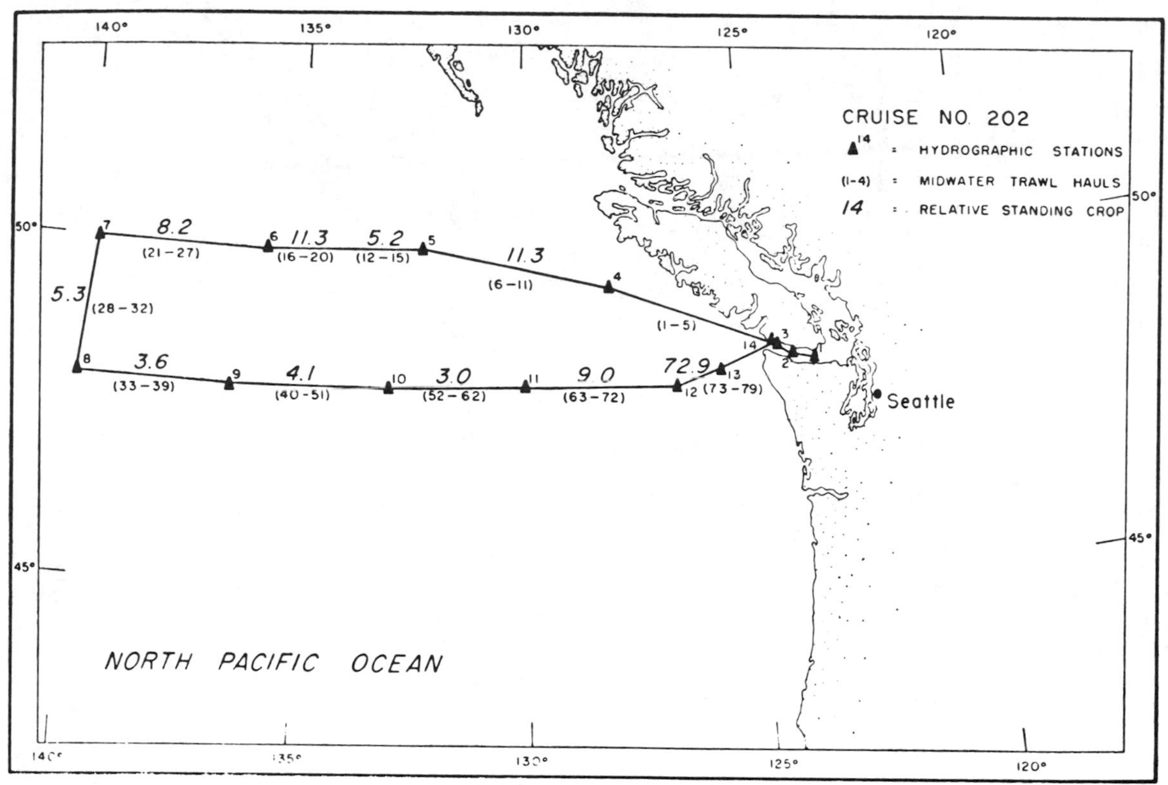

Figure 8.1—2. Cruise tracks showing position of hydrographic stations, position of hauls, and relative standing crop of plankton, in the eastern north Pacific.

(From Aron, W., The distribution of animals in the eastern north Pacific and its relationship to physical and chemical conditions, *J. Fish. Res. Bd. Canada,* 19(2), 291, 1962. With permission.)

Figure 8.1—3

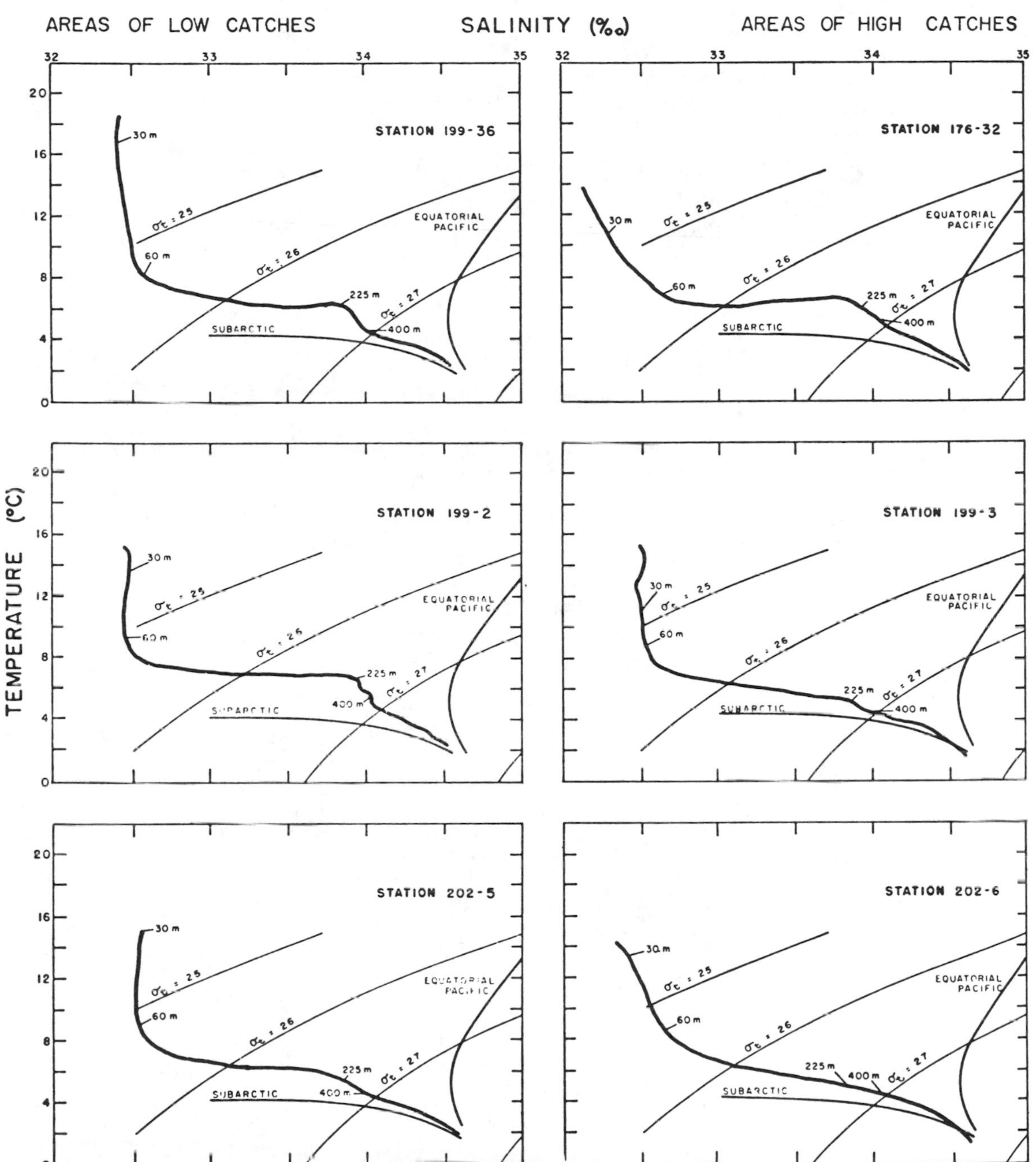

Figure 8.1—3. Comparison of temperature-salinity diagrams from oceanic areas of high and low plankton catches in the northeastern Pacific.

(From Aron, W., The distribution of animals in the eastern north Pacific and its relationship to physical and chemical conditions, *J. Fish. Res. Bd. Canada*, 19(2), 294, 1962. With permission.)

Figure 8.1—4

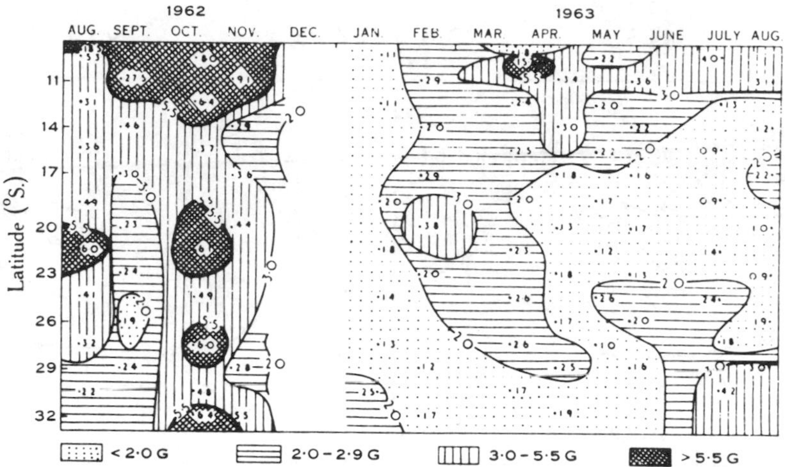

Figure 8.1—4. Contours of dry weight (g) per haul as a function of latitude and season for plankton catches in the central Indian Ocean.

(From Legand, M., Seasonal variations in the Indian Ocean along 110° E. VI. Macroplankton and micronekton biomass, *Aust. J. Mar. Freshwater Res.*, 20, 98, 1969. With permission.)

Figure 8.1—5

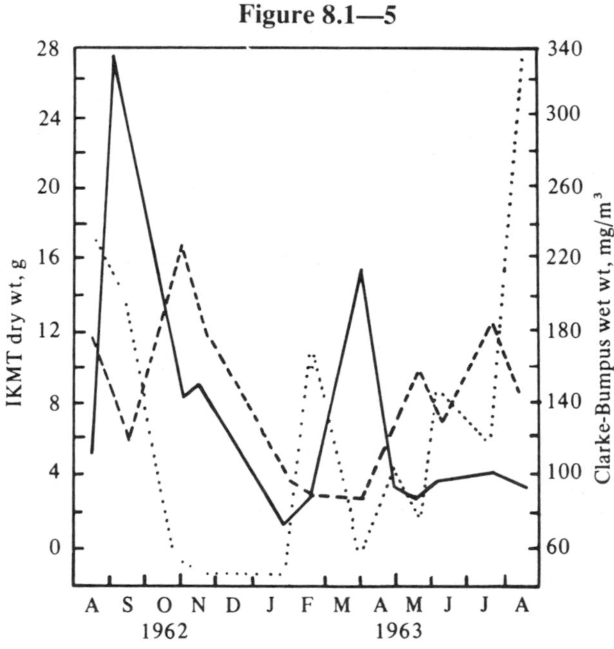

Figure 8.1—5. Comparison of the seasonal development of IKMT* plankton (● —— ●), IKMT macroplankton (X ----- X), and Clarke-Bumpus night plankton (Δ······· Δ) at stations of maximal abundance in the north central Indian Ocean.

* IKMT – Isaacs-Kidd midwater trawl.

(From Legand, M., Seasonal variations in the Indian Ocean along 110° E. VI. Macroplankton and micronekton biomass, *Aust. J. Mar. Freshwater Res.*, 20, 102, 1969. With permission.)

Table 8.1—2
DEPTH DISTRIBUTION OF PLANKTON CATCHES ACCORDING TO AREA IN THE EASTERN NORTH PACIFIC*

	Average volume of catch (*ml/min*)				
Area	**20–30**	**31–60**	**61–120**	**121–250**	**251–400**
A	43.7	12.7	4.3	5.8	No hauls
B	12.9	28.2	17.8	10.8	No hauls
C	18.3	7.7	No hauls	6.1	No hauls
D	0.6	6.1	No hauls	3.2	No hauls
E	28.6	22.1	No hauls	7.8	No hauls
F	1.8	3.4	6.2	5.1	No hauls
G	0.9	1.0	4.0	1.7	1.2
H	8.6	9.5	6.8	4.4	3.4
I	22.1	33.5	10.1	4.7	1.4
J	3.2	13.6	5.3	9.5	4.6
K	8.7	9.5	14.2	9.7	8.4

Note: Average volumes are 3 times average volumes caught by 3 ft net.

*Area code and description of area

Area	**Description**
A	Offshore area not influenced by Subarctic Current.
B	Offshore area where Subarctic Current is apparent south of 49°N latitude.
C	The Aleutian Island Area.
D	Area just off the Washington Coast with considerable Intermediate water present.
E	Offshore region composed mostly of Subarctic water.
F	Offshore region strongly influenced by the presence of Intermediate water and some Central and Equatorial water.
G	Offshore region similar in most respects to the previous region, but with greater quantities of Equatorial water.
H	Offshore region, oceanographically intermediate between E and F.
I	Offshore region, mostly Subarctic water, but Intermediate water is present along the cruise track.
J	Offshore region between 49°N, 135°W, and some Subarctic water
K	Offshore region (49°N, 135°W) with considerable Intermediate water present.

(From Aron, W., The distribution of animals in the eastern north Pacific and its relationship to physical and chemical conditions, *J. Fish. Res. Board Can.*, 19(2), 297, 299, 1962. With permission.)

Table 8.1—3
ESTIMATES OF ZOOPLANKTON PRODUCTION

Organism or group	Area and period	Production mgC/m² /day	Daily production ratios x/y standing crop	Production net primary production ratios x/y
Calanus finmarchicus	E. Barents Sea; year	7.8	0.002	0.03
Calanus cristatus	N.W. Pacific; summer	5.6	0.012	–
Calanus plumchrus	N.W. Pacific; summer	4.6	0.010	–
Eucalanus bungii	N.W. Pacific; summer	3.5	0.014	–
Calanus glacialis	N. Bering Sea; year	0.7	–	0.005
Calanus plumchrus	W. Bering Sea; year	3.1	–	0.012
Calanus cristatus	W. Bering Sea; year	3.8	–	0.015
Eucalanus bungii	W. Bering Sea; year	7.3	–	0.03
Diaptomus salinus	Aral Sea; year	0.66	0.007	–
Acartia clausi	Black Sea; year	0.38	0.035	0.001
Centropages kröyeri	Black Sea; summer	0.19	0.077	0.0002
Euphasia pacifica	N.E. Pacific; year	0.9	0.008	0.0048
Acartia tonsa	Chesapeake Bay estuary; summer	77	0.50	0.05
Acartia clausi	Black Sea bay; June	15	0.17	–
Acartia clausi	Black Sea, open sea; June	6.6	0.23	0.08
Calanus helgolandicus	Black Sea, open sea; June	28	0.15	0.07
zooplankton	Georges Bank; year	200	0.03	0.25
zooplankton	English Channel, year	75	0.10	0.30
zooplankton	Long Island Sound; year	166	0.17	0.30
zooplankton	N. North Sea; Apr.–Sept.	180	0.048	0.58
herbivorous copepods	North Sea; Jan.–June	4.9	0.08	0.14
copepods (mainly *Calanus*)	North Sea; Mar.–June	46 (author's calculation)	0.10 (author's calculation)	0.20 (author's calculation)
zooplankton	Gulf of Panama; Jan.–April	70 or 234	0.29 or 0.98	0.09 or 0.31

(From Mullin, M., Production of zooplankton in the ocean: the present status and problems, *Oceanogr. Mar. Biol. Annu. Rev.*, 7, 308, 1970. With permission by George Allen and Unwin Ltd., London.)

Figure 8.1—6

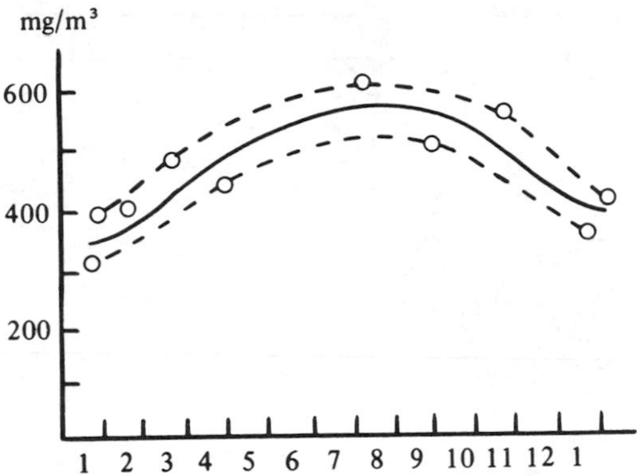

Figure 8.1—6. Seasonal changes in plankton biomass on the southwest part of the Florida shelf.

Figure 8.1—7

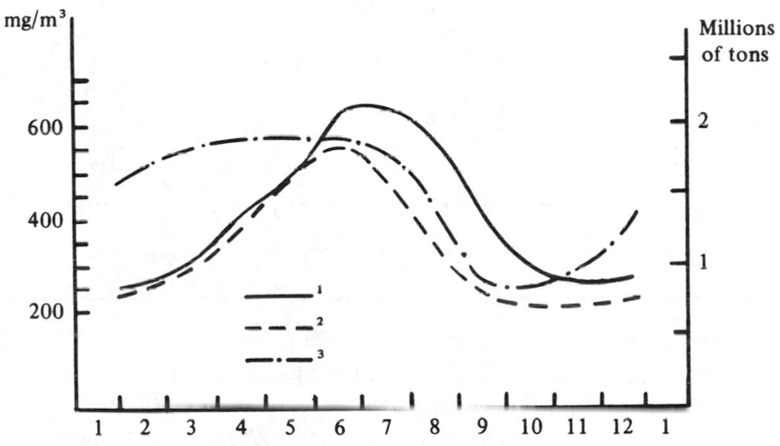

[1] Total quantity of plankton (millions of tons) on Bank.
[2] Average plankton biomass on eastern part of Bank.
[3] Average plankton biomass on western part of Bank.

Figure 8.1—7. Seasonal changes in biomass and quantity of plankton on the Campeche Bank.

Figure 8.1—8

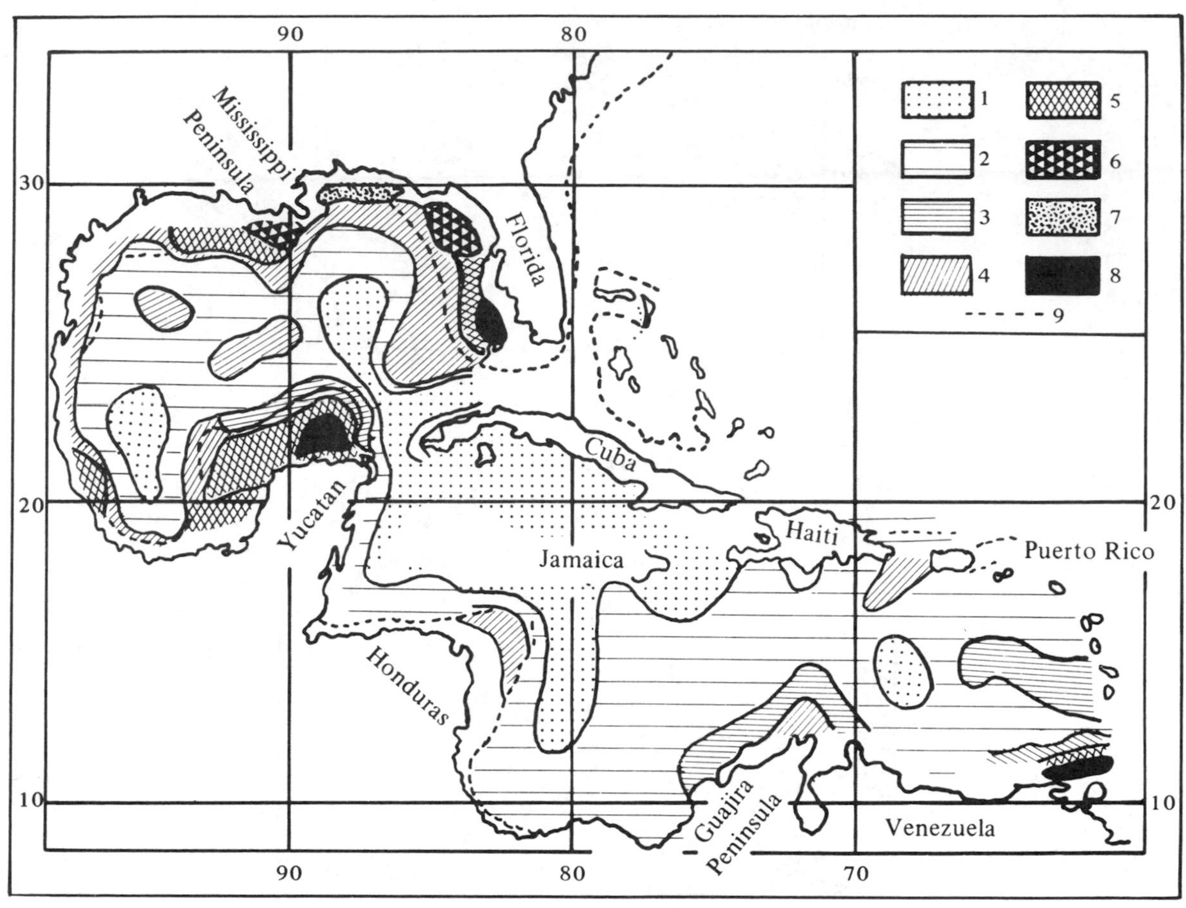

1) 30–100; 2) 50–150; 3) 100–200; 4) 100–300; 5) 200–600; 6) 200–1000;
7)100–3000; 8) 300–1000; 9) edge of shelf.

Figure 8.1—8. Mean plankton distributions in the upper 100 meter layer of the Caribbean and Gulf of Mexico in 1962 to 1966. Values are in mg/m^3.

Table 8.1—4

RELATIVE RICHNESS OF VARIOUS POPULATIONS OFF THE AFRICAN COAST, IN LONG ISLAND SOUND, AND IN THE SARGASSO SEA

Populations	Richest	Less rich	Least rich
Primary production	Africa	> Long Island	> Sargasso Sea
Zooplankton (surface layer)	Africa	> Long Island	> Sargasso Sea
Zooplankton (total)	Africa	> Sargasso Sea	> Long Island
Ultraplankton (surface layer)	Long Island	> Africa	> Sargasso Sea
Ultraplankton (total)	Africa	> Long Island	> Sargasso Sea

Table 8.1—5

BALANCE SHEET OF ESTIMATED PRODUCTION AND CONSUMPTION IN THE SARGASSO SEA OFF BERMUDA

Process	Depth (m)	Component	Process rate ($mgC/m^2/day$)
Production	0–100	Uncorrected C^{14} uptake	200
		Estimated total production	350–400
Consumption	0–300	Zooplankton	135
		Ultraplankton	200
	300–900	Zooplankton + ultraplankton	40
	900–bottom	Zooplankton + ultraplankton	6

Table 8.1—6

AMOUNT OF PLANKTON ($cm^3/1,000\ m^3$) AT VARIOUS DEPTHS AND IN DIFFERENT PARTS OF THE NORTH ATLANTIC

Depth (m)	Gulf Stream 40–43°N	Continental slope of North America 38–41°N	Gulf Stream 35–37°N	Sargasso Sea 20–37°N	North Equatorial Current 16–19°N	African Littoral 15–22°N	Canaries Current 27–36°N	32°29′N 20°09′W	32°34′N 16°19′W
0–50	74.5	199.0	54.9	64.7	89.6		78.6		
50–100	42.3	94.0	35.9	59.9	80.7	214.0	55.0		
100–200	26.2	35.1	23.3	32.1	30.6	94.7	27.0	26.0	28.0
200–500	25.4	30.8	11.1	10.2	15.9	59.9	15.6		
500–1,000	15.7	19.0	6.0	4.1	5.4	27.2	6.5		
1,000–2,000	5.0	7.8	2.8	1.2	1.6	–	2.1	5.0	4.0
2,000–3,000	–	–	2.1	0.6	1.0	–	0.8	1.0	0.7
3,000–4,000	–	–	–	0.3	0.3	–	0.2	0.6	–
4,000–5,000	–	–	–	0.1	–	–	–	–	–

Table 8.1—7
COMPARISONS OF DISPLACEMENT VOLUMES AND WET WEIGHTS OF ZOOPLANKTON IN VARIOUS REGIONS OF THE NORTH ATLANTIC

(1 = Summer; 2 = Fall; 3 = Winter; 4 = Spring; 5 = Yearly Mean)

Region	Displ. vol. ml/1,000 m^3	Wet wt. mg/m^3	Net mesh aperture (mm)	Depth range (m)
I. *Boreal North Atlantic and Neritic Waters*				
Iceland Coast				
(1)	250	–	stramin	0–50
(4)	450	–	stramin	0–25
Southern Norwegian Sea				
(1)	340	–	0.366	0–100
(3)	10	–	0.366	0–100
(4)	230	–	0.366	0–100
Labrador Current				
(2) 43° N	–	1072	0.17	0–200
Norwegian Sea (5)	–	>500	0.17	0–100
Cold-temperate Subarctic waters (5)	>100	>100	0.202	0–300
Western North Atlantic				
Coastal	8100	–	0.158	0–25
Slope water (4)	4300	–	0.158	0–50
Slope water (1)	–	430–1600	0.158	0–400
Coastal (5)	540	–	10 strands/cm	0–85
Offshore (5)	400	–	–	–
Gulf of Maine				
(3)	120	–	Front: 29–38 meshes/inch	variable
(1)	260	–	Rear: 48–54 meshes/inch	variable
Cape Cod Chesapeake Bay coastal shelf				
(1)	700–800	–	–	variable
(3)	400	–	–	variable
Georges Bank				
(1)	1500	–	0.366	0–25
(3)	200	–	0.366	0–25
Cape Hatteras-Cape Fear				
(1)	280	–	0.360	Variable over shelf

Table 8.1—7 (continued)
COMPARISONS OF DISPLACEMENT VOLUMES AND WET WEIGHTS OF ZOOPLANKTON IN VARIOUS REGIONS OF THE NORTH ATLANTIC

Region	Displ. vol. ml/1,000 m³	Wet wt. mg/m³	Net mesh aperture (mm)	Depth range (m)
Continental Slope 38°–41°N (2)	328	–	0.170	0–200
New York-Bermuda Coastal water (5)	1070	–	0.230	0–200
Slope water (5)	270	–	0.230	or less 0–200
II. *Central Waters (Sargasso Sea)*				
Sargasso Sea (5)	20	–	0.230	0–200
Sargasso Sea Bermuda (5)	28	–	0.203–0.366	0–500
S. W. Sargasso Sea	–	156.7	0.17	0–200
20°–37°N (2)				
Sargasso Sea (1)	–	45	0.158	0–400
Sargasso Sea (5)	–	50–100	0.170	0–100
Sargasso Sea (5)	–	<50–100	–	0–300
Sargasso Sea (5)	10–25	<10–25	0.202	0–300
III. *Boundary Currents*				
Florida Strait (4)	20	–	0.158	0–150
Florida Strait (4)	20	–	0.158	150–300
Gulf Stream off Florida (4)	50	–	0.158	0–100
Gulf Stream off Georgia (4)	70	–	0.158	0–150
Gulf Stream		137	0.158	0–400
Gulf Stream	–	250–500	0.170	0–100
Gulf Stream betw. N.Y.-Bermuda	30	–	0.230	0–200
Gulf Stream (north) 40°–43°N	–	143	0.17	0–200
Gulf Stream (south) 35°–37°N	–	114	0.17	0–200
N. Equat. Current 16°–19°N	–	201	0.17	0–200

Table 8.1—7 (continued)
COMPARISONS OF DISPLACEMENT VOLUMES AND WET WEIGHTS OF ZOOPLANKTON IN VARIOUS REGIONS OF THE NORTH ATLANTIC

Region	Displ. vol. ml/1,000 m³	Wet wt. mg/m³	Net mesh aperture (mm)	Depth range (m)
Canary Current 27° –36° N	–	161	0.17	0–200
Equat. Current (1)	100–>300	–	0.336	0–~50
Equat. Current region (5)	–	100–500	0.17	0–100
Equat. Current region (5)	–	100–300	–	0–300
All boundary Currents around Sargasso Sea (5)	25–100	25–100	0.202	0–300

Figure 8.1—9

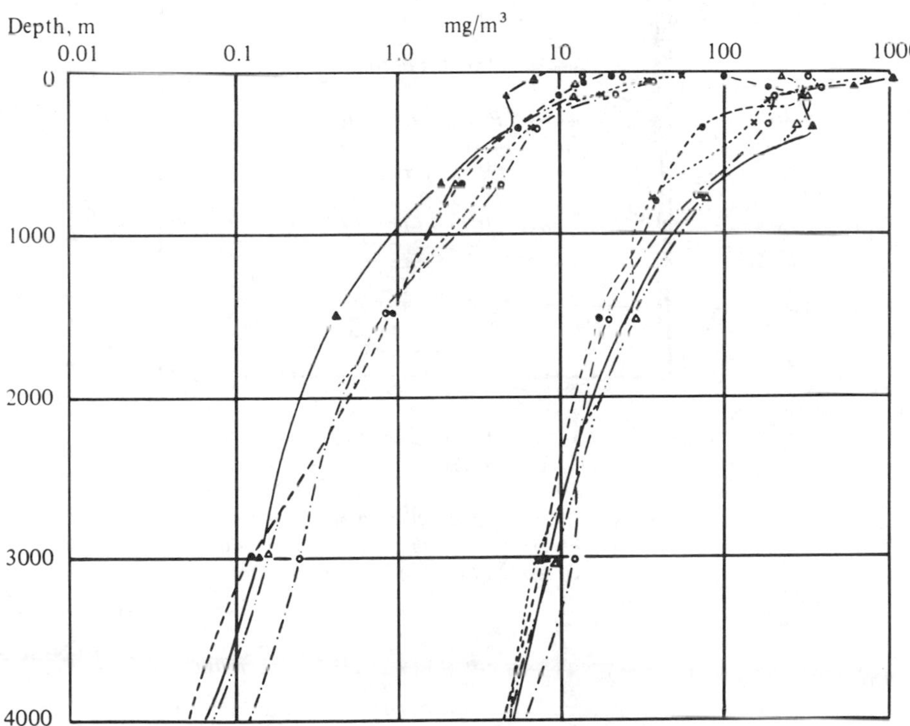

Figure 8.1—9. Vertical distribution of net plankton biomass (wet weight) in the Pacific, from divided hauls. The right family of curves is for stations off the Kurile Islands. The left family is for tropical stations (– – ● – – = average of two stations).

(From Banse, K., On the vertical distribution of zooplankton in the sea, in *Progress in Oceanography,* Vol. 2, Sears, M., Ed., Pergamon Press, New York, 1964. 82. With permission.)

Figure 8.1—10

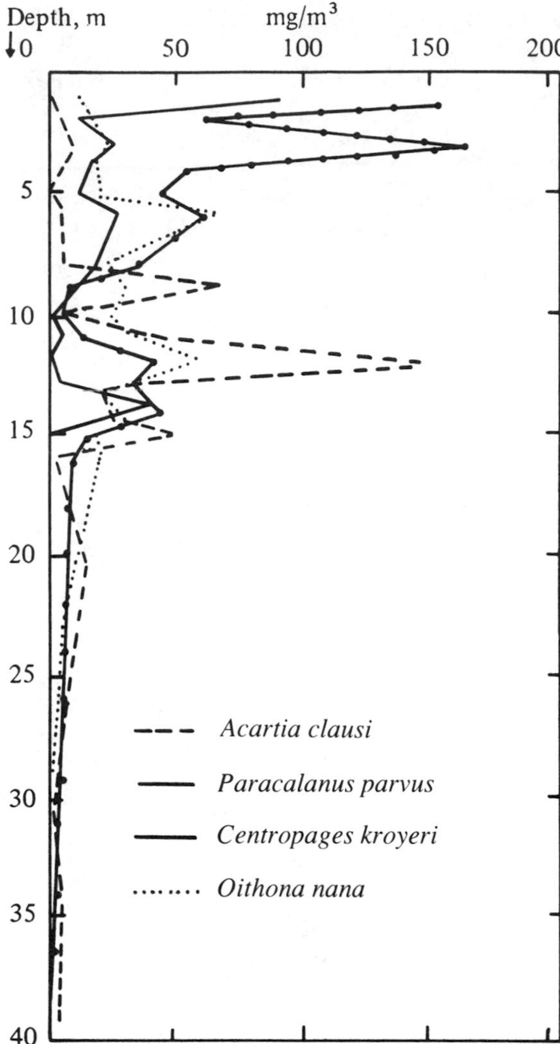

Figure 8.1—10. Vertical distribution of zooplankton biomass at two stations in the Black Sea.

(From Banse, K., On the vertical distribution of zooplankton in the sea, in *Progress in Oceanography,* Vol. 2, Sears, M., Ed., Pergamon Press, New York, 1964, 69. With permission.)

8.2. TAXON DIVERSITY AS A FUNCTION OF AREA AND DEPTH

Table 8.2—1
CHANGES IN THE RELATIVE NUMBERS OF CALANOIDA WITH DEPTH IN VARIOUS REGIONS OF THE OCEAN

	Open ocean			Isolated basin				
						Mediterranean Sea (Ionian Sea)		
Depth (m)	Northwestern Pacific Ocean (spring)	Tropical zone Pacific Ocean	Tropical zone Indian Ocean	Norwegian Sea (annual average)	Central North Polar Basin	Winter	Summer	Sea of Japan (winter)
0–50								
50–100	63.5	93.1	–	29.2	92.5	91.5	53.0	58.4
100–200	30.6	67.0	64.5	33.6	105	71.5	49.3	35.0
200–500	80.5	52.3	31.0	33.6	27.1	37.1	26.1	137.0
500–1000	16.2	40.0	61.0	160	27.1	8.8	27.0	56.4
1000–2000	59.0	28.5	18.7	5.1	5.8	3.5	4.7	3.1

(Numbers in each layer are expressed in % of numbers in the overlying layer.)

Table 8.2—2
MEAN ANNUAL TOTAL IN NUMBERS/m³ (A) AND MEAN PERCENTAGES (B) OF MAJOR ZOOPLANKTON GROUPS IN THE SARGASSO SEA

	0–500 m		500–1000 m		1000–1500 m		1500–2000 m		0–2000 m
	A	B	A*	B	A*	B	A*	B	A
Calanoids	89.8	42.1	12.0	43.1	5.4	45.7	2.8	48.8	27.5
Other copepods	60.9	28.0	10.4	37.7	4.2	38.3	2.0	36.4	19.4
Total copepods	150.7	70.1	22.4	80.8	9.6	84.0	4.8	85.2	46.9
Ostracods	14.8	7.0	2.2	7.7	*530*	4.8	*190*	2.8	4.4
Other Crustacea (including larval forms)	7.1	3.3	1.1	3.5	*370*	3.4	*270*	4.4	2.2
Total Crustacea	172.6	80.4	25.7	92.0	10.5	92.2	5.2	92.4	53.5
Tunicates	15.1	6.5	*185*	0.6	*20*	0	*20*	0.5	3.8
Chaetognaths	6.5	3.0	*312*	1.1	*25*	2.2	*35*	0.6	1.8
Coelenterates	6.8	3.1	*309*	1.0	*100*	0.8	*13*	0.4	1.8
Larval forms (noncrustacean)	4.4	2.1	*540*	1.9	*115*	1.0	*32*	0.6	1.3
Protozoa	5.1	2.6	*690*	2.5	*320*	3.4	*160*	5.0	1.6
Miscellaneous	4.9	2.3	*103*	0.9	*20*	0.4	0		1.3
Total No./m³	215.3		27.9		11.3		5.5		65.0
Total No./m²	107,650		13,952.5		5672.5		2750		130,025

*Italicized numbers are $\times 10^{-3}$, i.e., No./1000 m³.

(From Deevey, G. B. and Brooks, A. L., The annual cycle in quantity and composition of the zooplankton of the Sargasso Sea off Bermuda II Surface to 2000 m, *Limnol. Oceanogr.*, 16 (6), 933, 1971. With permission.)

Table 8.2—3
ANALYSES OF CONTINENTAL SHELF ZOOPLANKTON SOUTH OF NEW YORK

Organism	Water (wet weight)	Organic weight (ash free dry weight)	Constituent % of dry weight			
			Ash weight	C (Total)	N (Kjeldahl)	P (Total)
Cnidaria						
Pelagia noctiluca	95.9	31.0	69.0	8.2–9.9	1.4	0.14–0.16
Aequorea vitrina	99.1	50.9	49.1	17.8–22.5	0.5	0.021
Cyanea capillata	95.4	37.0	63.0	11.6	0.4–1.4	0.0–0.4
Ctenophora						
Beroe cucumis	95.3	29.7	70.3	–	1.1	0.16
Mnemiopsis sp.	95.0	25.0	75.0	6.4	0.2	0.12
Mollusca (Pteropoda)						
Limacina retroversa and sp.	81.3	35.8	64.2	28.3	4.1	0.58
Clione limacina	91.0	66.7	33.3	26.3	2.2–5.0	0.26–0.35
Corolla sp.	96.5	75.7	24.3	–	–	–
Arthropoda						
Calanus finmarchicus	89.8	82.4	17.6	35.7–41.7	4.7–5.9	0.39–0.68
Centropages typicus and *hamatus*	84.2	77.2	22.8	32.5–38.5	5.2–7.1	0.72–0.84
Euphausia krohnia	80.0	81.4	18.6	35.8	6.8	0.94
Meganyctiphanes norvegica	81.0	77.6	22.4	33.4–37.0	5.2–7.1	1.16
Lophogaster sp.	–	–	–	46.8	–	–
Idotea metallica	–	–	–	33.2	6.0	4.07
Chaetognatha						
Sagitta elegans	89.4	78.4	21.6	–	7.8	0.20–0.57
Tunicata						
Salpa fusiformis	96.0	22.9	77.1	7.2–10.6	0.5–1.5	0.19–0.28
Pyrosoma sp.	95.9	31.0	71.2	9.4	0.3–0.4	0.14

Table 8.2—4
NUMBER OF EPIPLANKTONIC SPECIES IN THE CALIFORNIA CURRENT REGION

Functional Group	Approx no. of species found in regions between the surface and 140 m depth
Amphipoda	⩾41
Chaetognatha	24
Cladocera	5
Copepoda	⩾235
Crustacean larvae	No estimate available
Ctenophora	⩾4
Decapoda	No estimate available
Euphausiacea	28
Heteropoda	13
Larvacea	26
Medusae	24
Mysidacea	⩾14
Ostracoda	⩾16
Pteropoda	29
Radiolaria, Tripylea	⩾20
Thaliacea	⩾23
Siphonophora	44
All Taxa Combined	⩾546

Table 8.2—5
NUMBER OF SPECIES OF CALANOIDA LIVING AT VARIOUS DEPTHS IN THE NORTHEAST

Atlantic Ocean (From 50° to 30°N)

Depth	0–200 m	200–500 m	500–1000 m	1000–2000 m	2000–3000 m	3000–4000 m	4000–5000 m
Total number of species	91	123	120	70	57	27	14
Number of appearing species	–	61	29	19	26	5	1
In % of total number of species at this depth	–	49	24	27	46	18	7
Number of disappearing species	29	32	69	39	36	14	–
In % of total number of species at this depth	32	26	58	56	61	52	–

Table 8.2—6
NUMBER OF SPECIES OF CALANOIDA LIVING AT VARIOUS DEPTHS IN THE NORTH ATLANTIC

(North of 50°N)

Depth	0–50 m	50–100 m	100–200 m	200–500 m	500–1000 m	1000–2000 m	Deeper than 2000 m
Total number of species	22	24	34	48	105	95	55
Number of disappearing species	–	7	10	17	59	21	3
Number of appearing species	5	0	3	2	31	43	–

Table 8.2—7
NUMBER OF SPECIES OF CALANOIDA AT VARIOUS DEPTHS OF THE ANTARCTIC AND SUBANTARCTIC REGIONS

Depth	0–50 m	50–100 m	100–200 m	250–500 m	500–750 m	750–1000 m
Total number of species	20	30	37	56	76	80
Number of appearing species	–	11	8	18	22	16
Number of disappearing species	0	1	0	2	12	–

Table 8.2—8
VERTICAL CHANGES IN THE DIVERSITY OF CALANID FAUNA IN THE EASTERN DEPRESSION OF THE NORTH POLAR BASIN

Depth (m)	No. of species	% of total calanid fauna	No. of appearing species	% of fauna of layer	No. of disappearing species	% of fauna of layer	Diversity index (α)	No. of specimens/m^3
0–50	17	37	–	–	0	0	2.8	96.9
50–100	25	54	8	32	1	4	4.1	89.6
100–200 (300)	34	74	10	29	4	12	4.8	94.2
200 (300)–800 (1000)	40	87	10	25	9	22	6.1	26.0
800 (1000)–bottom	31	68	1	3	–	–	4.9	1.5

Table 8.2—9
VARIATION WITH DEPTH OF THE NUMBERS ($SPEC/m^3$) OF COPEPODS (CALANOIDA) IN TROPICAL AREAS OF THE PACIFIC OCEAN[a] AND THE INDIAN OCEAN[b]

Hauls Made with BR 113/140 Nets

Depth (m)	Pacific Ocean (open waters)	Indian Ocean
0–50	12.4	18.6
50–100	11.3	–
100–200	6.9	12.0
200–500	3.5	3.7
500–1000	1.4	2.25
1000–2000	0.40	0.42
2000–3000	–	0.19
2000–4000	0.18	–
3000–4000	–	0.11
4000–6000	0.08	–
6000–8000	0.05	–

[a] Average of eight stations.
[b] Average of five stations.

Figure 8.2—1

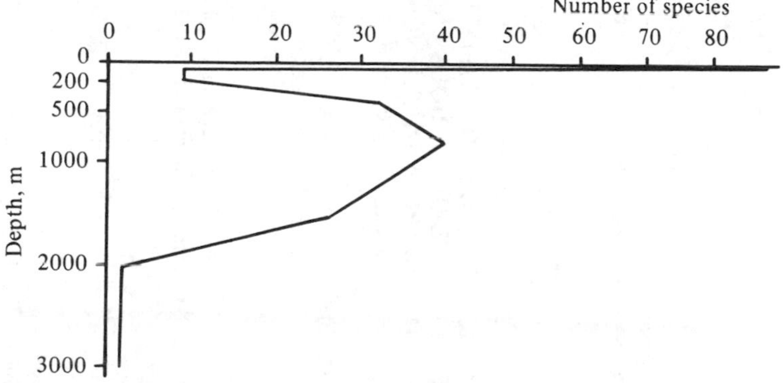

Figure 8.2—1. Number of species of copepods (Calanoida) appearing in each of the layers from which collections were taken in the tropical regions of the Indian Ocean.

Table 8.2—10
NUMBER OF SPECIES OF CALANOIDA OCCUPYING VARIOUS DEPTHS IN THE BAY OF BISCAY, AND VERTICAL CHANGES IN THE DIVERSITY INDEX (α)

Depth	0–180 m	180–275 m	275–550 m	550–730 m	730–920 m	920–1370 m	1370–1830 m	1830–2300 m	2300–2750 m	2750–3700 m
Total number of species	44	57	68	67	65	70	67	24	14	11
Number of "appearing" species		24	19	10	4	12	8	1	0	0
Number of "disappearing" species	10	8	11	6	6	11	36	10	3	
Diversity index (α)		7.6	8.8	11.9	11.8	12.2	14.0	5.3	3.3	

Table 8.2—11
VARIATION WITH DEPTH IN THE SIGNIFICANCE OF VARIOUS TROPHIC GROUPS OF COPEPODS IN THE NORTHWEST PACIFIC OCEAN

In % of the Total Mass of Copepoda, Except for Aphages

Depth (m)	Filter feeders	Predaceous carnivores	Species with mixed type of feeding
0–50	99.5	0	0.5
50–100	98.0	0	2.0
100–200	99.1	0.4	0.5
200–500	68.1	11.6	20.3
500–1000	15.1	36.9	48.0
1000–2000	4.0	68.0	28.0
2000–4000	9.7	66.0	22.3
4000–6000	0.34	21.3	78.3
6000–8000	>0.1	25.0	75.0

Table 8.2—12
VARIATION WITH DEPTH OF THE NUMBERS OF COPEPODS (CALANOIDA) IN THE NORTH PACIFIC OCEAN*

Hauls Made with Br 80/113 and BR 113/140 Nets

Depth (m)	Spec/m^3
0–50	270
50–100	142
100–200	37
200–500	42
500–1000	9.5
1000–2000	3.6
2000–4000	1.1
6000–8500	0.13

* Average of five stations.

Table 8.2—13
BIOMASS OF COPEPODA AND ITS PART IN THE TOTAL MASS OF PLANKTON IN SUBARCTIC[a] AND TROPICAL[b] REGIONS OF THE PACIFIC OCEAN

Subarctic Region

Depth	0–50 m	50–100 m	100–200 m	200–300 m	300–500 m	500–750 m	750–1000 m	1000–1500 m
Mg/m^3	461	78.3	38.9	202.6	152.2	57.4	33.5	17.6
% of total amount of plankton	82.8	57.7	39.8	76.8	70.3	61.1	66.1	65.6

Depth	1500–2000 m	2000–2500 m	2500–3000 m	3000–4000 m	4000–5000 m	5000–6000 m	6000–7000 m	7000–8700 m
Mg/m^3	8.6	4.63	1.84	0.78	0.41	0.18	0.15	0.065
% of total amount of plankton	48.9	32.3	33.6	58.0	42.9	28.4	25.1	27.6

Tropical Region

Depth	0–100 m	100–200 m	200–500 m	500–1000 m	1000–2000 m	2000–4000 m
Mg/m^3	20.2	10.4	2.6	1.3	0.31	0.1
% of total amount of plankton	42.5	62.2	40.1	43.1	52.3	51.0

[a] Average of nine stations.
[b] Average of five stations.

Table 8.2—14
BIOMASS OF CHAETOGNATHA AND THEIR ROLE IN THE TOTAL PLANKTONIC MASS IN THE SUBARCTIC[a] AND TROPICAL[b] REGIONS OF THE PACIFIC OCEAN

Subarctic Region

Depth	0–50 m	50–100 m	100–200 m	200–300 m	300–500 m	500–750 m	750–1000 m	1000–1500 m	1500–2000 m	2000–2500 m	2500–3000 m	3000–4000 m	4000–5000 m	5000–6000 m	6000–7000 m	7000–8000 m
Mg/m^3	36.8	33.6	38.7	33.1	28.4	15.4	6.4	1.4	5.6	7.8	2.0	0.058	0.012	0.002	<0.002	<0.002
% of total amount of plankton	8.7	28.7	43.9	14.5	13.2	15.3	12.7	5.4	30.1	43.6	37.1	4.5	0.9	0.6	0.4	<0.1

Tropical Region

Depth	0–100 m	100–200 m	200–500 m	500–1000 m	1000–2000 m	2000–4000 m	4000–8000 m
Mg/m^3	12.6	4.8	0.29	0.44	0.20	0.015	0
% of total amount of plankton	2.6	14	5.9	13.4	19.6	7.4	0

[a] Average of nine stations.
[b] Average of five stations.

Table 8.2—15
DISTRIBUTION OF CHAETOGNATHA IN THE WESTERN SUBTROPICAL ATLANTIC

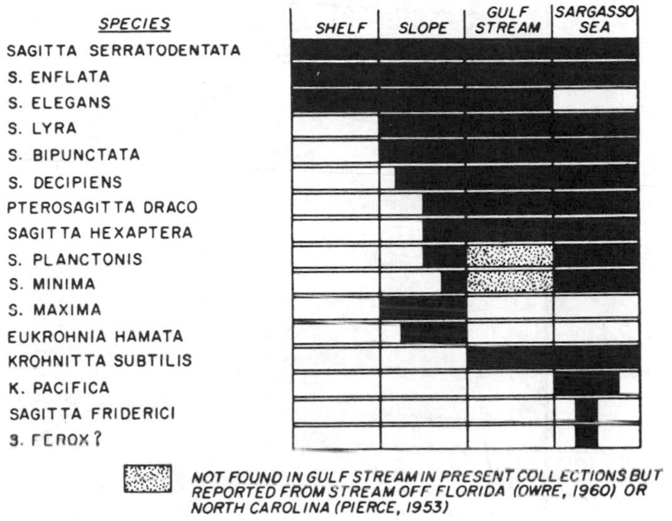

SPECIES | SHELF | SLOPE | GULF STREAM | SARGASSO SEA

SAGITTA SERRATODENTATA
S. ENFLATA
S. ELEGANS
S. LYRA
S. BIPUNCTATA
S. DECIPIENS
PTEROSAGITTA DRACO
SAGITTA HEXAPTERA
S. PLANCTONIS
S. MINIMA
S. MAXIMA
EUKROHNIA HAMATA
KROHNITTA SUBTILIS
K. PACIFICA
SAGITTA FRIDERICI
S. FEROX ?

NOT FOUND IN GULF STREAM IN PRESENT COLLECTIONS BUT REPORTED FROM STREAM OFF FLORIDA (OWRE, 1960) OR NORTH CAROLINA (PIERCE, 1953)

From The abundance, seasonal occurrence, and distribution of the epizooplankton between New York and Bermuda, by Grice, G. D. and Hart, A. D., *Ecol. Monogr.*, 32(4), 297, 1962. With permission.)

Table 8.2—16
NUMBER OF SPECIES OF EUPHAUSIACEA LIVING AT VARIOUS DEPTHS OF THE PACIFIC OCEAN

Depth	0– 200 m	200– 500 m	500– 1000 m	1000– 2000 m	2000– 4000 m	4000– 8000 m
Total number of species	45	46	43	17	7	4
Number of appearing species	–	5	3	0	0	0
Number of disappearing species	4	6	26	10	3	–

Figure 8.2—2

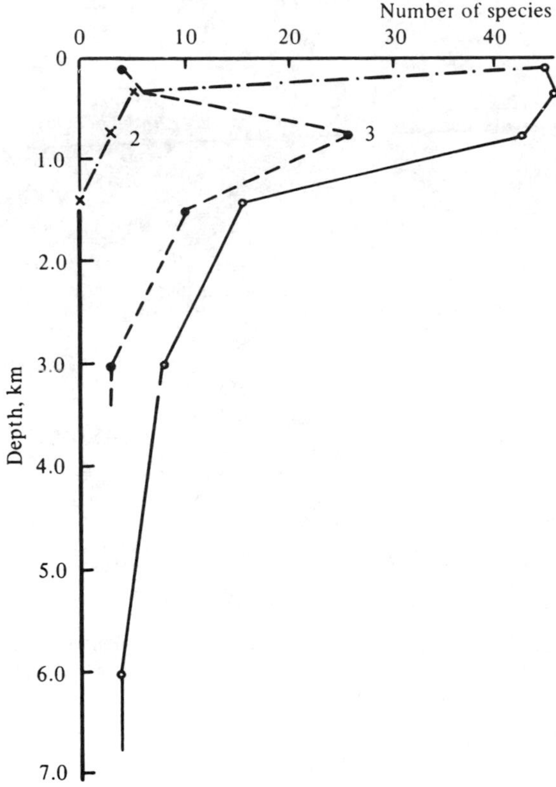

Figure 8.2—2. Variation with depth in numbers of species of euphausids in the central Pacific Ocean.

Table 8.2—17
BIOMASS OF EUPHAUSIACEA AND THEIR PART IN THE TOTAL MASS OF PLANKTON IN THE SUBARCTIC[a] AND TROPICAL[b] REGIONS OF THE PACIFIC OCEAN

Subartic Region

Depth	0–50 m	50–100 m	100–200 m	200–300 m	300–500 m	500–750 m	750–1000 m	1000–1500 m
Mg/m³	14.8	4.6	4.6	3.5	2.1	0.2	~0	0.2
% of total amount of plankton in the layer	3.6	7.0	5.0	1.3	0.9	0.2	<0.1	0.1

Depth	1500–2000 m	2000–2500 m	2500–3000 m	3000–4000 m	4000–5000 m	5000–6000 m	6000–7000 m	Deeper than 7000 m
Mg/m³	0.06	0.02	0.70	~ 0	0.03	0	0	0
% of total amount of plankton in the layer	0.3	<0.1	10.5	<0.1	2.3	0	0	0

Tropical Region

Depth	0–200 m	200–500 m	500–1000 m	1000–2000 m	2000–4000 m
Mg/m³	1.9	2.4	0.08	0.025	0.03
% of total amount of plankton in the layer	7.2	25.0	2.4	2.4	13.4

[a] Average of nine stations.
[b] Average of five stations.

Table 8.2—18
VERTICAL DISTRIBUTION OF PACIFIC EUPHAUSIDS BY FAUNAL ZONES

Epipelagic Zone	
Subarctic epipelagic:	
Thysanoessa raschii	
T. inermis	
T. spinifera	0–280 m
T. longipes	
Euphausia pacifica	
Transition-zone epipelagic:	
Nematoscelis difficilis	
Euphausia pacifica	0–280 m
Thysanoessa gregaria	
Euphausia gibboides	0–700 m
Thysanopoda acutifrons	
Central epipelagic:	
Thysanopoda obtusifrons	
T. aequalis, T. subaequalis	
Euphausia brevis	
E. mutica	
E. recurva	
E. hemigibba (North Pacific)	
E. gibba (South Pacific)	0–700 m
Nematoscellis atlantica	
N. microps	
Stylocheiron carinatum	
S. abbreviatum	
S. suhmii	
S. affine, Central Form	
Nematobrachion fexipes	
Equatorial epipelagic:	
Euphausia tenera	
E. distinguenda	0–280 m
Stylocheiron microphthalma	
Thysanopoda tricuspidata	
Euphausia diomedige	
E. eximia	
E. lamelligera	
E. fallax	
Nematoscelis gracilis	0–700 m
Stylocheiron affine	
W. Equatorial Form	
E. Equatorial Form	
Indo-Australian Form	

Table 8.2—18 (continued)
VERTICAL DISTRIBUTION OF PACIFIC EUPHAUSIDS BY FAUNAL ZONES

Mesopelagic Zone	
Cosmopolitan mesopelagic	
Stylocheiron maximum	140–1000 m
Subarctic mesopelagic	
Tessarabrachion oculatus	0–1000 m
Central-Equatorial mesopelagic (40° N.–40° S.)	
Stylocheiron longicorne	140–700 m
S. elongatum	
Thysanopoda pectinata	
T. orientalis	140–1000 m
T. monachantha	
Nematoscelis tenella	
Nematobrachion boopis	
Central mesopelagic:	
Thysanoessa parva	
Nematobrachion sexspinosus	280–1000 m
Stylocheiron robustum	
Thysanopoda cristata	
Bathypelagic Zone	
Thysanopoda cornuta	
T. egregia	
T. spinicaudata	
Bentheuphausia amblyops	

(From Brinton, E., The distribution of Pacific, euphausiids, *Bull. Scripps Inst. Oceanogr., Univ. Calif.* 8, 196, 1962. With permission.)

Table 8.2—19
STATISTICS FOR THE FIVE SPECIES OF ZOOPLANKTON THAT COMPRISE THE SUBARCTIC GROUP

Tessarabracnion oculatus, Thysanoessa longipes, Euphausia pacifica, Sagitta elegans, and *Limacina helicina*

Species	Frequency[a]	Abundance[b]	Average rank[c]	Dominance[d]	Dispersion (aggregated)[e]	Fidelity[f]	Vitality[g]
T. oculatus	25/62	2–744 (8)	1.24	0/62	430	Aleutians S to 40° N in Pacific	All stages present
Th. longipes	59/62	4–9784 (413)	3.35	13/62	2275	Arctic Ocean S to 40° N in Pacific	All stages present
E. pacifica	55/62	4–22,278 (345)	3.36	23/62	6950	60° N in Bering Sea, S to 25° N in Calif. Current to 35° N in Central Pacific	All stages present
S. elegans	60/62	17–7900 (1405)	4.15	29/62	1830	N. Atlantic, Arctic Ocean, S to 40° N in Pacific	All stages present
L. helicina	58/62	20–75,700 (360)	2.90	7/62	52,400	N. Atlantic, Arctic Ocean, S to 30° N in Calif. Current to 35° N in Central Pacific	All stages present

[a] Proportion of samples in which the species was found (total number of samples, 62). *T. oculatus* occurred in deeper tows at nine additional stations.
[b] Range and median (in parentheses) of numbers of individuals/1000 cu m in samples in which species was found.
[c] Species were ranked within each sample on the basis of numbers of individuals. Ranks for each species were averaged over the 62 samples (1: least abundant; 5: most abundant).
[d] Proportion of samples in which the species was among those making up 50% of the individuals; summation in each sample was begun with the most abundant species.
[e] The ratio of variance to mean; the expected value for a random (Poisson) distribution is 1.0.
[f] Degree of restriction to the Subarctic Water Mass.
[g] 2 Proportion of life lived in the Subarctic Water Mass.

(From Fager, E. W. and McGowan, J. A., Zooplankton Species groups in the north Pacific, Science, 140(3536), 458, © 1963, American Association for the Advancement of Science. With permission.)

8.3. CHEMICAL COMPOSITIONS

Table 8.3—1
TOTAL CARBON, CARBONATE CARBON, AND ORGANIC CARBON CONCENTRATION IN PLANKTONIC MARINE ORGANISMS[a]

Organism		Total carbon (°/₀ DW)	Carbonate carbon (°/₀ DW)[b]	Organic Carbon (°/₀ DW)	Organic Carbon (°/₀ a-f DW)
Cnidaria					
7b	*Cyanea capillata*	13.8	–	13.8	36.0
11B	*Physalia physalis*	31.4	–	31.4	62.8
22b	*Pelagia noctiluca*	12.9	–	12.9	26.0
28a	*Pelagia noctiluca*	15.9	–	15.9	31.2
57	*Aequorea vitrina*	26.8	–	26.8	52.5
	Average	17.5	–	17.5	41.7
Ctenophora					
14	*Mnemiopsis* sp.	6.4	–	6.4	20.6
Arthropoda					
1a	*Euphausia krohnii*	35.8	–	35.8	43.9
2a	*Centropages hamatus* *C. typicus* 1:1	36.3	–	36.3	46.2
4	*Calanus finmarchicus*	41.7	–	41.7	50.5
28b	*Meganyctiphanes norvegica*	42.0	–	42.0	51.6
29c	*Lophogaster* sp.	46.8	–	46.8	57.4
29	*Centropages* sp.	38.5	–	38.5	49.7
30	*Centropages* sp. *Sagitta elegans* 1:1	38.7	–	38.7	50.0
36a	*Idotea metallica*	33.2	2.36	30.8	48.0
36c	*Calanus finmarchicus*	39.8	–	39.8	48.3
40	Mixed copepods	35.6	–	35.6	46.0
56a	*Calanus finmarchicus* (a small admixture of euphausiids and shell-less pteropods)	37.8	–	37.8	46.0
	Average	38.3	–	38.0	48.9
Mollusca					
7a	*Limacina* sp.	28.3	2.74	25.6	56.0
8	*Ommastrephes* sp.	45.1	–	45.1	48.8
24b	*Sthenoteuthis* sp.	37.2	–	37.2	40.4
32	*Clione limacina*	26.3	–	26.3	39.4
38	*Illex illecebrosus*	39.2	–	39.2	42.6
39	Squid eggs (*Loligo*)	21.7	–	21.7	45.0
	Average	33.1	–	32.7	45.4
Chordata					
5	*Salpa* sp.	10.6	–	10.6	46.1
6b	*Salpa* sp.	9.6	–	3.6	39.0
27	*Salpa fusiformis*	7.8	–	7.8	33.9
55	*Pyrosoma* sp.	9.4	–	9.4	41.0
	Average	9.4	–	9.4	40.0

[a] All figures are % of dry weight except last column, which is ash-free dry weight.
[b] Where no results are listed, the inorganic carbonate was not detectable.

Table 8.3—1 (continued)
TOTAL CARBON, CARBONATE CARBON, AND ORGANIC CARBON CONCENTRATION IN PLANKTONIC MARINE ORGANISMS[a]

Organism		Total carbon (°/₀ DW)	Carbonate carbon (°/₀ DW)[b]	Organic Carbon	
				(°/₀ DW)	(°/₀ a-f DW)
Mixed Samples					
6a	Mixed copepods and phytoplankton	29.8	–	29.8	38.5
19	Copepods and phytoplankton	25.2	–	25.2	48.0
33b	Phytoplankton and fish	4.8	1.54	3.3	32.7
34a	Phytoplankton and copepods	6.6	–	6.6	56.0
34b	Copepods and phytoplankton	14.3	–	14.3	48.0
59	Mixed zooplankton	28.4	–	28.4	48.6
	Average	18.2	–	17.9	38.8

[a] All figures are % of dry weight except last column, which is ash-free dry weight.
[b] Where no results are listed, the inorganic carbonate was not detectable.

(From Curl, H. S., Jr., Analyses of carbon in marine plankton organisms, *J. Mar. Res.*, 20(3), 185, 1962. With permission.)

Table 8.3—2
ORGANIC CONTENT OF COPEPODS AND SAGITTAE BASED ON DRY WEIGHTS

	Protein %	Fat %	Carbohydrate %	Ash %	P_2O_5 %	Nitrogen %
Copepods	70.9–77.0	4.6–19.2	0–4.4	4.2–6.4	0.9–2.6	11.1–12.0
Sagittae	69.6	1.9	13.9	16.3	3.6	10.9

Table 8.3—3
TOTAL CARBON, NITROGEN, HYDROGEN AND ASH CONTENTS (% DRY WEIGHT), AND CARBON:NITROGEN RATIO OF NORTH PACIFIC ZOOPLANKTON

Sample number	Method of preservation	Total dry wt. analyzed (mg)	Carbon (%)	Nitrogen (%)	Hydrogen (%)	Ash (%)	C/N (ratio)
Dinoflagellida							
1	Drying	13.419	43.9	5.8	6.4	2.8	7.6
Pteropoda							
2	Drying	26.813	20.3	2.9	2.1	42.8	7.0
3	Drying	24.847	22.0	3.5	2.4	39.8	6.3
4	Drying	7.904	17.0	1.5	1.1	46.6	11.1
5	Drying	7.540	29.0	6.0	3.8	28.6	4.9
Average	–	–	22.4	3.5	2.4	39.3	7.3
Copepoda							
6a	Freezing	20.385	60.9	6.3	9.8	2.4	9.7
6b	Freezing	12.385	39.9	7.6	7.0	3.4	5.1
6c	Freezing	48.805	59.0	5.9	10.1	2.9	10.0
7	Drying	9.456	61.8	7.0	9.7	1.9	8.8
8a	Freezing	8.810	46.4	11.2	7.0	4.4	4.1
8b	Freezing	13.637	58.4	7.8	9.2	2.9	7.5
9	Freezing	4.206	48.0	12.7	7.6	4.3	3.8
10	Drying	7.189	49.9	7.6	7.7	3.9	6.5
11	Drying	7.749	52.8	9.9	8.3	3.4	5.3
12	Drying	19.327	58.4	7.1	9.6	2.1	8.2
13	Freezing	7.748	66.6	5.1	10.3	2.1	13.2
14	Drying	6.521	63.5	5.8	10.0	2.7	11.0
15	Freezing	4.071	47.4	13.1	7.3	3.3	3.6
16	Freezing	0.214	51.0	10.7	8.0	2.8	4.8
17	Freezing	5.422	46.6	12.6	7.2	3.7	3.7
18	Drying	6.745	46.6	11.2	7.2	3.3	4.2
19	Freezing	7.526	45.8	12.9	7.2	5.7	3.5
20	Drying	8.180	44.3	12.2	6.7	6.4	3.6
Average	–	–	53.3	9.4	8.4	3.4	6.5
Amphipoda							
21	Drying	7.796	48.4	8.2	7.5	13.4	5.9
22	Drying	11.690	25.9	4.4	4.4	37.7	6.0
23	Drying	19.390	45.9	6.1	7.1	10.0	7.5
Average	–	–	40.0	6.2	6.3	20.4	6.5
Mysidacea							
24	Drying	9.539	42.4	11.0	6.7	10.2	3.9
Euphausiacea							
25a	Drying	42.000	38.7	10.7	7.3	8.0	3.6
25b	Drying	8.969	39.6	10.1	6.7	8.5	3.9
26	Drying	44.472	47.2	10.0	7.6	8.1	4.7
Average	–	–	41.8	10.3	7.2	8.2	4.1

Table 8.3—3 (continued)
TOTAL CARBON, NITROGEN, HYDROGEN AND ASH CONTENTS (% DRY WEIGHT), AND CARBON:NITROGEN RATIO OF NORTH PACIFIC ZOOPLANKTON

Sample number	Method of preservation	Total dry wt. analyzed (mg)	Carbon (%)	Nitrogen (%)	Hydrogen (%)	Ash (%)	C/N (ratio)
Decapoda							
27	Drying	5.844	41.1	9.3	6.7	11.9	4.4
Insecta							
28	Freezing	10.947	52.6	9.7	7.8	5.6	5.4
Chaetognatha							
29	Drying	13.542	47.7	10.7	7.6	4.8	4.4
30	Drying	14.706	43.5	11.1	7.2	4.2	3.9
Average	–	–	45.6	10.9	7.4	4.5	4.2
Pisces							
31	Drying	14.302	41.5	11.2	7.0	8.9	3.7
32	Freezing	34.108	37.9	9.8	5.8	12.9	3.9
33	Drying	10.296	46.5	12.6	7.2	6.8	3.7
Average	–	–	42.0	11.2	6.7	9.5	3.8

(From Omori, M., Weight and chemical composition of some important oceanic zooplankton in the north Pacific Ocean, *Mar. Biol.*, 3, 8, 1969. With permission.)

Table 8.3—4
WET AND DRY WEIGHT OF NORTH PACIFIC ZOOPLANKTON

Sample number	Average wet wt./individual (mg)	Average dry wt./individual (mg)	Dry wt./wet wt. (%)
Dinoflagellida			
1	0.10	0.0011	1.1
Pteropoda			
2	16.00	4.97	31.1
3	–	3.98	–
4	0.22	0.08	36.4
5	0.56	0.14	25.0
Copepoda			
6a	20.81	3.47	16.7
6b	16.59	2.57	15.5
6c	26.20	8.87	33.9
7	4.55	1.26	27.7
8a	1.28	0.25	19.7
8b	1.46	0.31	21.2
9	1.00	0.13	13.0
10	9.08	1.09	12.0
11	1.04	0.14	13.5
12	17.47	3.23	18.5
13	–	0.10	–
14	2.72	0.51	18.8
15	1.88	0.23	12.2
16	5.64	0.52	9.2
17	0.42	0.06	14.3
18	3.10	0.41	13.2
19	2.21	0.25	11.3
20	0.20	0.03	16.5
Amphipoda			
21	2.56	0.47	18.4
22	5.99	2.19	36.6
23	14.22	3.13	22.0
Mysidacea			
24	5.25	0.98	18.7
Euphausiacea			
25a	7.63	1.54	20.2
25b	61.81	14.00	20.7
26	67.69	14.45	21.3
Decapoda			
27	1.20	0.16	13.3
Insecta			
28	2.91	0.81	27.8
Chaetognatha			
29	10.14	1.43	14.1
30	11.49	1.33	11.6
Pisces			
31	31.80	5.93	18.6
32	96.97	18.60	21.1
33	10.40	2.19	21.1

(From Omori, M., Weight and chemical composition of some important oceanic zooplankton in the north Pacific Ocean, *Mar. Biol.*, 3, 7, 1969. With permission.)

8.4. GEOGRAPHICAL DISTRIBUTION OF CHARACTERISTIC FAUNAL GROUPS

Table 8.4—1
THE AREAS OF THE NORTH PACIFIC IN WHICH THE LISTED SPECIES HAVE BEEN SHOWN TO OCCUR

Organism	Subarctic	Transitional	Central	Equatorial	Eastern Tropic Pacific	Warm water cosmopolites	Comments
PROTOZOA							
Foraminifera							
Globigerina quinqueloba	+						
Globigerinoides minuta	+						
Globigerina pachyderma	+						
Globorotalia truncatulinoides			+				
Pulleniatina obliquiloculata				+			
Sphaeroidinella dehiscens				+			
Globigerina conglomerata				+			
Globorotalia tumida				+			
Globorotalia hirsuta				+			
Globigerinella aequilateralis						+	Pure
Globigerinoides conglobata						+	Pure
Globigerinoides rubra						+	Pure
Orbulina universa						+	Pure
Globigerinoides sacculifera						+	Peak at equator
Globorotalia menardii						+	Peak at equator
Globigerina eggeri						+	Edge effect
Hastigerina pelagica						+	Edge effect
Radiolaria							
Castanidium apsteini	+						
Castanidium variabile	+						Doubtful, may be deep central too
Haeckeliana porcellana	+						Doubtful, may be deep central too
Castanea amphora			+				
Castanissa brevidentata			+				
Castanella thomsoni			+				T. zone w/upwelled water?
Castanea henseni			+				T. zone w/upwelled water?
Castanea globosa			+				T. zone w/upwelled water?
Castanidium longispinum				+			
Castanella aculeata				+			

Table 8.4—1 (continued)
THE AREAS OF THE NORTH PACIFIC IN WHICH THE LISTED SPECIES HAVE BEEN SHOWN TO OCCUR

Organism	Subarctic	Transitional	Central	Equatorial	Eastern Tropic Pacific	Warm water cosmopolites	Comments
CHAETOGNATHA							
Sagitta elegans	+						
Eukrohnia hamata	+						
Sagitta scrippsae		+					
Sagitta pseudoserratodentata			+				
Sagitta californica			+				Crossing W.T.P.
Sagitta ferox				+			
Sagitta robusta				+			Patchy
Sagitta regularis				+			
Sagitta hexaptera						+	
Sagitta enflata						+	Peak at equator
Pterosagitta draco						+	Peak at equator
Sagitta pacifica						+	Peak at equator
Sagitta minima						+	Edge effect
ANNELIDA							
Tomopteris septentrionalis						+	
Tomopteris pacifica	+						
Poeobius meseres	+						
ARTHROPODA							
Copepoda							
Calanus pacificus	+						May be T. zone
Calanus plumchrus	+						
Calanus tonsus	+						
Calanus cristatus	+						
Eucalanus bungii bungii	+						
Eucalanus elongatus hyalinus		+					South Pacific also
Eucalanus bungii californicus		+					
Clausocalanus pergens		+					
Clausocalanus lividus			+				
Eucalanus subcrassus				+			

Table 8.4—1 (continued)
THE AREAS OF THE NORTH PACIFIC IN WHICH THE LISTED SPECIES HAVE BEEN SHOWN TO OCCUR

Organism	Subarctic	Transitional	Central	Equatorial	Eastern Tropic Pacific	Warm water cosmopolites	Comments
Rhincalanus cornutus				+			
Eucalanus inermis					+		
Eucalanus crassus						+	Patchy, 'pure'
Rhincalanus nasutus						+	Very patchy, almost pure equatorial
Eucalanus attenuatus						+	Peak at equator; some edge effect
Eucalanus subtenuis						+	Patchy, peak at equator; some edge effect
Clausocalanus arcuicornis						+	
Eucalanus longiceps							
Rhincalanus gigas							
Clausocalanus laticeps							
Euphausiacea							
Thysanoessa longipes	+						
Euphausia pacifica	+						
Thysanopoda acutifrons		+					
Thysanoessa gregaria		+					
Euphausia gibboides		+					
Nematoscelis difficilismegalops		+					
Nematoscelis atlantica			+				
Euphausia brevis			+				
Euphausia hemigibba			+				
Euphausia gibba			+				
Euphausia mutica			+				
Stylocheiron suhmii			+				Crossing in W.T.P.
Euphausia diomediae				+			
Euphausia distinguenda				+			
Nematoscelis gracilis				+			
Euphausia distinguenda					+		
Euphausia eximia					+		
Euphausia lamelligera					+		
Euphausia tenera						+	Peak at equator
Stylocheiron abbreviatum						+	Avoids E.T.P.
Euphausia superba							

Table 8.4—1 (continued)
THE AREAS OF THE NORTH PACIFIC IN WHICH THE LISTED SPECIES HAVE BEEN SHOWN TO OCCUR

Organism	Subarctic	Transitional	Central	Equatorial	Eastern Tropic Pacific	Warm water cosmopolites	Comments
Amphipoda							
Parathimisto pacifica	+						
MOLLUSCA							
Pteropoda							
Limacina helicina	+						
Clio polita	+						
Corolla pacifica		+					
Clio balantium		+					
Cavolinia inflexa			+				
Clio pyramidata			+				Crossing in W.T.P.
Styliola subula			+				Crossing in W.T.P.
Limacina lesueuri			+				Crossing in W.T.P.
Clio n.sp.				+			
Cavolinia uncinata				+			Very patchy
Limacina trochiformis					+		
Limacina inflata						+	
Cavolinia longirostris						+	Very patchy; almost pure equatorial
Cavolinia gibbosa						+	Very patchy, avoids E.T.P.
Hyalocylix striata						+	
Creseis virgula						+	Edge effect
Creseis acicula						+	
Cavolinia tridentata						+	Peak at equator
Diacria trispinosa						+	Peak at equator
Limacina bulimoides						+	Avoids E.T.P.
Clio antarctica						+	
Heteropoda							
Caranaria japonica		+					
Gymnosomata							
Clione limacina	+						

Note: E.T.P. — Eastern Tropical Pacific; W.T.P. — Western Tropical Pacific

(From McGowan, J. A., "Oceanic Biogeography of the Pacific," in *The Micropalaeontology of Oceans*, Cambridge University Press (Eng.), 1971, p. 14 to 17. With permission.)

Table 8.4—2
TYPICAL COSMOPOLITAN OCEANIC SPECIES

Siphonophora

Physophora hydrostatica
Agalma elegans
Dimophyes arctica
Lensia conoidea
Chelopheys appendiculata
Sulculeolaria biloba

Medusae

Cosmetira pilosella
Laodicea undulata
Halicreas sp.
Periphylla periphylla

Mollusca

Euclio pyramidata
Euclio cuspidata
Diacria trispinosa
Pneumodermopsis ciliata
Taonidium pfefferi
Tracheloteuthis risei

Polychaeta

Travisiopsis lanceolata
Vanadis formosa
Rhynchonerella angelini
Tomopteris septentrionalis

Chaetognatha

Sagitta serratodentata
f. tasmanica
Sagitta hexaptera

Thaliacea

Salpa fusiformis
Dolioletta gegenbauri

Copepoda

Rhincalanus nasutus
Eucalanus elongatus
Pleuromamma robusta
Euchirella rostrata
Euchirella curticaudata
Oithona spinirostris

Other Crustacea

Lepas sp.
Munnopsis murrayi
Brachyscelus crusulum
Meganyctiphane norvegica

Euphausia krohni
Anchialus agilis

Table 8.4—3
SOME PLANKTONIC SPECIES TYPICAL OF DEEP WATER

Gaetanus pileatus
Arietellus plumifer
Pontoptilus muticus
Centraugaptilus rattrayi
Augaptilus megalaurus
and many other copepods

Amalopenaeus elegans
Hymenodora elegans
Boreomysis microps
Eucopia unguiculata
Cyphocaris anonyx
Scina sp.

Sagitta macrocephala
S. zetesios
Eukrohnia fowleri
Nectonemertes miriabilis
Spiratella helicoides
Histioteuthis boneltiana

Table 8.4—4
SPECIES GROUPS AND ASSOCIATED SPECIES OF EASTERN AUSTRALIAN SLOPE ZOOPLANKTON

Section I (Mean temperature [°C] >12 <17.6; mean chlorinity [°/₀₀] >19.40 <19.60)		Section II (Mean temperature [°C] >17.6 <21.6; mean chlorinity [°/₀₀] >19.50 <19.65)		
Group A (>12.4°C <17.6°C; >19.41 <19.60°/₀₀)	Group B (>13.4°C <17.0°C; >19.45 <19.53°/₀₀)	Group C (>17.6°C <20.8°C; >19.53 <9.59°/₀₀)	Group D (>19.40°C <20.6°C; >19.54 <19.60°/₀₀)	Group E (>20.0°C <21.6°C; >19.50 <19.61°/₀₀)
S. planctonis	*E. spinifera*	*O. cophocerca*	*D. denticulatum*	*S. robusta*
S. hamata	*T. gregaria*	*O. fusiformis*	*T. multitentaculata*	*S. ferox*
S. lyra	*E. recurva*	*O longicauda*	*M. huxleyi*	*S. enflata*
K. subtilis	*N. difficilis*		*S. magnum*	*S. bipunctata*
			O. rufescens	*S. s. pacifica*
				S. regularis
				P. draco
Ungrouped Species Occurring with One or More Grouped Species				
S. s. tasmanica	*I. zonaria*	*O. intermedia*	*P. macropus*	*S. hexaptera*
S. decipiens	*S. carinatum*	*G. cornutogastra*	*R. amboinensis*	*S. carinatum*
		G. parva	*C. pinnata*	*S. s. atlanticum*
		F. borealis sargassi	*D. gegenbauri*	*S. neglecta*
		F. pellucida	*O. cornutogastra*	*P. macropus*
			O. albicans	*C. acicula*
			S. minima	*C. virgula conica*
				S. abbreviatum
				S. minima
				D. gegenbauri
				O. albicans
Species from Other Groups Occurring with One or More Grouped Species				
S. s. pacifica (E)	*S. planctonis* (A)	*D. denticulatum* (D)	*S. enflata* (E)	*D. denticulatum* (D)
P. draco (E)	*S. lyra* (A)	*T. multitentaculata* (D)	*S. s. pacifica* (E)	*M. huxleyi* (D)
S. bipunctata (E)		*M. huxleyi* (D)	*S. bipunctata* (E)	*S. magnum* (D)
S. enflata (E)		*S. magnum* (D)	*S. regularis* (E)	*S. planctonis* (A)
S. magnum (D)		*O. rufescens* (D)	*S. planctonis* (A)	*S. lyra* (A)
T. gregaria (B)			(and each of Group C)	*K. subtilis* (A)

Probability (P ≤0.01) of species occurring as species groups or in association with other species.

Table 8.4—5
SPECIES COMPOSITION OF THE FIVE WORLD DISTRIBUTIONAL ZONES OF PLANKTONIC FORAMINIFERA

Northern and Southern Cold-water Regions

1. Artic and antarctic zones:
   Globigerina pachyderma (Ehrenberg): Left-coiling variety; right-coiling in subarctic and subantarctic zones.
2. Subarctic and subantarctic zones:
   Globigerina quinqueloba (Natland)
   Globigerina bulloides (d'Orbigny)
   Globigerinita bradyi (Wiesner)
   Globorotalia scitula (Brady)

Transition Zones

3. Northern and south transition zones between cold-water and warm-water regions:
   Globorotalia inflata (d'Orbigny): With mixed occurrences of subpolar and tropical-subtropical species.

Warm-water Region

4. Northern and southern subtropical zones:

Globigerinoides ruber (d'Orbigny): Pink variety in Atlantic Ocean only.	
Globigerinoides conglobatus (Brady): Autumn species.	
Hastigerina pelagica (d'Orbigny)	
Globigerinita glutinata (Egger)	
Globorotalia truncatulinoides (d'Orbigny)	
Globorotalia hirsuta (d'Orbigny)	Winter species
Globigerina rubescens (Hofker)	Winter species
Globigerinella aequilateralis (Brady)	Prefer outer margins of subtropical central water masses and into transitional zone.
Orbulina universa (d'Orbigny)	
Globoquadrina dutertrei (d'Orbigny)	
Globigerina falconensis (Blow)	
Globorotalia crassaformis (Galloway and Wissler)	

5. Tropical Zone:

Globigerinoides sacculifer (Brady): Including *Sphaeroidinella dehiscens* (Parker and Jones).	
Globorotalia menardii (d'Orbigny)	
Globorotalia tumida (Brady)	
Pulleniatina obliquiloculata (Parker and Jones)	
Candeina nitida (d'Orbigny)	
Hastigerinella digitata (Rhumbler)	
Globoquadrina conglomerata (Schwager)	Restricted to Indo-Pacific.
Globigerinella adamsi (Banner and Blow)	Restricted to Indo-Pacific.
Globoquadrina hexagona (Natland)	Restricted to Indo-Pacific.

The species are listed under the zone where their highest concentrations are observed, but they are not necessarily limited to these areas.
Most species listed under the Subtropical Zones are also common in the tropical waters.
* Usually located in central water masses between 20°N and 40°N, or between 20°S and 40°S latitude.

Table 8.4—6
THE WORLDWIDE DISTRIBUTION OF EIGHT SPECIES OF MESO- AND BATHYPELAGIC CHAETOGNATHS AS ELEMENTS OF COSMOPOLITAN, ARCTIC AND ANTARCTIC FAUNAS

Salinity (°/₀₀) Range, Above in Italics, and Temperature (°C) Range, Below, Are Given for Each Region

Region and Authority	*Eukrohnia fowleri*	*E. hamata*	*Sagitta decipiens*	*S. macrocephala*	*S. marri*	*S. maxima*	*S. planctonis*	*S. zetosios*
Arctic								
Greenland, Davis Strait Kramp, 1917	–	*33.6–34.0* ~0.5	–	–	–	*33.3–34.7* ~0.5–3.0	–	–
Umanak Fjord Kramp, 1917	–	*34.1–34.5* 0.5–1.0	–	–	–	*>34.0*	–	–
Disko Bay Kramp, 1917	–	~0.4–0.9	–	–	–	–	–	–
Near Egedesminde Kramp, 1917	–	*33.75* 0.52	–	–	–	–	–	–
S. Northern Storø Bredefjord Kramp, 1917	–	*33.6* ~0.87–0.4 *34.4* 3.2	–	–	–	–	–	–
Skovfjord Kramp, 1917	–	*33.9* 1.5	–	–	–	*33.87* 1.56	–	–
Greenland Baffin Bay and Labrador Sea Kramp, 1939	*34.9–35.1* 3.1–5.5	*33.0–34.9* 1.7–5.2	–	–	–	*33.4–35.0* ~1.2–6.0	–	*34.9–35.1* 3.1–5.5
Atlantic								
Gulf of Maine region Bigelow, 1926	–	*32.0–35.0* 1.3–9.0	–	–	–	*32.36–34.9* 1.63–9.0	–	–
Scottish region Fraser, 1952	–	*34.75–35.5*	–	–	–	*34.75–35.5* -1.0–12.0	–	*34.75–35.5* 8.0–9.0
Western Mediterranean Furnestin, 1957a	–	–	*37.47* 15.25–24.30	–	–	–	–	–
Eastern Mediterranean and entrance to Black Sea Furnestin, 1957a	–	–	25.9	–	–	–	–	–
Union of South Africa: (Tafel Bay–Lambert's Bay) Heydorn, 1959	–	*34.85–35.35* 11.92–14.97	–	–	–	–	*35.39* 14.94	–
Florida (off Miami) Owre, 1960	13.4–14.5	7.4	12.8–22.4	–	–	–	–	–

Table 8.4—6 (continued)
THE WORLDWIDE DISTRIBUTION OF EIGHT SPECIES OF MESO- AND BATHYPELAGIC CHAETOGNATHS AS ELEMENTS OF COSMOPOLITAN, ARCTIC AND ANTARCTIC FAUNAS

Region and Authority	*Eukrohnia fowleri*	*E. hamata*	*Sagitta decipiens*	*S. macrocephala*	*S. marri*	*S. maxima*	*S. planctonis*	*S. zetosios*
Pacific								
Chile Fagetti, 1958a	–	12.1–14.2	13.5	–	–	–	–	12.4
British Columbia Lea, 1955		*27.0–28.0* 8.5–10.0	–	–	–	–	–	–
Eastern tropical (off Mexico and Central America) Sund, 1961b	*34.65* 8.9	*34.41–34.82* 10.3–11.9	*34.40–34.92* 8.9–18.0	*34.65* 8.9	–	*34.78* 10.8	–	*34.65*
Eastern Australia-Tasmania Thomson, 1957	–	8.0–16.5	11.0–22.0	–	–	–	6.0–24.0	8.9
Kurile-Kamchatka Trench region Tchindonova, 1955	–	*33.1–34.5*	–	–	–	–	–	–
Indian								
Bay of Bengal (off Visakhapatnam) Rao and Ganapati, 1958	–	–	*22.03–34.61* 26.5–28.5	–	–	–	–	–
Antarctic								
Hoces Straits (Drake Passage) Balech, 1962	–	*33.77–34.09* 2.68–7.16	–	–	–	*34.17* 5.2	–	–
Antarctic and Subantarctic waters David, 1958b	–	−0.5–7.75	–	–	0.1–1.0	0.3–7.3	–	–
Antarctic-Subantarctic waters Fagetti, 1959	–	0.5–3.5	–	–	–	–	–	–
Antarctic Jameson, 1914	–	−2.7–3.8	–	–	–	–	0.3	–
Antarctic Convergence to Subtropical Convergence David, 1958b	–	*33.7–34.7*	–	–	*34.67–34.71*	*33.4–35.0*	–	–

Table 8.4—7
FAUNAL GROUPS OF CHAETOGNATHS IN THE INDIAN OCEAN

a) Cosmopolitan (common to Atlantic, Indian, and Pacific oceans): *S. lyra, S. enflata, S. hexaptera, S. minima, S. bipunctata, K. subtilis, K. pacifica, P. draco, S. gazellae, S. tasmanica.*

b) Cold-water representants: *S. gazellae, S. tasmanica, E. hamata.*

c) Tropical-equatorial, and restricted to the Indo-Pacific waters: *S. ferox, S. robusta, S. pacifica, S. pulchra, S. neglecta, S. bedoti, S. regularis.*

d) Mesoplanktonic: *S. decipiens, S. planctonis, S. zetesios.*

e) Deep water: *E. hamata* (in low latitudes), *E. fowleri, E. bathypelagica.*

Table 8.4—8
NUMERICALLY IMPORTANT SPECIES OF COPEPODS IN THE WESTERN ATLANTIC BY REGIONS

Neritic (14 samples)

	Freq.	Mean no./m³
Pseudocalanus minutus	13	559
Centropages typicus	14	450
Oithona similis	13	151
Temora longicornis	8	59
Paracalanus parvus	8	39
Calanus finmarchicus	11	32
Metridia lucens	12	16
Candacia armata	9	9

Slope (15 samples)

	Freq.	Mean no./m³
Centropages typicus	11	76
Pseudocalanus minutus	8	16
Oithona similis	6	14
Metridia lucens	11	15
Clausocalanus pergens	6	19
C. arcuicornis	7	13
Pleuromamma borealis	8	6
Oithona atlantica	12	6

Gulf Stream (3 samples)

	Freq.	Mean no./m³
Clausocalanus furcatus	3	27
Lucicutia flavicornis	3	9
Oithona plumifera	3	9
O. setigera	3	7
Calocalanus pavo	2	9
Farranula gracilis	3	4
Mecynocera clausi	3	2

Sargasso Sea (11 samples)

	Freq.	Mean no./m³
Clausocalanus furcatus	9	7
Oithona setigera	11	6
Lucicutia flavicornis	11	4
Ctenocalanus vanus	6	3
Farranula gracilis	6	2
Mecynocera clausi	9	2

From The abundance, seasonal occurrence, and distribution of the epizooplankton between New York and Bermuda, by Grice, G. D. and Hart, A. D., *Ecol. Monogr.*, 32(4), 297, 1962. With permission.)

Table 8.4—9
ANTARCTIC AND SUBANTARCTIC PELAGIC COPEPODS AND THEIR FAUNAL ASSOCIATIONS

Species	Type	Neritic	Subantarctic	Antarctic	Abyssal
Calanoida					
Calanus australis (Brodsky)	S		Farran, 1929 Vervoort, 1957 Brodsky, 1959		
Calanus propinquus (Brady)	S–B*			Vervoort, 1957	
Calanus tonsus (Brady)	S		Vervoort, 1957		
Calanus simillimus (Giesbrecht)	S*		Vervoort, 1957		
Calanoides acutus(Giesbrecht)	V*			Vervoort, 1957	
Megacalanus princeps	A				
(Wolfenden)	A				Vervoort, 1957
Bathycalanus bradyi (Wolfenden)	A				Vervoort, 1957
Eucalanus elongatus (Dana)	S–B		Farran, 1929		
Eucalanus longiceps (Matthews)	S–B		Vervoort, 1957	Hardy and Gunther, 1935	
Rhincalanus gigas (Brady)	V			Vervoort, 1957	
Rhincalanus nasutus (Giesbrecht)				Vervoort, 1957	
Microcalanus pygmaeus (G. O. Sars)	S–B			Vervoort, 1957	
Gaidius affinis (G. O. Sars)	B		Vervoort, 1957		
Gaidius intermedius (Wolfenden)	A				Vervoort, 1957
Gaidius tenuispinus (G. O. Sars)	B–A			Vervoort, 1957	Vervoort, 1957
Gaetanus antarcticus (Wolfenden)	B–A			Vervoort, 1957	Vervoort, 1957
Gaetanus latifrons (G. O. Sars)	A				Vervoort, 1957
Gaetanus minor (Farran)	B		Vervoort, 1957		

S – Surface.
B – Bathypelagic.
A – Abyssal.
V – Show vertical seasonal migration.
* – Species characteristic of Antarctic waters.

Table 8.4—9 (continued)
ANTARCTIC AND SUBANTARCTIC PELAGIC COPEPODS AND THEIR FAUNAL ASSOCIATIONS

Species	Type	Neritic	Subantarctic	Antarctic	Abyssal
Euchirella latirostris (Farran)	B*		Vervoort, 1957		
Euchirella rostrata (Claus)	B		Vervoort, 1957		
Euchirella rostromagna (Wolfenden)	B*			Vervoort, 1957	
Pseudochirella elongata (Wolfenden)	B*			Vervoort, 1957	
Pseudochirella hirsuta (Wolfenden)	A				Vervoort, 1957
Pseudochirella mawsoni (Vervoort)	B–A		Vervoort, 1957	Vervoort, 1957	Vervoort, 1957
Pseudochirella notacantha (G. O. Sars)	A				Vervoort, 1957
Pseudochirella pustulifera (G. O. Sars)	A				Hardy and Gunther, 1935
Pseudeuchaeta brevicauda (G. O. Sars)	A				Vervoort, 1957
Undeuchaeta major (Giesbrecht)	B–A		Vervoort, 1957		Vervoort, 1957
Ctenocalanus vanus (Giesbrecht)	B			Vervoort, 1957	
Clausocalanus arcuicornis (Dana)	S		Farran, 1929		
Clausocalanus laticeps (Farran)	S*		Vervoort, 1957		
Farrania frigida (Wolfenden)	A				Vervoort, 1957
Drenanopus pectinatus (Brady)	S	Vervoort, 1957			
Spinocalanus abyssalis (Giesbrecht)	A				Vervoort, 1957
Spinocalanus magnus (Wolfenden)	A				Vervoort, 1957
Spinocalanus spinosus (Farran)	A				Farran, 1929
Mimocalanus cultrifer (Farran)	A				Vervoort, 1957
Stephus longipes (Giesbrecht)	S			Tanaka, 1960	
Aetideus armatus (Boeck)	B		Vervoort, 1957		

S – Surface.
B – Bathypelagic.
A – Abyssal.
V – Show vertical seasonal migration.
* – Species characteristic of Antarctic waters.

Table 8.4—9 (continued)
ANTARCTIC AND SUBANTARCTIC PELAGIC COPEPODS AND THEIR FAUNAL ASSOCIATIONS

Species	Type	Neritic	Subantarctic	Antarctic	Abyssal
Euaetideus australis (Vervoort)	B		Vervoort, 1957		
Aetideopsis antarcticus (Wolfenden)	B*			Farran, 1929	
Aetideopsis minor (Wolfenden)	B*			Vervoort, 1957	
Chiridius polaris (Wolfenden)	B*			Vervoort, 1957	
Undeuchaeta minor (Giesbrecht)	S–B		Farran, 1929		
Euchaeta antarctica (Giesbrecht)	B*			Vervoort, 1957	
Euchaeta austrina (Giesbrecht)	B*			Vervoort, 1957	
Euchaeta biloba (Farran)	S–B		Vervoort, 1957	Vervoort, 1957	
Euchaeta erebi (Farran)	B*			Farran, 1929	
Euchaeta exigua (Wolfenden)	A				Vervoort, 1957
Euchaeta farrani (With)	A				Vervoort, 1957
Euchaeta rasa (Farran)	B–A			Vervoort, 1957	Vervoort, 1957
Euchaeta scotti (Farran)	B–A			Hardy and Gunther, 1935	Hardy and Gunther, 1935
Euchaeta similis (Wolfenden)	B*			Vervoort, 1957	
Valdiviella insignis (Farran)	A				Vervoort, 1957
Onchocalanus magnus (Wolfenden)	B*			Vervoort, 1957	
Onchocalanus wolfendeni (Vervoort)	B*			Vervoort, 1957	
Cornucalanus robustus (Vervoort)	B–A			Vervoort, 1957	Vervoort, 1957
Cephalophanes frigidus	A				Vervoort, 1957
Amallophora altera	B*			Vervoort, 1957	
Undinella brevipes (Farran)	A				Vervoort, 1957
Racovitzanus antarcticus (Giesbrecht)	B*			Vervoort, 1957	
Racovitzanus erraticus (Vervoort)	B*			Vervoort, 1957	

S – Surface.
B – Bathypelagic.
A – Abyssal.
V – Show vertical seasonal migration.
*** – Species characteristic of Antarctic waters.**

Table 8.4—9 (continued)
ANTARCTIC AND SUBANTARCTIC PELAGIC COPEPODS AND THEIR FAUNAL ASSOCIATIONS

Species	Type	Neritic	Subantarctic	Antarctic	Abyssal
Scolecithricella glacialis (Giesbrecht)	S*			Vervoort, 1957	
Scolecithricella dentipes (Vervoort)	B*			Vervoort, 1957	
Scolecithricella emarginata (Farran)	B			Hardy and Gunther, 1936	
Scolecithricella incisa (Farran	S–B*			Farran, 1929	
Scolecithricella minor (Brady)	B			Hardy and Gunther, 1936	
Scolecithricella ovata (Farran)	B		Vervoort, 1957	Vervoort, 1957	
Scolecithricella polaris (Wolfenden)	B*			Vervoort, 1957	
Scolecithricella robusta (T. Scott)	B			Vervoort, 1957	
Scolecithricella valida (Farran)	A				Vervoort, 1957
Scaphocalanus affinis (G. O. Sars)	B–A			Vervoort, 1957	Vervoort, 1957
Scaphocalanus brevicornis (G. O. Sars)	B			Hardy and Gunther, 1935	
Scaphocalanus echinatus (Farran)	B–A		Farran, 1929		
Scaphocalanus magnus (T. Scott)	A				Vervoort, 1957
Scaphocalanus subbrevicornis (Wolfenden)	B*			Vervoort, 1957	
Temorities brevis (G. O. Sars)	A				Vervoort, 1957
Metridia curticauda (Giesbrecht)	B–A			Vervoort, 1957	Vervoort, 1957

S – Surface.
B – Bathypelagic.
A – Abyssal.
V – Show vertical seasonal migration.
* – Species characteristic of Antarctic waters

Table 8.4—9 (continued)
ANTARCTIC AND SUBANTARCTIC PELAGIC COPEPODS AND THEIR FAUNAL ASSOCIATIONS

Species	Type	Neritic	Subantarctic	Antarctic	Abyssal
Metridia gerlachei (Giesbrecht)	S–B*			Vervoort, 1957	
Metridia lucens (Boeck)	S–B		Vervoort, 1957	Vervoort, 1957	
Metridia princeps (Giesbrecht)	B–A				Vervoort, 1957
Pleuromamma borealis (F. Dahl)	B		Farran, 1929		
Pleuromamma g.·acilis (Claus)	B		Farran, 1929		
Pleuromamma roᴌ·ısta (F. Dahl) *f. antarctica* (Steuer)	B–A		Vervoort, 1957	Vervoort, 1957	Vervoort, 1957
Lucicutia curta (Farran)	A				Vervoort, 1957
Lucicutia frigida (Wolfenden)	B–A			Vervoort, 1957	Vervoort, 1957
Lucicutia grandis (Giesbrecht)	A				Vervoort, 1957
Lucicutia macrocera (G. O. Sars)	A				Vervoort, 1957
Lucicutia magna (Wolfenden)	A				Farran, 1929
Lucicutia maxima (Steuer)	B			Hardy and Gunther, 1935	
Lucicutia wolfendeni (Sewell)	A				Vervoort, 1957
Disseta palumboi (Giesbrecht)	A				Hardy and Gunther, 1935
Heterorhabdus austrinus (Giesbrecht)	B*			Vervoort, 1957	
Heterorhabdus compactus (G. O. Sars)	A			Hardy and Gunther, 1935	Farran, 1929; Hardy and Gunther, 1935
Heterorhabdus farrani (Brady)	B*			Vervoort, 1957	
Heterorhabdus pustulifer (Farran)	B*			Vervoort, 1957	
Heterostylites major (F. Dahl)	A				Vervoort, 1957
Haloptilus fons (Farran)	B			Hardy and Gunther, 1935	

S – Surface.
B – Bathypelagic.
A – Abyssal.
V – Show vertical seasonal migration.
* – Species characteristic of Antarctic waters.

Table 8.4—9 (continued)
ANTARCTIC AND SUBANTARCTIC PELAGIC COPEPODS AND THEIR FAUNAL ASSOCIATIONS

Species	Type	Neritic	Subantarctic	Antarctic	Abyssal
Haloptilus ocellatus (Wolfenden)	B*			Vervoort, 1957	
Haloptilus oxycephalus (Giesbrecht)	B			Vervoort, 1957	
Augaptilus glacialis (G. O. Sars)	B			Vervoort, 1957	
Augaptilus megalurus (Giesbrecht)	B			Hardy and Gunther, 1935	
Euaugaptilus laticeps (G. O. Sars)	A				Vervoort, 1957
Euaugaptilus magnus (Wolfenden)	A				Vervoort, 1957
Centraugaptilus rattrayi (T. Scott)	B			Hardy and Gunther, 1935	
Pseudaugaptilus longiremis (G. O. Sars)	A				Vervoort, 1957
Pontoptilus ovalis (G. O. Sars)	A				Vervoort, 1957
Pachyptilus eurygnathus (G. O. Sars)	A				Vervoort, 1957
Arietellus simplex (G. O. Sars)	A				Vervoort, 1957
Phyllopus bidentatus (Brady)	B		Vervoort, 1957	Vervoort, 1957	
Candacia cheirura (Cleve)	S–B		Vervoort, 1957		
Candacia falcifera (Farran)	B			Vervoort, 1957	
Candacia maxima (Vervoort)	B		Vervoort, 1957	Vervoort, 1957	
Paralabidocera antarctica (I. C. Thompson)	S			Vervoort, 1957	
Cyclopoida					
Pseudocyclopina belgicae (Giesbrecht)	S			Giesbrecht, 1902	

S – Surface.
B – Bathypelagic.
A – Abyssal.
V – Show vertical seasonal migration.
* – Species characteristic of Antarctic waters.

Table 8.4—9 (continued)
ANTARCTIC AND SUBANTARCTIC PELAGIC COPEPODS AND THEIR FAUNAL ASSOCIATIONS

Species	Type	Neritic	Subantarctic	Antarctic	Abyssal
Mormonilla phasma (Giesbrecht)	B–A			Hardy and Gunther, 1935	Hardy and Gunther, 1935
Oithona frigida (Giesbrecht)	S		Vervoort, 1957	Vervoort, 1957	
Oithona similis (Claus)	S–B		Vervoort, 1957	Vervoort, 1957	
Ratania atlantica (Farran)	S–B			Vervoort, 1957	
Oncaea conifera (Giesbrecht)	S–B			Vervoort, 1957	
Oncaea curvata (Giesbrecht)	B			Vervoort, 1957	
Oncaea mediterranea (Giesbrecht)	S–B		Vervoort, 1957		
Oncaea notopus (Giesbrecht)	B			Vervoort, 1957	
Oncaea venuta (Philippi)	B			Tanaka, 1960	
Conea rapax (Giesbrecht)	B–A			Hardy and Gunther, 1935	
Lubbockia aculeata (Giesbrecht)	B–A			Vervoort, 1957	Vervoort, 1957

S – Surface.
B – Bathypelagic.
A – Abyssal.
V – Show vertical seasonal migration.
* – Species characteristic of Antarctic waters.

(From Vervoort, W., Biogeography and ecology in Antarctica: Notes on the biogeography and ecology of free-living copepoda, *Monographiae Biologicae,* 15, 394, 1965. With permission.)

Table 8.4—10
THE WORLDWIDE DISTRIBUTION OF EUPHAUSIIDS

The approximate latitudinal range of each species is stated. The Atlantic and Pacific Oceans are divided latitudinally into the region north of 40°N, the eastern (E), central (C), and western (W) areas of the region between 40°N and 40°S, and the region south of 40°S. Species occurring in the Mediterranean (Med.) are detailed separately. The Indian Ocean is divided into the eastern (E), central (C), and western (W) areas of the region north of 40°S, and the region south of 40°S, with species found in the Red Sea (Red) detailed separately.

		Atlantic						Pacific					Indian				
			40°N–40°S						40°N–40°S				N of 40°S				
Species	Latitudinal range	N of 40°N	E	C	W	Med.	S of 40°S	N of 40°N	E	C	W	S of 40°S	E	C	W	Red	S of 40°S
							Bentheuphausia										
B. amblyops	50°N 50°S	X	X	X	X			X	X	X	X	X	X	X	X		
							Thysanopoda										
T. monacantha	40°N–10°S		X		X				X	X	X		X	X	X		
T. cristata	35°N–40°S			X	X					X	X				X		
T. tricuspidata	35°N–30°S		X	X	X					X	X		X	X	X		
T. aequalis	40°N–40°S		X	X	X					X	X		X	X	X		
T. subaequalis	40°N–40°S				X	X				X					X		
T. obtusifrons	35°N–35°S		X	X	X					X	X		X	X	X		
T. pectinata	35°N–35°S		X		X				X	X	X		X	X	X		
T. orientalis	40°N–40°S				X				X	X	X			X	X		
T. microphthalma	40°N–40°S		X	X	X									X	X		
T. acutifrons	{70°N–40°N 40°S–60°S	X					X	X				X					X
T. cornuta	55°N–40°S		X		X			X	X	X	X				X		
T. ergregia	40°N–50°S		X						X	X	X	X					
T. spinicaudata	30°N–30°S								X	X							
							Meganyctiphanes										
M. norvegica	70°N–30°N	X				X											

Table 8.4—10 (continued)
THE WORLDWIDE DISTRIBUTION OF EUPHAUSIIDS

Species	Latitudinal range	Atlantic						Pacific					Indian				
			40°N–40°S						40°N–40°S				N of 40°S				
		N of 40°N	E	C	W	Med.	S of 40°S	N of 40°N	E	C	W	S of 40°S	E	C	W	Red	S of 40°S
Nyctiphanes																	
N. couchii	60°N–30°N	X				X											
N. australis	35°S–50°S											X					
N. capensis	30°S–40°S		X														
N. simplex	30°N–20°S				X				X								
Pseudeuphausia																	
P. latifrons	40°N–35°S										X		X	X	X	X	
P. sinica	30°N–15°N										X						
Euphausia																	
E. americana	40°N–10°S		X	X	X												
E. eximia	40°N–30°S		X			X			X								
E. krohnii	65°N–0°S	X	X	X	X	X											
E. mutica	40°N–40°S		X	X	X				X	X	X		X	X	X	X	
E. brevis	40°N–40°S		X	X	X	X			X	X	X		X	X	X	X	
E. diomedeae	25°N–25°S								X	X	X		X	X	X	X	
E. recurva	40°N–40°S		X						X		X		X	X	X		
E. superba	55°S–75°S						X					X					X
E. vallentini	45°S–60°S						X					X					X
E. lucens	35°S–50°S						X					X					X
E. frigida	50°S–65°S						X					X					X
E. pacifica	50°N–35°N							X									
E. nana	35°N–25°N										X						
E. crystallorophias	65°S–75°S						X					X					X
E. tenera	40°N–30°S		X	X	X				X	X	X		X	X	X		
E. similis	40°N–50°S						X				X	X	X	X	X		X
E. similis var. *armata*	15°N–50°S						X				X	X	X				X
E. mucronata	0°–10°S								X								
E. sibogae	0°–20°S										X						

Table 8.4—10 (continued)
THE WORLDWIDE DISTRIBUTION OF EUPHAUSIIDS

		Atlantic						Pacific					Indian				
			40°N–40°S						40°N–40°S				N of 40°S				
Species	Latitudinal range	N of 40°N	E	C	W	Med.	S of 40°S	N of 40°N	E	C	W	S of 40°S	E	C	W	Red	S of 40°S
E. distinguenda	30°N–20°S								X				X	X	X	X	
E. lamelligera	25°N–10°S								X								
E. gibba	20°S–40°S								X	X	X						
E. gibboides	40°N–40°S		X	X	X				X	X	X			?		?	
E. fallax	35°N–20°S										X						
E. sanzoi	20°N														?	X	
E. pseudogibba	30°N–30°S		X	X	X					X	X		X	X	X		
E. paragibba	20°N–20°S								X	X	X		X	X	X		
E. hemigibba	40°N–40°S		X	X	X	X			X	X	X		X	X	X		
E. spinifera	30°S–45°S						X					X					X
E. hanseni	25°N–40°S		X														
E. longirostris	40°S–55°S						X					X					X
E. triacantha	50°S–65°S						X					X					X
							Tessarabrachion										
T. oculatum	50°N–35°N							X									
							Thysanoëssa										
T. spinifera	60°N–25°N							X	X								
T. longipes	60°N–45°N							X									
T. inspinata	55°N–35°N							X									
T. inermis	75°N–40°N	X						X									
T. longicaudata	75°N–40°N	X															
T. parva	{40°N–20°N 25°S–40°S		X		X				X	X	X		X				
T. gregaria	{50°N–10°N 20°S–50°S	X	X	X	X	X	X	X				X	X				
T. vicina	50°S–75°S						X					X					X
T. macrura	50°S–75°S						X					X					X
T. raschii	75°N–40°N	X						X									

Table 8.4—10 (continued)
THE WORLDWIDE DISTRIBUTION OF EUPHAUSIIDS

Species	Latitudinal range	Atlantic						Pacific					Indian				
		N of 40°N	40°N–40°S			Med.	S of 40°S	N of 40°N	40°N–40°S			S of 40°S	N of 40°S			Red	S of 40°S
			E	C	W				E	C	W		E	C	W		
Nematoscelis																	
N. difficilis	45°N–20°N							X									
N. megalops	60°N–10°N; 20°S–55°S	X	X	X	X	X	X					X	X	X	X		X
N. tenella	40°N–40°S		X	X	X				X	X	X		X	X	X		
N. microps	40°N–40°S		X	X	X				X	X	X		X	X	X		
N. atlantica	40°N–40°S		X	X	X	X			X	X	X			?			
N. lobata	15°N–5°N										X						
N. gracilis	30°N–30°S								X	X	X		X	X	X		
Nematobrachion																	
N. flexipes	40°N–40°S		X	X	X				X	X	X		X	X	X		
N. sexspinosum	30°N–30°S				X					X	X				?		
N. boöpis	60°N–55°S	X	X	X	X				X	X	X	X	X	X	X		
Stylocheiron																	
S. carinatum	40°N–40°S		X	X	X				X	X	X		X	X	X	X	
S. affine	40°N–40°S		X		X				X	X	X		X	X	X	X	
S. suhmii	50°N–40°S		X	X	X	X			X	X	X		X	X	X	X	
S. microphthalma	35°N–25°S		X						X	X	X		X	X	X		
S. insulare	10°N–10°S				X						X						
S. elongatum	60°N–40°S	X	X	X	X				X	X	X		X	X	X	X	
S. indicum														X			
S. longicorne	60°N–50°S	X	X	X	X	X			X	X	X		X	X	X		
S. abbreviatum	50°N–40°S		X	X	X	X			X	X	X		X	X	X	X	
S. maximum	60°N–60°S	X	X	X	X	X		X	X	X	X	X	X	X	X		
S. robustum	30°N–30°S		X						X	X	X						

Figure 8.4—1

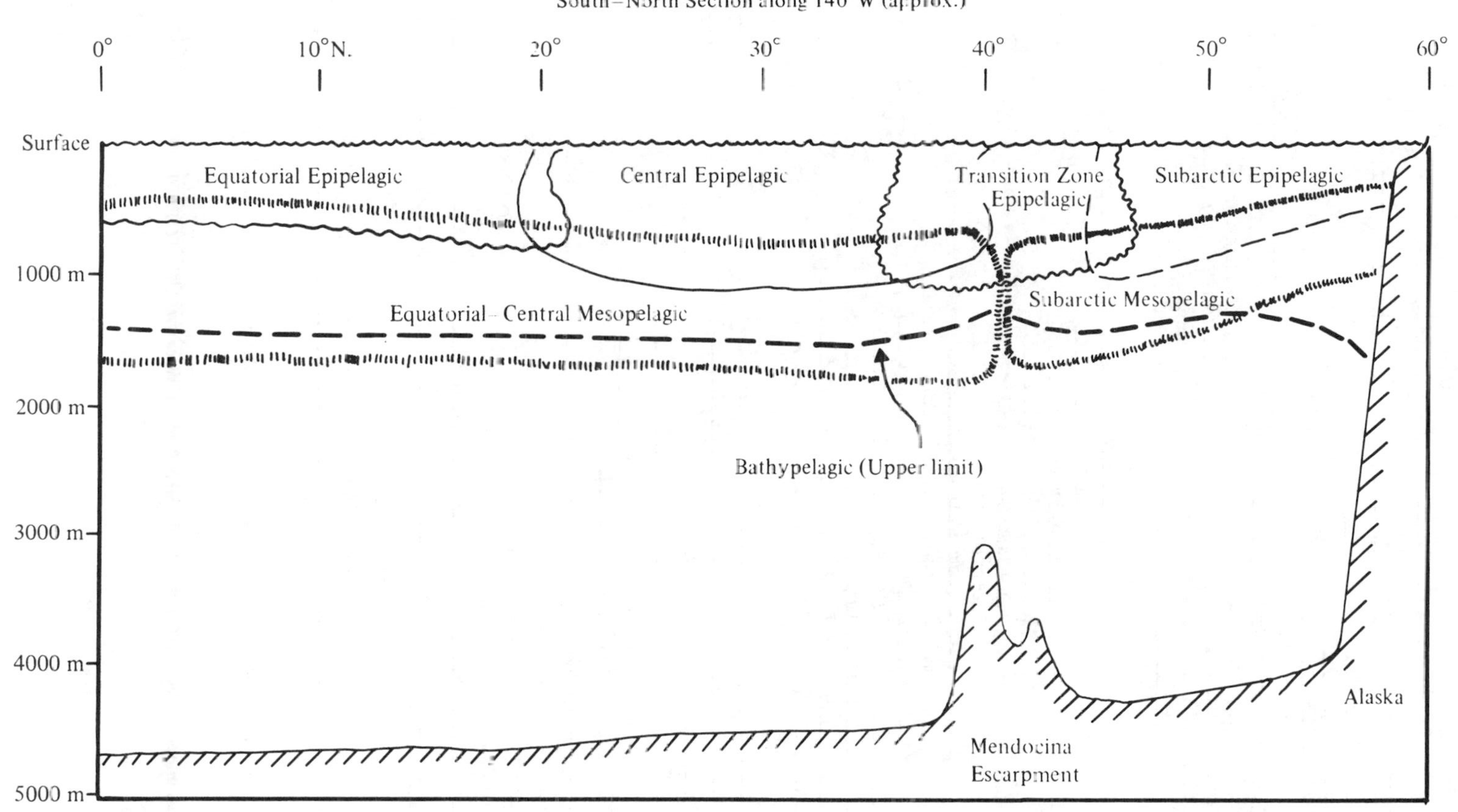

Figure 8.4—1. Bathymetric and latitudinal zonation of associations of euphausiid species in mid-oceanic profile in the North Pacific.

(From Brinton, E., The distribution of Pacific euphausiids. *Bull. Scripps Inst. Oceanogr. Univ. Calif.*, 8, 195, 1962. With permission.)

Figure 8.4—2

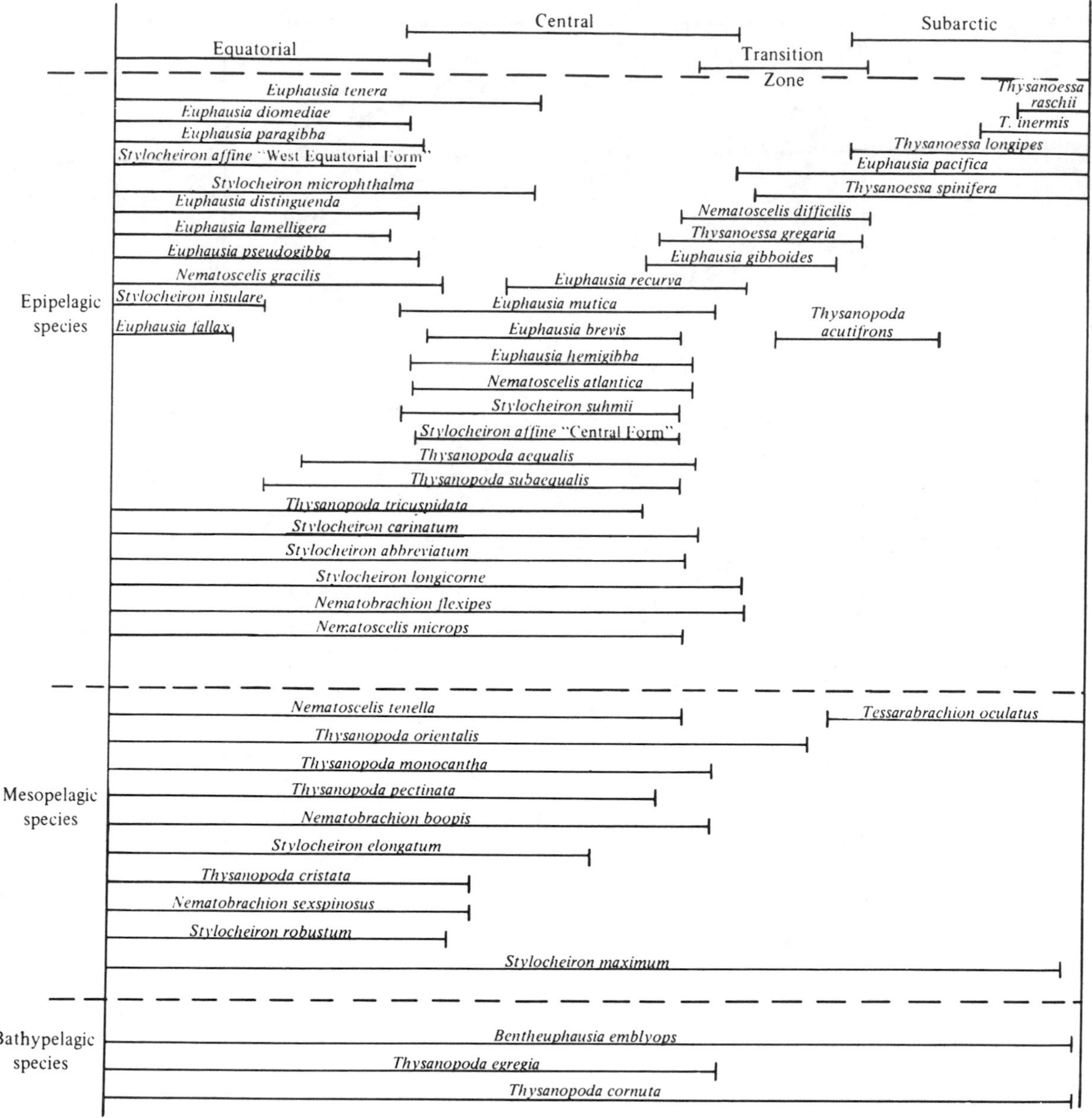

Figure 8.4—2. Species composition of bathymetric euphausiid associations in the North Pacific. Relationships to horizontal zones are also indicated.

(From Brinton, E., The distribution of Pacific euphausiids. *Bull. Scripps Inst. Oceanogr. Univ. Calif.*, 8, 194, 1962. With permission.)

Table 8.4—11
THE EUPHAUSIID FAUNA OF THE HIGH SEAS, AS COMPARED TO THOSE OF LITTORAL AND SUBLITTORAL PROVINCES IN REGARD TO LATITUDINAL EXTENTS

High Seas Euphausiid Groups	Littoral Fauna (Ekman)	Euphausiid Boundary Species
Thysanoessa longipes "Spined," *T. inermis*	Arctic	*Thysanoessa rashchii*
Subarctic group	East Asiatic Temperate, Northwest American Temperate	*T. spinifera*
Transition-zone group, North	Transition	*Nyctiphanes simplex*
East equatorial group	Pacific Tropical American	*E. lamelligera*
West equatorial group	Tropical Indo West Pacific	*P. latifrons*
Trans-equatorial group	–	–
	Peruvian North Chilean	*E. mucronata*
Transition-zone group, South	New Zealand	*N. australis*
Subantarctic group	Antiboreal South American	–
Antarctic group	Antarctic	*E. crystallorophias*

Figure 8.4—3
COMPOSITE DISTRIBUTION PATTERNS OF PACIFIC EUPHAUSIIDS

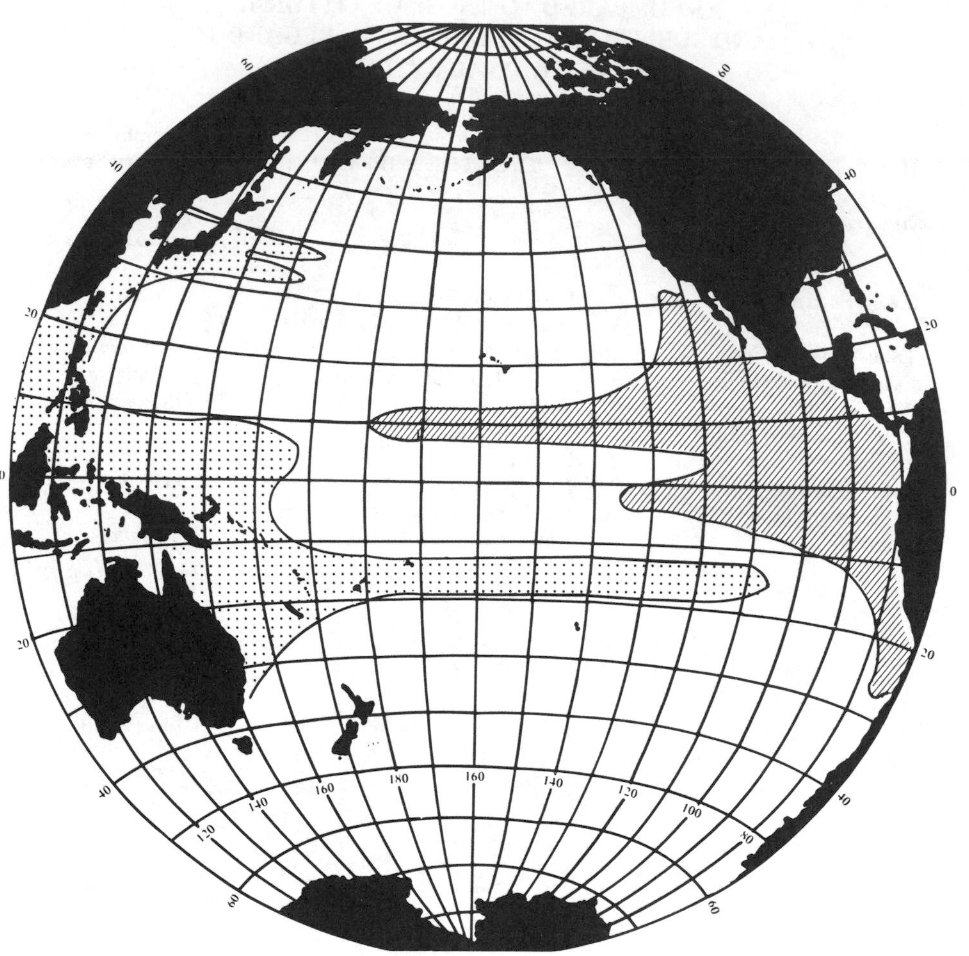

A. The western equatorial (Indo-Australian) group which includes *Euphausia pseudogibba, E. fallax, E. sibogae, Pseudeuphausia latifrons, Nematoscelis lobata, Stylocheiron insulare,* and *S. affine* "Indo-Australian Form." The eastern equatorial group includes *Euphausia lamelligera, E. distinguenda, E. eximia, Stylocheiron affine* "East Equatorial Form," and *Nyctiphanes simplex.*

(From Brinton, E., The distribution of Pacific euphausiids, *Bull. Scripps Inst. Oceanogr. Univ. Calif.*, 8, 212, 1962. With permission.)

Figure 8.4—3 (continued)

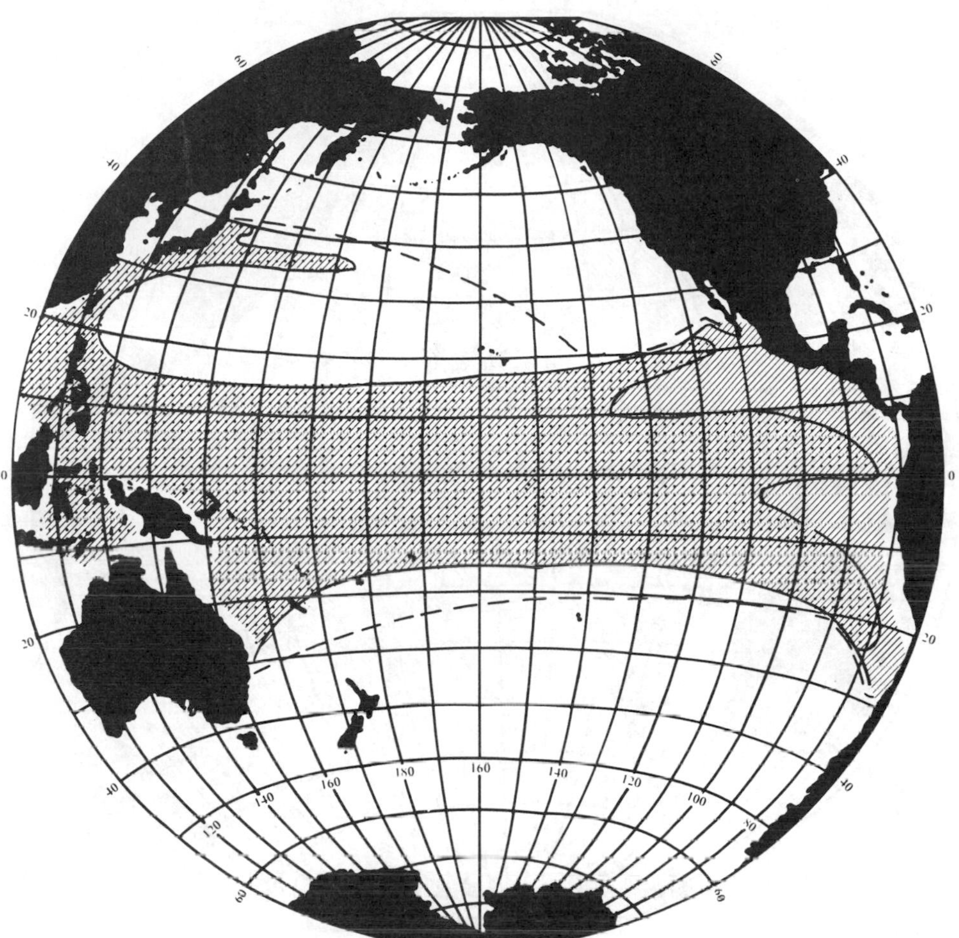

B. Composite distributions of groups of epipelagic trans-equatorial euphausiid species. Cross-hatched part: *Euphausia diomediae* and *Nematoscelis gracilis.* Strippled part: *Thysanopoda tricuspidata, Euphausia paragibba,* and *Stylocheiron microphthalma.* The dashed line indicates the limits of range of an equatorial-west central species, *Euphausia tenera.*

(From Brinton, E., The distribution of Pacific euphausiids, *Bull. Scripps Inst. Oceanogr. Univ. Calif.,* 8, 210, 1962. With permission.)

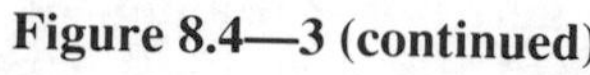

Figure 8.4—3 (continued)

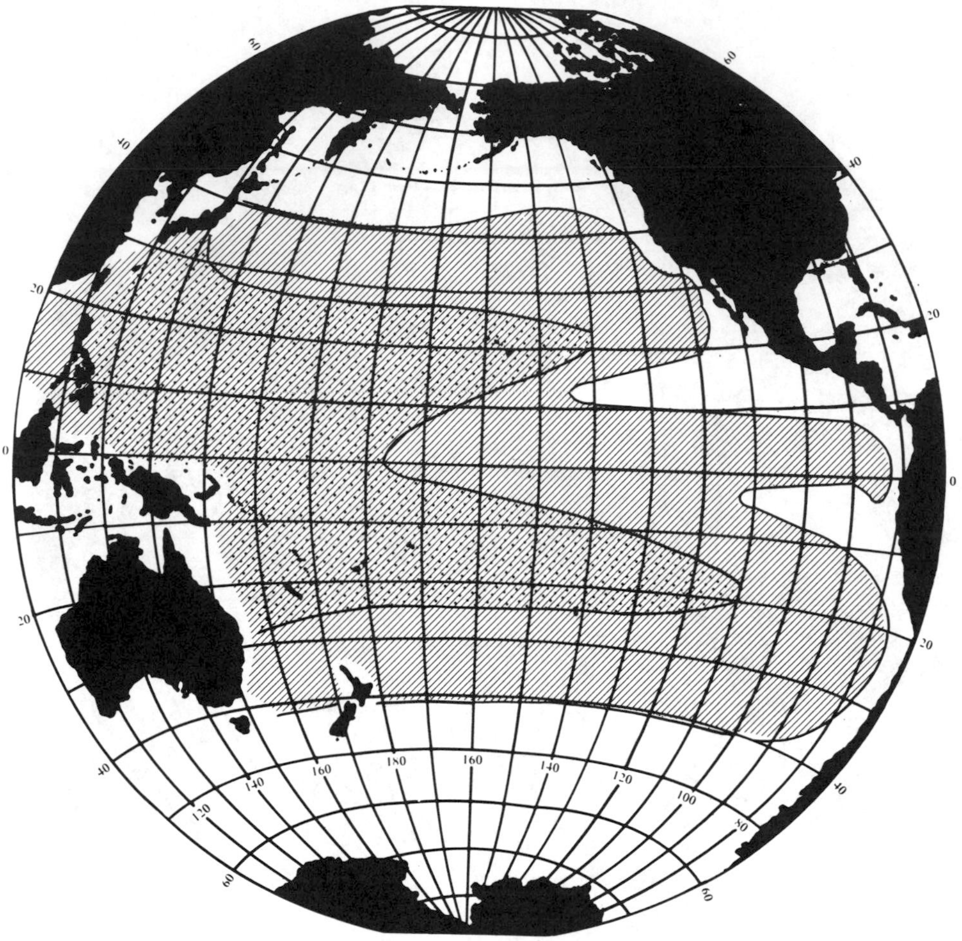

C. Composite distribution patterns of the central-equatorial mesopelagic euphausiid species. Cross-hatched part: *Thysanopoda orientalis, T. monocantha, T. pectinata, Nematoscelis tenella, Nematobrachion boopis,* and *Stylocheiron elongatum.* Stippled part: *Nematobrachion sexspinosus, Thysanopoda cristata,* and *Stylocheiron robustum.*

(From Brinton, E., The distribution of Pacific euphausiids, *Bull. Scripps Inst. Oceanogr. Univ. Calif.*, 8, 209, 1962. With permission.)

Figure 8.4—4

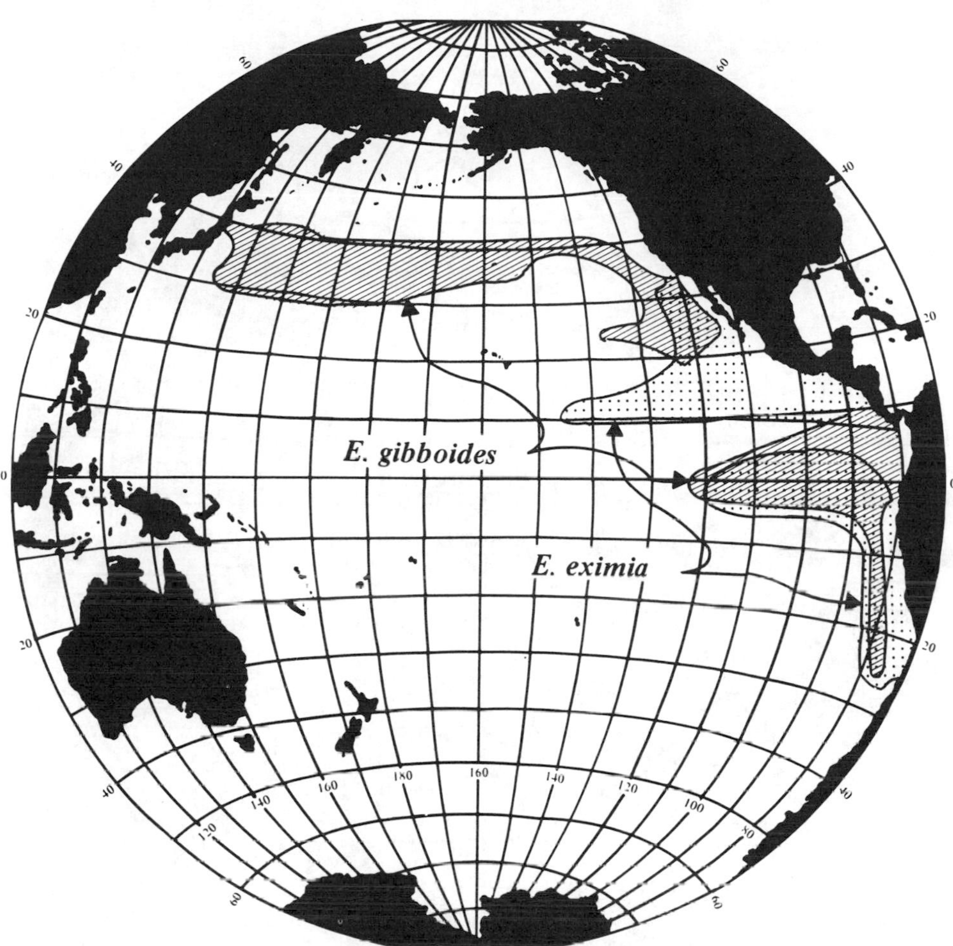

Figure 8.4—4. Distributions of the large transition-zone—equatorial-zone *Euphausia* species *E. gibboides* and *E. eximia*, showing differences in ranges in the North Pacific and similarities in the South Pacific.

(From Brinton, E., The distribution of Pacific euphausiids. *Bull. Scripps Inst. Oceanogr. Univ. Calif.*, 8, 204, 1962. With permission.)

Figure 8.4—5

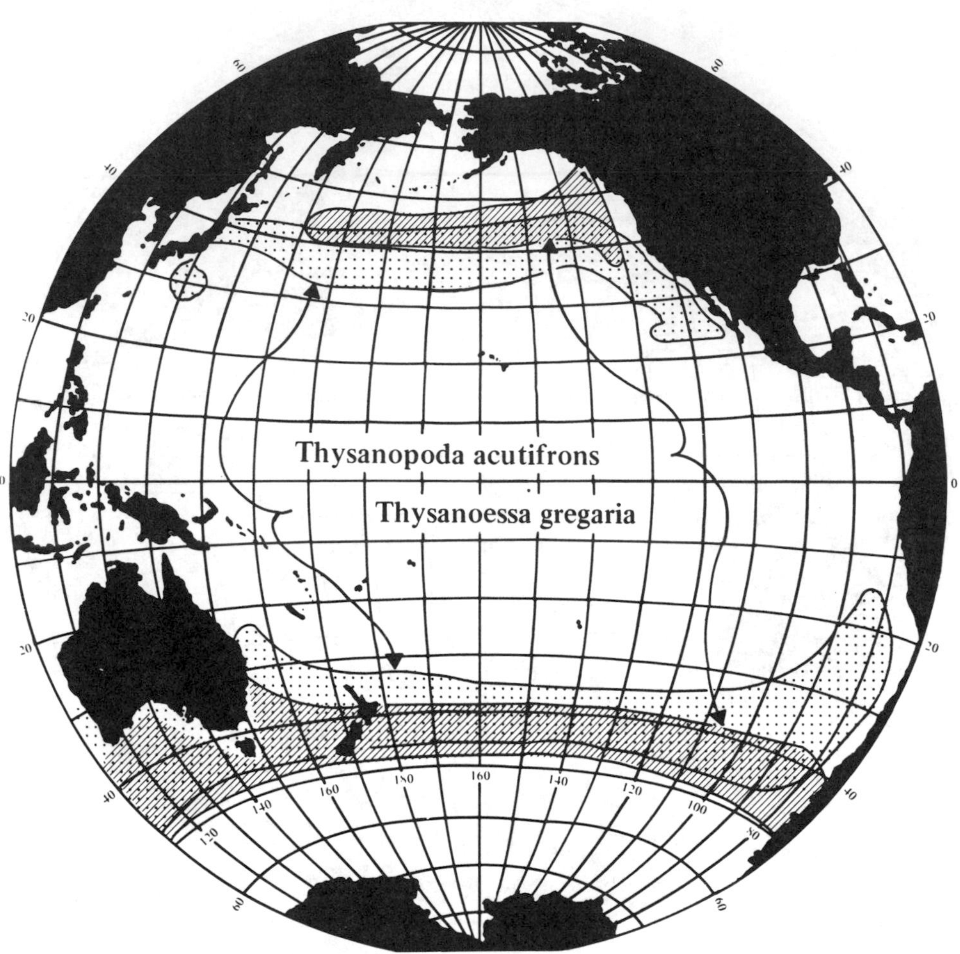

Figure 8.4—5. Distributions of the transition-zone species *Thysanopoda acutifrons* and *Thysanoessa gregaria*. Species of this zone, which lies between subarctic (or subantarctic) and central waters, are all antitropical (bipolar).

(From Brinton, E., The distribution of Pacific euphausiids. *Bull. Scripps Inst. Oceanogr. Univ. Calif.*, 8, 203, 1962. With permission.)

8.5. INDICATOR SPECIES AND THEIR ASSOCIATED WATER MASSES

Table 8.5—1
WATER MASSES OF THE OCEANS

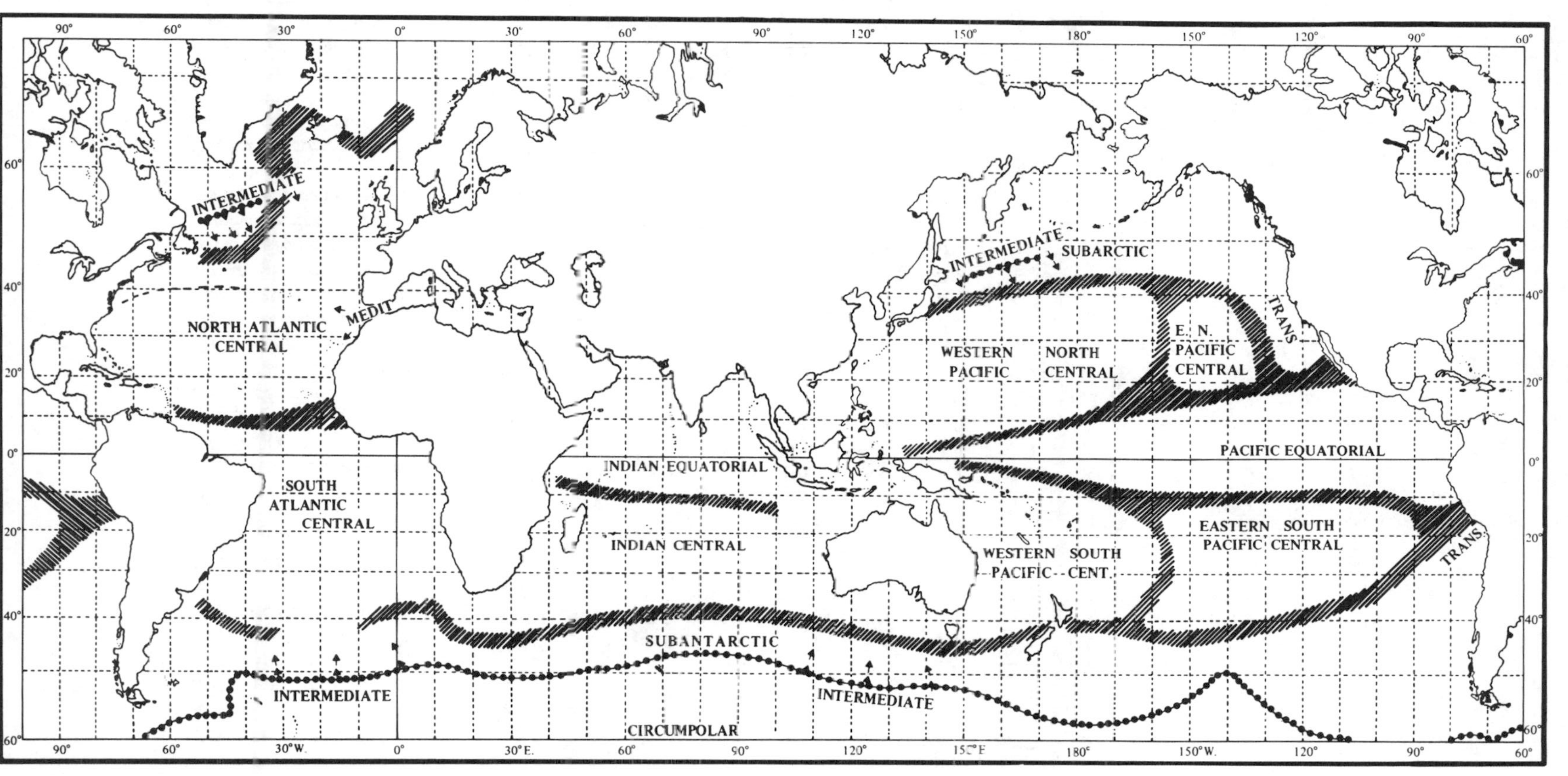

Note: Hatched bands delimit the major water masses; dotted lines indicate convergences where intermediate waters are formed. The dashed line in the north Atlantic approximates the southern boundary of "Gulf Stream water."

Table 8.5—2
INCIDENCES OF OCCURRENCE OF MAJOR PLANKTONIC SPECIES IN SEVEN WATER MASSES OF THE PACIFIC

	Calanus helgolandicus	138.1	21.8	31.9	8.1	0.9	140.5	28.0	83.7	1.8	105.9	60.7	3.1	4.4	5.2
	Acartia tonsa	59.8	10.1	4.6	1.6	0.1	2.8	0.6	–	–	0.1	0.3	–	–	–
	Nyctiphanes simplex	16.3	4.6	0.7	0.2	+	–	4.3	2.0	–	2.6	7.7	+	–	1.2
	Sagitta euneritica	7.2	3.6	0.3	0.8	–	14.8	8.6	3.6	0.3	1.3	0.2	–	0.6	–
UW	*Rhincalanus nasutus*	1.3	–	2.0	0.3	+	6.4	3.1	0.9	0.9	8.2	11.5	2.2	3.6	7.8
	Heterorhabdus papilliger	–	0.5	0.3	0.1	0.1	1.4	0.4	–	0.1	–	0.2	–	0.5	–
	Mysid gen. et sp.	0.8	+	0.6	+	–	0.4	+	0.2	0.1	–	0.1	–	–	–
	Candacia bipinnata	0.1	–	+	+	–	0.4	–	–	+	–	–	–	–	–
	Paracalanus sp.	0.1	–	–	–	–	0.4	–	–	–	–	–	–	–	–
	Aglaura hemistoma	0.1	0.9	14.7	–	4.0	0.7	0.7	2.4	3.1	11.4	0.2	0.8	0.8	3.2
	Pleuromamma abdominalis	+	1.0	1.7	0.6	1.0	0.9	2.1	8.7	1.9	5.4	1.1	–	0.5	0.8
	Sagitta bierii	0.8	1.4	4.3	1.2	1.2	3.1	3.0	5.9	5.9	15.7	4.3	0.4	–	0.8
	Diphyes dispar	0.2	0.8	1.7	1.5	0.1	2.4	2.0	6.7	1.5	6.7	2.8	1.0	0.3	1.4
	Muggiaea atlantica	0.1	0.7	–	–	+	2.0	2.0	3.4	0.3	1.1	0.2	0.6	+	0.1
CCW	*Eucalanus bungii californicus*	–	3.4	0.3	0.2	0.1	0.1	0.3	0.5	+	–	0.5	–	–	–
	Pleuromamma borealis	+	0.9	0.5	0.2	–	1.4	0.5	0.2	0.3	1.4	0.8	–	–	–
	Conchoecia striola	–	0.2	0.1	+	+	–	0.8	+	+	–	–	–	–	–
	Sagitta pseudoserratodentata	–	0.1	–	0.2	–	–	0.7	–	–	–	–	–	–	–
	Labidocera trispinosa	0.3	0.1	+	+	–	–	0.3	–	–	–	0.1	–	–	–
	Stylocheiron affine	+	0.1	0.1	0.2	+	0.1	0.1	0.1	0.2	–	–	–	–	–
	Euphausia gibboides	–	+	–	–	–	–	0.1	–	0.1	–	–	–	–	–
	Conchoecia alata	–	+	–	–	+	–	+	–	–	–	–	–	–	–
	Sagitta enflata	0.2	1.6	4.3	3.7	5.8	5.4	3.9	3.7	4.8	47.2	51.1	20.2	6.3	7.2
	Clausocalanus spp.	1.0	0.3	7.2	15.6	4.6	8.6	4.7	7.5	5.8	39.2	15.8	0.1	–	+
	Nannocalanus minor	–	–	–	8.4	5.4	1.2	0.1	0.4	0.8	19.5	5.8	8.4	5.8	3.2
	Eucalanus attenuatus	–	–	–	0.4	0.9	1.8	1.0	0.1	0.1	14.3	10.3	7.4	2.4	1.6
	Oikopleura longicanda	1.4	3.8	3.3	6.1	3.5	13.2	7.2	6.9	3.7	13.9	7.2	6.7	0.2	1.6
	Labidocera acutifrons	–	–	–	–	–	0.4	–	+	0.1	1.8	16.6	–	–	–
	Euchaeta marina/E. tenuis	0.6	0.1	1.4	0.8	1.3	1.4	–	2.7	–	9.5	1.8	1.0	–	1.0
	Eucalanus subtenuis	+	0.2	2.2	0.3	0.6	0.5	–	0.1	0.1	8.5	4.5	3.8	2.9	2.5
	Euaetideus bradyi	0.3	0.1	0.9	0.3	0.2	0.8	0.8	0.8	0.4	8.6	2.5	–	0.5	0.4
	Candacia curta	–	+	0.2	2.1	0.2	0.5	0.4	1.5	0.6	3.9	1.6	–	0.5	0.2
	Desmopterus papilio	–	0.2	+	–	+	0.1	0.1	0.1	–	3.9	0.8	0.3	0.6	0.4

UW – Upwelled Water.
CCW – California Current Water.

Table 8.5—2 (continued)
INCIDENCES OF OCCURRENCE OF MAJOR PLANKTONIC SPECIES IN SEVEN WATER MASSES OF THE PACIFIC

PCW	*Limacina inflata*	+	0.1	0.9	0.1	+	0.1	1.7	2.3	2.1	2.1	0.2	–	+	+
TSW	*Doliolum denticulatun*	–	+	0.4	0.1	0.2	0.4	0.1	0.8	0.2	3.4	1.2	19.4	5.4	0.2
	Salpa democratica	–	–	–	–	0.1	0.3	–	1.8	2.2	1.1	0.2	1.4	–	1.0
	Temora discaudata	–	+	0.3	0.1	+	0.2	+	–	+	1.0	1.8	1.7	1.0	0.4
	Conchoecia magna	+	0.1	0.1	+	+	0.1	0.2	0.2	0.1	1.6	0.5	–	+	0.1
	Fritillaria formica	+	0.2	0.2	+	+	0.3	0.8	–	0.5	1.5	0.1	+	–	–
	Candacia aethiopica	–	+	–	0.1	–	–	0.1	–	0.1	–	1.5	0.1	–	–
	Conchoecia giesbrechti	–	0.1	+	+	–	0.1	0.3	0.1	0.2	1.4	0.2	–	–	0.1
	Liriope tetraphylla	–	–	–	–	–	0.1	0.3	0.2	0.1	0.3	1.4	–	–	–
	Euchaeta wolfendeni	–	–	–	1.6	–	–	–	–	–	–	–	–	–	–
	Sagitta pacifica	–	+	–	0.6	1.1	0.1	+	–	–	–	–	0.2	0.8	0.6
	Copila mirabilis	–	–	–	–	–	–	–	–	0.2	1.1	0.3	–	0.6	0.2
	Sapphirina spp.	–	+	+	–	+	0.1	0.2	–	0.1	1.0	0.1	1.1	1.6	0.2
	Euphausia eximia	–	+	0.1	–	0.2	0.1	0.2	0.3	0.2	–	0.8	–	–	–
	Amphogona apicata	–	–	–	–	–	–	–	–	0.1	–	0.7	1.6	0.4	1.1
	Eudoxoides spiralis	–	+	+	+	–	–	0.1	0.1	0.1	0.5	–	–	–	–
	Pleuromamma quadrungulata	–	–	0.2	0.3	0.1	+	0.2	0.1	0.2	0.4	–	–	–	–

PCW – Central Pacific Water.
TSW – Tropical Surface Water.

(From Longhurst, A. R., Diversity and trophic structure of zooplankton communities in the California Current, *Deep-Sea Res.*, 14, 402, © 1967 Pergamon Press. With permission.)

Table 8.5—3
WATER MASS PREFERENCES OF UNGROUPED SPECIES IN THE PACIFIC

Central Tasman and Southwest Tasman Water Masses

Tunicata:	*"Ihlea magalhanica"*
	Pegea confederata
	Pyrosoma atlanticum
	Iasis zonaria
	Oikopleura parva
	Fritillaria borealis
Chaetognatha:	*Sagitta serratodentata tasmanica*
Euphausiacea:	*Euphausia similis*
	E. similis var. *armata*
Hyperiidae:	*Brachyscelus crustulum*

South Equatorial and Coral Sea Water Masses

Tunicata:	*Cyclosalpa pinnata*
	Brooksia rostrata
	Rittierella amboinensis
	Oikopleura intermedia
	O. cornutogastra
	Fritillaria formica
Chaetognatha:	*Sagitta neglecta*
Euphausiacea:	*Nematoscelis microps*
	Stylocheiron abbreviatum
Hyperiidae:	*Phrosina semilunata*
	Primno macropus
	Anchylomera blossevillei
Sergestidae:	*Lucifer hanseni*
	L. typus
Pteropoda:	*Creseis acicula*
	C. virgula conica

Table 8.5—4
SPECIES CHARACTERISTIC OF ARCTIC OR BOREAL WATER FOUND AT LOWER LATITUDES IN THE ATLANTIC

Calanus hyperboreus	*Dimophyes arctica*	*Spiratella helicina*
Metridia longa	*Sagitta maxima*	*Sergestes arcticus*
Pareuchaeta norvegica	*Eukrohnia hamata*	
Pareuchaeta barbata		

Table 8.5—5
SPECIES OF THE LUSITANIAN STREAM IN THE EASTERN NORTH ATLANTIC

Siphonophora

Rosacea plicata
R. cymbiformis
Nectopyramis diomedeae
N. thetis
Bassia bassensis
Vogtia (all species)
Hippopodius hippopus
Muggiaea spp.
Eudoxoides spiralis
Chuniphyes multidentata
Lensia – all species except *L. conoidea*
Stephanomia bijuga
Velella velella

Medusae

Rhopalonema velatum
Nausithoe punctata
Pelagia noctiluca

Chaetognatha

Sagitta lyra
S. serratodentata atlantica
S. bipunctata
Krohnitta subtilis

Polychaeta

Travisiopsis lobifera
Lagisca hubrechti
Sagitella kowalewskii

Mollusca

Euclio polita
Janthina britannica

Crustacea

Nematoscelis megalops
Stylocheiron spp.
Vibilia spp.
Phronima spp.
Sapphirina spp.
Phyllosoma larvae

Thaliacea

Cyclosalpa spp.
Ritteriella spp.
Thalia democratica
Thetys vagina
Iasis zonaria
Ihlea asymmetrica
Salpa maxima
Doliolina mulleri
Doliolum nationalis

Table 8.5—6
OCEANIC SPECIES FOUND IN THE MIXED WATER OF THE SCOTTISH SHELF IN THE EASTERN NORTH ATLANTIC

Frequent	Intermediate	Rare
	Coelenterata	
Dimophyes arctica	*Chelophyes appendiculata*	*Chuniphyes multidentata*
Lensia conoidea	*Lensia fowleri*	*Vogtia* spp.
Sulculeolaria biloba	*Hippopodius hippopus*	*Nectopyramis* spp.
Physophora hydrostatica	*Arachnactes larvae*	*Rosacea* spp.
Agalma elegans	*Staurophora mertonsii*	*Nausithoe* spp.
Laodicea undulata	*Pelagia noctiluca*	*Pantachogon haeckeli*
Cosmetira pilosella		*Halicreas* spp.
Phialidium hemisphericum		*Periphylla periphylla*
		Rhopalonema velatum
	Chaetognatha	
Sagitta serratodentata *f. tasmanica*	*Sagitta maxima*	*Sagitta serratodentata* *f. atlantia*
	Sagitta lyra	*Sagitta hexaptera*
	Eukrohnia hamata	*Sagitta zetesios*
	Polychaeta and Nemertea	
Tomopteris septentrionalis	*Nectonemertes mirabilis*	*Lagisca hubrechti*
		Vanadis formosa

Table 8.5—6 (continued)
OCEANIC SPECIES FOUND IN THE MIXED WATER OF THE SCOTTISH SHELF IN THE EASTERN NORTH ATLANTIC

Frequent	Intermediate	Rare
	Mollusca	
Clione limacina	*Clio pyramidata*	*Diacria trispinosa*
	Clio cuspidata	*Spiratella helcoides*
	Pneumodermopsis ciliata	*Spiratella helicina*
	Tracheloteuthis riseii	
	Taonidium pfefferi	
	Copepoda	
Rhincalanus nasutus	*Euchaeta hebes*	Other species of
Eucalanus elongatus	*Gaetanus pileatus*	*Euchaeta*
Pareuchaeta norvegica	*Gaidius tenuispinus*	*Pareuchaeta* and
Pleuromamma robusta	*Euchirella curticaudata*	*Euchirella*
Euchirella rostrata	*Metridia longa*	Most other bathypelagic species
Calanus hyperboreus	*Phaenna spinifera*	
Oithona spinirostris		
	Other Crustacea	
Thysanoessa longicaudata	*Euphausia krohni*	*Nematoscelis megalops*
Sergestes arcticus	*Stylocheiron longicorne*	*Thysanopoda acutifrons*
	Brachyscelus crusculum	*Nematobrachion boopis*
	Munnopsis murrayi	*Stylocheiron elongatum*
		Vibilia spp.
		Phronima sedentaria
		Scina spp.
		Ammalopeneus elegans
	Fish Larvae	
Maurolicus mulleri	*Bathylagus* spp.	*Nansenia groenlandica*
Myctophum glaciale	*Fierasfer* spp.	*Stomias boa*
Gadus poutassou	*Paralepis coregonoides*	*Argyropelecus hemigymnus*

Figure 8.5—1
ASSOCIATIONS OF MAJOR SPECIES AND SPECIES GROUPS OF EUPHAUSIIDS WITH WATER MASSES OF THE PACIFIC

Water mass designations are given in A; corresponding distributions of species are shown in B to S.

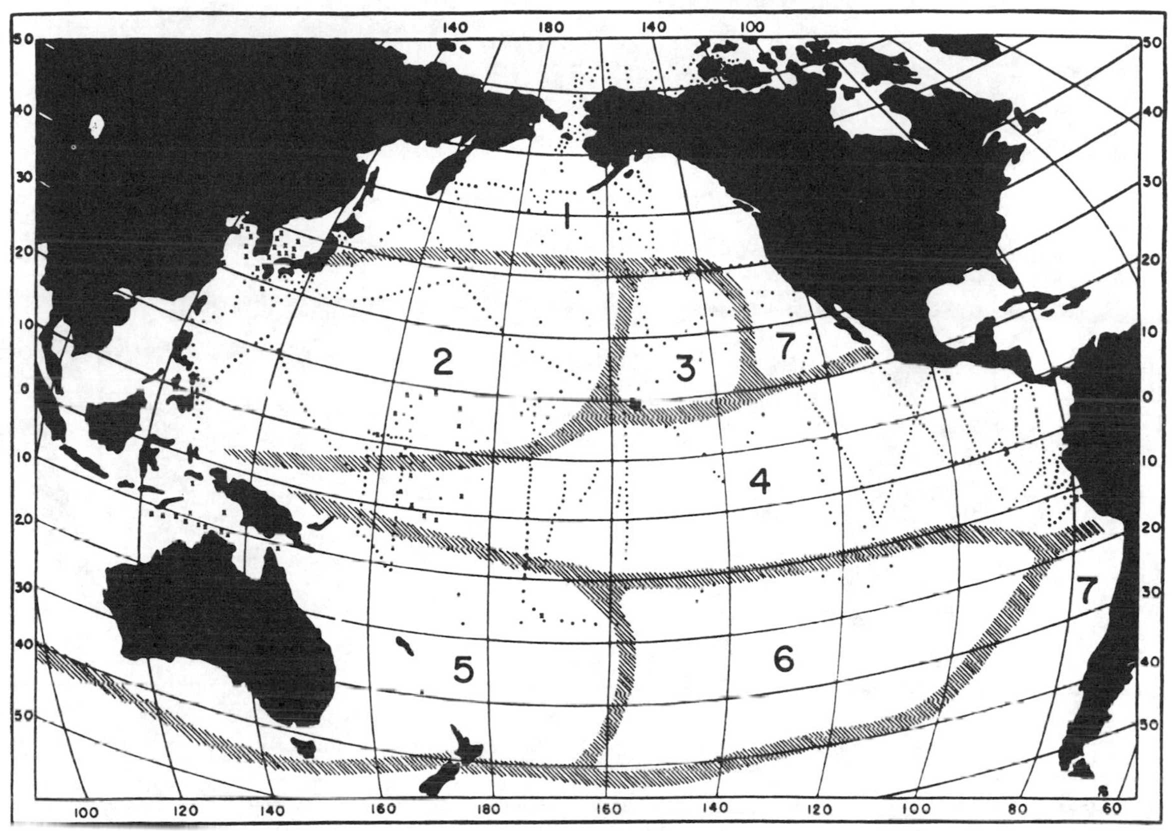

A. Water masses of the Pacific (100–500 m) according to Sverdrup, Johnson, and Fleming (1942). (1) Pacific Subarctic Water, (2) Western North Pacific Central Water, (3) Eastern North Pacific Central Water, (4) Pacific Equatorial Water, (5) Western South Pacific Central Water, (6) Eastern South Pacific Central Water, (7) Transition Water.

(From Bieri, R., The distribution of the planktonic Chaetognatha in the Pacific and their relationship to the water masses, *Limnol. Oceanogr.*, 4(1), 4, 1959.)

Figure 8.5—1 (continued)

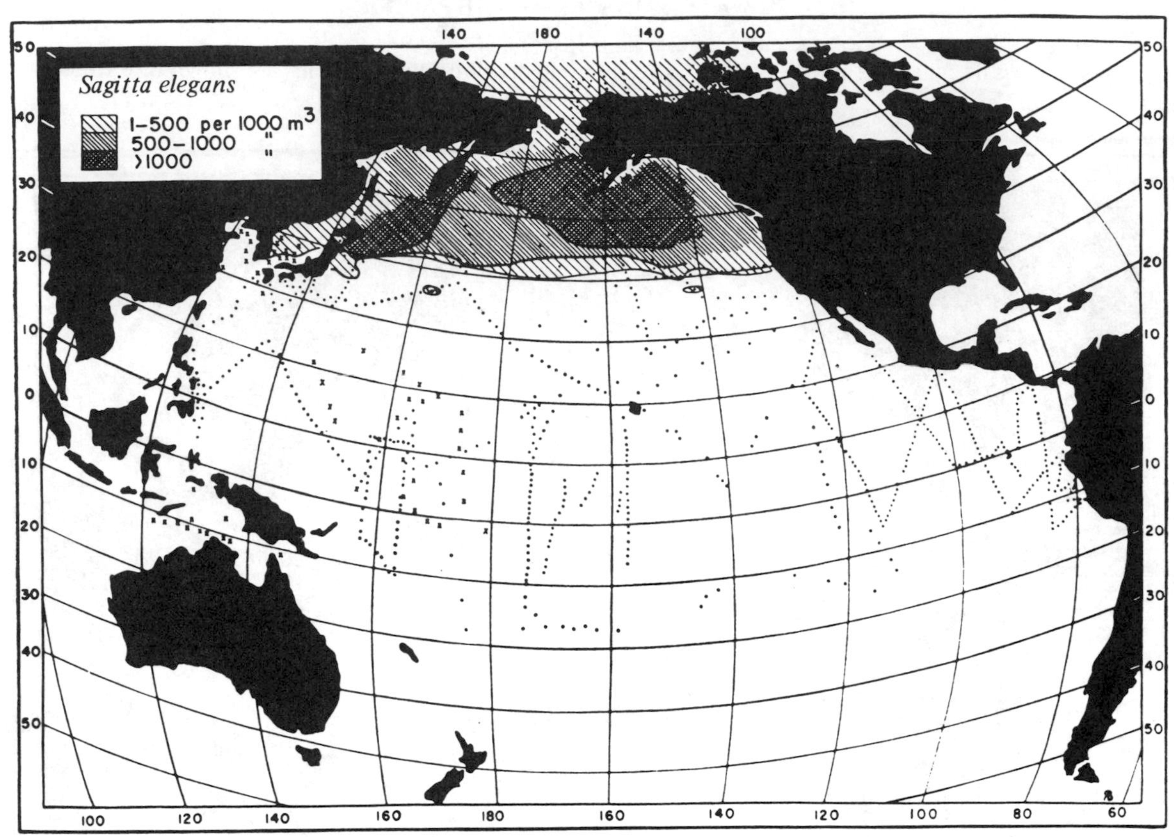

B. Distribution of *Sagitta elegans* in the Pacific and regions to the north.

(From Bieri, R., The distribution of the planktonic Chaetognatha in the Pacific and their relationship to the water masses, *Limnol. Oceanogr.*, 4(1), 4, 1959.)

Figure 8.5—1 (continued)

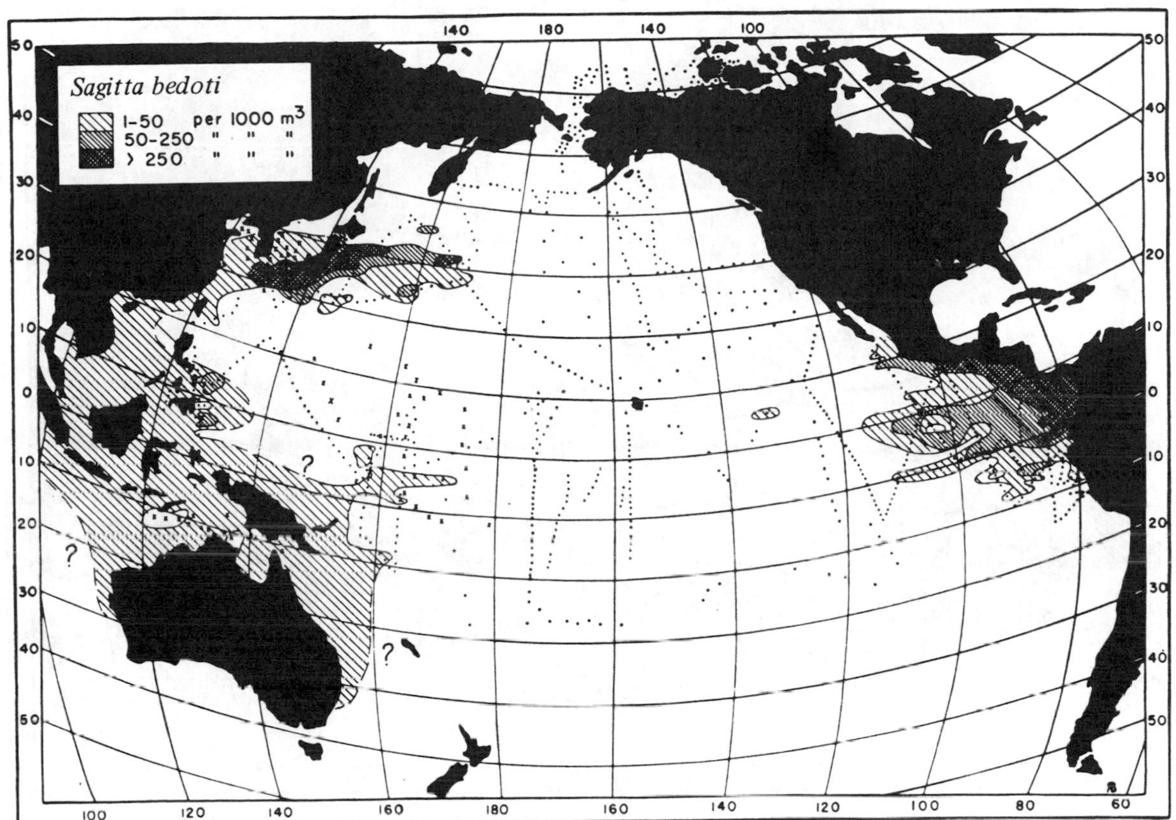

C. Known extent of the two *Sagitta bedoti* populations in the Pacific. The distribution of this species is reminiscent of the equatorial-west-central species such as *Sagitta robusta*, but it is apparently unable to cross the oceanic equatorial region.

(From Bieri, R., The distribution of the planktonic Chaetognatha in the Pacific and their relationship to the water masses, *Limnol. Oceanogr.*, 4(1), 11, 1959.)

Figure 8.5—1 (continued)

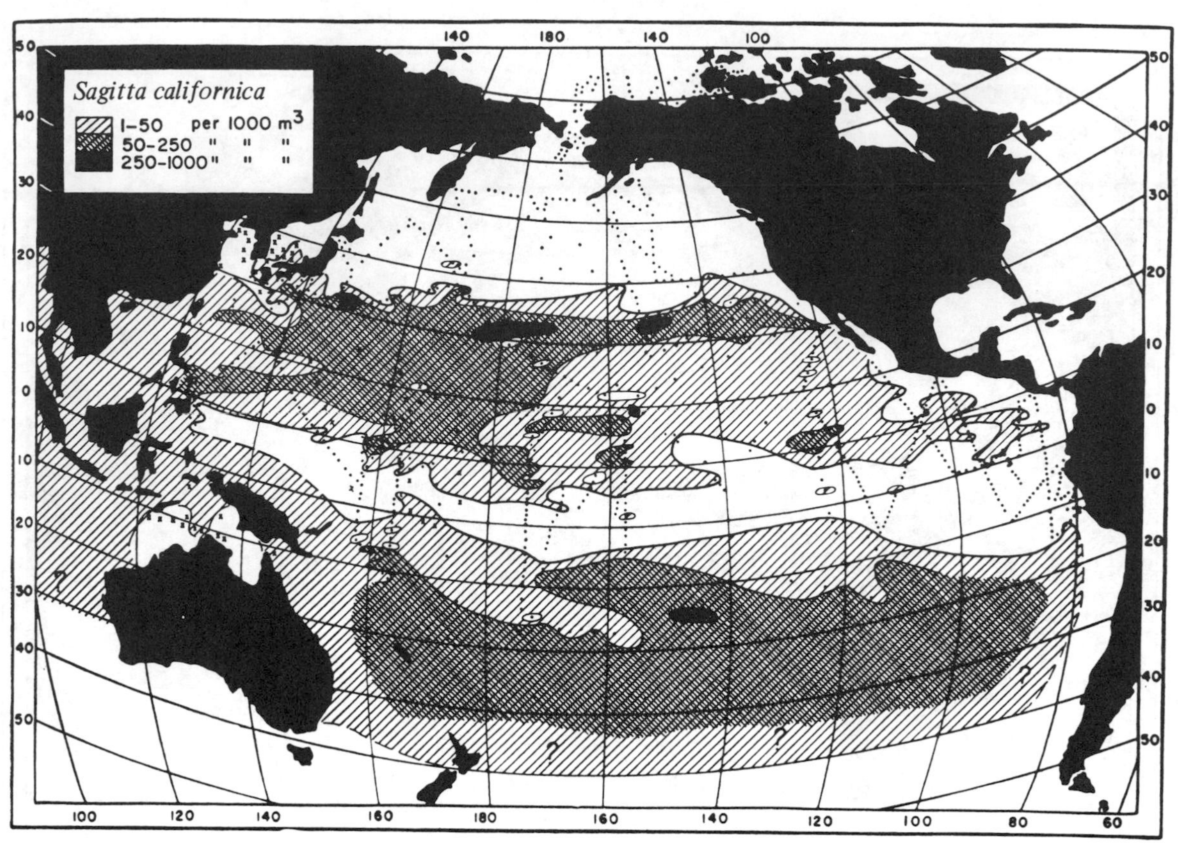

D. Distribution of *Sagitta californica* in the Pacific.

(From Bieri, R., The distribution of the planktonic Chaetognatha in the Pacific and their relationship to the water masses, *Limnol. Oceanogr.*, 4(1), 16, 1959.)

Figure 8.5—1 (continued)

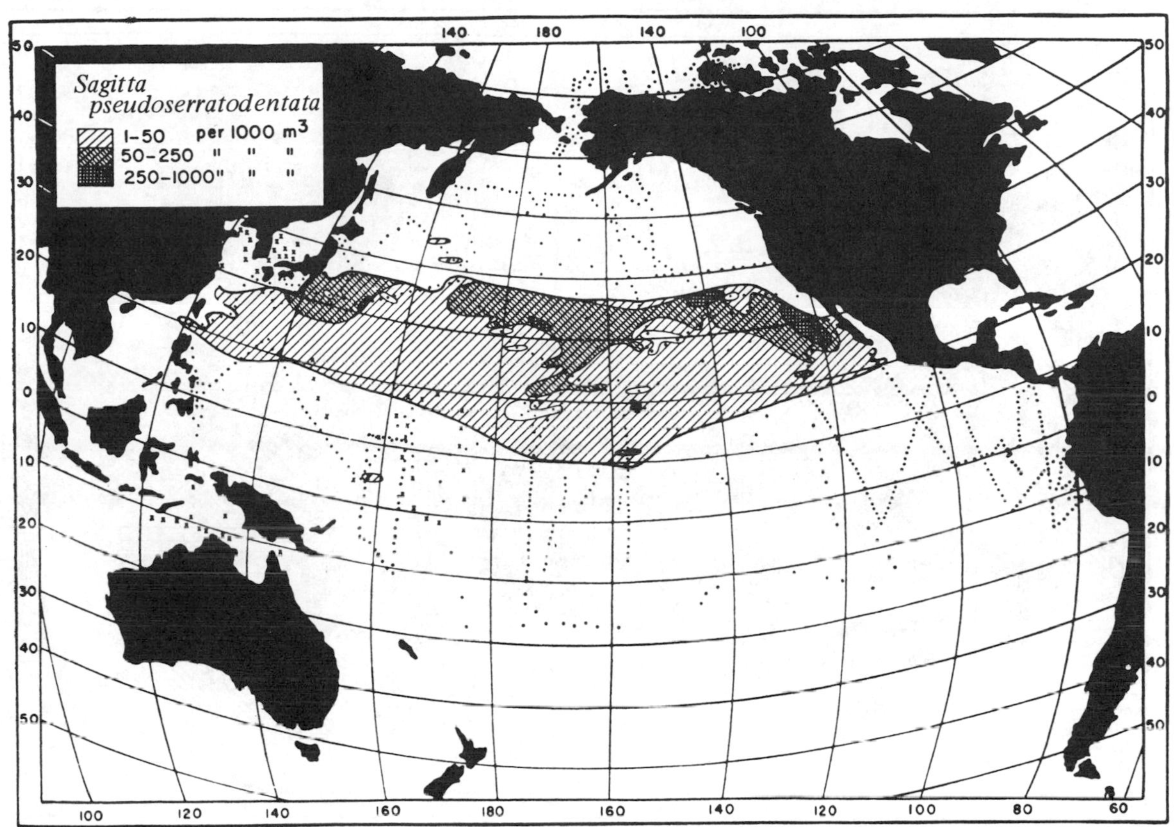

E. Distribution of *Sagitta pseudoserratodentata* in the Pacific. The limits of this species approximate fairly closely the limits of the Pacific Central Water.

(From Bieri, R., The distribution of the planktonic Chaetognatha in the Pacific and their relationship to the water masses, *Limnol. Oceanogr.*, 4(1), 10, 1959.)

Figure 8.5—1 (continued)

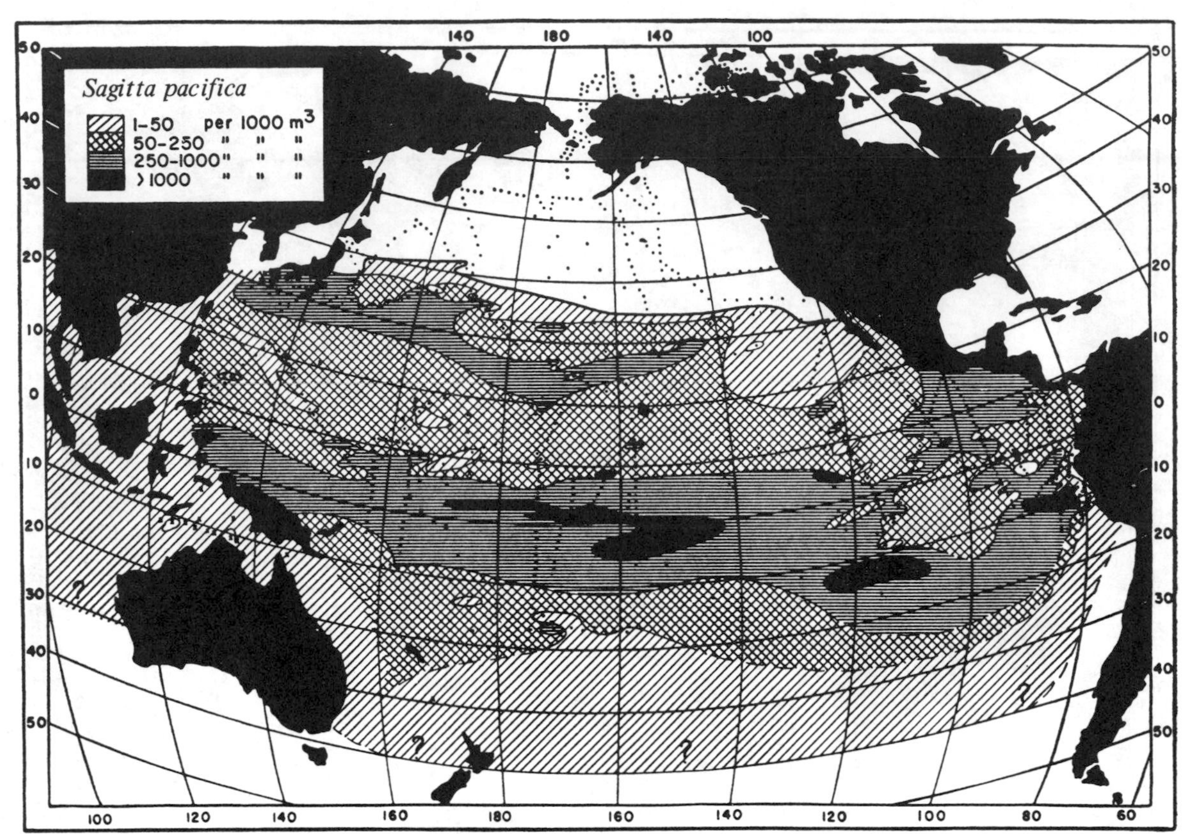

F. Occurrence of *Sagitta pacifica* (*serratodentata* group) in the Pacific.

(From Bieri, R., The distribution of the planktonic Chaetognatha in the Pacific and their relationship to the water masses, *Limnol. Oceanogr.*, 4(1), 11, 1959.)

Figure 8.5—1 (continued)

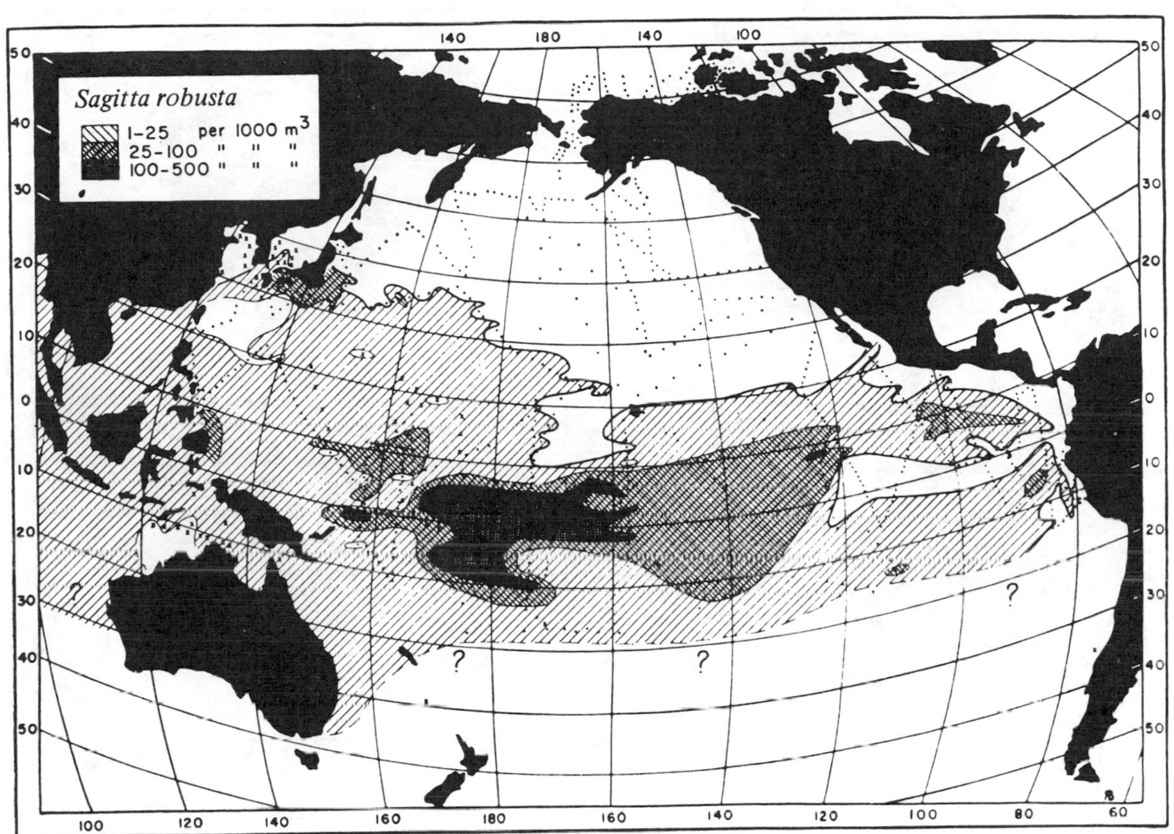

G. Occurrence of *Sagitta robusta* in the Pacific. This species is absent from the Eastern North Pacific Central Water.

(From Bieri, R., The distribution of the planktonic Chaetognatha in the Pacific and their relationship to the water masses, *Limnol. Oceanogr.*, 4(1), 13, 1959.)

Figure 8.5—1 (continued)

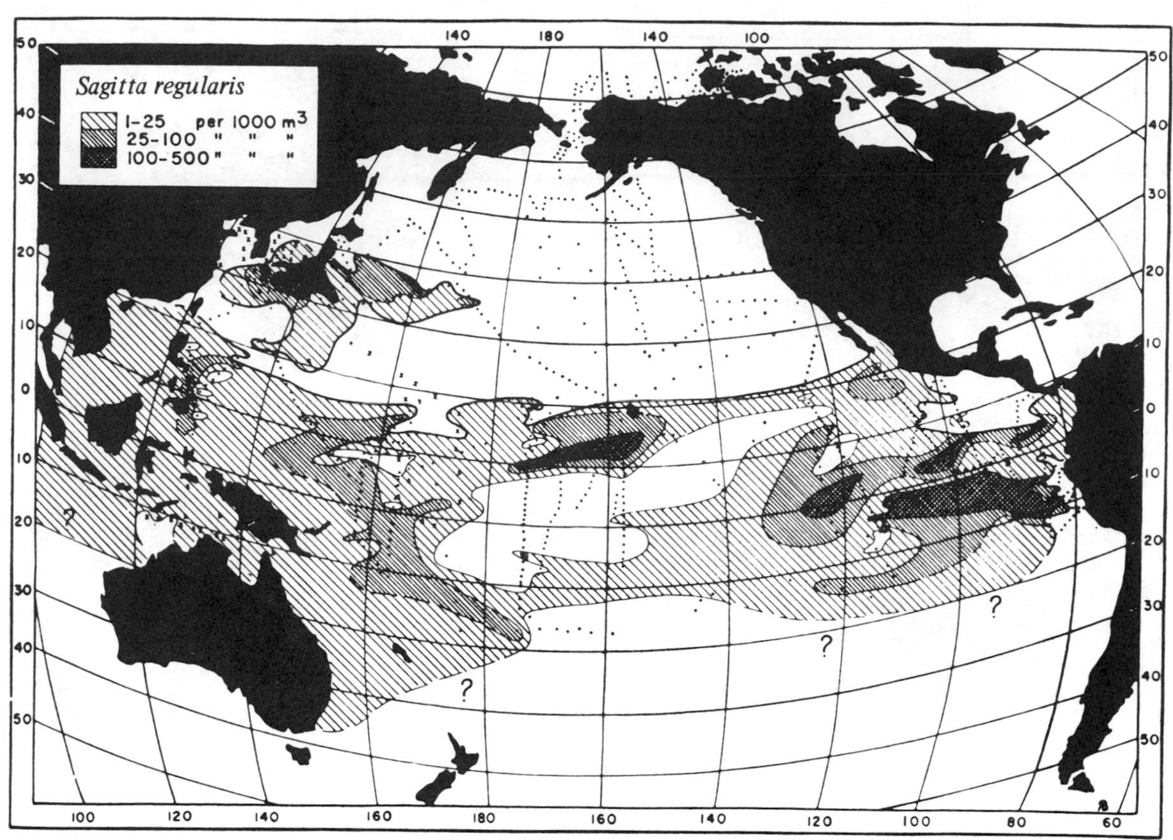

H. Distribution of *Sagitta regularis* in the Pacific. The distribution of this species is similar to that of *Sagitta robusta.*

(From Bieri, R., The distribution of the planktonic Chaetognatha in the Pacific and their relationship to the water masses, *Limnol. Oceanogr.*, 4(1), 13, 1959.)

Figure 8.5—1 (continued)

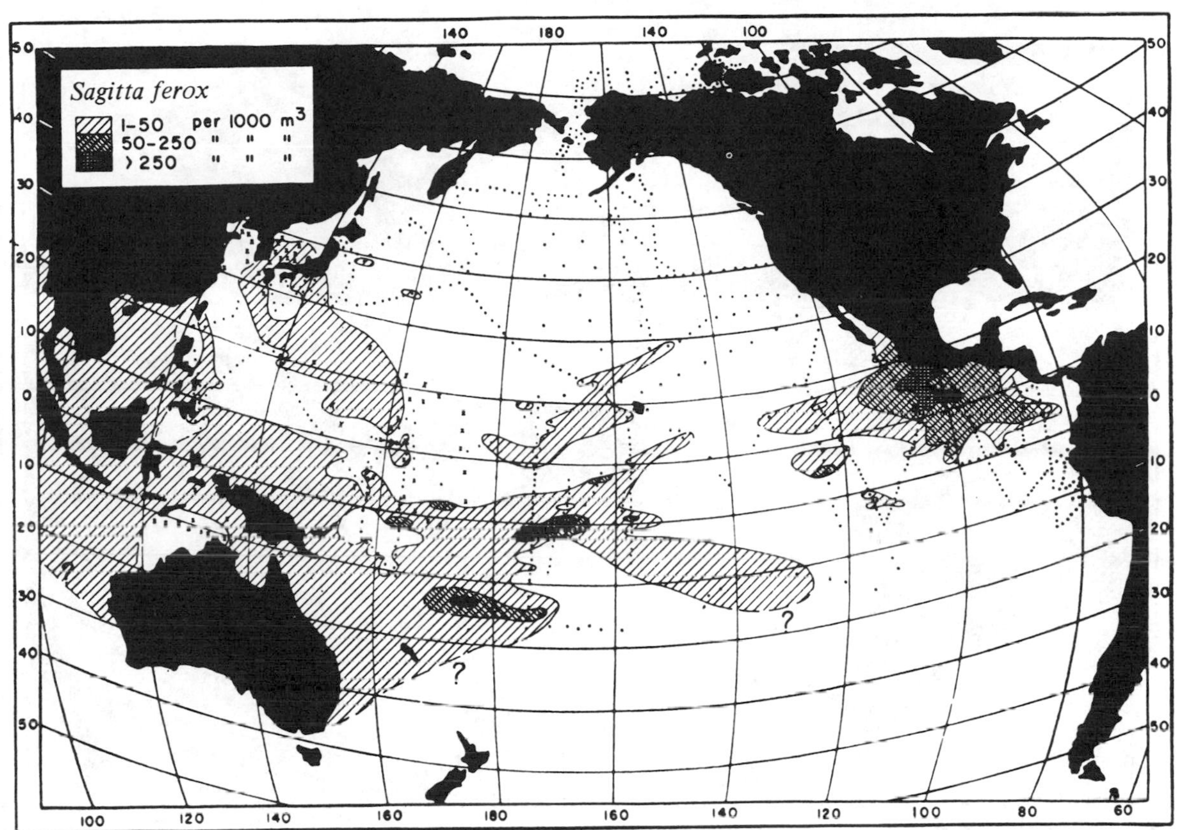

I. Distribution of *Sagitta ferox* in the Pacific. This species is similar in its distribution to *Krohnitta pacifica*.

(From Bieri, R., The distribution of the planktonic Chaetognatha in the Pacific and their relationship to the water masses, *Limnol. Oceanogr.*, 4(1), 15, 1959.)

Figure 8.5—1 (continued)

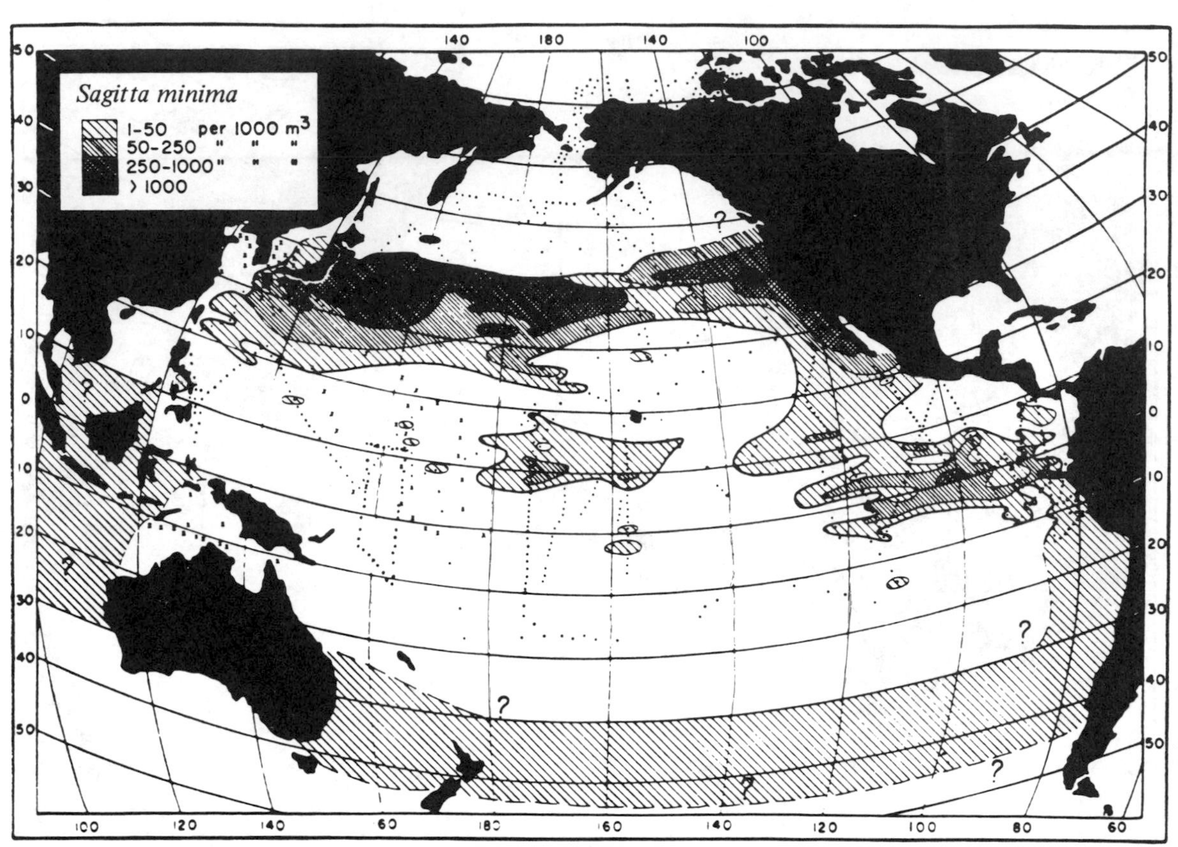

J. Distribution of *Sagitta minima* in the Pacific. This species is most common in the regions of mixing of water masses.

(From Bieri, R., The distribution of the planktonic Chaetognatha in the Pacific and their relationship to the water masses, *Limnol. Oceanogr.*, 4(1), 19, 1959.)

Figure 8.5—1 (continued)

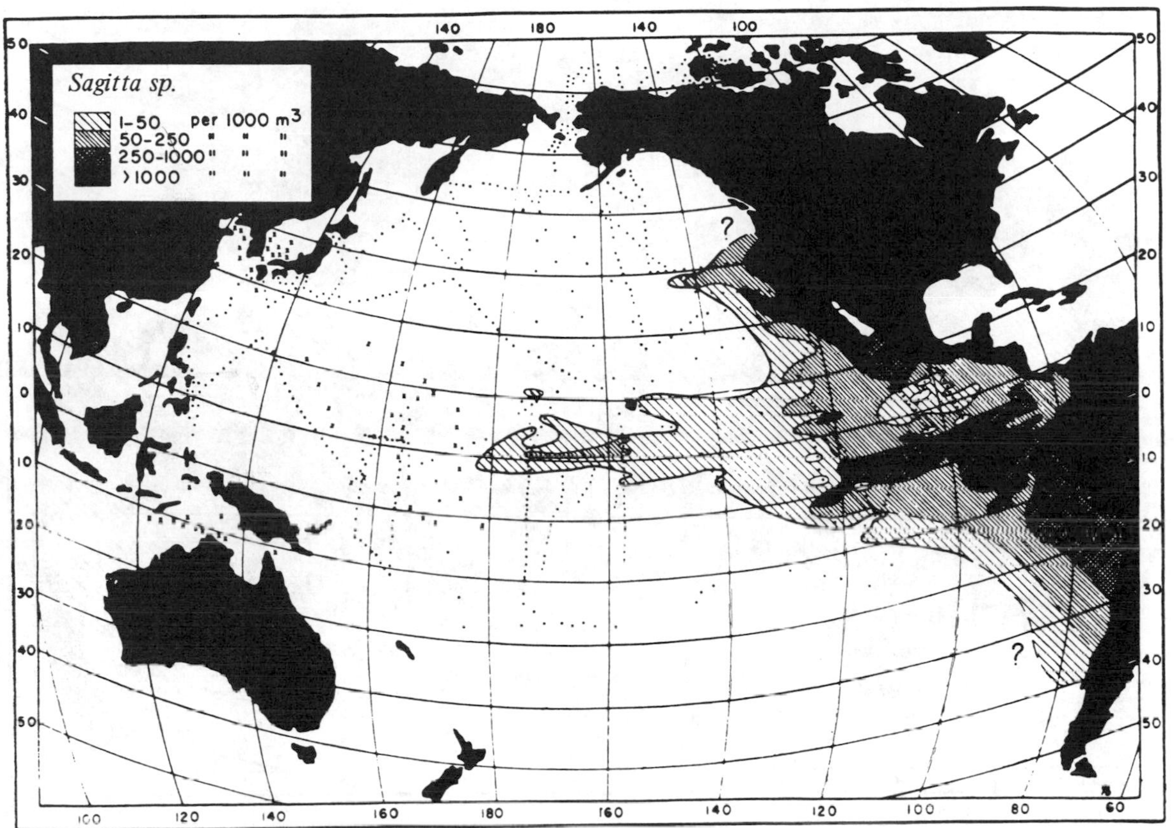

K. Distribution of *Sagitta* sp. (*serratodentata* group) in the Pacific.

(From Bieri, R., The distribution of the planktonic Chaetognatha in the Pacific and their relationship to the water masses, *Limnol. Oceanogr.*, 4(1), 10, 1959.)

Figure 8.5—1 (continued)

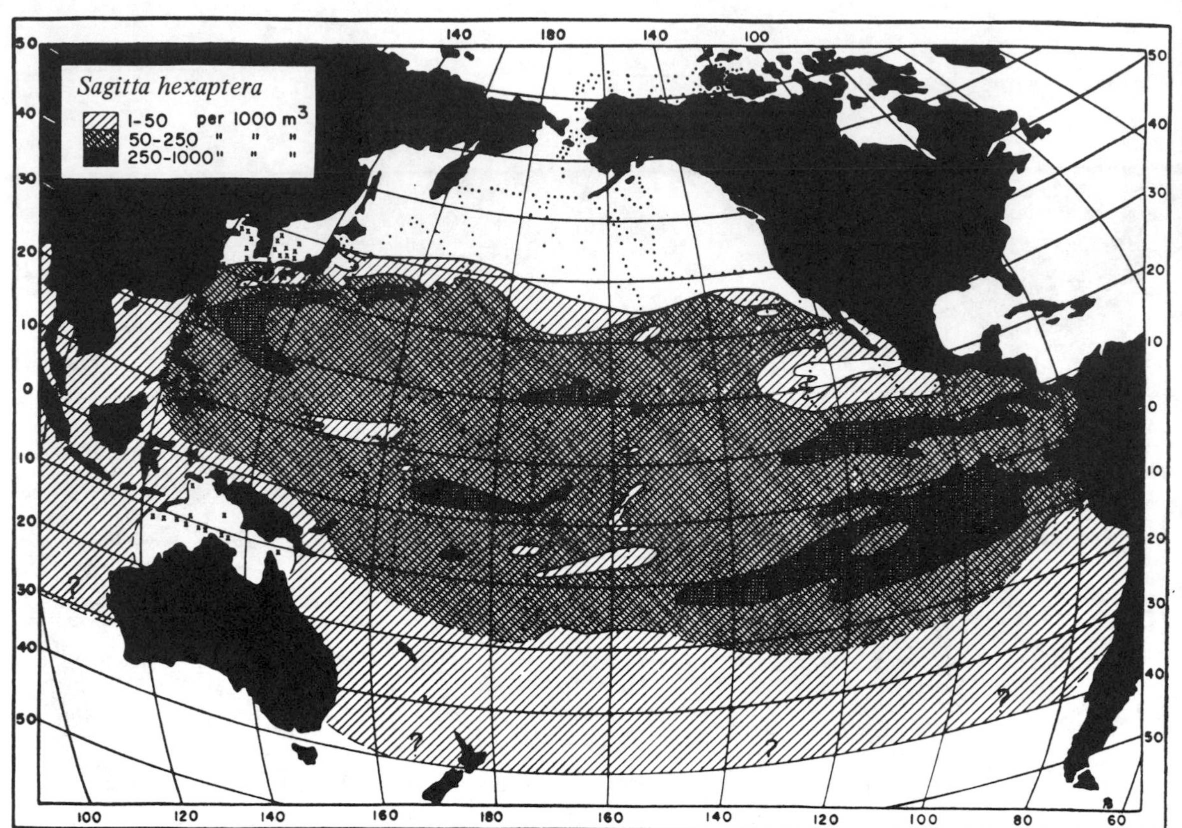

L. Distribution of *Sagitta hexaptera* in the Pacific.

(From Bieri, R., The distribution of the planktonic Chaetognatha in the Pacific and their relationship to the water masses, *Limnol. Oceanogr.*, 4(1), 19, 1959.)

Figure 8.5—1 (continued)

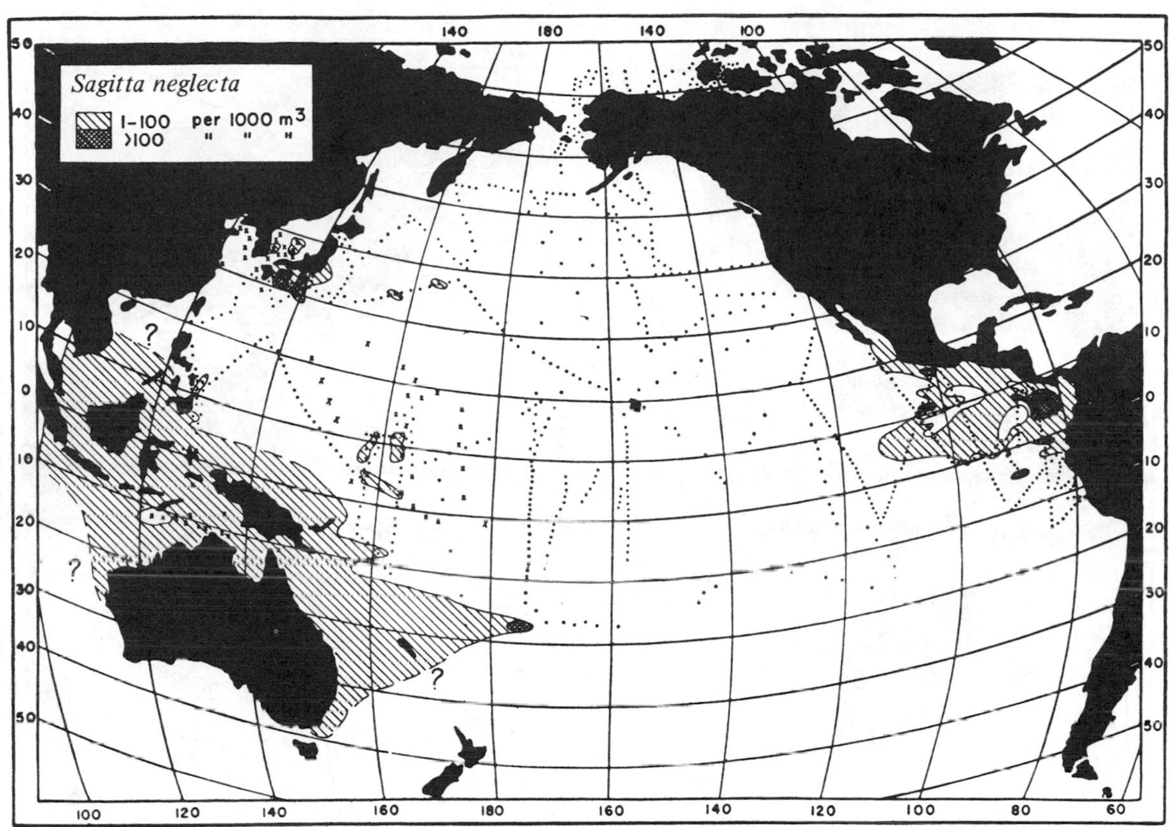

M. Known extent of *Sagitta neglecta* in the Pacific. It has essentially the same distribution as *Sagitta bedoti*, but is not so abundant as that species.

(From Bieri, R., The distribution of the planktonic Chaetognatha in the Pacific and their relationship to the water masses, *Limnol. Oceanogr.*, 4(1), 17, 1959.)

Figure 8.5—1 (continued)

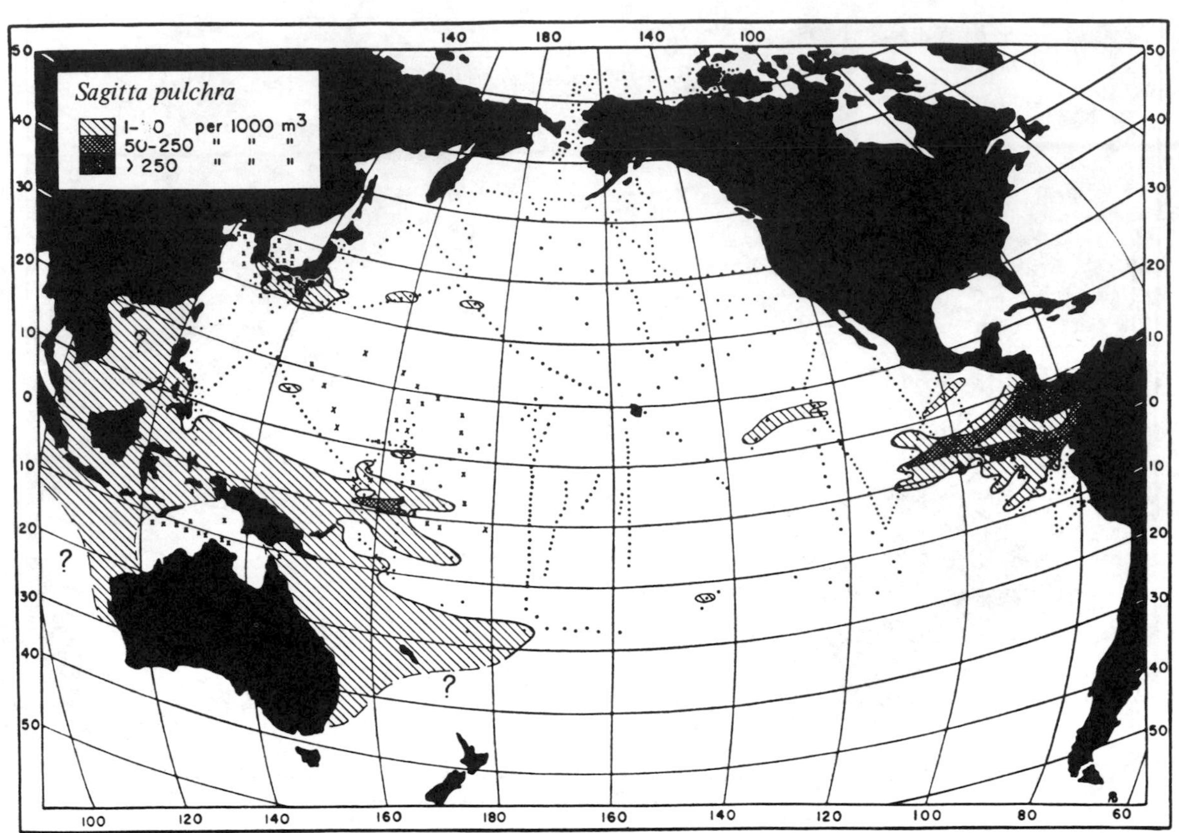

N. Known extent of *Sagitta pulchra* in the Pacific. It is similar in its distribution to *Sagitta bedoti* and *S. neglecta*.

(From Bieri, R., The distribution of the planktonic Chaetognatha in the Pacific and their relationship to the water masses, *Limnol. Oceanogr.*, 4(1), 17, 1959.)

Figure 8.5—1 (continued)

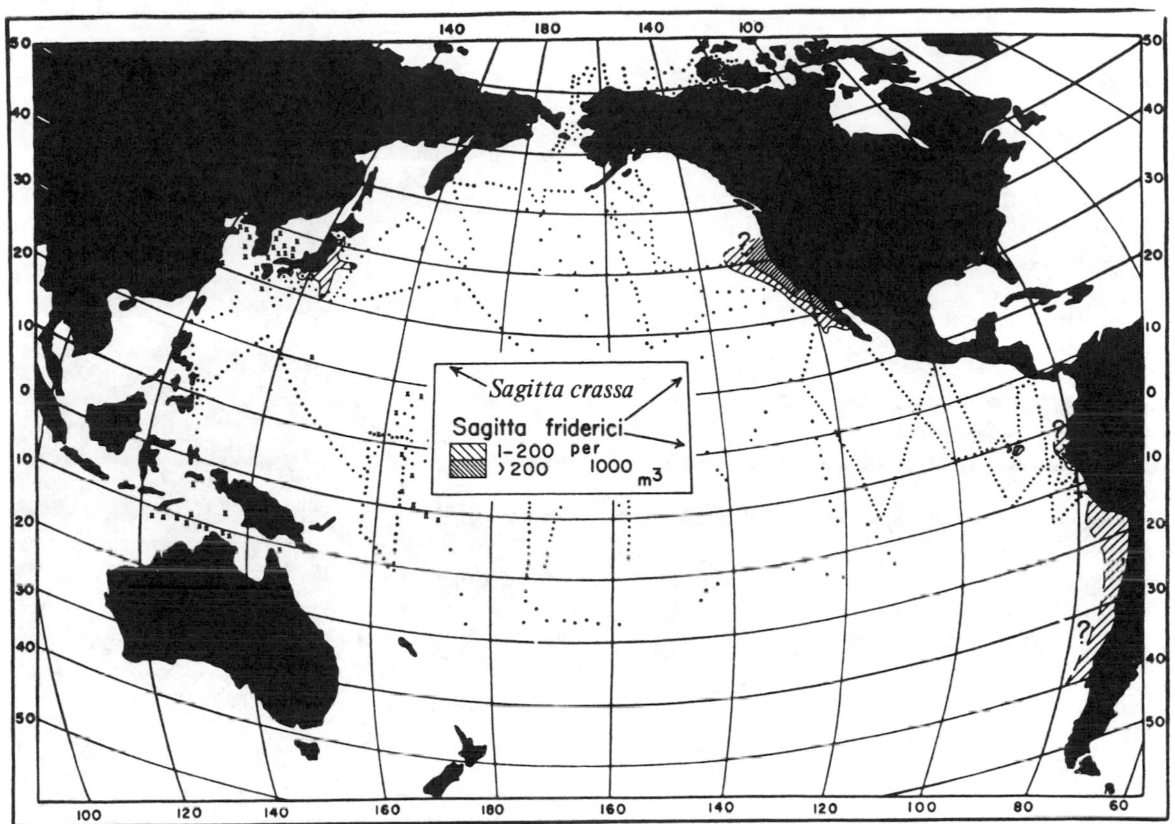

O. Known extent of the two *Sagitta friderici* populations in the Pacific. This species dominates the California and Peru Currents. *Sagitta crassa* off Japan is a closely related species.

(From Bieri, R., The distribution of the planktonic Chaetognatha in the Pacific and their relationship to the water masses, *Limnol. Oceanogr.*, 4(1), 8, 1959.)

Figure 8.5—1 (continued)

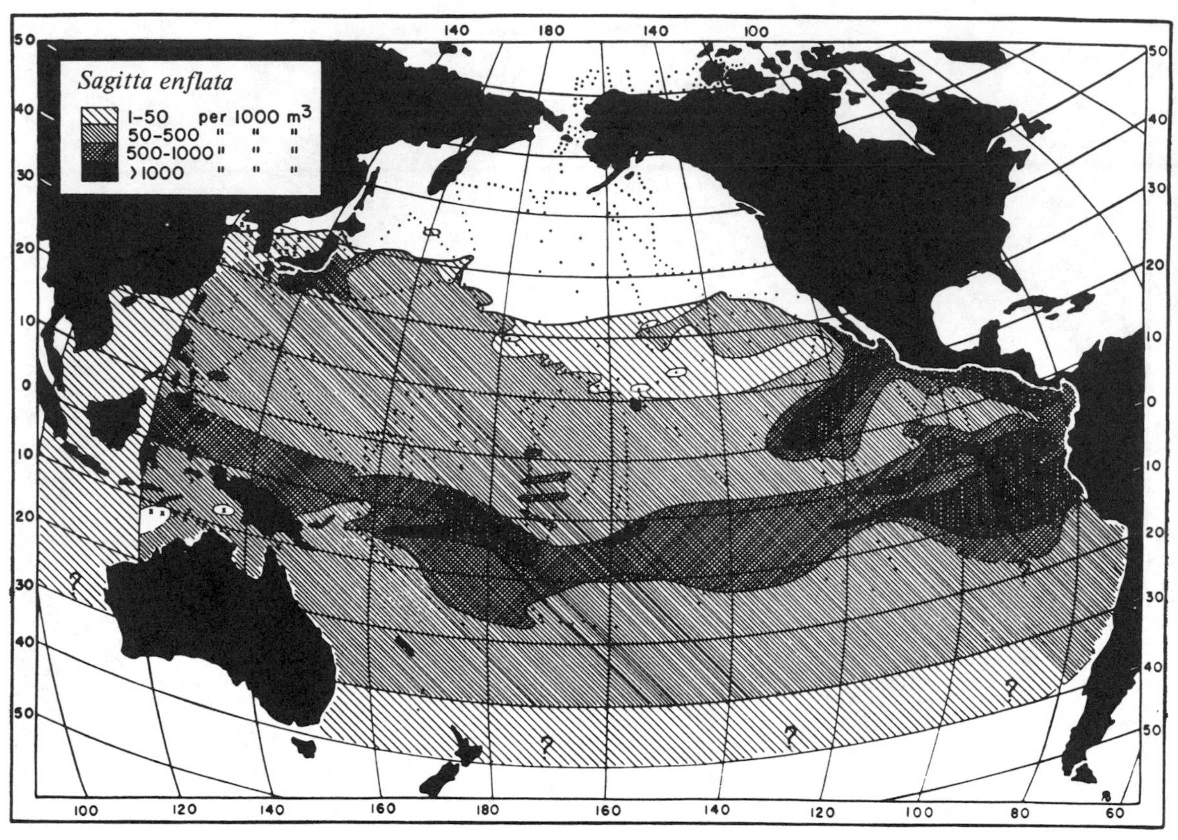

P. Occurrence of *Sagitta enflata* in the Pacific.

(From Bieri, R., The distribution of the planktonic Chaetognatha in the Pacific and their relationship to the water masses, *Limnol. Oceanogr.*, 4(1),16, 1959.)

Figure 8.5—1 (continued)

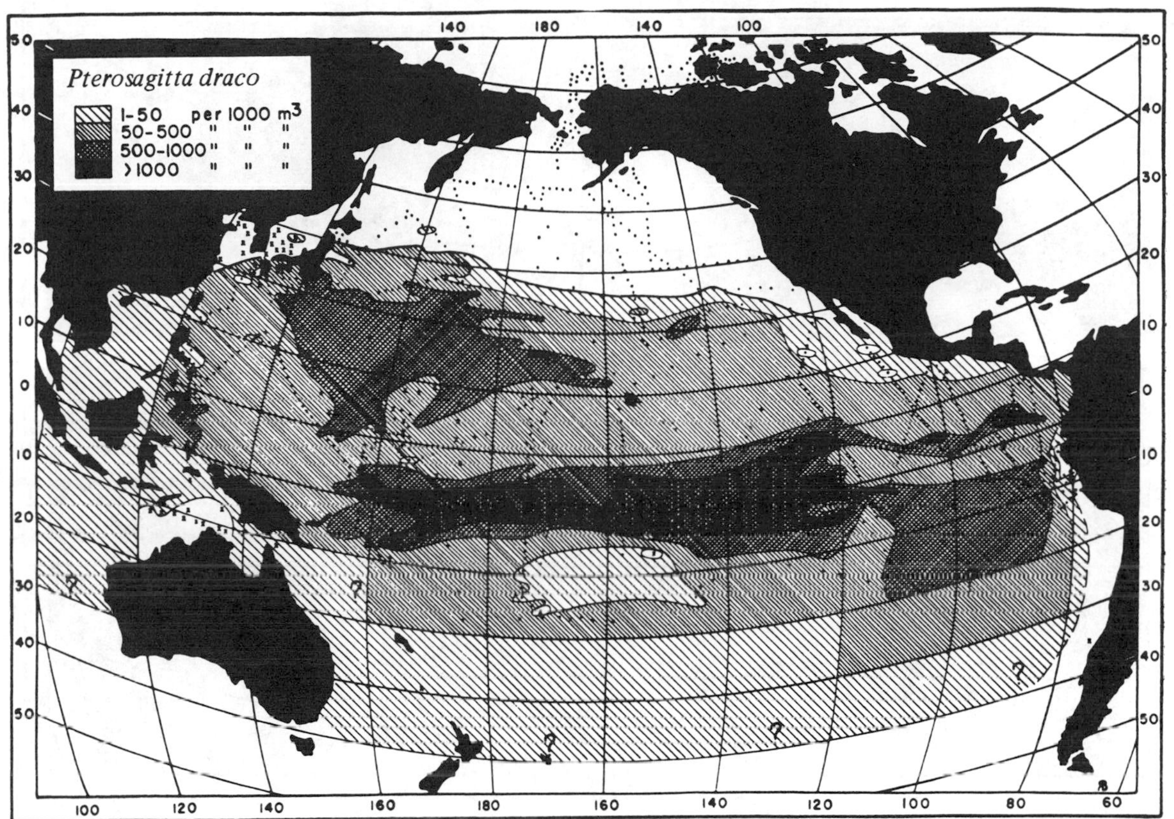

Q. Occurrence of *Pterosagitta draco* in the Pacific.

(From Bieri, R., The distribution of the planktonic Chaetognatha in the Pacific and their relationship to the water masses, *Limnol. Oceanogr.*, 4(1), 8, 1959.)

Figure 8.5—1 (continued)

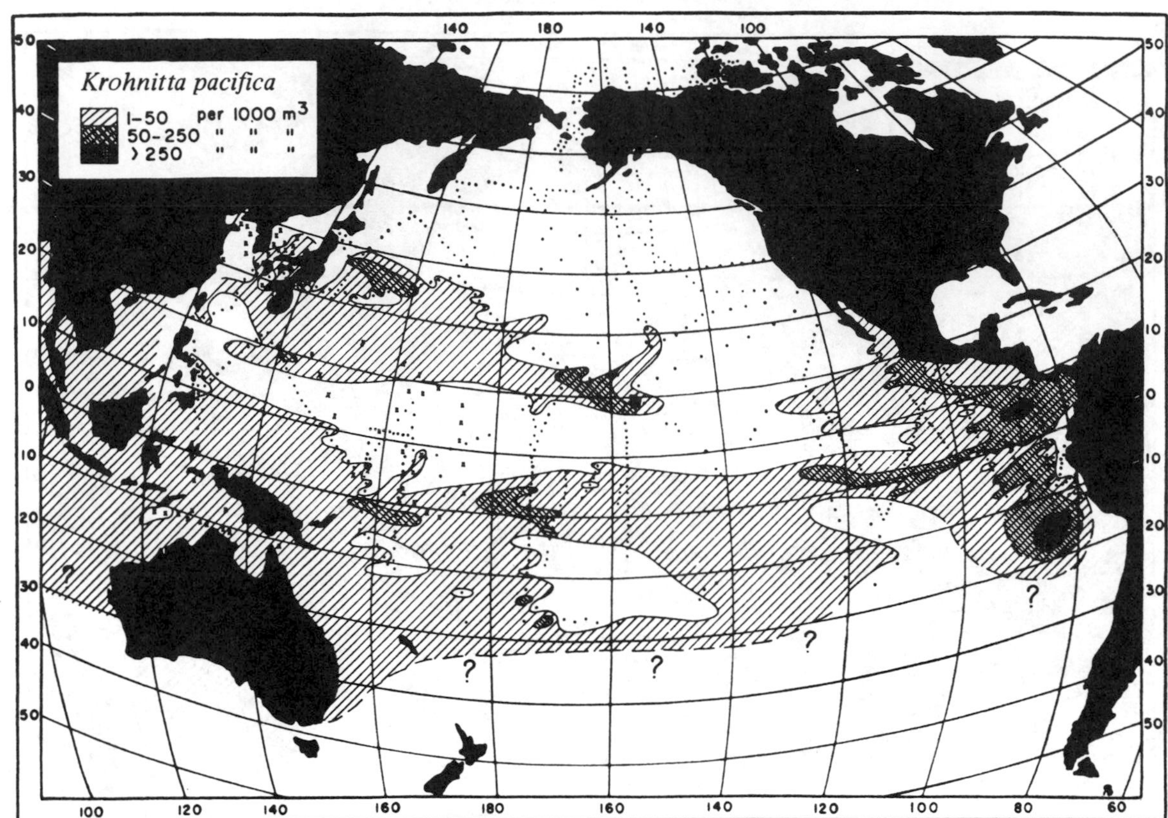

R. Occurrence of *Krohnitta pacifica* in the Pacific. Note that it extends slightly into the Eastern North Pacific Central Water northeast of Hawaii.

(From Bieri, R., The distribution of the planktonic Chaetognatha in the Pacific and their relationship to the water masses, *Limnol. Oceanogr.*, 4(1), 15, 1959.)

Figure 8.5—1 (continued)

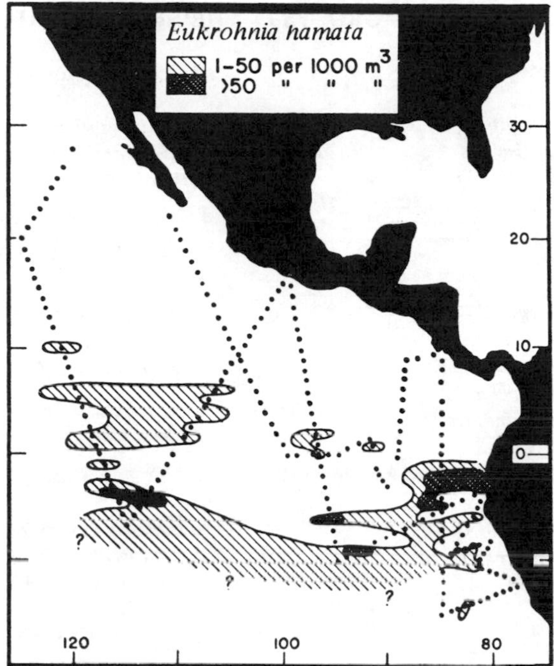

S. Concentration of *Eukrohnia hamata* in the upper 300 meters of water in the eastern equatorial Pacific.

(From Bieri, R., The distribution of the planktonic chaetognatha in the Pacific and their relationship to the water masses, *Limnol. Oceanogr.*, 4(1), 12, 1959. With permission.)

Table 8.5—7
SPECIES COMPOSITION OF RECURRENT ZOOPLANKTON GROUPS IN THE NORTH PACIFIC*

Group I

Euphausia hemigibba (E)
Euphausia mutica (E)
Euphausia recurva (E)

Euphausia tenera (E)
Nematoscelis atlantica (E)
Nematoscelis microps (E)
Nematoscelis tenella (E)
Stylocheiron carinatum (E)

Stylocheiron submii (E)
Pterosagitta draco (C)
Sagitta enflata (C)

Sagitta hexaptera (C)
Sagitta pacifica (C)

Creseis virgula (P)
Limacina bulimoides (P)

Limacina inflata (P)
Associated:
Euphausia brevis (E)
Hyalocylix striata (P)
Limacina trochiformis (P)

Group II

Euphausia pacifica (E)
Thysanoessa longipes (E)
Tessarabrachion oculatus (E)
Sagitta elegans (C)
Limacina helicina (P)
Associated:
Thysanoessa inermis (E)
Sagitta scrippsae (C)

Group III

Cavolinia inflexa (P)
Clio pyramidata (P)
Styliola subula (P)

Group IV

Euphausia diomediae (E)
Nematoscelis gracilis (E)
Sagitta robusta (C)

Group V

Euphausia gibboides (E)
Nematoscelis difficilis (E)
Thysanoessa gregaria (E)

Group VI

Limacina lesueuri (P)
Sagitta pseudoserratodentata (C)

Group VII

Atlanta lesueuri (H)
Atlanta turriculata (H)

No affinities

Euphausia paragibba (E)
Sagitta ferox (C)
Cavolinia longirostris (P)
Carinaria japonica (H)

Oxygyrus keraudreni (H)
Protatlanta souleveti (H)

Pterosoma planum (H)
Pterotrachea hippocamous (H)
Pterotrachea minuta (H)

*Major taxon identifications are: (C) chaetognath; (E) euphausiid; (H) heteropod; and (P) pteropod.

(From Fager, E. W. and McGowan, J. A., Zooplankton species groups in the north Pacific, *Science*, 140(3536), 455, © 1963, American Association for the Advancement of Science. With permission.)

Figure 8.5—2

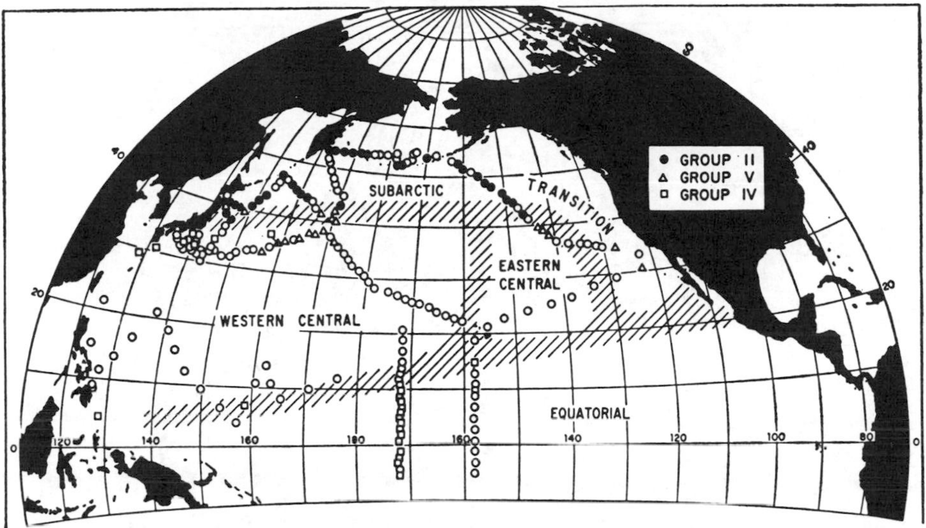

Figure 8.5—2. Distribution of zooplankton Groups II, IV, and V from Table 8.5—7 and their relationships to major water masses. Open circles are sampling stations where no members of the groups were found.

(From Fager, E. W. and McGowan, J. A., Zooplankton species groups in the north Pacific, *Science*, 140(3536), 456, © 1963, American Association for the Advancement of Science. With permission.)

Figure 8.5—3

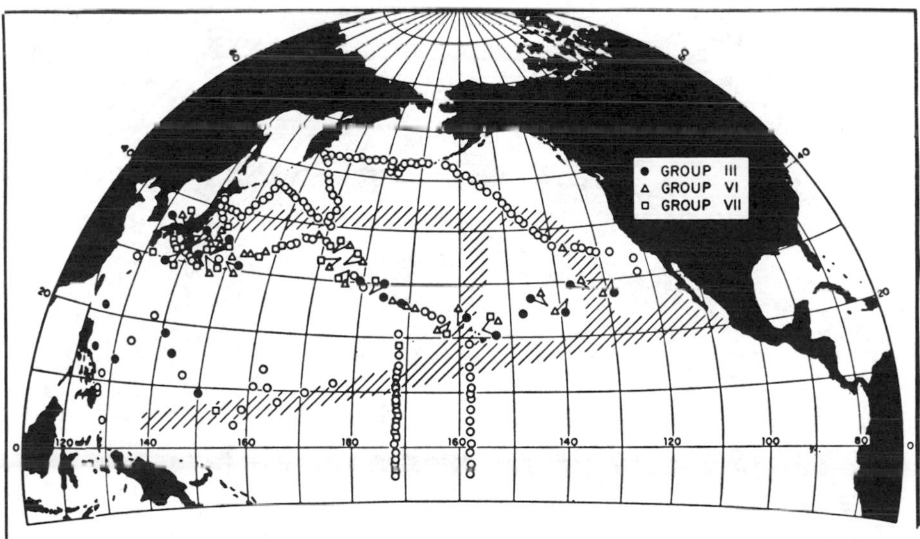

Figure 8.5—3. Distribution of zooplankton Groups III, VI, and VII from Table 8.5—7 and their relationships to major water masses. Open circles are sampling stations where no members of the groups were found.

(From Fager, E. W. and McGowan, J. A., Zooplankton species groups in the north Pacific, *Science*, 140(3536), 457, © 1963, American Association for the Advancement of Science. With permission.)

Figure 8.5—4

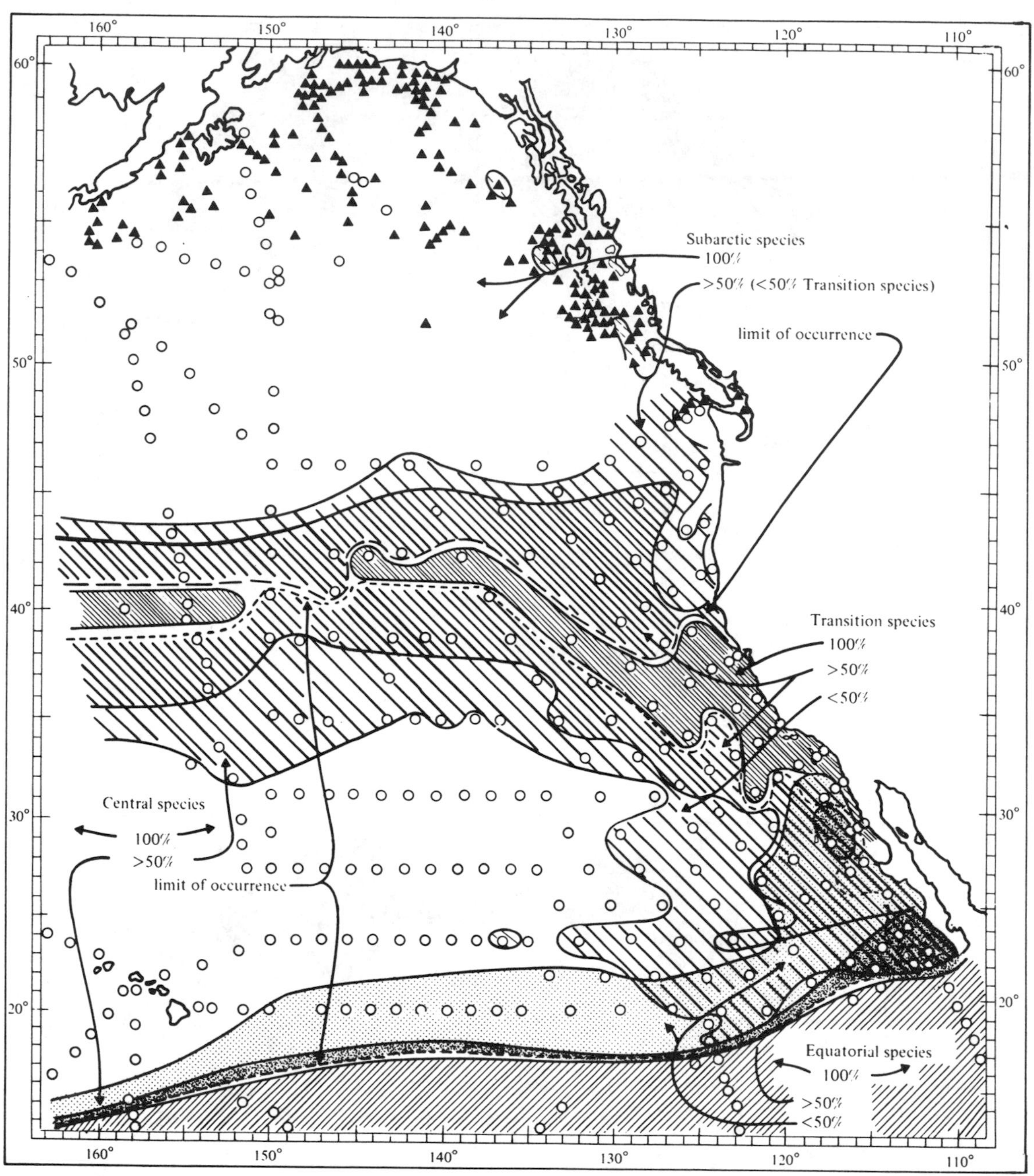

Figure 8.5—4. Associations of euphausiid species in the northeastern Pacific. The amount of a faunal group at a station is indicated as the percentage of the total euphausiid count (all species) made up by the summed counts of the species in that fauna.

(From Brinton, E., The distribution of Pacific euphausiids, *Bull. Scripps Inst. Oceanogr. Univ. Calif.* 8, 224, 1962. With permission.)

8.6. LENGTH, HEIGHT, WEIGHT, AND SAMPLE VOLUME RELATIONSHIPS IN MAJOR GROUPS

Table 8.6—1
RELATIONSHIPS OF LENGTH TO WEIGHT IN MAJOR ZOOPLANKTON GROUPS OF THE CALIFORNIA CURRENT REGION[a]

Ocular[b] μm units[b]	No/g
Amphipods	
10	1023
20	376
30	159
40	101
50	64
60	48
70	29
80	26
90	14
100	10
110	7
Chaetognaths	
20	6994
30	3808
40	2168
50	1522
60	762
70	614
80	425
90	330
100	253
110	200
120	142
130	110
140	75
150	61
160	44
170	40
180	34
190	31
200	27
210	23
220	18
230	16
240	14
250	13
260	11
270	9
280	6
290	5

Obcular[b] μm units[b]	No/g
Cladocerans	
3	11,731
6	3921
Copepods	
5	5814
10	1873
15	698
20	254
25	123
30	70
35	50
40	31
Crustacean larvae	
(a) Anomuran Zoea	
3	752
5	254
10	162
15	146
20	128
25	111
30	100
(b) Brachyuran Zoea	
4	3403
7	2762
10	949
13	500
15	200
25	65
(c) Megalopa	
15	500
20	235
30	105
40	68
50	39
60	23

[a] This table is based upon one gram samples of each size category of a functional group sorted from a variety of plankton collections characterized by pronounced diversity in species and in developmental stages.

[b] One ocular μm unit = 0.167 mm at 7X magnification.

Table 8.6—1 (continued) RELATIONSHIPS OF LENGTH TO WEIGHT IN MAJOR ZOOPLANKTON GROUPS OF THE CALIFORNIA CURRENT REGION[a]

Ocular[b] µm units[b]	No/g
Ctenophores	
Usually weighed	
Decapods	
(a) Galatheids	
30	286
40	125
50	36
60	28
70	18
80	15
90	10
(b) Hoplocarids	
8	1368
20	775
30	253
40	217
50	157
60	83
70	71
80	50
90	40
(c) Natantians	
5	7143
15	1744
25	765
35	539
45	379
55	147
65	54
75	42
85	29
95	20
105	18
115	15
125	13
135	11
145	8
155	7
165	6
175	5
185	4
195	3
205	2
215	1.5
250	1

Ocular[b] µm units[b]	No/g
Euphausiids	
5	11,111
10	6243
15	4200
20	1426
25	930
30	778
35	451
40	308
50	175
55	125
60	99
65	89
70	60
80	45
90	31
100	21
110	17
120	13
130	10
140	8
150	7
160	6
170	5
210	4
240	3
270	2
300	1
330	1
360	1
390	1
Heteropods	
5	7000
10	2147
15	941
20	525
30	375
40	222
50	185
60	158
70	105
80	83
90	61
100	48
(Larger forms usually weighed)	

[a] This table is based upon one gram samples of each size category of a functional group sorted from a variety of plankton collections characterized by pronounced diversity in species and in developmental stages.
[b] One ocular µm unit = 0.167 mm at 7X magnification.

Table 8.6—1 (continued)
RELATIONSHIPS OF LENGTH TO WEIGHT IN MAJOR ZOOPLANKTON GROUPS OF THE CALIFORNIA CURRENT REGION

Ocular[b] µm units[b]	No/g
Larvaceans	
10–15	13,200
15–20	7173
20–25	5200
25–30	3848
30–35	2564
35–40	2197
40–50	1923
50–60	405
60–70	267
Medusae	
Usually weighed	
Mysids	
15	5400
25	3420
35	2500
45	1075
55	625
65	300
75	154
85	133
95	48
110	15
Ostracods	
5	3465
10	1496

Ocular[b] µm units[b]	No/g
Ostracods (cont.)	
15	501
20	341
25	96
30	87
Pteropods	
5	3520
10	1444
15	910
20	327
25	185
30	140
35	122
40	100
45	62
50	42
55	33
Radiolarians	
4	5106
5	3998
6	2431
Thaliaceans	
Usually weighed	
Siphonophores	
Usually weighed	

[a] This table is based upon one gram samples of each size category of a functional group sorted from a variety of plankton collections characterized by pronounced diversity in species and in developmental stages.
[b] One ocular µm unit = 0.167 mm at 7X magnification.

Table 8.6—2
THE AMOUNT OF INTERSTITIAL LIQUID IN DRAINED "WET" PLANKTON SAMPLES

Total volume of drained "wet" plankton (ml)	Volume of interstitial liquid (ml)	Volume of organisms only (ml)	% interstitial liquid
35	12.0	23.0	34
20	5.5	14.5	28
10	4.0	6.0	40
32	11.0	21.0	34
32	5.5	26.5	17
11	5.0	6.0	45
23	7.5	15.5	33
12	5.0	7.0	42
24	8.0	16.0	33
12	4.5	7.5	38
10	4.0	6.0	40
27	9.0	18.0	33

Table 8.6—3
PERCENTAGE DECREASE IN VOLUME OF PRESERVED PLANKTON SAMPLES FROM ORIGINAL LIVE VOLUME, GIVEN FOR SELECTED TIME INTERVALS

		% decrease from original live plankton volume					
Original volume (ml)	Final volume (ml)	Imm. after pres.	1 day after pres.	10 days after pres.	1 month after pres.	1 year after pres.	2 years after pres.
13	11	92.3	92.3	84.6	84.6	84.6	84.6
12	10	83.3	83.3	83.3	83.3	83.3	83.3
(44)	35	(87)	(87)	(85)	(80)	(80)	(80)
50	32	84.0	80.0	70.0	68.0	64.0	64.0
48	24	70.8	52.1	52.1	52.1	50.0	50.0
25	12	76.0	60.0	52.0	52.0	48.0	48.0
84	32	75.0	45.2	41.6	39.3	38.1	38.1
57	20	82.4	52.6	42.1	42.1	35.1	35.1
77	27	62.4	49.4	40.3	39.0	36.4	35.1
42	10	61.9	38.1	33.3	30.1	23.8	23.8
119	23	47.3	31.8	20.2	20.2	17.8	17.8
93	12	64.4	39.7	22.6	18.3	14.0	12.9

8.7. SAMPLING EQUIPMENT AND COMPARATIVE EFFICIENCIES

Table 8.7—1
SOME CONTEMPORARY PLANKTON SAMPLING NETS AND THEIR CHARACTERISTICS

	Mouth area, m^3	Form	Mesh width, mm	Porosity	Open area ratio
Low-speed Nets (< 3 Knots)					
Coarse gauze (> 0.4 mm)					
Bongo net	0.38	Cone	0.51	0.51	6.8
FAO-Larval tuna	0.79	Cyl-cone[1]	0.51	0.51	4.8
WP-3 (Interim)	1.00	Cyl-cone	1.00	0.58	3.7
CalCOFI Standard	0.79	Cyl-cone	0.55	0.36	3.2
Tropical Juday–large	1.00	Red-cone[2]	0.45	0.40	3.1
Medium gauze (0.2–0.4 mm)					
CalCOFI Anchovy Egg	0.20	Cyl-cone	0.33	0.46	7.8
Australian Clarke-Bumpus	0.012	Cone	0.27	0.44	5.3
Indian Ocean Standard	1.00	Cyl-cone	0.33	0.46	4.3
NORPAC net	0.16	Cone	0.35	0.46	3.7
ICITA		Cone	0.28	0.42	3.1
Marunaka	0.28	Cone	0.33	0.45	2.4
Hensen Egg	0.42	Red-cone	0.30	0.44	2.1
Marutoku A	0.16	Cone	0.33	0.45	1.7
Fine gauze (< 0.2 mm)					
WP-2	0.25	Cyl-cone	0.20	0.45	6.0
Tropical Juday-Reg	0.50	Red-cone	0.17	0.32	4.2
Kitahara	0.05	Red-cone	0.11	0.32	4.2
Flowmeter	0.38	Cyl-cone	0.17	0.32	3.2
Bé net MPS	0.25	Pyramid	0.20	0.45	2.7
Marutoku A	0.16	Cone	0.11	0.32	1.2

[1] Cylinder-cone.
[2] Reduction cone-cone.

Table 8.7—1 (continued)
SOME CONTEMPORARY PLANKTON SAMPLING NETS AND THEIR CHARACTERISTICS

	Mouth area, m^3	Form	Mesh width, mm	Porosity	Open area ratio
Mixed gauzes					
International Standard	0.20	Cone	0.23	0.36	
			0.08	0.20	2.6
			10.00		
N70	0.37	Cone	0.37	0.34	2.4
			0.17	0.32	
High-speed Nets (> 3 Knots)					
Coarse gauze (> 0.4 mm)					
Miller high speed I	0.0081	Red-cone[2]	0.95	0.57	28.0
Jet net	0.0110	Red-cone-Enc[3]	0.44	0.48	27.0
Miller high speed II	0.0081	Red-cone	0.53	0.52	26.0
Medium gauze (0.2–0.4 mm)					
Miller high speed II	0.0081	Red-cone	0.26	0.44	22.0
Gulf III modified	0.0320	Red-cone-Enc	0.38	0.44	13.1
Hardy recorder	0.00016	Red-disc-Enc	0.22	0.37	11.8
Isaacs high speed	0.0005	Red-cyl-Enc	0.24	0.30	11.2
Catcher	0.0410	Red-cone-Enc	0.46	0.55	11.0
Hardy sampler	0.00029	Red-cone-Enc	0.22	0.37	8.1
Gulf III	0.1300	Red-cone-Enc	0.38	0.44	3.2
Small Hardy indicator	0.00013	Red-disc-Enc	0.22	0.37	2.3
Standard Hardy indicator	0.0011	Red-disc-Enc	0.22	0.37	1.4
Fine gauze (< 0.2 mm)					
Jashnov high speed	0.25	Pyramid	0.17	0.32	1.9

[3] Encased gauze section.

Table 8.7—2
INITIAL FILTRATION PERFORMANCE OF SOME CONTEMPORARY SAMPLERS

Plankton sampler	Towing velocity: 100 cm/sec					Towing velocity which will give an approach velocity of: (cm/sec)			
	Filtration efficiency (F)	Mesh velocity (v′) (cm/sec)	Approach velocity (v) (cm/sec)	Dynamic pressure at v cm/sec ($\frac{1}{2}qv^2$) (g/cm²)	Filtration pressure (K, $\frac{1}{2}qv^2$) (g/cm²)	v = 5	v = 10	v = 15	v = 20
Bongo	0.96	14	7	0.027	0.097	69	138	207	276
FAO larval tuna	0.97	20	10	0.054	0.176	49	97	146	194
WP-3 interim	0.98	26	15	0.120	0.239	33	65	98	130
CalCOFI standard	0.91	29	10	0.054	0.383	48	97	146	194
CalCOFI anchovy egg	0.95	12	6	0.016	0.089	90	179	269	358
Australian Clarke-Bumpus	0.88	17	7	0.027	0.158	52	104	156	208
Indian Ocean standard	0.96	22	10	0.053	0.244	49	98	147	196
NORPAC	0.96	26	12	0.072	0.309	42	84	126	168
ICITA	0.95	31	13	0.084	0.456	39	78	117	156
Marunaka	0.88	37	16	0.138	0.553	30	61	91	121
Marutoku A (medium)	0.80	47	21	0.228	0.817	24	47	71	94
WP-2	0.94	16	7	0.025	0.161	71	142	213	284
Flowmeter	0.90	28	9	0.041	0.476	56	112	167	223
Bé multiple	0.88	33	15	0.110	0.534	34	68	102	136
Marutoku A (fine)	0.59	49	16	0.125	1.237	32	64	96	128
Jashnov	0.78	41	13	0.089	0.871	38	76	113	151

Table 8.7—3
BOLTING CLOTH NUMBERS AND THEIR RESPECTIVE MESH APERTURES

No. of cloth	Mesh opening in microns	Mesh opening in in.	Open area %	No. of cloth	Mesh opening in microns	Mesh opening in in.	Open area %
1800	1800	.0709	61%	280	280	.0110	45%
1558	1558	.0614	60%	275	275	.0108	45%
1340	1340	.0528	59%	263	263	.0104	45%
1179	1179	.0464	59%	253	253	.0100	44%
1050	1050	.0413	59%	243	243	.0096	45%
947	947	.0373	58%	233	233	.0092	44%
860	860	.0339	57%	223	223	.0088	43%
782	782	.0308	56%	210	210	.0083	43%
760	760	.0299	56%	202	202	.0080	47%
706	706	.0278	53%	183	183	.0072	46%
656	656	.0258	53%	165	165	.0065	45%
630	630	.0248	52%	153	153	.0060	45%
602	602	.0237	51%	130	130	.0051	43%
571	571	.0225	50%	116	116	.0046	43%
526	526	.0207	50%	110	110	.0043	42%
505	505	.0199	50%	102	102	.0040	39%
471	471	.0185	50%	93	93	.0037	38%
452	452	.0178	48%	86	86	.0034	36%
423	423	.0167	47%	80	80	.0031	35%
405	405	.0159	47%	73	73	.0029	34%
390	390	.0154	47%	64	64	.0025	30%
363	363	.0143	47%	53	53	.0021	21%
351	351	.0138	46%	44	44	.0017	17%
333	333	.0131	46%	35	35	.0014	16%
316	316	.0124	46%	28	28	.0011	18%
308	308	.0121	45%	25	25	.0010	17%
295	295	.0116	45%	20	20	.0008	16%

Table 8.7—4
AVERAGE APERTURE SIZE OF STANDARD GRADE DUFOUR BOLTING SILK

Silk no.	Meshes/ in.	Size of aperture (mm)	Silk no.	Meshes/ in.	Size of aperture (mm)
0000	18	1.364	10	109	0.158
000	23	1.024	11	116	0.145
00	29	0.752	12	125	0.119
0	38	0.569	13	129	0.112
1	48	0.417	14	139	0.099
2	54	0.366	15	150	0.094
3	58	0.333	16	157	0.086
4	62	0.318	17	163	0.081
5	66	0.282	18	166	0.079
6	74	0.239	19	169	0.077
7	82	0.224	20	173	0.076
8	86	0.203	21	178	0.069
9	97	0.168	25	200	0.064

(From Sverdrup, H. V., Johnson, M. W., and Fleming, R. H., Observations and collections at sea, *The Oceans, Their Physics, Chemistry, and General Biology*, © 1942, renewed 1970, 378. By permission of Prentice-Hall, Inc., Englewood Cliffs, New Jersey.)

Table 8.7—5
SOME CHARACTERISTICS OF 200 μ BOLTING MATERIAL USED IN PLANKTON NETS

Gauze	Mesh openings of sample (μ)	Sifting surface (%)	Diameter of filaments (μ)		Whether tested
			Warp	Woof	
Nytal DIN 1171 30–200 μ	190–220	34.5	140	140	No
HD 200 μ	180–200	41.5	110	110	No
7xxx 200 μ	200–215	43	110	100	Yes
7 200 μ	190–215	47.5	80 + 2 × 60	80	Yes
200 μ	200–215	47.5	90	90	No
7P 200 μ	190–210	55	70	70	Yes
Estal Mono P.E. 200 μ	205–215	44.5	108	108	Yes
Monodur 200	200–250	40	127	127	Yes
Nitex 202	195–210	45	88 + 2 × 64	88	Yes

Table 8.7—6
THE COMBINED EFFECTS OF MESH SIZE AND POROSITY ON FILTRATION EFFICIENCY

Mesh width (mm)	Porosity	Open area ratio	Reduction in efficiency %
1.17	0.60	5.3	0
0.66	0.54	4.8	0
0.33	0.48	4.2	0
0.27	0.44	3.9	3
0.14	0.36	3.2	4
0.08	0.31	2.7	8
0.06	0.26	2.3	12

Table 8.7—7
DECREASE IN MESH WIDTH OF SILK ON IMMERSION IN WATER

Material		Mean mesh width, mm	Standard error
	Dry	0.312	± 0.0034
New silk	Wet	0.261	± 0.0029
	Dry	0.236	± 0.0040
Used silk	Wet	0.221	± 0.0073

Table 8.7—8
CHEMICAL RESISTANCE OF SOME MATERIALS USED IN PLANKTON NET GAUZE

Chemical	Nylon	Perlon	Polyester	Silk
Concentrated acids	Medium	Medium	Good	–
Dilute acids	Good	Good	Good	Good
Alkalis	Good	Good	Medium	Poor
Alcohols	Good	Good	–	–
Oxidizing agents	Medium	Medium	Good	–
Bleaching agents	Medium	Medium	Good	–
Sunlight	Medium	Medium	Good	Medium
Organic solvents	Good	Good	Good	Good
Soap	Good	Good	Good	Good
Petrol	Good	Good	–	–
Formic acid	Poor	Poor	–	–
Phenol	Poor	Poor	–	–

Section 9
Compounds from Marine Organisms

Section 9

COMPOUNDS FROM MARINE ORGANISMS

Compounds derived from marine flora and fauna are chronicled in Section 9. To facilitate rapid referral to the 901 compounds incorporated into this section, two tables have been developed, with the compounds tabulated in order of increasing molecular complexity. The formula, structure, name, source, activity, original references, and reported availability of each compound by synthesis or from nonmarine sources are specified. This compilation of chemical compounds has three main aims: (1) to draw attention to the marine environment as a source of novel organic substances, in many cases displaying structural types differing from those available from nonmarine sources; (2) to present an accurate record of organic compounds derived from marine organisms; and (3) to indicate where biological, microbiological, or pharmacological activity has been reported.[1,2]

REFERENCES

1. **Baker, J. T. and Murphy, V.,** *Handbook of Marine Science, Compounds from Marine Organisms*, Vol. 1, CRC Press, Cleveland, Ohio, 1976.
2. **Baker, J. T. and Murphy, V.,** *Handbook of Marine Science, Compounds from Marine Organisms*, Vol. 2, CRC Press, Boca Raton, Fla., 1981.

9.1. COMPOUNDS FROM MARINE ORGANISMS

Table 1

Molecular formula	Structural formula	Name	Source of compound	Available by synthesis	Available non-marine origin	Activity	References
C_1							
1. CH_5N	$MeNH_2$	Methylamine	Algae	+	+		284 IIb 375
2. CH_5N_3	$HN{=}C(NH_2)NH_2$	Guanidine	Octopus: *Octopus vulgaris* Sea anemone: *Alcyonium digitatum* Sponges: *Geodia gigas* *Hippospongia equina*	+	+		16 17 19 284 Ik
C_2							
3. $C_2H_5NO_2$	$HOOC\diagdown\diagup NH_2$	Glycine	Sponges Zoanthids	+	+	Gastric antacid	56, 88, 159, 284, Ij, 414
4. C_2H_7N	Me_2NH	Dimethylamine	Algae Soft coral: *Alcyonium digitatum* Sponge: *Hippospongia equina*	+	+		17 19 284 z 375
5. C_2H_7N	$\diagdown\diagup NH_2$	Ethylamine	Algae Sponge: *Hippospongia* *equina*	+	+		19 284 Ie 375
6. C_2H_7NO	$HO\diagup\diagdown\diagup NH_2$	Ethanolamine	Sponges	+	+		56 284 Id

Table 1 (continued)

Molecular formula	Structural formula	Name	Source of compound	Available by synthesis	Available non-marine origin	Activity	References
C_2 ...							
7. $C_2H_7NO_2S$	HO_2S NH_2	Hypotaurine	Sea anemones Sponges				56 87
8. $C_2H_7NO_3S$	HO_3S NH_2	Taurine	Algae Gorgonians Jellyfish Mollusks Sea anemones Sea pansy Sponges	+	+		16 20 22 56 87 264 284 IIx 336
9. $C_2H_7N_3$	HN= NH_2 NMe H	Methylguanidine	Soft coral: *Alcyonium digitatum*	+	+		17 284 IIc
10. $C_2H_8NO_3P$	H_2O_3P NH_2	2-Aminoethylphosphonic acid	Sea anemones: *Anthopleura elegantissima* *Metridium dianthus* *Tealia felina*	+			87 245 326
C_3							
11. $C_3H_7NO_2$	HOOC NH_2	α-Alanine	Sea worm: *Arenicola marina* Sponges	+	+		11 56 284 d
12. $C_3H_7NO_2$	HOOC NH_2	β-Alanine	Sponges	+	+		56 284 d

Table 1 (continued)

Molecular formula	Structural formula	Name	Source of compound	Available by synthesis	Available non-marine origin	Activity	References
C_3 . . .							
13. $C_3H_7NO_2$	$HOOC-CH_2-NHMe$	Sarcosine	Sea urchins Sponges Starfish	+	+		56 284 IIq
14. $C_3H_7NO_2S$	$HOOC-CH(NH_2)-CH_2-SH$	Cysteine	Sponges	+	+		56 284 v
15. $C_3H_7NO_3$	$HOOC-CH(NH_2)-CH_2-OH$	Serine	Sea worm: *Arenicola marina* Sponges	+	+		11 56 284 IIr
16. $C_3H_7NO_5S$	$HOOC-CH(NH_2)-CH_2-SO_3H$	Cysteic acid	Sponges	+	+		56 284 v
17. $C_3H_7N_3O_2$	$HN=C(NH_2)-NH-CH_2-COOH$	Glycocyamine	Gorgonians Sea anemone: *Anthopleura japonica* Sea worms Sponges: *Hippospongia equina* *Hymeniacidon caruncula*	+	+		19 84 274 284 Ij 327 336

Table 1 (continued)

Molecular formula	Structural formula	Name	Source of compound	Available by synthesis	Available non-marine origin	Activity	References
C_3 . . .							
18. $C_3H_8NO_5P$	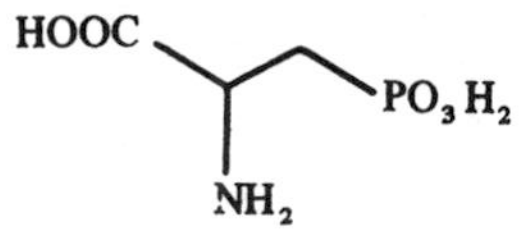 HOOC, PO_3H_2, NH_2	α-Amino-β-phosphonopropionic	Zoanthid: *Zoanthus sociatus*	+			246
19. C_3H_9N	 NH_2 NMe_3	Propylamine	Algae Sponge: *Hippospongia equina*	+			19 284 IIn 375
20. C_3H_9N	NMe_3	Trimethylamine	Algae Crab: *Limulus polyphemus* Jellyfish Sponge: *Hippospongia equina*	+	+		14 19 87 284 IIIc 375
21. $C_3H_9NO_3S$	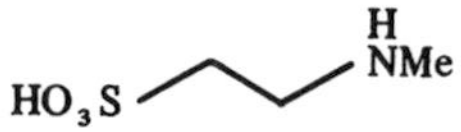 HO_3S, H NMe	N-Methyltaurine	Red algae Sponges: *Calyx nereis* *C. nicacensis*	+			20 22 264 284 IId
22. $C_3H_9N_3O_3S$	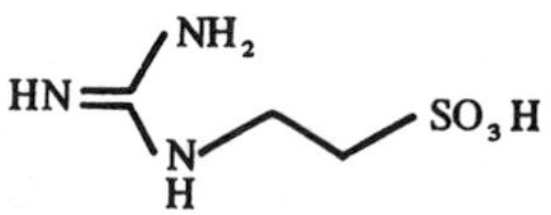 HN=, NH_2, N H, SO_3H	Taurocyamine	Sea anemones: *Actinia equina* *Anthopleura japonica* Sea worm: *Arenicola marina* Sponges				11 20 56 274 327

Table 1 (continued)

Molecular formula	Structural formula	Name	Source of compound	Available by synthesis	Available non-marine origin	Activity	References
C_3 ...							
23. $C_3H_{10}NO_3P$	 H_2O_3P … $\overset{H}{N}Me$	2-N-Methylamino-ethylphosphonic acid	Sea anemone: *Anthopleura xanthogrammica*	+			247
C_4							
24. $C_4H_4N_2$	 N, N	Pyrimidine	Sponge: *Cryptotethya crypta*	+			98 284 IIp
25. $C_4H_4N_2O_2$	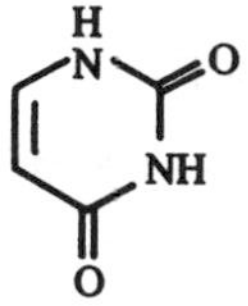 H, N, O, NH, O	Uracil	Sponge: *Cryptotethya crypta*	+	+		98 284 IIIf
26. $C_4H_7NO_4$	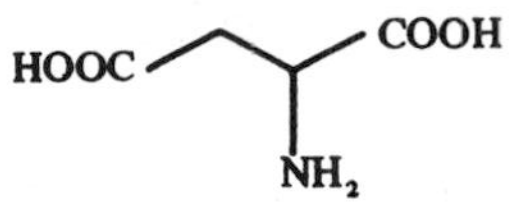 HOOC, COOH, NH_2	Aspartic acid	Sponges	+	+		56 284 f 336
27. $C_4H_8N_2O_3$	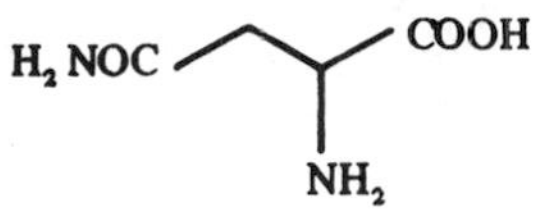 H_2NOC, COOH, NH_2	Asparagine	Sponges	+	+		56 284 f
28. $C_4H_9NO_2$	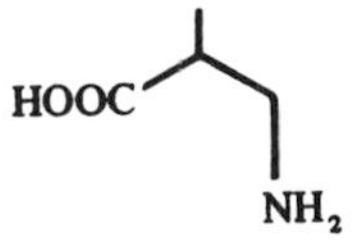 HOOC, NH_2	β-Aminoiso-butyric acid	Sponges				56

Table 1 (continued)

Molecular formula	Structural formula	Name	Source of compound	Available by synthesis	Available non-marine origin	Activity	References
C_4 . . .							
29. $C_4H_9NO_3$	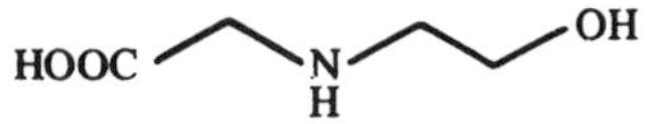	Petalonine	Alga: *Petalonia fascia*	+			385
30. $C_4H_9NO_3$	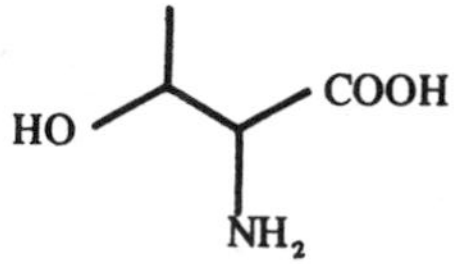	Threonine	Sea worm: *Arenicola marina* Sponges	+	+		11 56 284 IIz
31. $C_4H_9N_3O_2$	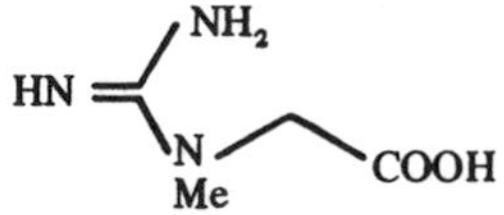	Creatine	Echinoderms Gorgonians Sea anemones Sea worms Soft coral Sponges	+	+		17 84 284 u 327 328
32. $C_4H_9N_3O_2$	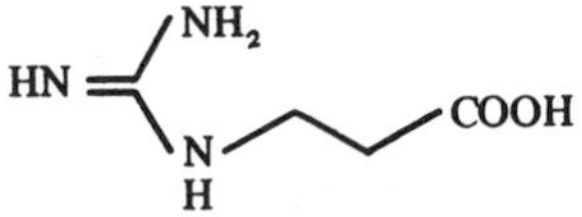	β-Guanidino-propionic acid	Sea anemone: *Anthopleura japonica*				274
33. $C_4H_{10}N_3O_5P$	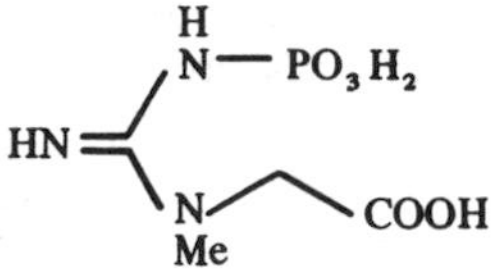	Phosphocreatine	Sponge: *Thetia lyncurium*	+	+		284 IIj 327 336

Table 1 (continued)

Molecular formula	Structural formula	Name	Source of compound	Available by synthesis	Available non-marine origin	Activity	References
C_4 ...							
34. $C_4H_{11}N$	NH_2	Isobutylamine	Algae	+	+		284 lu 375
35. $C_4H_{11}NO_3S$	HO_3S NMe_2	N,N-Dimethyl-taurine	Red algae Sponges: *Calyx nereis* *C. nicacensis*	+			20 22 263 264
36. $C_4H_{11}NS$	NH_2 MeS	2-Methylmercapto-propylamine	Algae		+		341 y 375
37. $C_4H_{12}N_2$	H_2N NH_2	Putrescine	Sponge: *Hippospongia equina*		+		19 336
38. $C_4H_{13}NO$	$[NMe_4]^+ HO^-$	Tetramine. Tetramethylammonium hydroxide	Gorgonian Jellyfish Mollusks Sea anemones			Paralytic agent	36 87 279 336 425 427
C_5							
39. $C_5H_2Br_2N_2$	Br Br N H CN	4,5-Dibromopyrrole-2-nitrile	Sponge: *Agelas oroides*				137

Table 1 (continued)

Molecular formula	Structural formula	Name	Source of compound	Available by synthesis	Available non-marine origin	Activity	References
C_5 . . .							
40. $C_5H_3Br_2NO_2$		4,5-Dibromopyrrole-2-carboxylic acid	Sponge: *Agelas oroides*				137
41. $C_5H_4Br_2N_2O$		4,5-Dibromopyrrole-2-carboxamide	Sponge: *Agelas oroides*				137
42. $C_5H_4N_4O$		Hypoxanthine	Soft coral: *Alcyonium digitatum*	+	+		17 284 Iq
43. $C_5H_4N_4O_2$		Xanthine	Soft coral: *Alcyonium digitatum* Sponge: *Calyx nicacensis*	+	+		17 20 284 IIIi
44. $C_5H_5N_5$		Adenine. Vitamin B_4	Crab: *Limulus polyphemus* Soft coral: *Alcyonium digitatum* Sponges: *Calyx nicacensis* *Geodia gigas*	+	+		14 16 17 20 284 a 336

Table 1 (continued)

Molecular formula	Structural formula	Name	Source of compound	Available by synthesis	Available non-marine origin	Activity	References
C_5 . . .							
45. $C_5H_5N_5O$	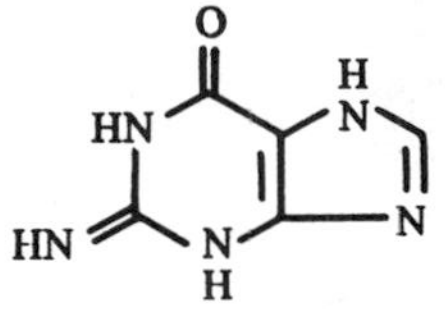	Guanine	Soft coral: *Alcyonium digitatum* Sponge: *Calyx nicacensis*	+	+		17 20 284 Ik 336
46. $C_5H_6N_2O_2$	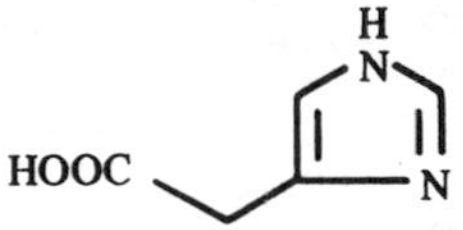	Imidazoleacetic acid	Sponge: *Hippospongia equina*		+		19
47. $C_5H_6N_2O_2$	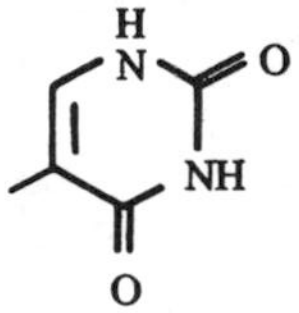	Thymine	Sponge: *Cryptotethya crypta*	+	+		98 284 IIIa
C_5							
48. $C_5H_9NO_2$	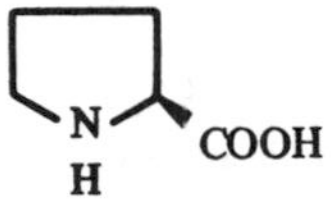	Proline	Sea worm: *Arenicola marina* Sponges	+	+		11 56 159 284 IIm 336
49. $C_5H_9NO_3$	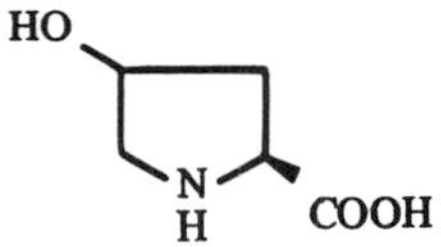	4-Hydroxyproline	Sponges	+	+		56 159 284 Ip 336

Table 1 (continued)

Molecular formula	Structural formula	Name	Source of compound	Available by synthesis	Available non-marine origin	Activity	References
C_5 . . .							
50. $C_5H_9NO_3S$	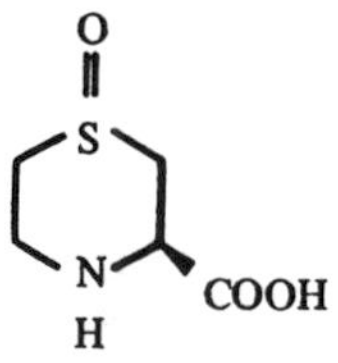	Chondrine. Yunaine	Algae: *Heterochordaria abien-tina* *Zonaria sinclairii*				272 391
51. $C_5H_9NO_4$	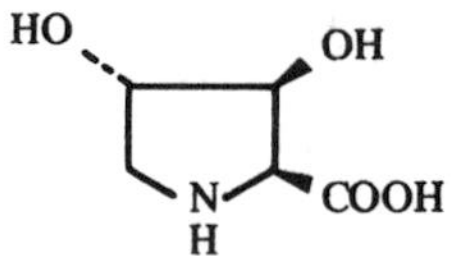	3,4-Dihydroxy-proline	Diatoms				232 304
52. $C_5H_9NO_4$	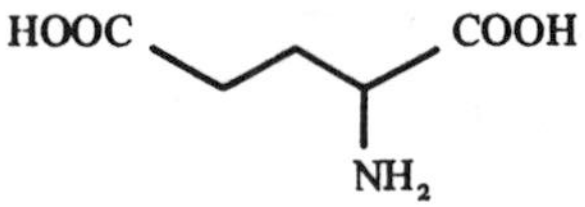	Glutamic acid	Sponges	+	+	Anti-epileptic	56 284 Ih 336
53. $C_5H_9N_3$	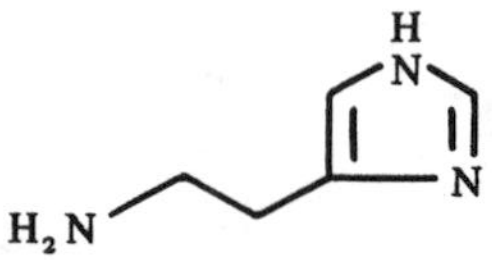	Histamine	Mollusk: *Neptunea arthritica* Sea anemones Soft coral: *Alcyonium digitatum* Sponges Zoanthid: *Palythoa* sp.	+	+		16 17 19 36 88 279 284 In 336 427

Table 1 (continued)

Molecular formula	Structural formula	Name	Source of compound	Available by synthesis	Available non-marine origin	Activity	References
C_5 . . .							
54. $C_5H_{10}N_2O_3$	H_2NOC COOH NH_2	Glutamine	Sponges	+	+		56 284 Ii
55. $C_5H_{10}O_2S$	 ^-OOC $\overset{+}{S}Me_2$	Dimethyl-β-propio-thetine	Dinoflagellate: *Gyrodinium cohnii*	+			221
56. $C_5H_{11}NO_2$	 ^-OOC $\overset{+}{N}Me_3$	Betaine	Crabs Jellyfish Sea worm Sponge	+	+		11 12 14 16 87 284 i 336
C_5							
57. $C_5H_{11}NO_2$	COOH NH_2	Valine	Sea worm: *Arenicola marina* Sponges	+	+		11 56 284 IIIh
58. $C_5H_{11}NO_2S$	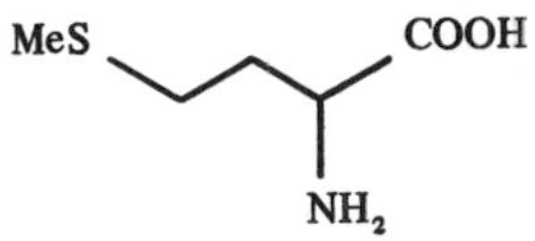 MeS COOH NH_2	Methionine	Sponges	+	+	Lipotropic agent	56 284 IIa

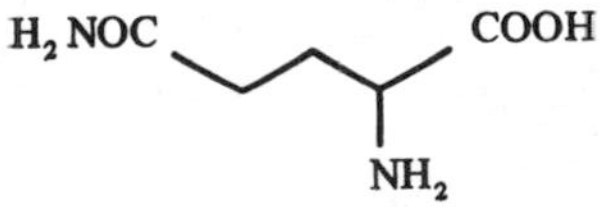

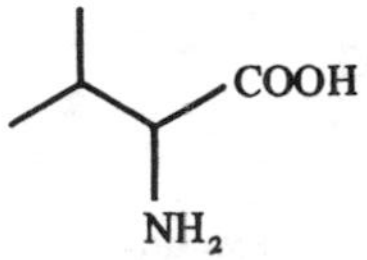

Table 1 (continued)

Molecular formula	Structural formula	Name	Source of compound	Available by synthesis	Available non-marine origin	Activity	References
C_5 . . .							
59. $C_5H_{11}NO_3S$	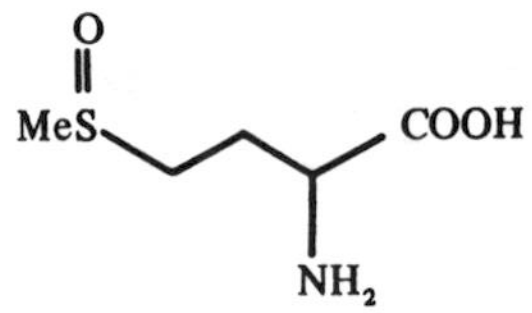 O, MeS, COOH, NH_2	Methionine sulfoxide	Sponges				56
60. $C_5H_{11}NO_4S$	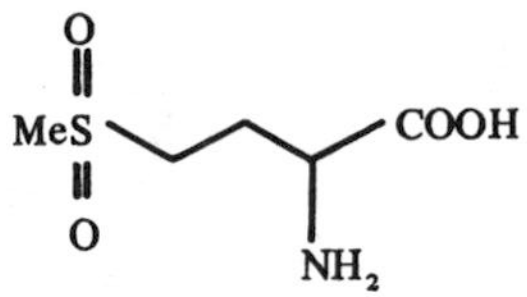 O, MeS, O, COOH, NH_2	Methionine sulfone	Sponges				56
61. $C_5H_{11}NS_2$	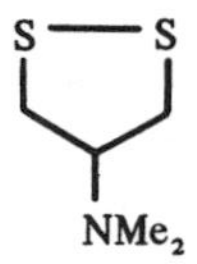 S—S, NMe_2	Nereistoxin	Sea worm: *Lumbriconereis heteropoda*	+		Insecticide. Neurotoxin	169 176 309 313
62. $C_5H_{11}N_3O_2$	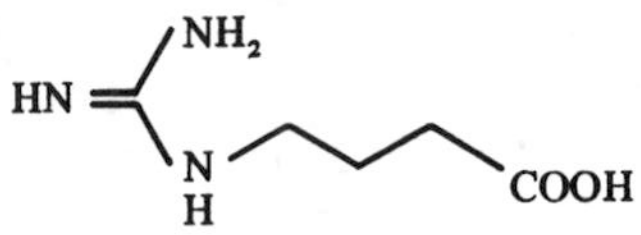 HN, NH_2, N H, COOH	γ-Guanidino-butyric acid	Sea anemone: *Anthopleura japonica*				274
63. $C_5H_{11}N_3O_3$	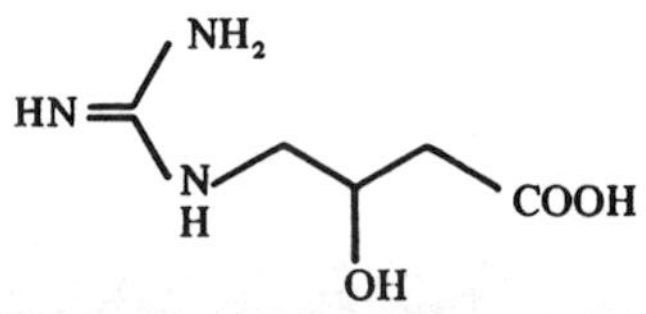 HN, NH_2, N H, COOH, OH	γ-Guanidino-β-hydroxybutyric acid	Sea anemone: *Anthopleura japonica*				274

Table 1 (continued)

Molecular formula	Structural formula	Name	Source of compound	Available by synthesis	Available non-marine origin	Activity	References
C_5 . . .							
64. $C_5H_{13}N$	NH_2	Isoamylamine	Algae Sponge: *Hippospongia equina*	+	+		19 284 It 375
65. $C_5H_{13}NO_3S$	$^{-}O_3S$ — $\overset{+}{N}Me_3$	Taurobetaine	Gorgonian: *Briareum asbestinum* Sponge: *Geodia gigas*	+			16 84 336
66. $C_5H_{13}NO_4S$	$^{-}O_3SO$ — $\overset{+}{N}Me_3$	Choline sulfate	Gorgonian: *Erythropodium caribaeorum* Red alga: *Gelidium cartilagineum*				88 264
67. $C_5H_{13}N_3O_3S$	$HN=C(NMe_2)$—NH — SO_3H	Asterubine	Starfish: *Asterias glacialis* *A. rubens*	+			2 3
68. $C_5H_{14}NO_3P$	$^{-}(HO_3P)$ — $\overset{+}{N}Me_3$	2-Trimethylamino-ethylphosphonic acid	Sea anemone: *Anthopleura xanthogrammica*	+			247

Table 1 (continued)

Molecular formula	Structural formula	Name	Source of compound	Available by synthesis	Available non-marine origin	Activity	References
C_5 . . .							
69. $C_5H_{14}N_4$		Agmatine	Crabs Mussel Octopus Scallop Sea anemone Soft coral Sponges Squid	+	+		4 16 17 19 254 274 284 c 336
70. $C_5H_{14}N_4O$		Hydroxyagmatine	Sea anemone: *Anthopleura japonica*				274
71. $C_5H_{15}NO_2$		Choline	Crab Mollusk Sea anemone Sea worm Soft coral Sponge	+	+		9 11 14 17 19 36 284 q 336
C_6							
72. $C_6H_4Br_2O$		2,6-Dibromophenol	Sea worm: *Balanoglossus biminiensis*	+			38

Table 1 (continued)

Molecular formula	Structural formula	Name	Source of compound	Available by synthesis	Available non-marine origin	Activity	References
C_6 . . .							
73. $C_6H_5N_5O_2$	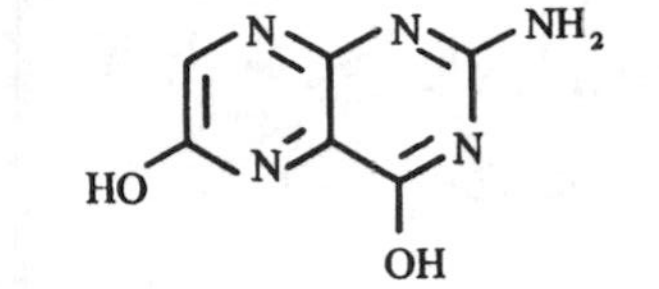	Xanthopterin	Crab: *Cancer pagurus*		+		138
74. $C_6H_6N_2O_2$	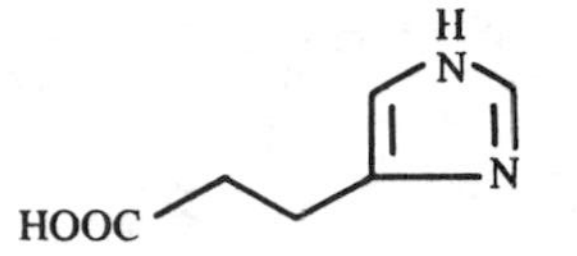	Urocanic acid. 4-Imidazolyl-acrylic acid	Sponge: *Hippospongia equina*	+	+		19 284 IIIg
75. $C_6H_7BrN_4O$	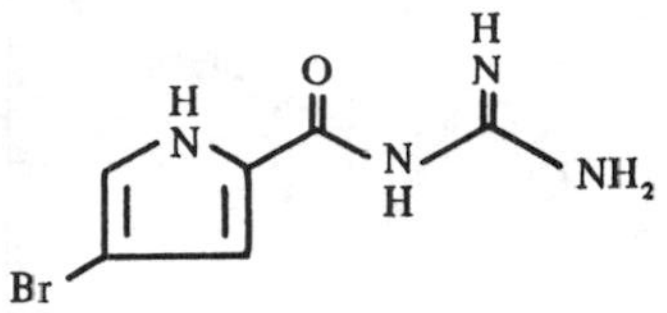	N-Amidino-4-bromo-pyrrole-2-carboxamide	Sponge: *Agelas* sp.			Antibiotic	130 376
76. $C_6H_7N_5$	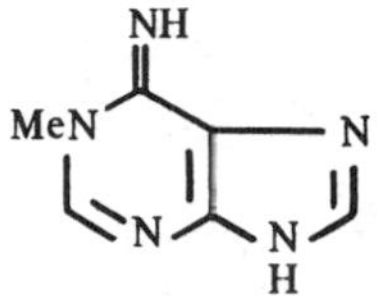	Spongopurine. 1-Methyladenine	Sponge: *Geodia gigas* Starfish: *Asterias amurensis*	+		Meiosis inducing	16 21 229 336
77. $C_6H_8N_2O_2$	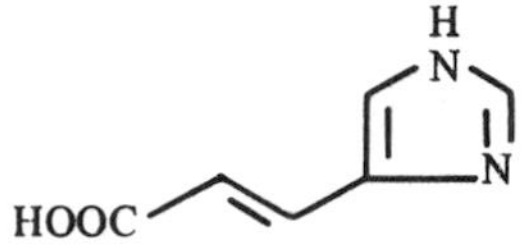	4-Imidazolyl-propionic acid	Sponge: *Hippospongia equina*		+		19

Table 1 (continued)

Molecular formula	Structural formula	Name	Source of compound	Available by synthesis	Available non-marine origin	Activity	References
C_6 . . . 78. $C_6H_8N_2O_2$	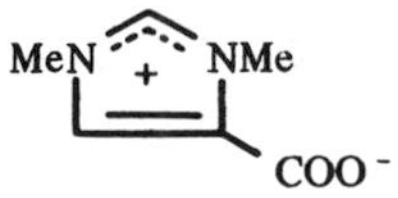	Norzooanemonin	Gorgonian: *Pseudopterogorgia americana*	+			423
79. $C_6H_8N_2O_3$	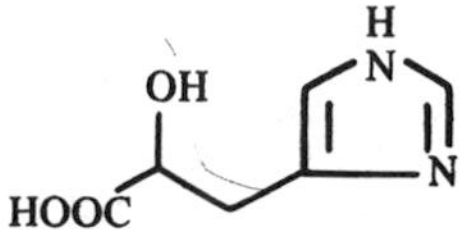	4-Imidazolyl-lactic acid	Sponge: *Hippospongia equina*				19
80. $C_6H_9NO_2$	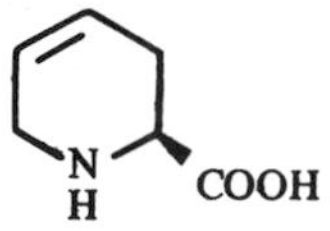	L-Baikiain	Red alga: *Corallina officinalis*		+		272
81. $C_6H_9N_3O_2$	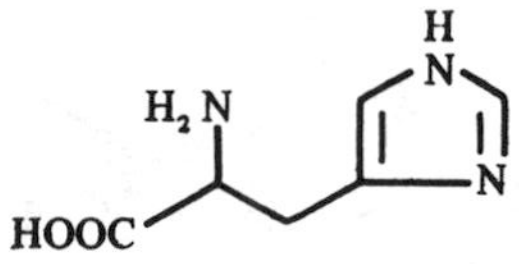	Histidine	Crab: *Limulus polyphemus* Soft coral: *Alcyonium digitatum* Sponges	+	+		14 17 19 56 284 In 336
82. $C_6H_{11}NO_2$	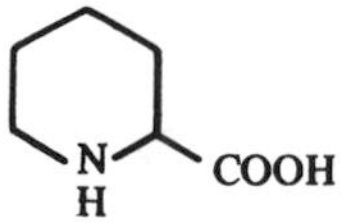	Pipecolic acid	Alga: *Petalonia fascia* Sponges	+	+		56 284 IIk 385

Table 1 (continued)

Molecular formula	Structural formula	Name	Source of compound	Available by synthesis	Available non-marine origin	Activity	References
C_6 . . .							
83. $C_6H_{12}N_4O_3$	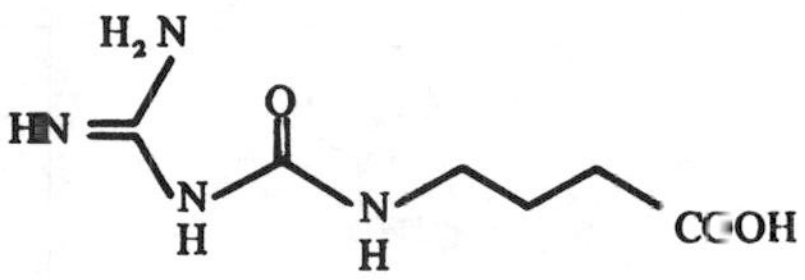	Gongrine	Red alga: *Gymnogongrus flabelliformis*	+			223 224 225
84. $C_6H_{13}NO_2$	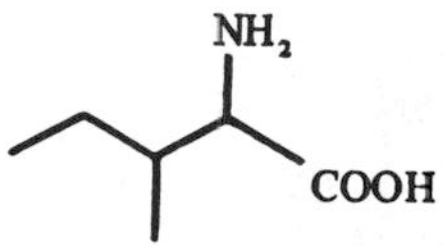	Isoleucine	Sea worm: *Arenicola marina* Sponges	+	+		11 56 284 Iv
85. $C_6H_{13}NO_2$	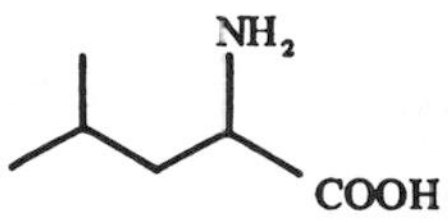	Leucine	Sea worm: *Arenicola marina* Sponges	+	+		11 56 284 Iy
86. $C_6H_{14}N_2O_2$	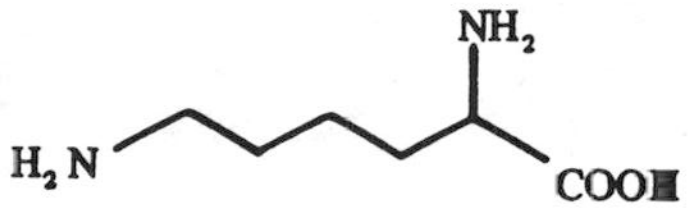	Lysine	Sea worm: *Arenicola marina* Sponges	+	+		11 19 56 284 Iz 336
87. $C_6H_{14}N_4O_2$	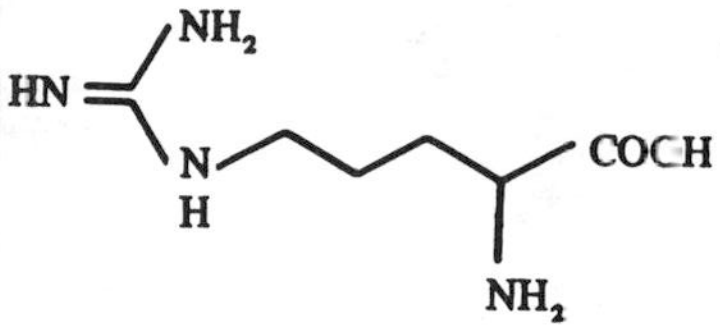	Arginine	Crustaceans Echinoderms Mollusks Sea anemones Sea worms Soft coral Sponges	+	+		4, 9, 14, 17, 56, 274, 284 e, 327, 328, 336

Table 1 (continued)

Molecular formula	Structural formula	Name	Source of compound	Available by synthesis	Available non-marine origin	Activity	References
C_6 . . .							
88. $C_6H_{14}N_4O_3$	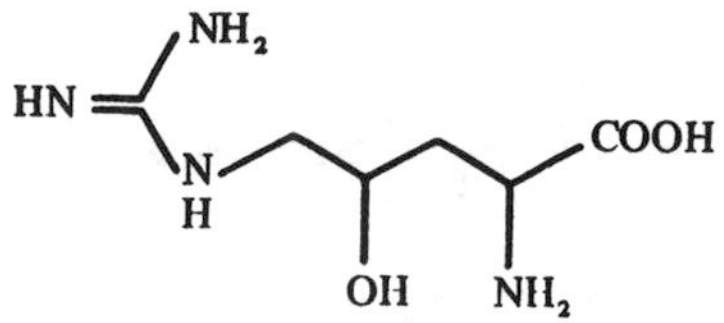	γ-Hydroxyarginine	Sea anemone: *Anthopleura japonica*				274
89. $C_6H_{15}N_4O_5P$	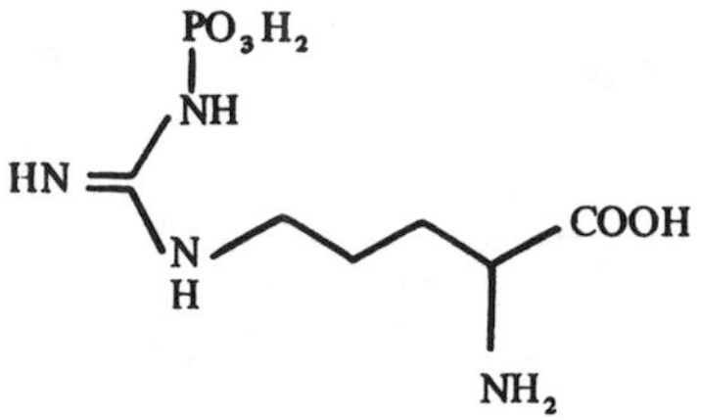	Phosphoarginine	Sponge: *Hymeniacidon caruncula*				327 336
C_7							
90. $C_7H_4Br_2K_2O_9S_2$	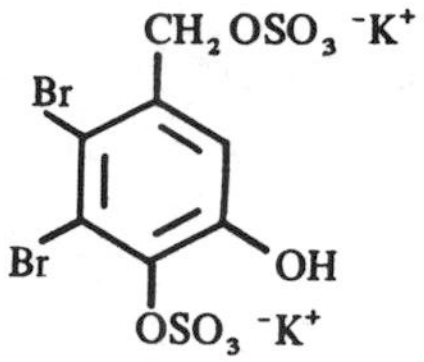	2,3-Dibromo-5-hydroxy-4-0-sulfate benzyl sulfate, dipotassium salt	Red algae: *Odonthalia corymbifera* *Polysiphonia lanosa* *P. nigrescens*	+			149 187 256 341 v
91. $C_7H_4Br_2O_3$	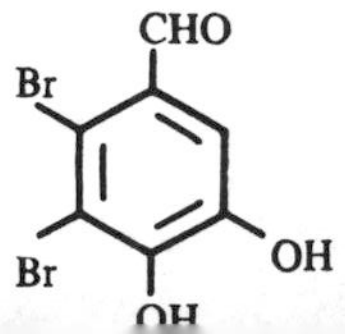	2,3-Dibromo-4,5-dihydroxybenz-aldehyde	Red algae: *Polysiphonia lanosa* *Rhodomela larix*	+			238 378

Table 1 (continued)

Molecular formula	Structural formula	Name	Source of compound	Available by synthesis	Available non-marine origin	Activity	References
C_7 . . .							
92. $C_7H_5BrO_3$	CHO, Br, OH, OH	3-Bromo-4,5-dihydroxybenzaldehyde	Red algae: *Polysiphonia lanosa* *P. morrowii*	+			337 341 u 378
93. $C_7H_6Br_2O_2$	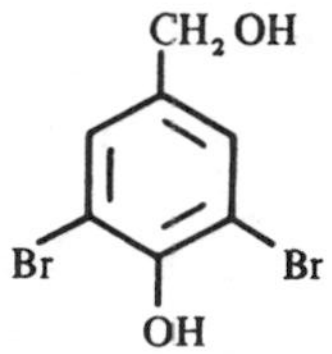 CH_2OH, Br, Br, OH	3,5-Dibromo-4-hydroxybenzyl alcohol	Red algae: *Odonthalia dentata* *Rhodomela confervoides*	+			101
94. $C_7H_6Br_2O_3$	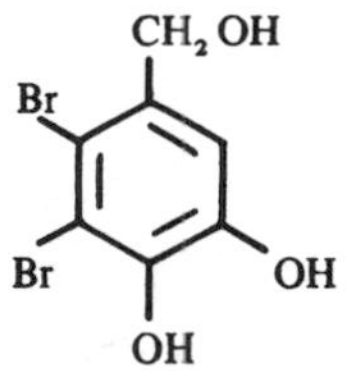 CH_2OH, Br, Br, OH, OH	Lanosol. 2,3-Dibromo-4,5-dihydroxybenzyl alcohol	Red algae: *Odonthalia corymbifera* *O. dentata* *Polysiphonia lanosa* *Rhodomela confervoides*				101 256 378
95. $C_7H_7NO_2$	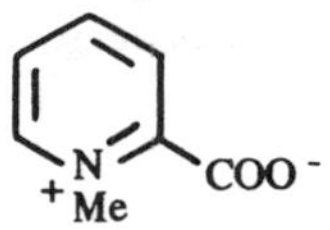 N, +Me, COO^-	Homarine	Crabs Gorgonian Jellyfish Lobster Sea anemones Sea urchin Sea worm Soft coral Sponge	+	+		9, 11, 12, 14, 17, 20, 188, 279, 284 Io, 336, 341 y, 425

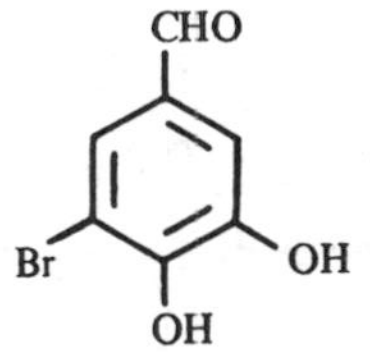

Table 1 (continued)

Molecular formula	Structural formula	Name	Source of compound	Available by synthesis	Available non-marine origin	Activity	References
C_7...							
96. $C_7H_7NO_2$		Trigonelline	Crabs Gorgonian Jellyfish Sea anemone Sea urchin Sponge	+	+		9 12 14 20 284 IIIb 336 423 425
97. $C_7H_9N_3O_2$		Spinacine	Crab: *Crangon vulgaris* Shark: *Acanthias vulgaris*	+			5 7 8 23 24
98. $C_7H_9N_5O$		Herbipoline	Sponge: *Geodia gigas*	+			13 16 18 336
99. $C_7H_{10}N_2O_2$		3,6-Dioxohexa-hydropyrrolo-(1,2-a)- pyrazine	Starfish: *Luidia clathrata*	+			316

Table 1 (continued)

Molecular formula	Structural formula	Name	Source of compound	Available by synthesis	Available non-marine origin	Activity	References
C_7 . . .							
100. $C_7H_{10}N_2O_2$	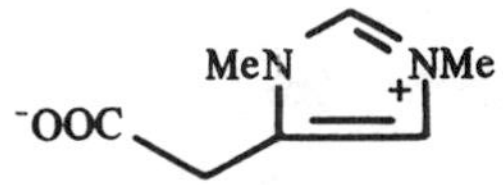	Zooanemonine	Sea anemone: *Anemonia sulcata* Sponge: *Hippospongia equina*	+			9 18 19 336
101. $C_7H_{13}N_3$	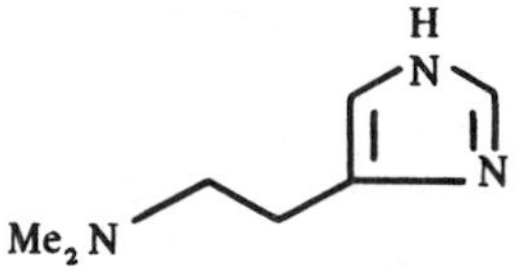	N, N-Dimethylhistamine	Sponges: *Geodia gigas* *Ianthella* sp.	+		Hypotensive	16 147 336
102. $C_7H_{15}N_5O_3$	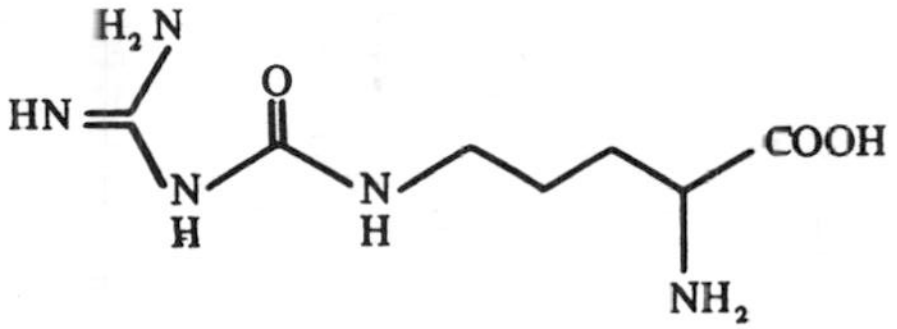	Gigartinine	Red alga: *Gymnogongrus flabelliformis*	+			223 224 225
103. $C_7H_{16}ClNO_3$	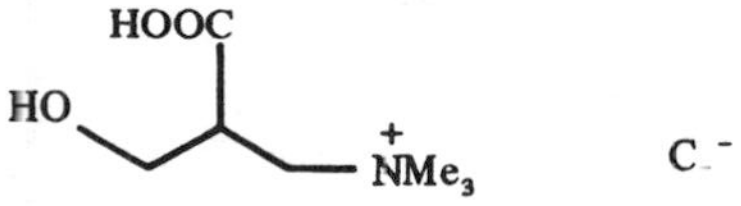	Atrinine	Mollusk: *Atrina pectinata japonica*				253
C_8							
104. $C_8H_5BrKNO_4S$	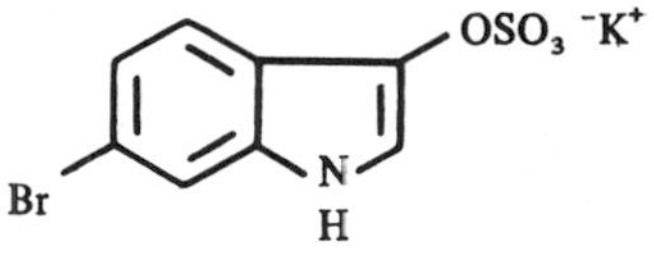	Potassium-6-bromoindoxyl sulfate	Mollusk: *Murex trunculus*				139

Table 1 (continued)

Molecular formula	Structural formula	Name	Available by synthesis	Available non-marine origin	Source of compound	Activity	References
C_8 ...							
105. $C_8H_6KNO_4S$	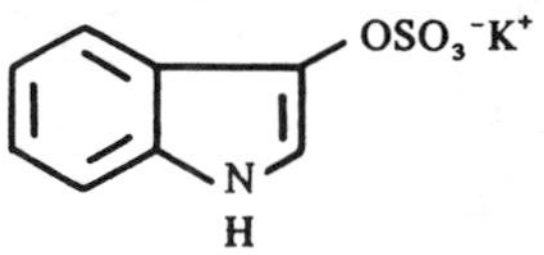	Potassium indoxyl sulfate	+	+	Mollusk: *Murex trunculus*		139 284 Ir
106. $C_8H_7Br_2NO_2$	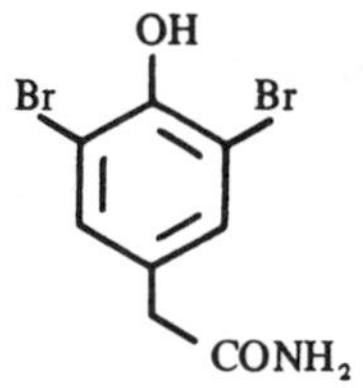	3,5-Dibromo-4-hydroxyphenyl-acetamide	+		Sponge: *Verongia archeri*	Antibiotic	377
107. $C_8H_7Br_2NO_3$	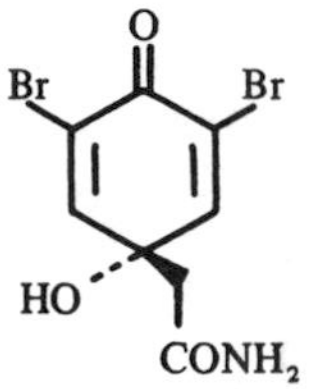 Artifact?	4-Acetamido-2,6-dibromo-4-hydroxy-cyclohexadienone	+		Sponges: *Verongia cauliformis* *V. fistularis* *V. thiona*	Antibiotic	31 349 350
108. $C_8H_8Br_2O_3$	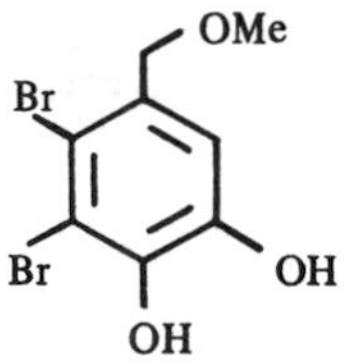	2,3-Dibromo-4,5-dihydroxybenzyl-methyl ether	+		Red algae: *Odonthalia corymbifera* *Rhodomela larix*		238 256

Table 1 (continued)

Molecular formula	Structural formula	Name	Source of compound	Available by synthesis	Available non-marine origin	Activity	References
C_8 . . .							
109. $C_8H_8O_4$	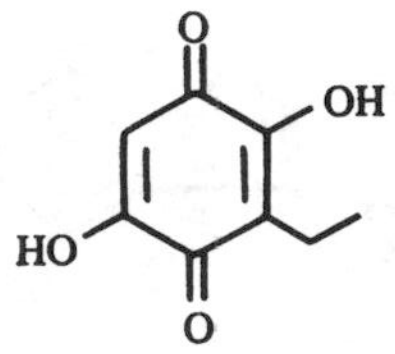	2,5-Dihydroxy-3-ethylbenzoquinone	Sea urchin: *Echinothrix diadema*	+			290 341 w
110. $C_8H_{11}N$	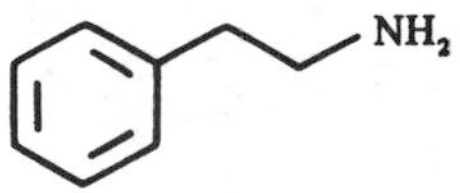	2-Phenylethyl-amine	Algae	+	+		284 IIh 375
111. $C_8H_{11}NO$	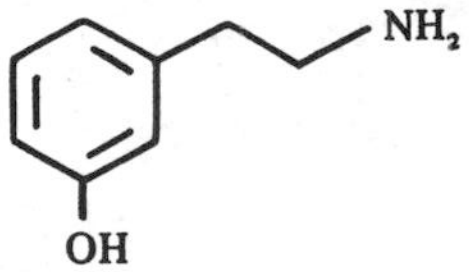	m-Tyramine	Zoanthid: *Palythoa* sp.				88
112. $C_8H_{11}NO_2$	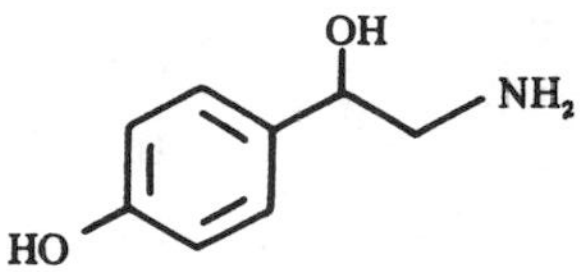	Octopamine	Octopuses: *Eledone moschata* *Octopus macropus* *O. vulgaris*	+	+	Sympathomi-metic	116 231 284 IIf
113. $C_8H_{14}ClN_3O_2$	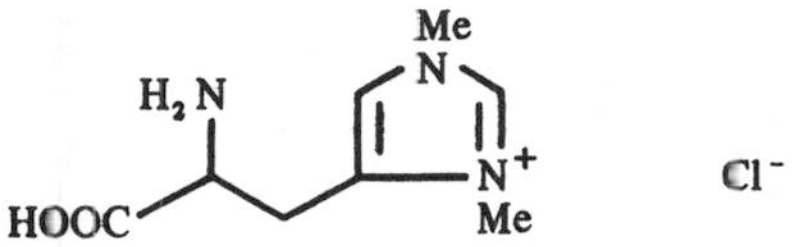	4-(β-Carboxy-β-aminoethyl)-1,3-dimethylimidazolium chloride	Red alga: *Gracilaria secundata*	+			272 273

Table 1 (continued)

Molecular formula	Structural formula	Name	Source of compound	Available by synthesis	Available non-marine origin	Activity	References
C_8 . . .							
114. $C_8H_{16}NO_2^+$		Choline acrylate	Mollusk: *Buccinum undatum*			Hypotensive. Neuromuscular blocking ability. Nicotinic action	309 429
C_9							
115. C_9H_6BrNOS		6-Bromo-2-methyl-thioindoleninone	Mollusks: *Dicathais orbita* *Mancinella bufo* *M. distinguenda* *M. keineri*				41 42
116. $C_9H_7BrKNO_6S_2$		6-Bromo-2-(methyl-sulfonyl)indol-3-yl potassium sulfate	Mollusk: *Murex trunculus*				139
117. $C_9H_7NO_2$		Indolyl-3-carbox-ylic acid	Alga: *Undaria pinnatifida*				1
118. $C_9H_8BrNO_4S_2$		Tyrindoxyl sulfate	Mollusk: *Dicathais orbita*				40

Table 1 (continued)

Molecular formula	Structural formula	Name	Source of compound	Available by synthesis	Available non-marine origin	Activity	References
C_9 . . .							
119. $C_9H_8Br_2O_4$	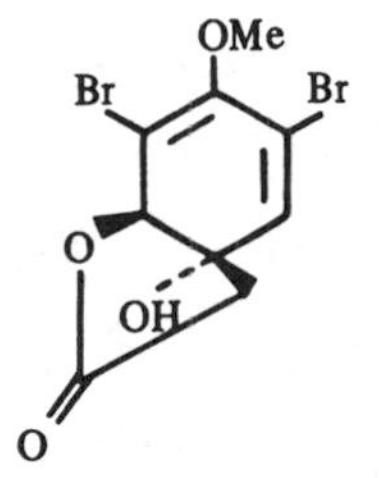	Aeroplysinin-2	Sponges: *Ianthella* sp. *Verongia aero-phoba*				287
120. $C_9H_8KNO_4S_2$	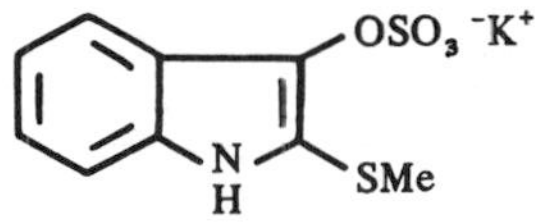	2-Methylthio-indo-xyl potassium sulfate	Mollusk: *Murex trunculus*				139
121. $C_9H_9BrClNO_3$	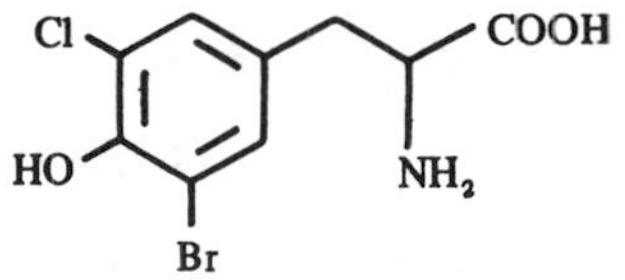	3-Bromo-5-chloro-tyrosine	Mollusk: *Buccinum undatum*				195
122. $C_9H_9Br_2NO_3$	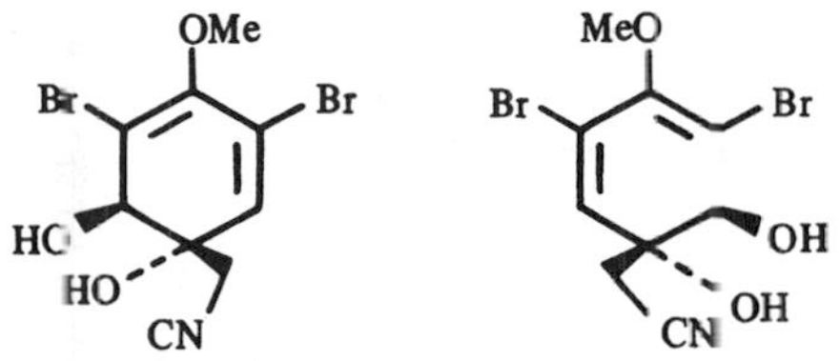 (+) (−)	(+)-Aeroplysinin-1 (−)-Aeroplysinin-1	Sponges: *Ianthella ardis* (correct name: *Pseudo-ceratina crassa*) *Verongia (Aplysina) aerophoba*	+		Antibacterial Antibacterial	100 120 125 142 377

Table 1 (continued)

Molecular formula	Structural formula	Name	Source of compound	Available by synthesis	Available non-marine origin	Activity	References
C_9 . . .							
123. $C_9H_9Br_2NO_3$		3,5-Dibromotyrosine	Gorgonian: *Primnoa lepadifera* Sponge: *Spongia officinalis obliqua* (bath sponge)	+		Antithyroid agent	6 267 284 x 297
124. $C_9H_9I_2NO_3$		3,5-Diiodotyrosine	Algae Gorgonian: *Gorgonia cavolinii* Sponges	+	+		6 109 267 284 y
125. $C_9H_{10}BrNO_3$		3-Bromotyrosine	Gorgonians Sponge: *Spongia officinalis obliqua*				267
126. $C_9H_{10}INO_3$		3-Iodotyrosine	Sponge: *Spongia officinalis obliqua*	+	+		267 284 Is
127. $C_9H_{11}NO_2$		Phenylalanine	Red alga: *Odonthalia corymbifera* Sponges	+	+		56 256 284 IIi

Table 1 (continued)

Molecular formula	Structural formula	Name	Source of compound	Available by synthesis	Available non-marine origin	Activity	References
C_9 . . .							
128. $C_9H_{11}NO_3$	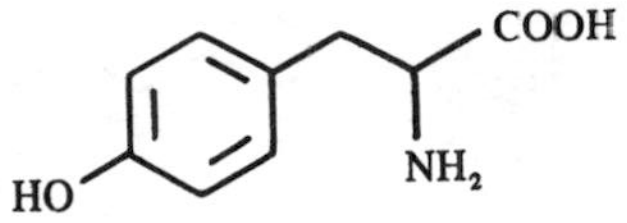 HO, COOH, NH_2	Tyrosine	Octopuses Sea worm: *Arenicola marina* Sponges	+	+		11 16 56 116 159 284 IIIe 336
129. $C_9H_{12}N_2O_6$	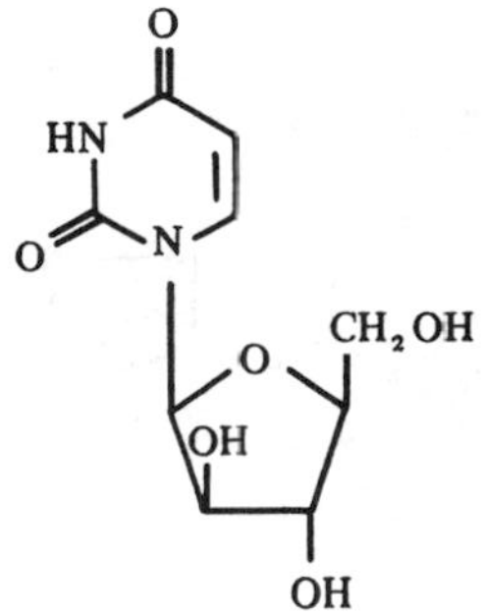 O, HN, O, N, O, CH_2OH, OH, OH	Spongouridine	Sponge: *Cryptotethya crypta*	+			54 98
130. $C_9H_{13}N_3O_5$	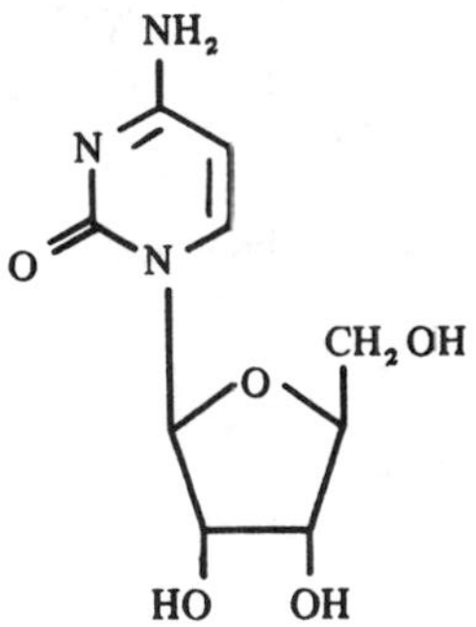 NH_2, N, O, N, O, CH_2OH, HO, OH	Cytidine	Sponge: *Cryptotethya crypta*	+	+		98 284 v

Table 1 (continued)

Molecular formula	Structural formula	Name	Source of compound	Available by synthesis	Available non-marine origin	Activity	References
C_9 . . .							
131. $C_9H_{15}N_3O_2$	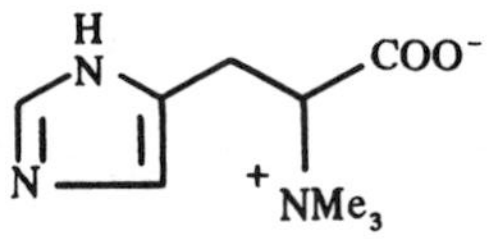 H, N, N, COO^-, $^+NMe_3$	Hercynine	Crab: *Limulus polyphemus* Sponge: *Hippospongia equina*	+	+		14 15 19 284 Im
132. $C_9H_{15}N_3O_2S$	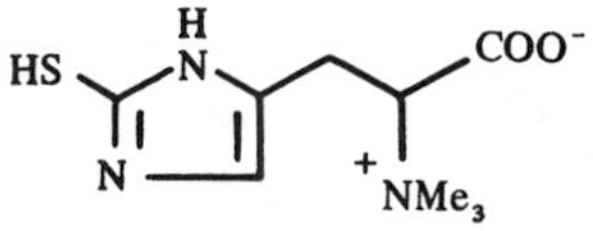 HS, H, N, N, COO^-, $^+NMe_3$	Ergothioneine	Crab: *Limulus polyphemus*	+	+		14 15 284 Ic
133. $C_9H_{18}N_4O_4$	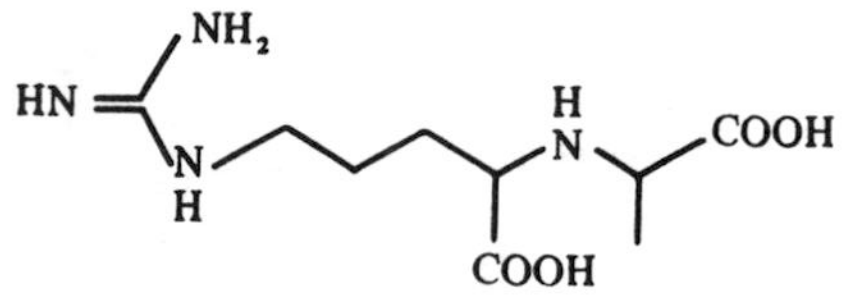 NH_2, HN, N, H, H, N, COOH, COOH	Octopin	Octopuses Scallops Squid: *Loligo pealii*	+			4 220 336
134. $C_9H_{22}N_2O_3$	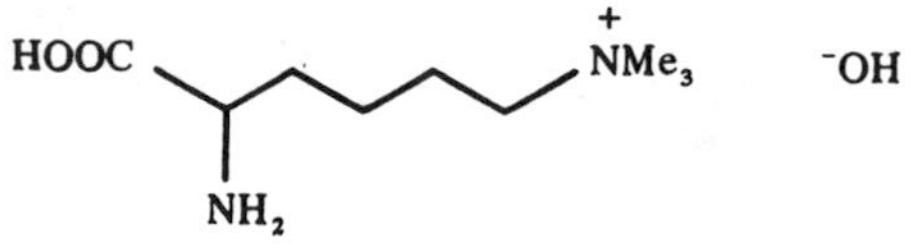 HOOC, NH_2, $^+NMe_3$, ^-OH	Laminine	Brown algae	+		Anthelmintic Antihypertensive	314 385 388 389 390

Table 1 (continued)

Molecular formula	Structural formula	Name	Source of compound	Available by synthesis	Available non-marine origin	Activity	References
C_{10}							
135. $C_{10}H_4Br_5NO$	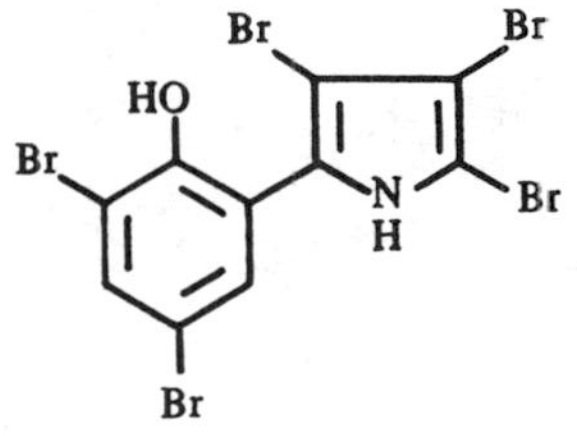	2-(2-Hydroxy-3,5-dibromophenyl)-3,4,5-tribromo-pyrrole	Bacterium: *Pseudomonas bromoutilis*	+		Antibiotic	68 175 266
136. $C_{10}H_6O_5$	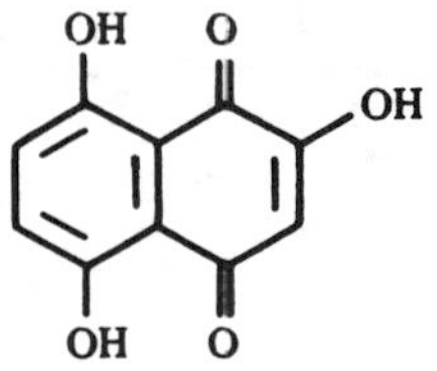	Naphthopurpurin	Sea urchins: *Echinothrix calamaris* *E. diadema*	+			290 397 h
137. $C_{10}H_6O_6$	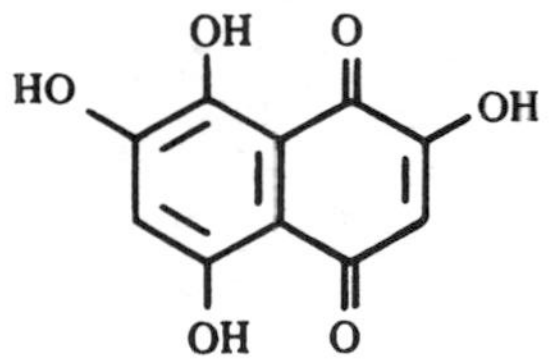	Mompain. 2,7-Dihydroxy-naphthazarin	Feather star: *Antedon* sp. Sea urchins	+			290 364 397 c
138. $C_{10}H_6O_6$	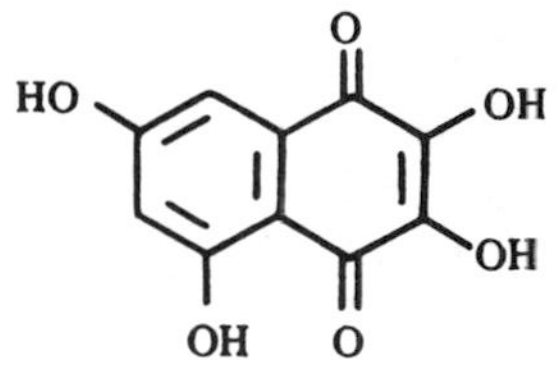	Spinochrome B	Sea urchins	+			32 156 290 397 d

Table 1 (continued)

Molecular formula	Structural formula	Name	Source of compound	Available by synthesis	Available non-marine origin	Activity	References
C_{10} . . .							
139. $C_{10}H_6O_7$	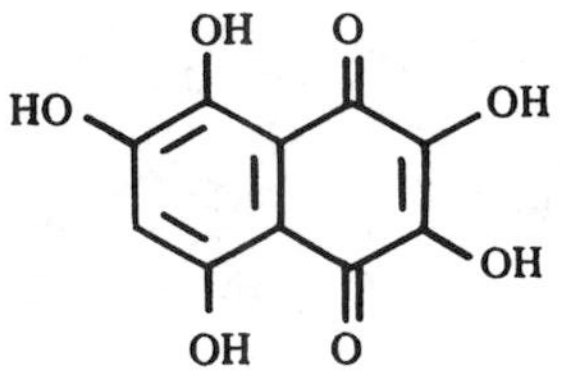	Spinochrome D	Sea urchins	+			32 290 366 397 d
140. $C_{10}H_6O_8$	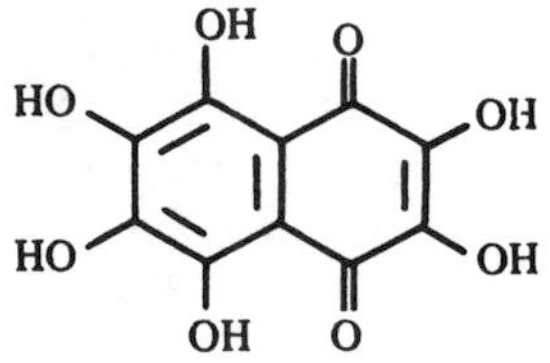	Spinochrome E	Sea urchins	+			32 341 x 366 369 370 397 e
141. $C_{10}H_7NO_4$	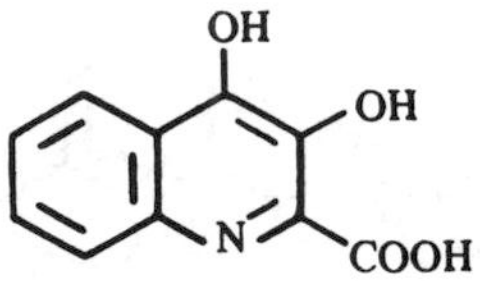	3,4-Dihydroxy-quinoline-2-carboxylic acid	Sponge: *Verongia aerophoba*	+			123
142. $C_{10}H_9NO_2$	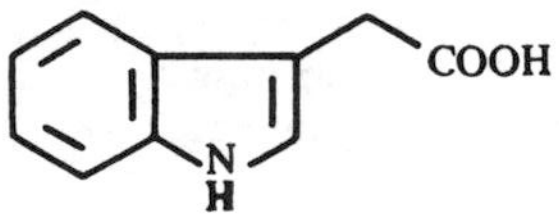	Indolyl-3-acetic acid	Alga: *Undaria pinnatifida*				1

Table 1 (continued)

Molecular formula	Structural formula	Name	Source of compound	Available by synthesis	Available non-marine origin	Activity	References
C_{10} . . .							
143. $C_{10}H_{10}BrNOS_2$	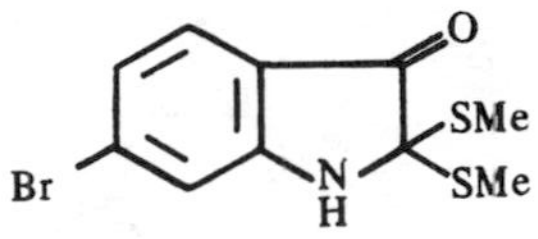	6-Bromo-2,2-dimethylthioindolin-3-one	Mollusks: *Dicathais orbita* *Mancinella bufo* *M. distinguenda* *M. keineri*				41 42
144. $C_{10}H_{10}Br_2N_2$	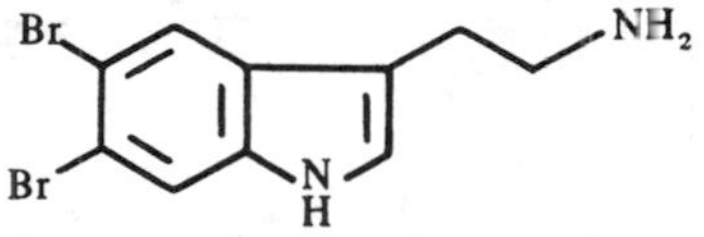	3-(2-Aminoethyl)-5,6-dibromoindole	Sponge: *Polyfibrospongia maynardii*			Antibacterial	413
145. $C_{10}H_{10}Br_2O_4$	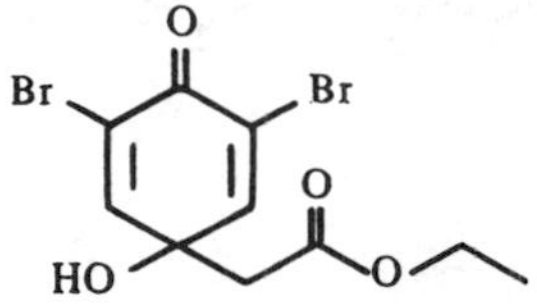	Ethyl-3,5-dibromo-1-hydroxy-4-oxo-2,5-cyclohexadiene-1-acetate	Opistobranch: *Tylodina fungina*				30
146. $C_{10}H_{12}Br_3Cl_3$	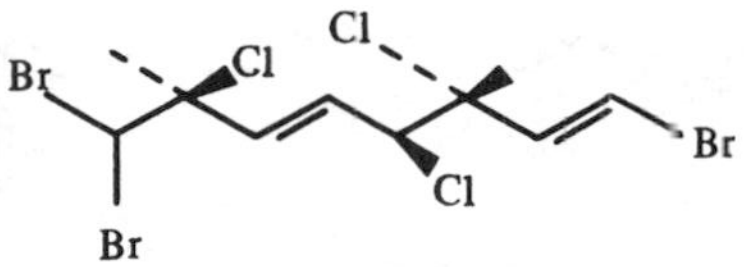	*3R,4S,7S-trans-trans*-3,7-Dimethyl-1,8,8-tribromo-3,4,7-trichloro-1,5-octadiene	Red alga: *Floccamium coccineum* Sea hare: *Aplysia californica*				127 128 373
147. $C_{10}H_{12}N_2$	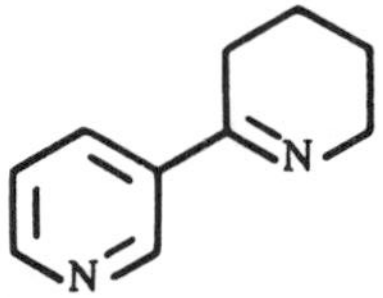	Anabaseine	Sea worm: *Paranemertes peregrina*	+			239

Table 1 (continued)

Molecular formula	Structural formula	Name	Source of compound	Available by synthesis	Available non-marine origin	Activity	References
C_{10} . . .							
148. $C_{10}H_{12}N_2O$	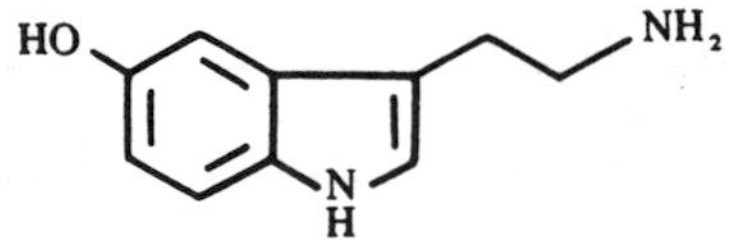	Serotonin	Mollusk: *Murex trunculus* Octopuses:- *Eledone moschata* *Octopus vulgaris* Sea anemones	+	+		116 278 279 284 IIr 336 426 427
149. $C_{10}H_{12}N_2O_4S$	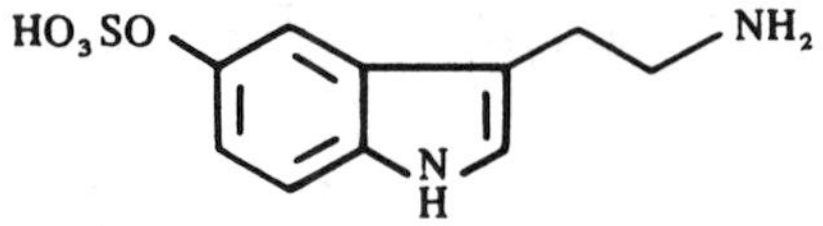	Serotonin sulfate	Zoanthids: *Palythoa caribaeorum* *P. mammilosa*				88
150. $C_{10}H_{13}Br_2NO_4$	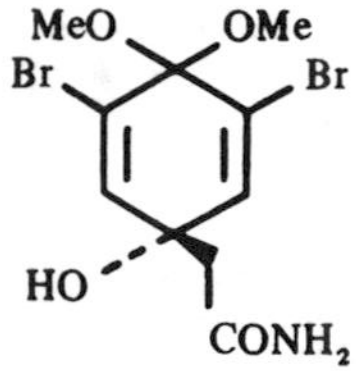 Artifact?	3-Acetamido-1,5-dibromo-6,6-dimethoxy-3-hydroxy-1,4-cyclohexadiene	Sponge: *Verongia fistularis*			Antibiotic	31 352
151. $C_{10}H_{13}N_5O_4$	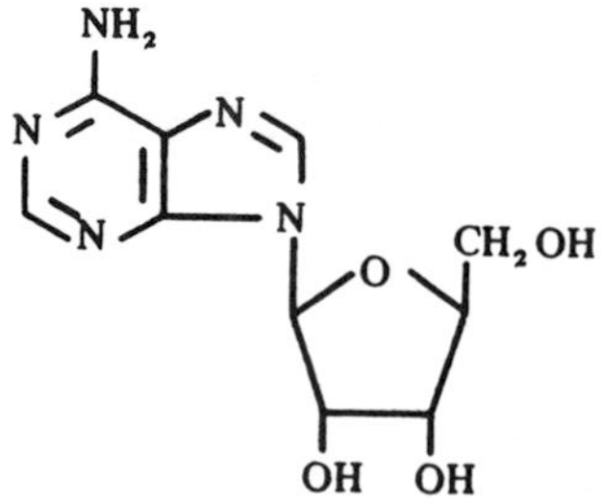	Adenosine	Sponge: *Cryptotethya crypta*	+	+		98 284 b

Table 1 (continued)

Molecular formula	Structural formula	Name	Source of compound	Available by synthesis	Available non-marine origin	Activity	References
C_{10} . . .							
152. $C_{10}H_{13}N_5O_5$	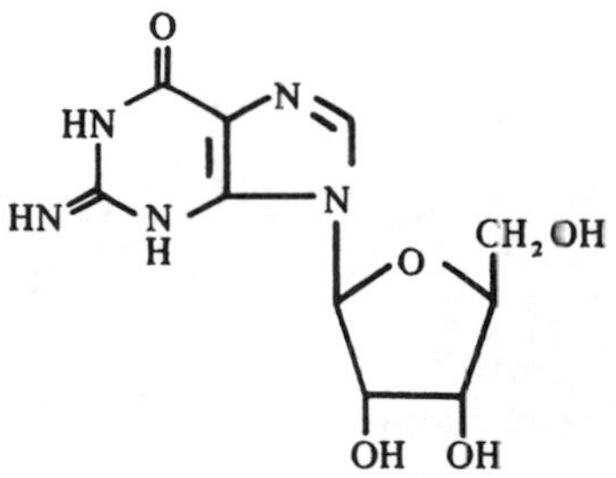	Guanosine	Sponge: *Crypto-tethya crypta*	+	+		98 284 II
153. $C_{10}H_{14}N_2O_6$	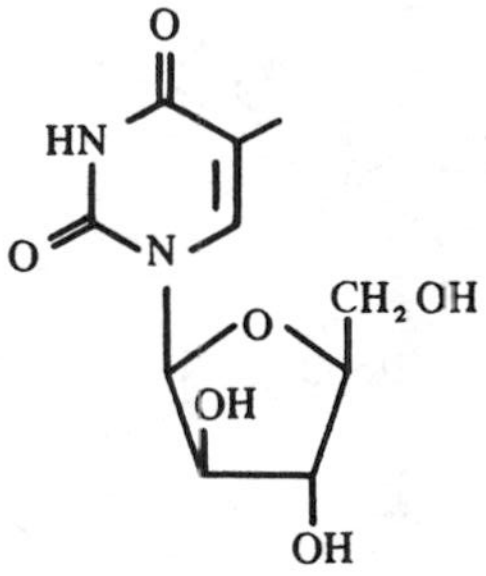	Spongothymidine	Sponge: *Crypto-tethya crypta*	+			54 98
154. $C_{10}H_{15}NO_4$	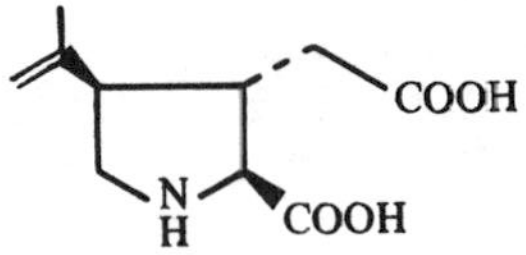	Allo-kainic acid	Red alga: *Digenea simplex*				130, 295, 296, 303, 341 z

Table 1 (continued)

Molecular formula	Structural formula	Name	Source of compound	Available by synthesis	Available non-marine origin	Activity	References
C_{10} . . .							
155. $C_{10}H_{15}NO_4$	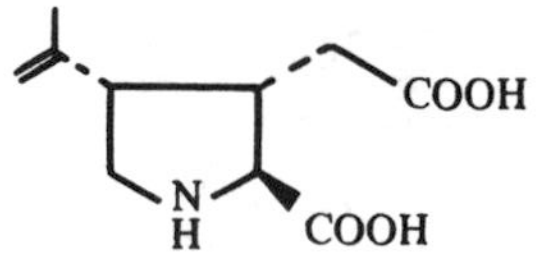	α-Kainic acid	Red alga: *Digenea simplex*	+		Anthelmintic	284 Iw, 288, 295, 296, 301, 302, 409, 410
156. $C_{10}H_{15}N_7O_3$	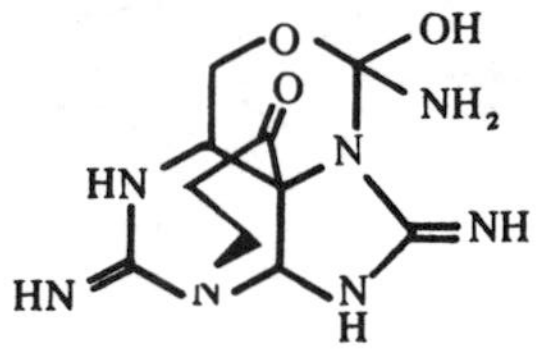 Isolated as dihydrochloride	Saxitoxin	Alaska butterclam: *Saxidomus giganteus* Dinoflagellate: *Gonyaulax catenella* Mussel: *Mytilus californianus*			Hypotensive. Paralytic poison	230 269 270 309 336 340 348 434 435
157. $C_{10}H_{16}Br_3ClO$	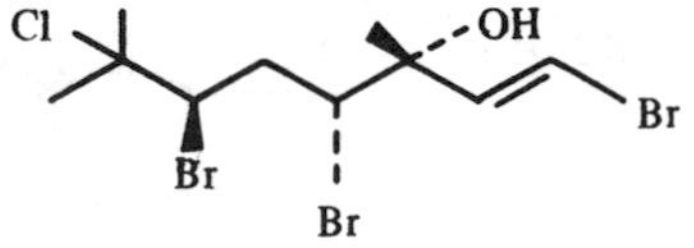	7-Chloro-3,7-dimethyl-1,4,6-tribromo-*trans*-1-octene-3-ol	Red alga: *Plocamium coccineum* Sea hare: *Aplysia californica*				128 373 432
158. $C_{10}H_{20}NO_2^+$	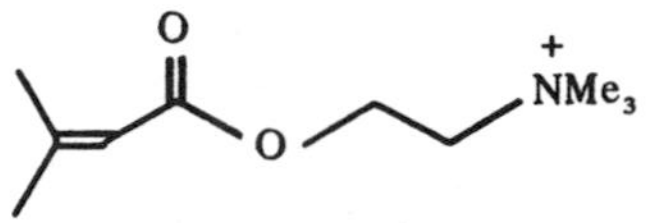	ββ-Dimethylacrylylcholine. Senecioylcholine	Mollusk: *Thais floridana*	+	+	Neuromuscular blocking agent	427 428 429

Table 1 (continued)

Molecular formula	Structural formula	Name	Source of compound	Available by synthesis	Available non-marine origin	Activity	References
C_{10} ...							
159. $C_{10}H_{26}N_4$	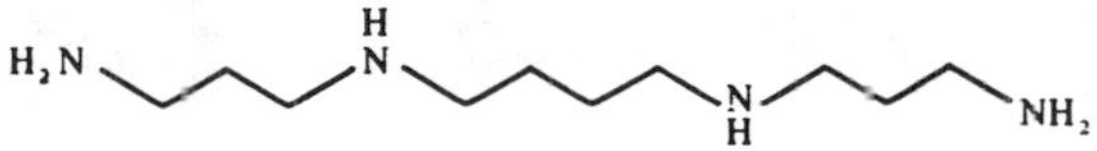	Spermine	Tunicate: *Ciona intestinalis*	+	+		10 222 284 IIt
C_{11}							
160. $C_{11}H_8O_8$	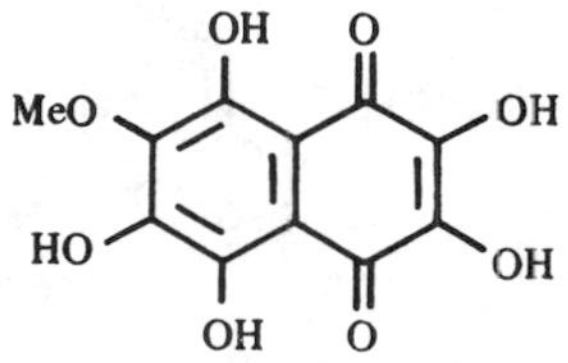	Namakochrome	Sea cucumber: *Polycheira rufescens*				364 397 j
161. $C_{11}H_{11}Br_2N_5O$	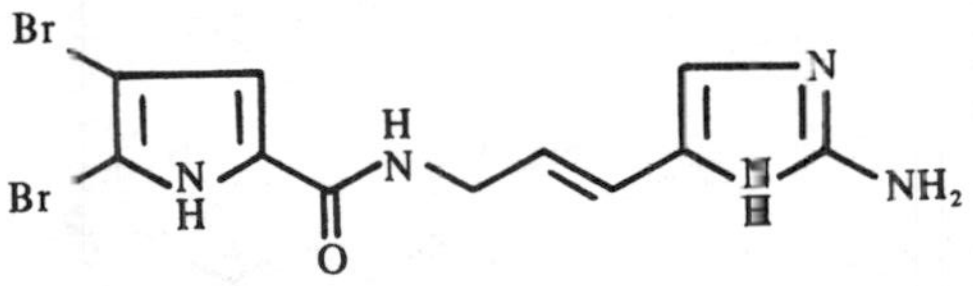	Oroidin	Sponge: *Agelas oroides*	+			137 146
162. $C_{11}H_{11}Br_2N_5O$	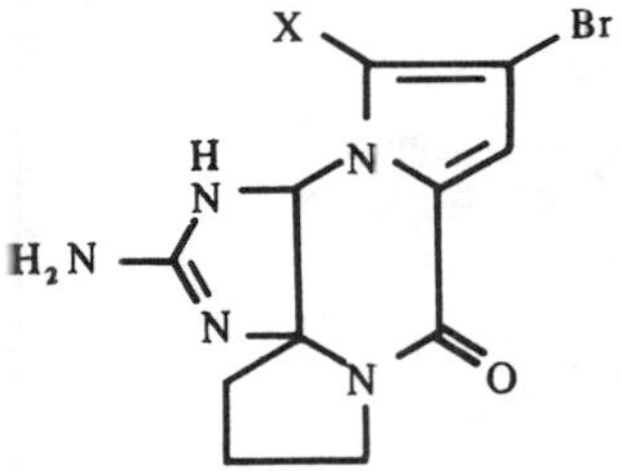 X = Br	4,5-Dibromophakellin	Sponge: *Phakellia flabellata*			Antimicrobial	351 353
163. $C_{11}H_{12}BrN_5O$	X = H	4-Bromophakellin				Antimicrobial	

Table 1 (continued)

Molecular formula	Structural formula	Name	Source of compound	Available by synthesis	Available non-marine origin	Activity	References
C_{11} . . .							
164. $C_{11}H_{12}Br_2N_2$	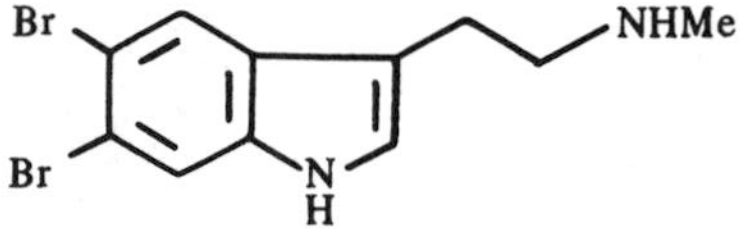	5,6-Dibromo-3-(2-methylaminoethyl)-indole	Sponge: *Polyfibrospongia maynardii*			Antibacterial	413
165. $C_{11}H_{12}N_2O_2$	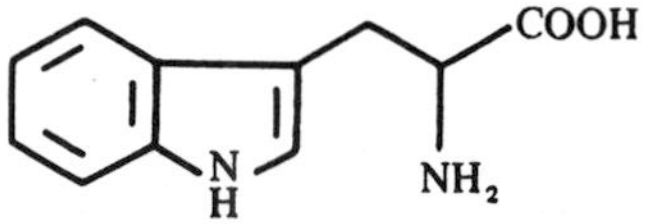	Tryptophan	Octopus: *Octopus vulgaris* Sponges	+	+		56 116 284 IIId 336
166. $C_{11}H_{15}Br_2NO_4$	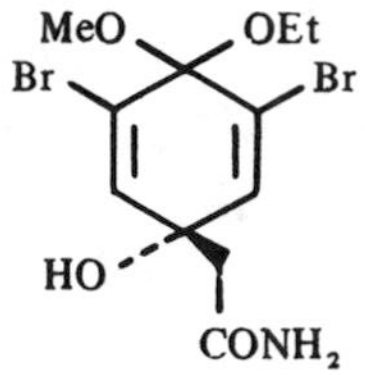 Artifact?	3-Acetamido-1,5-dibromo-6-ethoxy-3-hydroxy-6-methoxy-1,4 cyclohexadiene	Sponge: *Verongia* sp.			Antibiotic	30 31
167. $C_{11}H_{15}N_5O_5$	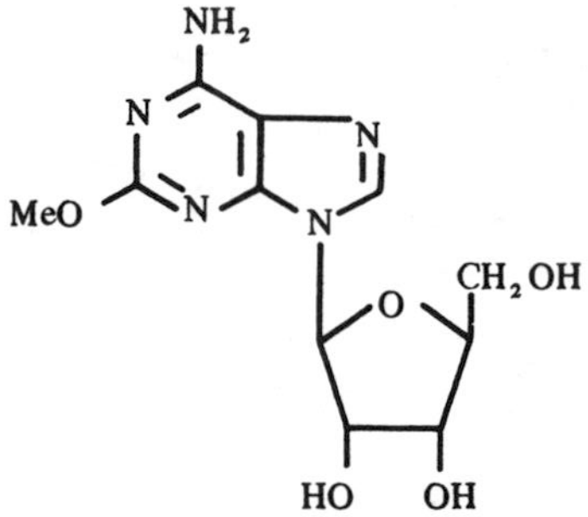	Spongosine	Sponge: *Cryptotethya crypta*				54

Table 1 (continued)

Molecular formula	Structural formula	Name	Source of compound	Available by synthesis	Available non-marine origin	Activity	References
C_{11}							
168. $C_{11}H_{16}$	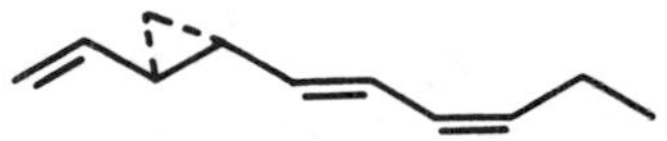	Dictyopterene B	Brown algae: *Dictyopteris* sp.	+			29 228 317 318
169. $C_{11}H_{16}$	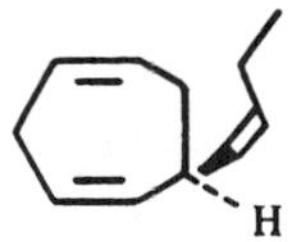 H	Dictyopterene D′ Ectocarpene	Brown algae: *Dictyopteris* sp. *Ectocarpus siliculosus*	+			227 228 300 318
170. $C_{11}H_{16}$		*trans, cis, cis*-1,3, 5,8-Undecatetraene	Brown algae: *Dictyopteris* sp.				228 317 318
171. $C_{11}H_{16}$		*trans,trans,cis*-1,3,5,8-Undecatetraene	Brown algae : *Dictyopteris* sp.				228 318
172. $C_{11}H_{17}N_3O_8$	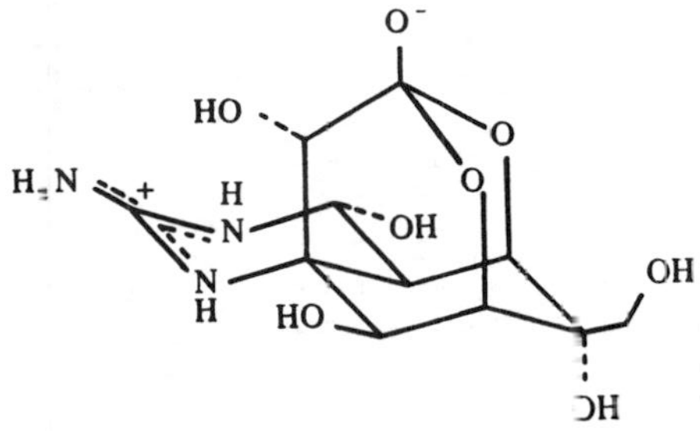 O^-, HO, O, O, H_2N, +, H, N, OH, N, H, HO, OH, OH	Tetrodotoxin	Pufferfish: *Sphoeroides (Fugu)* sp.	+		Hypotensive. Neurotoxin	64, 65, 158, 230, 243, 284 IIy, 298, 309, 336, 340, 402, 403, 436

Table 1 (continued)

Molecular formula	Structural formula	Name	Source of compound	Available by synthesis	Available non-marine origin	Activity	References
C_{11} . . .							
173. $C_{11}H_{18}$	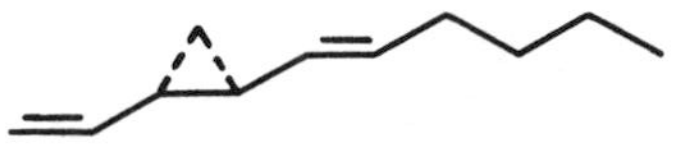	Dictyopterene A	Brown algae: *Dictyopteris australis* *D. plagiogramma*	+			228 292 312 318
174. $C_{11}H_{18}$	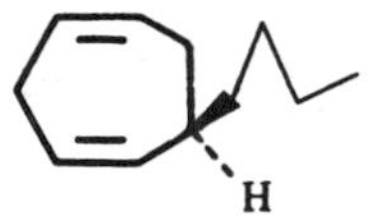	Dictyopterene C′	Brown algae: *Dictyopteris* sp.				228 318
175. $C_{11}H_{18}$		*trans,cis*-1,3,5-Undecatriene	Brown algae: *Dictyopteris* sp.				228 318
176. $C_{11}H_{18}$		*trans,trans*-1,3,5-Undecatriene	Brown algae: *Dictyopteris* sp.				228 318
177. $C_{11}H_{19}N_3O_3$	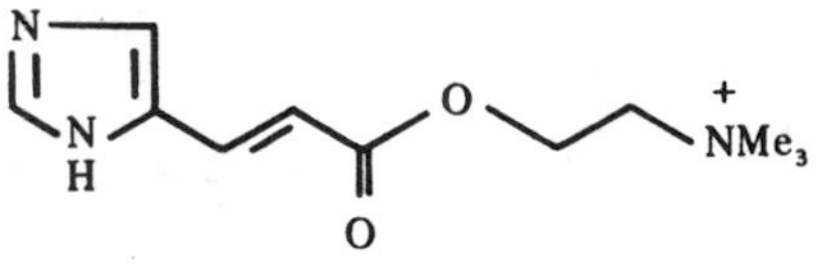 ^{-}OH	Murexine	Mollusks	+		Hypotensive. Muscle relaxant	117 284 IIe 315 331 340 427 429
178. $C_{11}H_{20}OS_2$	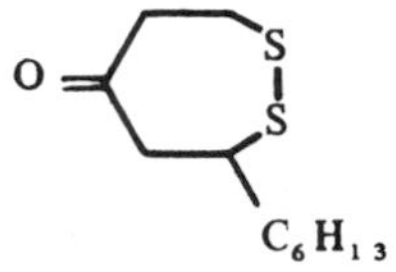	(−)-3-Hexyl-4,5-dithiacyclo-heptanone	Brown alga: *Dictyopteris plagiogramma*				228 330

Table 1 (continued)

Molecular formula	Structural formula	Name	Source of compound	Available by synthesis	Available non-marine origin	Activity	References
C_{11} . . . 179. $C_{11}H_{21}N_3O_3$	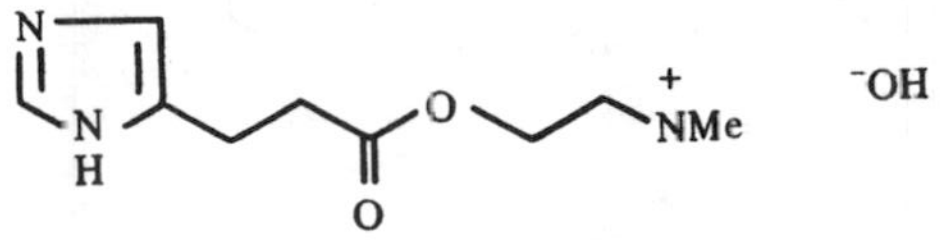	Dihydromurexine	Mollusk: *Thais haemostoma*	+			331
C_{12} . . . . 180. $C_{12}H_5Br_5O_2$	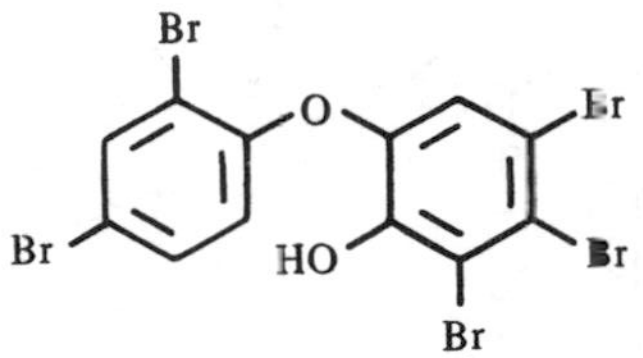	1-(2′,4′-Dibromo-phenoxy)-2-hydroxy-3,4,5-tribromo-benzene	Sponge: *Dysidea herbacea*	+		Antibacterial	351
181. $C_{12}H_5Br_5O_2$	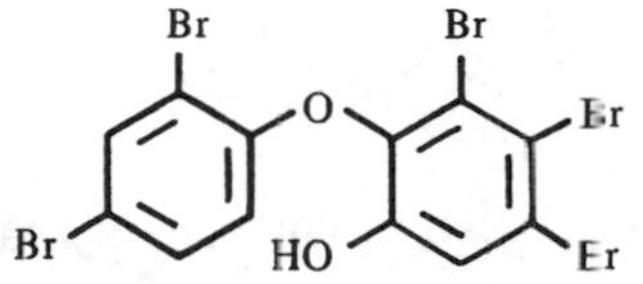	1-(2′,4′-Dibromo-phenoxy)-2-hydroxy-4,5,6-tribromo-benzene	Sponge: *Dysidea herbacea*	+		Antibacterial	351 354
182. $C_{12}H_6Br_4O_2$	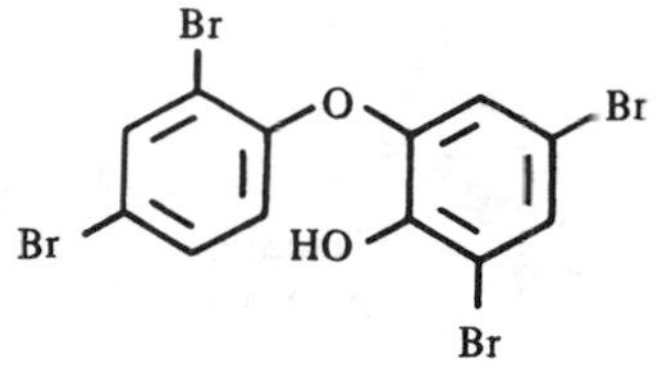	1-(2′,4′-Dibromo-phenoxy)-3,5-dibromo-2-hydroxybenzene	Sponge: *Dysidea herbacea*			Antibacterial	351

Table 1 (continued)

Molecular formula	Structural formula	Name	Source of compound	Available by synthesis	Available non-marine origin	Activity	References
C_{12} . . .							
183. $C_{12}H_7Br_3O_2$	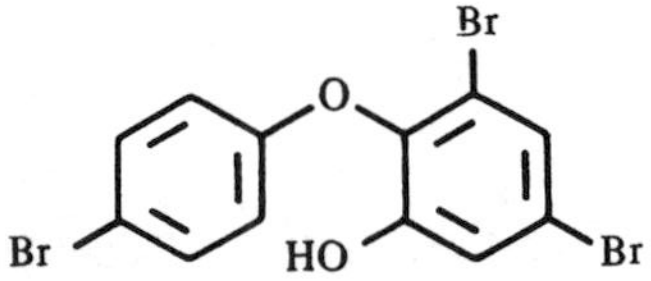	1-(4′-Bromo-phenoxy)-4,6-di-bromo-2-hydroxybenzene	Sponge: *Dysidea herbacea*			Antibacterial	351
184. $C_{12}H_8Br_2O_2$	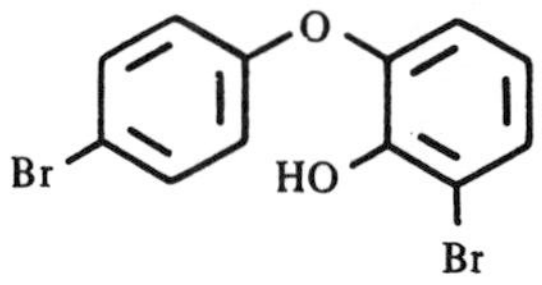	1-(4′-Bromo-phenoxy)-3-bromo-2-hydroxybenzene	Sponge: *Dysidea herbacea*	+		Antibacterial	351
185. $C_{12}H_8Br_2O_2$	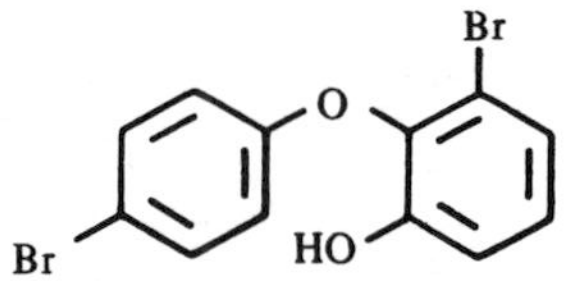	1-(4′-Bromo-phenoxy)-5-bromo-2-hydroxybenzene	Sponge: *Dysidea herbacea*			Antibacterial	354
186. $C_{12}H_8O_6$	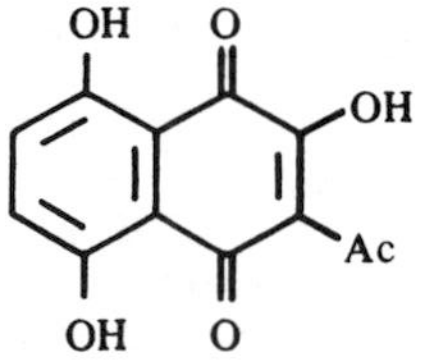	3-Acetyl-2-hydroxy-naphthazarin	Brittle stars: *Ophiocoma erinaceus* *O. insularia* Sea urchins: *Echinothrix calamaris* *E. diadema*				290 364 397 i

Table 1 (continued)

Molecular formula	Structural formula	Name	Source of compound	Available by synthesis	Available non-marine origin	Activity	References
C_{12} . . .							
187. $C_{12}H_8O_6$	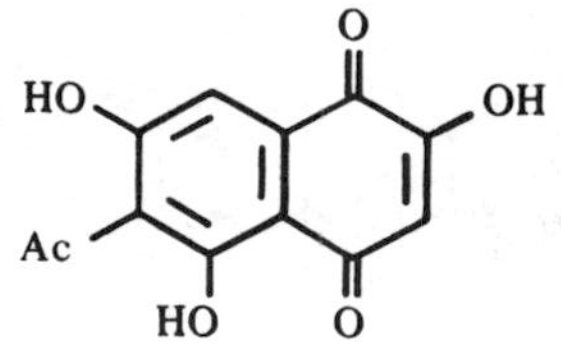	6-Acetyl-2,7-dihydroxy-juglone	Sea urchins: *Echinothrix calamaris* *E. diadema*	+			290 397 g
188. $C_{12}H_8O_6$	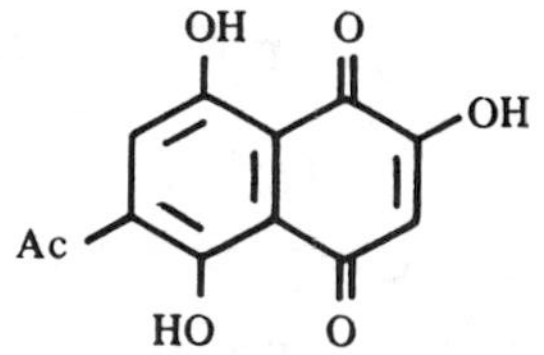	6-Acetyl-2-hydroxy-naphthazarin	Sea urchins: *Echinothrix calamaris* *E. diadema*	+			290 397 i
189. $C_{12}H_8O_7$	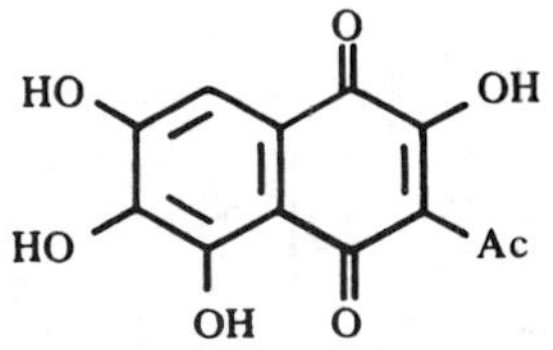	3-Acetyl-2,6,7-trihydroxyjuglone	Sea urchin: *Temnopleurus toreumaticus*	+			280 397 h
190. $C_{12}H_8O_7$	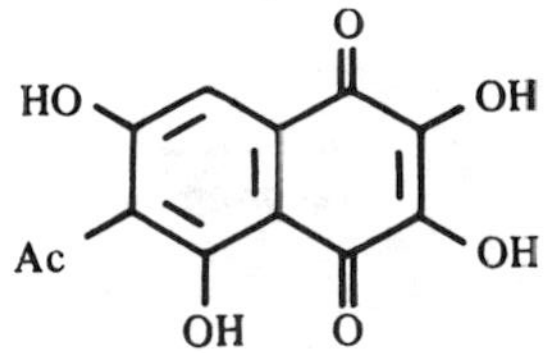	6-Acetyl-2,3,7-tri-hydroxyjuglone	Sea urchins: *Echinothrix calamaris* *E. diadema*	+			290 365 397 g

Table 1 (continued)

Molecular formula	Structural formula	Name	Source of compound	Available by synthesis	Available non-marine origin	Activity	References
C_{12} . . .							
191. $C_{12}H_8O_7$	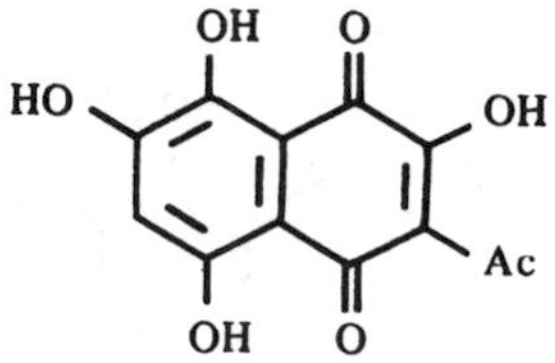	Spinochrome A	Brittle stars Feather stars Sea urchins	+			32 76 156 290 366 397 e
192. $C_{12}H_8O_8$	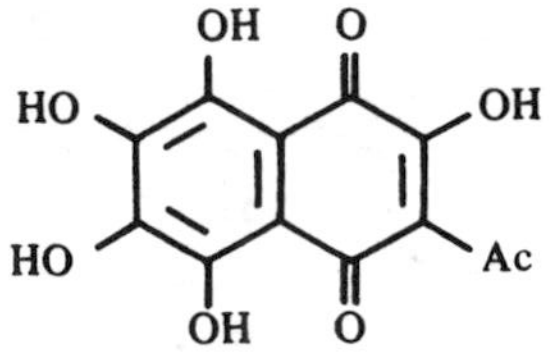	Spinochrome C	Sea urchins	+			32 76 290 366 397 f
193. $C_{12}H_{10}O_4$	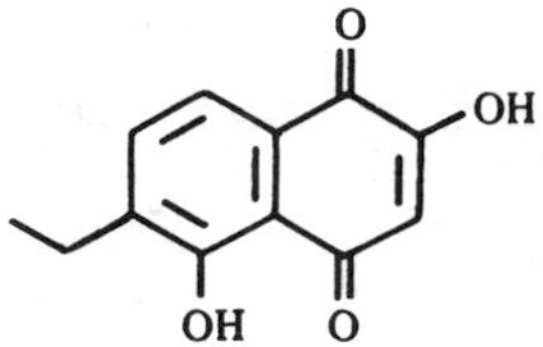	6-Ethyl-2-hydroxy-juglone	Sea urchin: *Echinothrix calamaris*				290 397 g
194. $C_{12}H_{10}O_5$	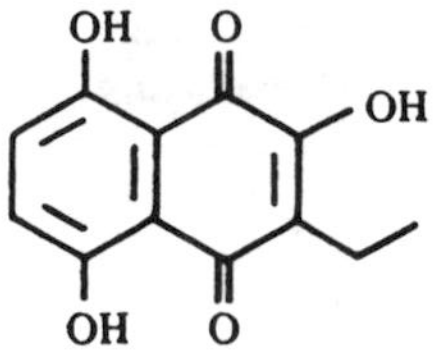	3-Ethyl-2-hydroxy-naphthazarin	Brittle stars: *Ophicoma erinaceus* *O. insularia*	+			364 397 i

Table 1 (continued)

Molecular formula	Structural formula	Name	Source of compound	Available by synthesis	Available non-marine origin	Activity	References
C_{12} . . .							
195. $C_{12}H_{10}O_5$	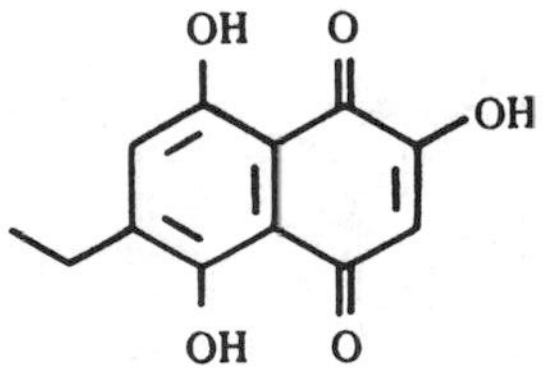	6-Ethyl-2-hydroxy-naphthazarin	Brittle stars: *Ophiocoma erinaceus* *O. insularia* Sea urchin: *Echinothrix calamaris*	+			290 364 397 i
196. $C_{12}H_{10}O_6$	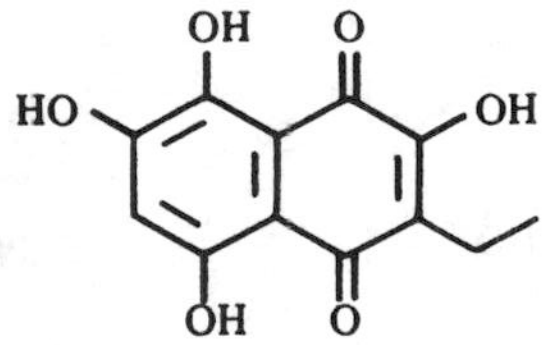	2,7-Dihydroxy-3-ethylnaphthazarin	Brittle star: *Ophicoma erinaceus* Sea urchins: *Echinothrix calamaris* *E. diadema*	+			290 364 397 i
197. $C_{12}H_{10}O_6$	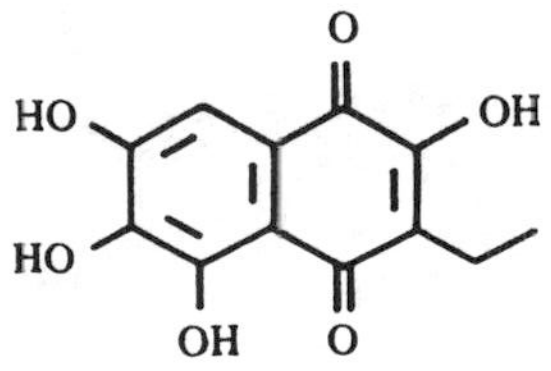	3-Ethyl-2,6,7-tri-hydroxyjuglone	Brittle stars: *Ophicoma erinaceus* *O. insularia*				364 397 h
198. $C_{12}H_{10}O_6$	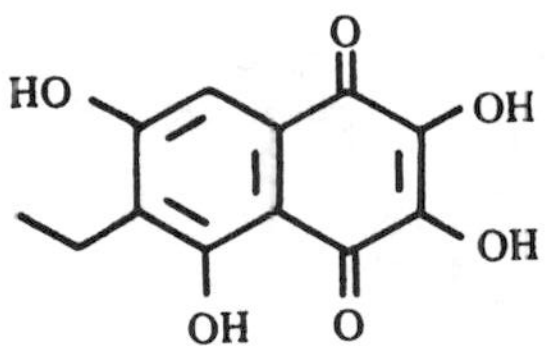	6-Ethyl-2,3,7-trihydroxyjuglone	Sea urchins: *Echinothrix calamaris* *E. diadema*				290 397 g

Table 1 (continued)

Molecular formula	Structural formula	Name	Source of compound	Available by synthesis	Available non-marine origin	Activity	References
C_{12} . . .							
199. $C_{12}H_{10}O_7$	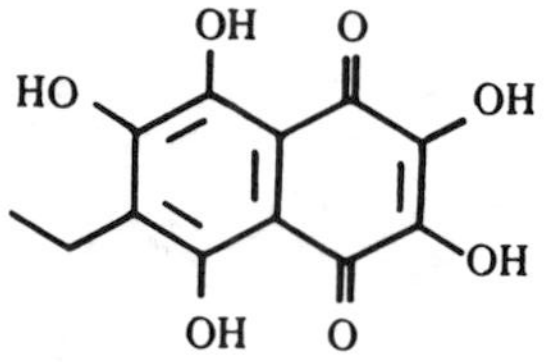	Echinochrome A	Brittle star: *Ophicoma erinaceus* Sea urchins	+			32 290 364 397 f 415
200. $C_{12}H_{10}O_8$	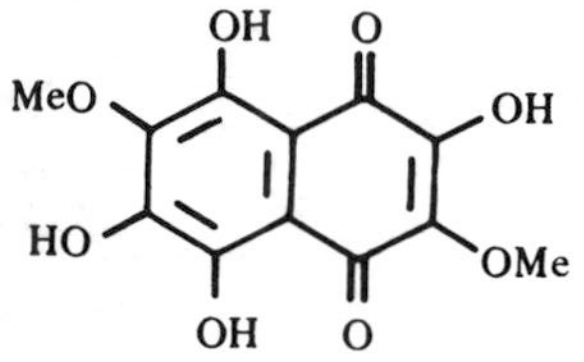	2,6-Dihydroxy-3,7-dimethoxy-naphthazarin	Starfish: *Acanthaster planci*	+			364 397 j
201. $C_{12}H_{10}O_8$	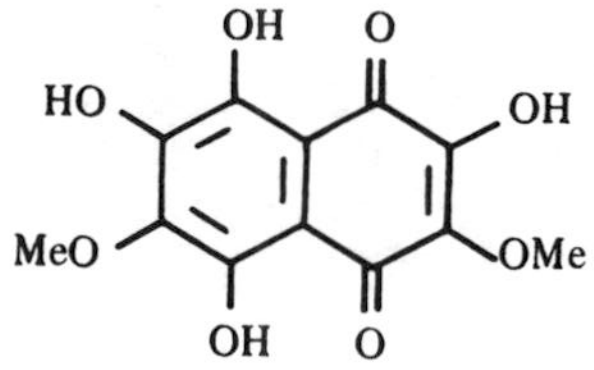	2,7-Dihydroxy-3,6-dimethoxy-naphthazarin	Starfish: *Acanthaster planci*	+			364 397 j
202. $C_{12}H_{14}N_6$	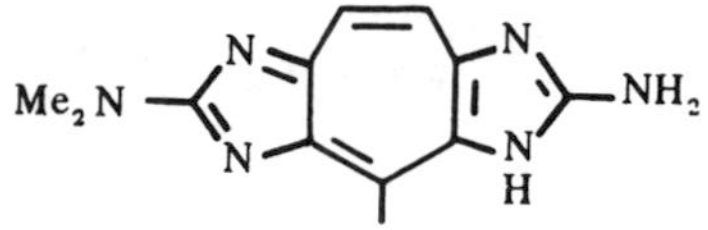	Parazoantho-xanthin D	Zoanthid: *Parazoanthus axinellae*				74 325

Table 1 (continued)

Molecular formula	Structural formula	Name	Source of compound	Available by synthesis	Available non-marine origin	Activity	References
C_{12} . . .							
203. $C_{12}H_{16}BrClO$	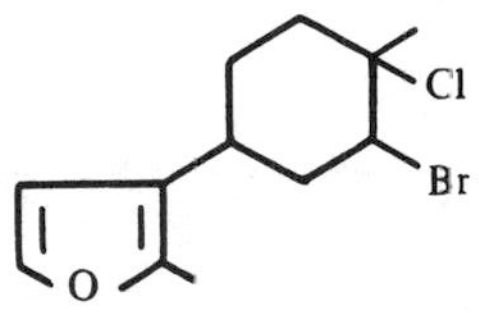	Furocaespitane	Red alga: *Laurencia caespitosa*				153
C_{13}. . . .							
204. $C_{13}H_{10}O_7$	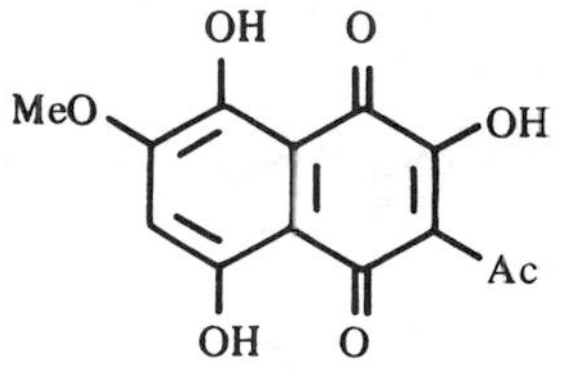	3-Acetyl-2-hydroxy-7-methoxynaphthazarin	Brittle stars: *Ophiocoma erinaceus* *O. insularia*				364 397 j
205. $C_{13}H_{12}O_7$	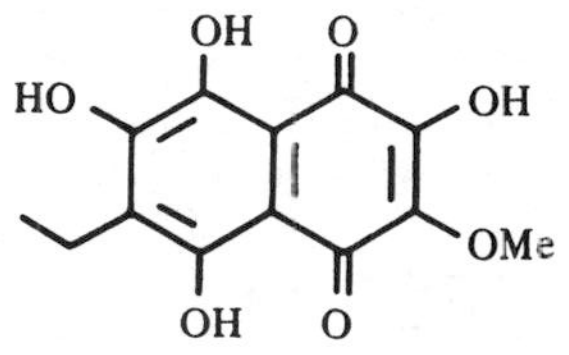	2,7-Dihydroxy-6-ethyl-3-methoxynaphthazarin	Sea urchin: *Diadema antillarum*				32 280 397 j
206. $C_{13}H_{12}O_7$	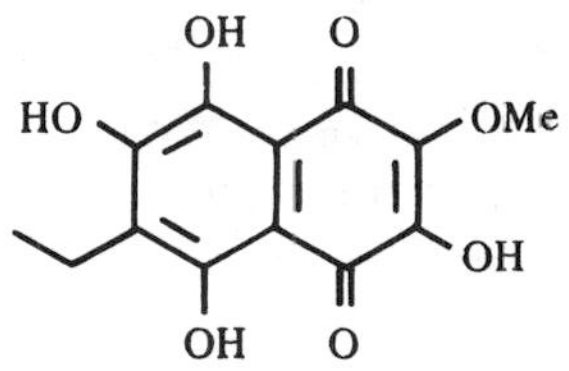	3,7-Dihydroxy-6-ethyl-2-methoxynaphthazarin	Sea urchin: *Diadema antillarum*				32 280 397 j

Table 1 (continued)

Molecular formula	Structural formula	Name	Source of compound	Available by synthesis	Available non-marine origin	Activity	References
C_{13} . . .							
207. $C_{13}H_{16}N_6$		Zoanthoxanthin	Zoanthid: *Parazoanthus axinellae*				73 75
208. $C_{13}H_{22}O_2S$		S-(*trans*-3-Oxo-4-undecaenyl) thioacetate	Brown alga: *Dictyopteris plagiogramma*				228 330
209. $C_{13}H_{24}O_2S$		S-(3-Oxoundecyl)-thioacetate	Brown alga: *Dictyopteris plagiogramma*				228 330
C_{14}							
210. $C_{14}H_{10}O_6$		2-Hydroxy-2′-methyl-2′H-pyrano (2,3-b) naphthazarin	Sea urchins: *Echinothrix calamaris* *E. diadema*				291 397 j
C_{15}							
211. $C_{15}H_{18}$		Cadalene	Brown alga: *Dictyopteris divaricata*	+	+		207 284 k 299 341 b

Table 1 (continued)

Molecular formula	Structural formula	Name	Source of compound	Available by synthesis	Available non-marine origin	Activity	References
C_{15} . . .							
212. $C_{15}H_{18}BrClO_2$	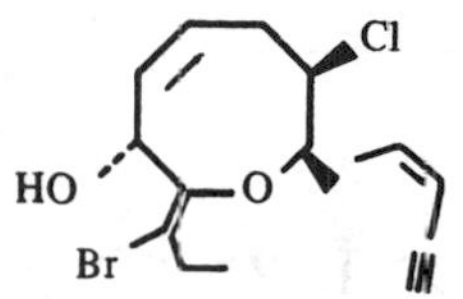	Chondriol	Red alga: *Chondria oppositiclada*			Antiviral	134 135
213. $C_{15}H_{18}O$	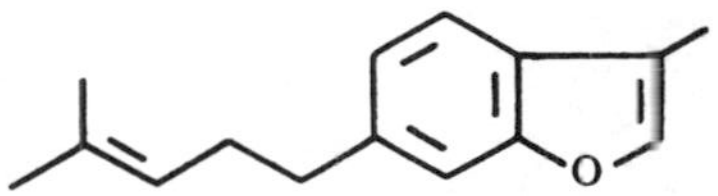	Furoventalene	Gorgonian: *Gorgonia ventalina*	+			420
214. $C_{15}H_{19}BrO$	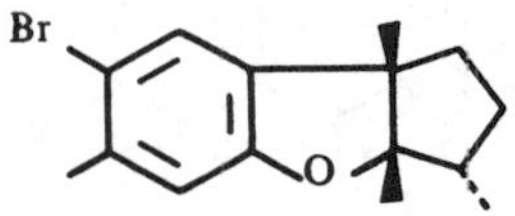	Aplysin	Red alga: *Laurencia okamurai* Sea hare: *Aplysia kurodai* *A. californica*	+			136 215 340 373 374 383 437 444
215. $C_{15}H_{19}BrO$	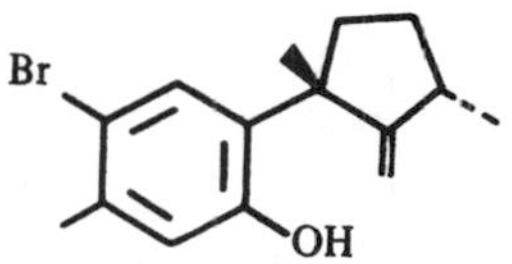	Isolaurinterol	Red alga: *Laurencia intermedia*				219

Table 1 (continued)

Molecular formula	Structural formula	Name	Source of compound	Available by synthesis	Available non-marine origin	Activity	References
C_{15} . . .							
216. $C_{15}H_{19}BrO$	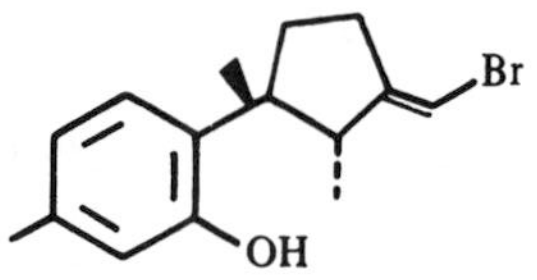	Laurenisol	Red alga: *Laurencia nipponica*				217
217. $C_{15}H_{19}BrO$	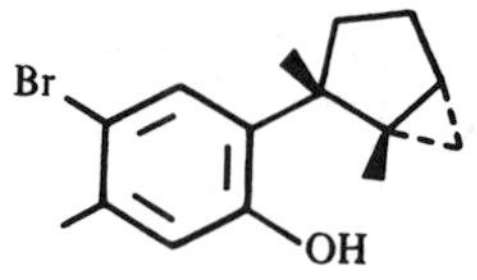	Laurinterol	Red algae: *Laurencia intermedia* *L. nipponica* *L. okamurai* Sea hare: *Aplysia californica*	+			136 210 215 219 373 374 383
218. $C_{15}H_{19}BrO_2$	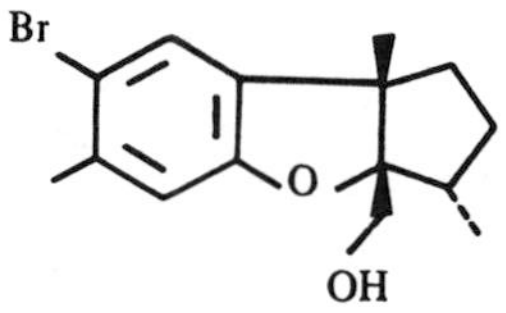	Aplysinol	Red alga: *Laurencia okamurai* Sea hare: *Aplysia kurodai*				215 444
219. $C_{15}H_{19}Br_2ClO$	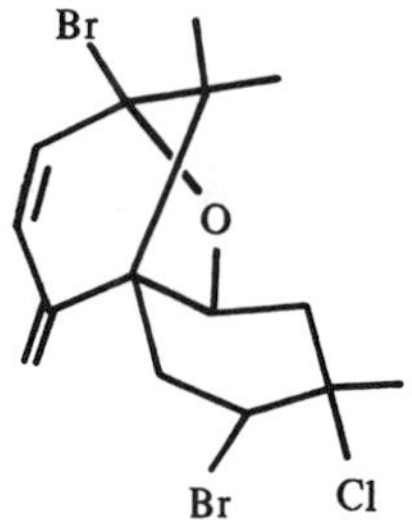	Pacifidiene	Sea hare: *Aplysia californica*				373 374

Table 1 (continued)

Molecular formula	Structural formula	Name	Source of compound	Available by synthesis	Available non-marine origin	Activity	References
C_{15} . . .							
220. $C_{15}H_{20}$	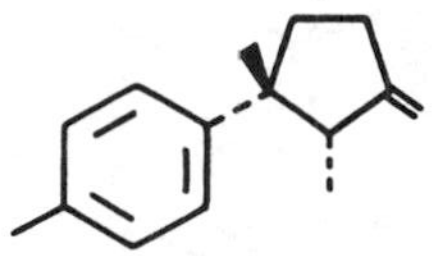	Laurene	Red algae: *Laurencia glandulifera* *L. nipponica*	+			209 211 216
221. $C_{15}H_{20}Br_2Cl_2O$	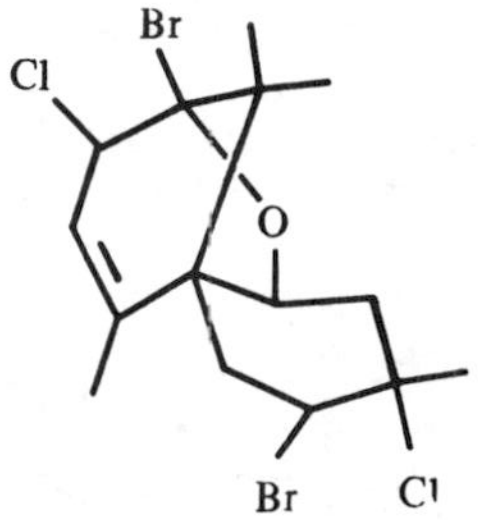	2,7-Dibromo-3,8-dichloro-2,3,7,8, 9.9a-hexahydro-5, 8,10,10-tetramethyl-6H-2,5a-methano-1-benzoxepin	Sea hare: *Aplysia californica*	+			374
222. $C_{15}H_{20}Br_2O_2$	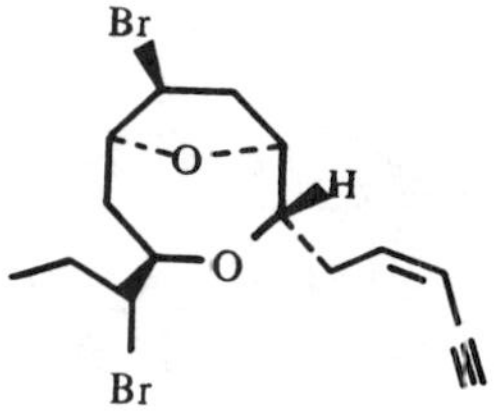	Isolaureatin	Red alga: *Laurencia nipponica*				214 218 258
223. $C_{15}H_{20}Br_2O_2$	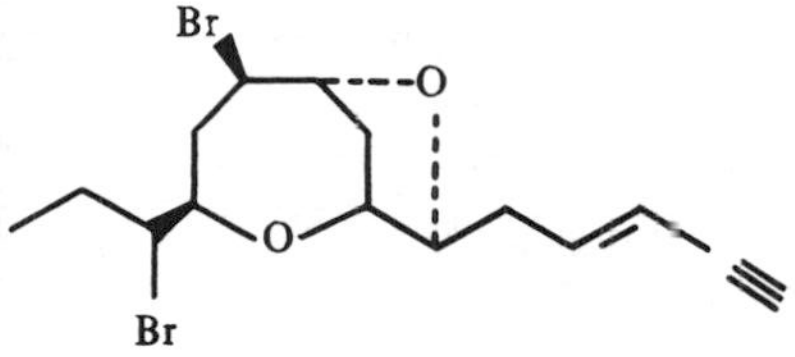	Isoprelaurefucin	Red alga: *Laurencia ripponica*				259

Table 1 (continued)

Molecular formula	Structural formula	Name	Source of compound	Available by synthesis	Available non-marine origin	Activity	References
C_{15} . . .							
224. $C_{15}H_{20}Br_2O_2$	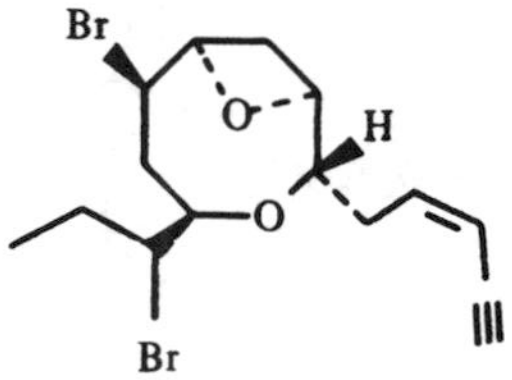	Laureatin	Red alga: *Laurencia nipponica*				213 218 258
225. $C_{15}H_{20}O$	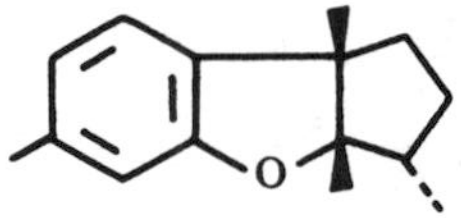	Debromoaplysin	Red alga: *Laurencia okamurai* Sea hares: *Aplysia californica* *A. kurodai*	+			136 215 373 383 437 444
226. $C_{15}H_{20}O$	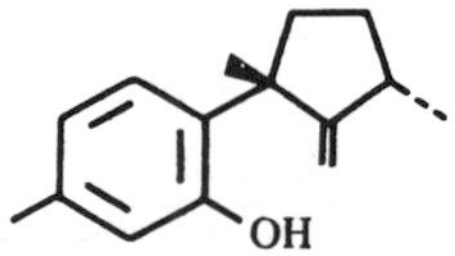	Debromoiso-laurinterol	Red algae: *Laurencia* sp.				136
227. $C_{15}H_{20}O$	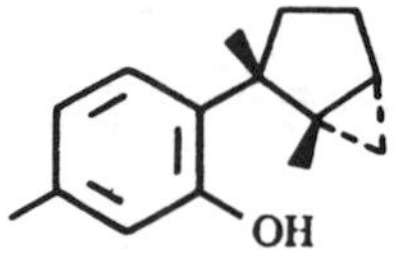	Debromolaurinterol	Red algae: *Laurencia intermedia* *L. okamurai*	+			136 210 215 219 383

Table 1 (continued)

Molecular formula	Structural formula	Name	Source of compound	Available by synthesis	Available non-marine origin	Activity	References
C_{15} . . .							
228. $C_{15}H_{20}O$	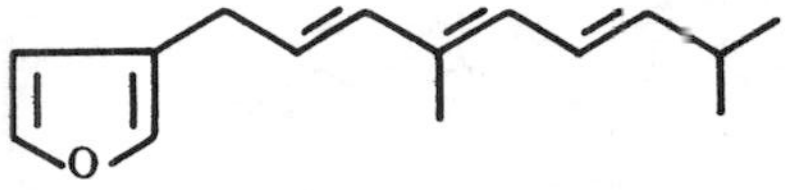	Dehydrodendrolasin	Sponge: *Pleraplysilla spinifera*				94
229. $C_{15}H_{20}O$	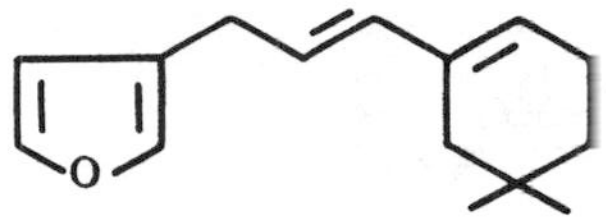	Pleraplysillin	Sponge: *Pleraplysilla spinifera*				94
230. $C_{15}H_{21}BrO_3$	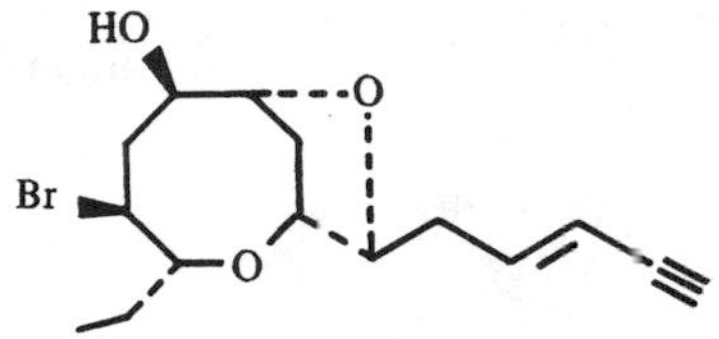	Laurefucin	Red alga: *Laurencia nipponica*				141 143
231. $C_{15}H_{21}Br_2ClO_2$	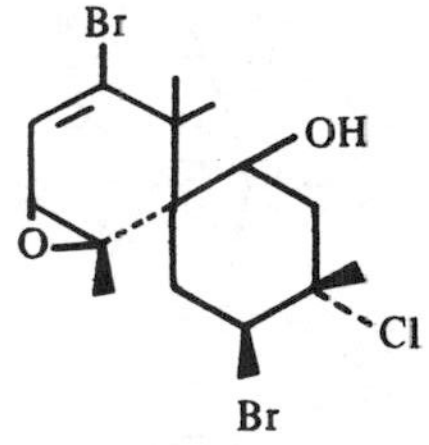	Prepacifenol	Red algae: *Laurencia filiformis* *L. pacifica*				363

Table 1 (continued)

Molecular formula	Structural formula	Name	Source of compound	Available by synthesis	Available non-marine origin	Activity	References
C_{15} . . .							
232. $C_{15}H_{21}Br_2ClO_2$	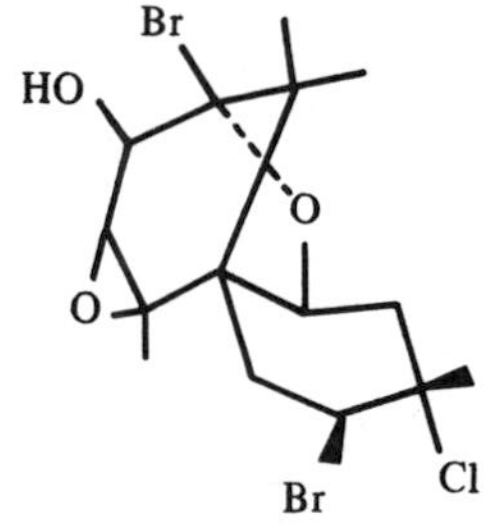	Johnstonol	Red algae: *Laurencia johnstonii* *L. okamurai* *L. pacifica* Sea hare: *Aplysia californica*				361 373
233. $C_{15}H_{21}NO_6$	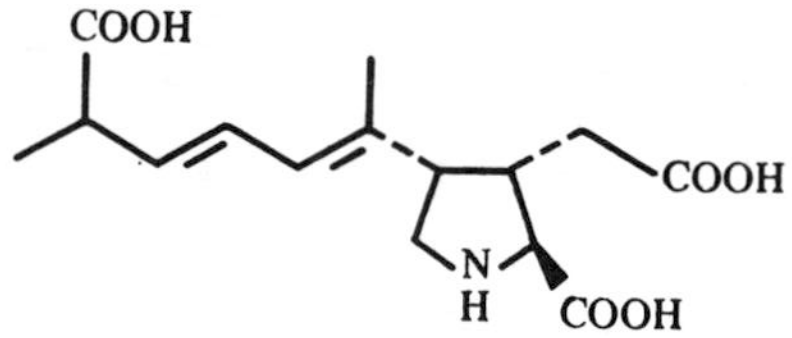	Domoic acid	Red alga: *Chondria armata*			Anthelmintic	386 387 392
234. $C_{15}H_{22}$	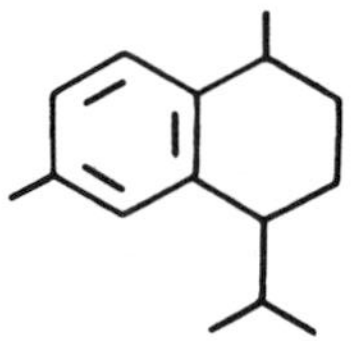	(+)-Calamenene	Gorgonian: *Pseudoplexaura porosa*				299 341 b 417
235. $C_{15}H_{22}$	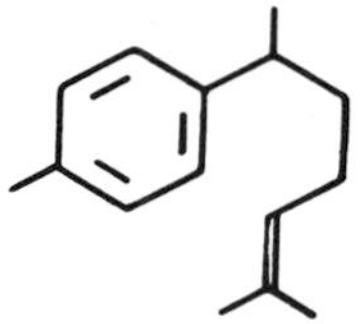	(−)-α-Curcumene	Gorgonians: *Plexaurella dichotoma* *P. fusifera* *P. grisea*				88

Table 1 (continued)

Molecular formula	Structural formula	Name	Source of compound	Available by synthesis	Available non-marine origin	Activity	References
C_{15} . . .							
236. $C_{15}H_{22}O$	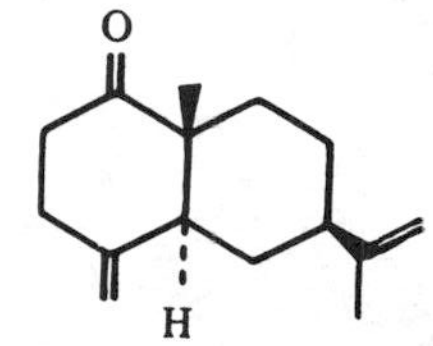	Dictyopterone	Brown alga: *Dictyopteris divaricata*				257
237. $C_{15}H_{23}BrO$	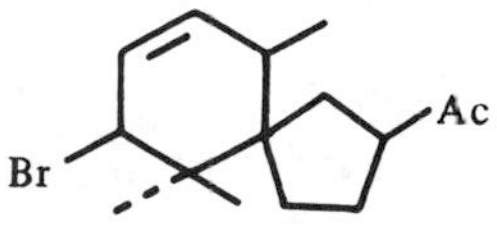	Spirolaurenone	Red alga: *Laurencia glandulifera*				384
238. $C_{15}H_{24}$	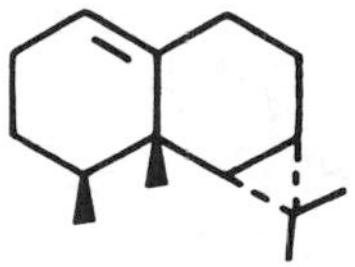	(−)-1(10)-Aristolene	Gorgonian: *Pseudopterogorgia americana*				341 d, e 419
239. $C_{15}H_{24}$	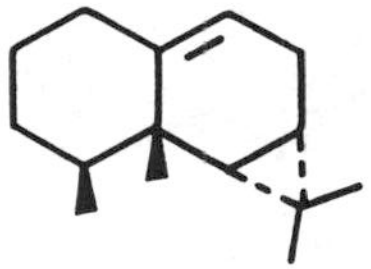	(+)-9-Aristolene	Gorgonian: *Pseudopterogorgia americana*		+		341 d 419
240. $C_{15}H_{24}$	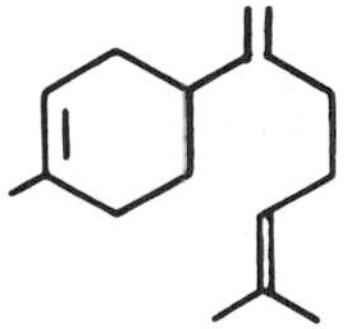	(+)-β-Bisabolene	Gorgonians: *Plexaurella dichotoma* *P. fusifera* *P. grisea*				88 417

Table 1 (continued)

Molecular formula	Structural formula	Name	Source of compound	Available by synthesis	Available non-marine origin	Activity	References
C_{15} . . .							
241. $C_{15}H_{24}$	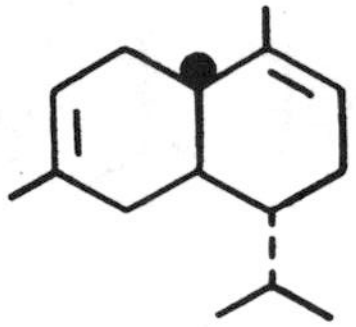	(+)-Cadinene	Gorgonian: *Plexaura crassa*		+		83 322
242. $C_{15}H_{24}$	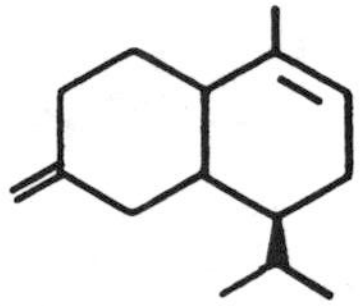	(−)-γ_1-Cadinene	Brown alga: *Dictyopteris divaricata*				207 341 b
243. $C_{15}H_{24}$	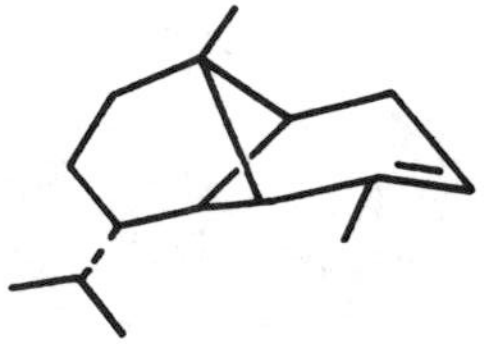	(−)-Copaene	Brown alga: *Dictyopteris divaricata*	+	+		207 284 s
244. $C_{15}H_{24}$	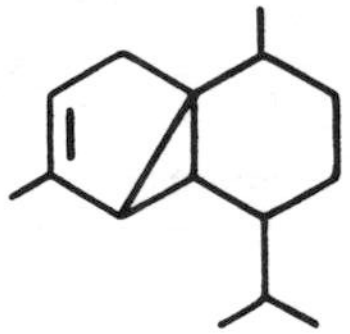	(+)-α-Cubebene	Gorgonian: *Pseudoplexaura porosa*	+			394 417

Table 1 (continued)

Molecular formula	Structural formula	Name	Source of compound	Available by synthesis	Available non-marine origin	Activity	References
C_{15} . . .							
245. $C_{15}H_{24}$	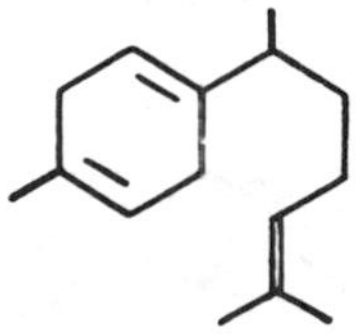	(+)-β-Curcumene	Gorgonians: *Plexaurella dichotoma* *P. fusifera* *P. grisea*				88
246. $C_{15}H_{24}$	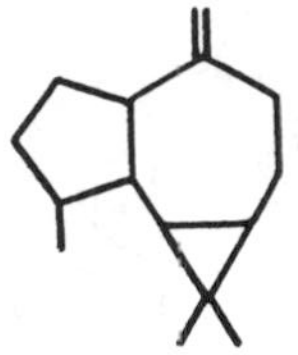	Decahydro-4-methylene -1,1,7-trimethyl-1H-cycloprop-(e)-azulene	Gorgonians: *Pseudoplexaura flagellosa* *P. porosa* *P. wagenaari*				88
247. $C_{15}H_{24}$	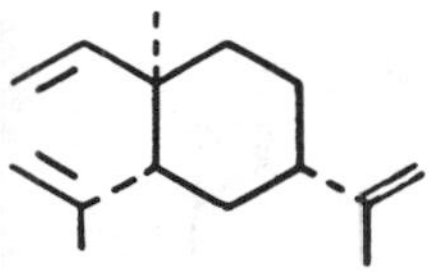 Artifact?	(+)-β-Elemene	Brown alga: *Dictyopteris divaricata* Gorgonian: *Eunicea mammosa*		+		207 341 d 417
248. $C_{15}H_{24}$	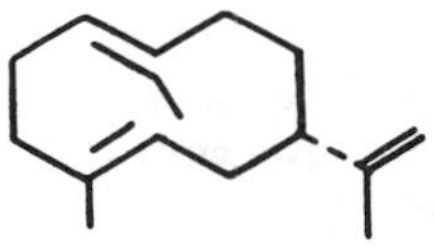	(−)-Germacrene-A	Gorgonian: *Eunicea mammosa*				417 422

Table 1 (continued)

Molecular formula	Structural formula	Name	Source of compound	Available by synthesis	Available non-marine origin	Activity	References
C_{15} . . .							
249. $C_{15}H_{24}$	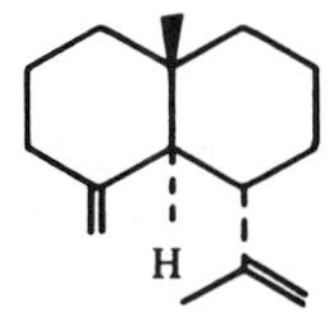	β-Gorgonene	Gorgonian: *Pseudopterogorgia americana*	+			59 193 419
250. $C_{15}H_{24}$	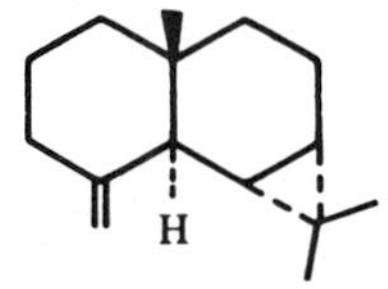	(+)-γ-Maaliene	Gorgonian: *Pseudopterogorgia americana*				419
251. $C_{15}H_{24}$	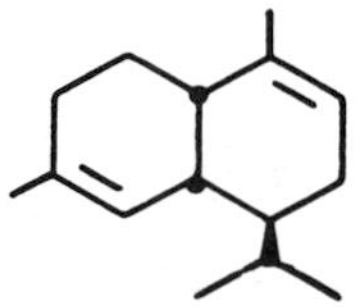	(+)-α-Muurolene	Gorgonians: *Eunicea mammosa* *E. palmeri* *Plexaurella dichotoma* *Pseudoplexaura porosa*				130 341 c 417
252. $C_{15}H_{24}$	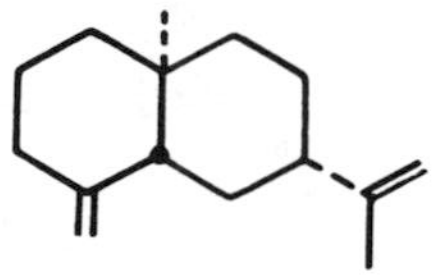	(−)-β-Selinene	Gorgonian: *Eunicea mammosa*				422

Table 1 (continued)

Molecular formula	Structural formula	Name	Source of compound	Available by synthesis	Available non-marine origin	Activity	References
C_{15} . . .							
253. $C_{15}H_{24}$	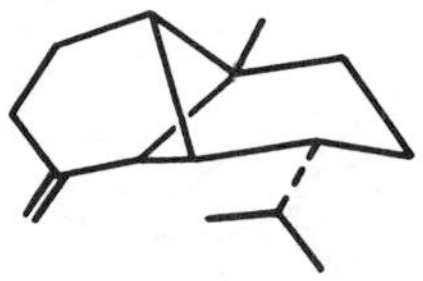	β-Ylangene	Gorgonians: *Eunicea palmeri* *Pseudoplexaura porosa*		+		341 e 417
254. $C_{15}H_{24}$	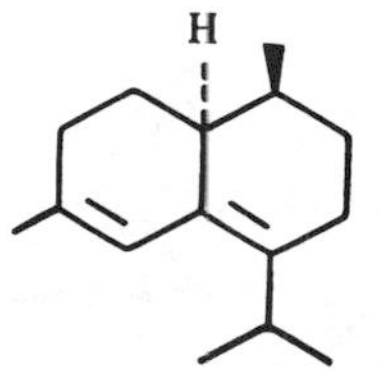	Zonarene	Brown alga: *Dictyopteris zonarioides*				132
255. $C_{15}H_{24}O$	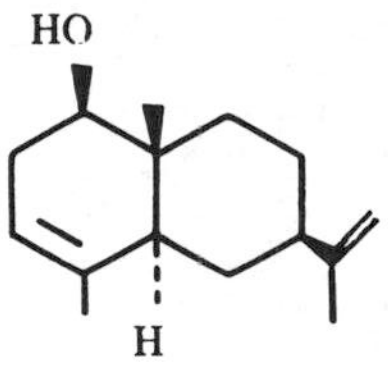	α-Selinen-1β-ol	Brown alga: *Dictyopteris divaricata*				257
256. $C_{15}H_{24}O$	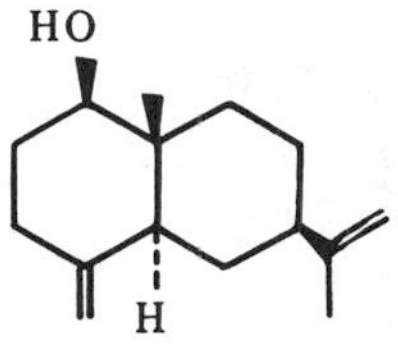	β-Selinen-1β-ol	Brown alga: *Dictyopteris divaricata*				257

Table 1 (continued)

Molecular formula	Structural formula	Name	Source of compound	Available by synthesis	Available non-marine origin	Activity	References
C_{15} ...							
257. $C_{15}H_{25}BrO$	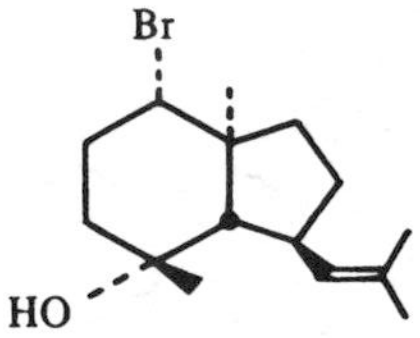	Oppositol	Red alga: *Laurencia subopposita*			Antibiotic	171
258. $C_{15}H_{25}Br_2ClO_2$	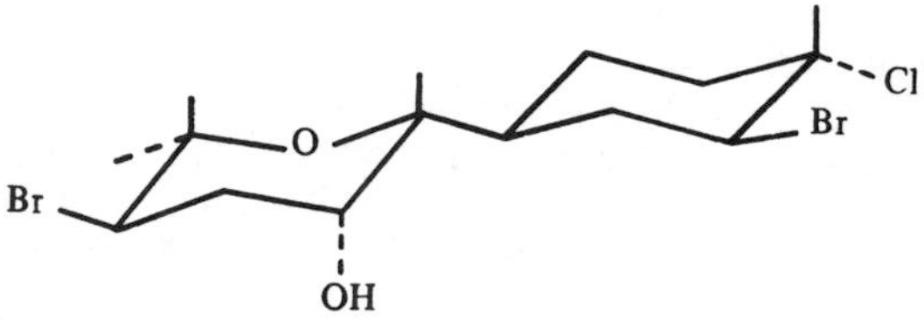	Caespitol	Red alga: *Laurencia caespitosa*				152 155
259. $C_{15}H_{26}$	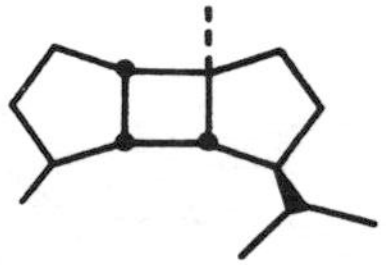	(+)-β-*epi*-Bourbonene	Gorgonians: *Pseudoplexaura flagellosa* *P. porosa* *P. wagenaari*				88
260 $C_{15}H_{26}O$	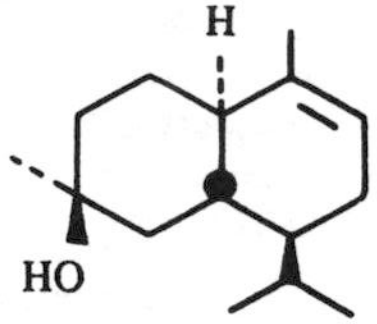	(–)-δ-Cadinol	Brown alga: *Dictyopteris divaricata*		+		207 341 a

Table 1 (continued)

Molecular formula	Structural formula	Name	Source of compound	Available by synthesis	Available non-marine origin	Activity	References
C_{15} ...							
261. $C_{15}H_{26}O_3S$	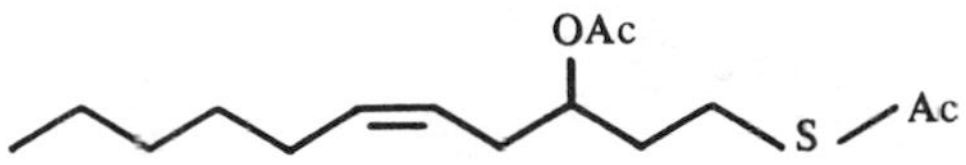	S-(–)-3-Acetoxyundec-5-enyl thioacetate	Brown algae: *Dictyopteris australis* *D. plagiogramma*				294
C_{16}							
262. $C_{16}H_8Br_2N_2O_2$	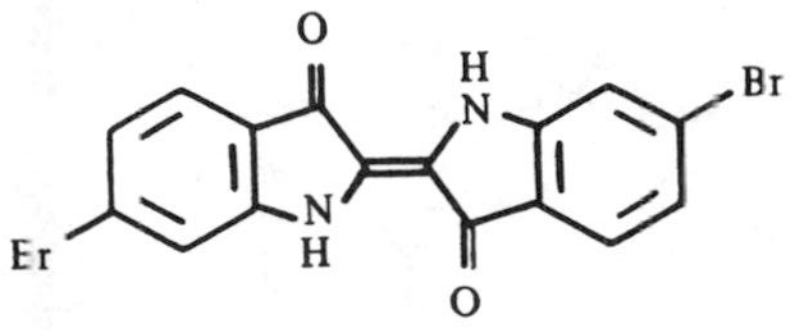	6,6′-Dibromo indigotin	Mollusks: *Murex brandaris* *M. trunculus* *Purpura aperta* *P. lapillus*	+			40
263. $C_{16}H_{10}O_7$	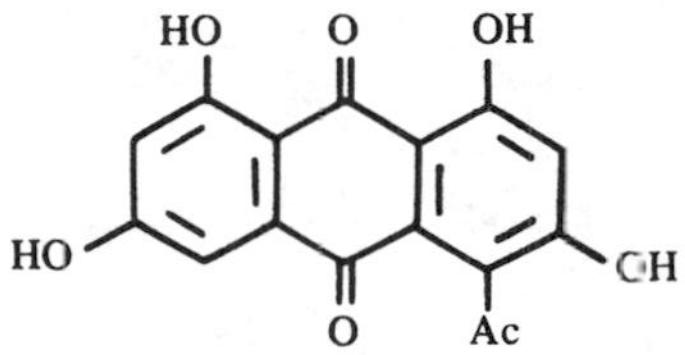	Rhodolamprometrin	Feather star: *Lamprometra klunzingeria*				114
264. $C_{16}H_{12}O_4$	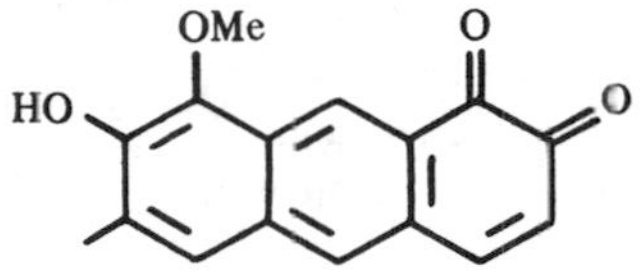	Hallachrome	Sea worms: *Halla parthenopeia* *Lumbriconereis impatiens*				323 324 397 m

Table 1 (continued)

Molecular formula	Structural formula	Name	Source of compound	Available by synthesis	Available non-marine origin	Activity	References
C_{16} . . .							
265. $C_{16}H_{13}NaO_8S$		Comantherin sodium sulfate ester	Feather star: *Comantheria perplexa*				160 241
266. $C_{16}H_{22}Cl_3NO_4$		5-Isopropyl-4-methoxy-1-(6,6,6-trichloro-3-methoxy-5-methyl-2-hexenoyl)-3-pyrrolin-2-one	Sponge: *Dysidea herbacea*				110
267. $C_{16}H_{25}N$		Axisonitrile-1	Sponge: *Axinella cannabina*				70

Table 1 (continued)

Molecular formula	Structural formula	Name	Source of compound	Available by synthesis	Available non-marine origin	Activity	References
C_{16} . . .							
268. $C_{16}H_{25}NS$	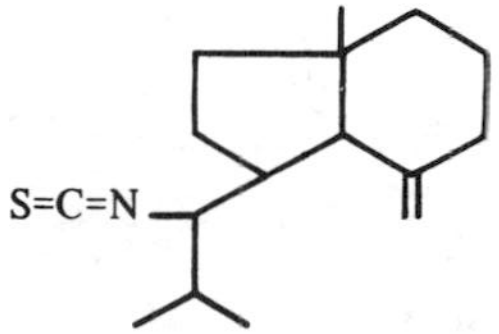	Axisothiocyanate-1	Sponge: *Axinella cannabina*				70
269. $C_{16}H_{32}O_2$	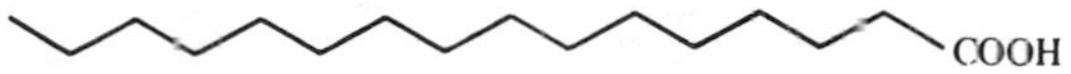	Palmitic acid	Green alga: *Caulerpa lamourouxii* (*C. racemosa* var. *lamourouxii*) Red alga: *Odonthalia corymbifera*		+		256 284 IIg 339
C_{17}							
270. $C_{17}H_{12}O_6$	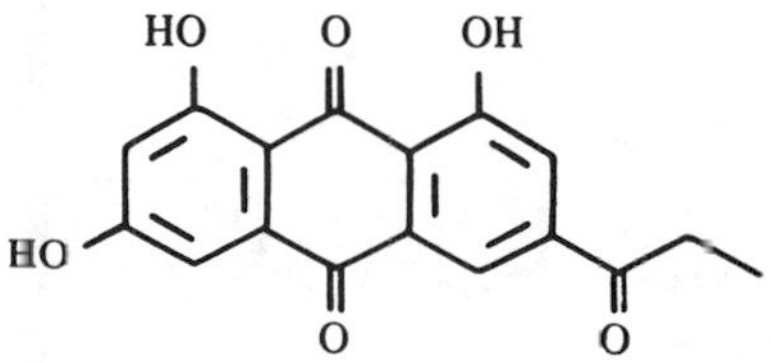	3-Propionyl-1,6,8-trihydroxy-9,10-anthraquinone	Feather star: *Comanthus bennetti*				44
271. $C_{17}H_{14}O_5$	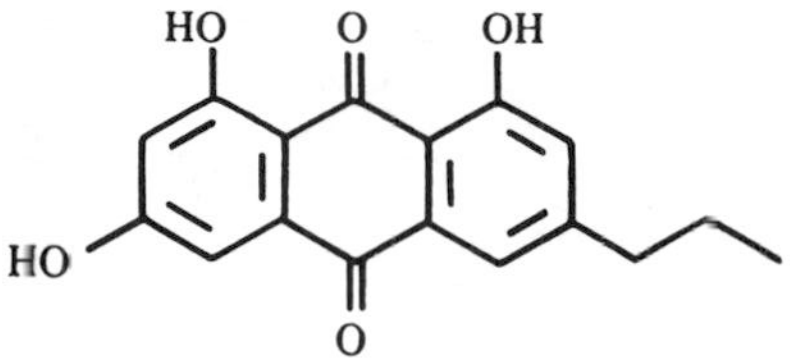	3-Propyl-1,6,8-trihydroxy-9,10-anthraquinone	Feather star: *Comanthus bennetti*				44

Table 1 (continued)

Molecular formula	Structural formula	Name	Source of compound	Available by synthesis	Available non-marine origin	Activity	References
C_{17} . . .							
272. $C_{17}H_{14}O_6$	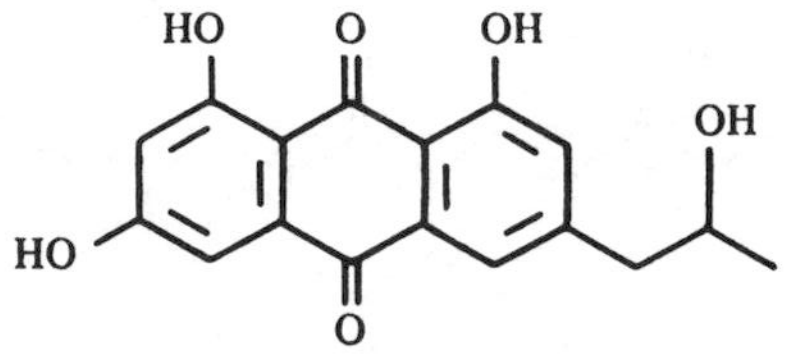	Isorhodoptilometrin	Feather star: *Ptilometra australis*				321 397 o
273. $C_{17}H_{14}O_6$	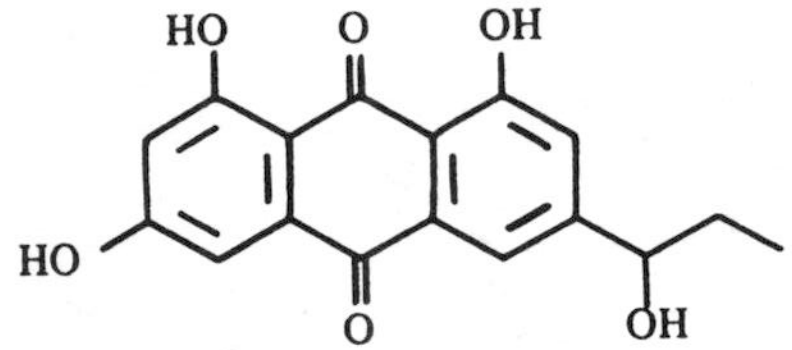	Rhodoptilometrin	Feather stars: *Comanthus bennetti* *Heterometra savignii* *Lamprometra klunzingeri* *Ptilometra australis*				44 114 321 397 n
274. $C_{17}H_{14}O_7$	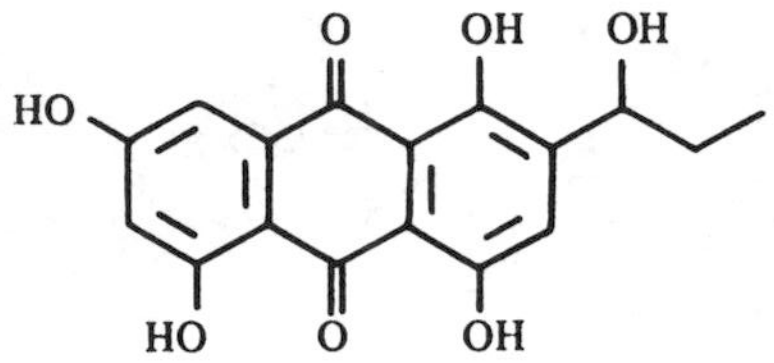	2-(1′-Hydroxy-propyl)-1,4,5,7-tetrahydroxy-9,10-anthraquinone	Feather star: *Comanthus bennetti*				44
275. $C_{17}H_{15}N_3O$	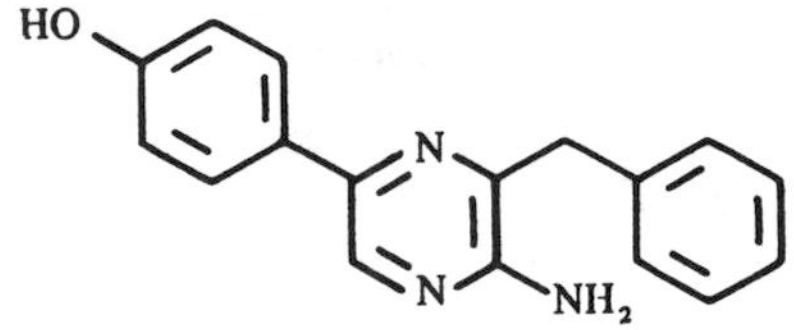	AF-350. 2-Amino-3-benzyl-5-(4′-hydroxy-phenyl)pyrazine (from the photoprotein Aequorin)	Jellyfish: *Aequorea* sp.	+			244

Table 1 (continued)

Molecular formula	Structural formula	Name	Source of compound	Available by synthesis	Available non-marine origin	Activity	References
C_{17} . . .							
276. $C_{17}H_{16}O_5$	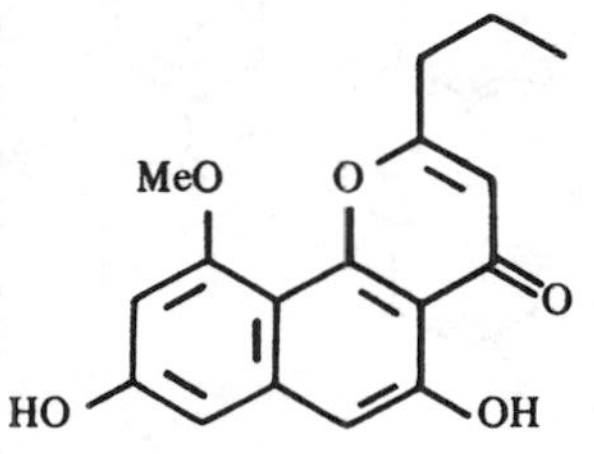 MeO, HO, OH, O (structure) Occurs as monosulfate	Comaparvin	Feather star: *Comanthus parvicirrus timorensis*				368
277. $C_{17}H_{23}BrO_3$	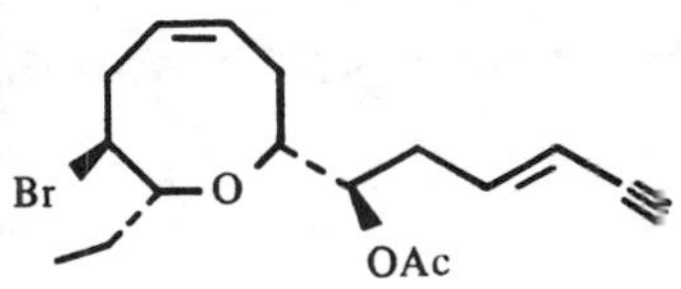 Br, O, OAc (structure)	Laurencin	Red alga: *Laurencia glandulifera*				71 208 212
278. $C_{17}H_{23}BrO_4$	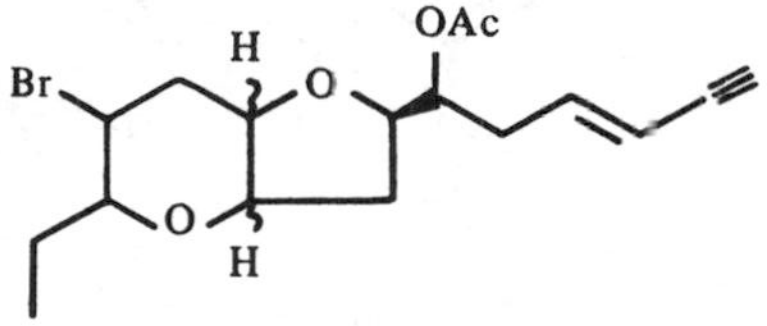 Br, H, O, OAc, O, H (structure)	Acetyllaurefucin	Red alga: *Laurencia nipponica*				141

Table 1 (continued)

Molecular formula	Structural formula	Name	Source of compound	Available by synthesis	Available non-marine origin	Activity	References
C_{18} 279. $C_{18}H_{14}O_7$	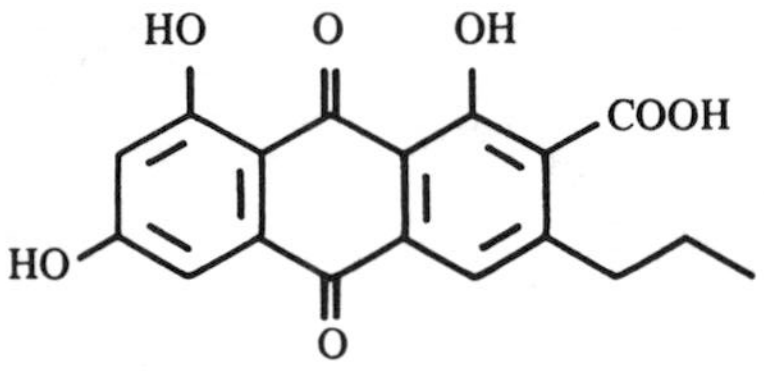	Ptilometric acid	Feather stars: *Ptilometra australis* *Tropiometra afra*				321
280. $C_{18}H_{17}NaO_8S$	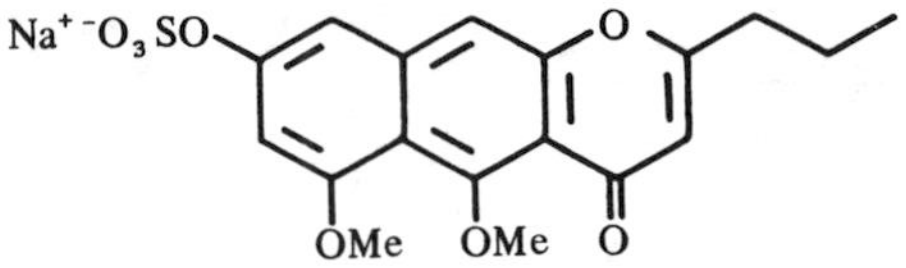	Neocomantherin sodium sulfate ester	Feather star: *Comantheria perplexa*				160 241
281. $C_{18}H_{18}O_6$	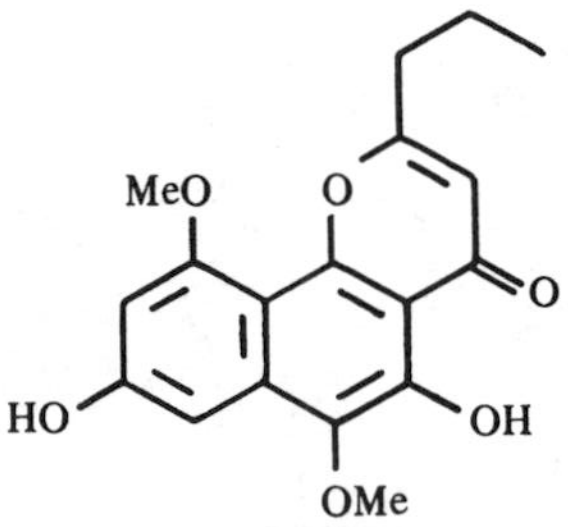 Occurs as monosulfate	6-Methoxy-comaparvin	Feather star: *Comanthus parvicirrus timorensis*				368

Table 1 (continued)

Molecular formula	Structural formula	Name	Source of compound	Available by synthesis	Available non-marine origin	Activity	References
C_{19}							
282. $C_{19}H_{16}O_7$	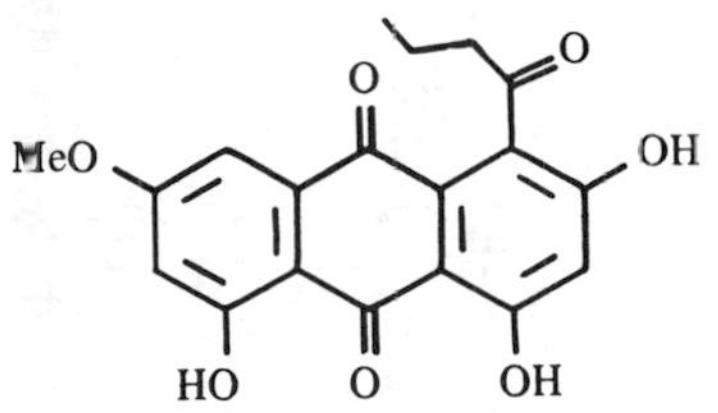	Rhodocomatulin 6-methyl ether	Feather stars: *Comatula cratera* *C. pectinata*				268 381 382 397 p
283. $C_{19}H_{16}O_8$	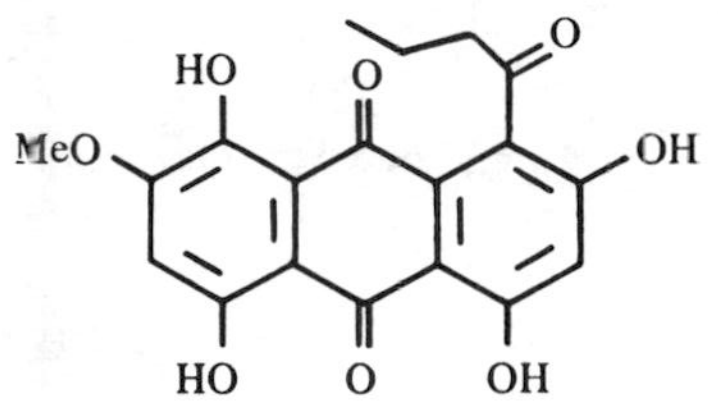	Rubrocomatulin 6-methyl ether	Feather stars: *Comatula cratera* *C. pectinata*				268 320 382 397 q
284. $C_{19}H_{18}O_6$	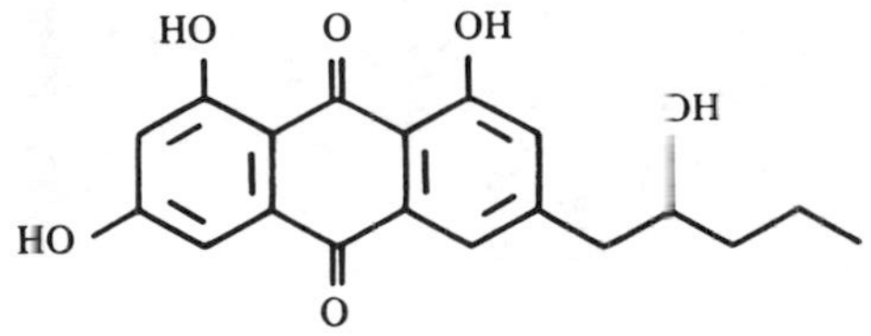	3-(2′-Hydroxy-pentyl)-1,6,8-trihydroxy-9,10-anthraquinone	Feather star: *Comanthus bennetti*				44

Table 1 (continued)

Molecular formula	Structural formula	Name	Source of compound	Available by synthesis	Available non-marine origin	Activity	References
C_{19} . . .							
285. $C_{19}H_{20}O_6$	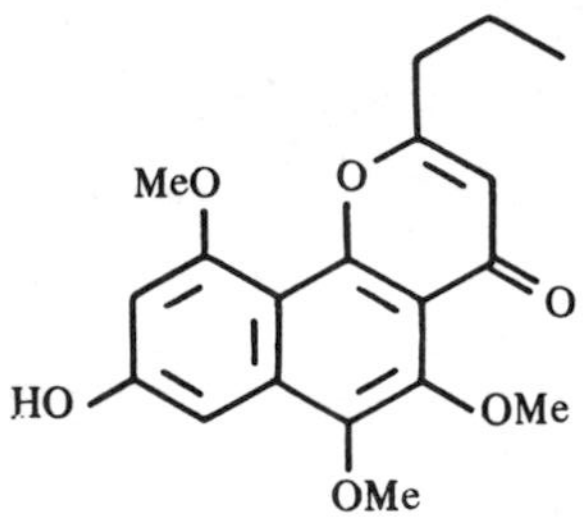 Occurs as monosulfate	6-Methoxy-comaparvin-5-methyl ether	Feather star: *Comanthus parvicirrus timorensis*				368
286. $C_{19}H_{38}$		Zamene	Shark liver oil				82 371
287. $C_{19}H_{40}$		Pristane	Brown alga: *Fucus vesiculosus* Shark liver oil	+	+		82 172 284 III 371
C_{20}							
288. $C_{20}H_{18}O_7$	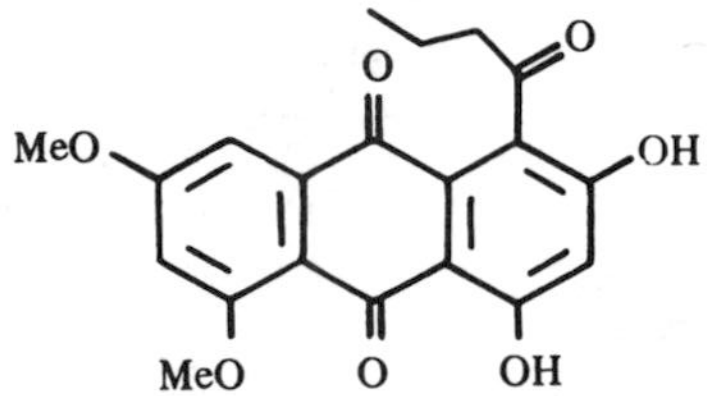	Rhodocomatulin 6,8-dimethyl ether	Feather stars: *Comatula cratera* *C. pectinata*				268 381 382 397 p

Table 1 (continued)

Molecular formula	Structural formula	Name	Source of compound	Available by synthesis	Available non-marine origin	Activity	References
C_{20} ...							
289. $C_{20}H_{24}O_2$	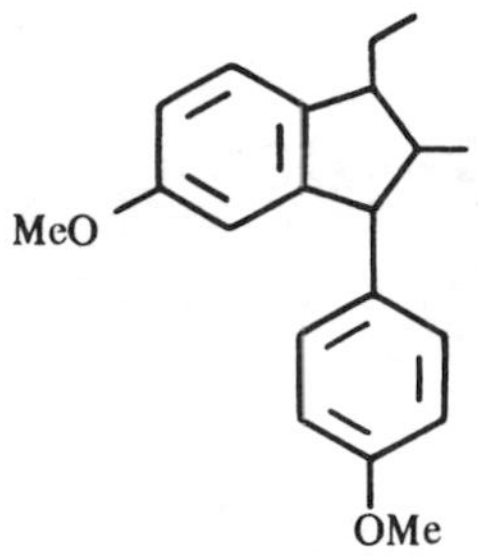 MeO, OMe	Metanethole	Sponge: *Spheciospongia vesparia*	+			53
290. $C_{20}H_{30}O_4$	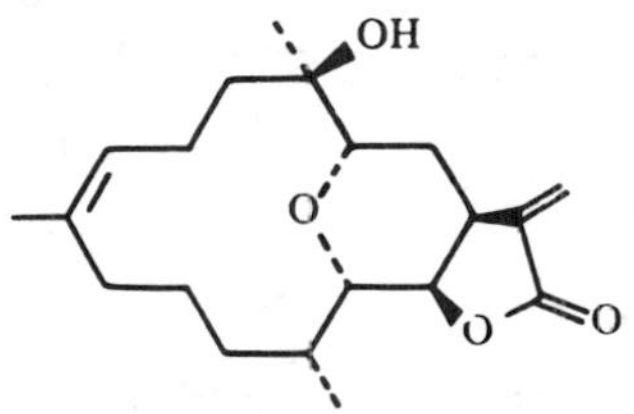 OH, O, O, O	Eunicin	Gorgonian: *Eunicea mammosa*			Antibacterial	192 417 418
291. $C_{20}H_{30}O_4$	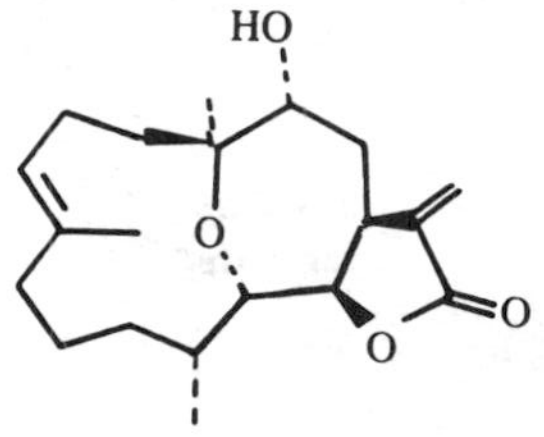 HO, O, O, O	Jeunicin	Gorgonian: *Eunicea mammosa*				88 417

Table 1 (continued)

Molecular formula	Structural formula	Name	Source of compound	Available by synthesis	Available non-marine origin	Activity	References
C_{20} ···							
292. $C_{20}H_{30}O_4$	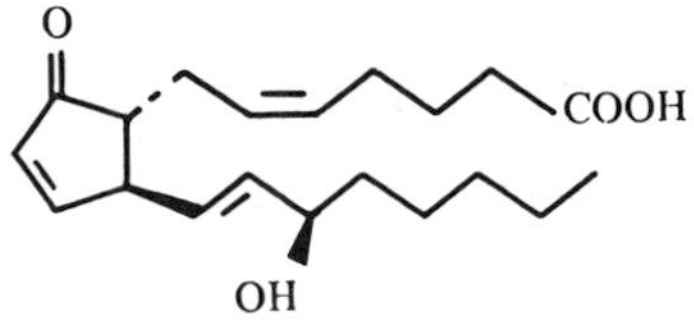	15*R*-PGA_2	Gorgonian: *Plexaura homomalla*				306 346 411 421
293. $C_{20}H_{30}O_4$	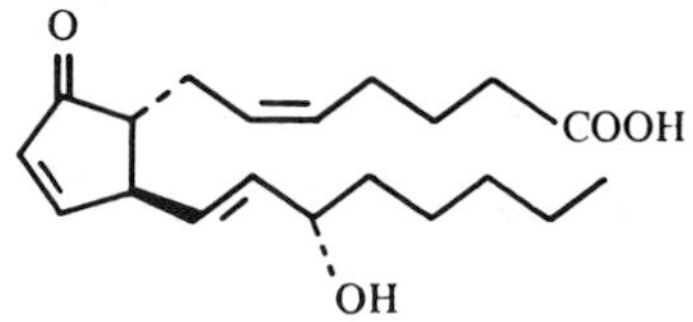	15*S*-PGA_2	Gorgonian: *Plexaura homomalla*		+		67 306 346 411
294. $C_{20}H_{30}O_4$	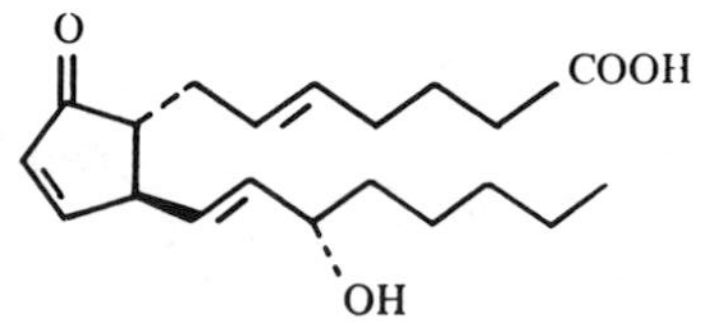	5-*trans*-15*S*-PGA_2	Gorgonian: *Plexaura homomalla*				67
295. $C_{20}H_{32}O$	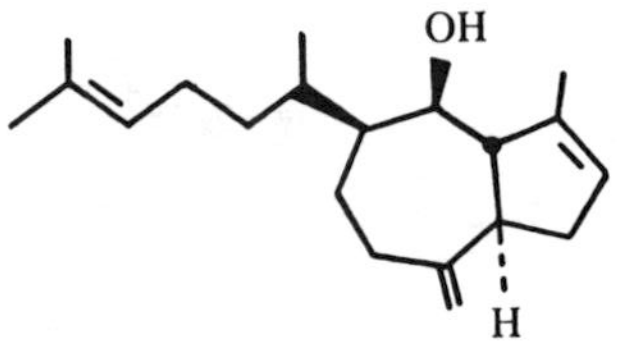	Pachydictyol A	Brown alga: *Pachydictyon coriaceum*			Antibiotic	186

Table 1 (continued)

Molecular formula	Structural formula	Name	Source of compound	Available by synthesis	Available non-marine origin	Activity	References
C_{20} . . .							
296. $C_{20}H_{32}O_5$	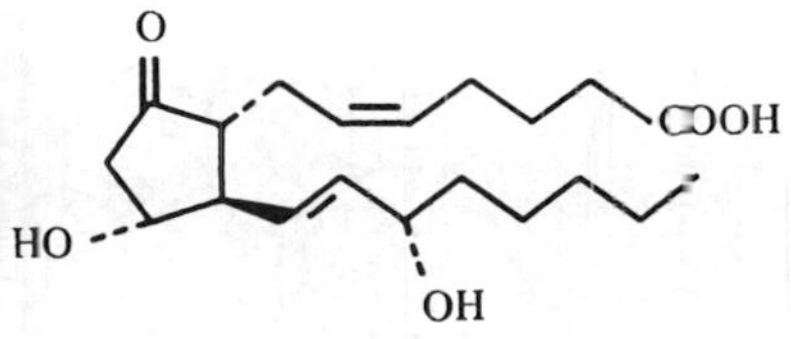 O, COOH, HO, OH	15*S*-PGE$_2$	Gorgonian: *Plexaura homomalla*	+	+	Cardiovascular. Hypotensive. Smooth muscle stimulating	66 284 IIo 306 346 347
297. $C_{20}H_{35}BrO_2$	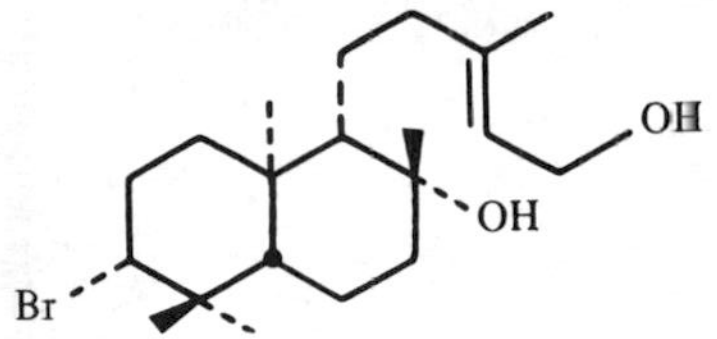 OH, OH, Br	Aplysin-20	Sea hare: *Aplysia kurodai*				281 340
298. $C_{20}H_{35}BrO_2$	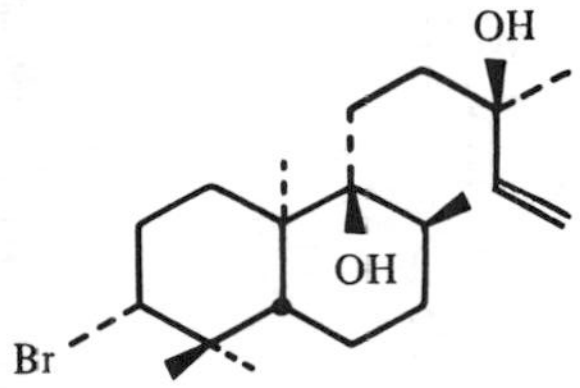 OH, OH, Br	Concinndiol	Red alga: *Laurencia concinna*				362

Table 1 (continued)

Molecular formula	Structural formula	Name	Source of compound	Available by synthesis	Available non-marine origin	Activity	References
C_{21}							
299. $C_{21}H_{22}O_4$	O; HO; OHC; O	Panicein-B_1	Sponge: *Halichondria panicea*				97
300. $C_{21}H_{22}O_4$	OH; HO; O; OHC	Panicein-B_2	Sponge: *Halichondria panicea*				97
301. $C_{21}H_{24}O_4$	O; O; O; O	Nitenin	Sponge: *Spongia nitens*				124

Table 1 (continued)

Molecular formula	Structural formula	Name	Source of compound	Available by synthesis	Available non-marine origin	Activity	References
C_{21} . . .							
302. $C_{21}H_{24}O_4$	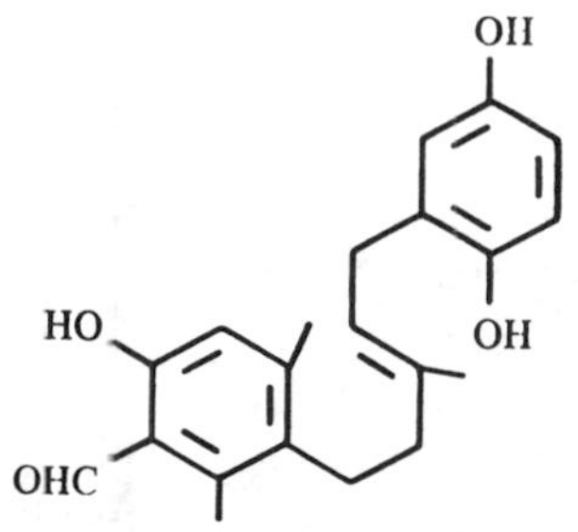 HO, OHC, OH, OH	Panicein-B_3	Sponge: *Halichondria panicea*				97
303. $C_{21}H_{24}O_5$	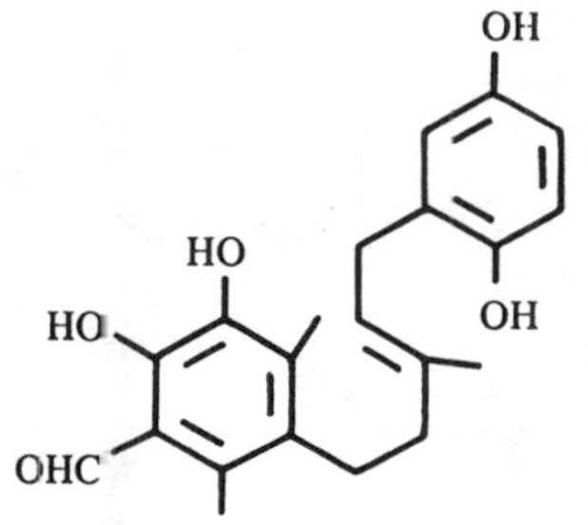 HO, HO, OHC, OH, OH	Panicein-C	Sponge: *Halichondria panicea*				97
304. $C_{21}H_{26}O_3$	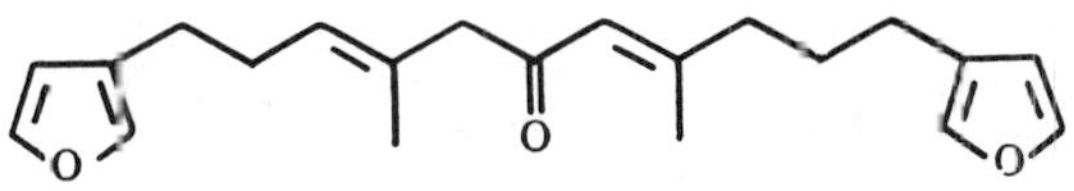 O, O, O	Furospongin-2	Sponges: *Hippospongia communis* *Spongia officinalis*				91
305. $C_{21}H_{26}O_3$	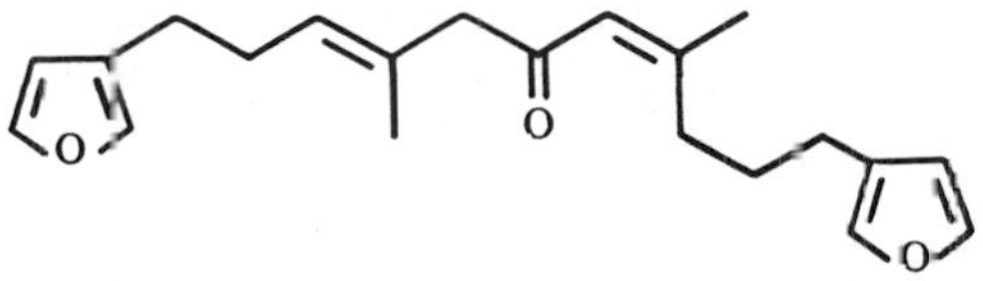 O, O, O	Isofurospongin-2	Sponges: *Hippospongia communis* *Spongia officinalis*				91

Table 1 (continued)

Molecular formula	Structural formula	Name	Source of compound	Available by synthesis	Available non-marine origin	Activity	References
C_{21} . . .							
306. $C_{21}H_{26}O_4$	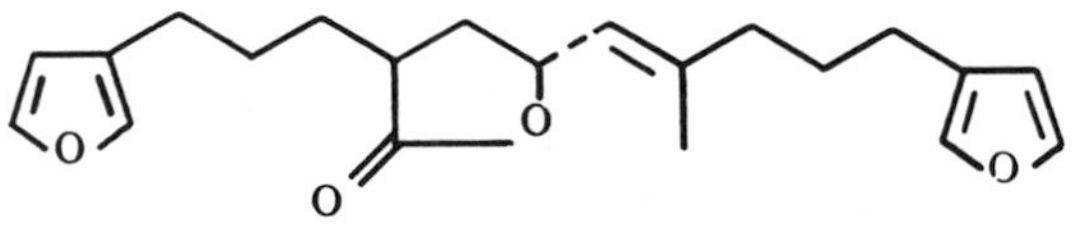	Dihydronitenin	Sponge: *Spongia nitens*				124
307. $C_{21}H_{28}O_2$	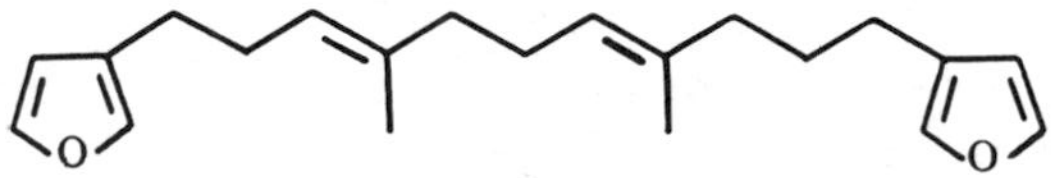	Anhydrofuro-spongin-1	Sponges: *Hippospongia communis* *Spongia officinalis*				91
308. $C_{21}H_{28}O_3$	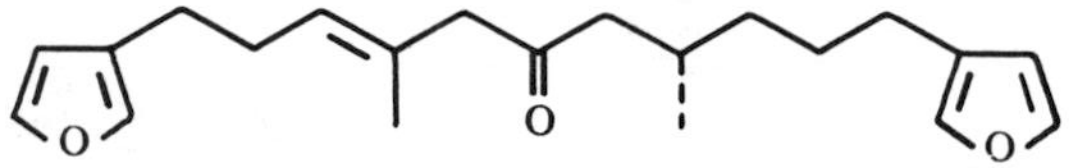	Dihydrofuro-spongin-2	Sponges: *Hippospongia communis* *Spongia officinalis*				91
309. $C_{21}H_{28}O_4$	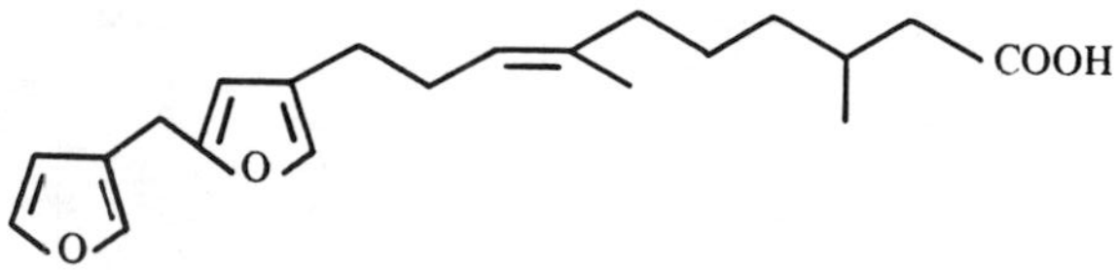	Ircinin-3	Sponge: *Ircinia oros*				95
310. $C_{21}H_{28}O_4$	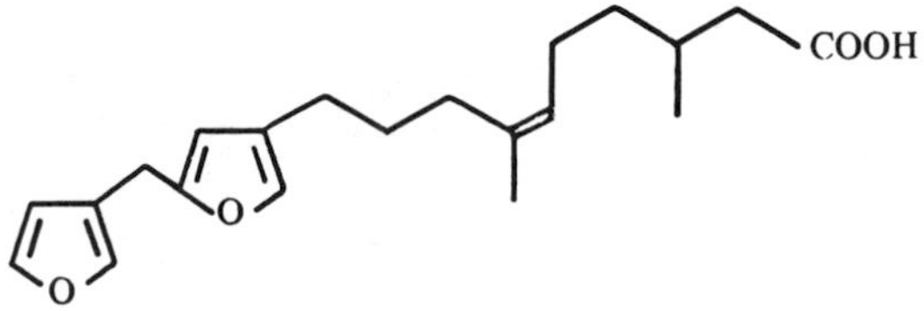	Ircinin-4	Sponge: *Ircinia oros*				95

Table 1 (continued)

Molecular formula	Structural formula	Name	Source of compound	Available by synthesis	Available non-marine origin	Activity	References
C_{21} . . .							
311. $C_{21}H_{30}O_2$	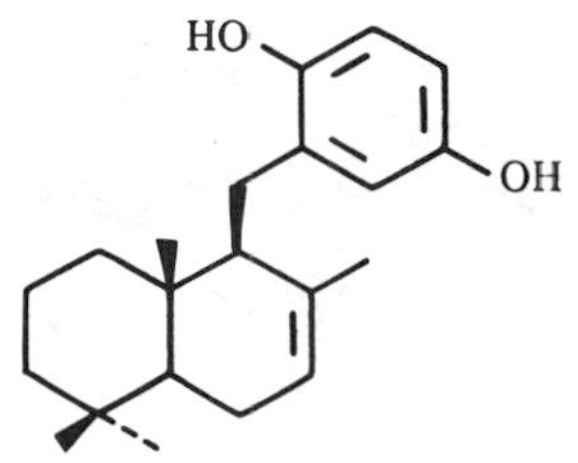 HO, OH	Isozonarol	Brown alga: *Dictyopteris zonarioides*			Fungicidal	133
312. $C_{21}H_{30}O_2$	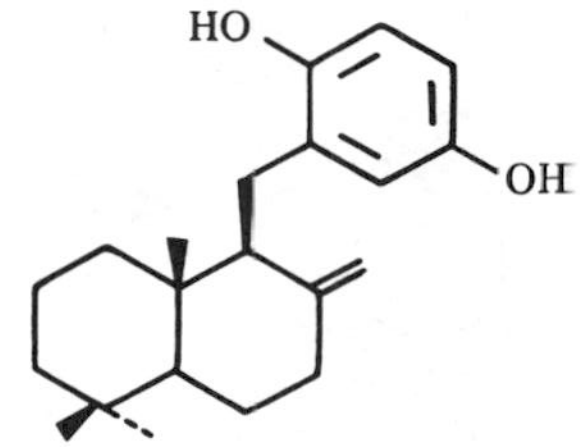 HO, OH	Zonarol	Brown alga: *Dictyopteris zonarioides*			Fungicidal	133
313. $C_{21}H_{30}O_3$	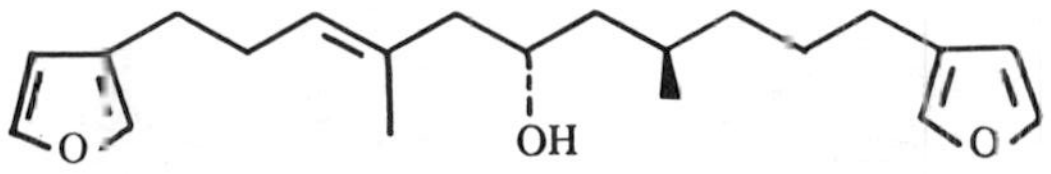 O, OH, O	Furospongin-1	Sponges: *Hippospongia communis* *Spongia officinalis*				89 91
314. $C_{21}H_{30}O_3$	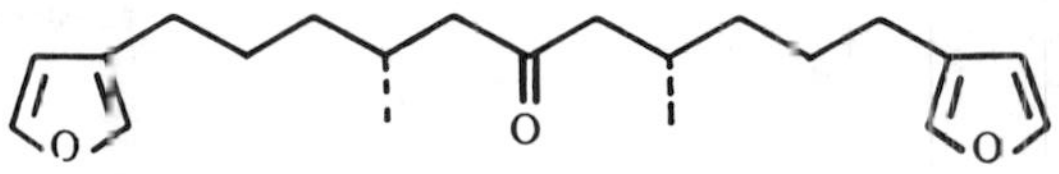 O, O, O	Tetrahydrofuro-spongin-2	Sponges: *Hippospongia communis* *Spongia officinalis*				91
315. $C_{21}H_{32}$		1,6,9,12,15,18-Henei-cosahexaene	Algae				172 445

Table 1 (continued)

Molecular formula	Structural formula	Name	Source of compound	Available by synthesis	Available non-marine origin	Activity	References
C_{21} . . .							
316. $C_{21}H_{32}$		3,6,9,12,15,18-Heneicosahexaene	Algae				58 261 262 445
317. $C_{21}H_{32}O_3$	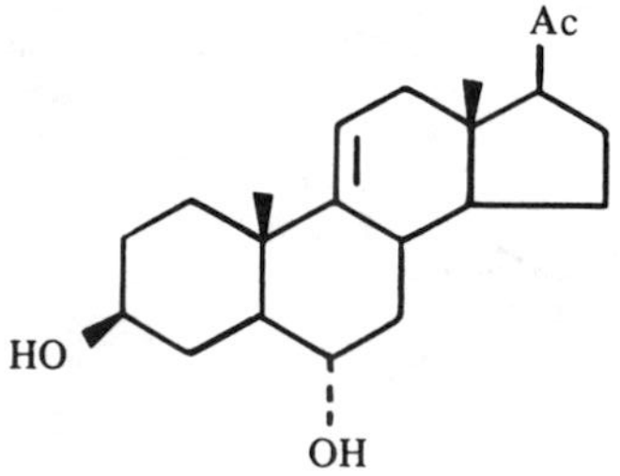 Ac, HO, OH	5α-Pregn-9(11)-ene-3β,6α-diol-20-one	Starfish: *Acanthaster planci* *Asterias amurensis* *A. forbesi* *A. rubens*				35 164 168 201 356 358
318. $C_{21}H_{32}O_4$	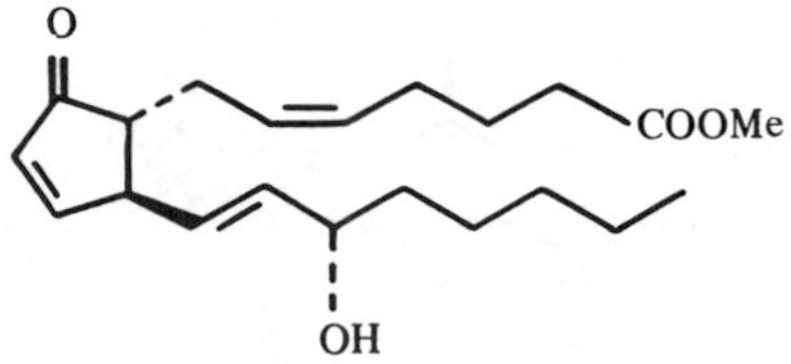 O, COOMe, OH	15*S*-PGA_2 methyl ester	Gorgonian: *Plexaura homomalla*				346 411
319. $C_{21}H_{32}O_6S$	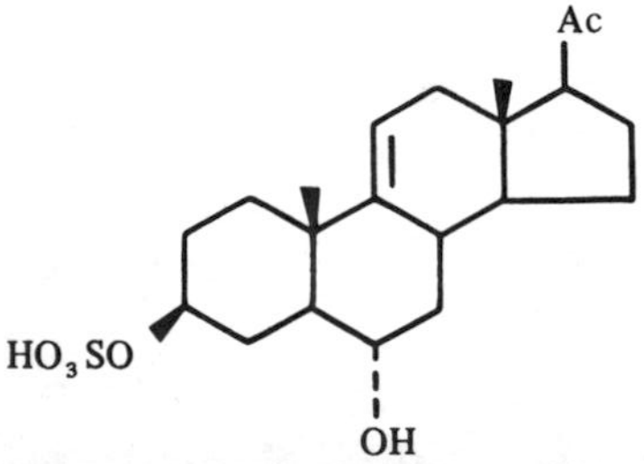 Ac, HO_3SO, OH	5α-Pregn-9(11)-ene-3β,6α-diol-20-one-3-sulfate	Starfish: *Asterias amurensis*				202 203

Table 1 (continued)

Molecular formula	Structural formula	Name	Source of compound	Available by synthesis	Available non-marine origin	Activity	References
C_{21} . . .							
320. $C_{21}H_{34}$		1,6,9,12,15-Heneicosa-pentaene	Algae: *Ascophyllum nodosum* *Fucus vesiculosus* *Laminaria digitata* *L. saccharina*				172 445
321. $C_{21}H_{34}O_3$		3β-6α-Dihydroxy-5α-pregnan-20-one	Starfish: *Asterias rubens*				168
322. $C_{21}H_{42}NO_2^+X^-$		O-(14-Methyl-4-pentadecenyl)-choline	Oyster				307
323. $C_{21}H_{44}O_3$		Batyl alcohol	Gorgonian: *Plexaura flexuosa* Shark liver oil	+	+		104 184 242 284 h

Table 1 (continued)

Molecular formula	Structural formula	Name	Source of compound	Available by synthesis	Available non-marine origin	Activity	References
C_{22}							
324. $C_{22}H_{12}O_{13}$	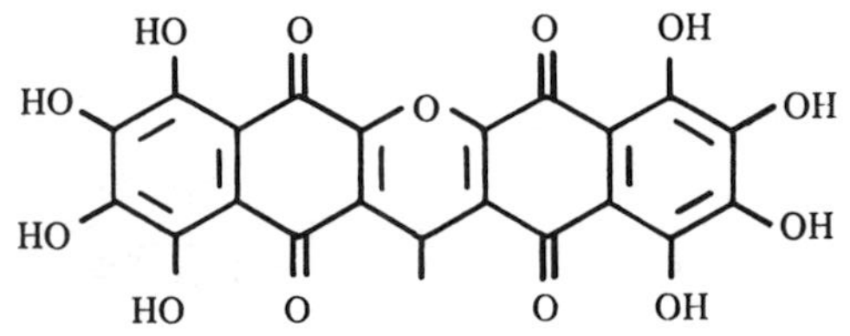	Anhydro-ethylidene -3,3′-bis (2,6,7-trihydroxy-naphthazarin)	Sea urchins: *Spatangus purpureus* *S. raschi* *Stronglyocentrotus drobachiensis*				280 397 l
325. $C_{22}H_{14}O_{14}$	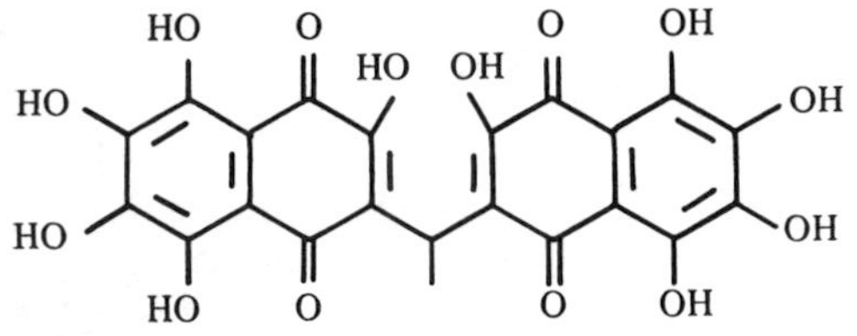	Ethylidene-3,3′-bis (2,6,7-tri-hydroxy-naphthazarin)	Sea urchin: *Spatangus purpureus*	+			280 397 k
326. $C_{22}H_{26}O_3$	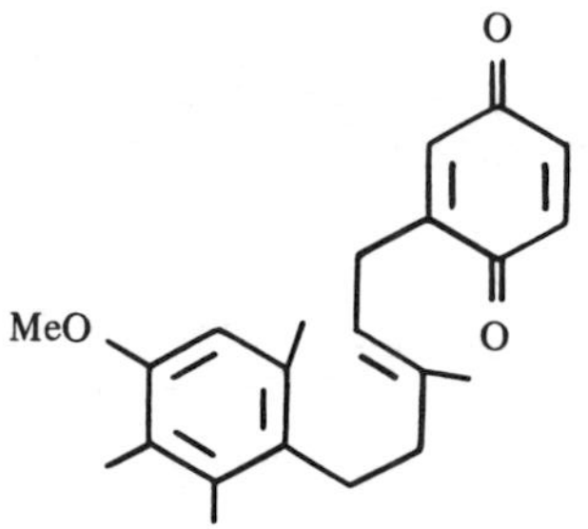	Panicein-A	Sponge: *Halichondria panicea*				97

Table 1 (continued)

Molecular formula	Structural formula	Name	Source of compound	Available by synthesis	Available non-marine origin	Activity	References
C_{22} . . .							
327. $C_{22}H_{32}N_2$		Hexadecahydro-1α, 2β,5β,8α-tetra-methyl-1,8-pyren-ediyl-diisocyanide	Sponge: *Adocia* sp.				43
328. $C_{22}H_{32}O_5$		Crassin acetate	Gorgonian: *Pseudoplexaura crassa*, *P. porosa*, *P. wagenaari*			Antibiotic	85, 88, 194, 417
329. $C_{22}H_{32}O_5$		Eupalmerin acetate	Gorgonians: *Eunicea mammosa*, *E. palmeri*, *E. succinea*				88
330. $C_{22}H_{34}O_4$		Ancepsenolide	Gorgonians: *Pterogorgia anceps*, *P. guadalupensis*				342, 345

Table 1 (continued)

Molecular formula	Structural formula	Name	Source of compound	Available by synthesis	Available non-marine origin	Activity	References
C_{22} . . .							
331. $C_{22}H_{36}O_5$	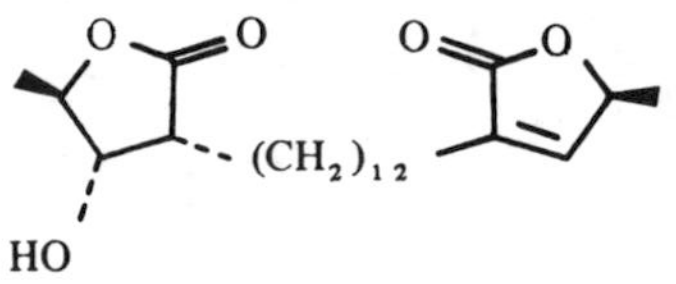	Hydroxy-ancepsenolide	Gorgonian: *Pterogorgia anceps*				343
332. $C_{22}H_{42}O_2S_2$	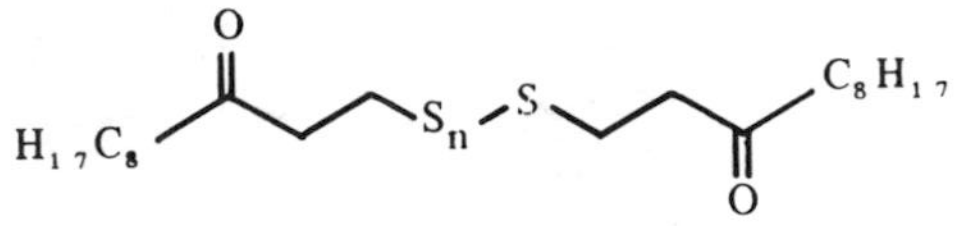 n = 1	Bis-(3-oxoundecyl) disulfide	Brown algae: *Dictyopteris australis* *D. plagiogramma*				228 293 330
333. $C_{22}H_{42}O_2S_3$	n = 2	Bis-(3-oxoundecyl) trisulfide	Brown algae: *Dictyopteris australis* *D. plagiogramma*				293
334. $C_{22}H_{42}O_2S_4$	n = 3	Bis-(3-oxoundecyl) tetrasulfide	Brown algae: *Dictyopteris australis* *D. plagiogramma*				293

Table 1 (continued)

Molecular formula	Structural formula	Name	Source of compound	Available by synthesis	Available non-marine origin	Activity	References
C_{23} . . .							
335. $C_{23}H_{34}O_5$	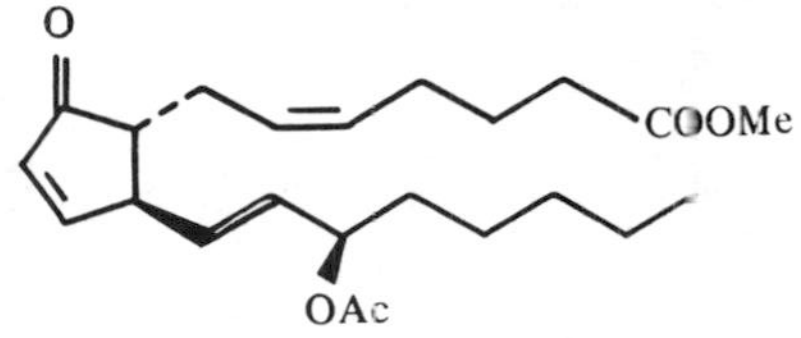	O-Acetyl (15*R*)-PGA_2 methyl ester	Gorgonian: *Plexaura homomalla*				346 411 421
336. $C_{23}H_{34}O_5$	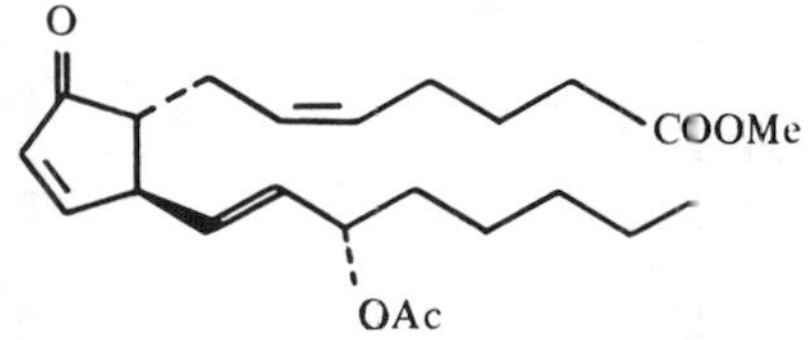	O-Acetyl(15*S*)-PGA_2 methyl ester	Gorgonian: *Plexaura homomalla*				411
337. $C_{23}H_{46}ClNO_4$	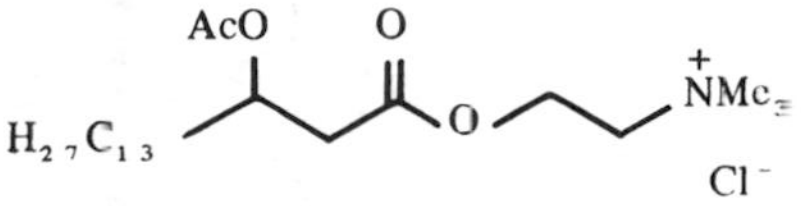	Pahutoxin	Boxfish: *Ostracion lentiginosus*	+			61 63 340
C_{24}							
338. $C_{24}H_{18}N_2O_4$	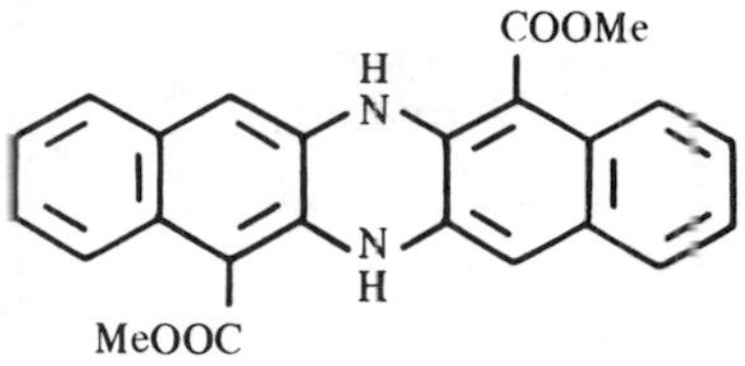	Caulerpin	Green algae: *Caulerpa lamourouxii* (*C. racemosa* var. *lamourouxii*) *C. racemosa* var. *clavifera* *C. serrulata* *C. sertularioides*				25 108 339

Table 1 (continued)

Molecular formula	Structural formula	Name	Source of compound	Available by synthesis	Available non-marine origin	Activity	References
C_{24} . . .							
339. $C_{24}H_{26}Br_4N_4O_8$	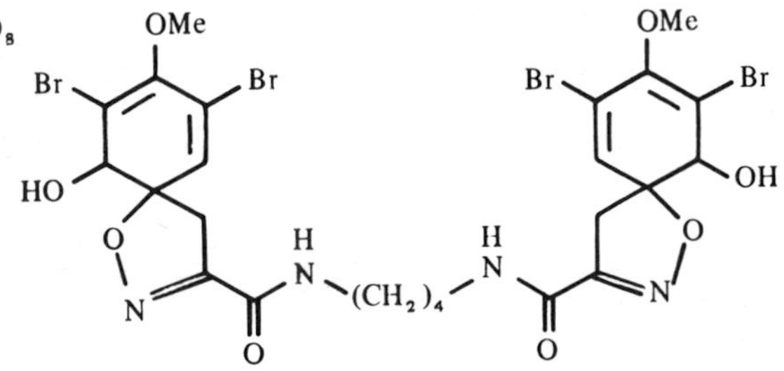	Aerothionin	Sponges: *Verongia aerophoba* *V. thiona*				121 289
340. $C_{24}H_{40}O_4$	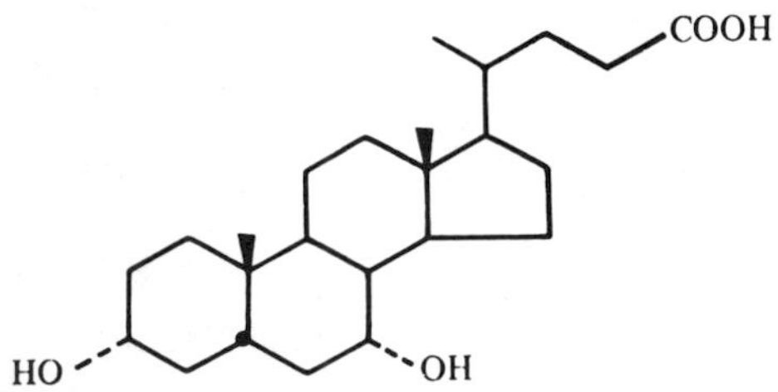	Chenodeoxycholic acid. 3α,7α-Dihydroxy-cholanic acid	Fish bile	+	+		179 182 284 o
341. $C_{24}H_{40}O_5$	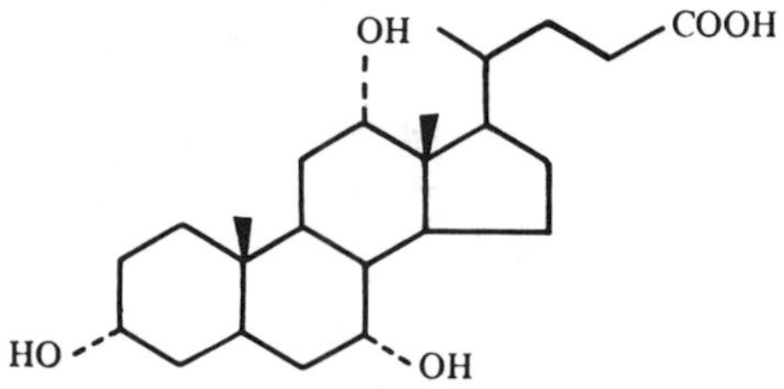	Allo-5α-cholic acid	Fish bile				182

Table 1 (continued)

Molecular formula	Structural formula	Name	Source of compound	Available by synthesis	Available non-marine origin	Activity	References
C_{24} . . .							
342. $C_{24}H_{40}O_5$	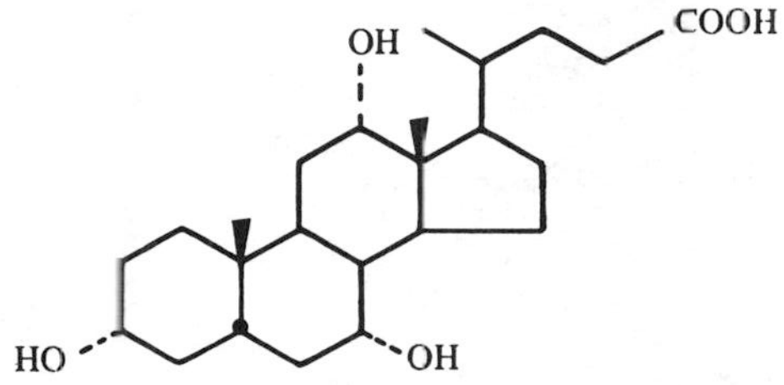 OH, COOH, HO, OH	Cholic acid. 3α,7α,12α-Trihydroxycholanic acid	Fish bile	+	+		178 179 182 284 q
343. $C_{24}H_{40}O_5$	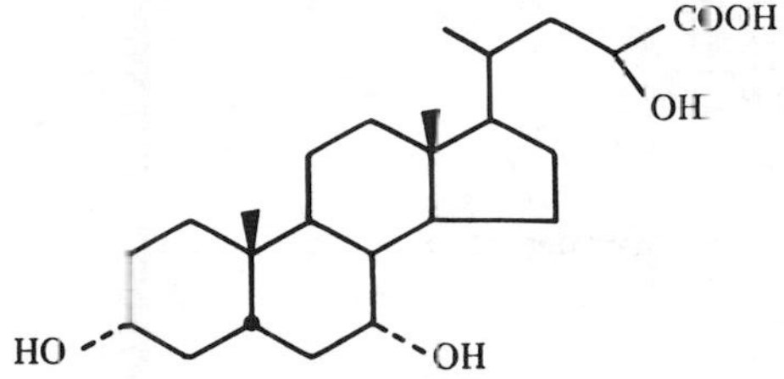 COOH, OH, HO, OH	β-Phocaecholic acid. 3α,7α,23-Trihydroxycholanic acid	Leopard seal: *Hydrurga leptonyx* Sea lion: *Zalophus californianus*				181
344. $C_{24}H_{40}O_7$	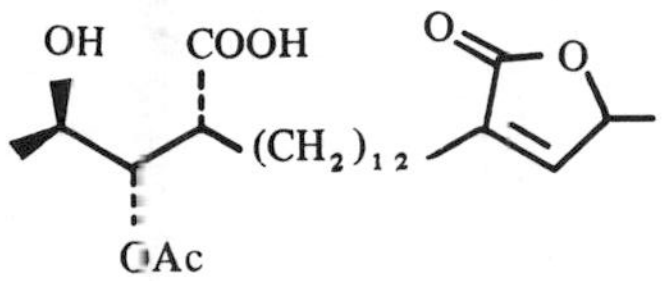 OH, COOH, O, O, $(CH_2)_{12}$, OAc	Guadalupensic acid	Gorgonian: *Pterogorgia guadalupensis*				130
345. $C_{24}H_{44}O_5$	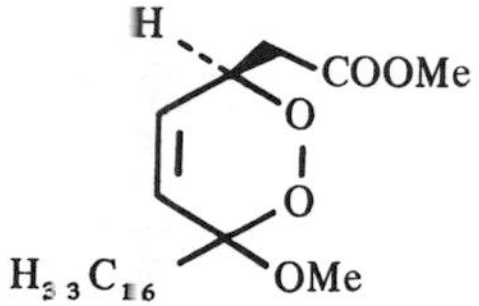 H, COOMe, O, O, $H_{33}C_{16}$, OMe	Chondrillin	Sponge: *Chondrilla* sp.				424

Table 1 (continued)

Molecular formula	Structural formula	Name	Source of compound	Available by synthesis	Available non-marine origin	Activity	References
C_{24} . . .							
346. $C_{24}H_{45}NO_3$	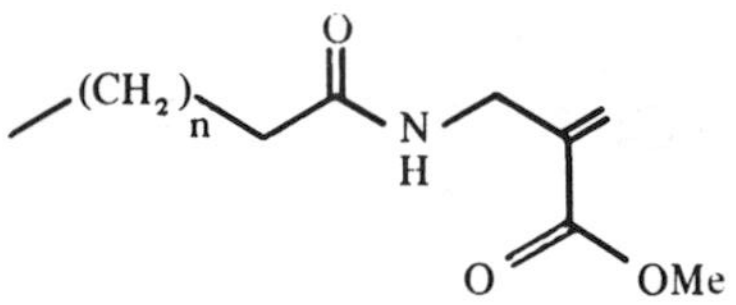 n = 16 (n = 13, 14, 15, 16, 17 detected)	N-Nonadecanoyl-2-methylene-β-alanine methyl ester	Sponge: *Fasciospongia cavernosa*				234
C_{25}							
347. $C_{25}H_{21}N_3O_3$	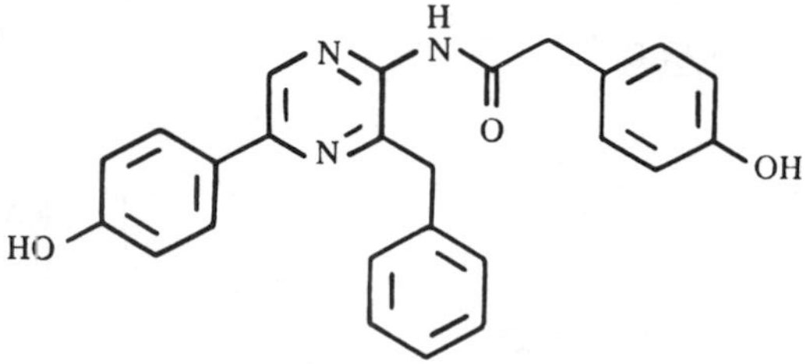	2-(N-*p*-Hydroxyphenyl ethanoyl) amino-3-benzyl-5-*p*-hydroxyphenyl pyrazine (from the photoprotein Aequorin)	Jellyfish: *Aequorea* sp.	+			359
348. $C_{25}H_{26}BrN_5O_{13}$	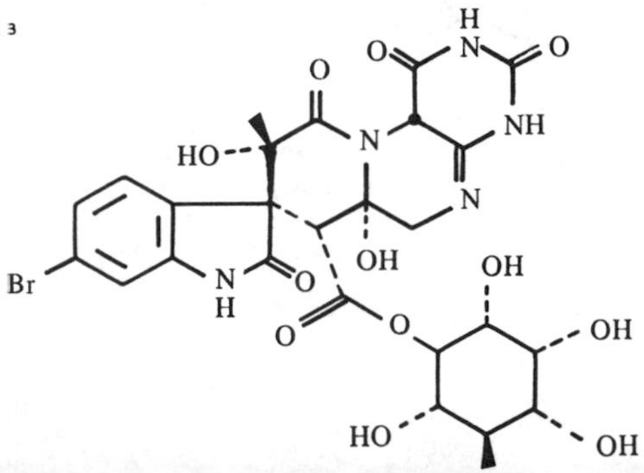	Surugatoxin	Mollusk: *Babylonia japonica*			Mydriatic. Parasympatholytic agent	177 255 305

Table 1 (continued)

Molecular formula	Structural formula	Name	Source of compound	Available by synthesis	Available non-marine origin	Activity	References
C_{25} ...							
349. $C_{25}H_{28}Br_4N_4O_8$	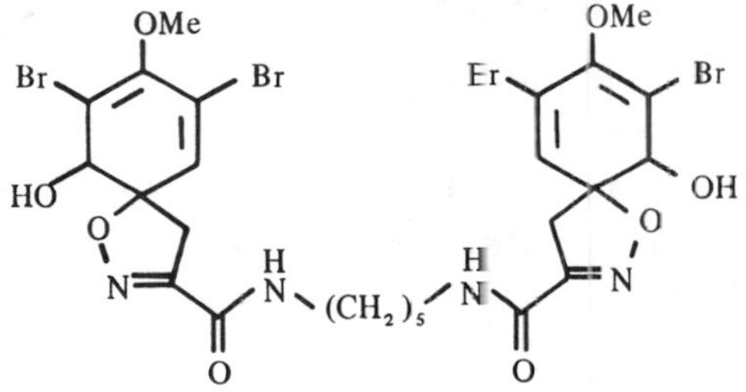	Homoaerothionin	Sponges: *Verongia aerophoba* *V. thiona*				122 289
350. $C_{25}H_{30}O_5$	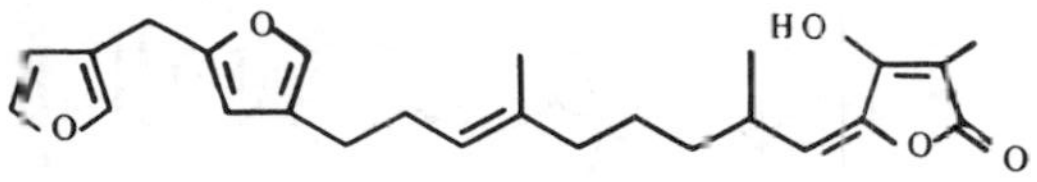	Ircinin-1	Sponge: *Ircinia oros*				92
351. $C_{25}H_{30}O_5$	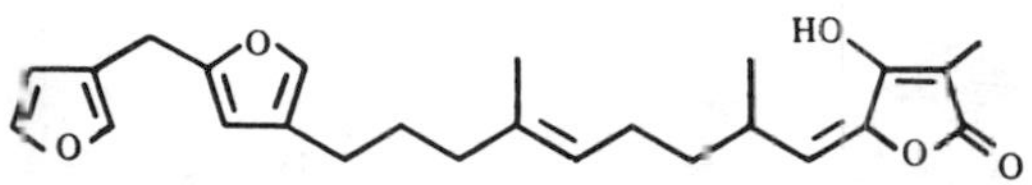	Ircinin-2	Sponge: *Ircinia oros*				92
352. $C_{25}H_{34}O_4$	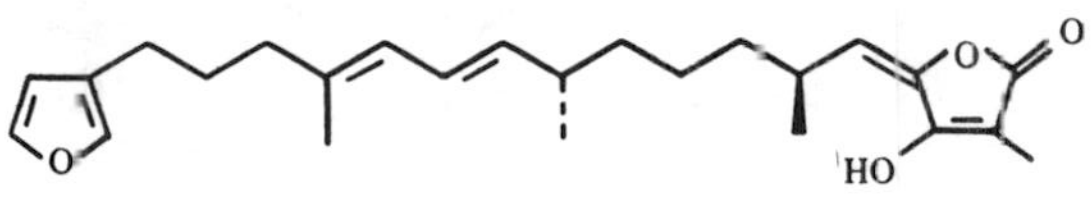	Fasciculatin	Sponge: *Ircinia fasciculata*				69
353. $C_{25}H_{34}O_4$	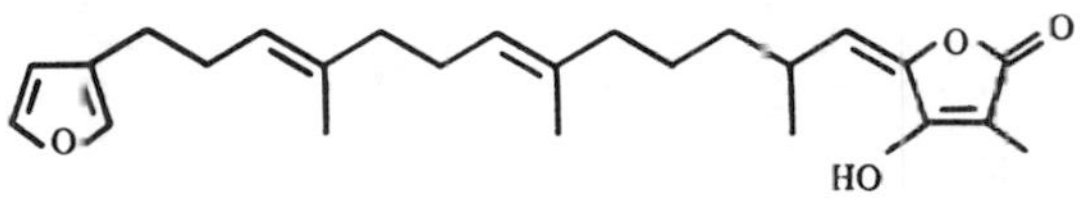	Variabilin	Sponge: *Ircinia variabilis*			Antibiotic	129

Table 1 (continued)

Molecular formula	Structural formula	Name	Source of compound	Available by synthesis	Available non-marine origin	Activity	References
C_{25} . . .							
354. $C_{25}H_{38}O$	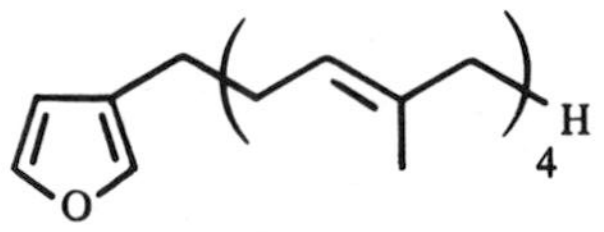	Furospinosulin-1	Sponge: *Ircinia spinosula*				93
C_{26}							
355. $C_{26}H_{36}O_2$	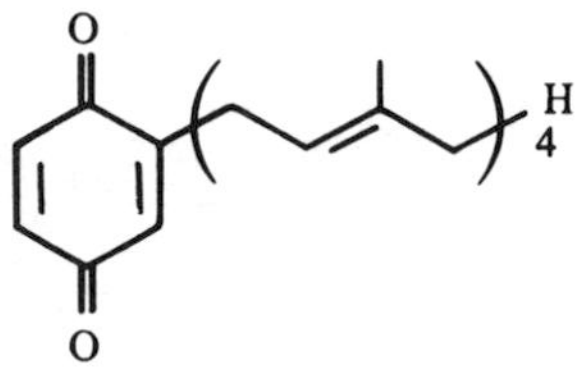	2-Tetraprenyl-1,4-benzoquinone	Sponge: *Ircinia muscarum*				90
356. $C_{26}H_{36}O_5$	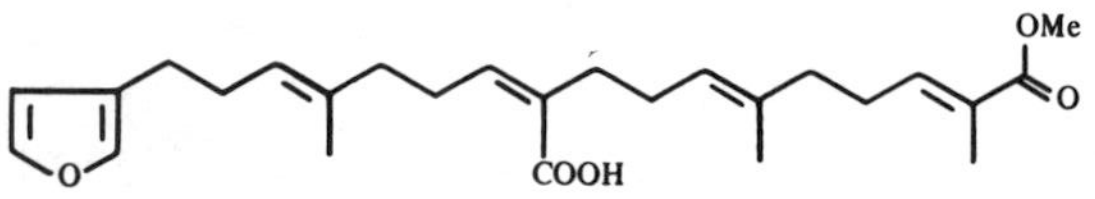	Furospongin-3	Sponge: *Spongia officinalis*				95
357. $C_{26}H_{36}O_5$	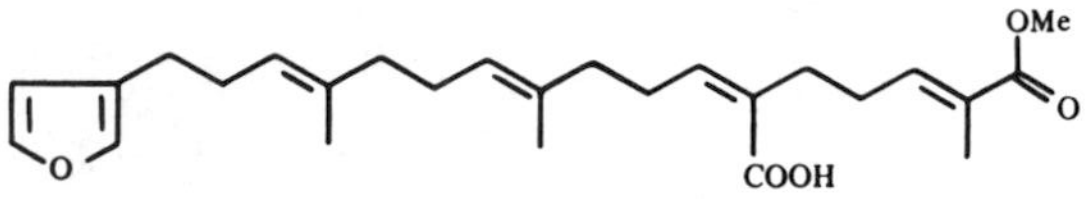	Furospongin-4	Sponge: *Spongia officinalis*				95
358. $C_{26}H_{38}O_2$	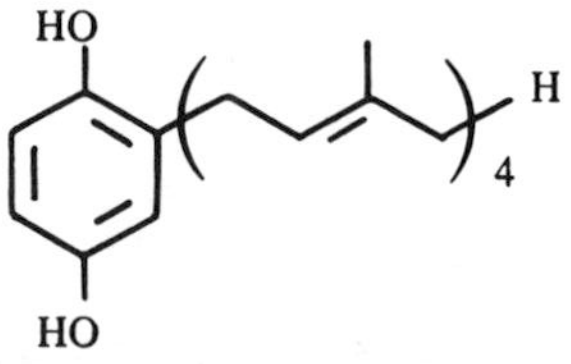	2-Tetraprenyl-1,4-benzoquinol	Sponge: *Ircinia muscarum*				90

Table 1 (continued)

Molecular formula	Structural formula	Name	Source of compound	Available by synthesis	Available non-marine origin	Activity	References
C_{26} . . .							
359. $C_{26}H_{42}O$	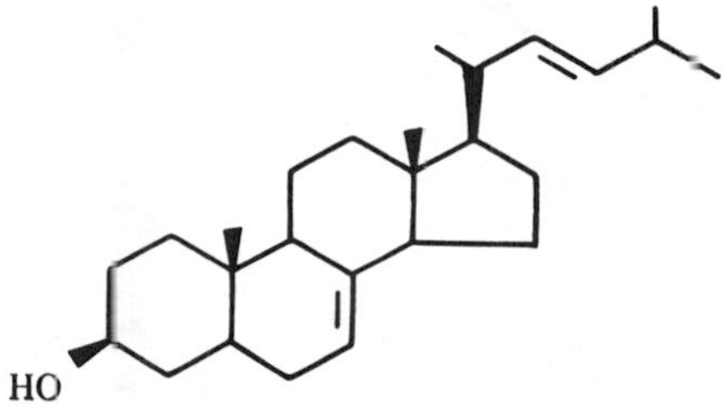 HO	Asterosterol. 22-*trans*-24-Nor-5α-cholesta-7,22-dien-3β-ol	Sea cucumber: *Stichopus japonicus* Starfish	+			251 252
360. $C_{26}H_{42}O$	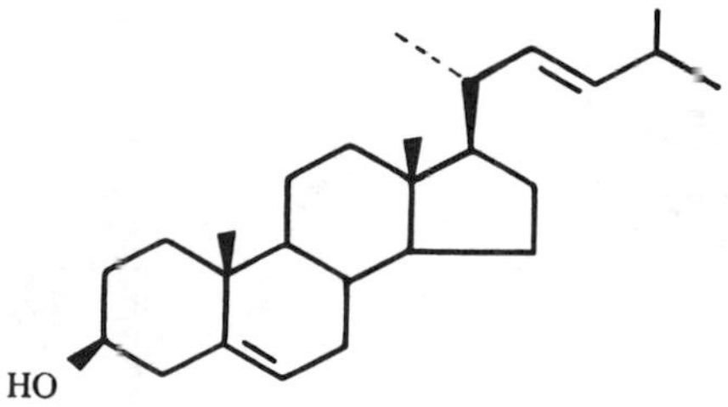 HO	24-Norcholesta-5,22-dien-3β-ol	Mollusks Sea cucumber Sea worm Sponge: *Cliona celata* Tunicate: *Halocynthia roretzi*	+			28 115 140 200 251
361. $C_{26}H_{42}O$	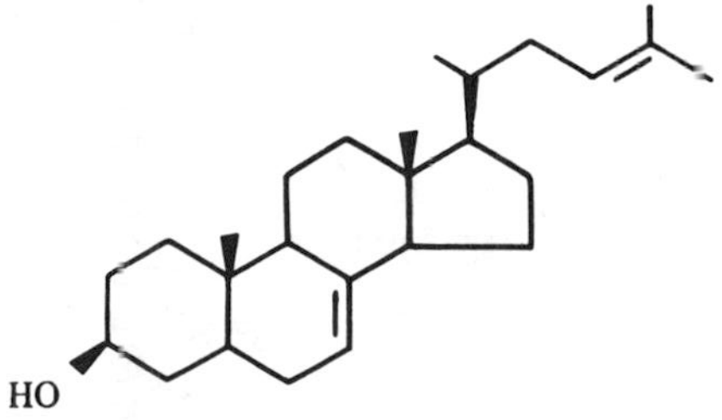 HO	24-Nor-5α-cholesta-7,23-dien-3β-ol	Tunicate *Halocynthia roretzi*				28

Table 1 (continued)

Molecular formula	Structural formula	Name	Source of compound	Available by synthesis	Available non-marine origin	Activity	References
C_{26} . . .							
362. $C_{26}H_{42}O_8$	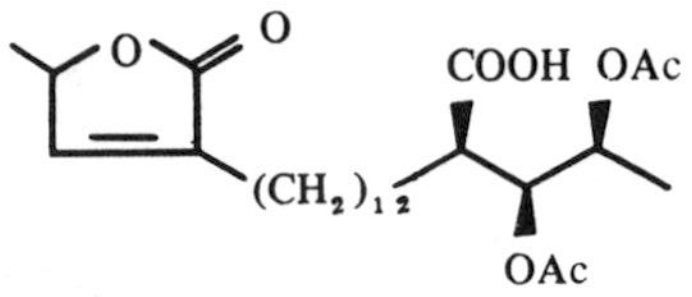	2-(13-Carboxy-14, 15-diacetoxy-hexadecanyl)-2-penten-4-olide	Gorgonian: *Pterogorgia guadalupensis*			Antibiotic	345
363. $C_{26}H_{44}O$	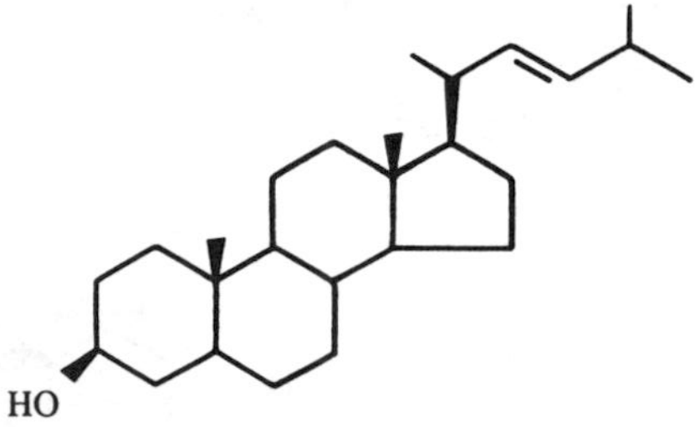	24-Norcholest-22-en-3β-ol	Tunicate: *Halocynthia roretzi* Sponge: *Hymeniacidon perleve*	+			115 286
364. $C_{26}H_{44}O$	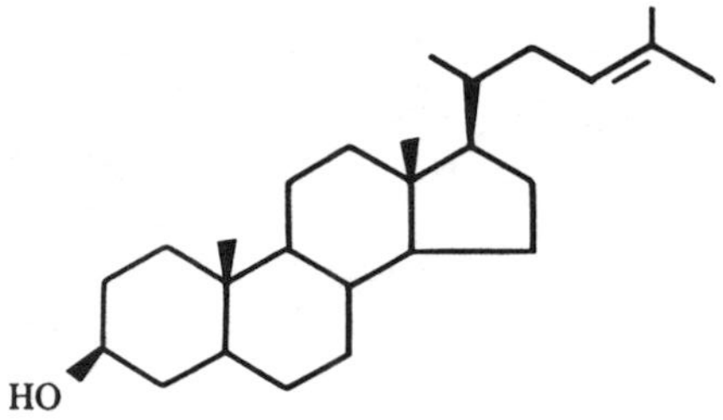	24-Norcholest-23-en-3β-ol	Tunicate: *Halocynthia roretzi*	+			28 285

Table 1 (continued)

Molecular formula	Structural formula	Name	Source of compound	Available by synthesis	Available non-marine origin	Activity	References
C_{26} ...							
365. $C_{26}H_{46}O_4S_2$		(–)-Bis-(3-acetoxy-undec-5-enyl) disulfide	Brown algae: *Dictyopteris* sp.				294
C_{27}							
366. $C_{27}H_{38}O_3$		4-Hydroxy-3-tetra-prenylbenzoic acid	Sponge: *Ircinia muscarum*				90
367. $C_{27}H_{40}O_3$		Taondiol	Brown alga: *Taonia atomaria*	+			150 151 154
368. $C_{27}H_{40}O_5$		Scalarin	Sponge: *Cacospongia scalaris*				126

Table 1 (continued)

Molecular formula	Structural formula	Name	Source of compound	Available by synthesis	Available non-marine origin	Activity	References
C_{27} . . . 369. $C_{27}H_{42}O_3$	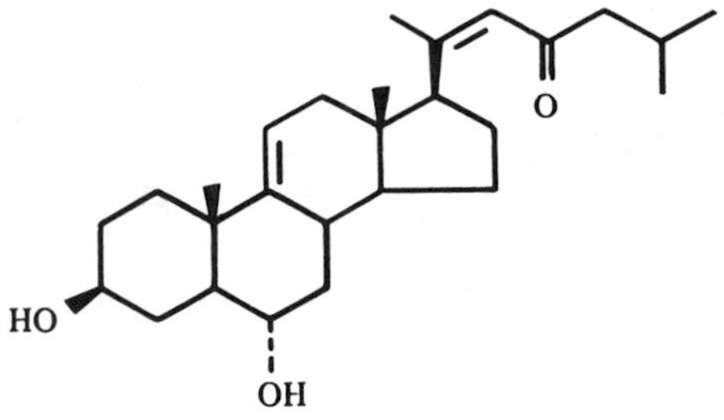	5α-Cholesta-9(11), 20(22)-diene-3β, 6α-diol-23-one	Starfish: *Acanthaster planci* *Asterias rubens*				168 356
370. $C_{27}H_{42}O_3$	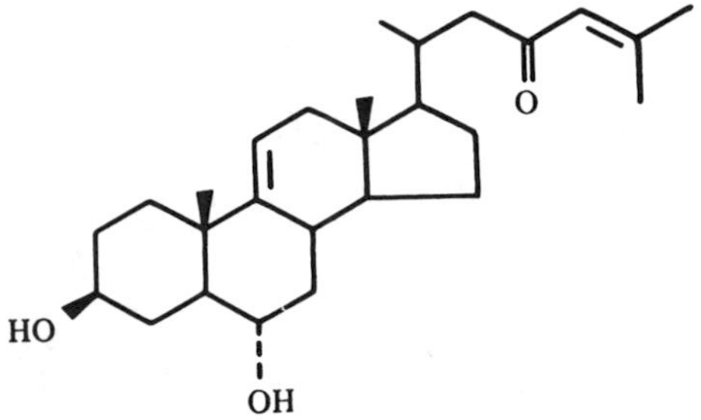	Marthasterone	Starfish: *Marthasterias glacialis*				271 367 406
371. $C_{27}H_{42}O_4$	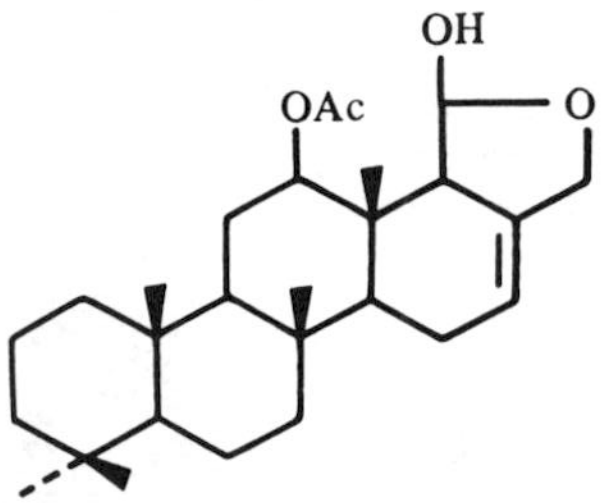	Deoxoscalarin	Sponge: *Spongia officinalis*				96

Table 1 (continued)

Molecular formula	Structural formula	Name	Source of compound	Available by synthesis	Available non-marine origin	Activity	References
C_{27} . . .							
372. $C_{27}H_{44}O$	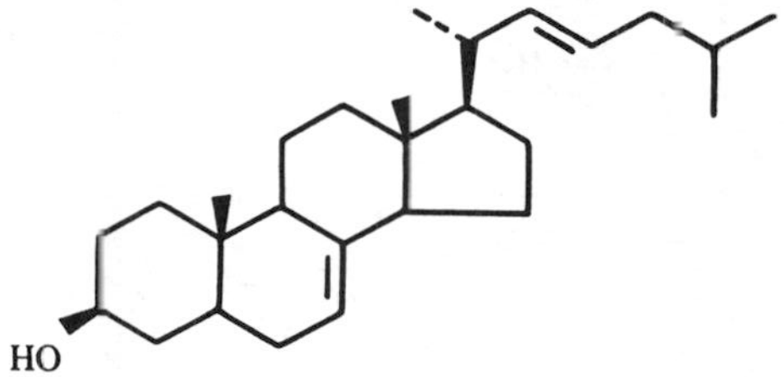 HO	22-*trans*-Cholesta-7,22-dien-3β-ol	Sea cucumber: *Holothuria atra* Starfish	+			162 168 251 252
373. $C_{27}H_{44}O$	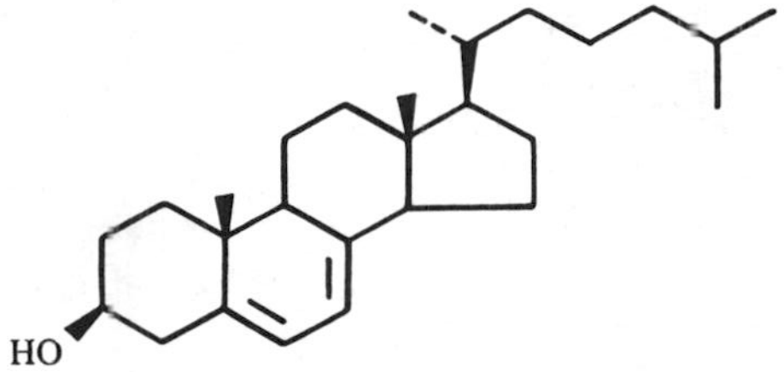 HO	7-Dehydrocholesterol. Provitamin D_3	Lobster: *Nephrops norvegicus* Mollusks	+	+		39 50 55 200 284 w
374. $C_{27}H_{44}O$	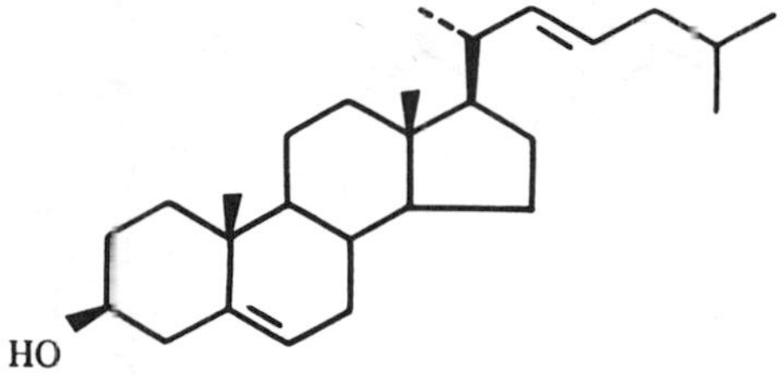 HO	22-Dehydrocholesterol (*cis* and *trans*)*	Crustaceans Feather star Mollusks* Red algae Sea cucumber Sponge*	+	+		39, 55, 115,* 162, 198, 199, 200,* 393, 395, 401

Table 1 (continued)

Molecular formula	Structural formula	Name	Source of compound	Available by synthesis	Available non-marine origin	Activity	References
C_{27} . . .							
375. $C_{27}H_{44}O$	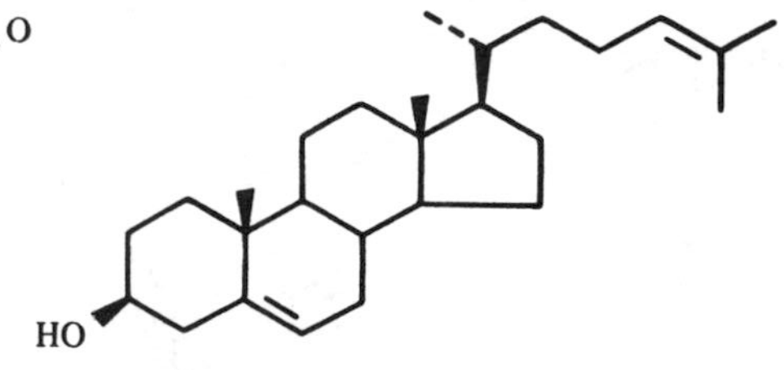	Desmosterol. 24-Dehydrocholesterol	Crustaceans Mollusks Red algae	+	+		39 118 198 199 200 341 m 395
376. $C_{27}H_{44}O_3$	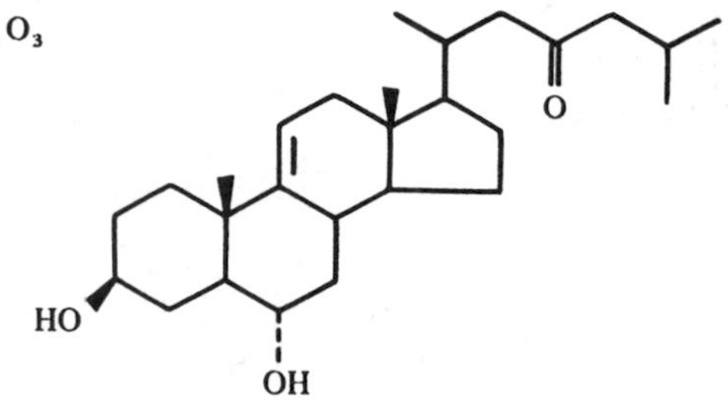	24,25-Dihydromarthasterone	Starfish: *Asterias rubens* *Marthasterias glacialis*				168 367 406
377. $C_{27}H_{44}O_6$	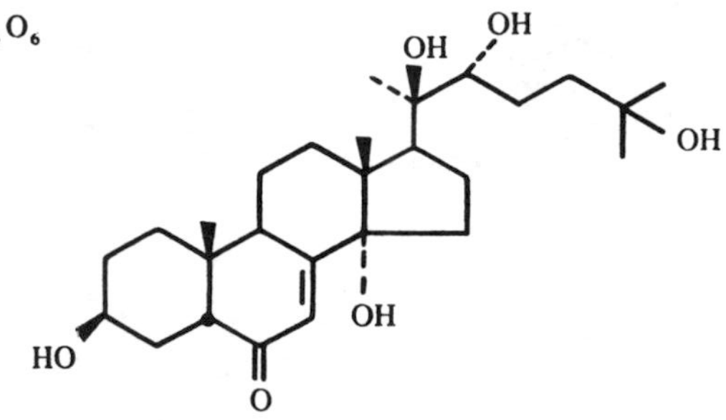	Deoxycrustecdysone	Crayfish: *Jasus lalandei*			Molting hormone	145

Table 1 (continued)

Molecular formula	Structural formula	Name	Source of compound	Available by synthesis	Available non-marine origin	Activity	References
C_{27} . . .							
378. $C_{27}H_{44}O_7$	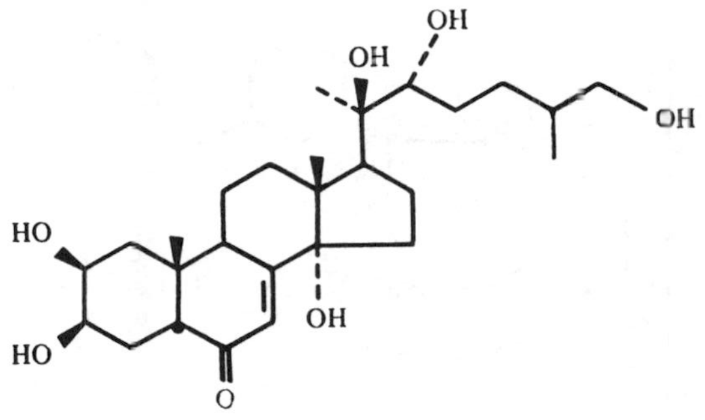	Callinecdysone A	Crayfish: *Callinectes sapidus*			Molting hormone	131
379. $C_{27}H_{44}O_7$	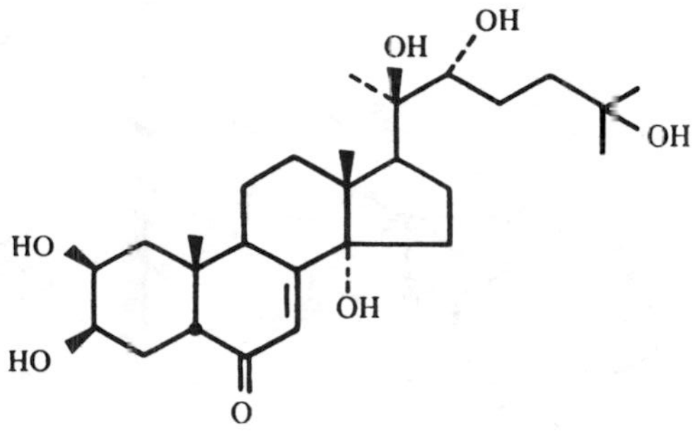	Crustecdysone	Crayfish: *Callinectes sapidus* *Jasus lalandei*	+	+	Molting hormone	131 174 189 190 196 341 n 360

Table 1 (continued)

Molecular formula	Structural formula	Name	Source of compound	Available by synthesis	Available non-marine origin	Activity	References
C_{27} . . .							
380. $C_{27}H_{46}O$	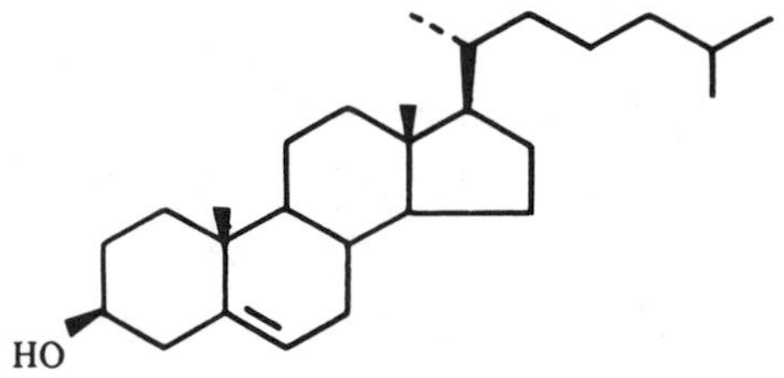	Cholesterol	Algae Coelenterates Crustaceans Echinoderms Mollusks Sea worms Sponges Tunicates Zoanthids	+	+		25, 39, 47, 50, 51, 55, 107, 115, 162, 163, 165, 198, 199, 200, 205, 251, 252, 256, 284 p, 319, 332, 338, 341 k, 395, 430, 431
381. $C_{27}H_{46}O$	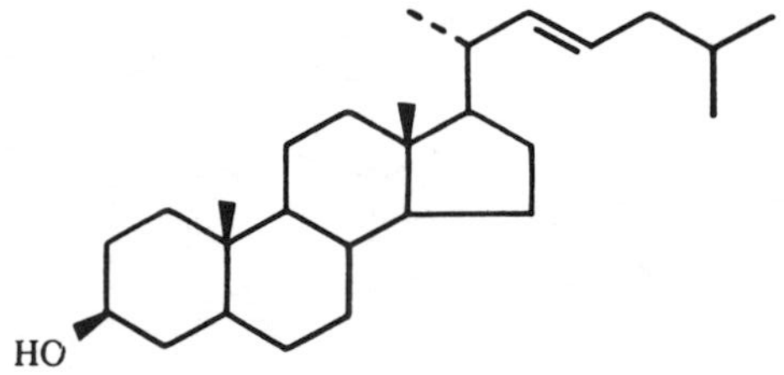	22,23-Dehydro-cholestanol	Sponge: *Hymeniacidon perleve*				115
382. $C_{27}H_{46}O$	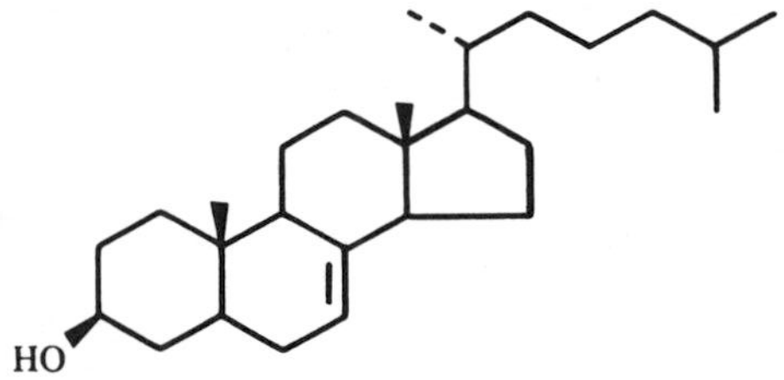	Lathosterol	Mollusks Sea cucumber Starfish	+			55 162 168 251 252 341 l

Table 1 (continued)

Molecular formula	Structural formula	Name	Source of compound	Available by synthesis	Available non-marine origin	Activity	References
C_{27} ...							
383. $C_{27}H_{48}O$	HO	Cholestanol	Mollusks Sponges Starfish	+	+		39 50 55 115 251 252
384. $C_{27}H_{48}O$	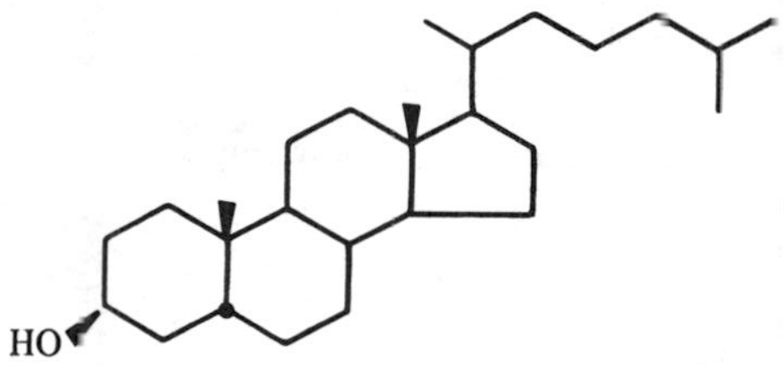 HO	Coprostanol	Whale oil	+	+		161 284 t
385. $C_{27}H_{48}O_3$	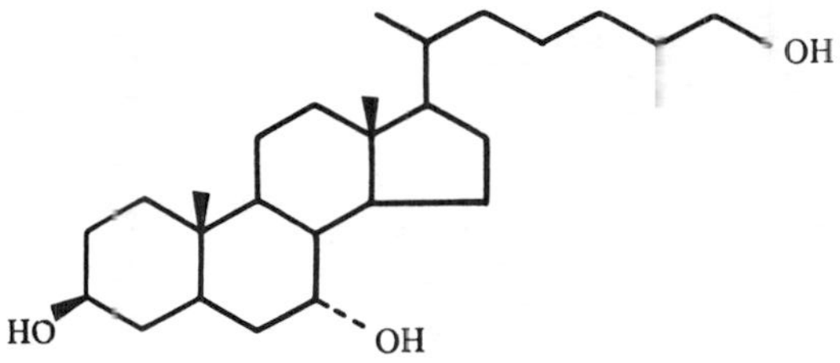 OH HO OH Probably occurs as a disulfate ester	16-Deoxymyxinol	Hagfish: *Eptatretus stoutii* *Myxine glutinosa*				33 34 398

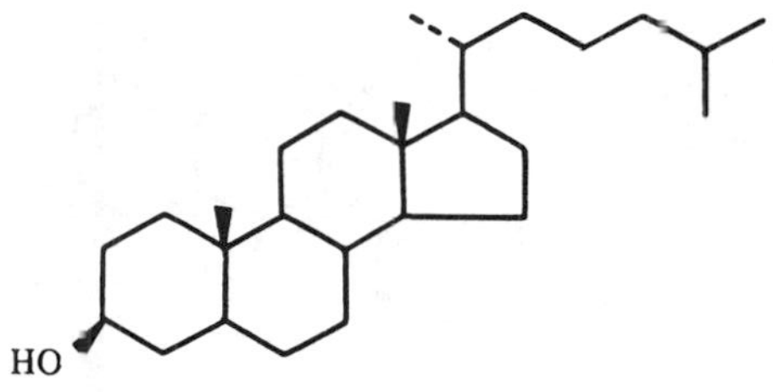

Table 1 (continued)

Molecular formula	Structural formula	Name	Source of compound	Available by synthesis	Available non-marine origin	Activity	References
C_{27} . . . 386. $C_{27}H_{48}O_4$	HO, OH, OH, OH Occurs as 3,26-disulfate ester	Myxinol	Hagfish: *Eptatretus stoutii* *Myxine glutinosa*				33 103 180 183 191
387. $C_{27}H_{48}O_6$	OH, HO, OH, OH, HO, OH Occurs as 26-sulfate ester	Scymnol	Fish				37, 46, 62, 99, 102, 178, 179, 182, 191, 433

Table 1 (continued)

Molecular formula	Structural formula	Name	Source of compound	Available by synthesis	Available non-marine origin	Activity	References
C_{28}							
388. $C_{28}H_{12}O_6$	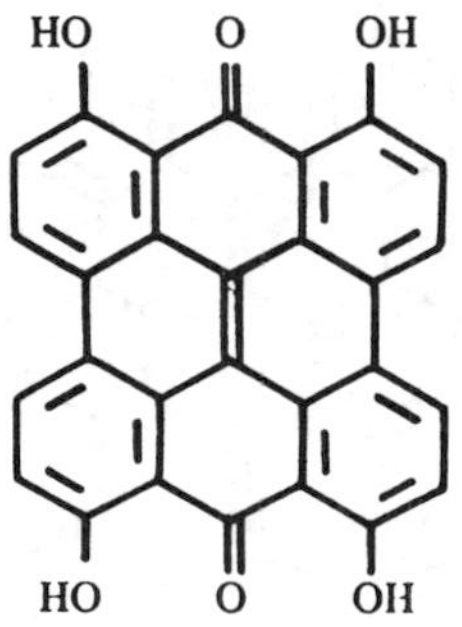	Fringelite H	Fossil feather stars: *Apiocrinus* sp.				57 397 r
389. $C_{28}H_{12}O_8$	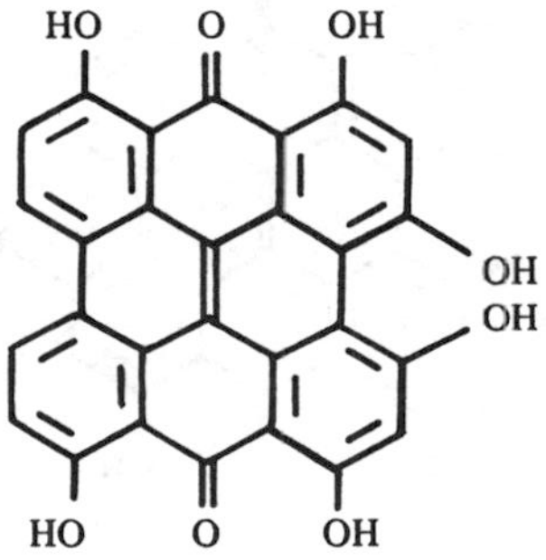	Fringelite F	Fossil feather stars: *Apiocrinus* sp.				57 397 r

Table 1 (continued)

Molecular formula	Structural formula	Name	Source of compound	Available by synthesis	Available non-marine origin	Activity	References
C_{28} . . .							
390. $C_{28}H_{12}O_9$	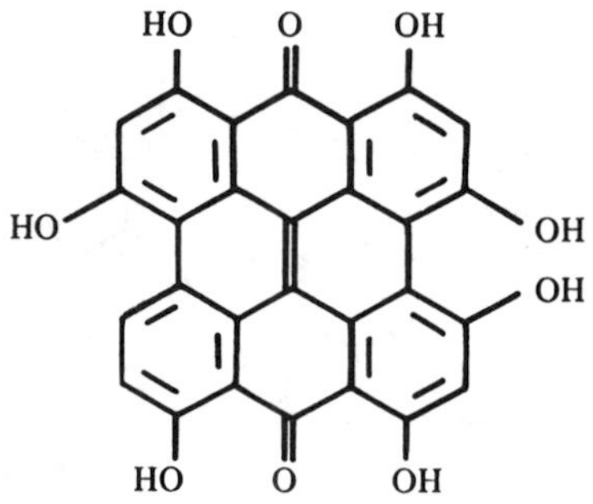	Fringelite E	Fossil feather stars: *Apiocrinus* sp.				57 397 r
391. $C_{28}H_{12}O_{10}$	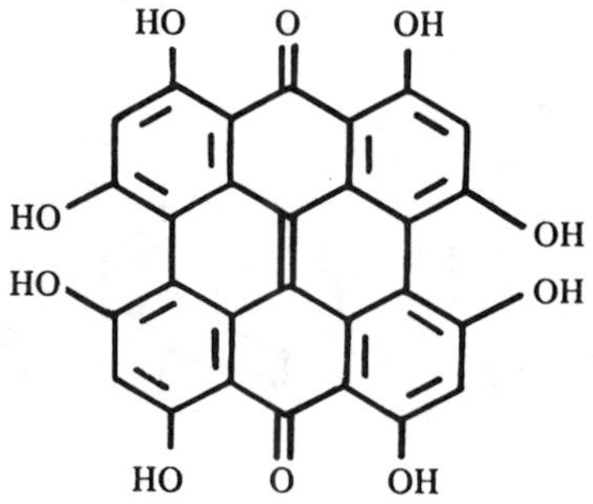	Fringelite D	Fossil feather stars: *Apiocrinus* sp.				57 397 r
392. $C_{28}H_{42}O_9$	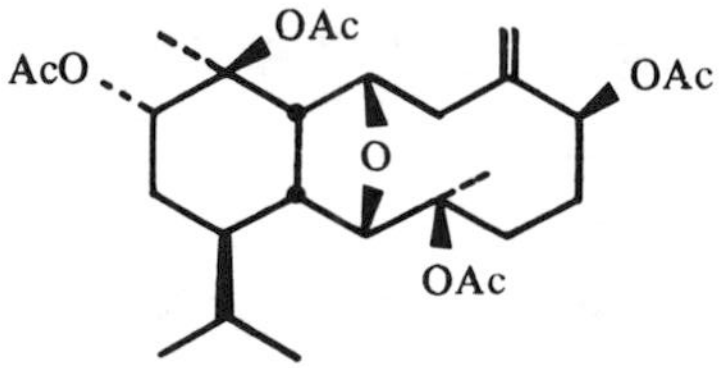	Eunicellin	Gorgonian: *Eunicella stricta*				240

Table 1 (continued)

Molecular formula	Structural formula	Name	Source of compound	Available by synthesis	Available non-marine origin	Activity	References
C_{28} . . . 393. $C_{28}H_{44}O$	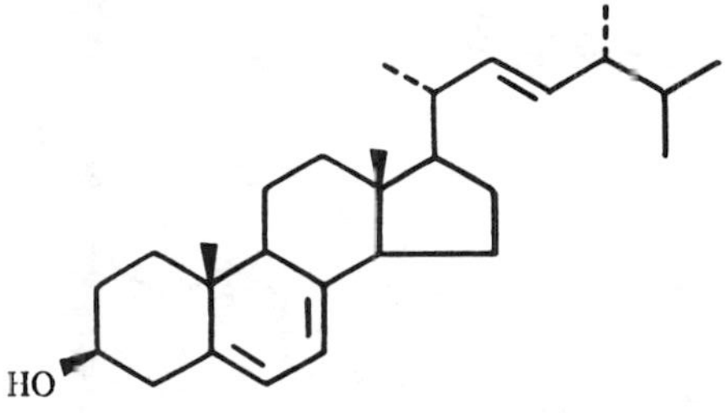	Ergosterol	Algae Crustaceans Marine yeast Mollusks		+		39 50 55 284 Ib 396
394. $C_{28}H_{46}O$	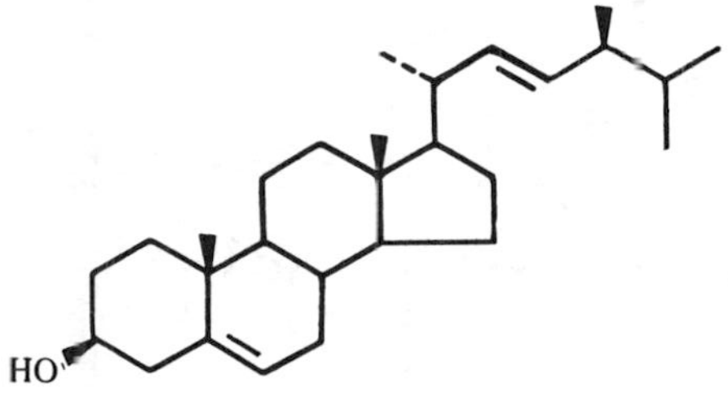	Brassicasterol	Brittle star Coral Crustaceans Feather star Green alga Mollusks Red algae Sponges Zoanthids		+		39, 50, 51, 55, 115, 162, 163, 198, 199, 200, 205, 395
395. $C_{20}H_{46}O$	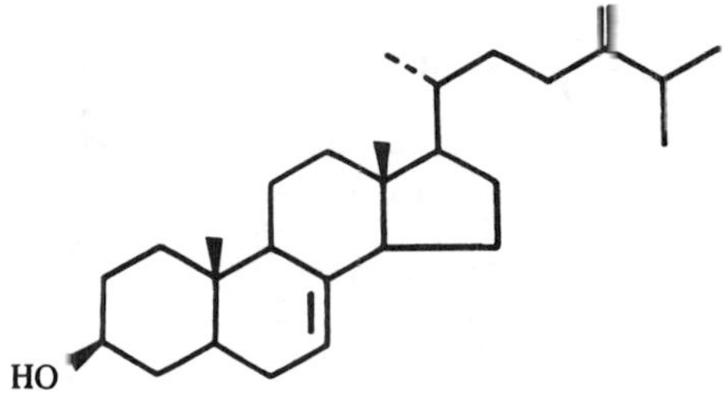	Episterol. 5α-Ergosta-7,24(28) -dien-3β-ol	Starfish		+		55 119 168 251 252 341 r

Table 1 (continued)

Molecular formula	Structural formula	Name	Source of compound	Available by synthesis	Available non-marine origin	Activity	References
C_{28} . . .							
396. $C_{28}H_{46}O$	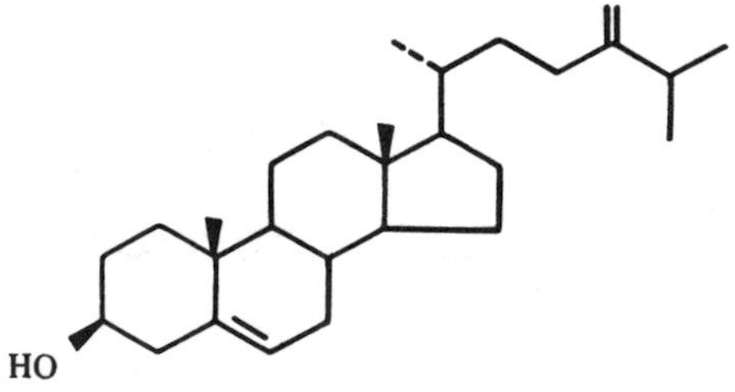	24-Methylene-cholesterol	Algae Crustaceans Mollusks Sponges Zoanthids	+	+		36, 55, 115, 163, 197, 198, 199, 200, 205, 338, 341 q, 395
397. $C_{28}H_{46}O$	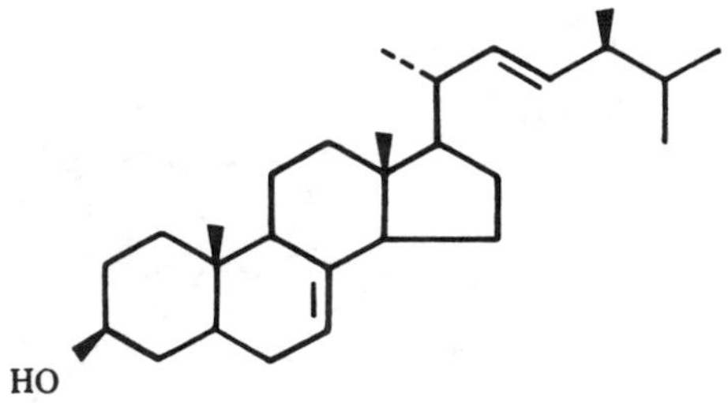	Stellasterol. 22-*trans*-(24*S*)-24-Methylcholesta-7,22-dien-3β-ol	Sea cucumber: *Holothuria atra* Starfish		+		50 55 130 162 168 251 252 341 q
398. $C_{28}H_{46}O_7$	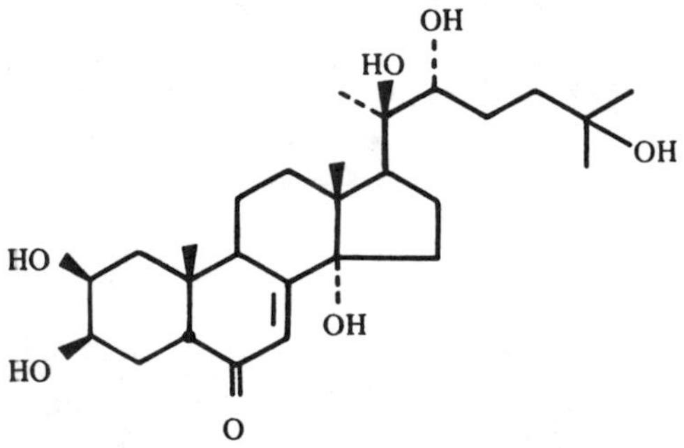	Callinecdysone B	Crayfish: *Callinectes sapidus*				131

Table 1 (continued)

Molecular formula	Structural formula	Name	Source of compound	Available by synthesis	Available non-marine origin	Activity	References
C_{28} . . .							
399. $C_{28}H_{48}O$	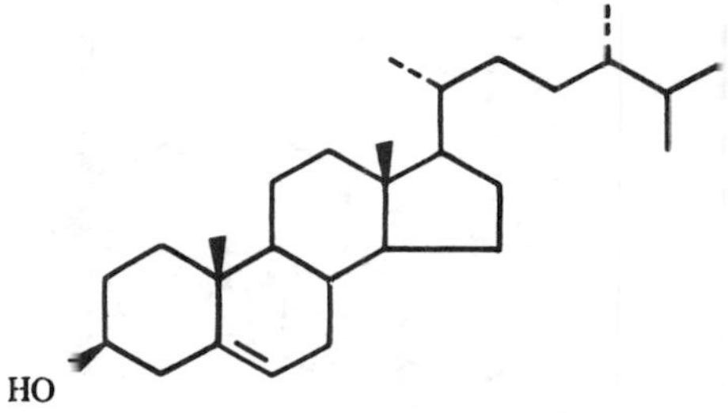 HO	Campesterol. 24α-Methyl-Δ^5-cholestenol	Brittle star Feather star Green alga Marine yeast Mollusks Sea urchin Zoanthid		+		50 55 162 163 200 205 284 l 341 o 396
400. $C_{28}H_{48}O$	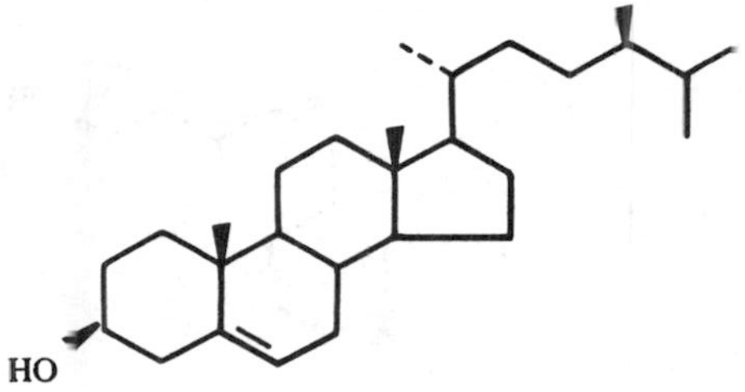 HO	22,23-Dihydrobrassica-sterol. 24β-Methyl-Δ^5-cholestenol	Crustaceans Sponge: *Cliona celata* Zoanthid: *Palythoa tuberculosa*				115 163 199
401. $C_{28}H_{48}O$	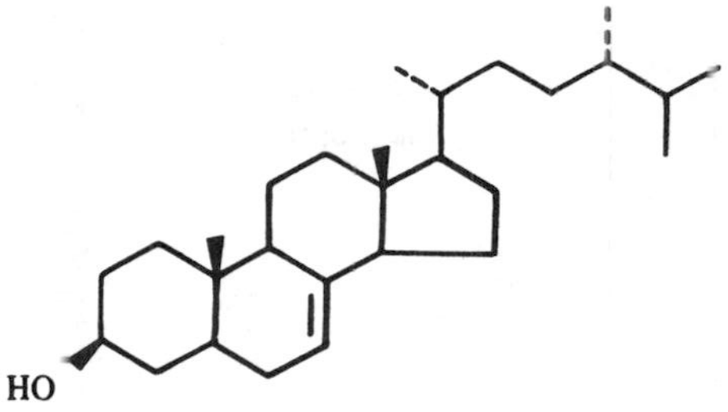 HO	Δ^7-Ergostenol	Sea cucumber: *Holothuria atra* Starfish		+		50 55 162 168 251 252 284 Ib 341 p

Table 1 (continued)

Molecular formula	Structural formula	Name	Source of compound	Available by synthesis	Available non-marine origin	Activity	References
C_{28} ... 402. $C_{28}H_{48}O$	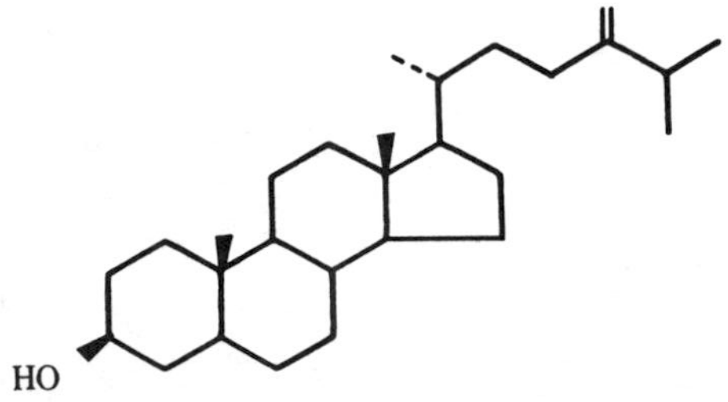	24-Methylene-cholestanol	Sponge: *Hymeniacidon perleve*				115
403. $C_{28}H_{48}O$	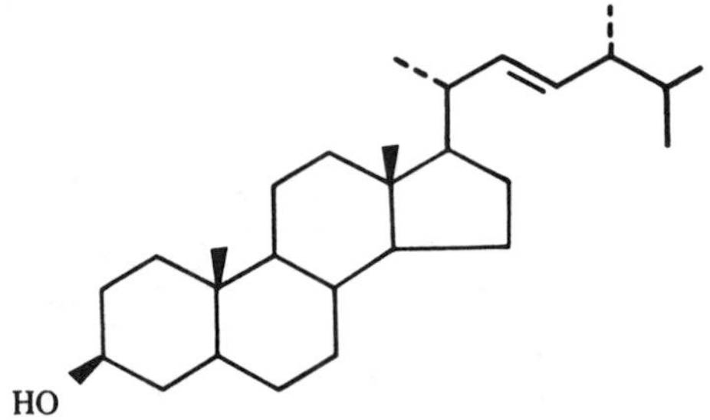	Neospongosterol	Sponges: *Hymeniacidon perleve* *Suberites compacta* *S. domuncula* *S. suberea*				39 48 50 55 115
C_{29} 404. $C_{29}H_{44}O$	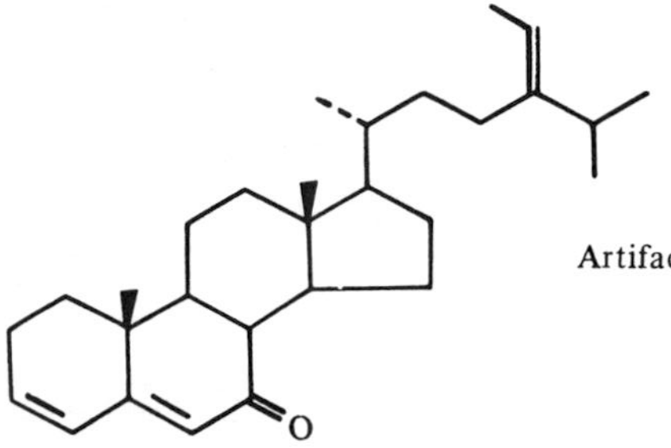 Artifact?	3,5,(E)-24(28)-Stigmastatrien-7-one	Brown alga: *Fucus evanescens*				206

Table 1 (continued)

Molecular formula	Structural formula	Name	Source of compound	Available by synthesis	Available non-marine origin	Activity	References
C_{29} ...							
405. $C_{29}H_{48}O$	HO	Chondrillasterol	Sponge: *Chondrilla nucula*				49 50 55 284 r
406. $C_{29}H_{48}O$	HO	23-Demethyl-gorgosterol	Gorgonians: *Gorgonia flabellum* *G. ventilina*				113 344
407. $C_{29}H_{48}O$	HO	24,28-Didehydroaply-sterol	Sponge: *Verongia cerophoba*	+			105 106

Table 1 (continued)

Molecular formula	Structural formula	Name	Source of compound	Available by synthesis	Available non-marine origin	Activity	References
C_{29} . . .							
408. $C_{29}H_{48}O$	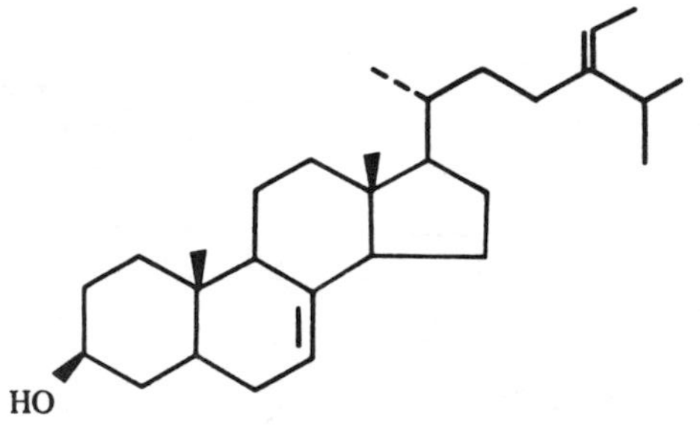 HO	24-Ethylidene-5α-cholest-7-en-3β-ol	Sea cucumber: *Holothuria atra* Starfish				162 168 251 252
409. $C_{29}H_{48}O$	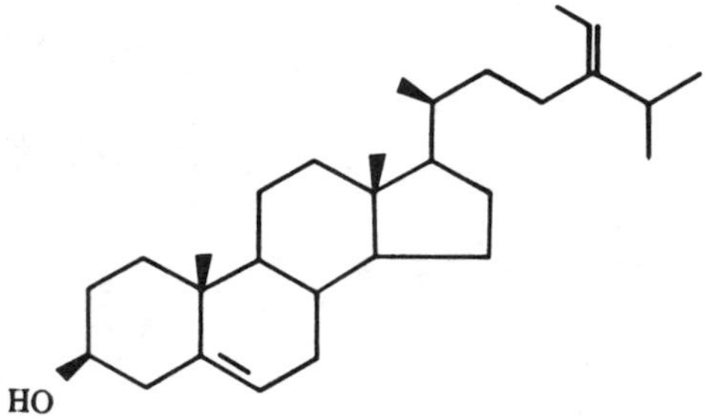 HO	Fucosterol	Algae Brittle star: *Ophiocoma insularia* Mollusks Sponge: *Hymeniacidon perleve*	+			39, 55, 115, 162, 198, 200, 205, 206, 284 Ig, 399
410. $C_{29}H_{48}O$	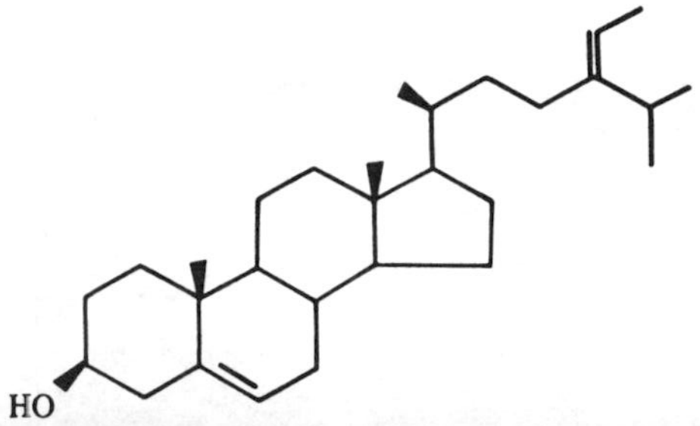 HO	28-Isofucosterol	Crab: *Geryon quinquedens* Green algae Mollusks Shrimp: *Pandalus borealis*	+	+		111 148 199 200 341 s 400

Table 1 (continued)

Molecular formula	Structural formula	Name	Source of compound	Available by synthesis	Available non-marine origin	Activity	References
C_{29} . . .							
411. $C_{29}H_{48}O$	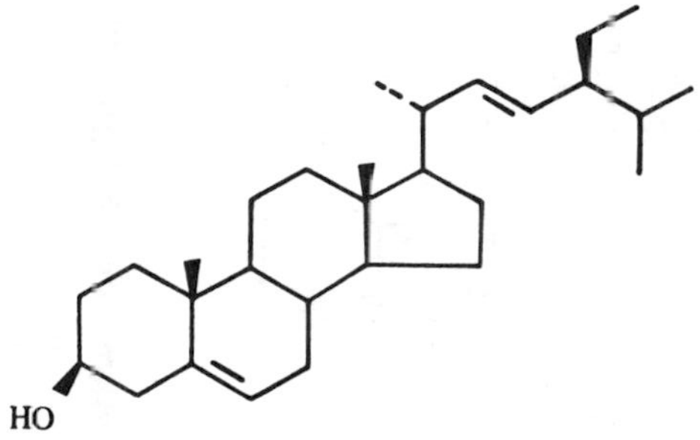	Poriferasterol	Mollusks Sponges Tunicate: *Cynthia roretzi*	+			39 50 52 55 115 200 380 412
412. $C_{29}H_{48}O$	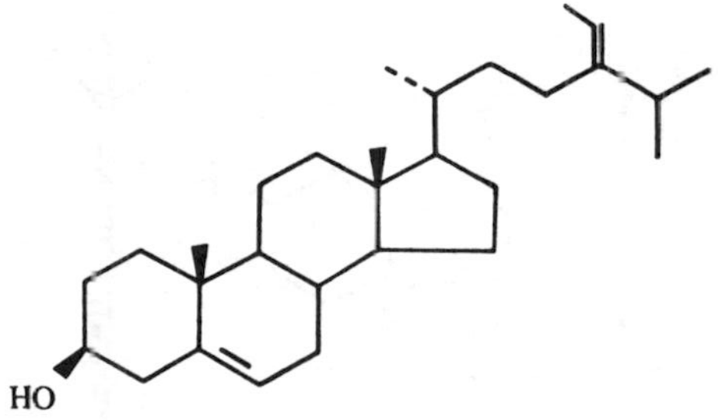	Sargasterol	Brown algae: *Eisenia bicyclis* *Sargassum ringgoldianum*	+	+		130 399
413. $C_{29}H_{48}O$	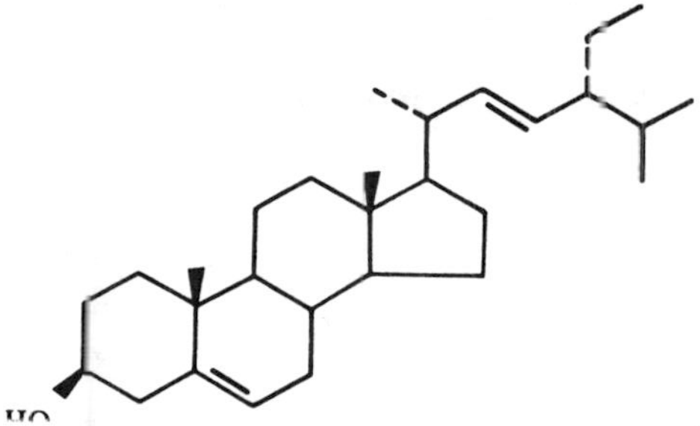	Stigmasterol	Brittle star: *Ophiocoma insularia* Feather star: *Antedon* sp. Mollusks Prawn: *Penaeus japonicus* Red algae	+	+		39 50 162 198 284 IIv 380 395

Table 1 (continued)

Molecular formula	Structural formula	Name	Source of compound	Available by synthesis	Available non-marine origin	Activity	References
C_{29} . . . 414. $C_{29}H_{48}O_2$	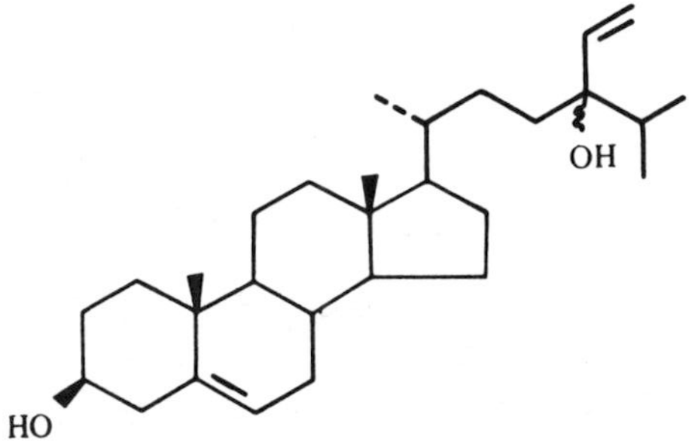	Saringosterol	Algae	+			204 205 206
415. $C_{29}H_{48}O_2$	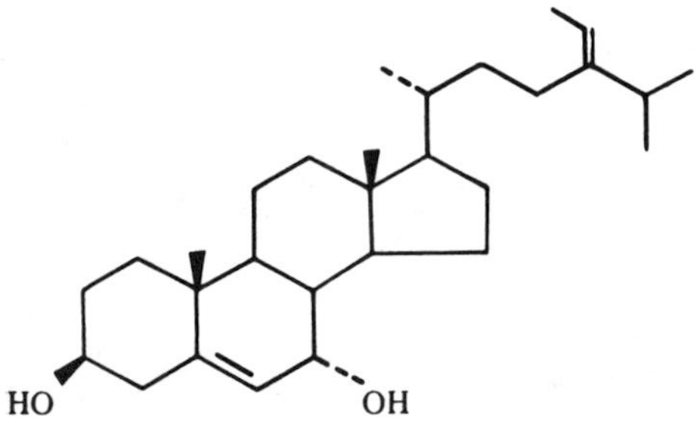	5,(E)-24,(28)-Stigmastadien-3β, 7α-diol	Brown alga: *Fucus evanescens*				206
416. $C_{29}H_{50}O$	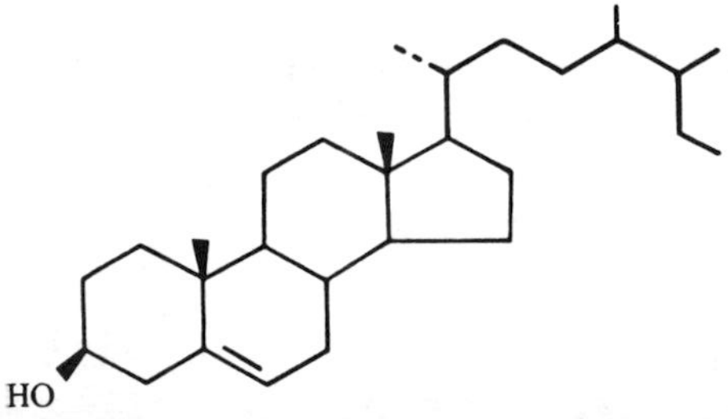	Aplysterol	Sponge: *Verongia aerophoba*				105 106

Table 1 (continued)

Molecular formula	Structural formula	Name	Source of compound	Available by synthesis	Available non-marine origin	Activity	References
C_{29} . . . 417. $C_{29}H_{50}O$	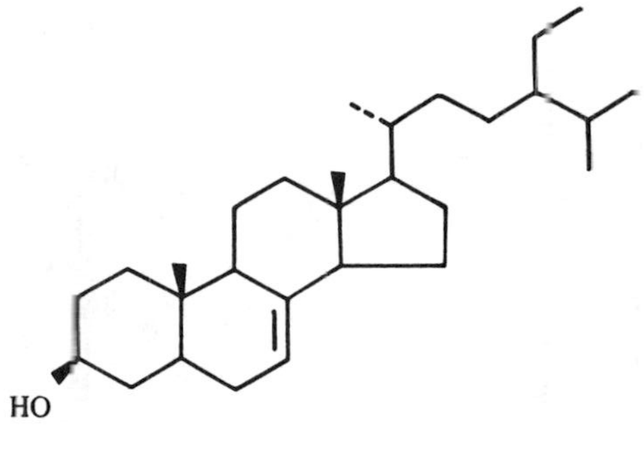 HO	24ξ-Ethylcholest-7-en-3β-ol	Sea cucumber: *Holothuria atra* Starfish				162 168 251 252
418. $C_{29}H_{50}O$	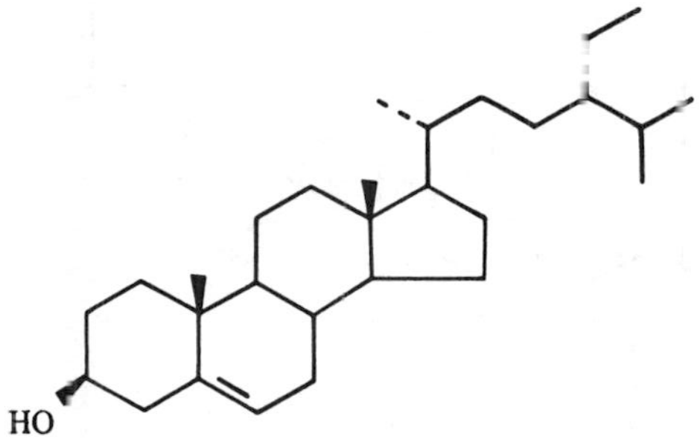 HO	β-Sitosterol	Algae Crustaceans Echinoderms Mollusks Sea anemones Sponge Zoanthids		+		25, 39, 55, 115, 162, 163, 198, 199, 200, 205, 284 IIs, 339, 395

Table 1 (continued)

Molecular formula	Structural formula	Name	Source of compound	Available by synthesis	Available non-marine origin	Activity	References
C_{29} . . . 419. $C_{29}H_{50}O$	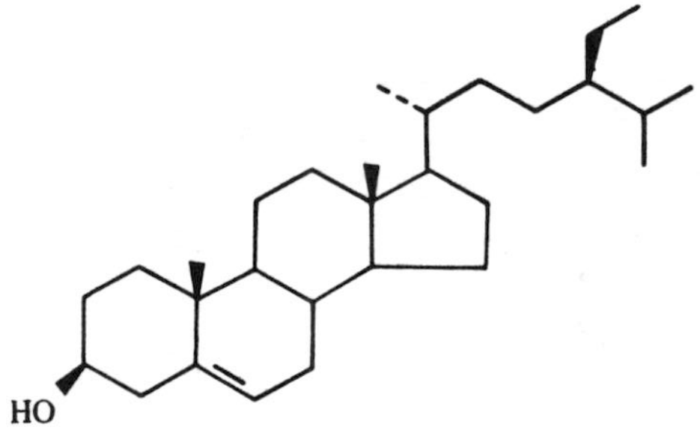	Clionasterol. γ-Sitosterol	Crab: *Limulus polyphemus* Mollusks Sponges Tunicate: *Cynthia roretzi*		+		39 50 52 55 115 284 IIs 412
420. $C_{29}H_{50}O_3$	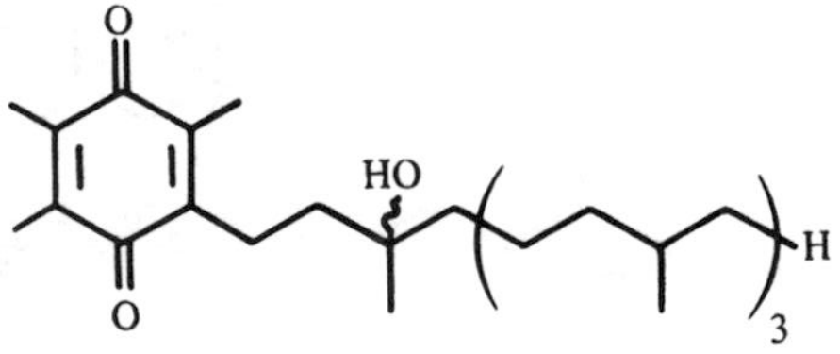	α-Tocopherol-quinone	Algae Limpets Shrimps	+			397 a
421. $C_{29}H_{52}O$	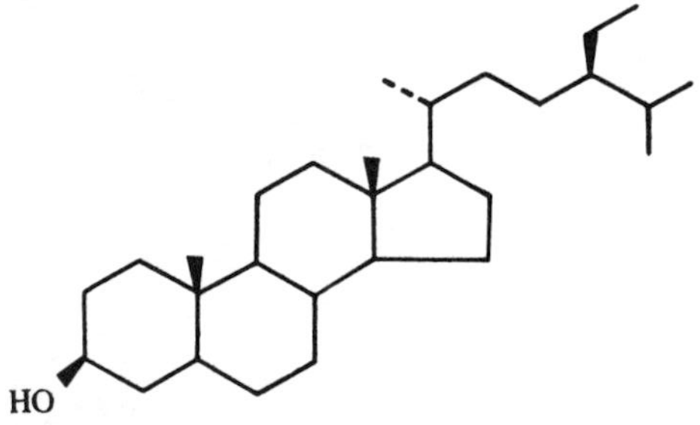	Poriferastanol	Sponges		+		50 322

Table 1 (continued)

Molecular formula	Structural formula	Name	Source of compound	Available by synthesis	Available non-marine origin	Activity	References
C_{30} 422. $C_{30}H_{44}O_4$	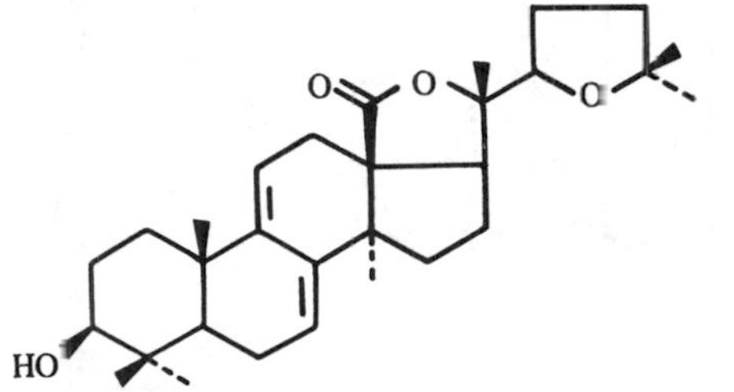	17-Desoxy-22,25 oxidoholothurin-ogenin	Sea cucumbers: *Actinopyga agassizi* *Holothuria forskali* *H. polii* *H. tubulosa*				77 165 166 167
423. $C_{30}H_{44}O_4$	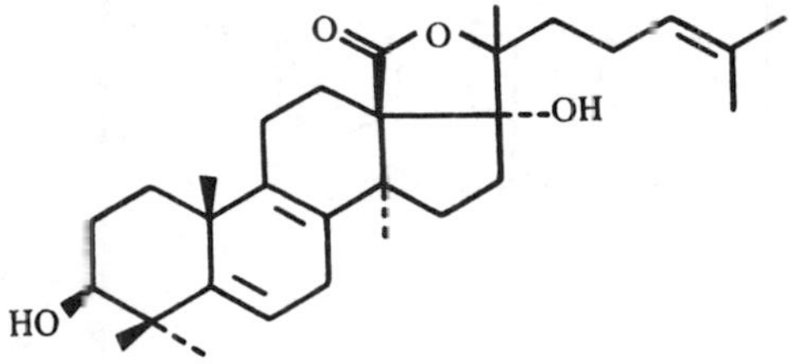	Stichopogenin A_2	Sea cucumber: *Stychopus japonicus*				112
424. $C_{30}H_{44}O_5$	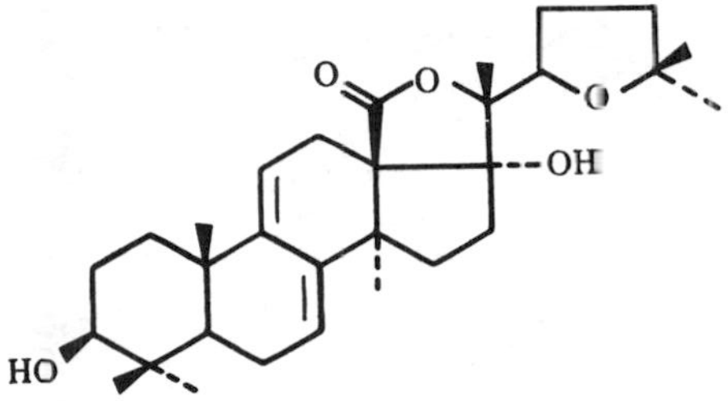	22,25-Oxidoholo-thurinogenin	Sea cucumbers: *Actinopyga agassizi* *Halodeima grisea* *Holothuria forskali* *H. polii* *H. tubulosa*				77 165 166 167 340 407

Table 1 (continued)

Molecular formula	Structural formula	Name	Source of compound	Available by synthesis	Available non-marine origin	Activity	References
C_{30} . . .							
425. $C_{30}H_{46}O$	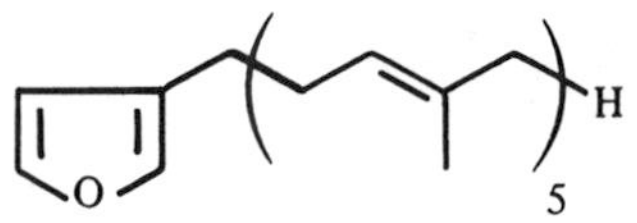	Furospinosulin-2	Sponge: *Ircinia spinosula*				93
426. $C_{30}H_{46}O_3$	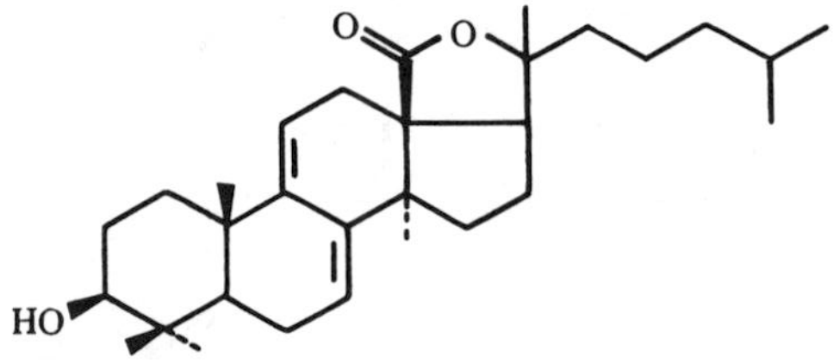	Seychellogenin	Sea cucumber: *Bohadschia koellikeri*				329 408
427. $C_{30}H_{46}O_4$	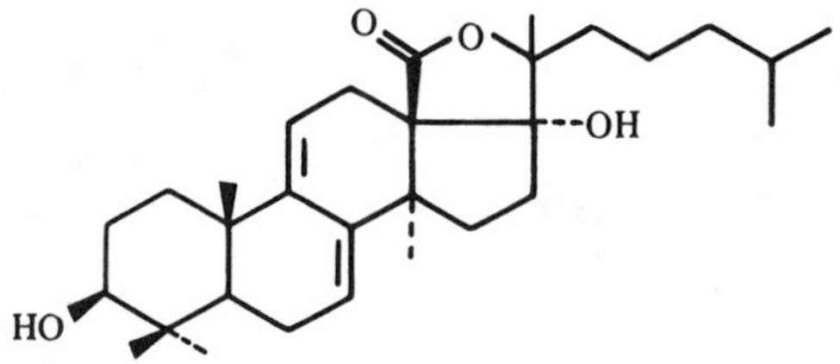	Holothurinogenin	Sea cucumber: *Holothuria polii*				166 340
428. $C_{30}H_{46}O_4$	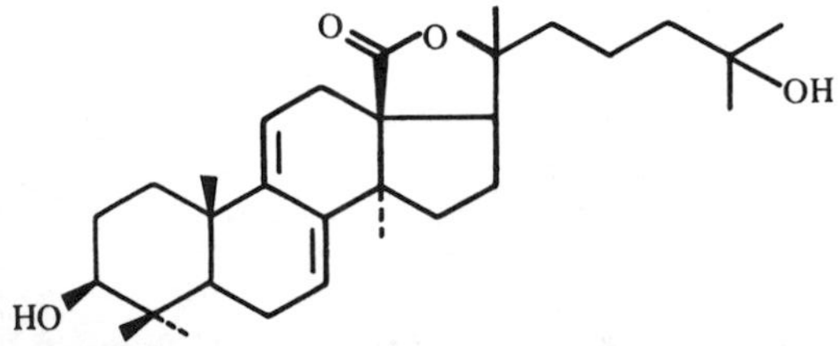	Koellikerigenin	Sea cucumber: *Bohadschia koellikeri*				329 408

Table 1 (continued)

Molecular formula	Structural formula	Name	Source of compound	Available by synthesis	Available non-marine origin	Activity	References
C_{30} . . .							
429. $C_{30}H_{46}O_5$	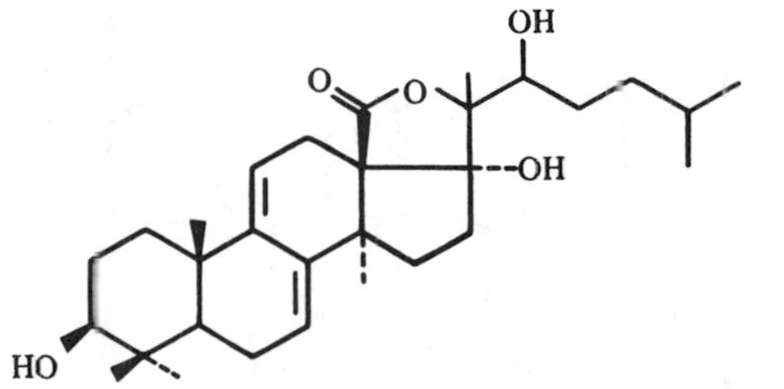	Griseogenin	Sea cucumbers: *Halodeima grisea* *Holothuria polii*				166 407
430. $C_{30}H_{46}O_5$	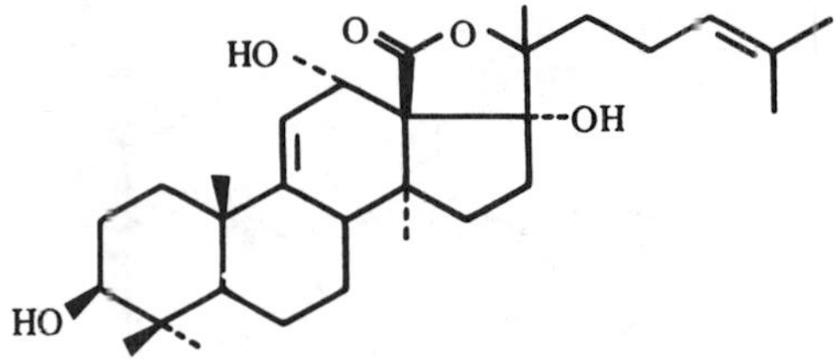	12α-Hydroxy-7,8-dihydro-24,25-dehydroholothurinogenin	Sea cucumber: *Actinopyga agassizi*				79 167
431. $C_{30}H_{46}O_5$	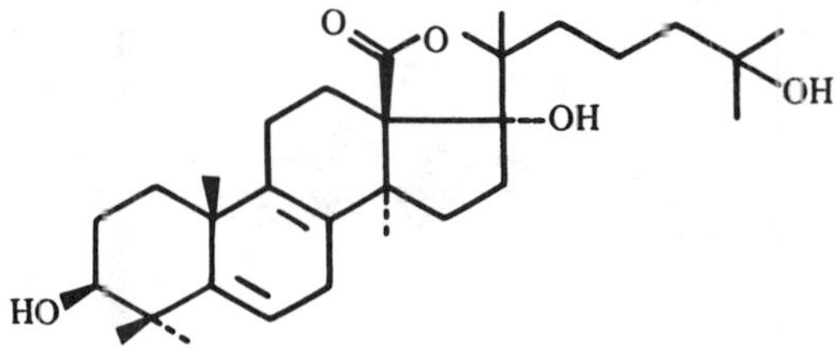	Stichopogenin A_4	Sea cucumber: *Stychopus japonicus*				112

Table 1 (continued)

Molecular formula	Structural formula	Name	Source of compound	Available by synthesis	Available non-marine origin	Activity	References
C_{30} . . .							
432. $C_{30}H_{46}O_6$	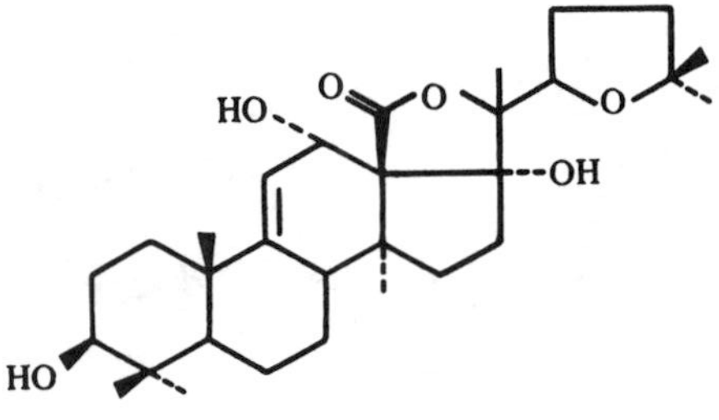	12α-Hydroxy-7,8-dihydro-22,25-oxidoholothurinogenin	Sea cucumber: *Actinopyga agassizi*				333
433. $C_{30}H_{50}$	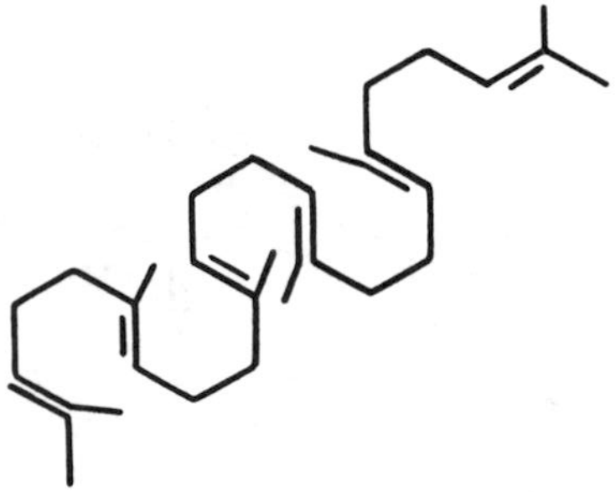	Squalene	Brown alga: *Fucus vesiculosus* Sea cucumber: *Stichopus japonicus* Shark liver oils	+	+	Bactericide	172 233 284 IIu 308 311 404
434. $C_{30}H_{50}O$	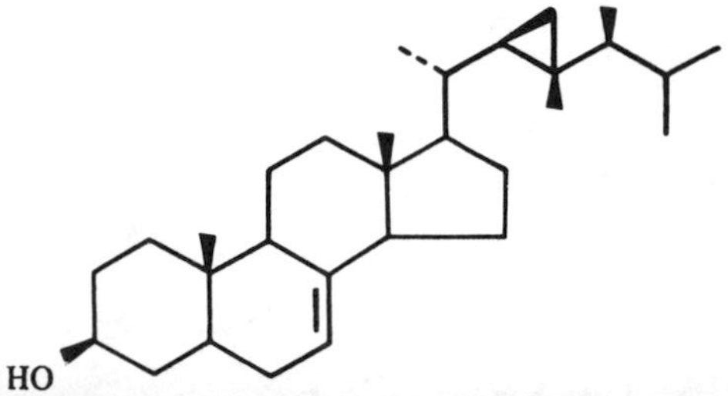	Acanthasterol	Starfish: *Acanthaster planci*				162 355

Table 1 (continued)

Molecular formula	Structural formula	Name	Source of compound	Available by synthesis	Available non-marine origin	Activity	References
C_{30} . . .							
435. $C_{30}H_{50}O$	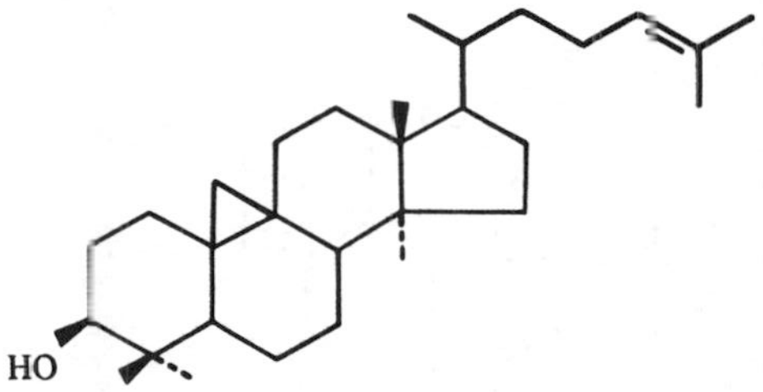	Cycloartenol	Red algae Sea cucumbers: *Holothuria tubulosa* *Stichopus japonicus*		+		39 310 311 341 f
436. $C_{30}H_{50}O$	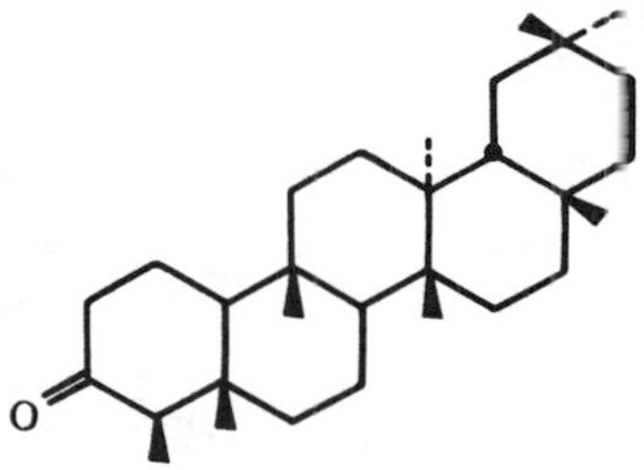	Friedelin	Algae: *Fucus evanescens* *Monostroma nitidum*		+		206 284 If 341 g 400
437. $C_{30}H_{50}O$	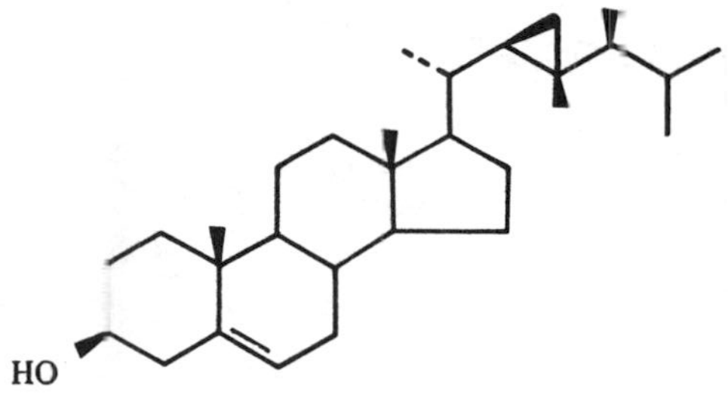	Gorgosterol	Gorgonians Zoanthids				39 47 55 86 163 170 265 344

Table 1 (continued)

Molecular formula	Structural formula	Name	Source of compound	Available by synthesis	Available non-marine origin	Activity	References
C_{30} . . .							
438. $C_{30}H_{50}O$	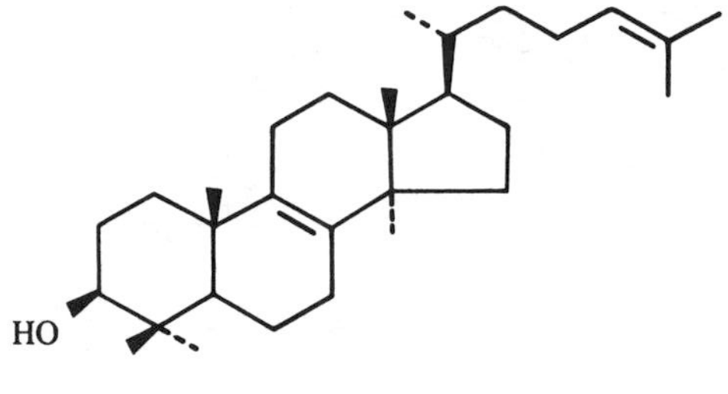 HO	Lanosterol	Red algae Sea cucumbers	+	+		39 165 284 Ix 310 311
439. $C_{30}H_{50}O$	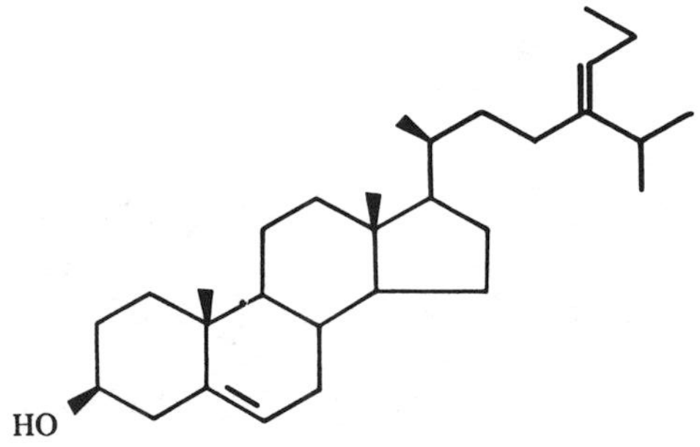 HO	29-Methyliso-fucosterol	Scallop: *Placopecten magellanicus*				200 341 t
440. $C_{30}H_{50}O$	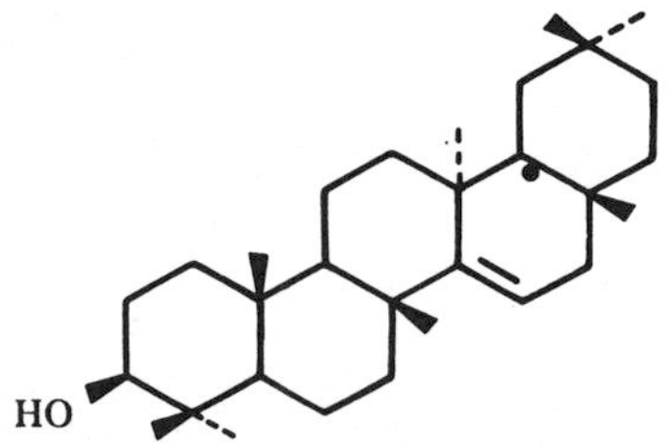 HO	Taraxerol	Green alga: *Caulerpa lamourouxii* (*C. racemosa* var. *lamourouxii*)		+		45 284 IIw 339

Table 1 (continued)

Molecular formula	Structural formula	Name	Source of compound	Available by synthesis	Available non-marine origin	Activity	References
C_{30} . . .							
441. $C_{30}H_{50}O_3$	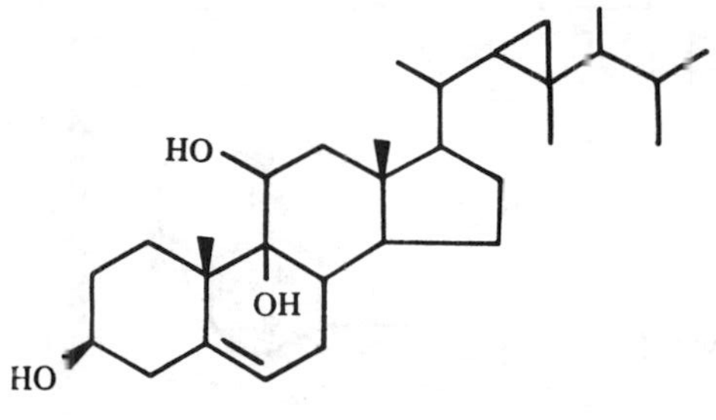	Gorgost-5-en-3,9, 11-triol one	Gorgonian: *Pseudopterogorgia americana*				88
442. $C_{30}H_{50}O_3$	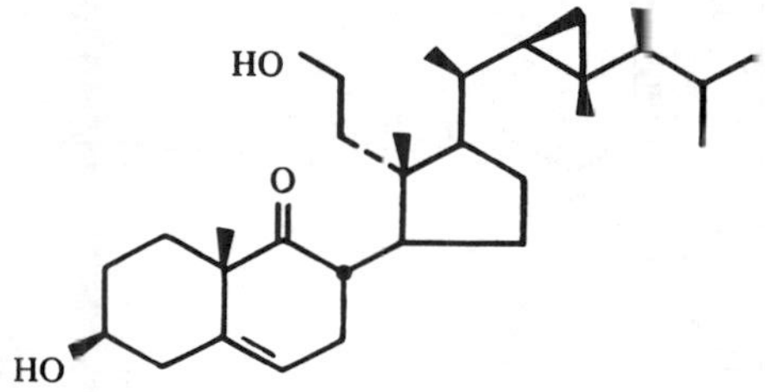	9,11-Secogorgost-5-en-3,11-diol-9-one	Gorgonian: *Pseudopterogorgia americana*				113 130
443. $C_{30}H_{50}O_4$	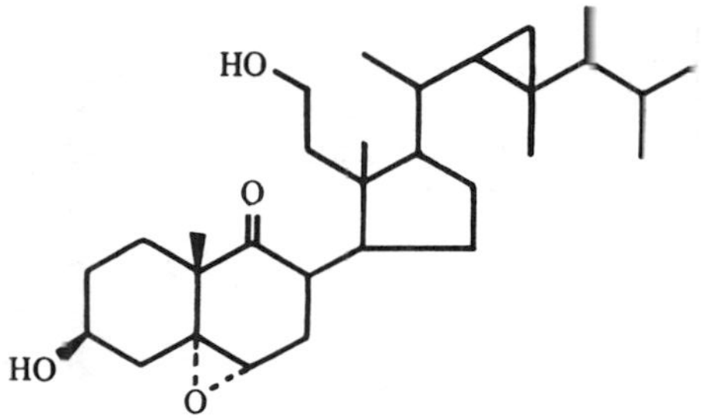	5,6-Epoxy-9,11-secogorgosten-3,11-diol-9-one	Gorgonian: *Pseudopterogorgia americana*				88

Table 1 (continued)

Molecular formula	Structural formula	Name	Source of compound	Available by synthesis	Available non-marine origin	Activity	References
C_{31}							
444. $C_{31}H_{42}O_3$	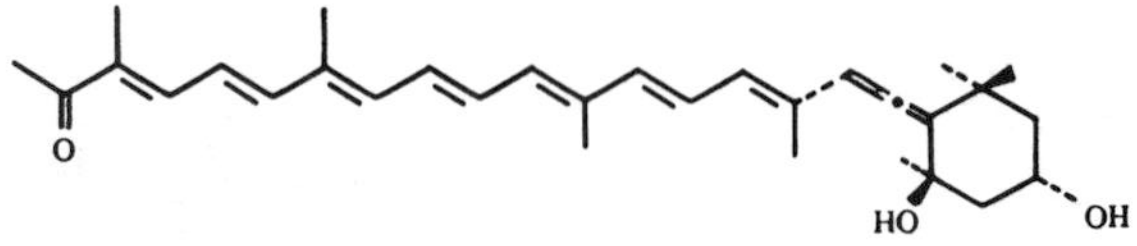	Paracentrone	Sea urchin: *Paracentrotus lividus*				144
445. $C_{31}H_{44}O_2$	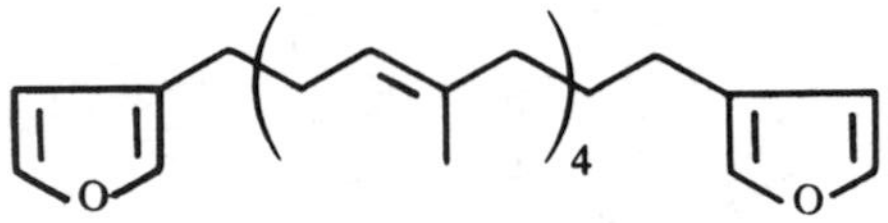	Difurospinosulin	Sponge: *Ircinia spinosula*				93
446. $C_{31}H_{48}O_4$	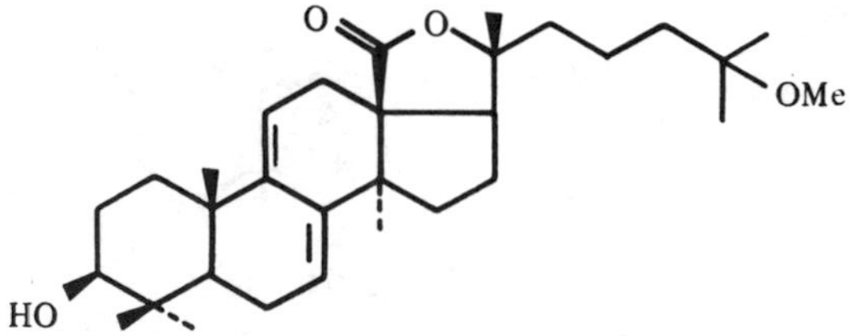	Ternaygenin	Sea cucumbers: *Bohadschia koellikeri* *Holothuria polii*				166 329 408
447. $C_{31}H_{48}O_5$	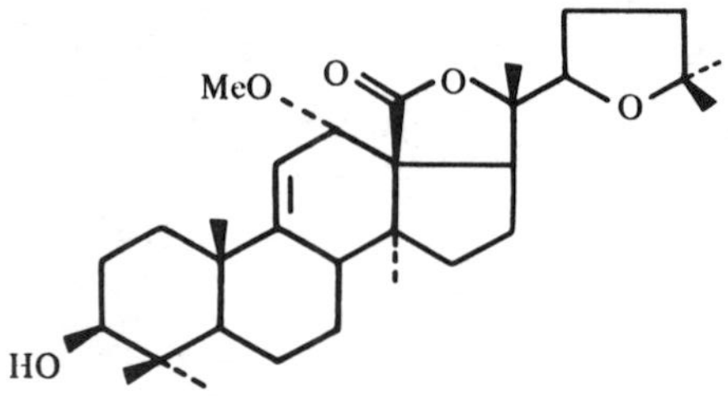	12α-Methoxy-17-desoxy-7,8-dihydro-22,25-oxido-holothurinogenin	Sea cucumber: *Actinopyga agassizi*				79 167

Table 1 (continued)

Molecular formula	Structural formula	Name	Source of compound	Available by synthesis	Available non-marine origin	Activity	References
C_{31} . . .							
448. $C_{31}H_{48}O_5$	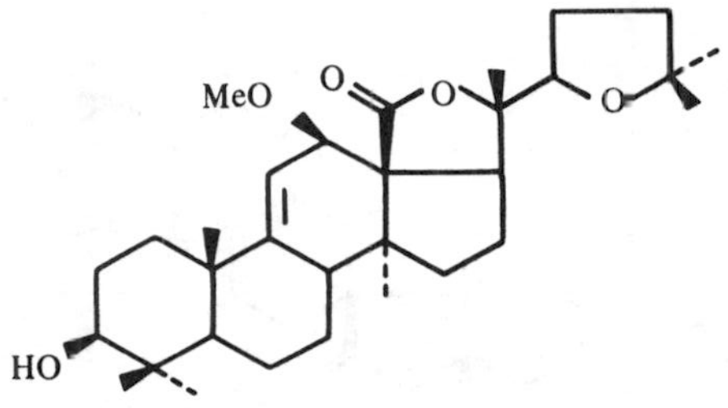 MeO, O, O, O, HO	12β-Methoxy-17-desoxy -7,8-dihydro-22,25-oxidoholothurinogenin	Sea cucumber: *Actinopyga agassizi*				78 167
449. $C_{31}H_{48}O_5$	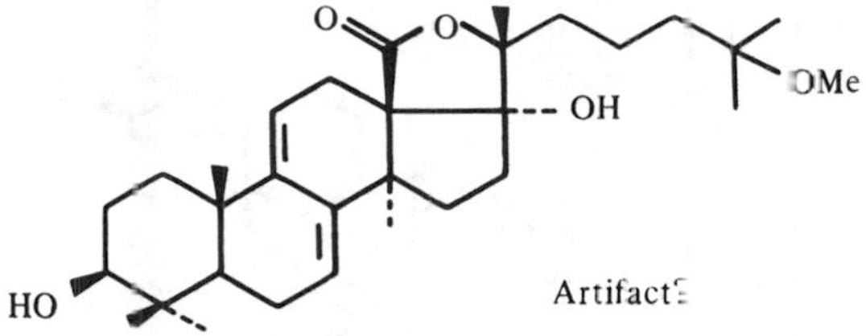 O, O, OH, OMe, HO, Artifact?	Praslinogenin	Sea cucumbers: *Bohadschia koellikeri* *Holothuria polii*				166 408
450. $C_{31}H_{48}O_6$	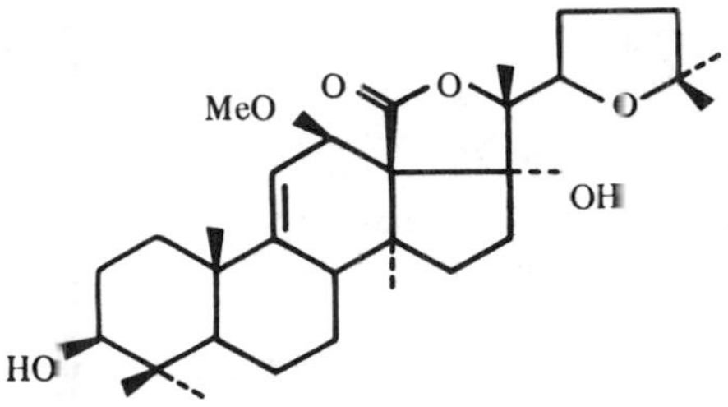 MeO, O, O, O, OH, HO	12β-Methoxy-7,8-dihydro-22,25-oxido -holothurinogenin	Sea cucumber: *Actinopyga agassizi*				78 79 167 333

Table 1 (continued)

Molecular formula	Structural formula	Name	Source of compound	Available by synthesis	Available non-marine origin	Activity	References
C_{31} . . .							
451. $C_{31}H_{50}O_6$	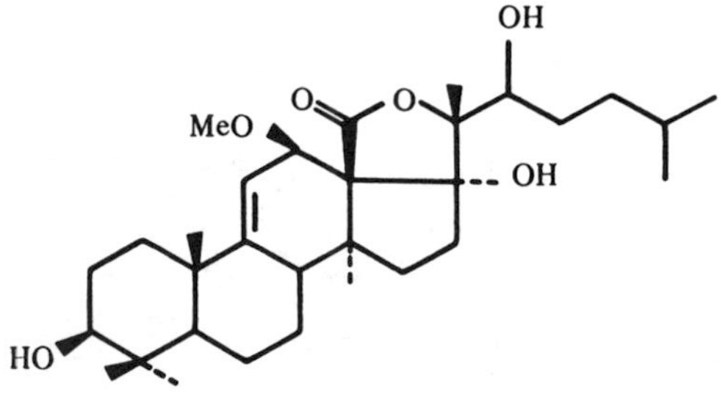	12β-Methoxy-7,8-dihydro-22-hydroxy-holothurinogenin	Sea cucumber: *Actinopyga agassizi*				78 79 167
C_{32}							
452. $C_{32}H_{50}O_5$	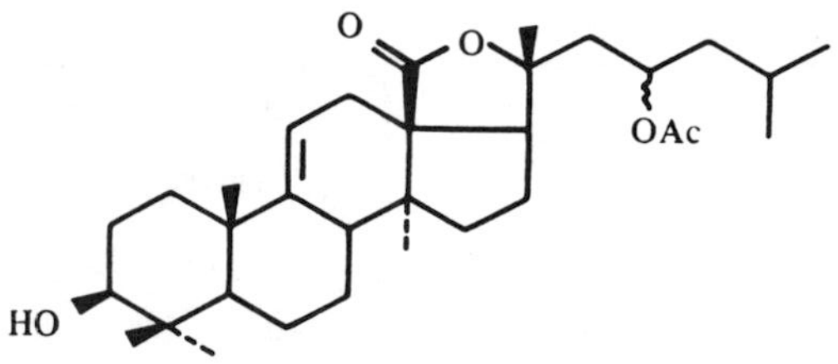	23ξ-Acetoxy-17-desoxy-7,8-dihydro-holothurinogenin	Sea cucumber: *Stichopus chloronotus*				333
453. $C_{32}H_{64}O_2$	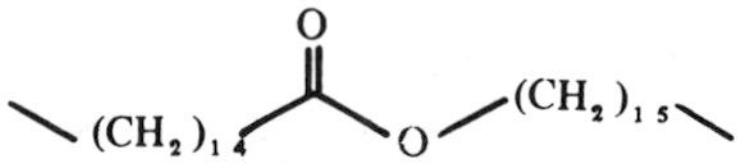 $CH_3(CH_2)_{14}COO(CH_2)_{15}CH_3$	Cetyl palmitate	Gorgonians Hard corals Soft coral Zoanthid				88 242

Table 1 (continued)

Molecular formula	Structural formula	Name	Source of compound	Available by synthesis	Available non-marine origin	Activity	References
C_{33} 454. $C_{33}H_{34}N_4O_6$	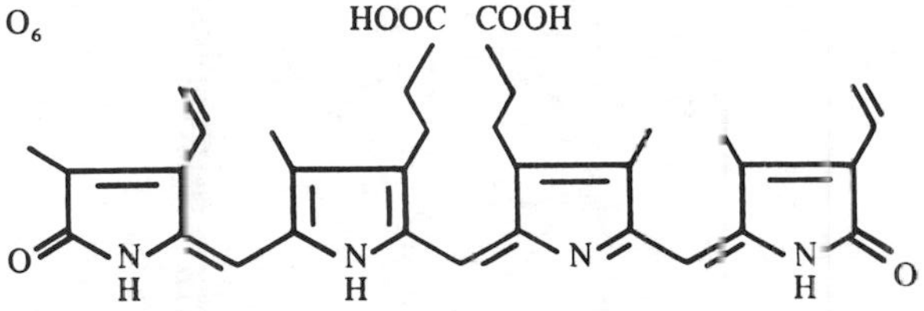	Biliverdin IXα. Helioporobilin	Coral: *Heliopora coerulea* Fish		+		284 j 335
455. $C_{33}H_{52}O_7$	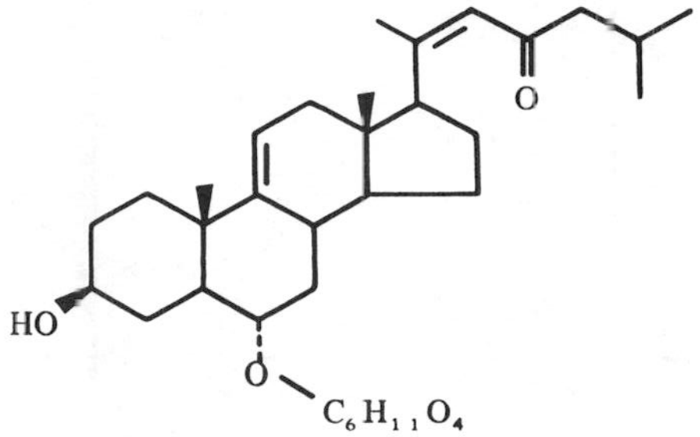	3β-Hydroxy-5α-cholesta-9(11),20(22)-dien-23-one-6α-yl-β-D-6′-deoxyglucoside	Starfish: *Acanthaster planci*				357
C_{34} 456. $C_{34}H_{40}N_4O_6$	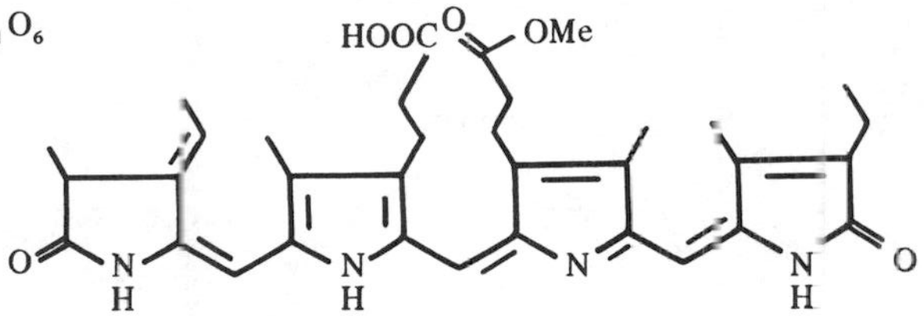	Aplysioverdin	Sea hare: *Aplysia* sp.				335

Table 1 (continued)

Molecular formula	Structural formula	Name	Source of compound	Available by synthesis	Available non-marine origin	Activity	References
C_{34} . . .							
457. $C_{34}H_{40}N_4O_6$	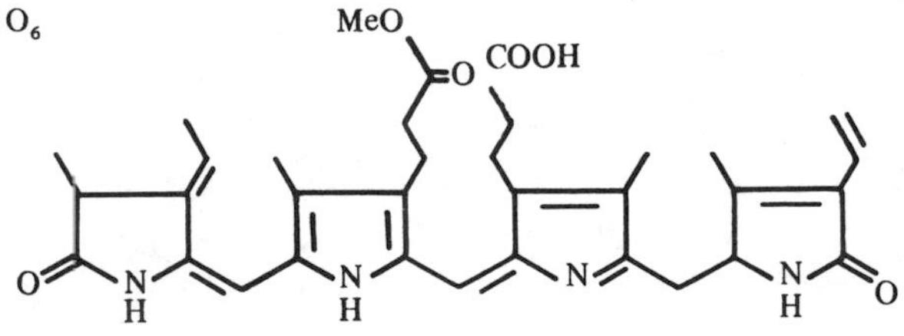	Aplysioviolin	Sea hare: *Aplysia limacina*				334 335
C_{35}							
458. $C_{35}H_{54}O$		Furospinosulin-3	Sponge: *Ircinia spinosula*				93
C_{36}							
459. $C_{36}H_{52}O_2$	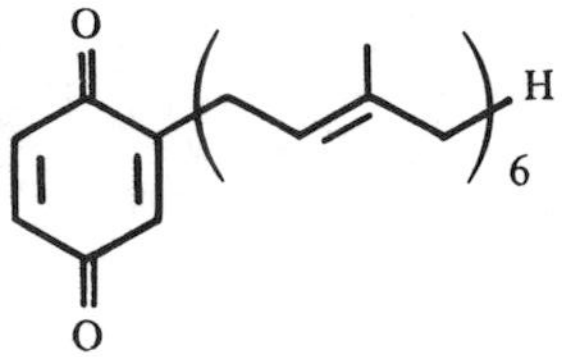	2-Hexaprenyl-1,4-benzoquinone	Sponge: *Ircinia spinosula*				93
460. $C_{36}H_{54}O_2$	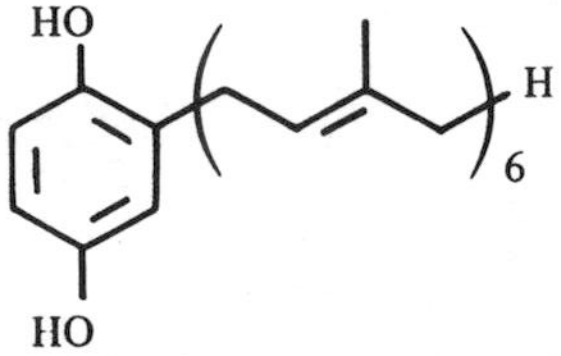	2-Hexaprenyl-1,4-benzoquinol	Sponge: *Ircinia spinosula*				93

Table 1 (continued)

Molecular formula	Structural formula	Name	Source of compound	Available by synthesis	Available non-marine origin	Activity	References
C_{36} . . .							
461. $C_{36}H_{56}O_9$	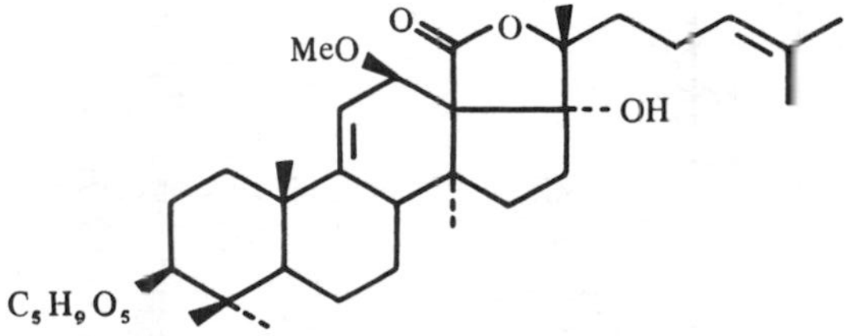 MeO, O, O, OH, $C_5H_9O_5$	12β-Methoxy-7,8-dihydro-24,25-dehydroholothur-inogenin-3β-xyloside	Sea cucumbers				79
C_{38}							
462. $C_{38}H_{48}O_4$	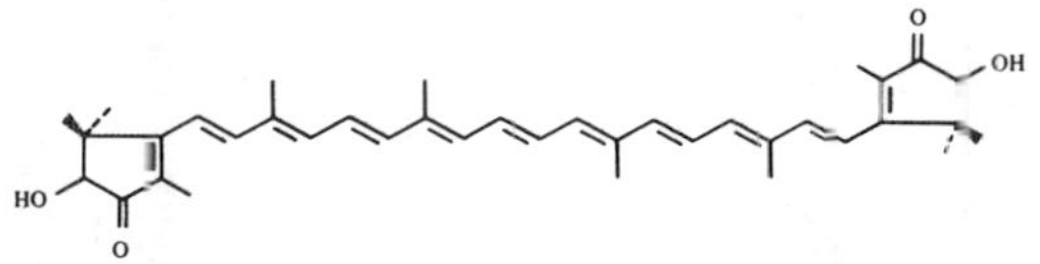 HO, O, O, OH Occurs esterified with C_{10}, C_{11} and C_{12} fatty acids. (Actinioerythrin)	Actinioerythrol	Sea anemone: *Actinia equina*				185
C_{39}							
463. $C_{39}H_{50}O_7$	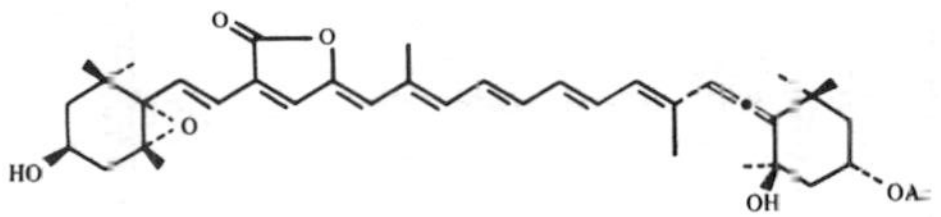 O, O, O, HO, OH, OAc	Peridinin	Dinoflagellates Sea anemones				379
464. $C_{39}H_{58}O_4$	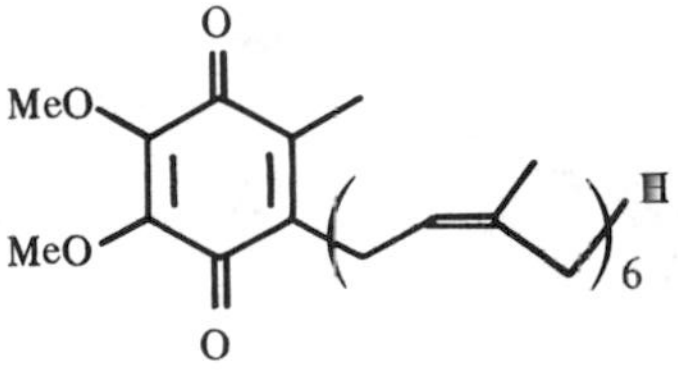 O, MeO, MeO, O, H, 6	Ubiquinone Q-6. 2,3-Dimethoxy-5-(3,7,11,15,19,23-hexamethyl-2,6,10,14,18,22-tetracosahexaenyl)-6-methyl-1,4-benzoquinone	Fish	+	+		284 IIIe 322 397 b

Table 1 (continued)

Molecular formula	Structural formula	Name	Source of compound	Available by synthesis	Available non-marine origin	Activity	References
C_{40}							
465. $C_{40}H_{46}$	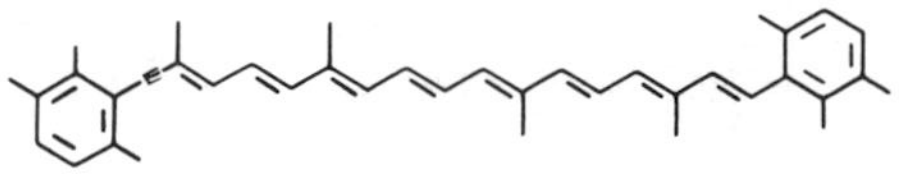	7,8-Didehydroiso-renieratene	Sponge: *Reniera japonica*				173
466. $C_{40}H_{46}$	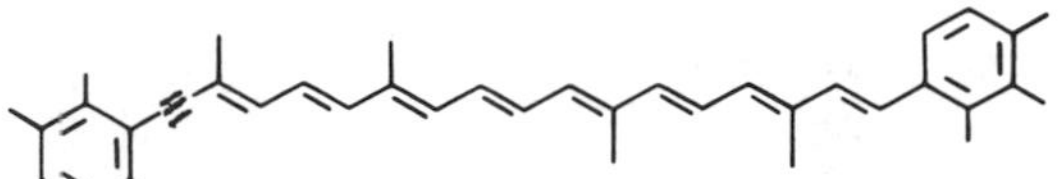	7,8-Didehydro-renieratene	Sponge: *Reniera japonica*				173
467. $C_{40}H_{48}$	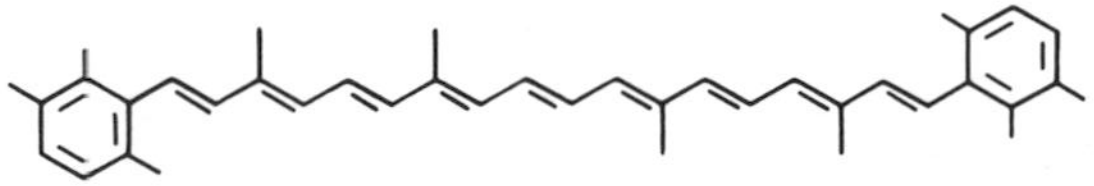	Isorenieratene	Sponge: *Reniera japonica*	+	+		440 442
468. $C_{40}H_{48}$	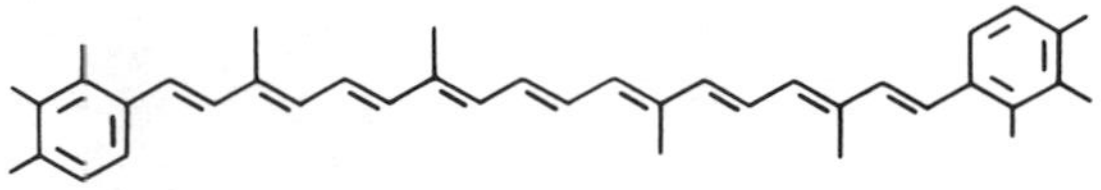	Renierapurpurin	Sponge: *Reniera japonica*	+			438 443
469. $C_{40}H_{48}$	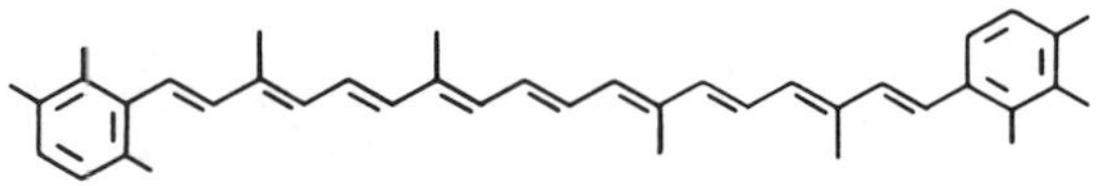	Renieratene	Sponge: *Reniera japonica*	+			438 439 441 443
470. $C_{40}H_{48}O_4$	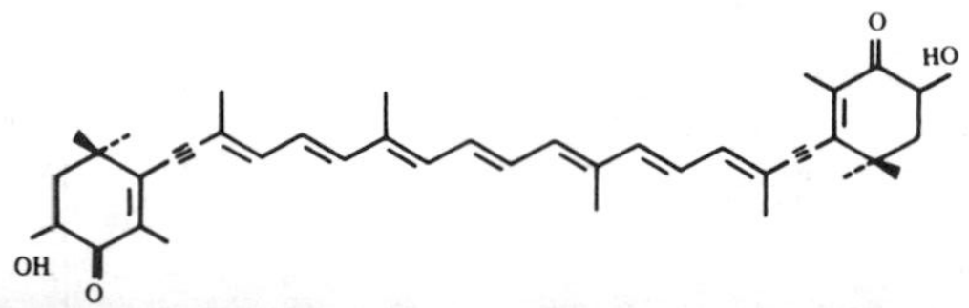	7,8,7′,8′-Tetrade-hydroastaxanthin	Starfish: *Asterias rubens*				372

Table 1 (continued)

Molecular formula	Structural formula	Name	Source of compound	Available by synthesis	Available non-marine origin	Activity	References
C_{40}							
471. $C_{40}H_{50}O_4$	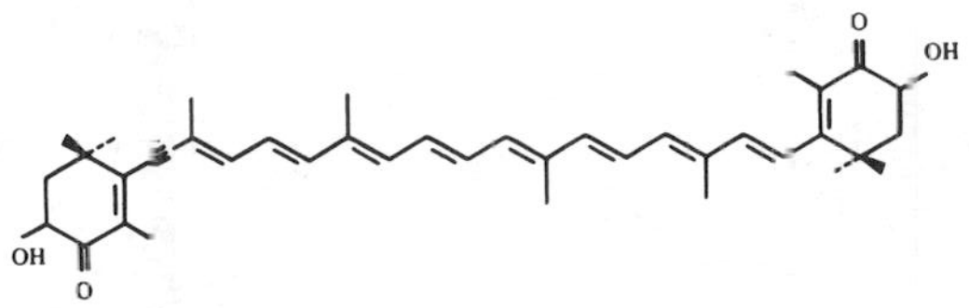	7,8-Didehydroastaxanthin	Starfish: *Asterias rubens*				372
472. $C_{40}H_{52}O_2$	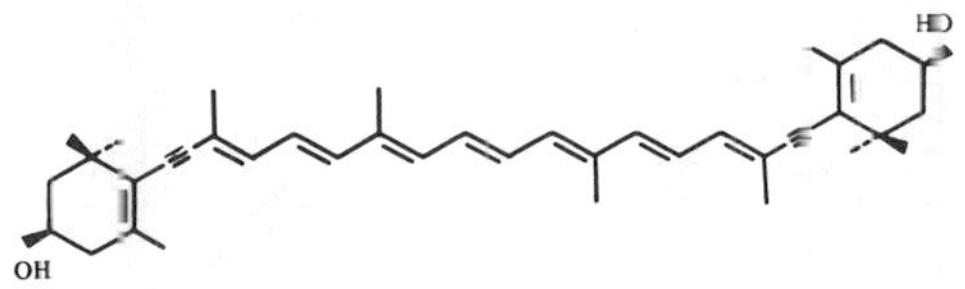	Alloxanthin. Cynthiaxanthin. Pectenoxanthin	Algae Mollusks Tunicate: *Halocynthia papillosa*				72 81 275
473. $C_{40}H_{52}O_2$	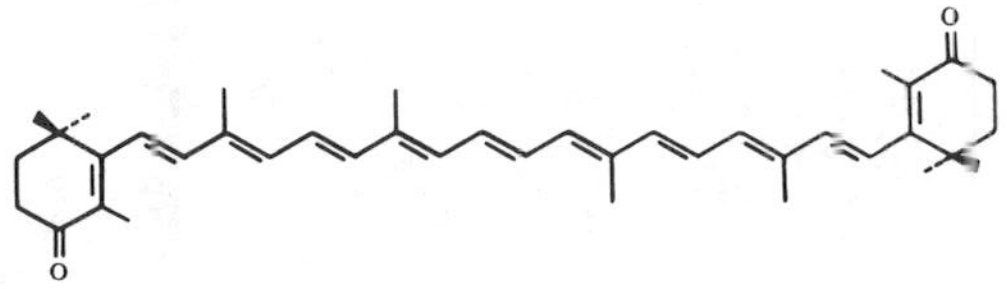	Canthaxanthin	Crab Isopods Sea cucumbers Shrimp		+		157 235 283
474. $C_{40}H_{52}O_3$	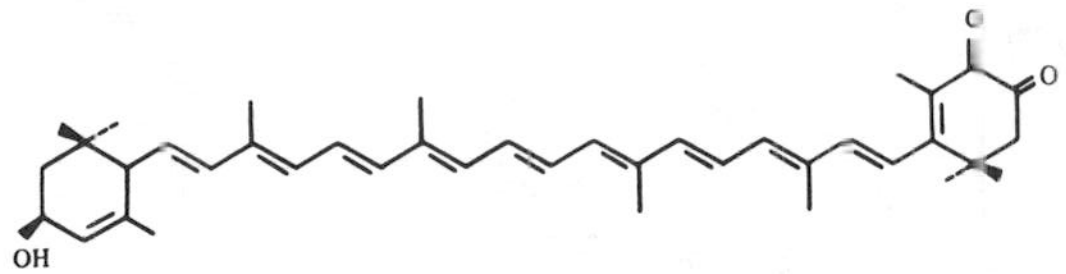	α-Doradecin	Sea bream				236
475. $C_{40}H_{52}O_3$	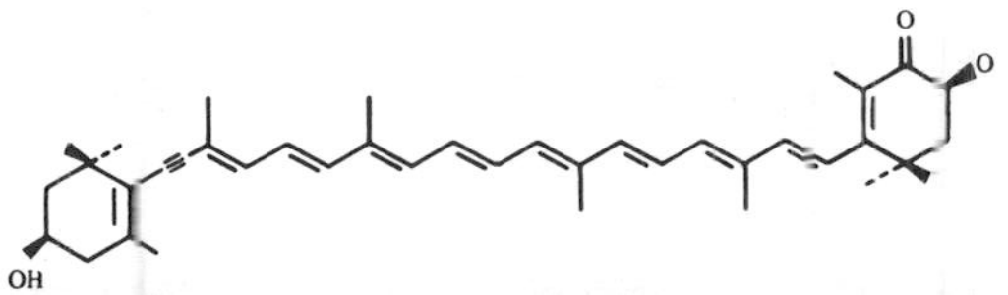	Pectenolone	Scallop: *Pecten maximus* Tunicate: *Halocynthia papillosa*				72

Table 1 (continued)

Molecular formula	Structural formula	Name	Source of compound	Available by synthesis	Available non-marine origin	Activity	References
C_{40} . . .							
476. $C_{40}H_{52}O_4$	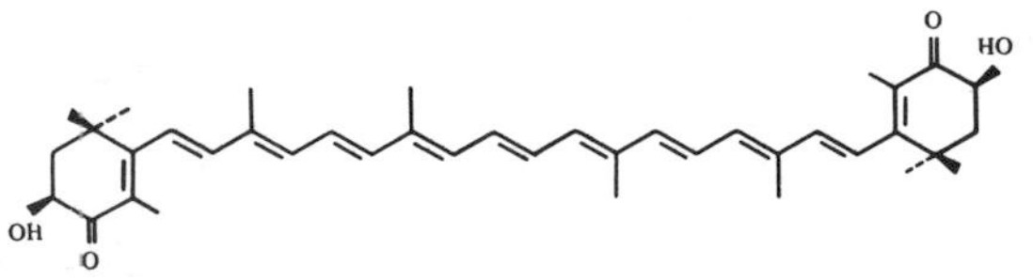	Astaxanthin	Crustaceans Fish Scallop Sea cucumbers Starfish Tunicate	+	+		72, 157, 235, 236, 237, 282, 283, 284 g, 341 h, i, 405
477. $C_{40}H_{54}O$	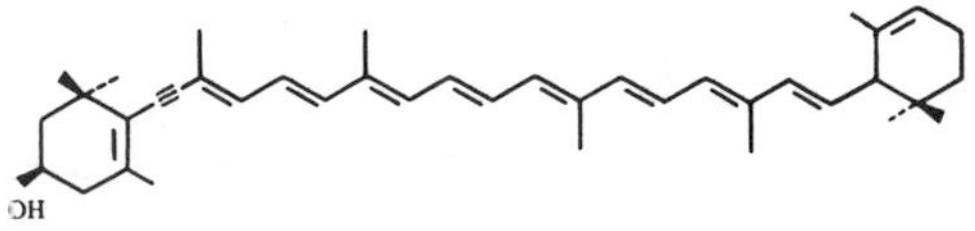	Crocoxanthin	Algae: *Cryptomonas ovata* *Hemiselmis virescens* *Rhodomonas* sp.				81 275
478. $C_{40}H_{54}O$	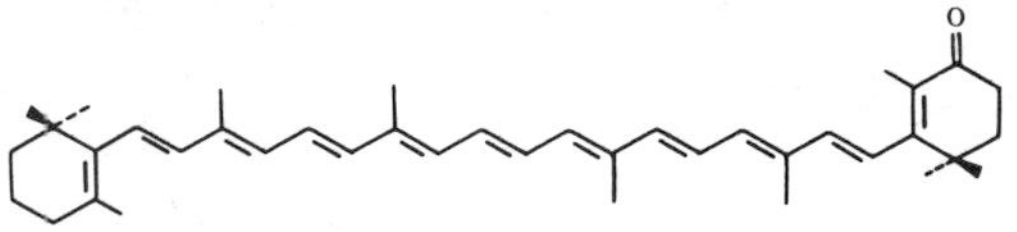	Echinenone	Crustaceans Mollusks Sea cucumbers Sea urchins Sponges Starfish	+	+		157 235 260 283 284 Ia
479. $C_{40}H_{54}O_2$	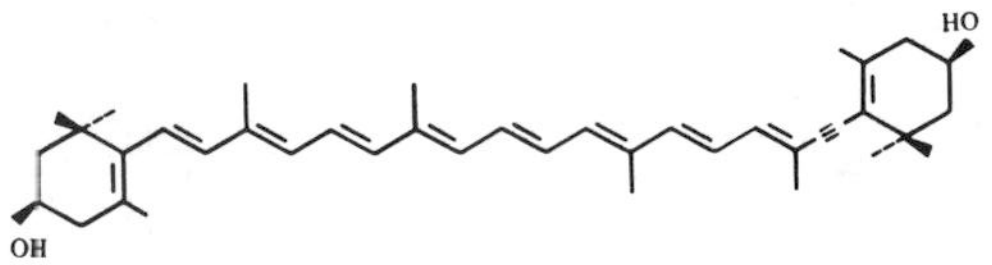	Diatoxanthin	Diatoms: *Isochrysis galbana* *Isthmia nervosa* *Navicula torquatum*				81 157 275 341 j
480. $C_{40}H_{54}O_2$	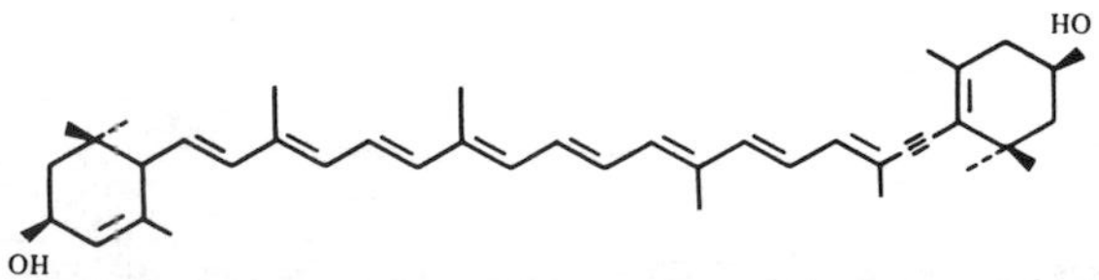	Monadoxanthin	Algae: *Cryptomonas ovata* *Hemiselmis virescens* *Rhodomonas* sp.				81 275

Table 1 (continued)

Molecular formula	Structural formula	Name	Source of compound	Available by synthesis	Available non-marine origin	Activity	References
C_{40} .... 481. $C_{40}H_{54}O_3$	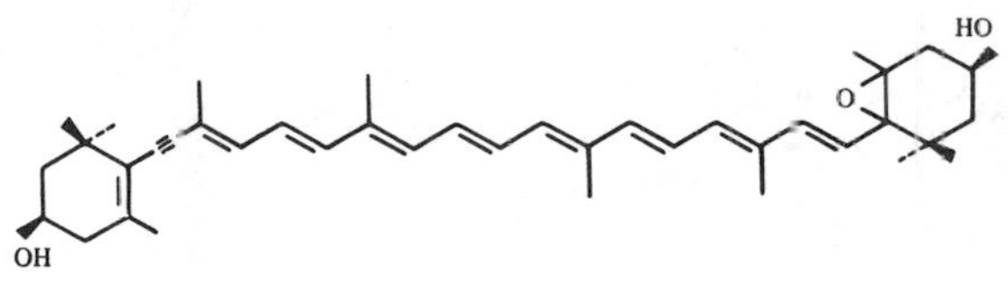	Diadinoxanthin	Diatom: *Nitzschia closterium* f. *minutissima* Dinoflagellate: *Glenodinium foliaceum*				26 157 277
482. $C_{40}H_{56}$	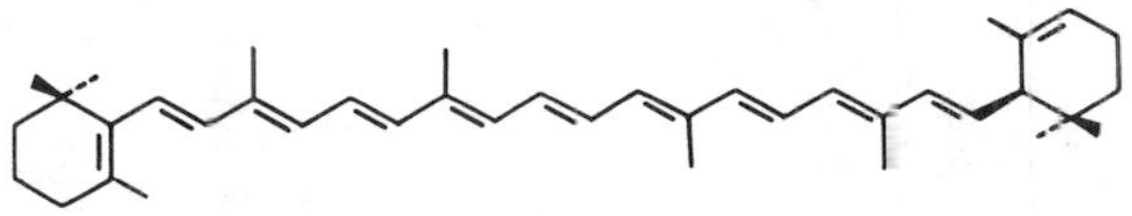	α-Carotene	Algae Sponge: *Reniera japonica*	+	+		157 173 249 284 m 438
483. $C_{40}H_{56}$	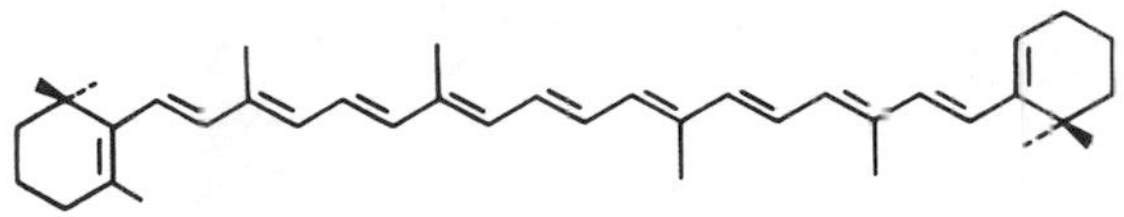	β-Carotene	Algae Crustaceans Fish Mollusks Sea cucumbers Sea urchins Sponges Starfish	+	+		157, 172, 173, 235, 237, 249, 277, 283, 284 n, 438
484. $C_{40}H_{56}$	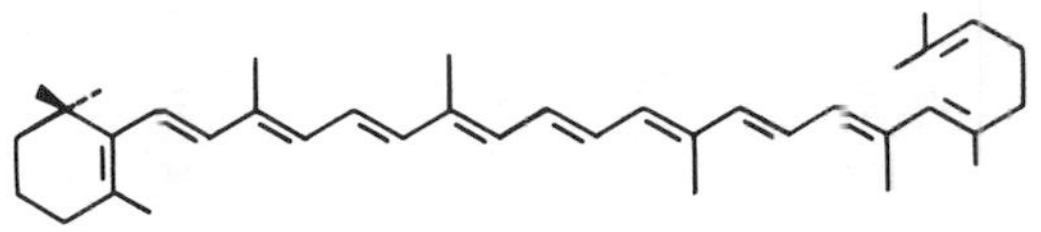	δ-Carotene	Crab: *Carcinus maenas*	+	+		157 235 276 284 n

Table 1 (continued)

Molecular formula	Structural formula	Name	Source of compound	Available by synthesis	Available non-marine origin	Activity	References
C_{40} . . .							
485. $C_{40}H_{56}$		ϵ-Carotene	Green alga: *Bryopsis corticulans* Diatom: *Navicula torquatum* Sea bream	 +	 +		80 157 236 276
486. $C_{40}H_{56}O_2$		Lutein. Xanthophyll	Algae Crabs Fish Isopods		+		157 235 236 249 284 IIIj
487. $C_{40}H_{56}O_2$		Zeaxanthin	Crab Fish Sea cucumbers Starfish	 +	 +		157 235 283 284 IIIk 322
488. $C_{40}H_{56}O_3$		Loroxanthin	Algae: *Cladophora ovoidea* *C. trichotoma* *Ulva rigida*				27
489. $C_{40}H_{56}O_4$		Siphonaxanthin	Green algae				248 249 250 416

Table 1 (continued)

Molecular formula	Structural formula	Name	Source of compound	Available by synthesis	Available non-marine origin	Activity	References
C_{40} ...							
490. $C_{40}H_{56}O_5$	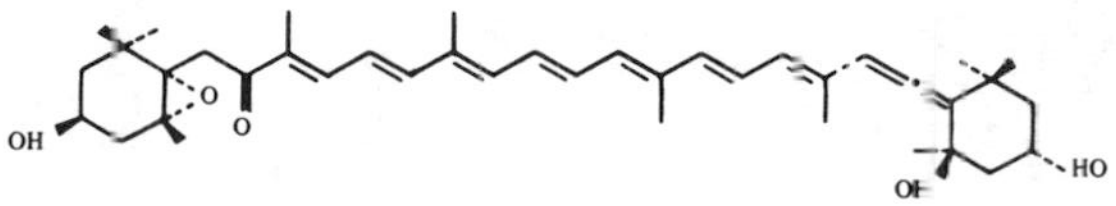	Fucoxanthinol	Sea urchin: *Paracentrotus lividus*				144
C_{41}							
491. $C_{41}H_{60}O_2$,	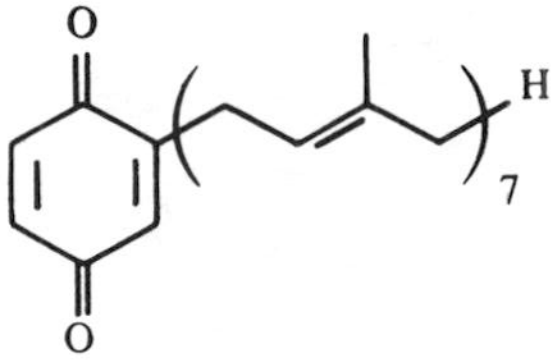	2-Heptaprenyl-1,4-benzoquinone	Sponge: *Ircinia spinosula*				93
492. $C_{41}H_{62}O_2$	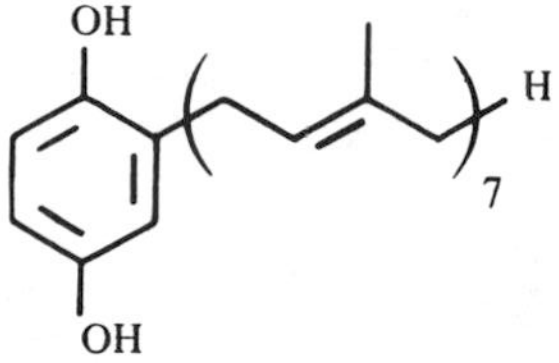	2-Heptaprenyl-1,4-benzoquinol	Sponge: *Ircinia spinosula*	+			93
493. $C_{41}H_{83}NO_2$	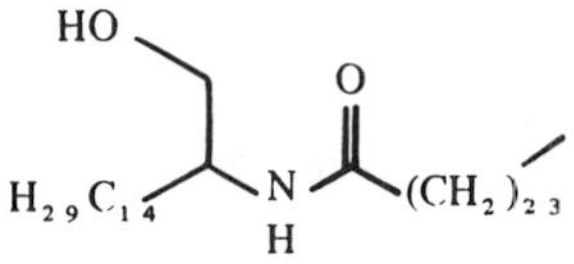	Caulerpicin*	Green algae: *Caulerpa lamourouxii* (*C. racemosa* var. *lamourouxii*) *C. racemosa* var. *clavifera* *C. serrulata* *C. sertularioides*				25 108 339

*(Mixture of C_{41}, C_{42}, C_{43} homologues).

Table 1 (continued)

Molecular formula	Structural formula	Name	Source of compound	Available by synthesis	Available non-marine origin	Activity	References
C_{42} 494. $C_{42}H_{58}O_6$	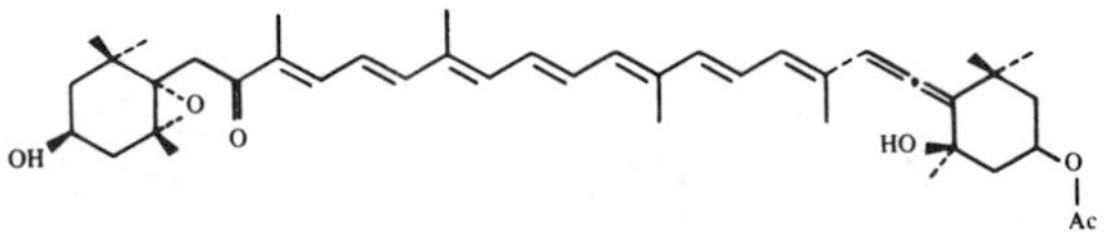 OH, O, O, HO, O, Ac	Fucoxanthin	Brown algae: *Fucus* sp. Dinoflagellate: *Glenodinium foliaceum* Sea urchin: *Paracentrotus lividus*				60 130 144 277 284 lg
495. $C_{42}H_{85}NO_2$	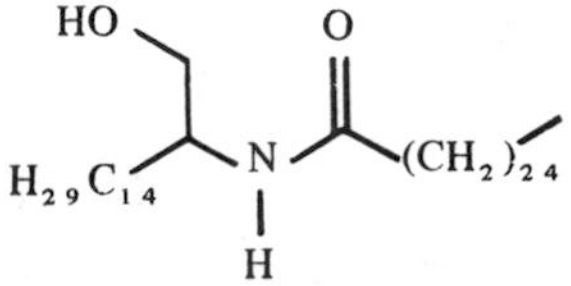 HO, O, $H_{29}C_{14}$, N, H, $(CH_2)_{24}$	Caulerpicin*	Green algae: *Caulerpa lamourouxii* *(C. racemosa* var. *lamourouxii)* *C. racemosa* var. *clavifera* *C. serrulata* *C. sertularioides*				25 108 339
C_{43} 496. $C_{43}H_{87}NO_2$	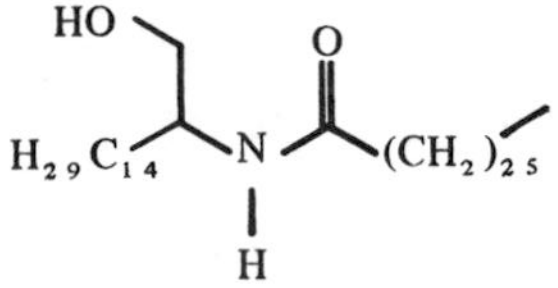 HO, O, $H_{29}C_{14}$, N, H, $(CH_2)_{25}$	Caulerpicin *	Green algae: *Caulerpa lamourouxii* *(C. racemosa* var. *lamourouxii)* *C. racemosa* var. *clavifera* *C. serrulata* *C. sertularioides*				25 108 339

*(Mixture of C_{41}, C_{42}, C_{43} homologues).

Table 1 (continued)

Molecular formula	Structural formula	Name	Source of compound	Available by synthesis	Available non-marine origin	Activity	References
C_{46}							
497. $C_{46}H_{68}O_2$	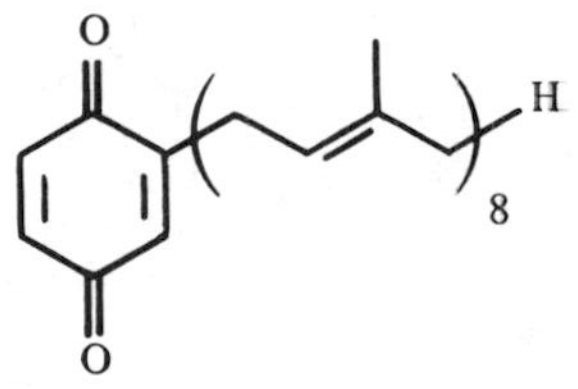	2-Octaprenyl-1,4-benzoquinone	Sponge: *Ircinia spinosula*				93
498. $C_{46}H_{70}O_2$	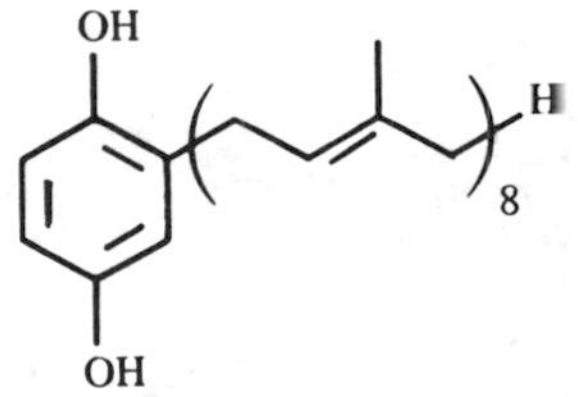	2-Octaprenyl-1,4-benzoquinol	Sponge: *Ircinia spinosula*				93
499. $C_{46}H_{70}O_3$	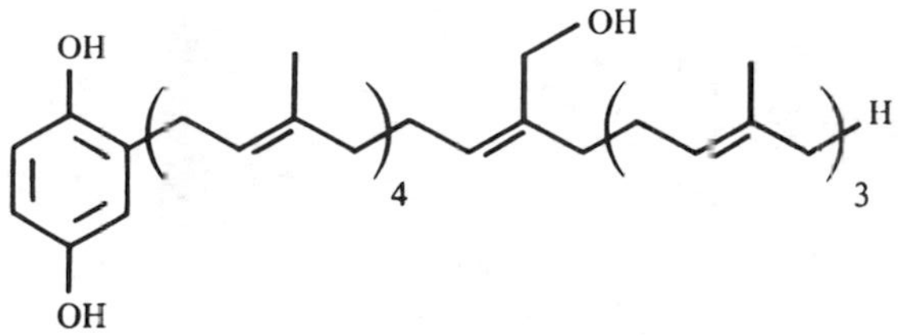	[19-(Hydroxymethyl)-3,7,11,15,23,27,31-heptamethyl-2,6,10,14,18,22,26,30-dotriaconta-octaenyl]-1,4-benzoquinol	Sponge: *Ircinia spinosula*				93

Table 1 (continued)

Molecular formula	Structural formula	Name	Source of compound	Available by synthesis	Available non-marine origin	Activity	References
C_{48} 500. $C_{48}H_{54}N_4O_{16}$	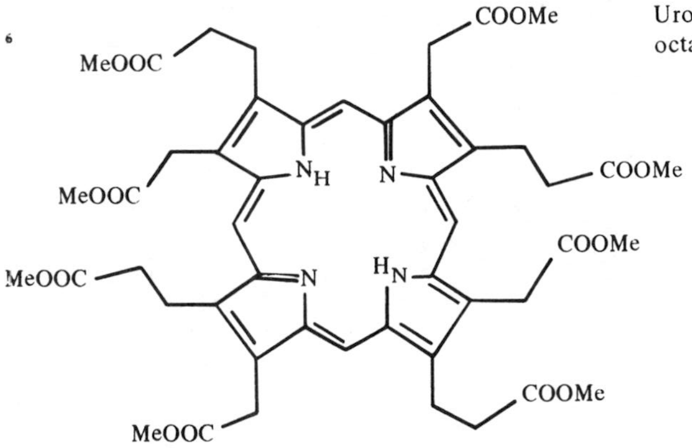 MeOOC, COOMe, NH, N	Uroporphyrin I octamethyl ester	Oyster: *Pinctada martensii*				226
C_{49} 501. $C_{49}H_{74}O_4$	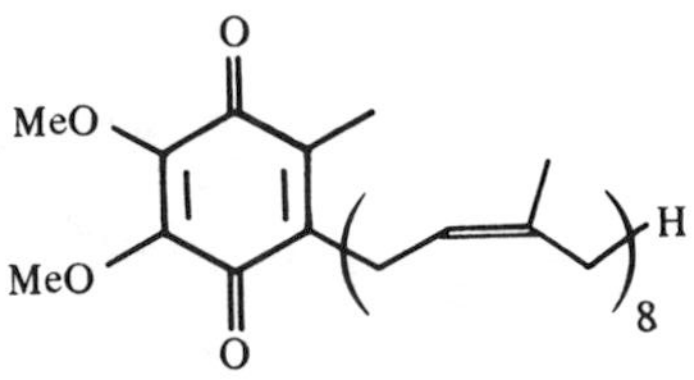 MeO, O, H, 8	Ubiquinone Q-8. 2,3-Dimethoxy-5-(3,7, 11,15,19,23,27,31-octamethyl-2,6,10,14, 18,22,26,30-dotria-conta-octaenyl)-6-methyl-1,4-benzoquin-one	Protozoans	+	+		284 IIIe 322 397 b
C_{52} 502. $C_{52}H_{76}O_5$	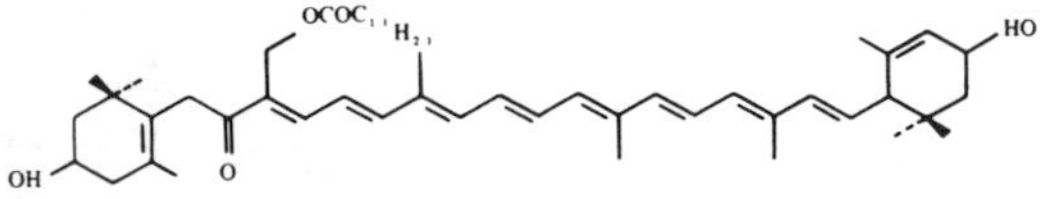 OH, O, $OCOC_{11}H_{23}$, HO	Siphonein	Green algae				248 249 416

Table 1 (continued)

Molecular formula	Structural formula	Name	Source of compound	Available by synthesis	Available non-marine origin	Activity	References
C_{54} 503. $C_{54}H_{82}O_4$	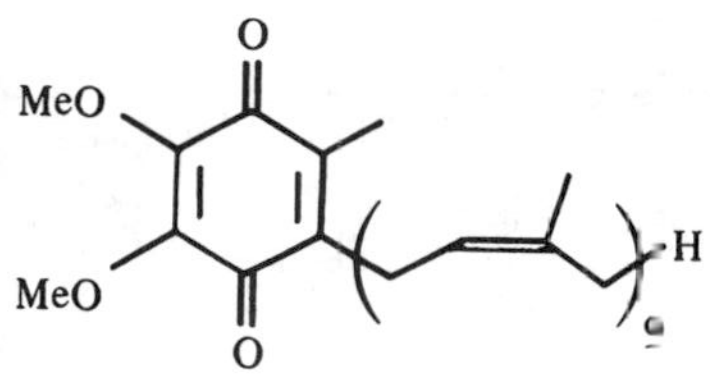 MeO, MeO, O, O, H, 9	Ubiquinone Q-9. 2,3-Dimethoxy-5-(3,7,11,15,19,23,27,31,35-nonamethyl-2,6,10,14,18,22,26,30,34-hexatriacontanonaenyl)-6-methyl-1,4-benzoquinone	Fish	+	+		284 IIIe 322 397 b
C_{59} 504. $C_{59}H_{90}O_4$	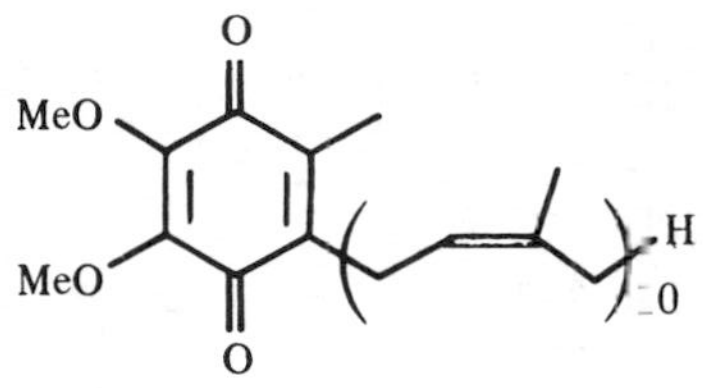 MeO, MeO, O, O, H, 10	Ubiquinone Q-10. 2,3-Dimethoxy-5-(3,7,11,15,19,23,27,31,35,39-decamethyl-2,6,10,14,18,22,26,30,34,38-tetracontadecaenyl)-6-methyl-1,4-benzoquinone	Crab Echinoderms Fish Sea worm	+	+		284 IIIe 322 397 b

COMPOUND NAME INDEX FOR TABLE 1

Name	Compound number
Acanthasterol	434
3-Acetamido-1,5-dibromo-6,6-dimethoxy-3-hydroxy-1,4-cyclohexadiene	150
3-Acetamido-1,5-dibromo-6-ethoxy-3-hydroxy-6-methoxy-1,4-cyclohexadiene	166
4-Acetamido-2,6-dibromo-4-hydroxycyclohexadienone	107
23ξ-Acetoxy-17-desoxy-7,8-dihydroholothurinogenin	452
S-(−)-3-Acetoxyundec-5-enyl thioacetate	261
6-Acetyl-2,7-dihydroxyjuglone	187
3-Acetyl-2-hydroxy-7-methoxynaphthazarin	204
3-Acetyl-2-hydroxynaphthazarin	186
6-Acetyl-2-hydroxynaphthazarin	188
Acetyllaurefucin	278
O-Acetyl(15R)-PGA$_2$ methyl ester	335
O-Acetyl(15S)-PGA$_2$ methyl ester	336
3-Acetyl-2,6,7-trihydroxyjuglone	189
6-Acetyl-2,3,7-trihydroxyjuglone	190
Actinioerythrol	462
Adenine	44
Adenosine	151
(+)-Aeroplysinin-1	122
(−)-Aeroplysinin-1	122
Aeroplysinin-2	119
Aerothionin	339
AF-350	275
Agmatine	69
α-Alanine	11
β-Alanine	12
Allo-5α-cholic acid	341
Allo-kainic acid	154
Alloxanthin	472
N-Amidino-4-bromopyrrole-2-carboxamide	75
2-Amino-3-benzyl-5-(4′-hydroxyphenyl)pyrazine	275
3-(2-Aminoethyl)-5,6-dibromoindole	144
2-Aminoethylphosphonic acid	10
β-Aminoisobutyric acid	28
α-Amino-β-phosphonopropionic acid	18
Anabaseine	147
Ancepsenolide	330
Anhydro-ethylidene-3,3′-bis(2,6,7-trihydroxy-naphthazarin)	324
Anhydrofurospongin-1	307
Aplysin	214
Aplysin-20	297
Aplysinol	218
Aplysioverdin	456
Aplysioviolin	457
Aplysterol	416
Arginine	87
(−)-1(10)-Aristolene	238
(+)-9-Aristolene	239
Asparagine	27
Aspartic acid	26
Astaxanthin	476
Asterosterol	359

COMPOUND NAME INDEX FOR TABLE 1 (continued)

Name	Compound number
Asterubine	67
Atrinine	103
Axisonitrile-1	267
Axisothiocyanate-1	268
L-Baikiain	80
Batyl alcohol	323
Betaine	56
Biliverdin IXα	454
(+)-β-Bisabolene	240
(−)-Bis-(3-acetoxyundec-5-enyl) disulfide	365
Bis-(3-oxoundecyl) disulfide	332
Bis-(3-oxoundecyl) tetrasulfide	334
Bis-(3-oxoundecyl) trisulfide	333
(+)-β-*epi*-Bourbonene	259
Brassicasterol	394
3-Bromo-5-chlorotyrosine	121
3-Bromo-4,5-dihydroxybenzaldehyde	92
6-Bromo-2,2-dimethylthioindolin-3-one	143
6-Bromo-2-(methylsulphonyl)indol-3-yl potassium sulfate	116
6-Bromo-2-methylthioindoleninone	115
4-Bromophakellin	163
1-(4′-Bromophenoxy)-3-bromo-2-hydroxybenzene	184
1-(4′-Bromophenoxy)-5-bromo-2-hydroxybenzene	185
1-(4′-Bromophenoxy)-4,6-dibromo-2-hydroxybenzene	183
3-Bromotyrosine	125
Cadalene	211
(+)-Cadinene	241
(−)-γ_1-Cadinene	242
(−)-δ-Cadinol	260
Caespitol	258
(+)-Calamenene	234
Callinecdysone A	378
Callinecdysone B	398
Campesterol	399
Canthaxanthin	473
4-(β-Carboxy-β-aminoethyl)-1,3-dimethylimidazolium chloride	113
2-(13-Carboxy-14,15-diacetoxyhexadecanyl)-2-penten-4-olide	362
α-Carotene	482
β-Carotene	483
δ-Carotene	484
ε-Carotene	485
Caulerpicin	493, 495, 496
Caulerpin	338
Cetyl palmitate	453
Chenodeoxycholic acid	340
7-Chloro-3,7-dimethyl-1,4,6-tribromo-*trans*-1-octene-3-ol	157
5α-Cholesta-9(11),20(22)-diene-3β,6α-diol-23-one	369
22-*trans*-Cholesta-7,22-dien-3β-ol	372
Cholestanol	383

COMPOUND NAME INDEX FOR TABLE 1 (continued)

Name	Compound number
Cholesterol	380
Cholic acid	342
Choline	71
Choline acrylate	114
Choline sulfate	66
Chondrillasterol	405
Chondrillin	345
Chondrine	50
Chondriol	212
Clionasterol	419
Comantherin sodium sulfate ester	265
Comaparvin	276
Concinndiol	298
(−)-Copaene	243
Coprostanol	384
Crassin acetate	328
Creatine	31
Crocoxanthin	477
Crustecdysone	379
(+)-α-Cubebene	244
(−)-α-Curcumene	235
(+)-β-Curcumene	245
Cycloartenol	435
Cynthiaxanthin	472
Cysteic acid	16
Cysteine	14
Cytidine	130
Debromoaplysin	225
Debromoisolaurinterol	226
Debromolaurinterol	227
Decahydro-4-methylene-1,1,7-trimethyl-1H-cycloprop-(e)-azulene	246
22,23-Dehydrocholestanol	381
7-Dehydrocholesterol	373
22-Dehydrocholesterol (*cis* and *trans*)	374
24-Dehydrocholesterol	375
Dehydrodendrolasin	228
23-Demethylgorgosterol	406
Deoxoscalarin	371
Deoxycrustecdysone	377
16-Deoxymyxinol	385
Desmosterol	375
17-Desoxy-22,25-oxidoholothurinogenin	422
Diadinoxanthin	481
Diatoxanthin	479
2,7-Dibromo-3,8-dichloro-2,3,7,8,9,9a-hexahydro-5,8,10,10-tetramethyl-6H-2,5a-methano-1-benzoxepin	221
2,3-Dibromo-4,5-dihydroxybenzaldehyde	91
2,3-Dibromo-4,5-dihydroxybenzyl alcohol	94
2,3-Dibromo-4,5-dihydroxybenzyl-methyl ether	108
3,5-Dibromo-4-hdyroxybenzyl alcohol	93
3,5-Dibromo-4-hydroxyphenylacetamide	106
2,3-Dibromo-5-hydroxy-4-0-sulfate benzyl sulfate, dipotassium salt	90

COMPOUND NAME INDEX FOR TABLE 1 (continued)

COMPOUND NAME INDEX FOR TABLE 1 (continued)

Name	Compound number
2,3-Dimethoxy-5-(3,7,11,15,19,23,27,31-octamethyl-2,6,10,14,18,22,26,30-dotriacontaoctaenyl)-6-methyl-1,4-benzoquinone	501
ββ-Dimethylacrylylcholine	158
Dimethylamine	4
N,N-Dimethylhistamine	101
Dimethyl-β-propiothetine	55
N,N-Dimethyltaurine	35
3R,4S,7S-*trans,trans*-3,7-Dimethyl-1,8,8-tribromo-3,4,7-trichloro-1,5-octadiene	146
3,6-Dioxohexahydropyrrolo-(1,2-a)-pyrazine	99
Domoic acid	233
α-Doradecin	474
Echinenone	478
Echinochrome A	199
Ectocarpene	169
(+)-β-Elemene	247
Episterol	395
5,6-Epoxy-9,11-secogorgosten-3,11-diol-9-one	443
5α-Ergosta-7,24(28)-dien-3β-ol	395
Δ^7-Ergostenol	401
Ergosterol	393
Ergothioneine	132
Ethanolamine	6
Ethylamine	5
24ξ-Ethylcholest-7-en-3β-ol	417
Ethyl-3,5-dibromo-1-hydroxy-4-oxo-2,5-cyclohexadiene-1-acetate	145
6-Ethyl-2-hydroxyjuglone	193
3-Ethyl-2-hydroxynaphthazarin	194
6-Ethyl-2-hydroxynaphthazarin	195
Ethylidene-3,3′-bis(2,6,7-trihydroxynaphthazarin	325
24-Ethylidene-5α-cholest-7-en-3β-ol	408
3-Ethyl-2,6,7-trihydroxyjuglone	197
6-Ethyl-2,3,7-trihydroxyjuglone	198
Eunicellin	392
Eunicin	290
Eupalmerin acetate	329
Fasciculatin	352
Friedelin	436
Fringelite D	391
Fringelite E	390
Fringelite F	389
Fringelite H	388
Fucosterol	409
Fucoxanthin	494
Fucoxanthinol	490
Furocaespitane	203
Furospinosulin-1	354
Furospinosulin-2	425
Furospinosulin-3	458
Furospongin-1	313

COMPOUND NAME INDEX FOR TABLE 1 (continued)

Name	Compound number
Furospongin-2	304
Furospongin-3	356
Furospongin-4	357
Furoventalene	213
(−)-Germacrene-A	248
Gigartinine	102
Glutamic acid	52
Glutamine	54
Glycine	3
Glycocyamine	17
Gongrine	83
β-Gorgonene	249
Gorgost-5-en-3,9,11-triol	441
Gorgosterol	437
Griseogenin	429
Guadalupensic acid	344
Guanidine	2
γ-Guanidinobutyric acid	62
γ-Guanidino-β-hydroxybutyric acid	63
β-Guanidinopropionic acid	32
Guanine	45
Guanosine	152
Hallachrome	264
Helioporobilin	454
1,6,9,12,15,18-Heneicosahexaene	315
3,6,9,12,15,18-Heneicosahexaene	316
1,6,9,12,15-Heneicosapentaene	320
2-Heptaprenyl-1,4-benzoquinol	492
2-Heptaprenyl-1,4-benzoquinone	491
Herbipoline	98
Hercynine	131
Hexadecahydro-1α,2β,5β,8α-tetramethyl-1,8-pyrenediyl-diisocyanide	327
2-Hexaprenyl-1,4-benzoquinol	460
2-Hexaprenyl-1,4-benzoquinone	459
(−)-3-Hexyl-4,5-dithiacycloheptanone	178
Histamine	53
Histidine	81
Holothurinogenin	427
Homarine	95
Homoaerothionin	349
Hydroxyagmatine	70
Hydroxyancepsenolide	331
γ-Hydroxyarginine	88
3β-Hydroxy-5α-cholesta-9(11),20(22)-dien-23-one-6α-yl-β-D-6′-deoxyglucoside	455
2-(2-Hydroxy-3,5-dibromophenyl)-3,4,5-tribromopyrrole	135
12α-Hydroxy-7,8-dihydro-24,25-dehydroholothurinogenin	430
12α-Hydroxy-7,8-dihydro-22,25-oxidoholothurinogenin	432
[19-(Hydroxymethyl)-3,7,11,15,23,27,31-heptamethyl-2,6,10,14,18,22,26,30-dotriacontaoctaenyl]-1,4-benzoquinol	499

COMPOUND NAME INDEX FOR TABLE 1 (continued)

Name	Compound number
2-Hydroxy-2′-methyl-2′H-pyrano (2,3-b)naphthazarin	210
3-(2′-Hydroxypentyl)-1,6,8-trihydroxy-9,10-anthraquinone	284
2-(N-p-Hydroxyphenyl ethanoyl) amino-3-benzyl-5-p-hydroxyphenylpyrazine	347
4-Hydroxyproline	49
2-(1′-Hydroxypropyl)-1,4,5,7-tetrahydroxy-9,10-anthraquinone	274
4-Hydroxy-3-tetraprenylbenzoic acid	366
Hypotaurine	7
Hypoxanthine	42
Imidazoleacetic acid	46
4-Imidazolylacrylic acid	74
4-Imidazolyllactic acid	79
4-Imidazolylpropionic acid	77
Indolyl-3-acetic acid	142
Indolyl-3-carboxylic acid	117
3-Iodotyrosine	126
Ircinin-1	350
Ircinin-2	351
Ircinin-3	309
Ircinin-4	310
Isoamylamine	64
Isobutylamine	34
28-Isofucosterol	410
Isofurospongin-2	305
Isolaureatin	222
Isolaurinterol	215
Isoleucine	84
Isoprelaurefucin	223
5-Isopropyl-4-methoxy-1-(6,6,6-trichloro-3-methoxy-5-methyl-2-hexenoyl)-3-pyrrolin-2-one	266
Isorenieratene	467
Isorhodoptilometrin	272
Isozonarol	311
Jeunicin	291
Johnstonol	232
α-Kainic acid	155
Koellikerigenin	428
Laminine	134
Lanosol	94
Lanosterol	438
Lathosterol	382
Laureatin	224
Laurefucin	230
Laurencin	277
Laurene	220
Laurenisol	216
Laurinterol	217
Leucine	85
Loroxanthin	488
Lutein	486
Lysine	86
(+)-γ-Maaliene	250

COMPOUND NAME INDEX FOR TABLE 1 (continued)

Name	Compound number
Marthasterone	370
Metanethole	289
Methionine	58
Methionine sulfone	60
Methionine sulfoxide	59
6-Methoxycomaparvin	281
6-Methoxycomaparvin-5-methyl ether	285
12α-Methoxy-17-desoxy-7,8-dihydro-22,25-oxidoholothurinogenin	447
12β-Methoxy-17-desoxy-7,8-dihydro-22,25-oxidoholothurinogenin	448
12β-Methoxy-7,8-dihydro-24,25-dehydroholothurinogenin-3β-xyloside	461
12β-Methoxy-7,8-dihydro-22-hydroxyholothurinogenin	451
12β-Methoxy-7,8-dihydro-22,25-oxidoholothurinogenin	450
1-Methyladenine	76
Methylamine	1
2-N-Methylaminoethylphosphonic acid	23
22-*trans*-(24S)-24-Methylcholesta-7,22-dien-3β-ol	397
24α-Methyl-Δ^5-cholestenol	399
24β-Methyl-Δ^5-cholestenol	400
24-Methylenecholestanol	402
24-Methylenecholesterol	396
Methylguanidine	9
29-Methylisofucosterol	439
2-Methylmercaptopropylamine	36
O-(14-Methyl-4-pentadecenyl)-choline	322
N-Methyltaurine	21
2-Methylthioindoxyl potassium sulfate	120
Mompain	137
Monadoxanthin	480
Mureninе	177
(+)-α-Muurolene	251
Myxinol	386
Namakochrome	160
Naphthopurpurin	136
Neocomantherin sodium sulfate ester	280
Neospongosterol	403
Nereistoxin	61
Nitenin	301
N-Nonadecanoyl-2-methylene-β-alanine methyl ester	346
22-*trans*-24-Nor-5α-cholesta-7,22-dien-3β-ol	359
24-Norcholesta-5,22-dien-3β-ol	360
24-Nor-5α-cholesta-7,23-dien-3β-ol	361
24-Norcholest-22-en-3β-ol	363
24-Norcholest-23-en-3β-ol	364
Norzooanemonin	78
2-Octaprenyl-1,4-benzoquinol	498
2-Octaprenyl-1,4-benzoquinone	497

COMPOUND NAME INDEX FOR TABLE 1 (continued)

Name	Compound number
Octopamine	112
Octopin	133
Oppositol	257
Oroidin	161
22,25-Oxidoholothurinogenin	424
S-(*trans*-3-Oxo-4-undecaenyl) thioacetate	208
S-(3-Oxoundecyl)-thioacetate	209
Pachydictyol A	295
Pacifidiene	219
Pahutoxin	337
Palmitic acid	269
Panicein-A	326
Panicein-B_1	299
Panicein-B_2	300
Panicein-B_3	302
Panicein-C	303
Paracentrone	444
Parazoanthoxanthin D	202
Pectenolone	475
Pectenoxanthin	472
Peridinin	463
Petalonine	29
15R-PGA_2	292
15S-PGA_2	293
5-*trans*-15S-PGA_2	294
15S-PGA_2 methyl ester	318
15S-PGE_2	296
Phenylalanine	127
2-Phenylethylamine	110
β-Phocaecholic acid	343
Phosphoarginine	89
Phosphocreatine	33
Pipecolic acid	82
Pleraplysillin	229
Poriferastanol	421
Poriferasterol	411
Potassium-6-bromoindoxyl sulfate	104
Potassium indoxyl sulfate	105
Praslinogenin	449
5α-Pregn-9(11)-ene-3β,6α-diol-20-one	317
5α-Pregn-9(11)-ene-3β,6α-diol-20-one-3-sulfate	319
Prepacifenol	231
Pristane	287
Proline	48
3-Propionyl-1,6,8-trihydroxy-9,10-anthraquinone	270
Propylamine	19
3-Propyl-1,6,8-trihydroxy-9,10-anthraquinone	271
Provitamin D_3	373
Ptilometric acid	279
Putrescine	37
Pyrimidine	24
Renierapurpurin	468

COMPOUND NAME INDEX FOR TABLE 1 (continued)

COMPOUND NAME INDEX FOR TABLE 1 (continued)

Name	Compound number
2-Tetraprenyl-1,4-benzoquinol	358
2-Tetraprenyl-1,4-benzoquinone	355
Tetrodotoxin	172
Threonine	30
Thymine	47
α-Tocopherolquinone	420
Trigonelline	96
3α,7α,12α-Trihydroxycholanic acid	342
3α,7α,23-Trihydroxycholanic acid	343
Trimethylamine	20
2-Trimethylaminoethylphosphonic acid	68
Tryptophan	165
m-Tyramine	111
Tyrindoxyl sulfate	118
Tyrosine	128
Ubiquinone Q-6	464
Ubiquinone Q-8	501
Ubiquinone Q-9	503
Ubiquinone Q-10	504
trans,cis,cis-1,3,5,8-Undecatetraene	170
trans,trans,cis-1,3,5,8-Undecatetraene	171
trans,cis-1,3,5-Undecatriene	175
trans,trans-1,3,5-Undecatriene	176
Uracil	25
Urocanic acid	74
Uroporphyrin I octamethyl ester	500
Valine	57
Variabilin	353
Vitamin B_4	44
Xanthine	43
Xanthophyll	486
Xanthopterin	73
β-Ylangene	253
Yunaine	50
Zamene	286
Zeaxanthin	487
Zoanthoxanthin	207
Zonarene	254
Zonarol	312
Zooanemonine	100

REFERENCES

1. **Abe, H., Uchiyama, M., and Sato, R.,** Isolation and identification of native auxins in marine algae, *Agr. Biol. Chem.,* 36, 2259, (1972).
2. **Ackermann, D.,** Asterubin, eine schwefelhaltige Guanidinverbindung der belebten Natur, *Hoppe-Seyler's Z. Physiol. Chem.,* 232, 206 (1935).
3. **Ackermann, D. and Müller, E.,** Zweite Synthese des Asterubins, *Hoppe-Seyler's Z. Physiol. Chem.,* 235, 233 (1935).
4. **Ackermann, D. and Mohr, M.,** Über das Vorkommen von Octopin, Agmatin und Arginin in der Octopodenart *Eledone moschata, Hoppe-Seyler's Z. Physiol. Chem.,* 250, 249 (1937).
5. **Ackermann, D. and Müller, E.,** Spinacin, ein Bestandteil der Selachierleber, *Hoppe-Seyler's Z. Physiol. Chem.,* 268, 277 (1941).
6. **Ackermann, D. and Burchard, C.,** Zur Kenntnis der Spongine, *Hoppe-Seyler's Z. Physiol. Chem.,* 271, 183 (1941).
7. **Ackermann, D.,** Nachweis des Imidazolkernes im Spinacin, *Hoppe-Seyler's Z. Physiol. Chem.,* 276, 268 (1942).
8. **Ackermann, D. and Skraup, S.,** Endgültige Konstitutionsermittlung und Synthese des Spinacins, *Hoppe-Seyler's Z. Physiol. Chem.,* 284, 129 (1949).
9. **Ackermann, D.,** Über das Vorkommen von Homarin, Trigonellin und einer neuen Base Anemonin in der Anthozoe *Anemonia sulcata, Hoppe-Seyler's Z. Physiol. Chem.,* 295, 1 (1953).
10. **Ackermann, D. and Janka, R.,** Erstmalige Beobachtung von Spermin bei Avertebraten (*Cionia intestinalis*), *Hoppe-Seyler's Z. Physiol. Chem.,* 296, 279 (1954).
11. **Ackermann, D.,**Über das Vorkommen von Homarin, Taurocyamin, Cholin, Lysin und anderen Aminosäuren sowie Bernsteinsäure in dem Meereswurm *Arenicola marina, Hoppe-Seyler's Z. Physiol. Chem.,* 302, 80 (1955).
12. **Ackermann, D. and List, P. H.,** Über das Vorkommen von Trimethylaminoxyd, Homarin, Trigonellin und einer Base $C_4H_9O_2N$ in der Krabbe *(Crangon vulgaris), Hoppe-Seyler's Z. Physiol. Chem.,* 306, 260 (1957).
13. **Ackermann, D. and List, P. H.** Konstitutionsermittlung des Herbipolins, einer neuen tierischen Purinbase, *Hoppe-Seyler's Z. Physiol. Chem.,* 309, 286 (1957).
14. **Ackermann, D. and List, P. H.,** Über das Vorkommen von Herzynin, Ergothionein, Homarin, Trigonellin, Glykokollbetain, Cholin, Trimethylamin, Adenin und fast sämtlicher Aminosäuren des Eiweisses in *Limulus polyphemus* L., *Hoppe-Seyler's Z. Physiol. Chem.,* 313, 30 (1958).
15. **Ackermann, D. and List, P. H.,** Über das Vorkommen von Ergothionein und Herzynin in *Limulus polyphemus* L, *Naturwissenschaften,* 45, 131 (1958).
16. **Ackermann, D. and List, P. H.,** Über das Vorkommen von Taurobetain, Taurin und Inosit im Riesenkieselschwamm, *Hoppe-Seyler's Z. Physiol. Chem.,* 317, 78 (1959).
17. **Ackermann, D. and Menssen, H. G.,** Niedrigmolekulare, N-haltige Inhaltstoffe der Lederkoralle *Alcyonium digitatum, Hoppe-Seyler's Z. Physiol. Chem.,* 317, 144, (1959).
18. **Ackermann, D. and List, P. H.,** Zur Konstitution des Zooanemonins und des Herbipolins, *Hoppe-Seyler's Z. Physiol. Chem.,* 318, 281 (1960).
19. **Ackermann, D. and Menssen, H. G.,** N-haltige Inhaltsstoffe des Pferdeschwammes *Hippospongia equina, Hoppe-Seyler's Z. Physiol. Chem.,* 322, 198 (1960).
20. **Ackermann, D. and Pant, R.,** Inhaltsstoffe des Schwammes *Calix nicacensis, Hoppe-Seyler's Z. Physiol. Chem.,* 326, 197 (1961).
21. **Ackermann, D. and List, P. H.,** Spongopurin, eine in der Natur neue Purinbase, *Naturwissenschaften,* 48, 74 (1961).
22. **Ackermann, D. and Pant, R.,** Erstmaliges Vorkommen von Mono- und Dimethyltaurin in der Tierwelt, *Naturwissenschaften,* 48, 646 (1961).
23. **Ackermann, D.,** Über das Vorkommen von Spinacin in der Krabbe *Crangon vulgaris, Hoppe-Seyler's Z. Physiol. Chem.,* 328, 275 (1962).
24. **Ackermann, D. and Hoppe-Seyler, G.,** Vergleich zweier biologischer, vom Histidin und Histamin ableitbarer, isomerer Ringsysteme (Spinacin und Zapotidin), *Hoppe-Seyler's Z. Physiol. Chem.,* 336, 283 (1964).
25. **Aguilar-Santos, G. and Doty, M. S.,** Chemical studies on three species of the marine algal genus *Caulperpa, Drugs from the Sea, Trans.,* Marine Technology Society, 173 (1967).
26. **Aitzetmüller, K., Svec, W. A., Katz, J. J., and Strain, H. H.,** Structure and chemical identity of diadinoxanthin and the principal xanthophyll of *Euglena, Chem. Commun.,* 32 (1968).
27. **Aitzetmüller, K., Strain, H. H., Svec, W. A., Grandolfo, M., and Katz, J. J.,** Loroxanthin, a unique xanthophyll from *Scendesmus obliquus* and *Chlorella vulgaris, Phytochemistry,* 8, 1761 (1969).
28. **Alcaide, A., Viala, J., Pinte, F., Itoh, M., Nomura, T., and Barbier, M.,** Stérols à 26 atomes de carbone du Tunicier *Halocynthia roretzi, C. R. Acad. Sci. Paris,* 273, 1386 (1971).
29. **Ali, A., Sarantakis, D., and Weinstein, B.,** Synthesis of the natural product (±)- dictyopterene B, *Chem. Commun.,* 940, (1971).
30. **Andersen, R. J. and Faulkner, D. J.,** Antibiotics from marine organisms of the Gulf of California, *Food-Drugs from the Sea, Proceedings,* Marine Technology Society, 111 (1972).
31. **Andersen, R. J. and Faulkner, D. J.,** A novel antibiotic from a sponge of the genus *Verongia, Tetrahedron Lett.,* 1175 (1973).

32. **Anderson, H. A., Mathieson, J. W., and Thomson, R. H.,** Distribution of spinochrome pigments in echinoids, *Comp. Biochem. Physiol.*, 28, 333 (1969).
33. **Anderson, I. G., Haslewood, G. A. D., Cross, A. D., and Tökés, L.,** New evidence for the structure of myxinol, *Biochem. J.*, 104, 1061 (1967).
34. **Anderson, I. G. and Haslewood, G. A. D.,** Comparative studies of bile salts. 16-Deoxymyxinol, a second bile alcohol from hagfish, *Biochem. J.*, 112, 763 (1969).
35. **ApSimon, J. W., Buccini, J. A., and Badripersaud, S.,** Marine organic chemistry. I. Isolation of 3β,6α-dihydroxy-5α-pregn-9(11)-en-20-one from the saponins of the starfish *Asterias forbesi*. A rapid method for extracting starfish saponins, *Can. J. Chem.*, 51, 850 (1973).
36. **Asano, M. and Itoh, M.,** Salivary poison of a marine gastropod *Neptunea arthritica* Bernardi, and the seasonal variation of its toxicity, *Ann. N. Y. Acad. Sci.*, 90, 674 (1960).
37. **Ashikari, H.,** Über die Galle des "Akajei" – Fisches (*Dasyatis akajei*) und die Konstitution des Scymnols, *J. Biochem.*, (Tokyo), 29, 319 (1939).
38. **Ashworth, R. B. and Cormier, M. J.,** Isolation of 2,6-dibromophenol from the marine hemichordate *Balanoglossus biminiensis, Science,* 155, 1558, (1967).
39. **Austin, J.,** The sterols of marine invertebrates and plants, *Advances in Steroid Biochemistry and Pharmacology*, Briggs, M. H., Ed., Academic Press, New York, 1, 73 (1970).
40. **Baker, J. T. and Sutherland, M. D.,** Pigments of marine animals. VIII. Precursors of 6,6′-dibromoindigotin (tyrian purple) from the mollusc *Dicathais orbita Gmelin, Tetrahedron Lett.*, 43 (1968).
41. **Baker, J. T. and Duke, C. C.,** Chemistry of the indoleninones. II. Isolation from the hypobranchial glands of marine molluscs of 6-bromo-2,2-dimethylthioindolin-3-one and 6-bromo-2-methylthioindoleninone as alternative precursors to tyrian purple, *Aust. J. Chem.*, 26, 2153 (1973).
42. **Baker, J. T. and Duke, C. C.,** Isolation from the hypobranchial glands of marine molluscs of 6-bromo-2,2-dimethylthioindolin-3-one and 6-bromo-2-methylthioindoleninone as alternative precursors to tyrian purple, *Tetrahedron Lett.*, 2481 (1973).
43. **Baker, J. T., Hawes, G. B., Oberhänsli, W. E., and Wells, R. J.,** personal communication.
44. **Bartolini, G. L., Erdman, T. R., and Scheuer, P. J.,** Anthraquinone pigments from the crinoid *Comanthus bennetti, Tetrahedron,* 29, 3699 (1973).
45. **Beaton, J. M., Spring, F. S., Stevenson, R., and Stewart, J. L.,** The Constitution of taraxerol (Skimmiol): A new naturally occurring triterpenoid type, *Chem. Ind.*, 1454 (1954).
46. **Bergmann, W. and Pace, W. T.,** Scymnol, *J. Am. Chem. Soc.*, 65, 477 (1943).
47. **Bergmann, W., McLean, M. J., and Lester, D.,** Contributions to the study of marine products. XIII. Sterols from various marine invertebrates, *J. Org. Chem.*, 8, 271 (1943).
48. **Bergmann, W., Gould, D. H., and Low, E. M.,** Contributions to the study of marine products. XVII. Spongosterol, *J. Org. Chem.*, 10, 570 (1945).
49. **Bergmann, W. and McTigue, F. H.,** Contributions to the study of marine products. XXI. Chondrillasterol, *J. Org. Chem.*, 13, 738 (1948).
50. **Bergmann, W.,** Comparative biochemical studies on the lipids of marine invertebrates, with special reference to the sterols, *J. Mar. Res.* 8, 137 (1949).
51. **Bergmann, W. and Ottke, R. C.,** Contributions to the study of marine products. XXIV. The occurrence of brassicasterol in mollusks, *J. Org. Chem.*, 14, 1085 (1949).
52. **Bergmann, W., McTigue, F. H., Low, E. M., Stokes, W. M., and Feeney, R. J.,** Contributions to the study of marine products. XXVI. Sterols from sponges of the family Suberitidae, *J. Org. Chem.*, 15, 96 (1950).
53. **Bergmann, W. and McAleer, W. J.,** The isolation of metanethole from the sponge *Spheciospongia vesparia, J. Am. Chem. Soc.*, 73, 4969 (1951).
54. **Bergmann, W. and Burke, D. C.,** Contributions to the study of marine products. XL. The nucleosides of sponges. IV. Spongosine, *J. Org. Chem.*, 21, 226 (1956).
55. **Bergmann, W.,** Sterols: Their Structure and Distribution, *Comp. Biochem.*, Florkin, M. and Mason, H. S., Eds., Academic Press, New York, 3, 103 (1962).
56. **Bergquist, P. R. and Hartman, W. D.,** Free amino acid patterns and the classification of the Demospongiae, *Mar. Biol.*, 3, 247 (1969).
57. **Blumer, M.,**Organic pigments: Their long term-fate, *Science,* 149, 722 (1965).
58. **Blumer, M., Mullin, M. M. and Guillard, R. R. L.,** A polyunsaturated hydrocarbon (3,6,9,12,15, 18-heneicosahexaene) in the marine food web, *Mar. Biol.*, 6, 226 (1970).
59. **Boeckman, R. K., Jr. and Silver, S. M.,** Synthesis of the non-isoprenoid sesquiterpene, β-gorgonene, *Tetrahedron Lett.*, 3497 (1973).
60. **Bonnett, R., Mallams, A. K., Spark, A. A., Tee, J. L., Weedon, B. C. L., and McCormick, A.,** Carotenoids and related compounds. Part XX. Structure and reactions of fucoxanthin, *J. Chem. Soc. (C)*, 429 (1969).
61. **Boylan, D. B. and Scheuer, P. J.,** Pahutoxin: A fish poison, *Science,* 155, 52 (1967).
62. **Bridgwater, R. J., Briggs, T., and Haslewood, G. A. D.,** Comparative studies of "bile salts." 14. Isolation from shark bile and partial synthesis of scymnol, *Biochem. J.*, 82, 285 (1962).
63. **Brock, V. E.,** Possible production of substances poisonous to fishes by the box fish *Ostracion lentiginosus* Schneider, *Copeia,* 3, 195 (1956).

64. **Brown, M. S. and Mosher, H. S.,** Tarichatoxin: Isolation and purification, *Science,* 140, 295 (1963).
65. **Buchwald, H. D., Durham, L., Fischer, H. G., Harada, R., Mosher, H. S., Kao, C. Y., and Fuhrman, F. A.,** Identity of tarichatoxin and tetrodotoxin, *Science,* 143, 474 (1964).
66. **Bundy, G. L., Schneider, W. P., Lincoln, F. H., and Pike, J. E.,** The synthesis of prostaglandins E_2 and $F_2\alpha$ from (15R) - and (15S)-PGA_2. *J. Am. Chem. Soc.,* 94, 2123 (1972).
67. **Bundy, G. L., Daniels, E. G., Lincoln, F. H., and Pike, J. E.,** Isolation of a new naturally occurring prostaglandin, 5-*trans*-PGA_2. Synthesis of 5-*trans*-PGE_2 and 5-*trans*-$PGF_2\alpha$, *J. Am. Chem. Soc.,* 94, 2124 (1972).
68. **Burkholder, P. R., Pfister, R. M., and Leitz, F. H.,** Production of a pyrrole antibiotic by a marine bacterium, *Appl. Microbiol.,* 14, 649 (1966).
69. **Cafieri, F., Fattorusso, E., Santacroce, C., and Minale, L.,** Fasciculatin, a novel sesterterpene from the sponge *Ircinia fasciculata, Tetrahedron,* 28, 1579 (1972).
70. **Cafieri, F., Fattorusso, E., Magno, S., Santacroce, C., and Sica, D.,** Isolation and structure of axisonitrile-1 and axisothiocyanate-1, two unusual sesquiterpenoids from the marine sponge *Axinella cannabina, Tetrahedron,* 29, 4259 (1973).
71. **Cameron, A. F., Cheung, K. K., Ferguson, G., and Robertson, J. M.,** Laurencia natural products. Part 1. Crystal structure and absolute stereochemistry of laurencin, *J. Chem. Soc. (B),* 559 (1969).
72. **Campbell, S. A., Mallams, A. K., Waight, E. S., Weedon, B. C. L., Barbier, M., Lederer, E., and Salaque, A.,** Pectenoxanthin, cynthiaxanthin, and a new acetylenic carotenoid, pectenolone, *Chem. Commun.,* 941 (1967).
73. **Cariello, L., Crescenzi, S., Prota, G., Giordano, F., and Mazzarella, L.,** Zoanthoxanthin, a heteroaromatic base from *Parazoanthus cfr. axinellae* (zoantharia): Structure confirmation by X-ray crystallography, *Chem. Commun.,* 99 (1973).
74. **Cariello, L., Crescenzi, S., Prota, G., and Zanetti, L.,** New zoanthoxanthins from the Mediterranean zoanthid *Parazoanthus axinellae, Experientia,* 30, 849 (1974).
75. **Cariello, L., Crescenzi, S., Prota, G., Capasso, S., Giordano, F., and Mazzarella, L.,** Zoanthoxanthin, a natural 1,3,5,7-tetrazacyclopent- (f)-azulene from *Parazoanthus axinellae, Tetrahedron,* 30, 3281 (1974).
76. **Chang, C. W. J., Moore, R. E. and Scheuer, P. J.,** Spinochromes A(M) and C(F), *Tetrahedron Lett.,* 3557 (1964).
77. **Chanley, J. D., Mezzetti, T., and Sobotka, H.,** The holothurinogenins, *Tetrahedron,* 22, 1857 (1966).
78. **Chanley, J. D. and Rossi, C.,** The holothurinogenins - II. Methoxylated neo-holothurinogenins, *Tetrahedron,* 25, 1897 (1969).
79. **Chanley, J. D. and Rossi, C.,** The neo-holothurinogenins - III. Neo-holothurinogenins by enzymatic hydrolysis of desulfated holothurin A, *Tetrahedron,* 25, 1911 (1969).
80. **Chapman, D. J. and Haxo, F. T.,** Identity of ϵ-carotene and ϵ_1-carotene, *Plant Cell Physiol.,* 4, 57 (1963).
81. **Chapman, D. J.,** Three new carotenoids isolated from algae, *Phytochemistry,* 5, 1331 (1966).
82. **Christensen, P. K. and Sörensen, N. A.,** Studies related to pristane. V. The constitution of zamene, *Acta Chem. Scand.,* 5, 751 (1951).
83. **Ciereszko, L. S., Sifford, D. H., and Weinheimer, A. J.,** Chemistry of coelenterates. I. Occurrence of terpenoid compounds in gorgonians, *Ann. N.Y. Acad. Sci.,* 90, 917 (1960).
84. **Ciereszko, L. S., Odense, P. H., and Schmidt, R. W.,** Chemistry of coelenterates. II. Occurrence of taurobetaine and creatine in gorgonians, *Ann. N.Y. Acad. Sci.,* 90, 920 (1960).
85. **Ciereszko, L. S.,** Chemistry of coelenterates. III. Occurrence of antimicrobial terpenoid compounds in the zooxanthellae of alcyonarians, *Trans. N.Y. Acad. Sci.* Ser. II, 24, 502 (1962).
86. **Ciereszko, L. S., Johnson, M. A., Schmidt, R. W., and Koons, C. B.,** Chemistry of coelenterates. VI. Occurrence of gorgosterol, a C_{30} sterol, in coelenterates and their zooxanthellae, *Comp. Biochem. Physiol.,* 24, 899 (1968).
87. **Ciereszko, L. S.,** Nitrogen compounds in porifera and coelenterata, in *Comparative Biochemistry of Nitrogen Metabolism.* Vol. I. *The Invertebrates,* Campbell, J. W., Ed., Academic Press, New York, 1970, 57.
88. **Ciereszko, L. S. and Karns, T. K. B.,** Comparative biochemistry of coral reef coelenterates, in *Biology and Geology of Coral Reefs.* Vol. II.: *Biology 1,* Jones, O. A. and Endean, R., Eds., Academic Press, New York, 1973, 183.
89. **Cimino, G., De Stefano, S., Minale, L., and Fattorusso, E.,** Furospongin-1, a new C-21 furanoterpene from the sponges *Spongia officinalis* and *Hippospongia communis, Tetrahedron,* 27, 4673 (1971).
90. **Cimino, G., De Stefano, S., and Minale, L.,** Prenylated quinones in marine sponges: *Ircinia* sp., *Experientia,* 28, 1401 (1972).
91. **Cimino, G., De Stefano, S., Minale, L., and Fattorusso, E.,** Minor C-21 furanoterpenes from the sponges *Spongia officinalis* and *Hippospongia communis, Tetrahedron,* 28, 267 (1972).
92. **Cimino, G., De Stefano, S., Minale, L., and Fattorusso, E.,** Ircinin -1 and -2, linear sesterterpenes from the marine sponge *Ircinia oros, Tetrahedron,* 28, 333 (1972).
93. **Cimino, G., De Stefano, S., and Minale, L.,** Polyprenyl derivatives from the sponge *Ircinia spinosula, Tetrahedron,* 28, 1315 (1972).
94. **Cimino, G., De Stefano, S., Minale, L., and Trivellone, E.,** New sesquiterpenes from the marine sponge *Pleraplysilla spinifera, Tetrahedron,* 28, 4761 (1972).
95. **Cimino, G., De Stefano, S., and Minale, L.,** Further linear furanoterpenes from marine sponges, *Tetrahedron,* 28, 5983 (1972).

96. **Cimino, G., De Stefano, S., and Minale, L.,** Deoxoscalarin, a further sesterterpene with the unusual tetracyclic carbon skeleton of scalarin, from *Spongia officinalis, Experientia,* 29, 934 (1973).
97. **Cimino, G., De Stefano, S., and Minale, L.,** Paniceins, unusual aromatic sesquiterpenoids linked to a quinol or quinone system from the marine sponge *Halichondria panicea, Tetrahedron,* 29, 2565 (1973).
98. **Cohen, S. S.,** Sponges, cancer chemotherapy and cellular aging, *Perspect. Biol. Med.,* 6, 215 (1963).
99. **Cook, J. W.,** Bile acids of elasmobranch fish, *Nature,* 147, 388 (1941).
100. **Cosulich, D. B. and Lovell, F. M.,** An X-ray determination of the structure of an antibacterial compound from the sponge *Ianthella ardis, Chem. Commun.,* 397 (1971).
101. **Craigie, J. S. and Gruenig, D. E.,** Bromophenols from red algae, *Science,* 157, 1058 (1967).
102. **Cross, A. D.,** Scymnol sulphate and anhydroscymnol, *J. Chem. Soc.,* 2817 (1961).
103. **Cross, A. D.,** Nuclear-magnetic-resonance and mass-spectral study of myxinol tetra-acetate, *Biochem. J.,* 100, 238 (1966).
104. **Davies, W. H., Heilbron, I. M., and Jones, W. E.,** The unsaponifiable matter from the oils of elasmobranch fish. Part IX. The structure of batyl and selachyl alcohols, *J. Chem. Soc.,* 165 (1933).
105. **De Luca, P., De Rosa, M., Minale, L., and Sodano, G.,** Marine sterols with a new pattern of side-chain alkylation from the sponge *Aplysina (= Verongia) aerophoba, J. Chem. Soc. Perkin I,* 2132 (1972).
106. **De Luca, P., De Rosa, M., Minale, L., Puliti, R., Sodano, G., Giordano, F., and Mazzarella, L.,** Synthesis of 24,28-didehydroaplysterol and X-ray crystal structure of aplysterol: unusual marine sterols, *Chem. Commun.,* 825 (1973).
107. **Dorée, C.,** The occurrence and distribution of cholesterol and allied bodies in the animal kingdom, *Biochem. J.,* 4, 72 (1909).
108. **Doty, M. S. and Aguilar-Santos, G.,** Transfer of toxic algal substances in marine food chains, *Pac. Sci.* 24, 351 (1970).
109. **Drechsel, E.,** Beiträge zur Chemie einiger Seethiere, *Z. Biol.,* 33, 85 (1907).
110. **Dunstan, P. J., Hofheinz, W., and Oberhänsli, W. E.,** personal communication.
111. **Dusza, J. P.,** Contributions to the study of marine products. XLIX. Synthesis of 29-isofucosterol, *J. Org. Chem.,* 25, 93 (1960).
112. **Elyakov, G. B., Kuznetsova, T. A., Dzizenko, A. K., and Elkin, Yu, N.,** A chemical investigation of the trepang (*Stychopus japonicus* Selenka): The structure of triterpenoid aglycones obtained from trepang glycosides, *Tetrahedron Lett.,* 1151 (1969).
113. **Enwall, E. L., van der Helm, D., Hsu, I. N., Pattabhiraman, T., Schmitz, F. J., Spraggins, R. L., and Weinheimer, A. J.,** Crystal structure and absolute configuration of two cyclopropane containing marine steroids, *Chem. Commun.,* 215 (1972).
114. **Erdman, T. R. and Thomson, R. H.,** Naturally occurring quinones. Part XXI. Anthraquinones in the crinoids *Heterometra savignii* (J. Müller) and *Lamprometra klunzingeri* (Hartlaub), *J. Chem Soc. Perkin I,* 1291 (1972).
115. **Erdman, T. R. and Thomson, R. H.,** Sterols from the sponges *Cliona celata* Grant and *Hymeniacidon perleve* Montagu, *Tetrahedron,* 28, 5163 (1972).
116. **Erspamer, V.,** Wirksame Stoffe der hinteren Speicheldrüsen der Octopoden und der Hypobranchialdrüse der Purpurschnecken. *Arzneim. Forsch.,* 2, 253 (1952).
117. **Esrpamer, V. and Benati, O.,** Identification of murexine as β-[imidazolyl-(4)]-acryl-choline, *Science,* 117, 161 (1953).
118. **Fagerlund, U. H. M. and Idler, D. R.,** Marine sterols. IV. 24-Dehydrocholesterol: Isolation from a barnacle and synthesis by the Wittig reaction, *J. Am. Chem. Soc.,* 79, 6473 (1957).
119. **Fagerlund, U. H. M. and Idler, D. R.,** Marine sterols. V. Isolation of 7,24(28)-ergostadien-3β-ol from starfish, *J. Am. Chem. Soc.,* 81, 401 (1959).
120. **Fattorusso, E., Minale, L., and Sodano, G.,** Aeroplysinin-I, a new bromo-compound from *Aplysina aerophoba, Chem. Commun.,* 751 (1970).
121. **Fattorusso, E., Minale, L., Sodano, G., Moody, K., and Thomson, R. H.,** Aerothionin, a tetrabromo-compound from *Aplysina aerophoba* and *Verongia thiona, Chem. Commun.,* 752 (1970).
122. **Fattorusso, E., Minale, L., Moody, K., Sodano, G., and Thomson, R. H.,** Homo-aerothionin, a second tetrabromo-compound from *Aplysina aerophoba* and *Verongia thiona, Gazz. Chim. Ital.,* 101, 61 (1971).
123. **Fatorusso, E., Forenza, S., Minale, L. and Sodano, G.,** Isolation of 3,4-dihydroxyquinoline-2-carboxylic acid from the sponge *Aplysina aerophoba, Gazz. Chim. Ital.,* 101, 104 (1971).
124. **Fattorusso, E., Minale, L., Sodano, G., and Trivellone, E.,** Isolation and structure of nitenin and dihydronitenin, new furanoterpenes from *Spongia nitens, Tetrahedron,* 27, 3909 (1971).
125. **Fattorusso, E., Minale, L., and Sodano, G.,** Aeroplysinin-1, an antibacterial bromo-compound from the sponge *Verongia aerophoba, J. Chem. Soc. Perkin I,* 16 (1972).
126. **Fattorusso, E., Magno, S., Santacroce, C., and Sica, D.,** Scalarin, a new pentacyclic C-25 terpenoid from the sponge *Cacospongia scalaris, Tetrahedron,* 28, 5993 (1972).
127. **Faulkner, D. J., Stallard, M. O., Fayos, J., and Clardy, J.,** (3R, 4S, 7S) *trans, trans*-3,7-Dimethyl-1,8,8-tribromo-3,4,7-trichloro-1,5-octadiene, a novel monoterpene from the sea hare *Aplysia californica, J. Am. Chem. Soc.,* 95, 3413 (1973).

128. **Faulkner, D. J. and Stallard, M. O.,** 7-Chloro-3,7-dimethyl-1,4,6-tribromo-1-octen-3-ol, a novel monoterpene alcohol from *Aplysia californica, Tetrahedron Lett.* 1171 (1973).
129. **Faulkner, D. J.,** Variabilin, an antibiotic from the sponge *Ircinia variabilis, Tetrahedron Lett.*, 3821 (1973).
130. **Faulkner, D. J. and Andersen, R. J.,** Natural products chemistry of the marine environment, in *The Sea,* Vol. 5, *Marine Chemistry,* Goldberg, E. D., Ed., John Wiley & Sons, New York, 1974, 679.
131. **Faux, A., Horn, D. H. S., Middleton, E. J., Fales, H. M., and Lowe, M. E.,** Moulting hormones of a crab during ecdysis, *Chem. Commun.*, 175 (1969).
132. **Fenical, W., Sims, J. J., Wing, R. M., and Radlick, P. C.,** Zonarene, a sesquiterpene from the brown seaweed *Dictyopteris Zonarioides, Phytochemistry,* 11, 1161 (1972).
133. **Fenical, W., Sims, J. J., Squatrito, D., Wing, R. M., and Radlick, P.,** Zonarol and isozonarol, fungitoxic hydroquinones from the brown seaweed *Dictyopteris zonarioides, J. Org. Chem.*, 38, 2383 (1973).
134. **Fenical, W., Sims, J. J., and Radlick, P.,** Chondriol, a halogenated acetylene from the marine alga *Chondria oppositiclada, Tetrahedron Lett.*, 313 (1973).
135. **Fenical, W., Gifkins, K. B., and Clardy, J.,** X-ray determination of chondriol; a re-assignment of structure, *Tetrahedron Lett.*, 1507, (1974).
136. **Feutrill, G. I., Mirrington, R. N., and Nichols, R. J.,** The total synthesis of (±)-laurinterol and related compounds, *Aust. J. Chem.*, 26, 345 (1973).
137. **Forenza, S., Minale, L., Riccio, R., and Fattorusso, E.,** New bromo-pyrrole derivatives from the sponge *Agelas oroides, Chem. Commun.*, 1129 (1971).
138. **Forrest, H. S.,** Pteridines: Structure and metabolism, *Comp. Biochem.*, Florkin, M. and Mason, H. S., Eds., Academic Press, New York, 4, 615 (1962).
139. **Fouquet, H. and Bielig, H.-J.,** Biological precursors and genesis of tyrian-purple, *Angew. Chem. Int. Ed. Engl.*, 10, 816 (1971).
140. **Fryberg, M., Oehlschlager, A. C., and Unrau, A. M.,** Synthesis of a novel C_{26} marine sterol, *Chem. Commun.*, 1194 (1971).
141. **Fukuzawa, A., Kurosawa, E., and Irie, T.,** Laurefucin and acetyllaurefucin, new bromo compounds from *Laurencia nipponica* Yamada (1), *Tetrahedron Lett.*, 3 (1972).
142. **Fulmor, W., Van Lear, G. E., Morton, G. O., and Mills, R. D.,** Isolation and absolute configuration of the aeroplysinin I enantiomorphic pair from *Ianthella ardis, Tetrahedron Lett.*, 4551 (1970).
143. **Furusaki, A., Kurosawa, E., Fukuzawa, A., and Irie, T.,** The revised structure and absolute configuration of laurefucin from *Laurencia nipponica* Yamada, *Tetrahedron Lett.*, 4579 (1973).
144. **Galasko, G., Hora, J., Toube, T. P., Weedon, B. C. L., André, D., Barbier, M., Lederer, E., and Villanueva, V. R.,** Carotenoids and related compounds. Part XXII. Allenic carotenoids in sea urchins, *J. Chem. Soc. (C),* 1264 (1969).
145. **Galbraith, M. N., Horn, D. H. S., Middleton, E. J., and Hackney, R. J.,** Structure of deoxycrustecdysone, a second crustacean moulting hormone, *Chem. Commun.*, 83 (1968).
146. **Garcia, E. E., Benjamin, L. E., and Fryer, R. I.,** Reinvestigation into the structure of oroidin, a bromopyrrole derivative from marine sponge, *Chem. Commun.*, 78 (1973).
147. **German, V. F.,** Nα, Nα-Dimethylhistamine, the hypotensive principle of the sponge *Ianthella* sp., *J. Pharm. Sci.*, 60, 495 (1971).
148. **Gibbons, G. F., Goad, L. J., and Goodwin, T. W.,** The identification of 28 isofucosterol in the marine green algae *Enteromorpha intestinalis* and *Ulva lactuca, Phytochemistry,* 7, 983 (1968).
149. **Glombitza, K.-W. and Stoffelen, H.,** 2,3-Dibrom-5-hydroxybenzyl-1′,4-disulfat (Dikaliumsalz) aus Rhodomelaceen, *Planta Med.*, 22, 391 (1972).
150. **González, A. G., Darias, J., and Martin, J. D.,** Taondiol, a new component from *Taonia atomaria, Tetrahedron Lett.*, 2729 (1971).
151. **González, A. G., Darias, J., Martin, J. D., and Pascual, C.,** Marine natural products of the Atlantic Zone-V. The structure and chemistry of taondiol and related compounds, *Tetrahedron,* 29, 1605 (1973).
152. **González, A. G., Darias, J., and Martin, J. D.,** Caespitol, a new halogenated sesquiterpene from *Laurencia caespitosa, Tetrahedron Lett.*, 2381 (1973).
153. **González, A. G., Darias, J., and Martin, J. D.,** Furocaespitane, a new furan from *Laurencia caespitosa, Tetrahedron Lett.*, 3625 (1973).
154. **González, A. G., Martin, J. D., and Rodriguez, M. L.,** Stereospecific total synthesis of *dl*-taondiol methyl ether, *Tetrahedron Lett.*, 3657 (1973).
155. **González, A. G., Darias, J., Martin, J. D., and Pérez, C.,** Revised structure of caespitol and its correlation with isocaespitol, *Tetrahedron Lett.*, 1249 (1974).
156. **Goodwin, T. W. and Srisukh, S.,** A study of the pigments of the sea urchins *Echinus esculentus* L. and *Paracentrotus lividus* Lamarck, *Biochem. J.*, 47, 69 (1950).
157. **Goodwin, T. W.,** Carotenoids: Structure, distribution and function, *Comp. Biochem.*, Florkin, M. and Mason, H. S., Eds., Academic Press, New York, 4, 643 (1962).
158. **Goto, T., Kishi, Y., Takahashi, S., and Hirata, Y.,** Tetrodotoxin, *Tetrahedron,* 21, 2059 (1965).
159. **Gross, J., Sokal, Z., and Rougvie, M.,** Structural and chemical studies on the connective tissue of marine sponges, *J. Histochem. Cytochem.*, 4, 227 (1956).

160. **Grossert, J. S.,** Natural products from echinoderms, *Chem. Soc. Rev.*, 1, 1 (1972).
161. **Gupta, A. K. S.,** Vorkommen von Koprostanol in Walöl, *Hoppe-Seyler's Z. Physiol. Chem.*, 348, 1688 (1967).
162. **Gupta, K. C. and Scheuer, P. J.,** Echinoderm sterols, *Tetrahedron*, 24, 5831 (1968).
163. **Gupta, K. C. and Scheuer, P. J.,** Zoanthid sterols. *Steroids*, 13, 343 (1969).
164. **Gurst, J. E. Sheikh, Y. M., and Djerassi, C.,** Synthesis of corticosteroids from marine sources, *J. Am. Chem. Soc.*, 628 (1973).
165. **Habermehl, G. and Volkwein, G.,** Über Gifte der mittlemeerischen Holothurien, *Naturwissenschaften*, 55, 83 (1968).
166. **Habermehl, G. and Volkwein, G.,** Über Gifte der mittelmeerischen Holothurien, II. Die Aglyka der Toxine von *Holothuria polii, Liebig's Ann. Chem.*, 731, 53 (1970).
167. **Habermehl, G. and Volkwein, G.,** Aglycones of the toxins from the cuvierian organs of *Holothuria forskali* and a new nomenclature for the aglycones from holothurioideae, *Toxicon*, 9, 319 (1971).
168. **Habermehl, G. and Christ, B.,** Steroide aus *Asterias rubens, Z. Naturforsch.*, 28C, 225 (1973).
169. **Hagiwara, H., Numata, M., Konishi, K., and Oka, Y.,** Synthesis of nereistoxin and related compounds, *Chem. Pharm. Bull.* (Tokyo), 13, 253 (1965).
170. **Hale, R. L., Leclercq, J., Tursch, B., Djerassi, C., Gross, R. A., Jr., Weinheimer, A. J., Gupta, K., and Scheuer, P. J.,** Demonstration of a biogenetically unprecedented side chain in the marine sterol, gorgosterol, *J. Am. Chem. Soc.*, 92, 2179 (1970).
171. **Hall, S. S., Faulkner, D. J., Fayos, J., and Clardy, J.,** Oppositol, a brominated sesquiterpene alcohol of a new skeletal class from the red alga *Laurencia subopposita, J. Am. Chem. Soc.*, 95, 7187 (1973).
172. **Halsall, T. G. and Hills, I. R.,** Isolation of heneicosa-1,6,9,12,15,18-hexaene and −1,6,9,12,15-pentaene from the alga *Fucus vesiculosus, Chem. Commun.*, 448 (1971).
173. **Hamasaki, T., Okukado, N., and Yamaguchi, M.,** Two natural acetylenic aromatic carotenoids, *Bull. Chem. Soc. Jap.*, 46, 1884 (1973).
174. **Hampshire, F. and Horn, D. H. S.,** Structure of crustecdysone, a crustacean moulting hormone, *Chem. Commun.*, 37 (1966).
175. **Hanessian, S. and Kaltenbronn, J. S.,** Synthesis of a bromine-rich marine antibiotic, *J. Am. Chem. Soc.*, 88, 4509 (1966).
176. **Hashimoto, Y. and Okaichi, T.,** Some chemical properties of nereistoxin, *Ann. N.Y. Acad. Sci.*, 90, 667 (1960).
177. **Hashimoto, Y., Miyazawa, K., Kamiya, H., and Shibota, M.,** Toxicity of the Japanese ivory shell, *Bull. Jap. Soc. Sci. Fish.*, 33, 661 (1967).
178. **Haslewood, G. A. D.,** 6. Bile salts of fish, *Biochem. Soc. Symp.*, 6, 83 (1951).
179. **Haslewood, G. A. D.,** Recent developments in our knowledge of bile salts, *Physiol. Rev.*, 35, 178 (1955).
180. **Haslewood, G. A. D.,** A bile alcohol sulphate from the hagfish *Eptatretus stoutii, Biochem. J.*, 78, 30P (1961).
181. **Haslewood, G. A. D.,** Comparative studies of 'bile salts.' 13. Bile acids of the leopard seal *Hydrurga leptonyx*, and of two snakes of the genus *Bitis, Biochem. J.*, 78, 352 (1961).
182. **Haslewood, G. A. D.,** Bile salts: Structure, distribution and possible biological significance as a species character, *Comp. Biochem.*, Florkin, M. and Mason, H. S., Eds., Academic Press, New York, 3, 205 (1962).
183. **Haslewood, G. A. D.,** Comparative studies of bile salts. Myxinol disulphate, the principal bile salt of hagfish (Myxinidae), *Biochem. J.*, 100, 233 (1966).
184. **Heilbron, I. M. and Owens, W. M.,** The unsaponifiable matter from the oils of elasmobranch fish. Part IV. The establishment of the structure of selachyl and batyl alcohols as monoglyceryl ethers, *J. Chem. Soc.*, 942 (1928).
185. **Hertzberg, S. and Liaaen-Jensen, S.,** Animal carotenoids. 3. The carotenoids of *Actinia equina*. Structure determination of actinioerythrin and violerythrin, *Acta Chem. Scand.*, 23, 3290 (1969).
186. **Hirschfeld, D. R., Fenical, W., Lin, G. H. Y., Wing, R. M., Radlick, P., and Sims, J. J.,** Marine natural products. VIII. Pachydictyol A, an exceptional diterpene alcohol from the brown alga *Pachydictyon coriaceum, J. Am. Chem. Soc.*, 95, 4049 (1973).
187. **Hodgkin, J. H., Craigie, J. S., and McInnes, A. G.,** The occurrence of 2,3-dibromobenzyl alcohol 4,5-disulfate, dipotassium salt, in *Polysiphonia lanosa, Can. J. Chem.*, 44, 74 (1966).
188. **Hoppe-Seyler, F. A.,** Über das Homarin, eine bisher unbekannte tierische Base, *Hoppe-Seyler's Z. Physiol. Chem.*, 222, 105 (1933).
189. **Horn, D. H. S., Middleton, E. J., Wunderlich, J. A., and Hampshire, F.,** Identity of the moulting hormones of insects and crustaceans, *Chem. Commun.*, 339 (1966).
190. **Horn, D. H. S., Fabbri, S., Hampshire, F., and Lowe, M. E.,** Isolation of crustecdysone (20R-hydroxyecdysone) from a crayfish (*Jasus lalandei* H. Milne-Edwards), *Biochem. J.*, 109, 399 (1968).
191. **Hoshita, T. and Kazuno, T.,** Chemistry and metabolism of bile alcohols and higher bile acids, *Adv. Lipid Res.*, 6, 207 (1968).
192. **Hossain, M. B., Nicholas, A. F., and van der Helm, D.,** The molecular structure of eunicin iodoacetate, *Chem. Commun.*, 385 (1968).
193. **Hossain, M. B. and van der Helm, D.,** The crystal structure of β-gorgonene – silver nitrate, *J. Am. Chem. Soc.*, 90, 6607 (1968).
194. **Hossain, M. B. and van der Helm, D.,** The crystal structure of crassin p-iodobenzoate, *Rec. Trav. Chim. Pays-Bas*, 88, 1413 (1969).

195. **Hunt, S. and Breuer, S. W.,** Isolation of a new naturally occurring halogenated amino acid: Monochloromonobromotyrosine, *Biochim. Biophys. Acta,* 252, 401 (1971).
196. **Hüppi, G. and Siddall, J. B.,** Steroids: CCCXXXIII. Synthetic studies on insect hormones. V. The synthesis of crustecdysone (20-hydroxyecdysone), *J. Am. Chem. Soc.,* 89, 6790 (1967).
197. **Idler, D. R. and Fagerlund, U. H. M.,** Marine sterols. III. The synthesis of 24-methylenecholesterol and 25-dehyrocholesterol, *J. Am. Chem. Soc.,* 79, 1988 (1957).
198. **Idler, D. R., Saito, A., and Wiseman, P.,** Sterols in red algae (Rhodophyceae), *Steroids,* 11, 465 (1968).
199. **Idler, D. R. and Wiseman, P.,** Sterols of crustacea, *Int. J. Biochem.,* 2, 91 (1971).
200. **Idler, D. R. and Wiseman, P.,** Molluscan sterols: A review, *J. Fish. Res. Board Can.* 29, 385 (1972).
201. **Ikegami, S., Kamiya, Y., and Tamura, S.,** A new sterol from asterosaponins A and B, *Tetrahedron Lett.,* 1601 (1972).
202. **Ikegami, S., Kamiya, Y., and Tamura, S.,** Studies on asterosaponins. V. A novel steroid conjugate, 5α-pregn-9(11)-ene-3β,6α-diol-20-one-3-sulfate, from a starfish saponin, asterosaponin A, *Tetrahedron,* 29, 1807 (1973).
203. **Ikegami, S., Kamiya, Y., and Tamura, S.,** A new steroidal sulfate obtained from a starfish saponin, asterosaponin A, *Tetrahedron Lett.,* 731 (1973).
204. **Ikekawa, N., Tsuda, K., and Morisaki, N.,** Saringosterol: a new sterol from brown algae, *Chem. Ind.,* 1179 (1966).
205. **Ikekawa, N., Morisaki, N., Tsuda, K., and Yoshida, T.,** Sterol compositions in some green algae and brown algae, *Steroids,* 12, 41 (1968).
206. **Ikekawa, N., Morisaki, M., and Hirayama, K.,** Two new sterols from *Fucus evanescens, Phytochemistry,* 11, 2317 (1972).
207. **Irie, T., Yamamoto, K., and Masamune, T.,** Sesquiterpenes from *Dictyopteris divaricata. I, Bull. Chem. Soc. Jap.,* 37, 1053 (1964).
208. **Irie, T., Suzuki, M., and Masamune, T.,** Laurencin, a constituent from *Laurencia* species, *Tetrahedron Lett.,* 1091 (1965).
209. **Irie, T., Yasunari, Y., Suzuki, T., Imai, N., Kurosawa, E., and Masamune, T.,** A new sesquiterpene hydrocarbon from *Laurencia glandulifera, Tetrahedron Lett.,* 3619 (1965).
210. **Irie, T., Suzuki, M., Kurosawa, E., and Masamune, T.,** Laurinterol and debromolaurinterol, constituents from *Laurencia intermedia, Tetrahedron Lett.,* 1837 (1966).
211. **Irie, T., Suzuki, T., Itô, S., and Kurosawa, E.,** The absolute configuration of laurene and α-cuparenone (1), *Tetrahedron Lett.,* 3187 (1967).
212. **Irie, T., Suzuki, M., and Masamune, T.,** Laurencin, a constituent of *Laurencia glandulifera* Kützing, *Tetrahedron,* 24, 4193 (1968).
213. **Irie, T., Izawa, M., and Kurosawa, E.,** Laureatin, a constituent from *Laurencia nipponica* Yamada (1), *Tetrahedron Lett.,* 2091 (1968).
214. **Irie, T., Izawa, M., and Kurosawa, E.,** Isolaureatin, a constituent from *Laurencia nipponica* Yamada (1), *Tetrahedron Lett.,* 2735 (1968).
215. **Irie, T., Suzuki, M., and Hayakawa, Y.,** Isolation of aplysin, debromoaplysin and aplysinol from *Laurencia okamurai* Yamada, *Bull. Chem. Soc. Jap.,* 42, 843 (1969).
216. **Irie, T., Suzuki, T., Yasunari, Y., Kurosawa, E., and Masamune, T.,** Laurene, a sesquiterpene hydrocarbon from *Laurencia* species, *Tetrahedron,* 25, 459 (1969).
217. **Irie, T., Fukuzawa, A., Izawa, M., and Kurosawa, E.,** Laurenisol, a new sesquiterpenoid containing bromine from *Laurencia nipponica* Yamada (1), *Tetrahedron Lett.,* 1343 (1969).
218. **Irie, T., Izawa, M., and Kurosawa, E.,** Laureatin and isolaureatin, Constituents of *Laurencia nipponica* Yamada, *Tetrahedron,* 26, 851 (1970).
219. **Irie, T., Suzuki, M., Kurosawa, E., and Masamune, T.,** Laurinterol, debromolaurinterol and isolaurinterol, constituents of *Laurencia intermedia* Yamada, *Tetrahedron,* 26, 3271 (1970).
220. **Irvin, J. L. and Wilson, D. W.,** Studies on octopine. II. The nitrogenous extractives of squid and octopus muscle, *J. Biol. Chem.,* 127, 565 (1939).
221. **Ishida, Y. and Kadota, H.,** Isolation and identification of dimethyl-β-propiothetin from *Gyrodinium cohnii, Agr. Biol. Chem.,* 31, 756 (1967).
222. **Israel, M., Rosenfield, J. S., and Modest, E. J.,** Analogs of spermine and spermidine. I. Synthesis of polymethylenepolyamines by reduction of cyanoethylated α, ω-alkylenediamines, *J. Med. Chem.,* 7, 710 (1964).
223. **Ito, K. and Hashimoto, Y.,** Isolation of a new amino acid, 'gigartinine,' from a red alga *Gymnogongrus flabelliformis, Bull. Jap. Soc. Sci. Fish.,* 32, 274 (1966).
224. **Ito, K. and Hashimoto, Y.,** Gigartinine: a new amino-acid in red algae, *Nature,* 211, 417 (1966).
225. **Ito, K. and Hashimoto, Y.,** Syntheses of DL-gigartinine and gongrine, *Agr. Biol. Chem.,* 33, 237 (1969).
226. **Iwakiri, Y., Yamaguchi, M., and Tsumaki, T.,** A pigment of pearl-oyster shells, *Memoirs Kyushu Uni, Fac. Sci. Ser. C, Chem.,* 3, 161 (1960).
227. **Jaenicke, L., Akintobi, T., and Müller, D. G.,** Synthesis of the sex attractant of *Ectocarpus siliculosus, Angew. Chem. Int. Ed., Engl.,* 10, 492 (1971).
228. **Jaenicke, L. and Müller, D. G.,** Gametenlockstoffe bei niederen Pflanzen und Tieren, *Fortschr. Chem. Org. Naturst.,* 30, 61 (1973).

229. **Kanatani, H., Shirai, H., Nakanishi, K., and Kurokawa, T.,** Isolation and identification of meiosis inducing substance in starfish *Asterias amurensis, Nature,* 221, 273 (1969).
230. **Kao, C. Y.,** Pharmacology of tetrodotoxin and saxitoxin, *Fed. Proc.,* 31, 1117 (1972).
231. **Kappe, T. and Armstrong, M. D.,** Preparation of D- and L-octopamine, *J. Med. Chem.,* 7, 569 (1964).
232. **Karle, I. L., Daly, J. W., and Witkop, B.,** 2,3-*cis*-3,4-*trans*-3,4-Dihydroxy -L-proline: Mass spectrometry and X-ray analysis, *Science,* 164, 1401 (1969).
233. **Karrer, P. and Helfenstein, A.,** Synthese des Squalens, *Helv. Chim. Acta,* 14, 78 (1931).
234. **Kashman, Y., Fishelson, L., and Neeman, I.,** N-Acyl-2-methylene-β-alanine methyl esters from the sponge *Fasciospongia cavernosa, Tetrahedron,* 29, 3655 (1973).
235. **Katayama, T., Yokoyama, H., and Chichester, C. O.,** The biosynthesis of astaxanthin II. The carotenoids in Benibuna *Carassius auratus,* especially the existence of new keto carotenoids, α-doradecin and α-doradexanthin *Bull. Jap. Soc. Sci. Fish.,* 36, 702 (1970).
236. **Katayama, T., Hirata, K., Yokoyama, H., and Chichester, C. O.,** The Biosynthesis of astaxanthin. III. The carotenoids in sea breams, *Bull. Jap. Soc. Sci. Fish.,* 36, 709 (1970).
237. **Katayama, T., Yokoyama, H., and Chichester, C. O.,** The biosynthesis of astaxanthin. 1. The Structure of α-doradexanthin and β-doradexanthin, *Int. J. Biochem.,* 1, 438 (1970).
238. **Katsui, N., Suzuki, Y., Kitamura, S., and Irie, T.,** 5,6-Dibromoprotocatechualdehyde and 2,3-dibromo-4,5-dihydroxybenzyl methyl ether. New dibromophenols from *Rhodomela larix, Tetrahedron,* 23, 1185 (1967).
239. **Kem, W. R., Abbott, B. C., and Coates, R. M.,** Isolation and structure of a hoplonemertine toxin, *Toxicon,* 9, 15 (1971).
240. **Kennard, O., Watson, D. G., Riva di Sanseverino, L., Tursch, B., Bosmans, R., and Djerassi, C.,** Chemical studies of marine invertebrates. IV. Terpenoids LXII. Eunicellin, a diterpenoid of the gorgonian *Eunicella stricta.* X-ray diffraction analysis of eunicellin dibromide, *Tetrahedron Lett.* 2879 (1968).
241. **Kent, R. A., Smith, I. R., and Sutherland, M. D.,** Pigments of marine animals. X. Substituted naphthopyrones from the crinoid *Comantheria perplexa, Aust. J. Chem.,* 23, 2325 (1970).
242. **Kind, C. A. and Bergmann, W.,** Contributions to the study of marine products. XI. The occurrence of octadecyl alcohol, batyl alcohol and cetyl palmitate in gorgonias, *J. Org. Chem.,* 7, 424 (1942).
243. **Kishi, Y., Fukuyama, T., Aratani, M., Nakatsubo, F., Goto, T., Inoue, S., Tanino, H., Sugiura, S., and Kakoi, H.,** Synthetic studies on tetrodotoxin and related compounds. IV. Stereospecific total syntheses of DL-tetrodotoxin, *J. Am. Chem. Soc.,* 94, 9219 (1972).
244. **Kishi, Y., Tanino, H., and Goto, T.,** The structure confirmation of the light-emitting moiety of bioluminescent jellyfish *Aequorea, Tetrahedron Lett.,* 2747 (1972).
245. **Kittredge, J. S., Roberts, E., and Simonsen, D. G.,** The occurrence of free 2-aminoethylphosphonic acid in the sea anemone *Anthopleura elegantissima, Biochemistry,* 1, 624 (1962).
246. **Kittredge, J. S. and Hughes, R. R.,** The occurrence of α-amino-β-phosphonopropionic acid in the zoanthid *Zoanthus sociatus,* and in the ciliate *Tetrahymena pyriformis, Biochemistry,* 3, 991 (1964).
247. **Kittredge, J. S., Isbell, A. F., and Hughes, R. R.,** Isolation and characterization of the N-methyl derivatives of 2-aminoethyl-phosphonic acid from the sea anemone *Anthopleura xanthogrammica, Biochemistry,* 6, 289 (1967).
248. **Kleinig, H. and Egger, K.,** Zur Struktur von Siphonaxanthin und Siphonein, den Hauptcarotinoiden siphonoler Grünalgen, *Phytochemistry,* 6, 1681 (1967).
249. **Kleinig, H.,** Carotenoids of siphonous green algae: A chemotaxonomical study, *J. Phycol.,* 5, 281 (1969).
250. **Kleinig, H., Nitsche, H., and Egger, K.,** The structure of siphonaxanthin, *Tetrahedron Lett.,* 5139 (1969).
251. **Kobayashi, M., Tsuru, R., Todo, K., and Mitsuhashi, H.,** Asteroid sterols, *Tetrahedron Lett.,* 2935 (1972).
252. **Kobayashi, M., Tsuru, R., Todo, K., and Mitsuhashi, H.,** Marine sterols. II. Asterosterol, a new C_{26} sterol from *Asterias amurensis* Lütken, *Tetrahedron,* 29, 1193 (1973).
253. **Konosu, S., Chen, Y.-N., and Watanabe, K.,** Atrinine, a new betaine isolated from the adductor muscle of fan-mussel, *Bull. Jap. Soc. Sci. Fish.,* 36, 940 (1970).
254. **Kossel, A.,** Über das Agmatin, *Hoppe-Seyler's Z. Physiol. Chem.,* 66, 257 (1910).
255. **Kosuge, T., Zenda, H., Ochiai, A., Masaki, N., Noguchi, M., Kimura, S., and Narita, H.,** Isolation and structure determination of a new marine toxin, surugatoxin from the Japanese ivory shell *Babylonia japonica, Tetrahedron Lett.,* 2545 (1972).
256. **Kurata, K., Amiya, T., and Yabe, K.,** Studies on the constituents of a red marine alga *Odonthalia corymbifera, Bull. Jap. Soc. Sci. Fish,* 39, 973 (1973).
257. **Kurosawa, E., Izawa, M., Yamamoto, K., Masamune, T., and Irie, T.,** Sesquiterpenes from *Dictyopteris divaricata.* II. Dictyopterol and dictyopterone, *Bull. Chem. Soc. Jap.,* 39, 2509 (1966).
258. **Kurosawa, E., Furusaki, A., Izawa, M., Fukuzawa, A., and Irie, T.,** The absolute configurations of laureatin and isolaureatin, *Tetrahedron Lett.,* 3857 (1973).
259. **Kurosawa, E., Fukuzawa, A., and Irie, T.,** Isoprelaurefucin, new bromo compound from *Laurencia nipponica* Yamada, *Tetrahedron Lett.,* 4135 (1973).
260. **Lederer, E.,** Échinénone et pentaxanthine; deux nouveaux caroténoides trouvés dans l'oursin (*Echinus esculentus*), *C.R. Acad. Sci. Paris,* 201, 300 (1935).
261. **Lee, R. F., Nevenzel, J. C., Paffenhöfer, G.-A., Benson, A. A., Patton, S., and Kavanagh, T. E.,** A unique hexaene hydrocarbon from a diatom (*Skeletonema costatum*), *Biochim. Biophys. Acta,* 202, 386 (1970).

262. **Lee, R. F. and Loeblich III, A. R.,** Distribution of 21:6 hydrocarbon and its relationship to 22:6 fatty acid in algae, *Phytochemistry,* 10, 593 (1971).
263. **Lindberg, B.,** Low-molecular carbohydrates in algae. X. Investigation of *Furcellaria fastigiata, Acta Chem. Scand.,* 9, 1093 (1955).
264. **Lindberg, B.,** Methylated taurines and choline sulphate in red algae, *Acta Chem. Scand.,* 9, 1323 (1955).
265. **Ling, N. C., Hale, R. L., and Djerassi, C.,** The structure and absolute configuration of the marine sterol gorgosterol, *J. Am. Chem. Soc.,* 92, 5281 (1970).
266. **Lovell, F. M.,** The structure of a bromine-rich marine antibiotic, *J. Am. Chem. Soc.,* 88, 4510 (1966).
267. **Low, E. M.,** Halogenated amino acids of the bath sponge, *J. Mar. Res.,* 10, 239 (1951).
268. **Low, T. F., Park, R. J., Sutherland, M. D., and Vessey, I.,** Pigments of marine animals. III. The synthesis of some substituted polyhydroxyanthraquinones, *Aust. J. Chem.,* 18, 182 (1965).
269. **McFarren, E. F., Schantz, E. J., Campbell, J. E., and Lewis, K. H.,** Chemical determination of paralytic shellfish poison in clams, *J. Assoc. Offic. Analyt. Chemists,* 41, 168 (1958).
270. **McFarren, E. F., Schantz, E. J., Campbell, J. E., and Lewis, K. H.,** A modified Jaffe test for determination of paralytic shellfish poison, *J. Assoc. Offic. Analyt. Chemists,* 42, 399 (1959).
271. **Mackie, A. M. and Turner, A. B.,** Partial characterization of a biologically active steroid glycoside isolated from the starfish *Marthasterias glacialis, Biochem. J.,* 117, 543 (1970).
272. **Madgwick, J. C., Ralph, B. J., Shannon, J. S., and Simes, J. J.,** Non-protein amino acids in Australian seaweeds, *Arch. Biochem. Biophys.,* 141, 766 (1970).
273. **Madgwick, J. C. and Ralph, B. J.,** Enzymic synthesis of 1,3-dimethylhistidine in *Gracilaria secundata, Bot. Mar.,* 15, 170 (1972).
274. **Makisumi, S.,** Guanidino compounds from a sea-anemone, *Anthopleura japonica* Verrill, *J. Biochem.* (Tokyo), 49, 284 (1961).
275. **Mallams, A. K., Waight, E. S., Weedon, B. C. L., Chapman, D. J., Haxo, F. T., Goodwin, T. W., and Thomas, D. M.,** A new class of carotenoids, *Chem. Commun.,* 301 (1967).
276. **Manchand, P. S., Rüegg, R., Schwieter, U., Siddons, P. T., and Weedon, B. C. L.,** Carotenoids and related compounds. Part. XI. Syntheses of δ-carotene and ε-carotene, *J. Chem. Soc.,* 2019 (1965).
277. **Mandelli, E. F.,** Carotenoid pigments of the dinoflagellate *Glenodinium foliaceum* Stein, *J. Phycol.,* 4, 347 (1968).
278. **Mathias, A. P., Ross, D. M., and Schachter, M.,** Identification and distribution of 5-hydroxytryptamine in a sea anemone, *Nature,* 180, 658 (1957).
279. **Mathias, A. P., Ross, D. M., and Schachter, M.,** Distribution of histamine, 5-hydroxytryptamine, tetramethyl-ammonium and other substances in coelenterates possessing nematocysts, *J. Physiol.,* 142, 56P (1958).
280. **Mathieson, J. W. and Thomson, R. H.,** Naturally occurring quinones. Part XVIII. New spinochromes from *Diadema antillarum, Spatangus purpureus* and *Temnopleurus toreumaticus, J. Chem. Soc. (C).,* 153 (1971).
281. **Matsuda, H., (the late) Tomiie, Y., Yamamura S., and Hirata, Y.,** The structure of Aplysin-20, *Chem. Commun.,* 898 (1967).
282. **Matsuno, T., Ishida, T., Ito, T., and Sakushima, A.,** Gonadal pigment of sea-cucumber (*Holothuria leucospilota* Brandt), *Experientia,* 25, 1253 (1969).
283. **Matsuno, T. and Ito, T.,** Gonadal pigments of sea-cucumber *Stichopus japonicus* Selenka (Echinodermata), *Experientia,* 27, 509 (1971).
284. *Merck Index,* Eighth Ed., Merck & Co., Inc., New Jersey, 1968, (a) p. 19; (b) p. 20; (c) p. 24; (d) p. 27; (e) p. 99; (f) p. 106; (g) p. 108; (h) p. 123; (i) p. 145; (j) p. 148; (k) p. 185; (l) p. 198; (m) p. 212; (n) p. 213; (o) p. 229; (p) p. 253; (q) p. 254; (r) p. 255; (s) p. 283; (t) p. 284; (u) p. 291; (v) p. 317; (w) p. 325; (x) p. 346; (y) p. 369; (z) p. 374; (Ia) p. 402; (Ib) p. 416; (Ic) p. 417; (Id) p 426; (Ie) p. 431; (If) p. 471; (Ig) p. 473; (Ih) p. 496; (Ii) p. 497; (Ij) p. 500; (Ik) p. 510; (Il) p. 511; (Im) p. 525; (In) p. 532; (Io) p. 533; (Ip) p. 554; (Iq) p. 559; (Ir) p. 563; (Is) p. 573; (It) p. 579; (Iu) p. 581; (Iv) p. 585; (Iw) p. 597; (Ix) p. 607; (Iy) p. 616; (Iz) p. 633; (IIa) p. 675; (IIb) p. 680; (IIc) p. 687; (IId) p. 692; (IIe) p. 705; (IIf) p. 755; (IIg) p. 778; (IIh) p. 807; (IIi) p. 814; (IIj) p. 823; (IIk) p. 836; (IIl) p. 865; (IIm) p. 869; (IIn) p. 875; (IIo) p. 879; (IIp) p. 893; (IIq) p. 933; (IIr) p. 943; (IIs) p. 951; (IIt) p. 974; (IIu) p. 977; (IIv) p. 982; (IIw) p. 1013; (IIx) p. 1015; (IIy) p. 1030; (IIz) p. 1048; (IIIa) p. 1050; (IIIb) p. 1074; (IIIc) p. 1077; (IIId) p. 1086; (IIIe) p. 1091; (IIIf) p. 1092; (IIIg) p. 1097; (IIIh) p. 1100; (IIIi) p. 1119; (IIIj) p. 1120; (IIIk) p. 1126.
285. **Métayer, A., Viala, J., Alcaide, A., and Barbier, M.,** Synthèses de quelques stérols à 26 atomes de carbone, *C.R. Acad. Sci. Paris,* 274, 662 (1972).
286. **Métayer, A. and Barbier, M.,** Synthèse du 5α diméthyl-24 cholène-22 *trans* ol-3β,*C. R. Acad. Sci. Paris,* 276, 201 (1973).
287. **Minale, L., Sodano, G., Chan, W. R., and Chen, A. M.,** Aeroplysinin-2, a dibromolactone from marine sponges *Aplysina (Verongia) aerophoba* and *Ianthella* sp., *Chem. Commun.,* 674 (1972).
288. **Miyasaki, M.,** Studies on the components of *Digenea simplex* Ag. IV. Studies on the structure of kainic acid, the anthelmintic component. (4), *J. Pharm. Soc. Jap.,* 75, 692 (1955).
289. **Moody, K., Thomson, R. H., Fattorusso, E., Minale, L., and Sodano, G.,** Aerothionin and homoaerothionin: Two tetrabromo spirocyclohexadienylisoxazoles from *Verongia* sponges, *J. Chem. Soc. Perkin I,* 18, (1972).
290. **Moore, R. E., Singh, H., and Scheuer, P. J.,** Isolation of eleven new spinochromes from echinoids of the genus *Echinothrix, J. Org. Chem.,* 31, 3645 (1966).

291. **Moore, R. E., Singh, H., and Scheuer, P. J.**, A pyranonaphthazarin pigment from the sea urchin *Echinothrix diadema, Tetrahedron Lett.* 4581 (1968).
292. **Moore, R. E., Pettus, J. A. Jr., and Doty, M. S.**, Dictyopterene A, an odoriferous constituent from algae of the genus *Dictyopteris, Tetrahedron Lett.*, 4787 (1968).
293. **Moore, R. E.**, Bis-(3-oxoundecyl) polysulphides in *Dictyopteris, Chem. Commun.*, 1168 (1971).
294. **Moore, R. E., Mistysyn, J., and Pettus, J. A., Jr.**, (–)-Bis-(3-acetoxyundec-5-enyl) disulphide and S-(–)-3-acetoxy-undec-5-enyl thioacetate, possible precursors to undeca-1,3,5-trienes in *Dictyopteris, Chem. Commun.*, 326 (1972).
295. **Morimoto, H. and Nakamori, R.**, Studies on the active components of *Digenea simplex* Ag. and related compounds. XXXVI. Isomerization and isomers of kainic acid (15). Stereochemical configuration of kainic acid and its derivatives. (3), *J. Pharm. Soc. Jap.*, 76, 294 (1956).
296. **Morimoto, H. and Nakamori, R.**, Stereochemical structures of kainic acid and its isomers, *Proc. Jap. Acad. Sci.*, 32, 41 (1956).
297. **Mörner, C. Th.**, Zur Kenntnis der organischen Gerüstsubstanz des Anthozoënskeletts. IV. Mitteilung. Isolierung und Identifizierung der Bromgorgosäure, *Hoppe-Seyler's Z. Physiol. Chem.*, 88, 138 (1913).
298. **Mosher, H. S., Fuhrman, F. A., Buchwald, H. D., and Fischer, H. G.**, Tarichatoxin-tetrodotoxin: A potent neurotoxin, *Science*, 144, 1100 (1964).
299. **Motl, O., Romaňuk, M., and Herout, V.**, On terpenes. CLXXVIII. Composition of the oil from *Amorpha fruticosa* L. Fruits. structure of (–)-γ-amorphene. *Collect Czech. Chem. Commun.*, 31, 2025 (1966).
300. **Müller, D. G., Jaenicke, L., Donike, M., and Akintobi, T.**, Sex attractant in a brown alga: Chemical structure, *Science*, 171, 815, (1971).
301. **Murakami, S., Takemoto, T., Tei, Z., and Daigo, K.**, Studies on the effective principles of *Digenea simplex* Ag. VIII. Structure of kainic acid. (1), *J. Pharm. Soc. Jap.* 75, 866 (1955).
302. **Murakami, S., Takemoto, T., Tei, Z., and Daigo, K.**, Studies on the effective principles of *Digenea simplex* Ag. IX. Structure of kainic acid. (2), *J. Pharm. Soc. Jap.*, 75, 869 (1955).
303. **Murakami, S., Takemoto, T., Tei, Z., and Daigo, K.**, Studies on the effective principles of *Digenea simplex* Ag. XI. Structure of α-allokainic acid, *J. Pharm. Soc. Jap.* 75, 1255 (1955).
304. **Nakajima, T. and Volcani, B. E.**, 3,4-Dihydroxyproline: A new amino acid in diatom cell walls, *Science*, 164, 1400 (1969).
305. **Nakamura, T.**, Surugatoxin preparation from *Babylonia japonica*, for use as parasympatholytic agent, Derwent No. 76928T. Patent No. JA-4725309.
306. **Nakano, J.**, Cardiovascular effect of a prostaglandin isolated from a gorgonian *Plexaura homomalla, J. Pharm. Pharmacol.*, 21, 782 (1969).
307. **Nakazawa, Y.**, Studies on the new glycolipide in oyster. V. On the nitrogenous components and structure of the glycolipide, *J. Biochem.* (Tokyo), 46, 1579 (1959).
308. **Nicolaides, N. and Laves, F.**, The stereochemistry of squalene. A new method for the determination of *cis-trans* isomerism, *J. Am. Chem. Soc.*, 76, 2596 (1954).
309. **Nigrelli, R. F., Stempien, M. F., Jr., Ruggieri, G. D., Liguori, V. R., and Cecil, J. T.**, Substances of potential biomedical importance from marine organisms, *Fed. Proc.*, 26, 1197 (1967).
310. **Nomura, T., Tsuchiya, Y., André, D. and Barbier, M.**, Sur les Fractions Insaponifiables des Holothuries *Stichopus japonicus* et *Holothuria tubulosa, Bull. Jap. Soc. Sci. Fish.*, 35, 293 (1969).
311. **Nomura, T., Tsuchiya, Y., André, D., and Barbier, M.**, Sur la Biosynthèse des Stérols de l'Holothurie *Stichopus japonicus, Bull. Jap. Soc. Sci. Fish.*, 35, 299 (1969).
312. **Ohloff, G. and Pickenhagen, W.**, Synthese von (±)-dictyopteren A. *Helv. Chim. Acta*, 52, 880 (1969).
313. **Okaichi, T. and Hashimoto, Y.**, The structure of nereistoxin, *Agr. Biol. Chem.*, 26, 224 (1962).
314. **Ozawa, H., Gomi, Y., and Otsuki, I.**, Pharmacological studies on laminine monocitrate, *J. Pharm. Soc., Jap.*, 87, 935 (1967).
315. **Pasini, C., Vercellone, A., and Erspamer, V.**, Synthesis of murexine, *Liebig's Ann. Chem.*, 578, 6(1953).
316. **Pettit, G. R., von Dreele, R. B., Bolliger, G., Traxler, P. M., and Brown, P.**, Isolation and structural elucidation of 3,6-dioxo–hexahydro-pyrrolo(1,2-a)-pyrazine from the echinoderm *Luidia clathrata, Experientia*, 29, 521 (1973).
317. **Pettus, J. A., Jr. and Moore, R. E.**, Isolation and structure determination of an undeca-1,3,5,8-tetraene and dictyopterene B from algae of the genus *Dictyopteris, Chem. Commun.*, 1093 (1970).
318. **Pettus, J. A., Jr. and Moore, R. E.**, The isolation and structure determination of dictyopterenes C′ and D′ from *Dictyopteris.* Stereospecificity in the Cope rearrangement of dictyopterenes A and B, *J. Am. Chem. Soc.*, 93, 3087 (1971).
319. **Phleger, C. F. and Benson, A. A.**, Cholesterol and hyperbaric oxygen in swimbladders of deep sea fishes, *Nature*, 230, 122 (1971).
320. **Powell, V. H., Sutherland, M. D., and Wells, J. W.**, Pigments of marine animals. V. Rubrocomatulin monomethyl ether, an anthraquinoid pigment of the *Comatula* genus of crinoids, *Aust. J. Chem.*, 20, 535 (1967).
321. **Powell, V. H. and Sutherland, M. D.**, Pigments of marine animals. VI. Anthraquinoid pigments of the crinoids *Ptilometra australis* Wilton and *Tropiometra afra* Hartlaub, *Aust. J. Chem.*, 20, 541 (1967).
322. **Premuzic, E.**, Chemistry of natural products derived from marine sources, *Fortschr. Chem. Org. Naturst.* 29, 417 (1971).

Prota, G., D'Agostino, M., and Misuraca, G., Isolation and characterization of hallachrome, a red pigment from the sea worm *Halla parthenopeia, Experientia,* 27, 15 (1971).

Prota, G., D'Agostino, M., and Misuraca, G., The structure of hallachrome: 7-hydroxy-8-methoxy-6-methyl-1,2-anthraquinone, *J. Chem. Soc., Perkin I,* 1614, (1972).

Prota, G., personal communication.

Quin, L. D., 2-Aminoethylphosphonic acid in insoluble protein of the sea anemone *Metridium dianthus, Science,* 144, 1133 (1964).

Roche, J. and Robin, Y., Sur les phosphagènes des éponges, *C. R. Soc. Biol.,* 148, 1541 (1954).

Roche, J., Thoai, N.-v., and Robin, Y., Sur la présence de créatine chez les invertébrés et sa signification biologique, *Biochem. Biophys. Acta,* 24, 514 (1957).

Roller, P., Djerassi, C., Cloetens, R., and Tursch, B., Terpenoids. LXIV. Chemical studies of marine invertebrates. V. The isolation of three new holothurinogenins and their chemical correlation with lanosterol, *J. Am. Chem. Soc.,* 91, 4918 (1969).

Roller, P., Au, K., and Moore, R. E., Isolation of S-(3-oxoundecyl) thioacetate-bis-(3-oxoundecyl) disulphide, (–)-3-hexyl-4,5-dithiacycloheptanone, and S-(*trans*-3-oxoundec-4-enyl) thioacetate from *Dictyopteris, Chem. Commun.,* 503 (1971).

Roseghini, M., Occurrence of dihydromurexine (imidazolepropionylcholine) in the hypobranchial gland of *Thais (purpura) haemastoma, Experientia,* 27, 1008 (1971).

Rosenheim, O. and King, H., The ring system of sterols and bile acids, *Nature,* 130, 315 (1932).

Rothberg, I., Tursch, B. M., and Djerassi, C., Terpenoids. LXVIII. 23ξ-Acetoxy-17-deoxy-7,8-dihydroholothurinogenin, a new triterpenoid sapogenin from a sea cucumber, *J. Org. Chem.,* 38, 209 (1973).

Rüdiger, W., Aplysioviolin, ein neuartiger Gallenfarbstoff, *Naturwissenshaften.,* 53, 613 (1966).

Rüdiger, W., Gallenfarbstoffe und Biliproteide, *Fortschr. Chem. Org. Naturst.,* 29, 60 (1971).

Russell, F. E., Comparative pharmacology of some animal toxins, *Fed. Proc.,* 26, 1206 (1967).

Saito, T. and Ando, Y., Bromine compounds in seaweed. I. A bromophenolic compound from the red alga *Polysiphonia morrowii* Harv, *Chem. Abstr.,* 51, 17810i (1957).

Salaque, A., Barbier, M., and Lederer, E., Sur la biosynthese des stérols de l'huitre (*Ostrea gryphea*) et de l'oursin (*Paracentrotus lividus*), *Comp. Biochem. Physiol.,* 19, 45 (1966).

Santos, G. A. and Doty, M. S., Constituents of the green alga *Caulerpa lamourouxii, Lloydia,* 34, 88 (1971).

Scheuer, P. J., The chemistry of some toxins isolated from marine organisms, *Fortschr. Chem. Org. Naturst.,* 27, 322 (1969).

Scheuer, P. J., *Chemistry of Marine Natural Products,* Academic Press: New York, 1973. (a) p. 3; (b) p. 4; (c) p. 5; (d) p. 6; (e) p. 7; (f) p. 35; (g) p. 37; (h) p. 43; (i) p. 44 (j) p. 47; (k) p. 61; (l) p. 62; (m) p. 63; (n) p. 65; (o) p. 68; (p) p. 70; (q) p. 71; (r) p. 72; (s) p. 77; (t) p. 81; (u) p. 89; (v) p. 90; (w) p. 91; (x) p. 101; (y) p. 122; (z) p. 139.

Schmitz, F. J., Kraus, K. W., Ciereszko, L. S., Sifford, D. H., and Weinheimer, A. J., Ancepsenolide: A novel bisbutenolide of marine origin. Chemistry of coelenterates V, *Tetrahedron Lett.,* 97 (1966).

Schmitz, F. J., Lorance, E. D., and Ciereszko, L. S., Chemistry of coelenterates. XII. Hydroxyancepsenolide, a dilactone from the octocoral *Pterogorgia anceps, J. Org. Chem.,* 34, 1989 (1969).

Schmitz, F. J. and Pattabhiraman, T., New marine sterol possessing a side chain cyclopropyl group: 23-Demethylgorgosterol, *J. Am. Chem. Soc.,* 92, 6073 (1970).

Schmitz, F. J. and Lorance, E. D., Chemistry of coelenterates. XXI. Lactones from the gorgonian *Pterogorgia guadalupensis, J. Org. Chem.,* 36, 719 (1971).

Schneider, W. P., Hamilton, R. D., and Ruhland, L. E., Occurrence of esters of (15S)-prostaglandin A_2 and E_2 in coral, *J. Am. Chem. Soc.,* 94, 2122 (1972).

Schneider, W. P., Bundy, G. L., and Lincoln, F. H., Preparation of prostaglandin E_2 from *Plexaura homomalla, Chem. Commun.,* 254 (1973).

Schuett, W. and Rapoport, H., Saxitoxin, the paralytic shellfish poison. Degradation to a pyrrolopyrimidine, *J. Am. Chem. Soc.,* 84, 2266 (1962).

Sharma, G. M. and Burkholder, P. R., Studies on antimicrobial substances of sponges. I. Isolation, purification, and properties of a new bromine-containing antibacterial substance, *J. Antibiot. (Tokyo), Ser. A.,* 20, 200 (1967).

Sharma, G. M. and Burkholder, P. R., Studies on antimicrobial substances of sponges. II. Structure and synthesis of a bromine-containing antibacterial compound from a marine sponge, *Tetrahedron Lett.,* 4147 (1967).

Sharma, G. M., Vig, B., and Burkholder, P. R., Antimicrobial substances of marine sponges. IV. *Food-drugs from the sea, Proc.,* Marine Technol. Soc., 307 (1969).

Sharma, G. M., Vig, B., and Burkholder, P. R., Studies on the antimicrobial substances of sponges. IV. Structure of a bromine-containing compound from a marine sponge, *J. Org. Chem.,* 35, 2823 (1970).

Sharma, G. M. and Burkholder, P. R., Structure of dibromophakellin, a new bromine-containing alkaloid from the marine sponge *Phakellia flabellata, Chem. Commun.,* 151 (1971).

Sharma, G. M. and Vig, B., Studies on the antimicrobial substances of sponges. VI. Structures of two antibacterial substances isolated from the marine sponge *Dysidea herbacea, Tetrahedron Lett.,* 1715 (1972).

Sheikh, Y. M., Djerassi, C., and Tursch, B. M., Acansterol: A cyclopropane-containing marine sterol from *Acanthaster planci, Chem. Commun.,* 217 (1971).

356. **Sheikh, Y. M., Tursch, B. M. and Djerassi, C.,** 5α-Pregn-9(11)-ene-3β,6α-diol-20-one and 5α-cholesta-9(11),20(22)-diene-3β,6α-diol-23-one. Two novel steroids from the starfish *Acanthaster planci, J. Am. Chem. Soc.*, 94, 3278 (1972).
357. **Sheikh, Y. M. and Djerassi, C.,** Characterization of 3β-hydroxy-5α-cholesta-9(11), 20(22)-dien-23-one-6α-yl-β-D-6′-deoxy glucoside from the starfish *Acanthaster planci, Tetrahedron Lett.*, 2927 (1973).
358. **Shimizu, Y.,** Characterization of an acid hydrolysis product of starfish toxins as a 5α-pregnane derivative, *J. Am. Chem. Soc.*, 94, 4051 (1972).
359. **Shimomura, O. and Johnson, F. H.,** Chemical nature of the light emitter in bioluminescence of aequorin, *Tetrahedron Lett.*, 2963 (1973).
360. **Siddall, J. B., Horn, D. H. S., and Middleton, E. J.,** Synthetic studies on insect hormones. The synthesis of a possible metabolite of crustecdysone (20-hydroxyecdysone), *Chem. Commun.*, 899 (1967).
361. **Sims, J. J., Fenical, W., Wing, R. M., and Radlick, P.,** Marine natural products, III. Johnstonol, an unusual halogenated epoxide from the red alga *Laurencia johnstonii, Tetrahedron Lett.*, 195 (1972).
362. **Sims, J. J., Lin, G. H. Y., Wing, R. M., and Fenical, W.,** Marine natural products. Concinndiol, a bromo-diterpene alcohol from the red alga *Laurencia concinna, Chem. Commun.*, 470 (1973).
363. **Sims, J. J., Fenical, W., Wing, R. M., and Radlick, P.,** Marine natural products. IV. Prepacifenol, a halogenated epoxy sesquiterpene and precursor to pacifenol from the red alga *Laurencia filiformis, J. Am. Chem. Soc.*, 95, 972 (1973).
364. **Singh, H., Moore, R. E., and Scheuer, P. J.,** The distribution of quinone pigments in echinoderms, *Experientia*, 23, 624 (1967).
365. **Singh, H., Folk, T. L., and Scheuer, P. J.,** Synthesis of juglone derivatives: Hydroxy acetyl ethyl substituents, *Tetrahedron*, 25, 5301 (1969).
366. **Singh, I., Moore, R. E., Chang, C. W. J., and Scheuer, P. J.,** The synthesis of spinochromes A, C, D and E., *J. Am. Chem. Soc.*, 87, 4023 (1965).
367. **Smith, D. S. H., Turner, A. B., and Mackie, A. M.,** Marine steroids. Part I. Structures of the principal aglycones from the saponins of the starfish *Marthasterias glacialis, J. Chem. Soc. Perkin I*, 1745 (1973).
368. **Smith, I. R. and Sutherland, M. D.,** Pigments of marine animals. XI. Angular naphthopyrones from the crinoid *Comanthus parvicirrus timorensis, Aust. J. Chem.*, 24, 1487 (1971).
369. **Smith, J. and Thomson, R. H.,** Spinochrome E, *Tetrahedron Lett.*, 10 (1960).
370. **Smith, J. and Thomson, R. H.,** Naturally occurring quinones. Part V. Spinochromes E and N, *J. Chem. Soc.*, 1008 (1961).
371. **Sörensen, J. S. and Sörensen, N. A.,** Studies related to pristane. III. The identity of norphytane and pristane, *Acta Chem. Scand.*, 3, 939 (1949).
372. **Sörensen, N. A., Liaaen-Jensen, S., Bördalen, B., Haug, A., Enzell, C., and Francis, G.,** "Asterinsäure" - an acetylenic carotenoid, *Acta Chem. Scand.*, 22, 344 (1968).
373. **Stallard, M. O. and Faulkner, D. J.,** Chemical constituents of the digestive gland of the sea hare *Aplysia californica.* I. Importance of diet, *Comp. Biochem. Physiol.*, 49B, 25 (1974).
374. **Stallard, M. O. and Faulkner, D. J.,** Chemical constituents of the digestive gland of the sea hare *Aplysia californica.* II. Chemical transformations, *Comp. Biochem. Physiol.*, 49B, 37 (1974).
375. **Steiner, M. and Hartmann, T.,** Über Vorkommen und Verbreitung flüchtiger Amine bei Meeresalgen, *Planta*, 79, 113 (1968).
376. **Stempien, M. F., Jr., Nigrelli, R. F., and Chib, J. S.,** Isolation and synthesis of physiologically active substances from sponges of the genus *Agelas*, 164th ACS Meeting, Abstracts, MEDI 21 (1972).
377. **Stempien, M. F., Jr., Chib, J. S., Nigrelli, R. F., and Mierzwa, R. A.,** Physiologically active substances from marine sponges. II. Antimicrobial substances present in extracts of the sponge *Verongia archeri* and other species of the genus *Verongia,Food-Drugs from the Sea,* Proc., Marine Technol. Soc.,105 (1972).
378. **Stoffelen, H., Glombitza, K.-W., Murawski, U., Bielaczek, J., and Egge, H.,** Bromphenole aus *Polysiphonia lanosa* (L.) Tandy, *Planta Med.*, 22, 396 (1972).
379. **Strain, H. H., Svec, W. A., Aitzetmüller, K., Kiosen, H., Norgard, S., Liaaen-Jensen, S., Haxo, F. T., Wegfahrt, P., and Rapoport, H.,** The structure of peridinin, the characteristic dinoflagellate carotenoid, *J. Am. Chem. Soc.*, 93, 1823 (1971).
380. **Sucrow, W., Caldeira, P. P., and Slopianka, M.,** Die Synthese von (24S)-und (24R)-Aethylcholesta-5,22,25-trien-3β-ol sowie von Poriferasterin und Stigmasterin, *Chem. Ber.*, 106, 2236 (1973).
381. **Sutherland, M. D. and Wells, J. W.,** Anthraquinone pigments from the crinoid *Comatula pectinata, Chem. Ind.*, 291 (1959).
382. **Sutherland, M. D. and Wells, J. W.,** Pigments of marine animals. IV. The anthraquinoid pigments of the crinoids *Comatula pectinata* L. and *C. cratera* A. H. Clark, *Aust. J. Chem.*, 20, 515 (1967).
383. **Suzuki, M., Hayakawa, Y., and Irie, T.,** The acid-catalyzed rearrangement of laurinterol derivatives, *Bull. Chem. Soc. Jap.*, 42, 3342 (1969).
384. **Suzuki, M., Kurosawa, E., and Irie, T.,** Spirolaurenone, a new sesquiterpenoid containing bromine from *Laurencia glandulifera* Kützing, *Tetrahedron Lett.*, 4995 (1970).
385. **Takagi, N., Hsu, H. Y., and Takemoto, T.,** Studies on the hypotensive constituents of marine algae. V. Amino acid constituents of *Petalonia fascia, J. Pharm. Soc. Jap.*, 90, 899 (1970).

386. **Takemoto, T. and Daigo, K.,** Über die Inhaltsstoffe von *Chondria armata* und ihre pharmakologische Wirkung, *Arch. Pharm.,* 293, 627 (1960).

387. **Takemoto, T., Nakajima, T., and Daigo, K.,** Der fliegentötende Bestandteil von *Chondria armata, Jap. J. Pharm. Chem.,* 35, 21 (1963).

388. **Takemoto, T., Daigo, K., and Takagi, N.,** Studies on the hypotensive constituents of marine algae. I. A new basic amino acid "laminine" and the other basic constituents isolated from *Laminaria angustata, J. Pharm. Soc. Jap.,* 84, 1176 (1964).

389. **Takemoto, T., Daigo, K., and Takagi, N.,** Studies on the hypotensive constituents of marine algae. II. Synthesis of laminine and related compounds, *J. Pharm. Soc. Jap.,* 84, 1180 (1964).

390. **Takemoto, T., Daigo, K., and Takagi, N.,** Studies on the hypotensive constituents of marine algae. III. Determination of laminine in laminariaceen, *J. Pharm. Soc. Jap.,* 85, 37 (1965).

391. **Takemoto, T., Takagi, N., and Daigo, K.,** Studies on the hypotensive constituents of marine algae. IV. Amino acid constituents of *Heterochordaria abietina, J. Pharm. Soc. Jap.,* 85, 843 (1965).

392. **Takemoto, T., Daigo, K., Kondo, Y. and Kondo, K.,** Studies on the constituents of *Chondria armata.* VIII. On the structure of domoic acid. (1), *J. Pharm. Soc. Jap.,* 86, 874 (1966).

393. **Tamura, T., Wainai, T., Truscott, B., and Idler, D. R.,** Isolation of 22-dehydrocholesterol from scallop, *Can. J. Biochem.,* 42, 1331 (1964).

394. **Tanaka, A., Uda, H., and Yoshikoshi, A.,** Total synthesis of α-cubebene, β-cubebene and cubebol, *Chem. Commun.,* 308 (1969).

395. **Teshima, S.-i. and Kanazawa, A.,** Sterol compositions of marine crustaceans, *Bull. Jap. Soc. Sci. Fish.,* 37, 63 (1971).

396. **Teshima, S.-i. and Kanazawa, A.,** Sterol composition of marine occurring yeast, *Bull. Jap. Soc. Sci. Fish.,* 37, 68 (1971).

397. **Thomson, R. H.,** *Naturally Occurring Quinones,* second ed., Academic Press, London, 1971. (a) p. 171; (b) p. 173; (c) p. 256; (d) p. 261; (e) p. 268; (f) p. 270; (g) p. 271; (h) p. 272; (i) p. 273; (j) p. 274; (k) p. 275; (l) p. 276; (m) p. 387; (n) p. 443; (o) p. 445; (p) p. 492; (q) p. 507; (r) p. 594.

398. **Tökés, L.,** Nuclear-magnetic-resonance and mass-spectral study of 16-deoxymyxinol, *Biochem. J.,* 112, 765 (1969).

399. **Tsuda, K., Hayatsu, R., Kishida, Y., and Akagi, S.,** Steroid studies. VI. Studies on the constitution of sargasterol, *J. Am. Chem. Soc.,* 80, 921 (1958).

400. **Tsuda, K. and Sakai, K.,** Untersuchungen über Steroide. XX. Die Sterine aus grünen Meeres-Algen, *Chem. Pharm. Bull.* (Tokyo), 8, 554 (1960).

401. **Tsuda, K., Sakai, K., Tanabe, K., and Kishida, Y.,** Steroid studies. XVI. Isolation of 22-dehydrocholesterol from *Hypnea japonica, J. Am. Chem. Soc.,* 82, 1442 (1960).

402. **Tsuda, K., Ikuma, S., Kawamura, M., Tachikawa, R., Sakai, K., Tamura, C., and Amakasu, O.,** Tetrodotoxin. VII. On the structures of tetrodotoxin and its derivatives, *Chem. Pharm. Bull.* (Tokyo), 12, 1357 (1964).

403. **Tsuda, K.,** Über Tetrodotoxin, Giftstoff der Bowlfische, *Naturwissenschaften,* 53, 171 (1966).

404. **Tsujimoto, M.,** A highly unsaturated hydrocarbon in shark liver oil, *J. Ind. Eng. Chem.,* 8, 889 (1916).

405. **Tsukuda, N. and Amano, K.,** Studies on the discoloration of red fishes. I. Content of carotenoid pigments in eighteen species of red fishes, *Bull. Jap. Soc. Sci. Fish.,* 32, 334 (1966).

406. **Turner, A. B., Smith, D. S. H., and Mackie, A. M.,** Characterization of the principal steroidal saponins of the starfish *Marthasterias glacialis*: Structures of the aglycones, *Nature,* 233, 209 (1971).

407. **Tursch, B., de Souza Guimarães, I. S., Gilbert, B., Aplin, R. T., Duffield, A. M., and Djerassi, C.,** Chemical studies of marine invertebrates. II. Terpenoids. LVIII. Griseogenin, a new triterpenoid sapogenin of the sea cucumber *Halodeima grisea* L., *Tetrahedron,* 23, 761 (1967).

408. **Tursch, B., Cloetens, R., and Djerassi, C.,** Chemical studies of marine invertebrates. VI. Terpenoids. LXV. Praslinogenin, a new holothurinogenin from the Indian Ocean sea cucumber *Bohadschia koellikeri, Tetrahedron Lett.,* 467 (1970).

409. **Ueno, Y., Tanaka, K., Ueyanagi, J., Nawa, H., Sanno, Y., Honjo, M., Nakamori, R., Sugawa, T., Uchibayashi, M., Osugi, K., and Tatsuoka, S.,** 12. Studies on the active components of *Digenea simplex* Ag. and related compounds. V. Synthesis of α-kainic acid, *Proc. Jap. Acad. Sci.,* 33, 53 (1957).

410. **Ueyanagi, J., Nawa, H., Honjo, M., Nakamori, R., Tanaka, K., Ueno, Y., and Tatsuoka, S.,** Studies on the active components of *Digenea simplex* Ag. and related compounds. XLIX. Synthesis of α-kainic acid. (2), *J. Pharm. Soc. Jap.,* 77, 618 (1957).

411. Upjohn Co., Extraction of prostanoic acid derivatives from marine *Plexaura homomalla* (Esper) 1792, form S or R using frozen material. Prostaglandins with known activity, Derwent No. 20232T, Patent No. DT-2145018.

412. **Valentine, F. R., Jr. and Bergmann, W.,** Contributions to the study of marine products. VIII. The sterol of sponges: Clionasterol and poriferasterol, *J. Org. Chem.,* 6, 452 (1941).

413. **Van Lear, G. E., Morton, G. O., and Fulmor, W.,** New antibacterial bromoindole metabolites from the marine sponge *Polyfibrospongia maynardii, Tetrahedron Lett.,* 299 (1973).

414. **von Holt, C.,** Uptake of glycine and release of nucleoside-polyphosphates by zooxanthellae, *Comp. Biochem. Physiol.,* 26, 1071 (1968).

415. **Wallenfels, K. and Gauhe, A.,** Synthese von Echinochrom A, *Chem. Ber.,* 76, 325 (1943).

416. **Walton, T. J., Britton, G., Goodwin, T. W., Diner, B., and Moshier, S.,** The structure of siphonaxanthin, *Phytochemistry,* 9, 2545 (1970).
417. **Weinheimer, A. J., Schmitz, F. J., and Ciereszko, L. S.,** Chemistry of coelenterates. VII. The occurrence of terpenoid compounds in gorgonians, in *Drugs from the Sea,* Trans. Marine Technol. Soc., 135 (1967).
418. **Weinheimer, A. J., Middlebrook, R. E., Bledsoe, J. O., Jr., Marsico, W. E., and Karns, T. K. B.,** Eunicin, an oxa-bridged cembranolide of marine origin, *Chem. Commun.,* 384 (1968).
419. **Weinheimer, A. J., Washecheck, P. H., van der Helm, D., and Hossain, M. B.,** The sesquiterpene hydrocarbons of the gorgonian *Pseudopterogorgia americana,* the nonisoprenoid β-gorgonene, *Chem. Commun.,* 1070 (1968).
420. **Weinheimer, A. J. and Washecheck, P. H.,** The structure of the marine benzofuran, furoventalene, a non-farnesyl sesquiterpene. Chemistry of coelenterates. XIV, *Tetrahedron Lett.,* 3315 (1969).
421. **Weinheimer, A. J. and Spraggins, R. L.,** The occurrence of two new prostaglandin derivatives (15-*epi*-PGA_2 and its acetate, methyl ester) in the Gorgonian *Plexaura homomalla.* Chemistry of coelenterates. XV. *Tetrahedron Lett.,* 5185 (1969).
422. **Weinheimer, A. J., Youngblood, W. W., Washecheck, P. H., Karns, T. K. B., and Ciereszko, L. S.,** Isolation of the elusive (–)-germacrene-A from the gorgonian *Eunicea mammosa.* Chemistry of coelenterates. XVIII, *Tetrahedron Lett.,* 497 (1970).
423. **Weinheimer, A. J., Metzner, E. K., and Mole, M. L., Jr.,** A new marine betaine, norzooanemonin, in the gorgonian *Pseudopterogargia americana, Tetrahedron,* 29, 3135 (1973).
424. **Wells, R. J.,** personal communication.
425. **Welsh, J. H. and Prock, P. B.,** Quaternary ammonium bases in the coelenterates, *Biol. Bull.,* 115, 551 (1958).
426. **Welsh, J. H.,** 5-Hydroxytryptamine in coelenterates, *Nature,* 186, 811 (1960).
427. **Welsh, J. H.,** Composition and mode of action of some invertebrate venoms, *Annu. Rev. Pharmacol.,* 4, 293 (1964).
428. **Whittaker, V. P.,** The identity of natural and synthetic β-β-dimethylacrylylcholine, *Biochem. J.,* 71, 32 (1959).
429. **Whittaker, V. P.,** Pharmacologically active choline esters in marine gastropods, *Ann. N.Y. Acad. Sci.,* 90, 695 (1960).
430. **Wieland, H. and Dane, E.,** Untersuchungen über die Konstitution der Gallensäuren. XXXIX. Mitteilung. Zur Kenntnis der 12-Oxy-cholansäure, *Hoppe-Seyler's Z. Physiol. Chem.,* 210, 268 (1932).
431. **Wilber, C. G. and Bayors, W. M.,** A comparative study of the lipids in some marine annelids, *Biol. Bull. Woods Hole,* 93, 99 (1947).
432. **Willcott, M. R., Davis, R. E., Faulkner, D. J., and Stallard, M. O.,** The configuration and conformation of 7-chloro-1,6-dibromo-3,7-dimethyl-3,4-epoxy-1-octene, *Tetrahedron Lett.,* 3967 (1973).
433. **Windhaus, A., Bergmann, W. and König, G.,** Über einige Versuche mit Scymnol, *Hoppe-Seyler's Z. Physiol. Chem.,* 189, 148 (1930).
434. **Wong, J. L., Brown, M. S., Matsumoto, K., Oesterlin, R., and Rapoport, H.,** Degradation of saxitoxin to a pyrimido(2,1-b) purine, *J. Am. Chem. Soc.,* 93, 4633 (1971).
435. **Wong, J. L., Oesterlin, R., and Rapoport, H.,** The structure of saxitoxin, *J. Am. Chem. Soc.,* 93, 7344 (1971).
436. **Woodward, R. B.,** The structure of tetrodotoxin, *Pure Appl. Chem.,* 9, 49 (1964).
437. **Yamada, K., Yazawa, H., Toda, M., and Hirata, Y.,** The synthesis of (±)-aplysin and (±)-debromoaplysin, *Chem. Commun.,* 1432 (1968).
438. **Yamaguchi, M.,** On carotenoids of a sponge *Reniera japonica, Bull. Chem. Soc. Jap.,* 30, 111 (1957).
439. **Yamaguchi, M.,** Chemical constitution of renieratene, *Bull. Chem. Soc. Jap.,* 30, 979 (1957).
440. **Yamaguchi, M.,** Chemical constitution of isorenieratene, *Bull. Chem. Soc. Jap.,* 31, 51 (1958).
441. **Yamaguchi, M.,** Renieratene, a new carotenoid containing benzene rings, isolated from a sea sponge, *Bull. Chem. Soc. Jap.,* 31, 739 (1958).
442. **Yamaguchi, M.,** Total synthesis of isorenieratene, *Bull. Chem. Soc. Jap.,* 32, 1171 (1959).
443. **Yamaguchi, M.,** Total syntheses of renieratene and renierapurpurin, *Bull. Chem. Soc. Jap.,* 33, 1560 (1960).
444. **Yamamura, S. and Hirata, Y.,** Structures of aplysin and aplysinol, naturally occurring bromo-compounds, *Tetrahedron,* 19, 1485 (1963).
445. **Youngblood, W. W. and Blumer, M.,** Alkanes and alkenes in marine benthic algae, *Mar. Biol.,* 21, 163 (1973).

Table 2

Molecular formula	Structural formula	Name	Source of compound	Available by synthesis	Available nonmarine sources	Activity	References
C_1							
1. $CHBr_3$		Bromoform	Red alga: *Asparagopsis taxiformis*				85
C_2							
2. $C_2H_8NO_4P$		Phosphoethanolamine	Pufferfish: *Fugu vermiculare porphyreum*				175
C_3							
3. $C_3H_2Br_3ClO$		3-Chloro-1,1,3-tribromoacetone	Red alga: *Asparagopsis taxiformis*	+		Antimicrobial	82 85
4. $C_3H_2Br_4O$		1,1,3,3-Tetrabromoacetone	Red alga: *Asparagopsis taxiformis*	+		Antimicrobial	82 85
5. $C_3H_3Br_2ClO$		1-Chloro-1,3-dibromoacetone	Red alga: *Asparagopsis taxiformis*	+		Antimicrobial	82 85

Table 2 (continued)

Molecular formula	Structural formula	Name	Source of compound	Available by synthesis	Available nonmarine sources	Activity	References
C_3							
6. $C_3H_3Br_2ClO$	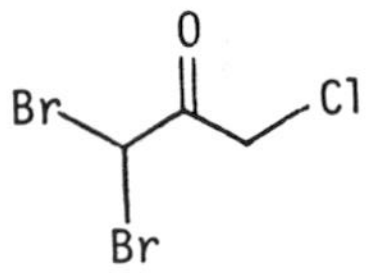 O, Br, Cl, Br	3-Chloro-1,1-dibromo-acetone	Red alga: *Asparagopsis taxiformis*	+		Antimicrobial	82 85
7. $C_3H_3Br_2IO$	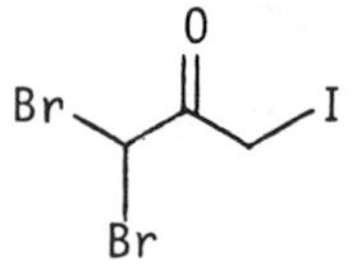 O, Br, I, Br	1,1-Dibromo-3-iodoacetone	Red alga: *Asparagopsis taxiformis*				85
8. $C_3H_3Br_3O$	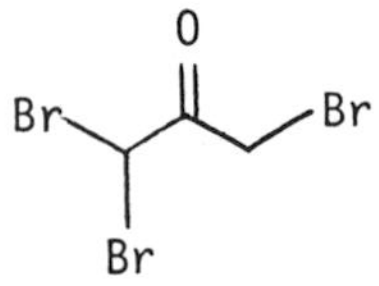 O, Br, Br, Br	1,1,3-Tribromoacetone	Red alga: *Asparagopsis taxiformis*	+		Antimicrobial	82 85
9. C_3H_4BrClO	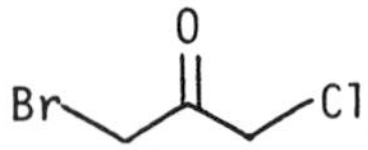 O, Br, Cl	1-Bromo-3-chloroacetone	Red alga: *Asparagopsis taxiformis*	+		Antimicrobial	82
10. $C_3H_4Br_2O$	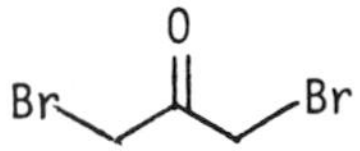 O, Br, Br	1,3-Dibromoacetone	Red alga: *Asparagopsis taxiformis*	+		Antimicrobial	82
11. $C_3H_4O_2$	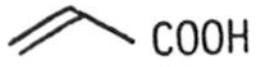 COOH	Acrylic acid	Yellow-brown alga: *Phaeocystis* sp.	+		Antibiotic	202 245

12. C_3H_5BrO	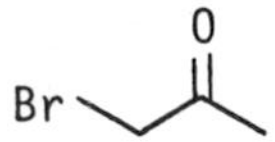	Bromoacetone	Red alga: *Asparagopsis taxiformis*				85
13. C_3H_5IO	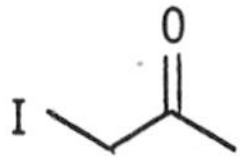	Iodoacetone	Red alga: *Asparagopsis taxiformis*				85
14. $C_3H_5N_3$	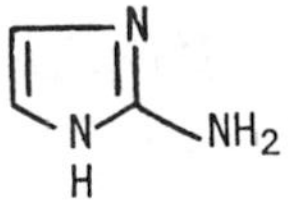	2-Aminoimidazole	Sponge: *Reniera cratera*				46
15. $C_3H_6O_3$	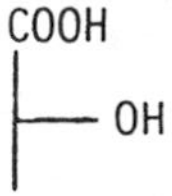	D(−)-Lactic acid	Octopus: *Octopus dofleini martini*	+			191 202
16. C_3H_9NO	$Me_3N \longrightarrow O$	Trimethylamine oxide	Algae Fish				92,116 175,202
C_4 17. C_4HBr_4N	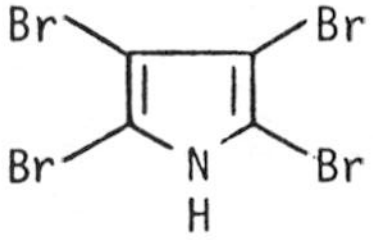	Tetrabromopyrrole	Bacterium: *Chromobacterium* sp.	+		Antibiotic	6
18. $C_4H_4O_4$	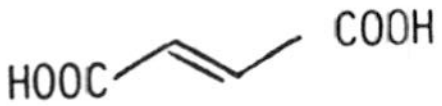	Fumaric acid	Clam: *Tapes japonica*	+	+		117 202
19. $C_4H_6O_4$	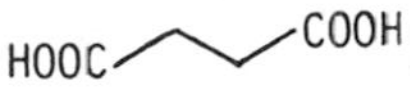	Succinic acid	Clam: *Tapes japonica* Sea worm: *Arenicola marina*	+	+	Laxative	3,117 202

Table 2 (continued)

Molecular formula	Structural formula	Name	Source of compound	Available by synthesis	Available nonmarine sources	Activity	References
C_4							
20. $C_4H_6O_5$	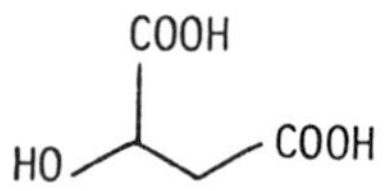	Malic acid	Clam: *Tapes japonica*	+	+		117 202
21. $C_4H_7N_3O$	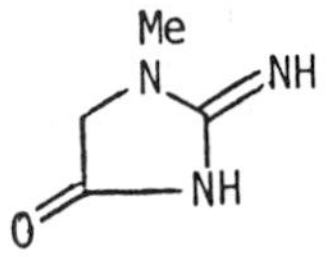	Creatinine	Fish	+	+		175 202 254
22. $C_4H_9NO_2$	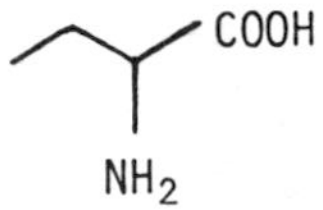	α-Amino-*n*-butyric acid	Fish: *Chrysophrys major* *Trachurus japonicus*	+			175 202
23. $C_4H_9NO_2$		γ-Amino-*n*-butyric acid	Algae Pufferfish: *Fugu vermiculare porphyreum*	+			175 190 202
24. $C_4H_9NO_3$	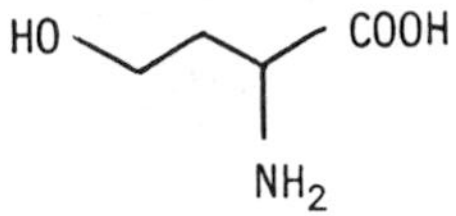	Homoserine	Gorgonians: *Eunicella cavolini* *E. stricta* *E. verrucosa*	+			39 202
C_5							
25. $C_5H_{11}NO_3S$	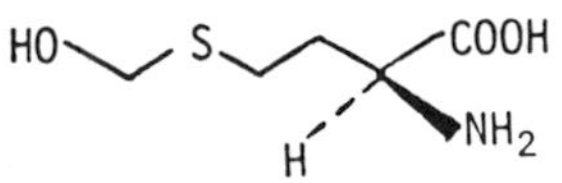	(−)-*S*-Hydroxymethyl-L-homocysteine	Red alga: *Chondrus ocellatus*				257

Formula	Compound	Source			Ref.
26. $C_5H_{11}NO_3S$	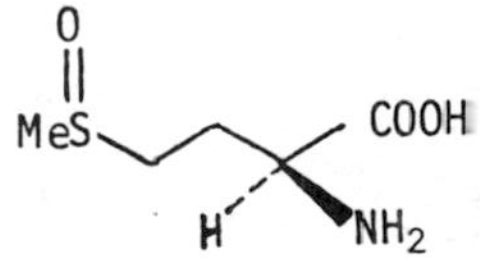 L(−)-Methionine-l-sulfoxide	Red alga: *Grateloupia turuturu*	+		208
27. $C_5H_{12}N_2O_2$	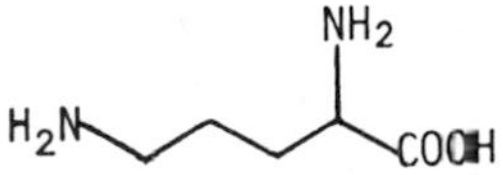 Ornithine	Algae Fish	+	+	175,190 202,208
C_6					
28. $C_6H_5N_5O_2$	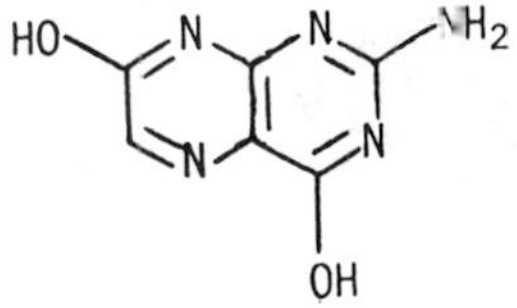 Isoxanthopterin	Copepods			210
29. $C_6H_6O_3$	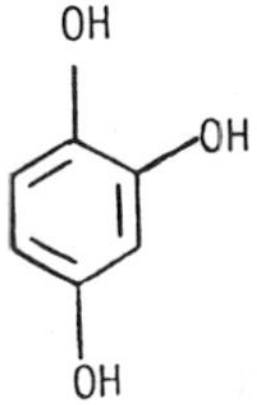 Hydroxyhydroquinone 1,2,4-Benzenetriol	Sponge *Axinella polypoides*	+		46 202
30. $C_6H_6O_3$	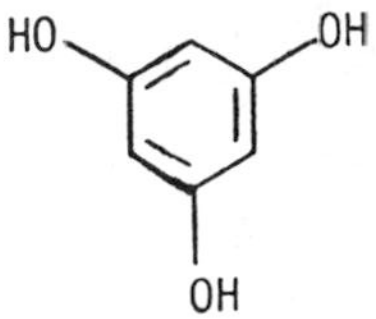 Phloroglucinol	Brown algae	+		94 202

Table 2 (continued)

Molecular formula	Structural formula	Name	Source of compound	Available by synthesis	Available nonmarine sources	Activity	References
C_6							
31. $C_6H_7N_5O_2$		7,8-Dihydroxanthopterin	Salmon: *Oncorhynchus kisutch*				179
32. $C_6H_{12}N_2O_4S_2$		Cystine	Algae Clam: *Tapes japonica*	+	+		117 190 202
33. $C_6H_{12}O_8S$		L-Fucose-4-sulfate	Sea urchins: *Hemicentrotus pulcherrimus* *Pseudocentrotus depressus*				132
34. $C_6H_{13}NO_3S$		*N*-Methylmethionine sulfoxide	Red alga: *Grateloupia turuturu*	+			208

C_7

35. $C_6H_{13}N_3O_3$	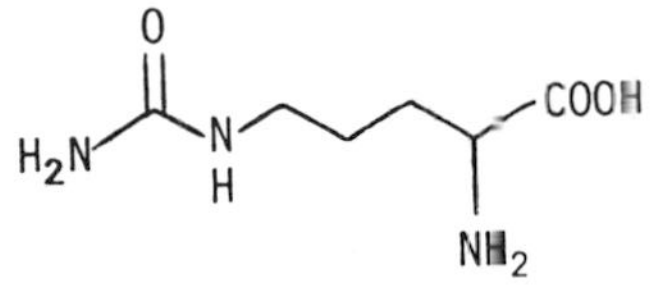	Citrulline	Algae	+	+		190 202 208
36. $C_7H_4Br_2O_2$	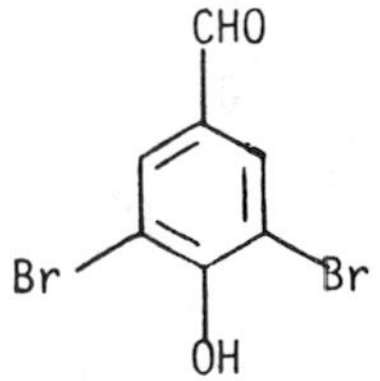	3,5-Dibromo-4-hydroxy-benzaldehyde	Sea worm: *Taelepus setosus*	+			119
37. $C_7H_5Br_3O_3$	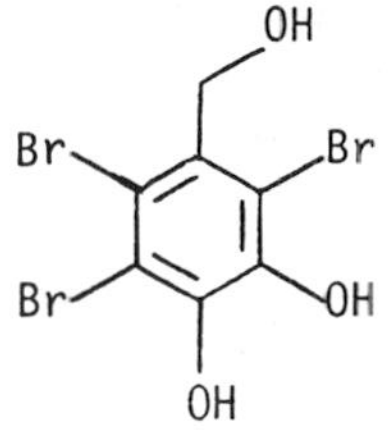	2,3,6-Tribromo-4,5-dihydroxybenzyl alcohol	Red algae: *Polysiphonia lanosa* *Rhodomela subfusca*			Antibiotic	97
38. $C_7H_5N_5O_3$	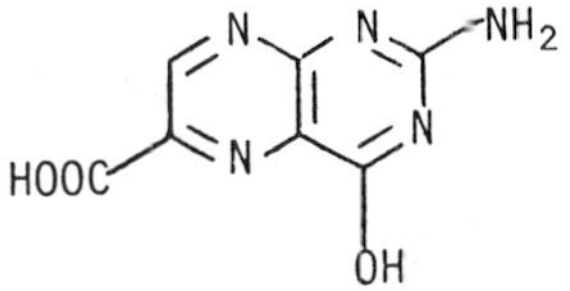	Pterin-6-carboxylic acid	Salmon: *Oncorhynchus kisutch*				179
39. $C_7H_5O_2$	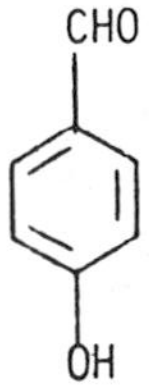	4-Hydroxybenzaldehyde	Bacterium: *Chromobacterium* sp.	+	+	Antibiotic	6 202

Table 2 (continued)

Molecular formula	Structural formula	Name	Source of compound	Available by synthesis	Available nonmarine sources	Activity	References
C_7							
40. $C_7H_7BrO_2$	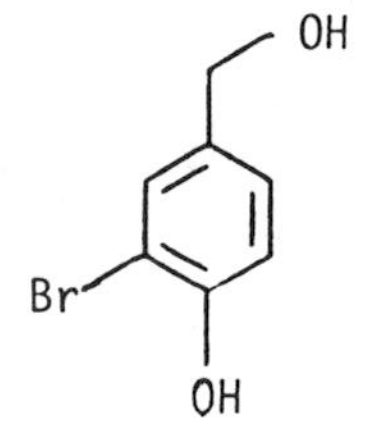	3-Bromo-4-hydroxybenzyl alcohol	Red algae: *Halopithys incurvus*, *Polysiphonia brodiaei*, *P. fruticulosa*, *P. nigra*, *P. urceolata*			Antibiotic	97
41. $C_7H_7BrO_3$	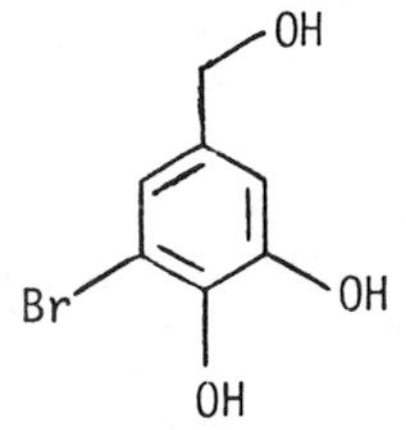	3-Bromo-4,5-dihydroxybenzyl alcohol	Red algae: *Halopithys incurvus*, *Polysiphonia lanosa*, *P. nigrescens*, *P. urceolata*	+		Antibiotic	97 252
42. $C_7H_{10}Br_4O$	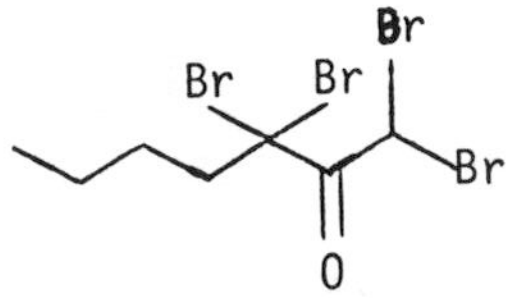	1,1,3,3-Tetrabromo-2-heptanone	Red alga: *Bonnemaisonia hamifera*	+			232 248

43. $C_7H_{11}Br_2IO$	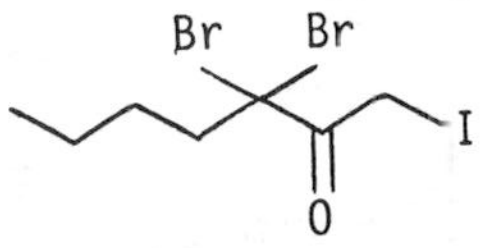 1-Iodo-3,3-dibromo-2-heptanone	Red alga: *Bonnemaisonia hamifera*		232 248
44. $C_7H_{11}Br_3O$	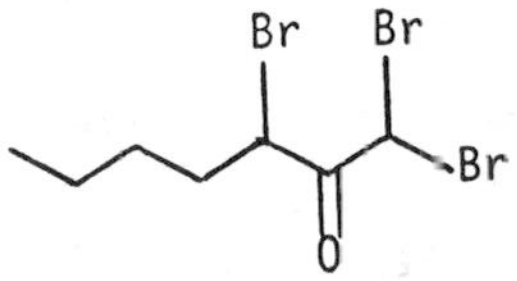 1,1,3-Tribromo-2-heptanone	Red alga: *Bonnemaisonia hamifera*		232 248
45. $C_7H_{11}Br_3O$	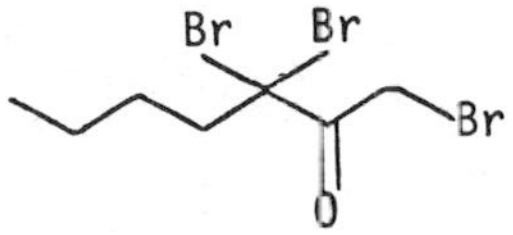 1,3,3-Tribromo-2-heptanone	Red alga: *Bonnemaisonia hamifera*		232 248
46. $C_7H_{11}N_3O_2$	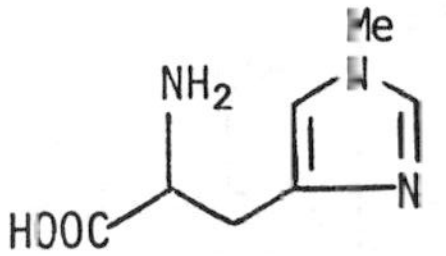 1-Methylhistidine	Brown algae: *Phyllospora comosa*, *Zonaria turneriana*; Red algae: *Corallina officinalis*, *Gracilaria secundata*	+	189 190
47. $C_7H_{11}N_3O_2$	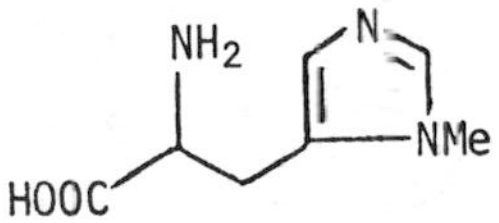 3-Methylhistidine	Pufferfish: *Fugu vermiculare porphyreum*		175
48. $C_7H_{12}Br_2O$	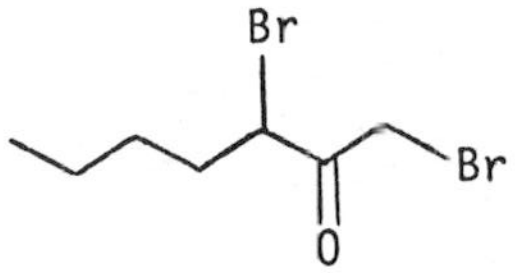 1,3-Dibromo-2-heptanone	Red alga: *Bonnemaisonia hamifera*		232 248

Table 2 (continued)

Molecular formula	Structural formula	Name	Source of compound	Available by synthesis	Available nonmarine sources	Activity	References
C_7							
49. $C_7H_{15}NO_3$	^-OOC, OH, $\overset{+}{N}Me_3$	L-Carnitine Vitamin B_T	Coelenterates Crustaceans Echinoderms Fish Molluscs Sea worms Sponge	+	+		33 89 202
50. $C_7H_{15}NO_3$	HO, COO^-, $^+NMe_3$	Homoserine betaine	Green alga: *Monostroma nitidum* Molluscs: *Callista brevisiphonata* *Turbo argyrostoma*	+			2 280
51. $C_7H_{17}NO_3$	O, O, $\overset{+}{N}Me_3$ ^-OH	Acetylcholine	Lobster: *Homarus americanus* Molluscs Starfish: *Asterias rubens*	+	+		58 69 164
52. $C_7H_{18}ClNS$	MeS, $\overset{+}{N}Me_3$ Cl^-	[3-(Methylthio)propyl] trimethylammonium chloride	Mollusc: *Turbo argyrostoma*	+		Cholinomimetic	277 278
C_8							
53. $C_8H_2Br_6N_2$	Br, Br, Br, Br, Br, Br, N H, N H	Hexabromo-2,2′-bipyrrole	Bacterium: *Chromobacterium* sp.	+		Antibiotic	6

54. $C_8H_6Br_2O_3$	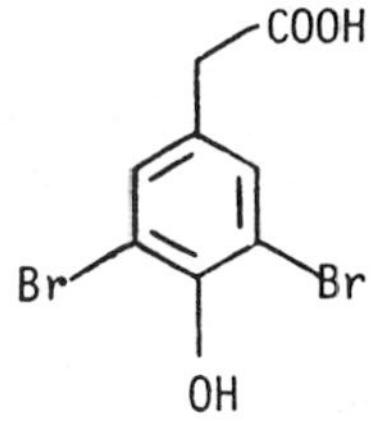 3,5-Dibromo-4-hydroxyphenylacetic acid	Red alga: *Halopithys incurvus*		43
55. $C_8H_8O_2$	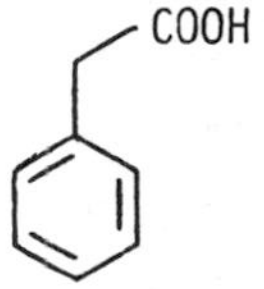 Phenylacetic acid	Brown alga: *Undaria pinnatifida*	+	1
56. $C_8H_8O_3$	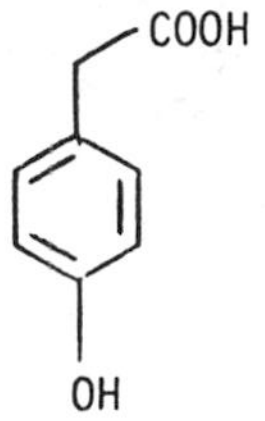 *p*-Hydroxyphenylacetic acid	Brown alga: *Undaria pinnatifida*	+	1
57. $C_8H_8O_4$	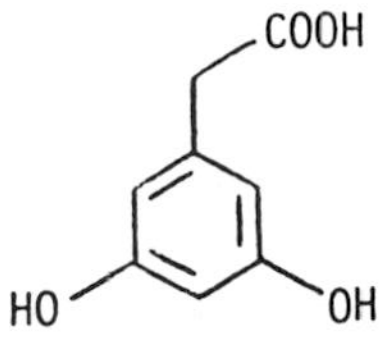 3,5-Dihydroxyphenylacetic acid	Octopus: *Octopus vulgaris* Squid: *Loligo vulgaris*	+	163
58. $C_8H_{10}N_2O_2$	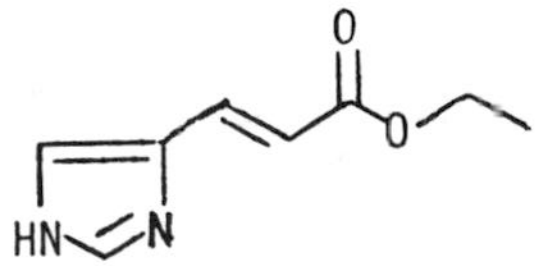 Ethyl urocanate	Molluscs: *Concholepas concholepas* *Murex trunculus*		233

Table 2 (continued)

Molecular formula	Structural formula	Name	Source of compound	Available by synthesis	Available nonmarine sources	Activity	References
C_8							
59. $C_8H_{11}NO$	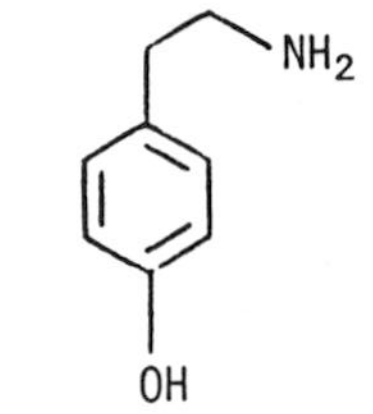 NH_2, OH	Tyramine	Octopus: *Hapalochlaena maculosa*	+	+	Sympathomimetic	125 202
60. $C_8H_{11}NO_2$	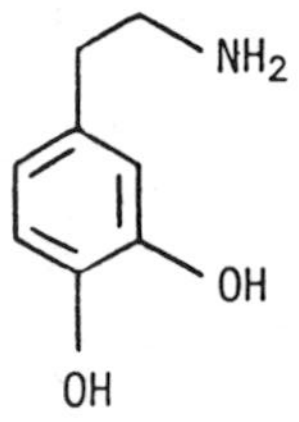 NH_2, OH, OH	Dopamine	Crab Lobster Molluscs Sea urchin Starfish	+	+	Investigative antihypotensive agent	58 163 202
61. $C_8H_{11}NO_3$	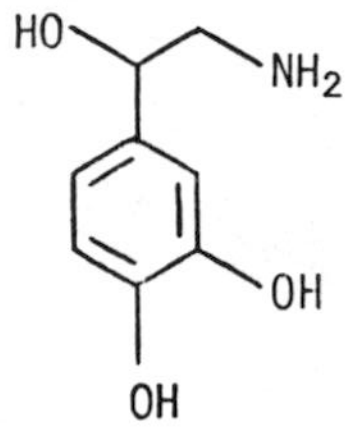 HO, NH_2, OH, OH	*l*-Noradrenaline	Clam Octopuses Sea urchin Squids Starfish	+	+	Sympathomimetic; Vasopressor	58 163 202
62. C_8H_{12}	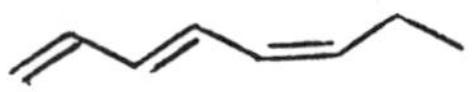	Fucoserratene	Brown alga: *Fucus serratus*	+			135 218

63. $C_8H_{17}NO_2$	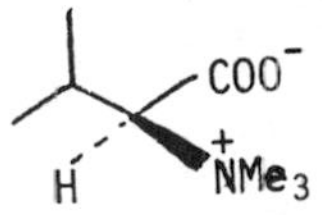	L-Valine betaine	Mollusc: *Callista brevisiphonata*	+			280
64. $C_8H_{21}Cl_2NS$	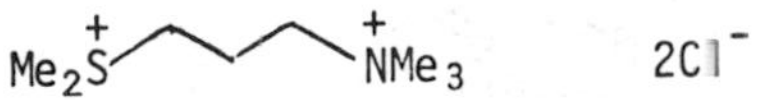	[3-(Dimethylsulfonio) propyl]-trimethylammonium dichloride	Mollusc: *Turbo argyrostoma*	+		Cholinomimetic	277 281
65. $C_8H_{22}NO_7P$	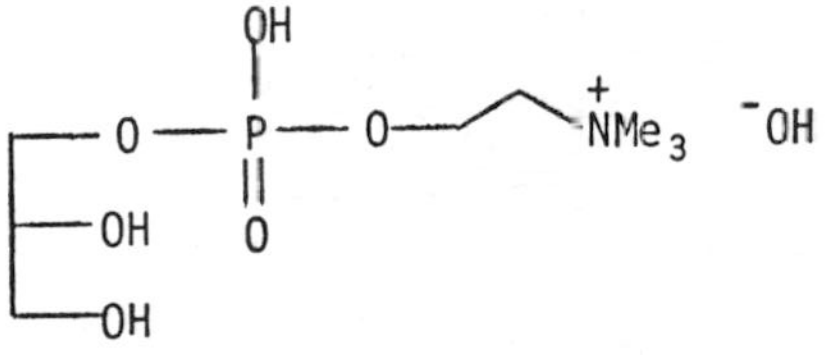	Glycerylphosphorylcholine	Octopus: *Octopus dofleini martini*		+		33
C_9							
66. $C_9H_6Br_2O_4$	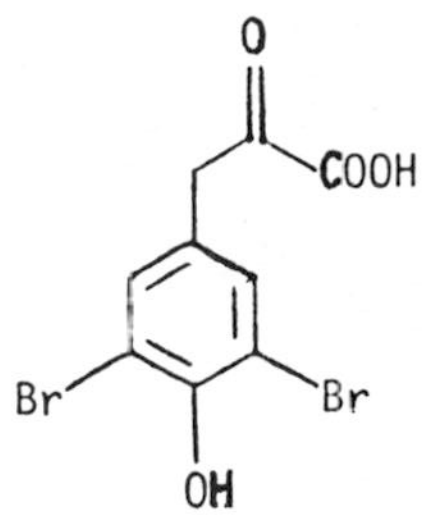	3,5-Dibromo-4-hydroxyphenylpyruvic acid	Red alga: *Halopithys incurvus*				43
67. $C_9H_7N_5O_5$	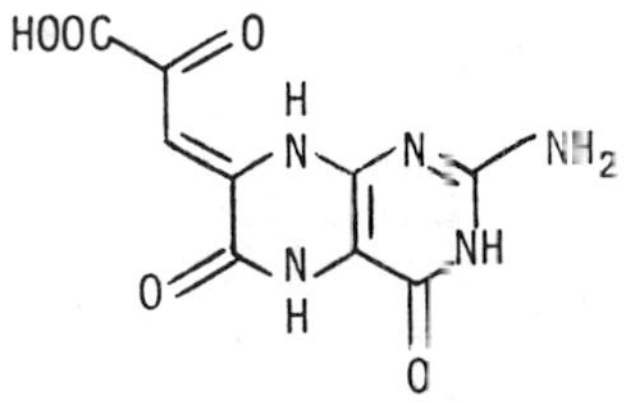	Erythropterin	Copepods	+	+		202 210

Table 2 (continued)

Molecular formula	Structural formula	Name	Source of compound	Available by synthesis	Available nonmarine sources	Activity	References
C_9							
68. $C_9H_9Br_3O_2$	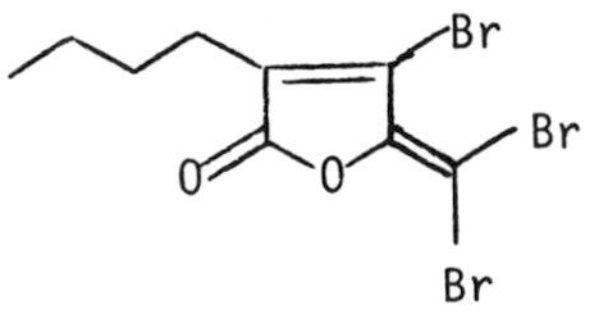	Fimbrolide C	Red alga: *Delisea fimbriata*				162
69. $C_9H_9Br_3O_3$	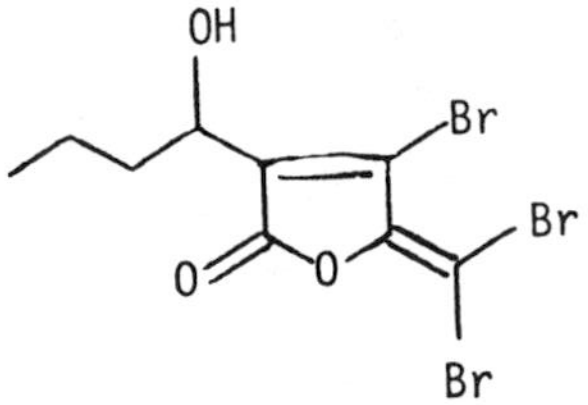	Hydroxyfimbrolide C	Red alga: *Delisea fimbriata*				162
70. $C_9H_{10}Br_2O_2$	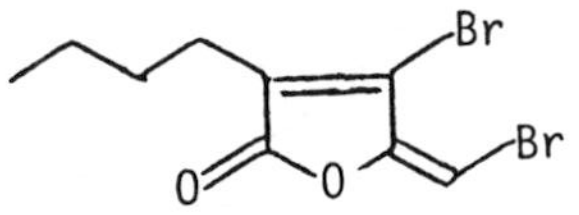	Fimbrolide A	Red alga: *Delisea fimbriata*				162
71. $C_9H_{10}Br_2O_2$	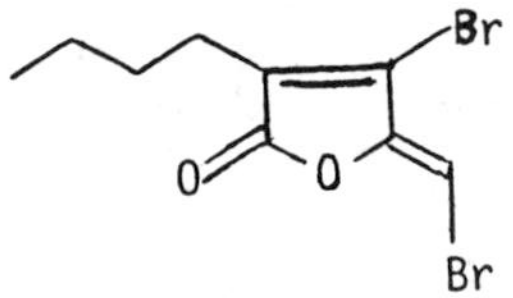	Fimbrolide B	Red alga: *Delisea fimbriata*				162

72. $C_9H_{10}Br_2O_3$	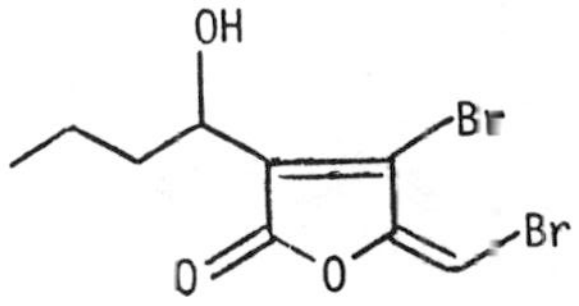 Hydroxyfimbrolide A	Red alga: *Delisea fimbriata*			162
73. $C_9H_{10}Br_2O_3$	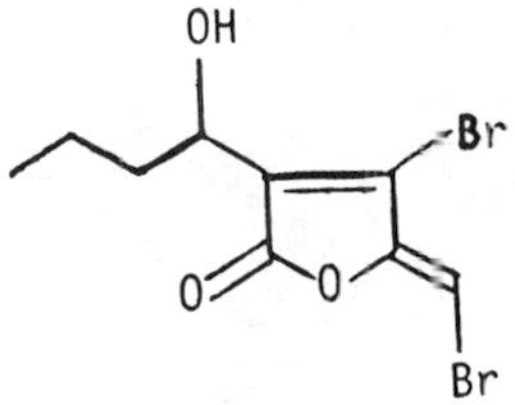 Hydroxyfimbrolide B	Red alga: *Delisea fimbriata*			162
74. $C_9H_{11}N_5O_4$	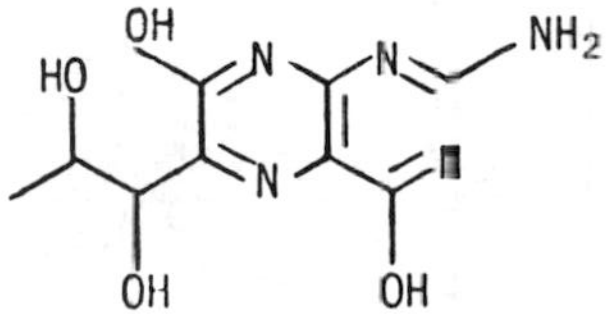 Ichthyopterin	Salmon: *Oncorhynchus kisutch*		+	179 202
75. $C_9H_{13}N_2O_9P$	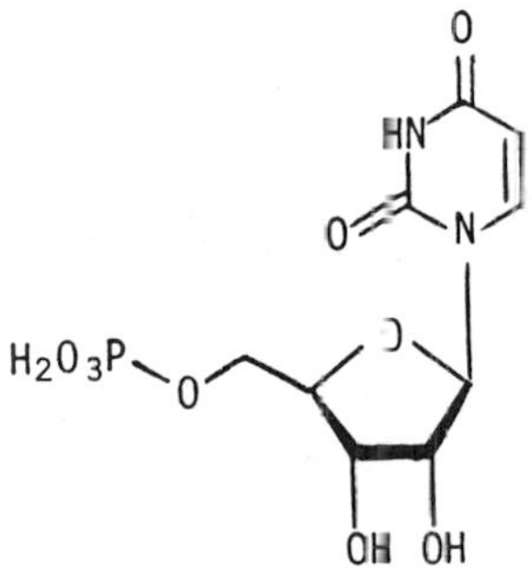 5′-Uridylic acid	Clam: *Tapes japonica*	+	+	117 202

Table 2 (continued)

Molecular formula	Structural formula	Name	Source of compound	Available by synthesis	Available nonmarine sources	Activity	References
C_9							
76. $C_9H_{14}N_4O_3$	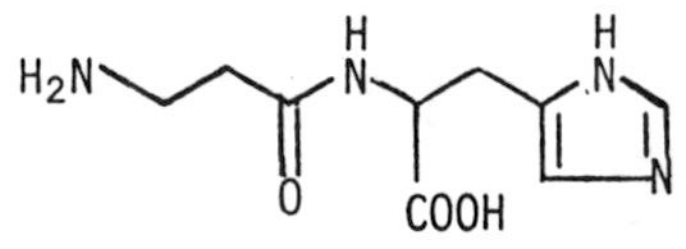	Carnosine	Fish	+	+		202 254
77. $C_9H_{18}O_8$	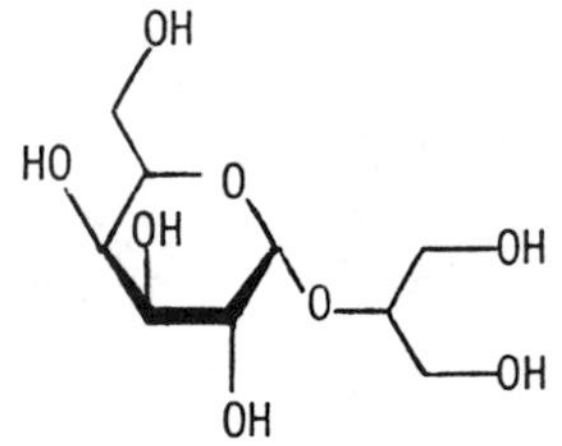	Floridoside	Red alga: *Plocamium costatum*				162
C_{10}							
78. $C_{10}H_{10}N_6$	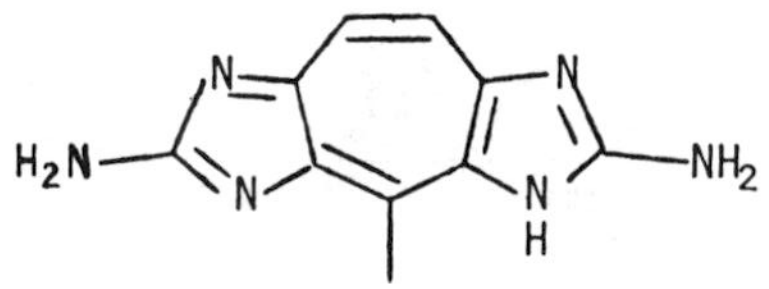	Parazoanthoxanthin A	Zoanthid: *Parazoanthus axinellae*				40
79. $C_{10}H_{11}Cl_3O$	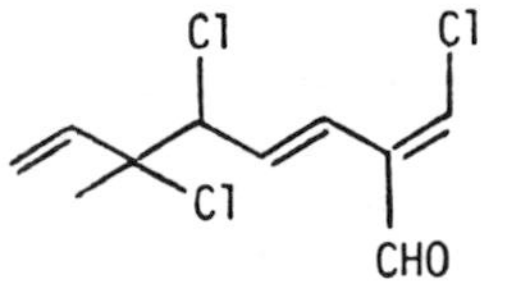	Cartilagineal	Red alga: *Plocamium cartilagineum*				59

80. $C_{10}H_{12}Br_2O_3$	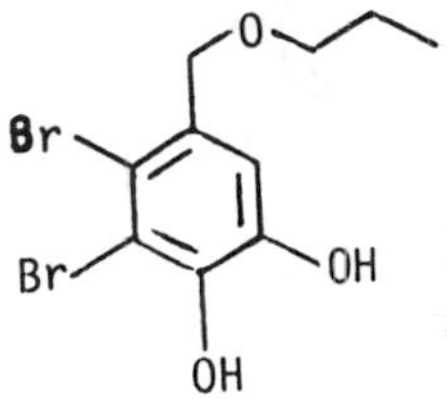	2,3-Dibromo-4,5-dihydroxybenzyl-*n*-propyl ether	Red algae: *Polysiphonia lanosa*, *P. nigrescens*			Antibiotic	97
81. $C_{10}H_{12}N_2O_4$	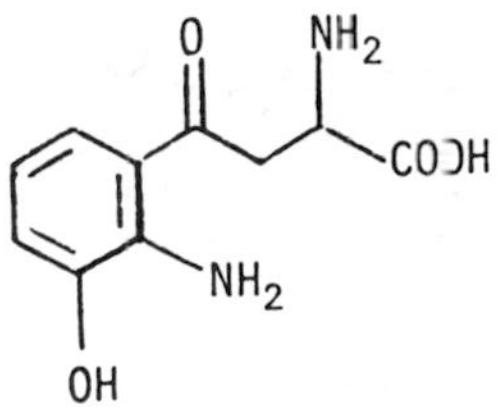	3-Hydroxy-L-kynurenine	Gorgonians: *Eunicella cavolini*, *E. stricta*, *E. verrucosa*		+		39 69
82. $C_{10}H_{12}N_4O_5$	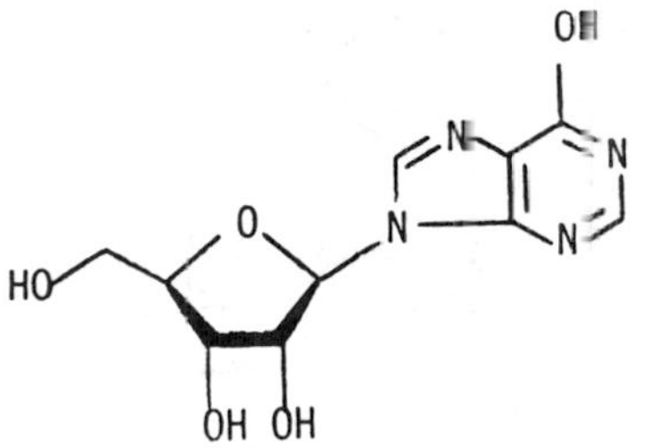	Inosine	Clam: *Tapes japonica* Fish	+	+		117 175 202
83. $C_{10}H_{12}O_3$	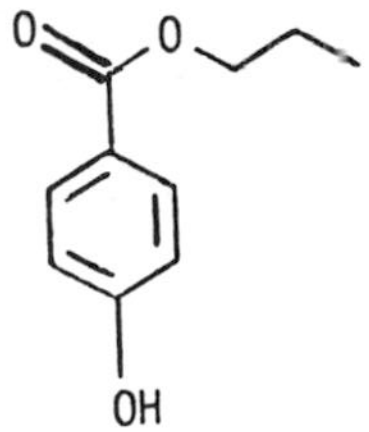	*n*-Propyl-4-hydroxybenzoate	Bacterium: *Chromobacterium* sp.	+		Antibiotic	6

Table 2 (continued)

Molecular formula	Structural formula	Name	Source of compound	Available by synthesis	Available nonmarine sources	Activity	References
C_{10}							
84. $C_{10}H_{13}BrCl_4$	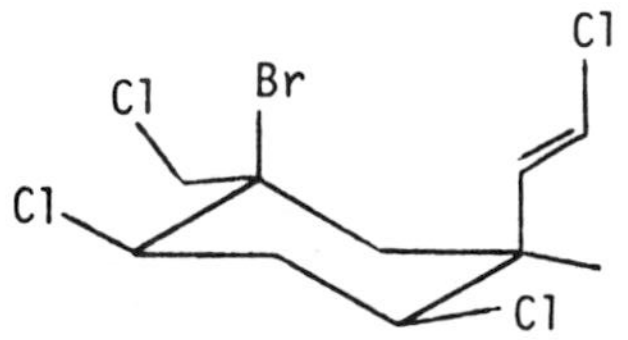	Violacene	Red alga: *Plocamium violaceum*				220
85. $C_{10}H_{13}Br_2Cl$	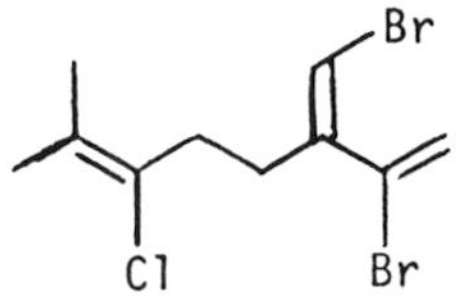	6-Chloro-2,(Z)-9-dibromomyrcene	Red alga: *Desmia hornemanni*				126
86. $C_{10}H_{13}Br_2Cl$	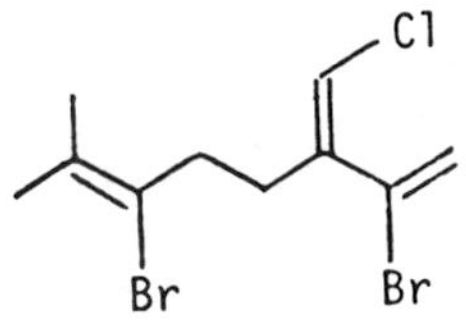	(Z)-9-Chloro-2,6-dibromomyrcene	Red alga: *Desmia hornemanni*				126
87. $C_{10}H_{13}N_4O_8P$	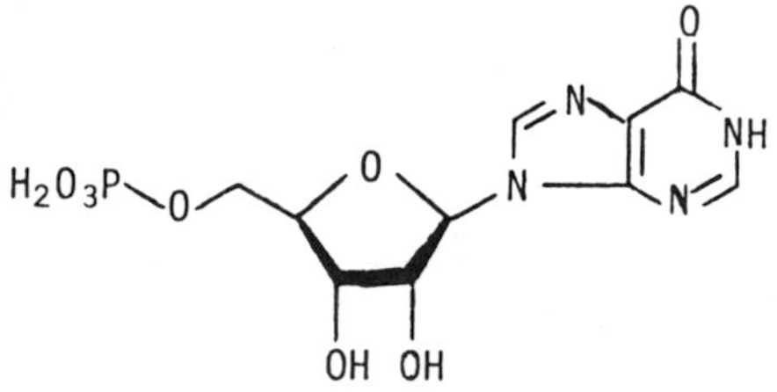	Inosinic acid	Clam: *Tapes japonica* Fish	+	+		117 175 179 202

88. $C_{10}H_{14}BrCl$	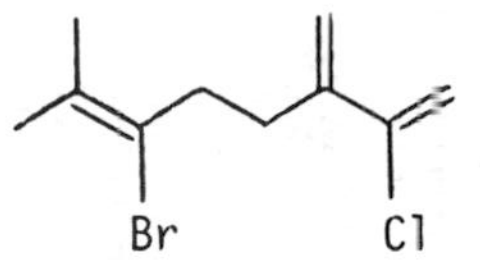 6-Bromo-2-chloromyrcene	Red alga: *Desmia hornemanni*			126
89. $C_{10}H_{14}BrCl$	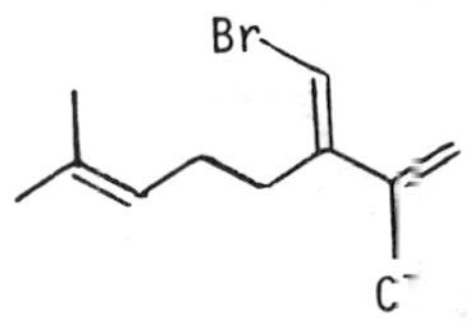 (E)-9-Bromo-2-chloromyrcene	Red alga: *Desmia hornemanni*			126
90. $C_{10}H_{14}BrCl$	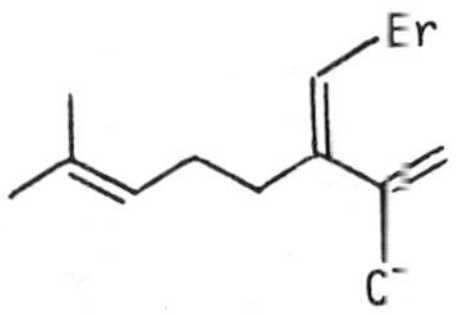 (Z)-9-Bromo-2-chloromyrcene	Red alga: *Desmia hornemanni*			126
91. $C_{10}H_{14}N_5O_7P$	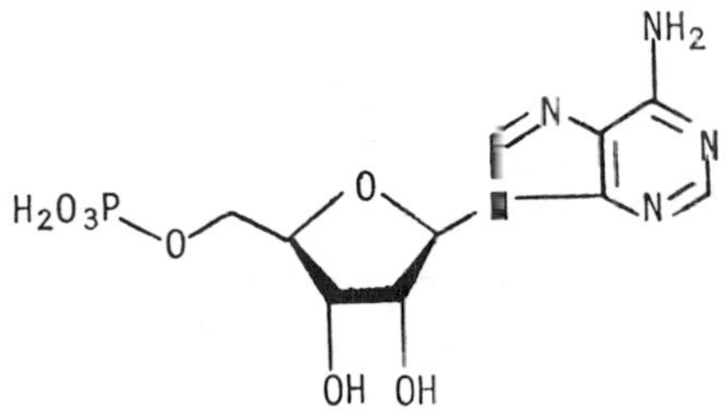 5′-Adenylic acid	Clam: *Tapes japonica* Fish	+	+	117 175 202
92. $C_{10}H_{15}Br$	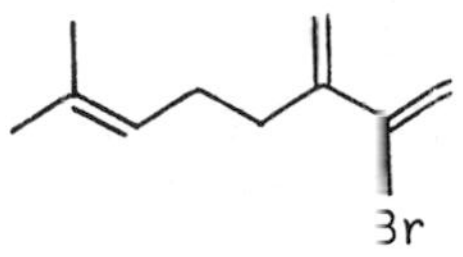 2-Bromomyrcene	Red alga: *Desmia hornemanni*			126

Table 2 (continued)

Molecular formula	Structural formula	Name	Source of compound	Available by synthesis	Available nonmarine sources	Activity	References
C_{10}							
93. $C_{10}H_{15}Br$	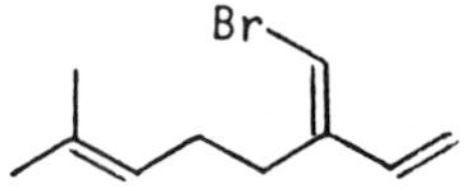 Br	(E)-9-Bromomyrcene	Red alga: *Desmia hornemanni*				126
94. $C_{10}H_{15}Br$	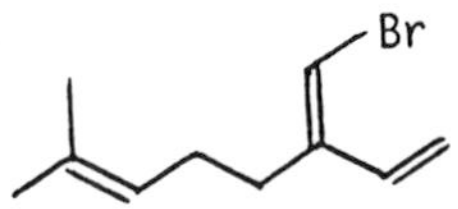 Br	(Z)-9-Bromomyrcene	Red alga: *Desmia hornemanni*				126
95. $C_{10}H_{15}Cl$	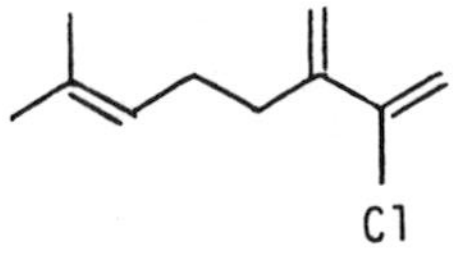 Cl	2-Chloromyrcene	Red alga: *Desmia hornemanni*				126
96. $C_{10}H_{15}N_5O_{10}P_2$	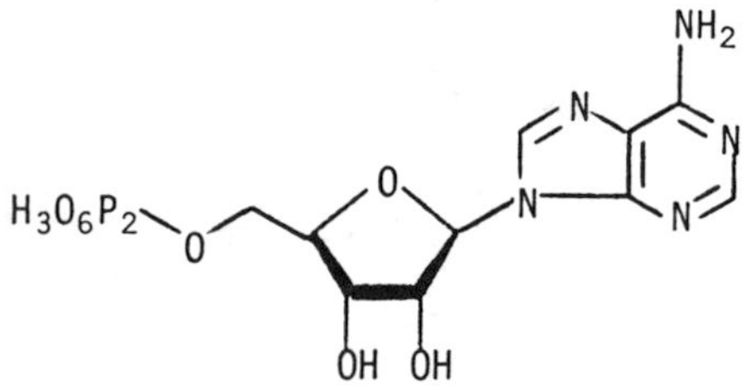 NH_2, N, N, N, N, O, $H_3O_6P_2$, O, OH, OH	Adenosine diphosphate	Clam: *Tapes japonica* Fish	+	+		117 175 202
97. $C_{10}H_{16}$	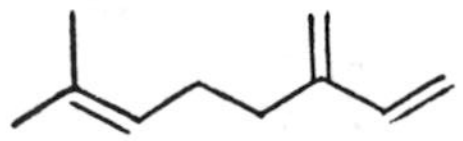	Myrcene	Red alga: *Desmia hornemanni*	+	+		126 202

Formula	Name	Source			Ref.
98. $C_{10}H_{16}N_4O_3$	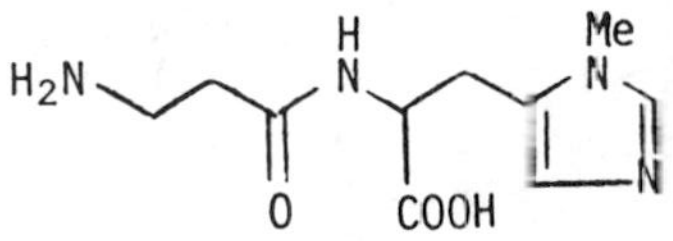 Anserine	Fish	+	+	202 254
99. $C_{10}H_{16}N_5O_{13}P_3$	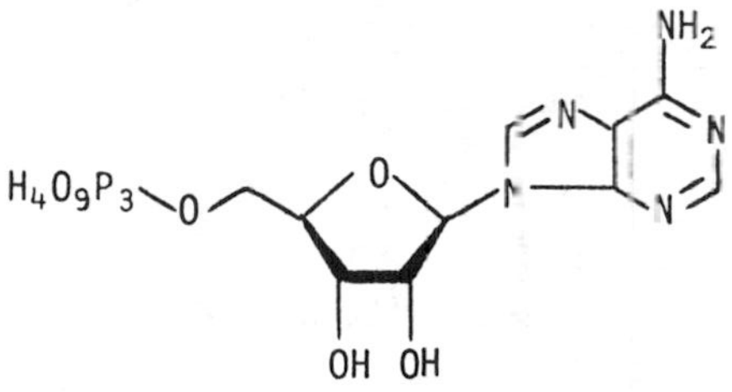 Adenosine triphosphate	Fish	+	+	175 202
100. $C_{11}H_{11}Br_3O_4$	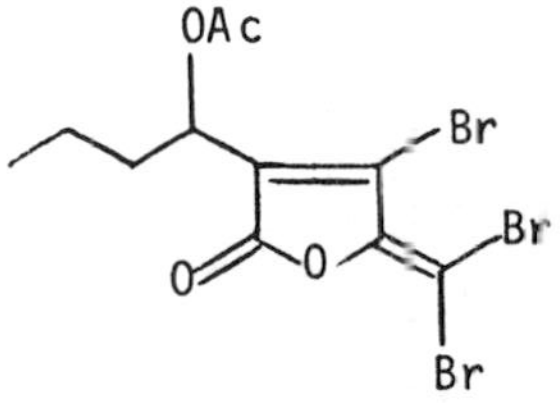 Acetoxyfimbrolide G	Red alga: *Delisea fimbriata*			162
101. $C_{11}H_{12}BrIO_4$	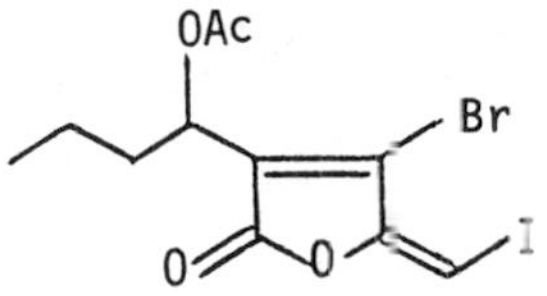 Acetoxyfimbrolide C	Red alga: *Delisea fimbriata*			162
102. $C_{11}H_{12}Br_2O_4$	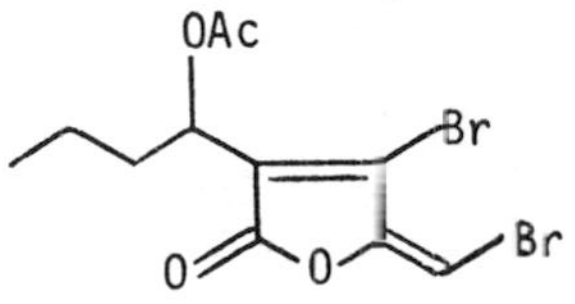 Acetoxyfimbrolide A	Red alga: *Delisea fimbriata*			162

Table 2 (continued)

Molecular formula	Structural formula	Name	Source of compound	Available by synthesis	Available nonmarine sources	Activity	References
C_{11}							
103. $C_{11}H_{12}Br_2O_4$	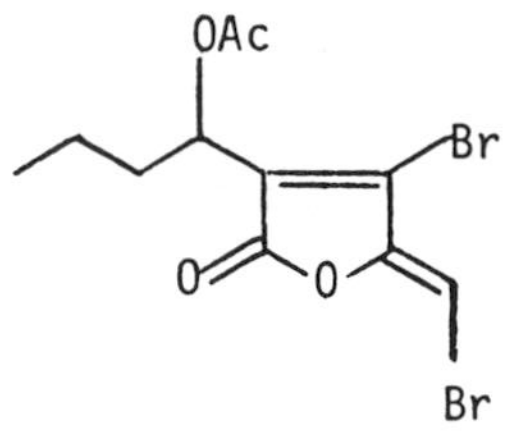	Acetoxyfimbrolide B	Red alga: *Delisea fimbriata*				162
104. $C_{11}H_{12}N_6$	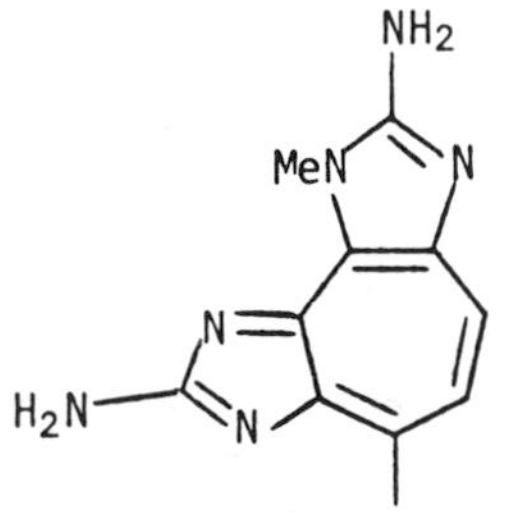	3-Norpseudozoantho-xanthin	Zoanthid: *Epizoanthus arenaceus*				42
105. $C_{11}H_{13}N_5O$	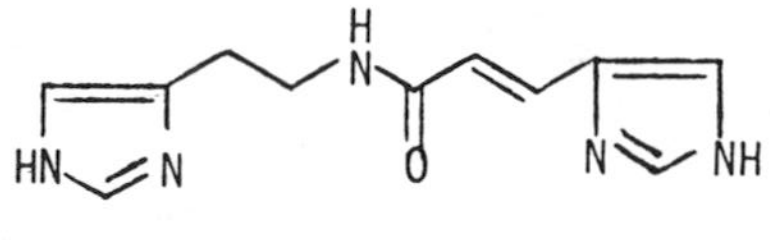	*N*-Urocanylhistamine	Mollusc: *Drupa concatenata*				236
106. $C_{11}H_{15}N_5O$	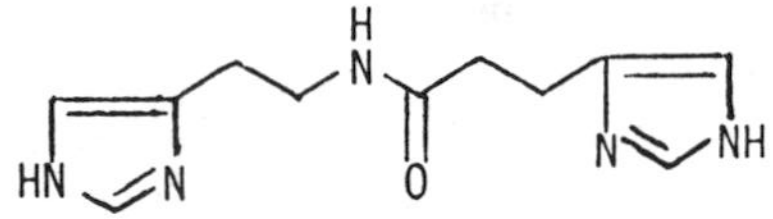	*N*(4-Imidazole-propionyl)histamine	Mollusc: *Drupa concatenata*	+			235

Compound	Structure	Name	Source	Activity	Reference
107. $C_{11}H_{16}$	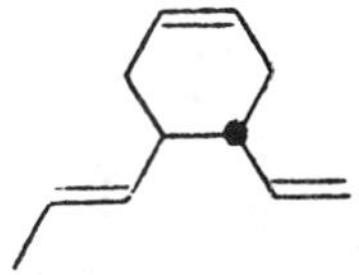	Aucantene	Brown alga: *Cutleria multifida*		134
108. $C_{11}H_{16}$	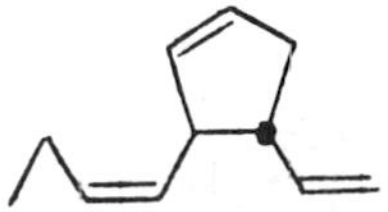	Multifidene	Brown alga: *Cutleria multifida*		134 219
109. $C_{11}H_{16}O$	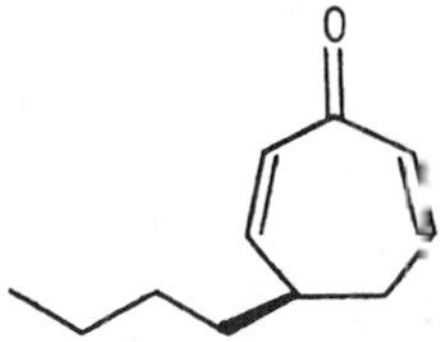	(+)-*R*-4-Butylcyclohepta-2,6-dienone	Brown algae: *Dictyopteris australis* *D. plagiogramma*	+	213
110. $C_{11}H_{16}O$	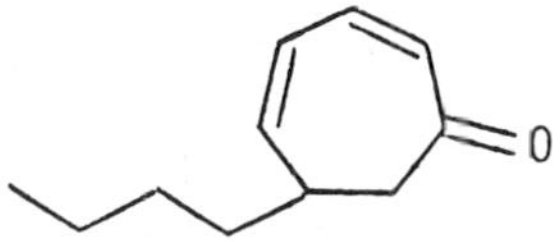	(+)-6-Butylcyclohepta-2,4-dienone	Brown algae: *Dictyopteris australis* *D. plagiogramma*		213
111. $C_{11}H_{18}$		*cis, trans*-1,3,5-Undecatriene	Brown algae: *Dictyopteris australis* *D. plagiogramma*		214
112. $C_{11}H_{18}$		*trans, trans, trans*-2,4,6-Undecatriene	Brown algae: *Dictyopteris australis* *D. plagiogramma*		214

Table 2 (continued)

Molecular formula	Structural formula	Name	Source of compound	Available by synthesis	Available nonmarine sources	Activity	References
C_{11}							
113. $C_{11}H_{18}ClNO$	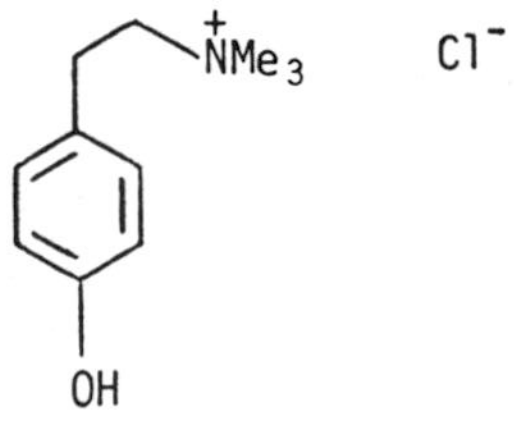	Candicine chloride	Mollusc: *Turbo argyrostoma*	+	+	Stimulating and paralyzing nicotinic and curariform actions	279
114. $C_{11}H_{19}NO_9$	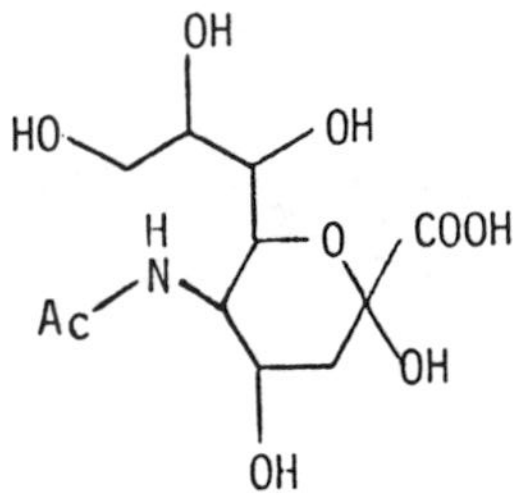	*N*-Acetylneuraminic acid	Sea urchins: *Anthocidaris crassispina* *Hemicentrotus pulcherrimus*	+	+		69 124 202
115. $C_{11}H_{19}NO_{10}$	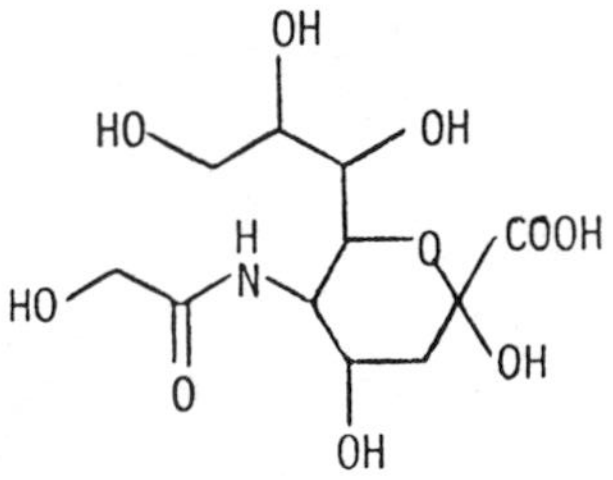	*N*-Glycolylneuraminic acid	Sea urchins: *Anthocidaris crassispina* *Hemicentrotus pulcherrimus* *Pseudocentrotus depressus*		+		124 202

Compound	Name	Source		Ref.
C_{12}				
116. $C_{12}H_{10}O_7$	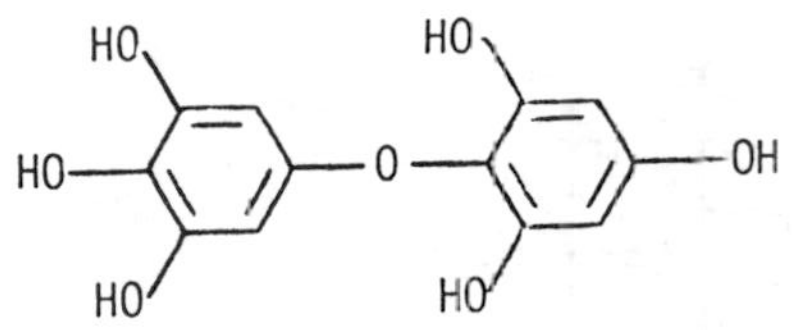 Bifuhalol	Brown alga: *Bifurcaria bifurcata*		96
117. $C_{12}H_{14}N_6$	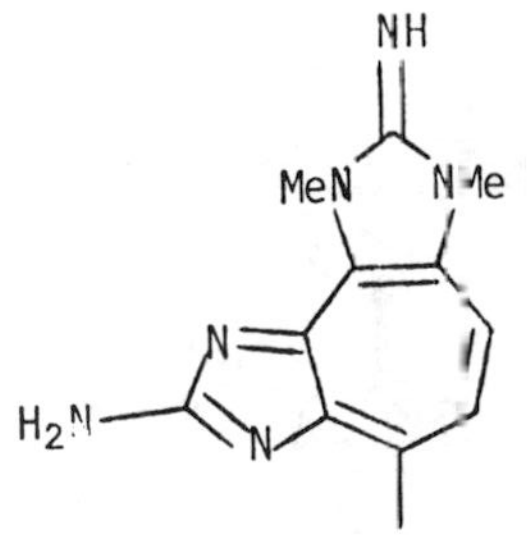 Pseudozoanthoxanthin	Zoanthid: *Epizoanthus arenaceus*		42
118. $C_{12}H_{21}N_3O_3$	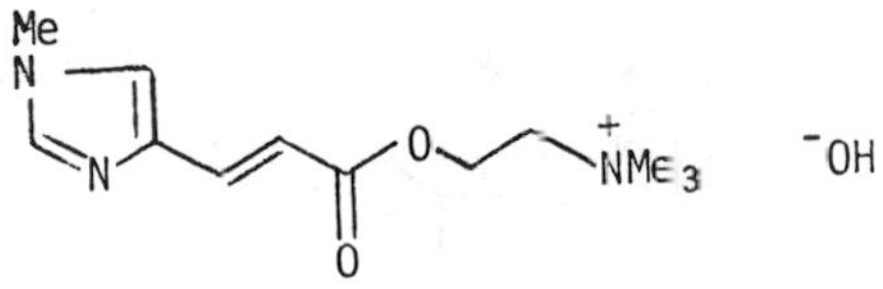 *N*-Methylmurexine	Mollusc: *Nucella emarginata*		15
C_{13}				
119. $C_{13}H_8Br_4O_2$	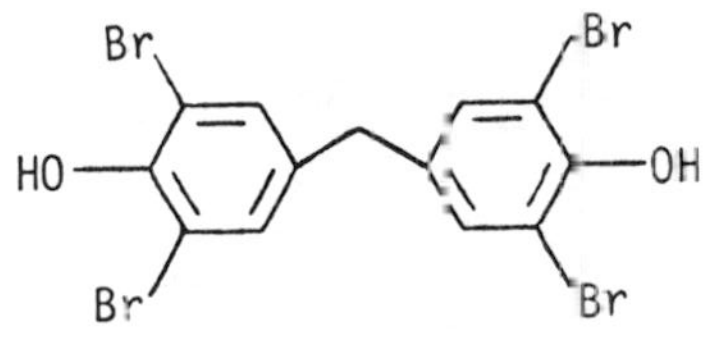 Bis(3,5-dibromo-4-hydroxyphenyl)methane	Sea worm: *Thelepus setosus*	+	119

Table 2 (continued)

Molecular formula	Structural formula	Name	Source of compound	Available by synthesis	Available nonmarine sources	Activity	References
C_{11}							
120. $C_{13}H_{12}Br_2N_2O_5$	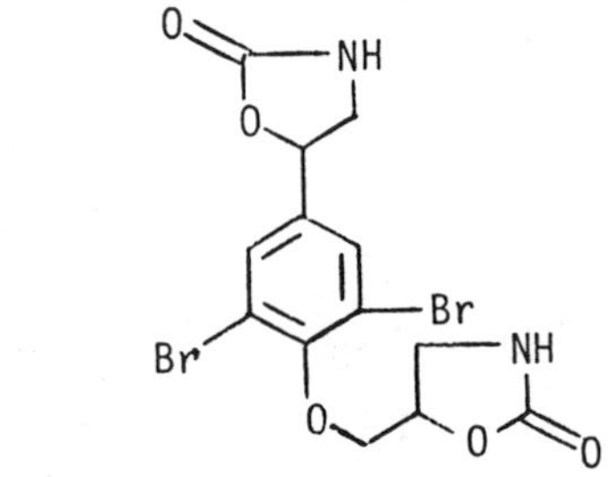	LL-PAA216	Sponge: *Verongia lacunosa*				29
121. $C_{13}H_{16}N_6$	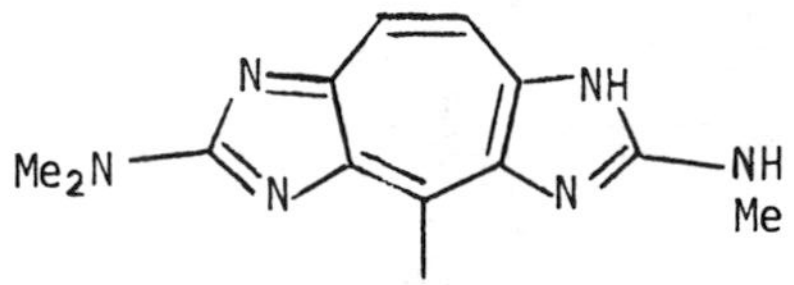	Epizoanthoxanthin A	Zoanthid: *Epizoanthus arenaceus*				41 42
122. $C_{13}H_{28}$		Tridecane	Brown algae: *Fucus distichus* *F. vesiculosus*				283
C_{14}							
123. $C_{14}H_9Br_3O_3$	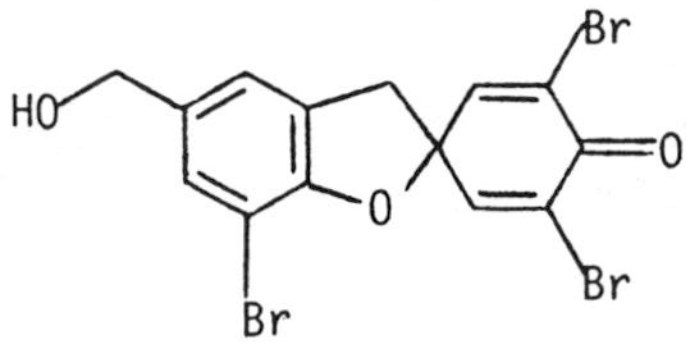	Thelepin	Sea worm: *Thelepus setosus*				119

No. / Formula	Structure	Compound	Source		Ref.
124. $C_{14}H_{11}Br_3O_3$	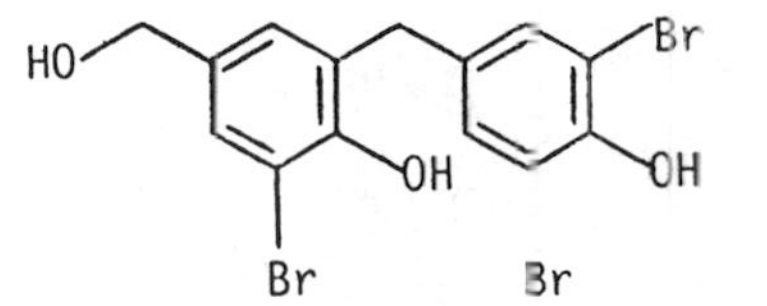	2,4′-Dihydroxy-5-hydroxymethyl-3,3′,5′-tribromodiphenyl methane	Sea worm: *Thelepus setosus*	+	119
125. $C_{14}H_{18}N_6$	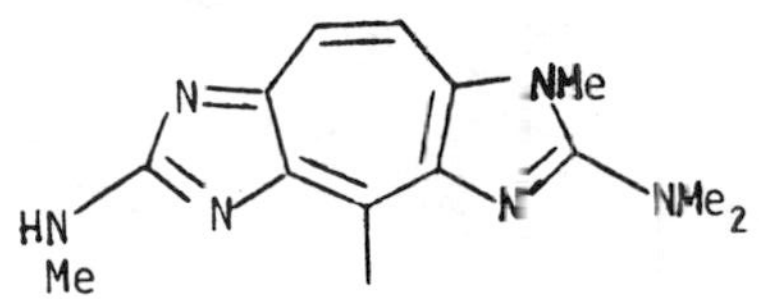	Epizoanthoxanthin B	Zoanthid: *Epizoanthus arenaceus*		42
126. $C_{14}H_{18}N_6$	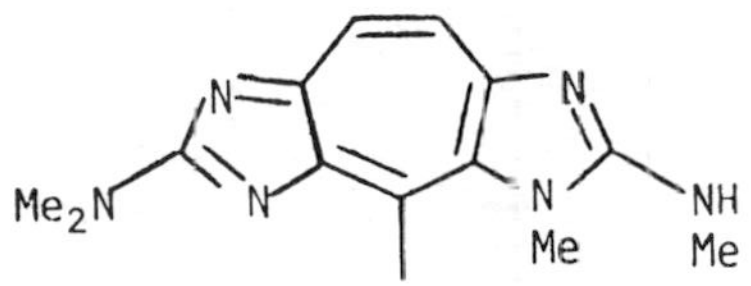	Parazoanthoxanthin E	Zoanthid: *Parazoanthus axinellae*		41
127. $C_{14}H_{18}N_6$	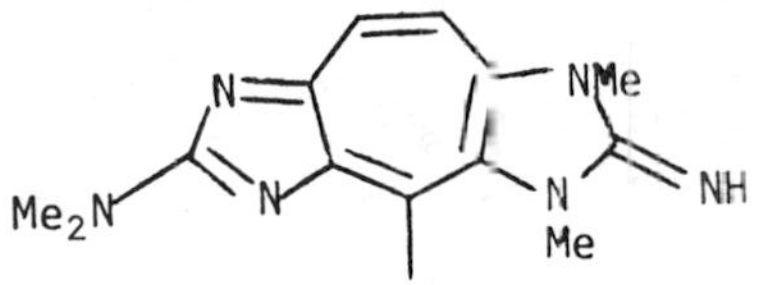	Parazoanthoxanthin F	Zoanthid: *Parazoanthus axinellae*		41
128. $C_{14}H_{23}NC_9$	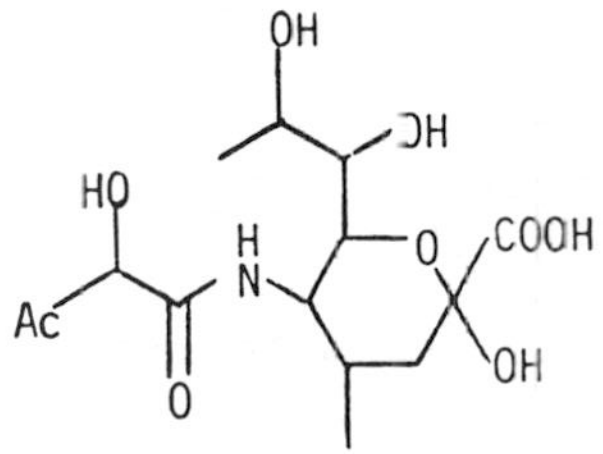	*N*-Acetoglycolyl-4-methyl-4,9-dideoxyneuraminic acid	Sea urchin: *Pseudocentrotus depressus*		124
129. $C_{14}H_{30}$	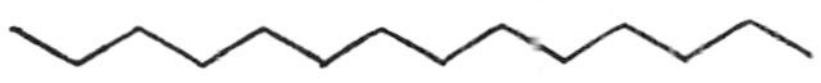	Tetradecane	Brown alga: *Fucus distichus*		283

Table 2 (continued)

Molecular formula	Structural formula	Name	Source of compound	Available by synthesis	Available nonmarine sources	Activity	References
130. $C_{14}H_{30}O$	 OH	Myristyl alcohol Tetradecanol	Coelenterates	+			194 202
C_{15} 131. $C_{15}H_{18}BrClO_2$	Cl, Br, O, O	Rhodophytin	Red alga: *Laurencia* sp.				80
132. $C_{15}H_{19}BrO$	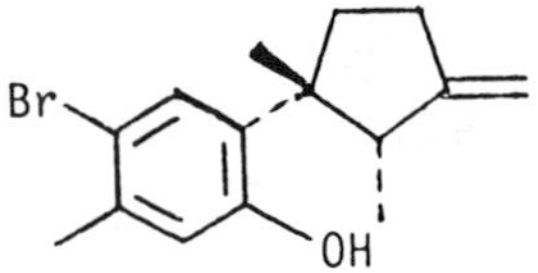 Br, OH	Allo-laurinterol	Red alga: *Laurencia filiformis* f. *dendritica*				162
133. $C_{15}H_{19}BrO$	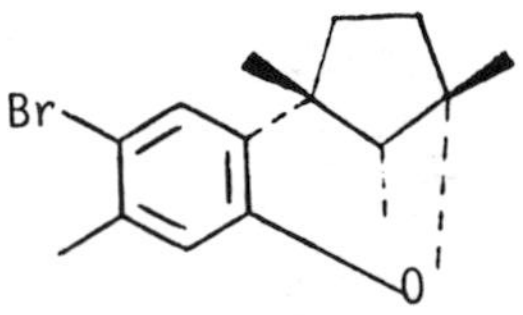 Br, O	Filiformin	Red alga: *Laurencia filiformis* f. *dendritica*				162

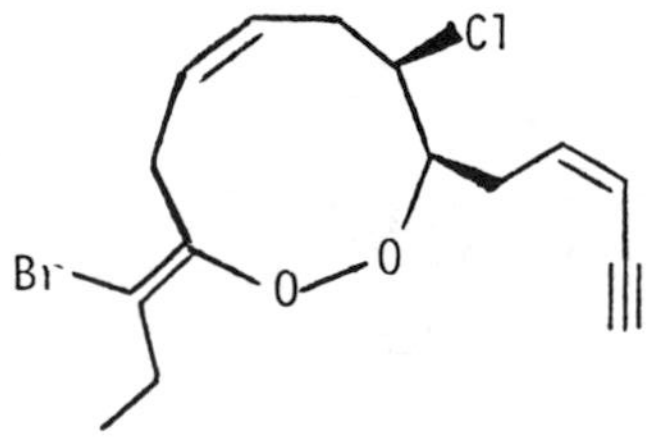

	Name	Source	Ref.
134. $C_{15}H_{19}BrO_2$	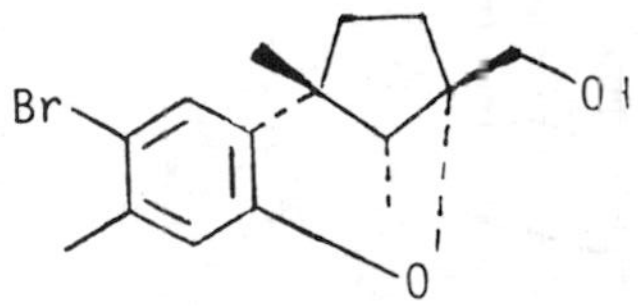 Filiforminol	Red alga: *Laurencia filiformis* f. *dentritica*	162
135. $C_{15}H_{19}Br_2ClO$	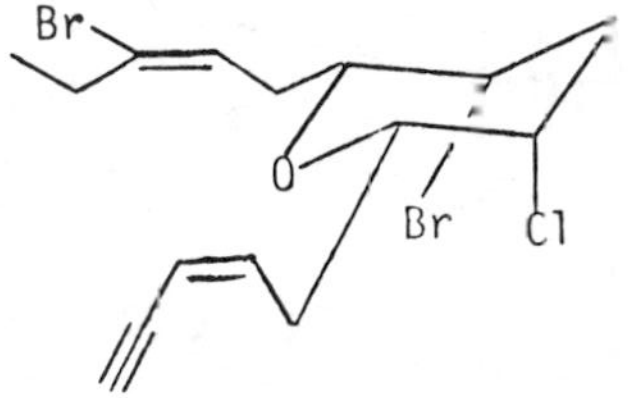 Dactylyne	Sea hare: *Aplysia dactylomela*	185 244
136. $C_{15}H_{20}Br_2O_2$	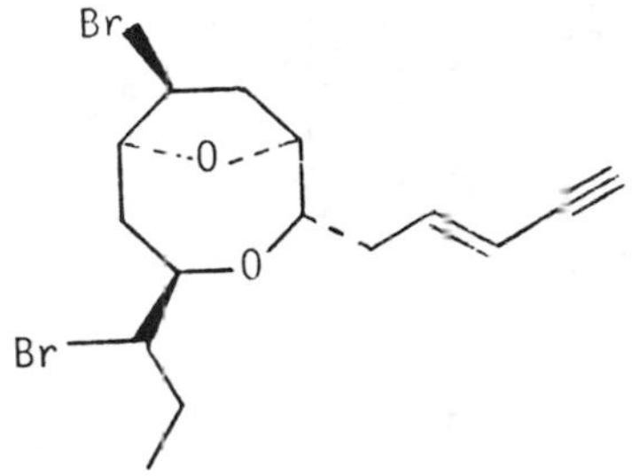 T-Isolaureatin 3-*trans*-Isolaureatin	Red alga: *Laurencia nipponica*	177
137. $C_{15}H_{20}Br_2O_2$	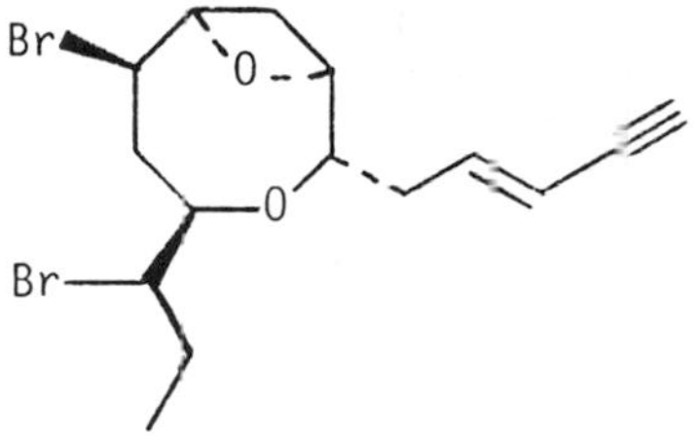 T-Laureatin 3-*trans*-Laureatin	Red alga: *Laurencia nipponica*	177

Table 2 (continued)

Molecular formula	Structural formula	Name	Source of compound	Available by synthesis	Available nonmarine sources	Activity	References
C_{15}							
138. $C_{15}H_{21}BrO_2$	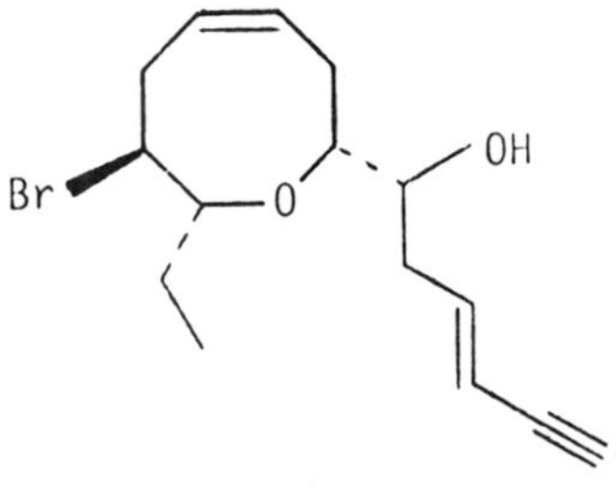	Deacetyllaurencin	Red alga: *Laurencia nipponica*				177
139. $C_{15}H_{21}Br_2ClO_2$	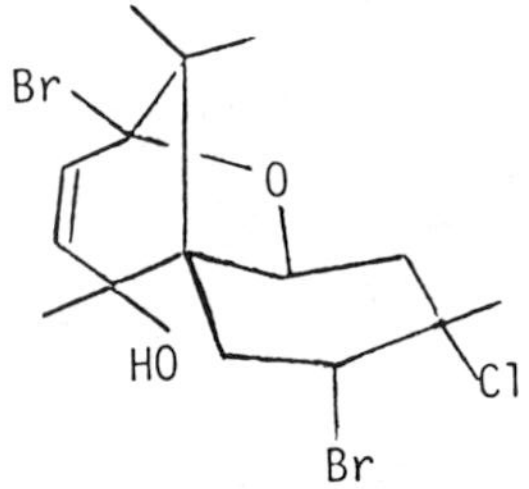	Pacifenol	Red algae: *Laurencia tasmanica* *L. nidifica* Sea hare: *Aplysia californica*				246 251 270
140. $C_{15}H_{21}Br_2ClO_3$	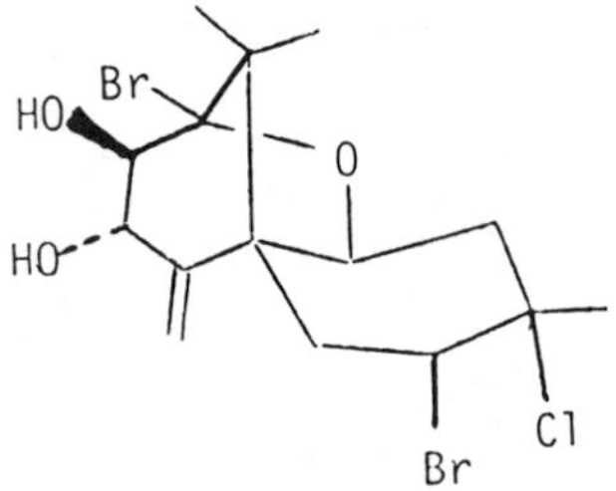	2,7-Dibromo-8-chloro-octahydro-8,10,10-trimethyl-5-methylene-6H-2,5a-methano-1-benzoxepin-3,4-diol	Sea hare: *Aplysia californica*				78

Compound	Name	Source	Ref.
141. $C_{15}H_{21}Br_2ClO_3$	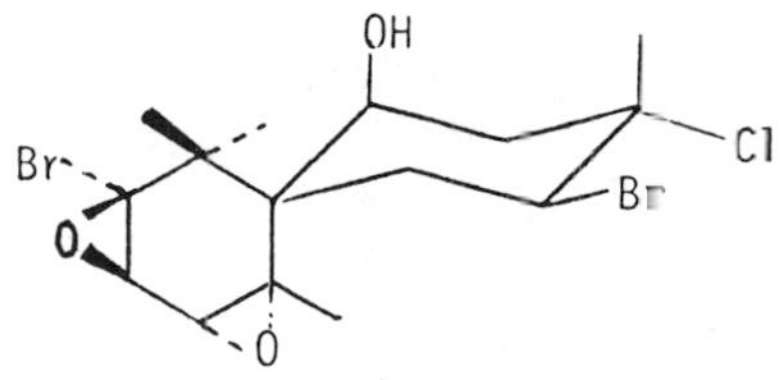 Prepacifenol epoxide	Red alga: *Laurencia johnstonii*; Sea hare: *Aplysia californica*	78
142. $C_{15}H_{22}$	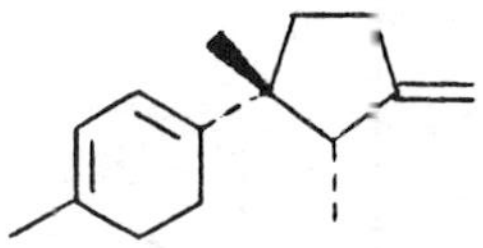 Dihydrolaurene	Red alga: *Laurencia filiformis* f. *dendritica*	162
143. $C_{15}H_{22}BrCl$	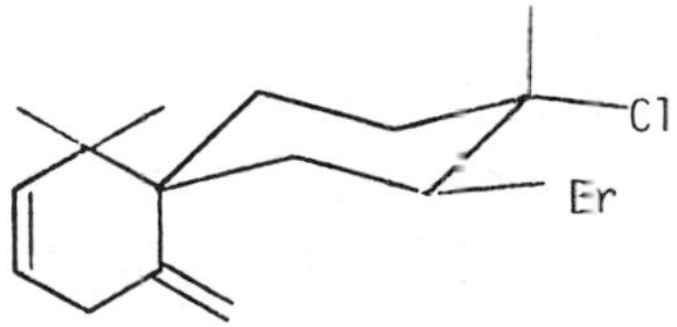 Nidificiene	Red alga: *Laurencia nidifica*	270
144. $C_{15}H_{22}BrClO$	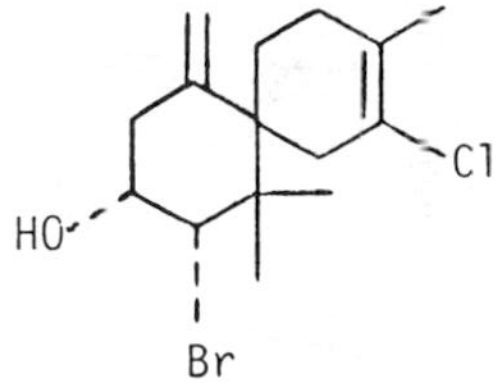 Elatol	Red alga: *Laurencia elata*	247
C_{15} 145. $C_{15}H_{22}O_2$	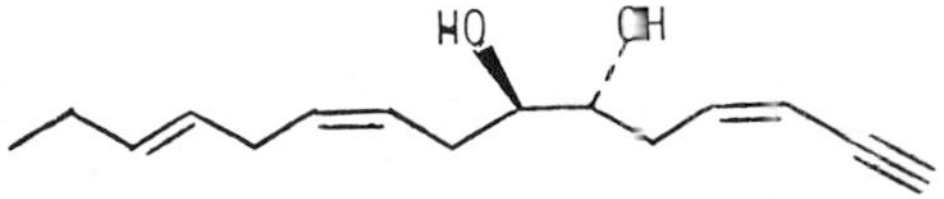 6*R*, 7*R* *cis*-Laurediol	Red alga: *Laurencia nipponica*	177

Table 2 (continued)

Molecular formula	Structural formula	Name	Source of compound	Available by synthesis	Available nonmarine sources	Activity	References
C_{15}							
146. $C_{15}H_{22}O_2$	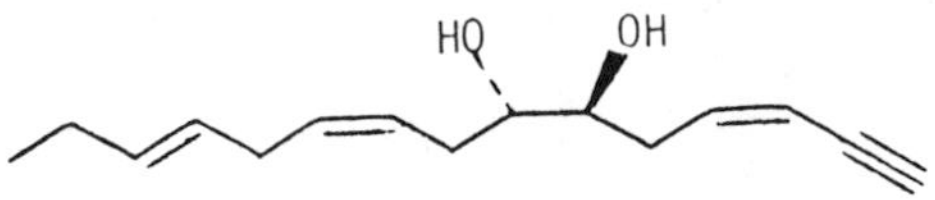	6*S*, 7*S* *cis*-Laurediol	Red alga: *Laurencia nipponica*				177
147. $C_{15}H_{22}O_2$	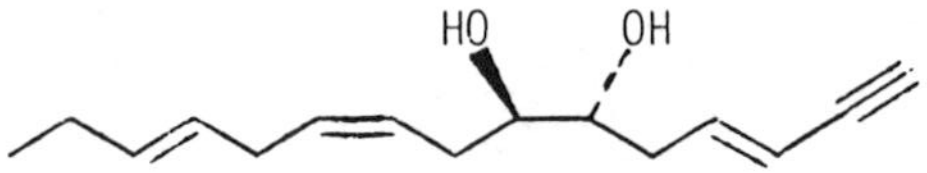	6*R*, 7*R* *trans*-Laurediol	Red alga: *Laurencia nipponica*				177
148. $C_{15}H_{22}O_2$	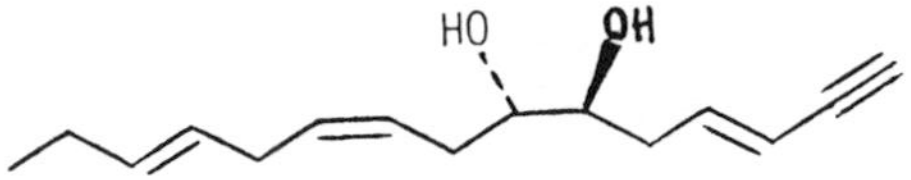	6*S*, 7*S* *trans*-Laurediol	Red alga: *Laurencia nipponica*				177
149. $C_{15}H_{23}BrO$	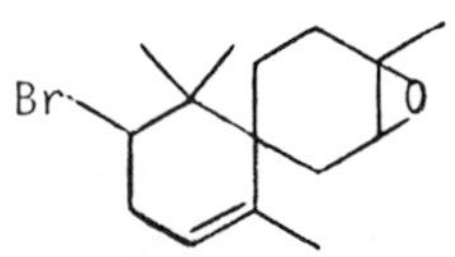	4-Bromo-α-chamigren-8,9-epoxide	Red alga: *Laurencia glanduli-fera*				255
150. $C_{15}H_{23}BrO$	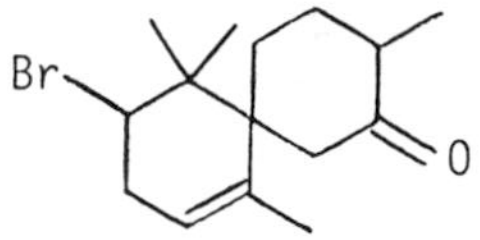	4-Bromo-α-chamigren-8-one	Red alga: *Laurencia glanduli-fera*				255

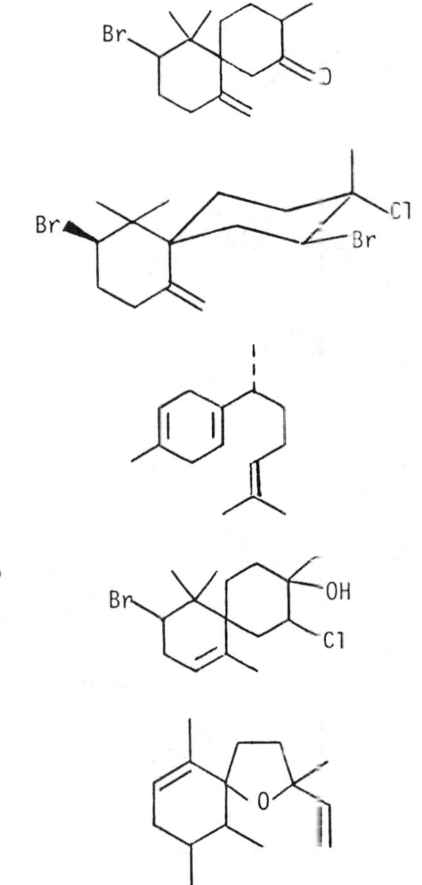

No.	Formula	Name	Source	Ref.
151.	$C_{15}H_{23}BrC$	4-Bromo-β-chamigren-8-one	Red alga: *Laurencia glanduli-fera*	255
152.	$C_{15}H_{23}Br_2Cl$	Nidificene	Red alga: *Laurencia nidifica*	270
153.	$C_{15}H_{24}$	(−)-β-Curcumene	Gorgonians: *Muricea elongata*, *Plexaurella nutans*	138
154.	$C_{15}H_{24}BrClO$	Glanduliferol	Red alga: *Laurencia glanduli-fera*	256
155.	$C_{15}H_{24}O$	Dactyloxene-B	Sea hare: *Aplysia dactylomela*	243

Table 2 (continued)

Molecular formula	Structural formula	Name	Source of compound	Available by synthesis	Available nonmarine sources	Activity	References
C_{15}							
156. $C_{15}H_{24}O_3$	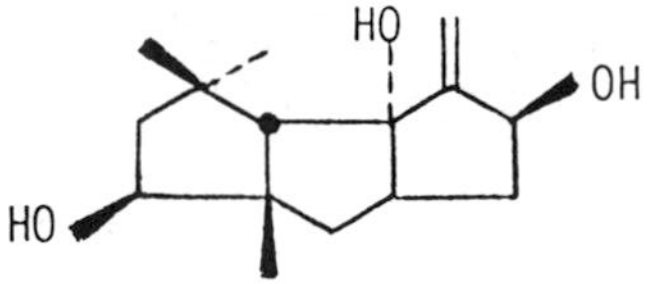	Δ9(12)-Capnellene-3β, 8β,10α-triol	Soft coral: *Capnella imbricata*				141
157. $C_{15}H_{25}Br_2ClO_2$	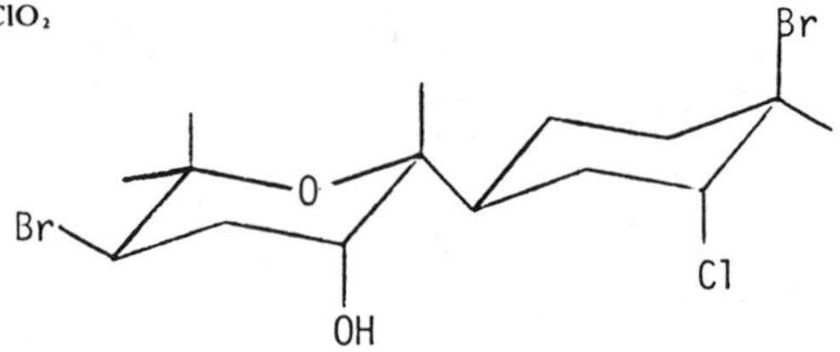	Isocaespitol	Red alga: *Laurencia caespitosa*				99 101
158. $C_{15}H_{26}O$	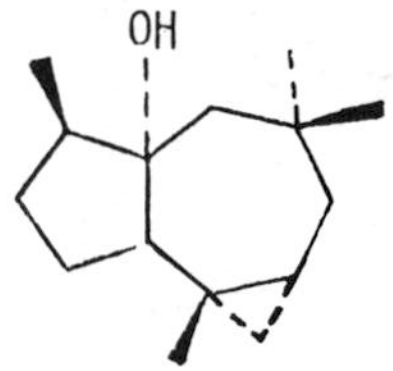	Africanol	Soft coral: *Lemnalia africana*				266
159. $C_{15}H_{26}O$	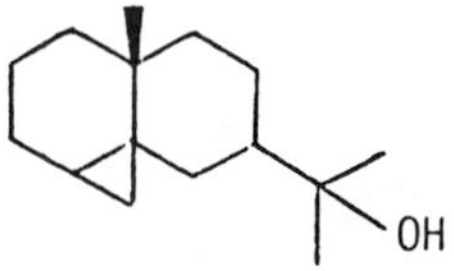	Cycloeudesmol	Red alga: *Chondria oppositiclada*			Antibiotic	81

Compound	Name	Source		Activity	Ref.
160. $C_{15}H_{32}$	 Pentadecane	Algae			283
161. $C_{15}H_{32}O$	 Pentadecanol	Coelenterates			194
162. $C_{15}H_{32}O$	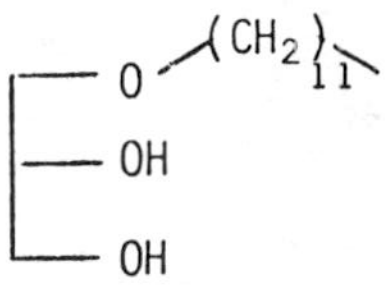 1-*O*-Dodecylglycerol	Cod liver oil Herring: *Clupea harengus* Mussel *Mytilus edulis*	+		114
C_{16}					
163. $C_{16}H_{22}O$	2-Diprenyl-1,4-benzo-quinol	Tunicates: *Aplidium crateri-ferum* *Aplidium* sp.			83 162
164. $C_{16}H_{25}N$	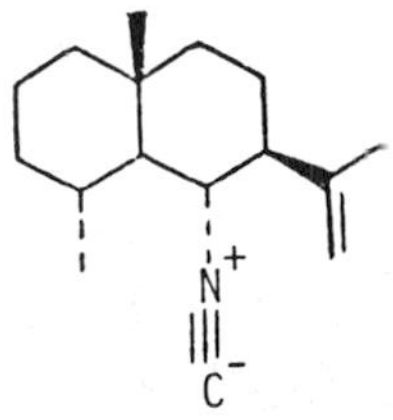 Acanthellin-1	Sponge: *Acanthella acuta*		Antibacterial	206
165. $C_{16}H_{25}N$	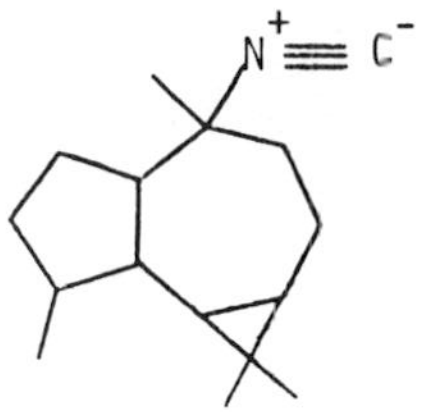 Axisonitrile-2	Sponge: *Axinella cannabina*			77

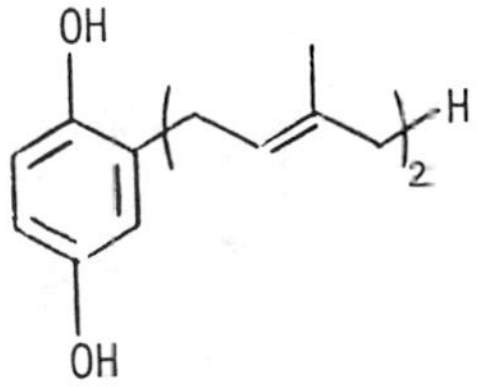

Table 2 (continued)

Molecular formula	Structural formula	Name	Source of compound	Available by synthesis	Available nonmarine sources	Activity	References
C_{16}							
166. $C_{16}H_{25}N$	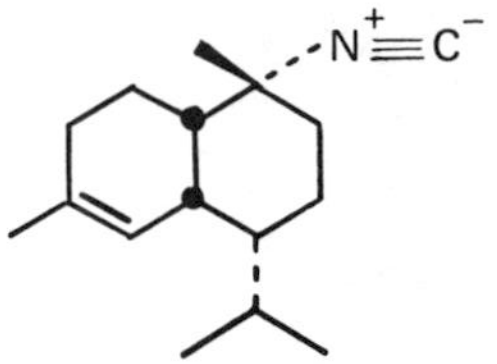	1,2,3,4,4a,7,8,8a-Octahydro-4-isopropyl-1,6-dimethyl-1-naphthyl isocyanide	Sponge: *Halichondria* sp.				36 37
167. $C_{16}H_{25}NS$	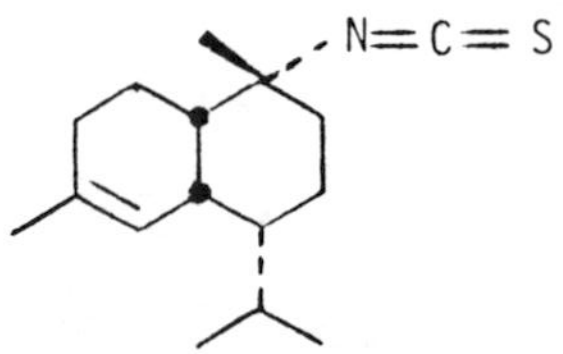	1,2,3,4,4a,7,8,8a-Octahydro-4-isopropyl-1,6-dimethyl-1-naphthyl isothiocyanate	Sponge: *Halichondria* sp.				36 37
168. $C_{16}H_{26}O_2$	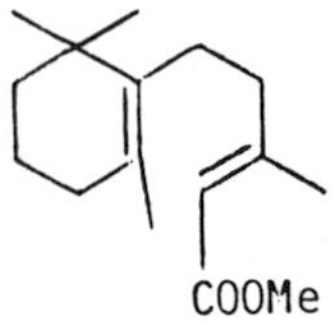 COOMe	Methyl-*trans*-monocyclofarnesate	Sponge: *Halichondria panicea*				45
169. $C_{16}H_{27}NO$	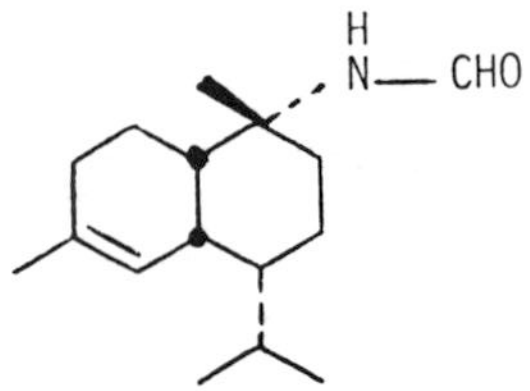 H N— CHO	*N*-(1,2,3,4,4a,7,8,8a-Octahydro-4-isopropyl-1,6-dimethyl-1-naphthyl)formamide	Sponge: *Halichondria* sp.				36 37

170. $C_{16}H_{34}$	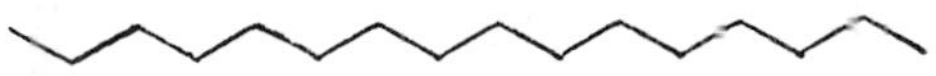 Hexadecane	Algae: *Ascophyllum nodosum* *Fucus vesiculosus* *Rhodomela confervoides*			283
171. $C_{16}H_{34}O$	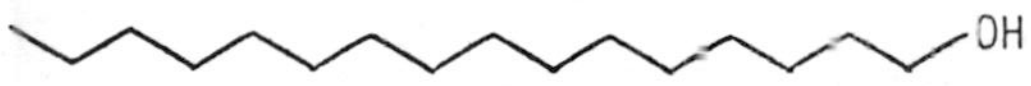 Cetyl alcohol Hexadecanol	Algae Coelenterates	+	+	194 202
C_{17} 172. $C_{17}H_{15}N_3O_4S$	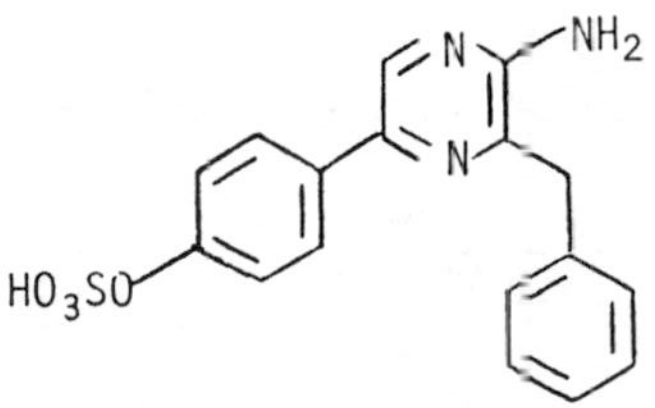 AF-350 monosulfate	Squid: *Watasenia scintillans*	+		109
173. $C_{17}H_{20}N_4O_6$	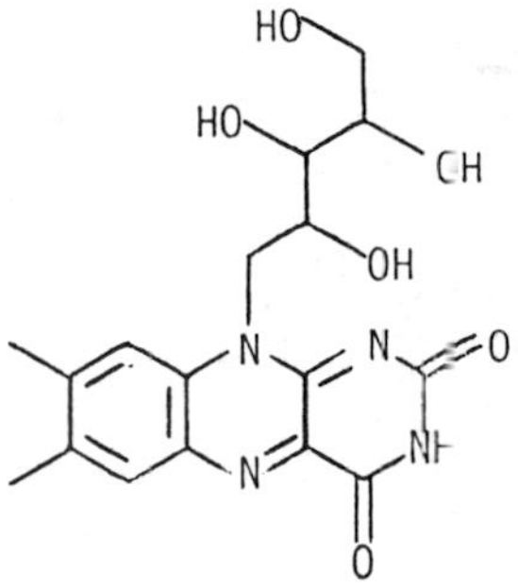 Riboflavine	Copepods Fish	+	+	60 202 210

Table 2 (continued)

Molecular formula	Structural formula	Name	Source of compound	Available by synthesis	Available nonmarine sources	Activity	References
C_{17}							
174. $C_{17}H_{27}Br_2ClO_3$	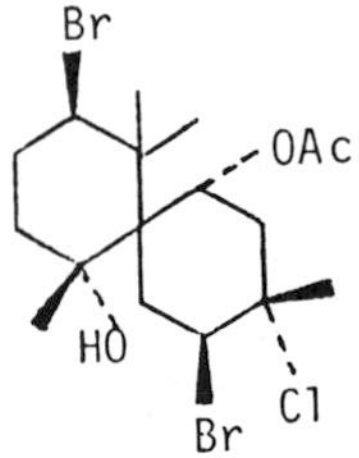	Acetoxyintricatol	Red alga: *Laurencia intricata*				187
175. $C_{17}H_{34}O_2$	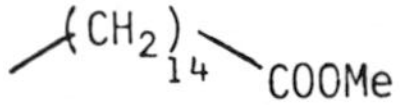	Methylpalmitate	Red alga: *Desmia hornemanni*				126
176. $C_{17}H_{34}O_4$	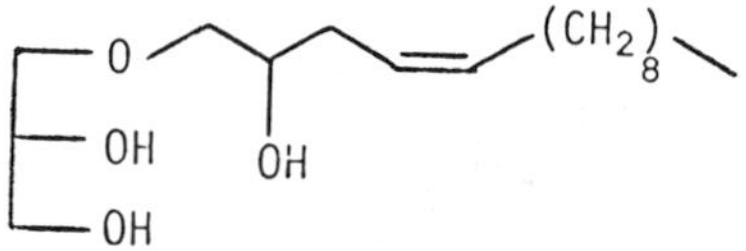	1-*O*-(2-Hydroxy-4-tetradecenyl)glycerol	Shark: *Somniosus microcephalus*				113 115
177. $C_{17}H_{36}$		Heptadecane	Algae				85,283
178. $C_{17}H_{36}O$	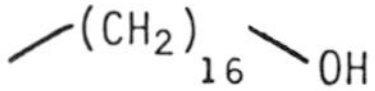	Heptadecanol	Coelenterates				194
179. $C_{17}H_{36}O_3$	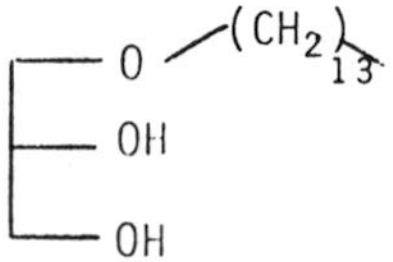	1-*O*-Tetradecylglycerol	Cod liver oil Crayfish Herrings Mackerel Mussel Shrimp		+		114

Compound	Name	Source			Ref.
180. $C_{17}H_{36}O_4$ 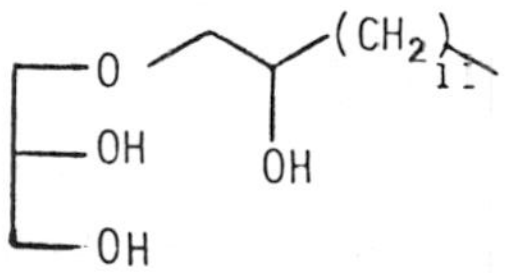	1-*O*-(2-Hydroxytetradecyl)glycerol	Shark: *Somniosus microcephalus*			115
C_{18} 181. $C_{18}H_{14}O_{10}$	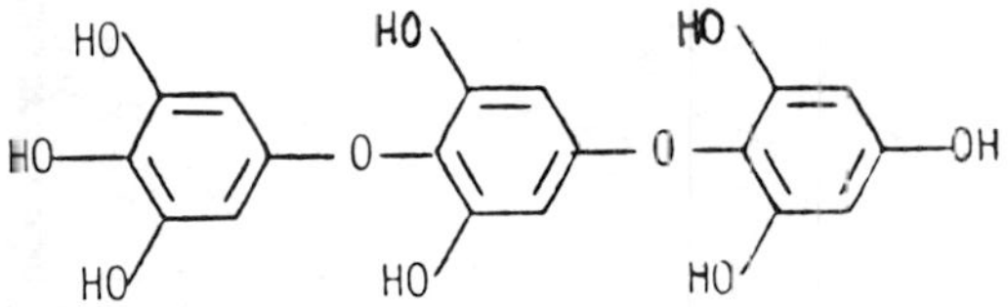 Trifuhalol	Brown alga: *Halidrys siliquosa*	+		95
182. $C_{18}H_{22}O_2$	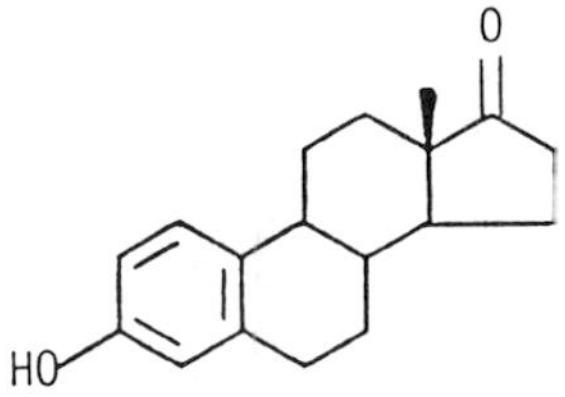 Estrone	Dogfish: *Squalus suckleyi* Mollusc: *Pecten hericius*	+	+	32 202 273
183. $C_{18}H_{24}O_2$	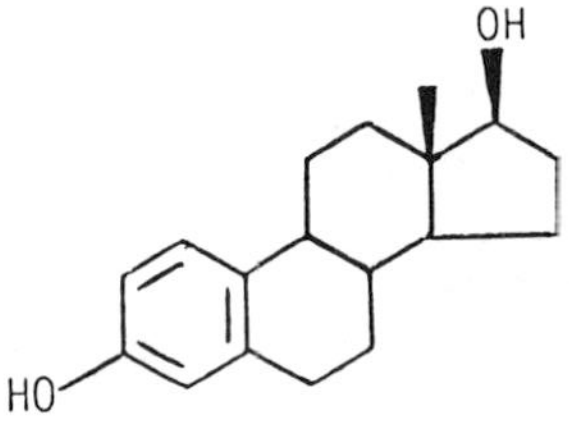 β-Estradiol	Dogfish: *Squalus suckleyi* Mollusc: *Pecten hericius* Sea urchin: *Strongylocentrotus franciscanus* Starfish: *Pisaster ochraceous*	+	+	31 32 202 273
184. $C_{18}H_{36}O$	 Oleyl alcohol *cis*-9-Octadecen-1-ol	Coelenterates	+		194 202

Table 2 (continued)

Molecular formula	Structural formula	Name	Source of compound	Available by synthesis	Available nonmarine sources	Activity	References
C_{18}							
185. $C_{18}H_{38}$	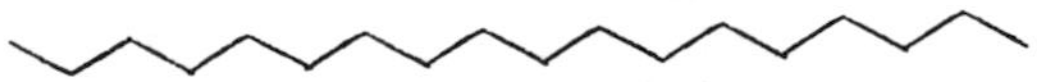	Octadecane	Algae				283
186. $C_{18}H_{38}O$	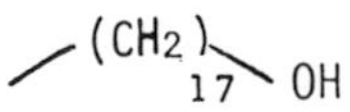 $(CH_2)_{17}$ OH	Stearyl alcohol Octadecanol	Algae Coelenterates	+			194 202
187. $C_{18}H_{38}O_3$	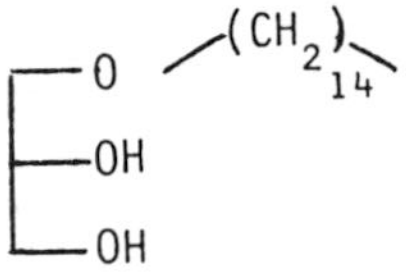 O $(CH_2)_{14}$; OH; OH	1-*O*-Pentadecylglycerol	Cod liver oil Crayfish Herrings Mackerel Mussel Shrimp		+		114
188. $C_{18}H_{38}O_4$	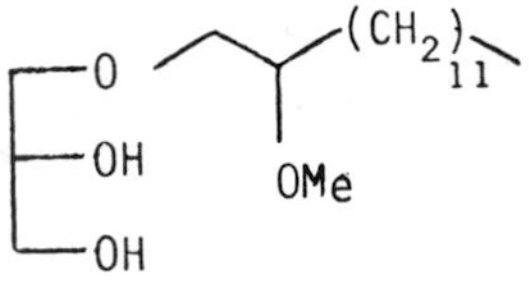 O $(CH_2)_{11}$; OMe; OH; OH	1-*O*-(2-Methoxytetradecyl)glycerol	Cod liver oil Crayfish: *Nephrops norvegicus* Herring: *Clupea harengus*	+			114
C_{19}							
189. $C_{19}H_{30}$	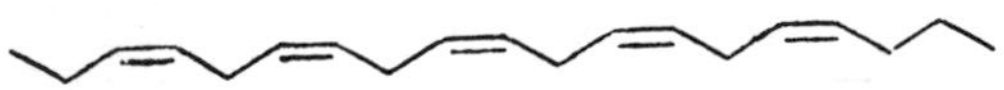	*cis, cis, cis, cis, cis*-3,6,9,12,15-Nonadecapentaene	Brown algae: *Porphyra leucostica* *Pylaiella littoralis* *Scytosiphon lomentaria*				283
190. $C_{19}H_{32}$		*cis, cis, cis, cis*-4,7,10,13-Nonadecatetraene	Algae: *Porphyra leucostica* *Scytosiphon lomentaria*				283

Formula	Name	Source			Ref.
191. $C_{19}H_{38}O_4$	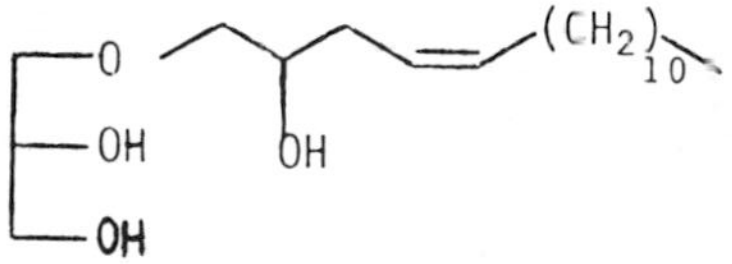 1-*O*-(2-Hydroxy-4-hexadecenyl)glycerol	Shark: *Somniosus microcephalus*	+		113 115
192. $C_{19}H_{40}$	 Nonadecane	Algae			283
193. $C_{19}H_{40}O$	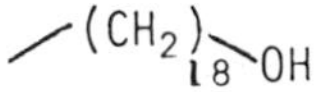 Nonadecanol	Coelenterates			194
194. $C_{19}H_{40}O_3$	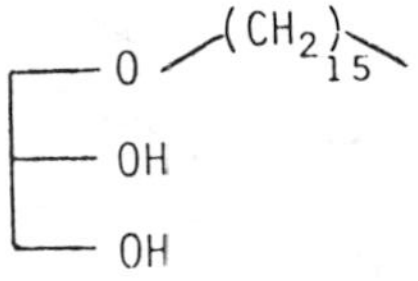 Chimyl alcohol 1-*O*-Hexadecylglycerol	Algae Cod liver oil Coelenterates Crayfish Herrings Mackerel Mussel Shrimp	+	+	114 194 202
195. $C_{19}H_{40}O_4$	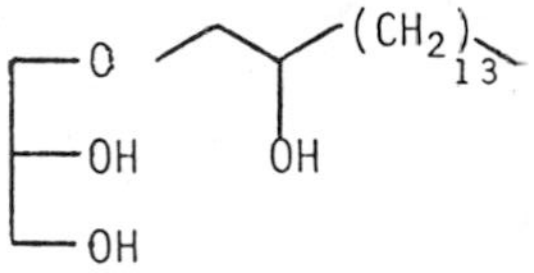 1-*O*-(2-Hydroxyhexadecyl)glycerol	Shark: *Somniosus microcephalus*	+		115
196. $C_{19}H_{40}O_4$	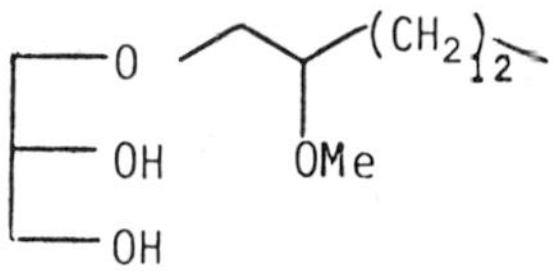 1-*O*-(2-Methoxypentadecyl)glycerol	Cod liver oil Crayfish Herrings Mackerel	+		114

Table 2 (continued)

Molecular formula	Structural formula	Name	Source of compound	Available by synthesis	Available nonmarine sources	Activity	References
C_{20}							
197. $C_{20}H_{13}N_3O_3$	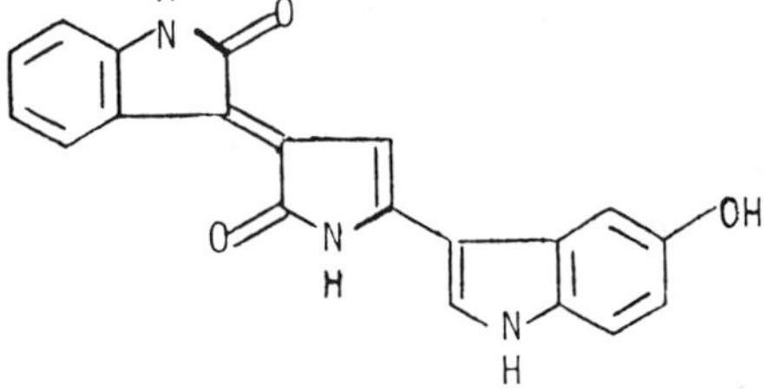	Violacein	Bacterium: *Chromobacterium* sp.	+			6 202
198. $C_{20}H_{17}N_3O_2$	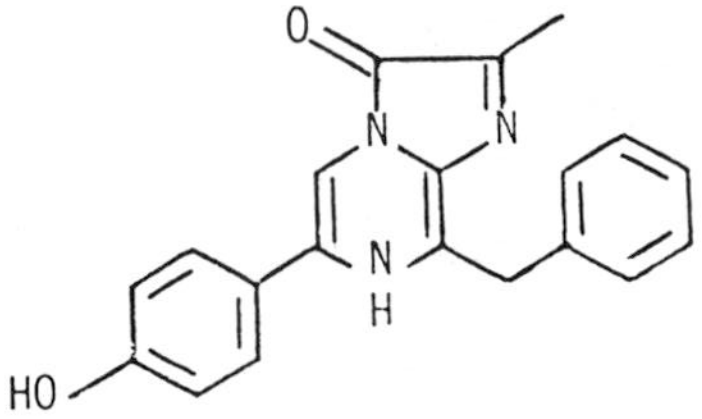	Renilla luciferin	Soft coral: *Renilla reniformis*	+			53,54 55,56 57,104 121,122 123,184
199. $C_{20}H_{24}O_4$	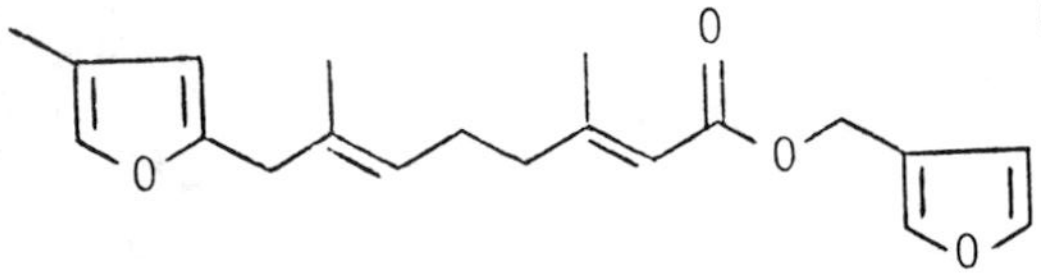	Pleraplysillin-2	Sponge: *Pleraplysilla spinifera*				48

No. / Formula	Name	Source	Activity	Ref.
200. $C_{20}H_{28}O_3$	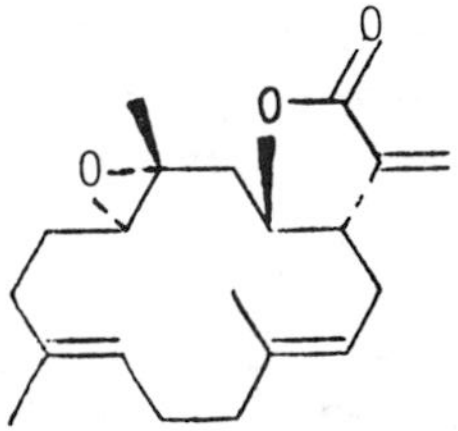 Lobophytolide	Soft coral: *Lobophytum cristagalli*		267
201. $C_{20}H_{28}O_3$	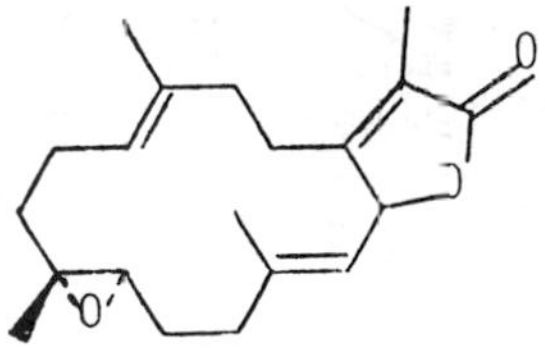 2,3,6,7,10a,13,14,14a-Octahydro-1a,5,8,12-tetramethyloxireno[9,10]cyclotetradeca[1,2-b]furan-9(1aH)-one	Soft coral: *Sarcophyton glaucum*		150
202. $C_{20}H_{28}O_3$	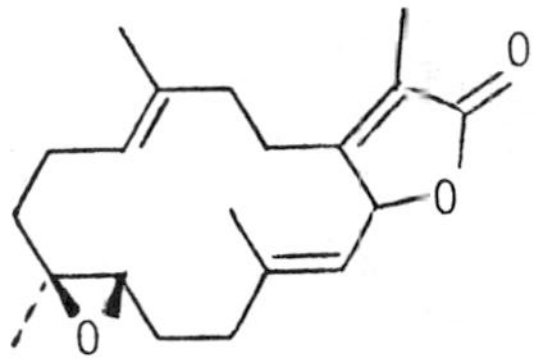 2,3,6,7,10a,13,14,14a-Octahydro-1a,5,8,12-tetramethyloxireno[9,10]cyclotetradeca[1,2-b]furan-9(1aH)-one	Soft coral: *Sarcophyton glaucum*		150
203. $C_{20}H_{28}O_3$	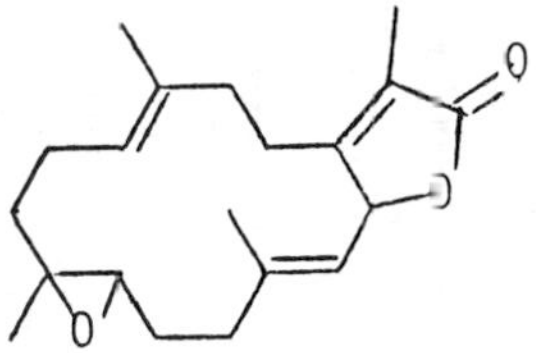 Sarcophine	Soft coral: *Sarcophyton glaucum* *S. trocheliophorum*	Antiacetylcholine Cholinesterase inhibitor	17 162 222

Table 2 (continued)

Molecular formula	Structural formula	Name	Source of compound	Available by synthesis	Available nonmarine sources	Activity	References
C_{20}							
204. $C_{20}H_{28}O_4$	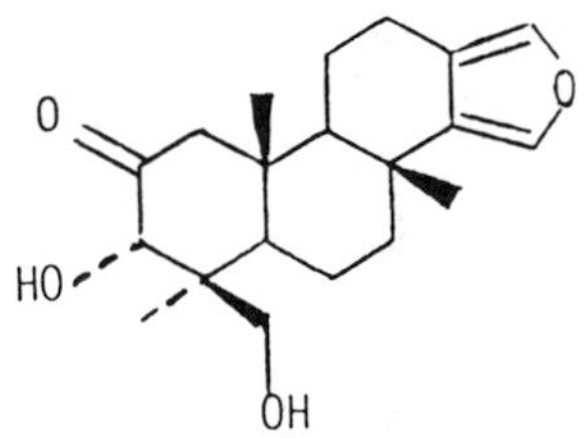	Spongiadiol	Sponges: *Spongia* sp.				162
205. $C_{20}H_{28}O_4$	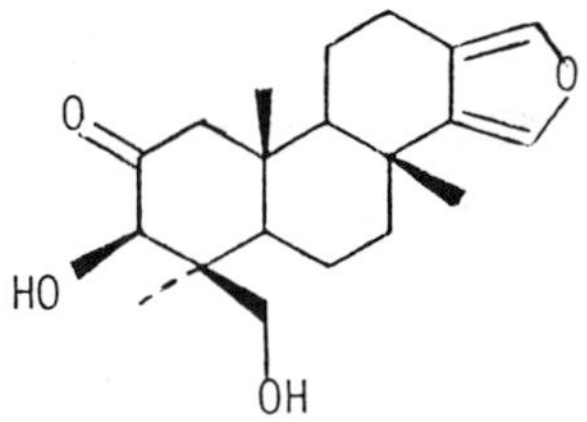	*epi*-Spongiadiol	Sponges: *Spongia* sp.				162
206. $C_{20}H_{28}O_5$	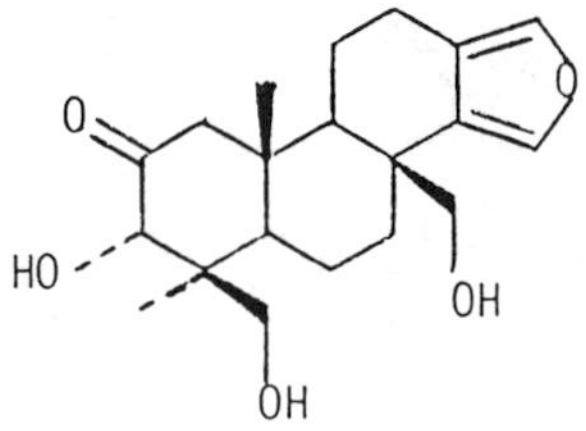	Spongiatriol	Sponges: *Spongia* sp.				162

No.	Name	Source			Ref.
207. $C_{20}H_{28}O_5$	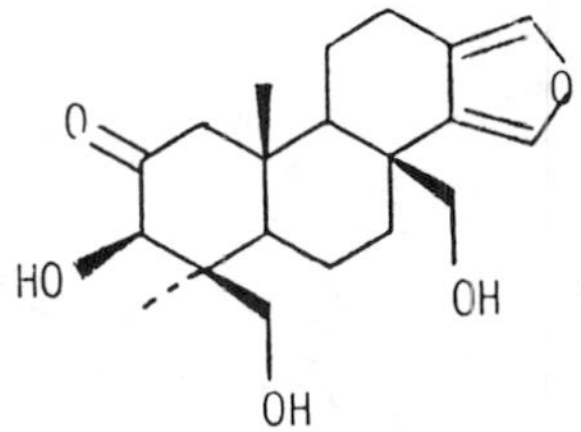 *epi*-Spongiatriol	Sponges *Spongia* sp.			162
208. $C_{20}H_{30}O$	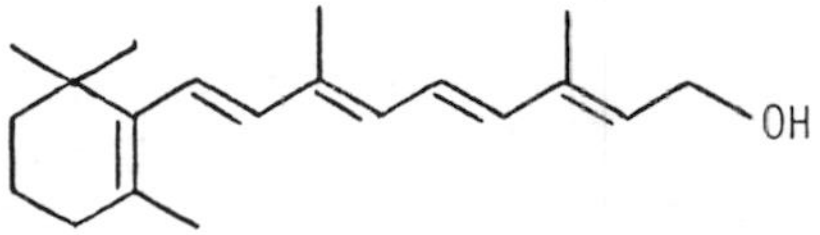 Retinol Vitamin A	Crustaceans Fish	+	+	14,62 86,87 202
209. $C_{20}H_{30}O_2$	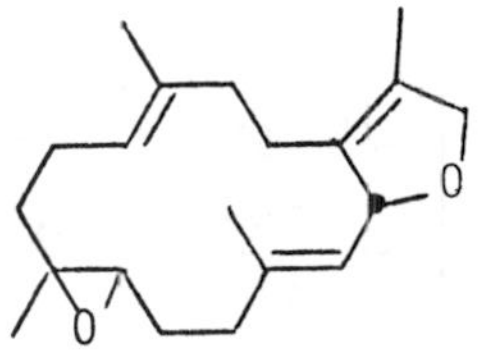 1a,2,3,6,7,9,10a,13,14,14a-Decahydro-1a,5,8,12-tetramethyloxireno-[9,10]cyclotetradeca-[1,2-b]furan	Soft coral: *Sarcophyton glaucum*			150
210. $C_{20}H_{30}O_2$	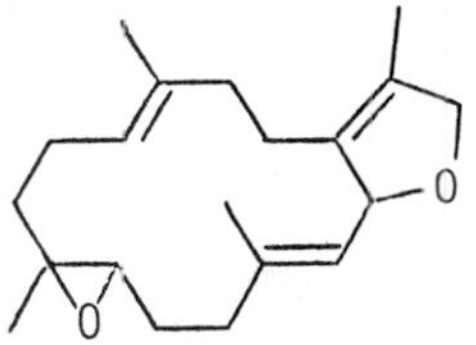 1a,2,3,6,7,9,10a,13,14,14a-Decahydro-1a,5,8,12-tetramethyloxireno-[9,10]cyclotetradeca-[1,2-b]furan	Soft coral: *Sarcophyton glaucum*			150
211. $C_{20}H_{30}O$	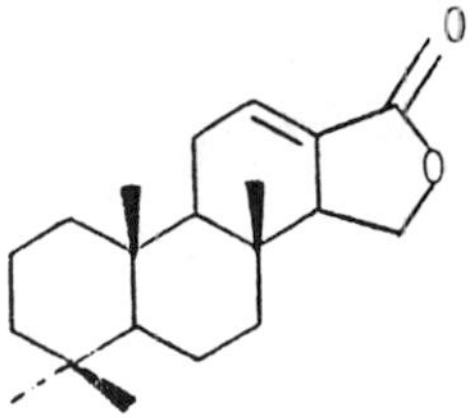 Isoagatholactone	Sponge: *Spongia officinalis*			50

Table 2 (continued)

Molecular formula	Structural formula	Name	Source of compound	Available by synthesis	Available nonmarine sources	Activity	References
C_{20}							
212. $C_{20}H_{30}O_4$	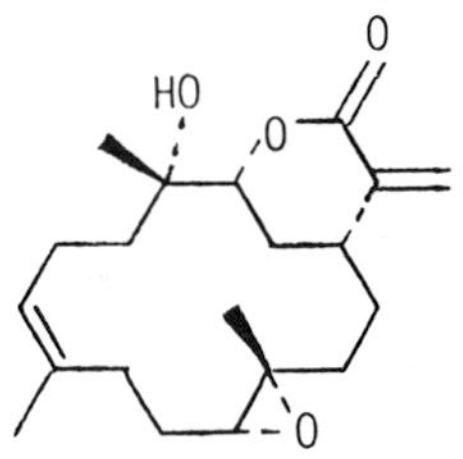	Flexibilide	Soft coral: *Sinularia flexibilis*				161
213. $C_{20}H_{32}O$	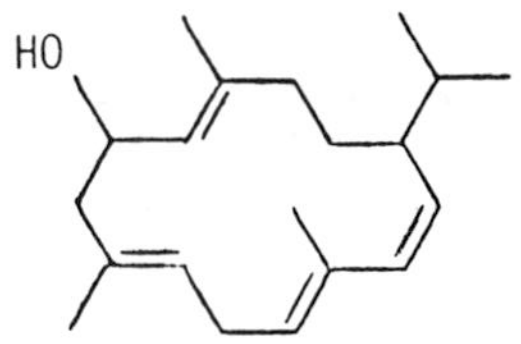	6-Isopropyl-3,9,13-trimethyl-2,7,9,12-cyclotetradecatetraen-1-ol	Soft coral: *Sarcophyton glaucum*				150
214. $C_{20}H_{32}O_2$	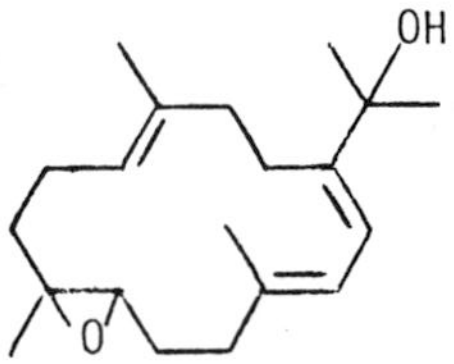	Dehydroepoxynephthenol	Soft coral: *Lobophytum* sp.				51

215. $C_{20}H_{32}O_4$ 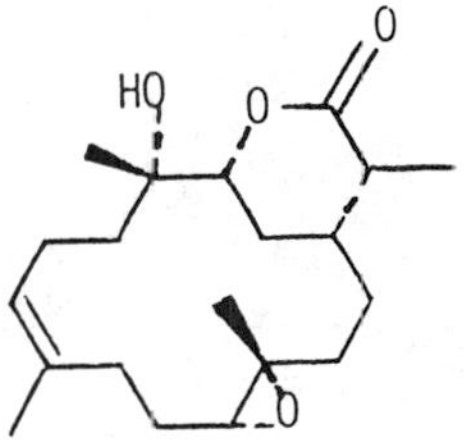	HO, O, O, O	Dihydroflexibilide	Soft coral: *Sinularia flexibilis*		161
216. $C_{20}H_{34}O$ 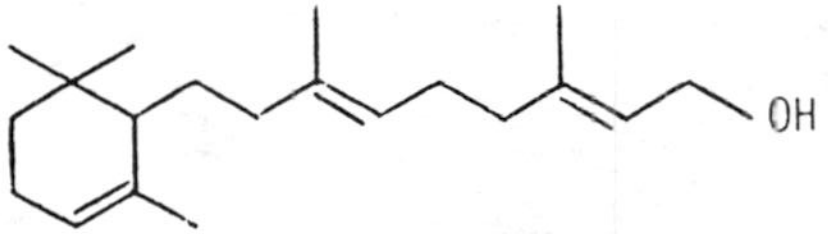	OH	Caulerpol	Green alga: *Caulerpa brownii*		21
217. $C_{20}H_{34}O$ 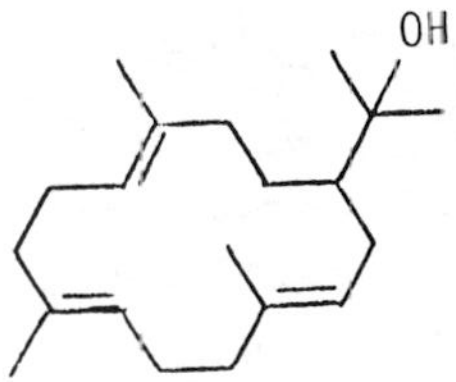	OH	Nephthenol	Soft coral: *Nephthea* sp.		242
218. $C_{20}H_{34}O_5$ 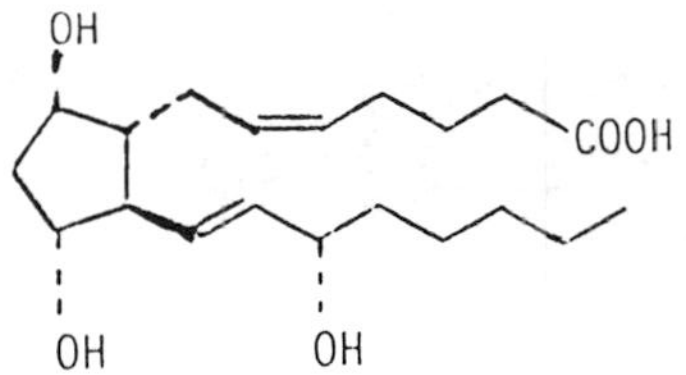	OH, COOH, OH, OH	$PGF_{2\alpha}$	Tuna: *Thunnus thynnus*	+	202 224

Table 2 (continued)

Molecular formula	Structural formula	Name	Source of compound	Available by synthesis	Available nonmarine sources	Activity	References
C_{20}							
219. $C_{20}H_{36}O_5$	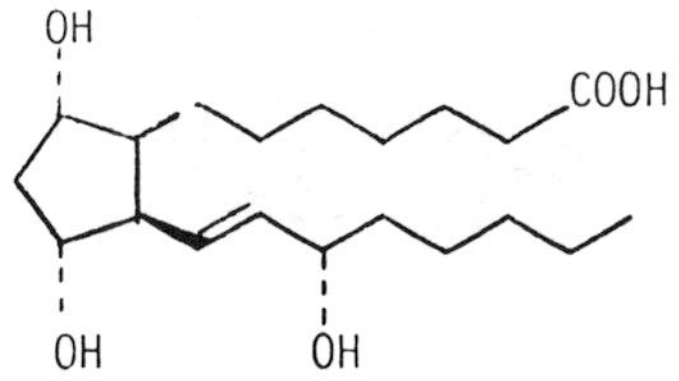 OH, COOH, OH, OH	$PGF_{1\alpha}$	Salmon: *Oncorhynchus keta*		+		202 224
220. $C_{20}H_{40}O$	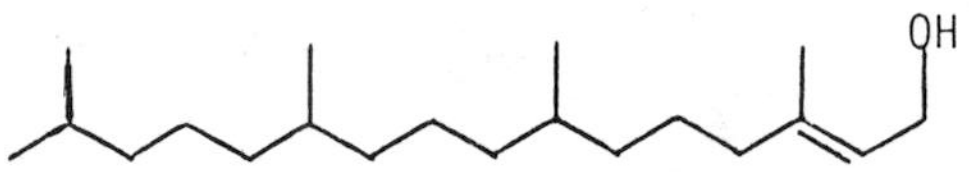 OH	Phytol	Brown alga: *Hizikia fusiformis*	+		Lipase-activator	174 202
221. $C_{20}H_{40}O_4$	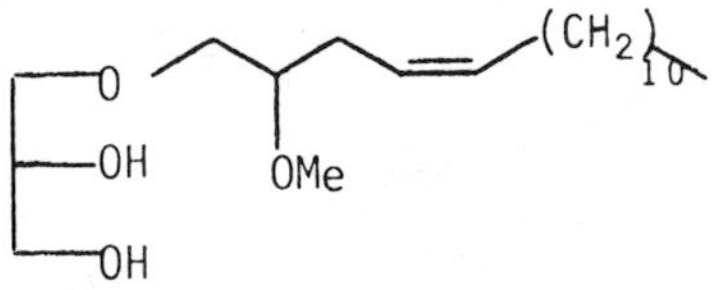 O, OH, OH, OMe, $(CH_2)_{10}$	1-*O*-(2-Methoxy-4-hexadecenyl)glycerol	Shark: *Somniosus microcephalus*	+			113
222. $C_{20}H_{42}$	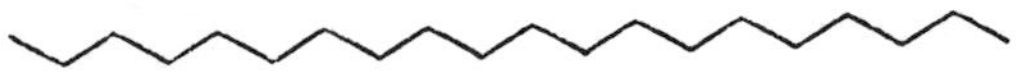	Eicosane	Algae				283
223. $C_{20}H_{42}O$	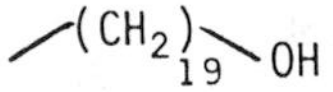 $(CH_2)_{19}$ OH	Eicosanol	Coelenterates				194
224. $C_{20}H_{42}O_3$	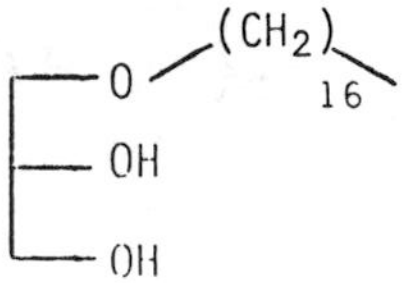 O, OH, OH, $(CH_2)_{16}$	1-*O*-Heptadecylglycerol	Cod liver oil Coelenterates Crayfish Herrings Mackerel Mussel Shrimp		+		114 194

No. / Formula	Compound	Source			Activity	Ref.
225. $C_{20}H_{42}O_4$	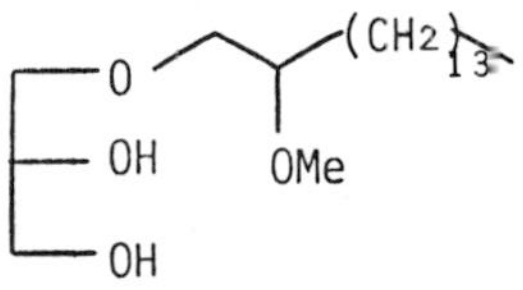 1-*O*-(2-Methoxyhexadecyl)glycerol	Cod liver oil Crayfish Herrings Mackerel Mussel Shark Shrimp	+	+	Antitumor	26 113 114
C_{21} 226. $C_{21}H_{26}O_3$	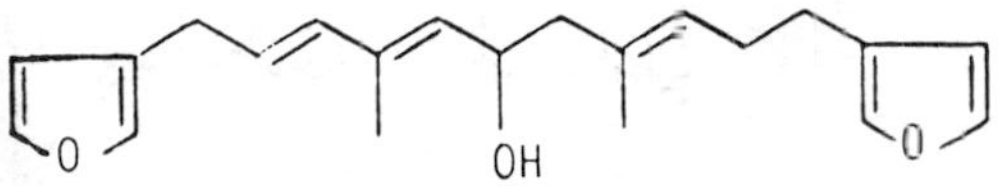 Tetradehydrofurospongin-1	Sponges: *Spongia* sp.				162
227. $C_{21}H_{28}O_2$	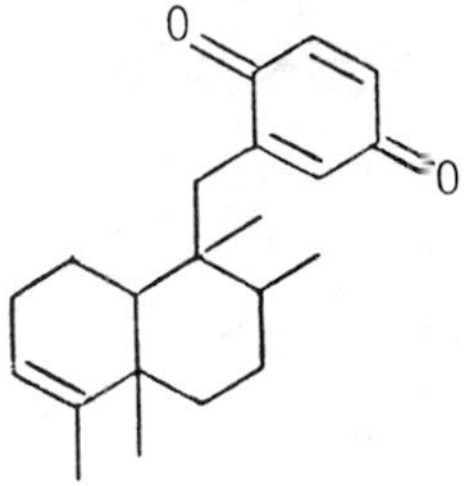 Avarone	Sponge: *Dysidea avara*				207
228. $C_{21}H_{28}O_3$	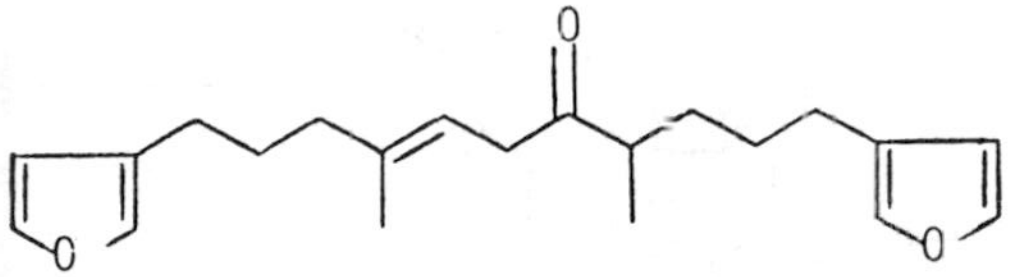 Furospongenone	Sponge: *Spongia* sp.				162
229. $C_{21}H_{30}O_2$	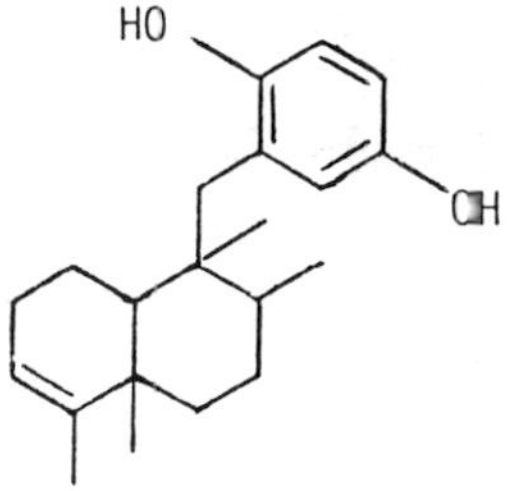 Avarol	Sponge: *Dysidea avara* *Dysidea* sp.				162 207

Table 2 (continued)

Molecular formula	Structural formula	Name	Source of compound	Available by synthesis	Available nonmarine sources	Activity	References
C_{21}							
230. $C_{21}H_{30}O_2$	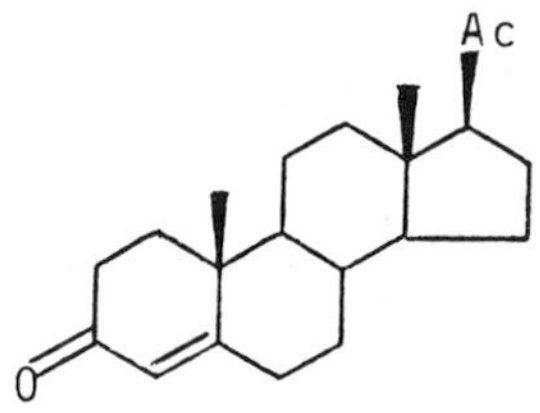	Progesterone	Dogfish: *Squalus suckleyi* Mollusc: *Pecten hericius* Sea urchin: *Strongylocentrotus franciscanus* Starfish: *Pisaster ochraceous*	+	+		31 32 202 273
231. $C_{21}H_{30}O_2$	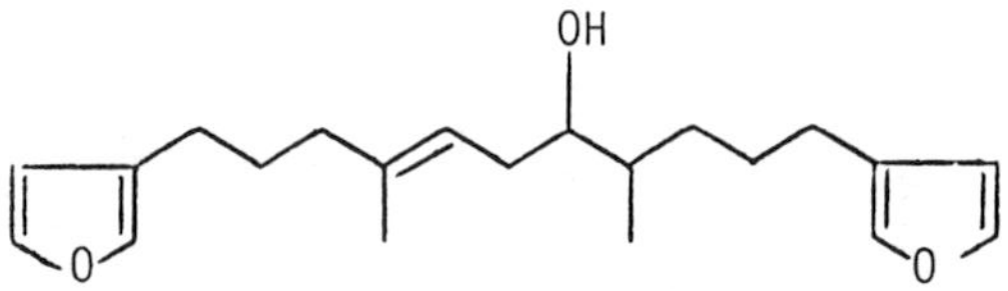	Furospongenol	Sponges: *Spongia* sp.				162
232. $C_{21}H_{30}O_5$	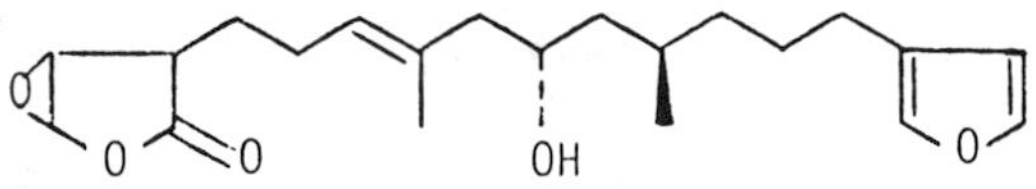	4-[11-(3-Furyl)-6-hydroxy-4,8-dimethyl-3-undecenyl]-2,6-dioxabicyclo[3.1.0]hexan-3-one	Sponge: *Spongia officinalis*				47
233. $C_{21}H_{30}O_5$	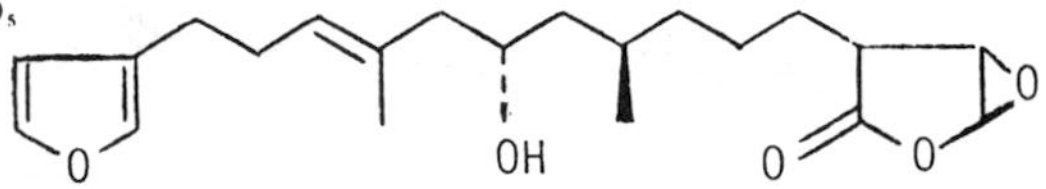	4-[11-(3-Furyl)-6-hydroxy-4,8-dimethyl-8-undecenyl]-2,6-dioxabicyclo[3.1.0]hexan-3-one	Sponge: *Spongia officinalis*				47
234. $C_{21}H_{30}O_5$	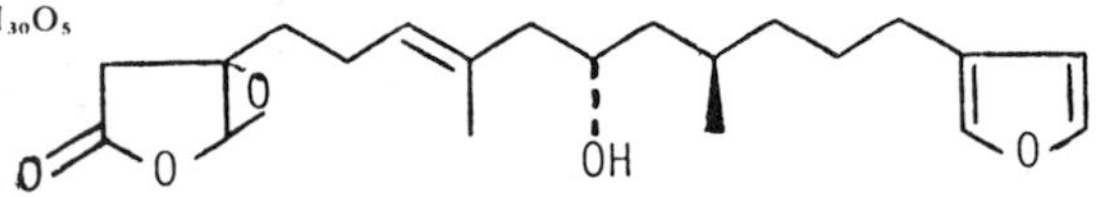	5-[11-(3-Furyl)-6-hydroxy-4,8-dimethyl-3-undecenyl]-2,6-dioxabicyclo[3.1.0]hexan-3-one	Sponge: *Spongia officinalis*				47

No.	Formula	Name	Source	Ref.
235.	$C_{21}H_{30}O_5$	5-[11-3-Furyl)-6-hydroxy-4,8-dimethyl-8-undecenyl]-2,6-dioxabicyclo[3.1.0] hexan-3-one	Sponge: *Spongia officinalis*	47
236.	$C_{21}H_{30}O_5$	3-[11-(3-Furyl)-6-hydroxy-4,8-dimethyl-3-undecenyl]-5-hydroxy-2(5H)-furanone	Sponge: *Spongia officinalis*	47
237.	$C_{21}H_{30}O_5$	3-[11-(3-Furyl)-6-hydroxy-4,8-dimethyl-8-undecenyl]-5-hydroxy-2(5H)-furanone	Sponge: *Spongia officinalis*	47
238.	$C_{21}H_{30}O_5$	4[11-(3-Furyl)-6-hydroxy-4,8-dimethyl-3-undecenyl]-5-hydroxy-2(5H)-furanone	Sponge: *Spongia officinalis*	47
239.	$C_{21}H_{30}O_5$	4-[11-(3-Furyl)-6-hydroxy-4,8-dimethyl-8-undecenyl]-5-hydroxy-2(5H)-furanone	Sponge: *Spongia officinalis*	47
240.	$C_{21}H_{32}O_2$	Δ^4-3-Keto-pregnen-20β-ol	Mollusc: *Pecten hericius*; Sea urchin: *Strongylocentrotus franciscanus*	32

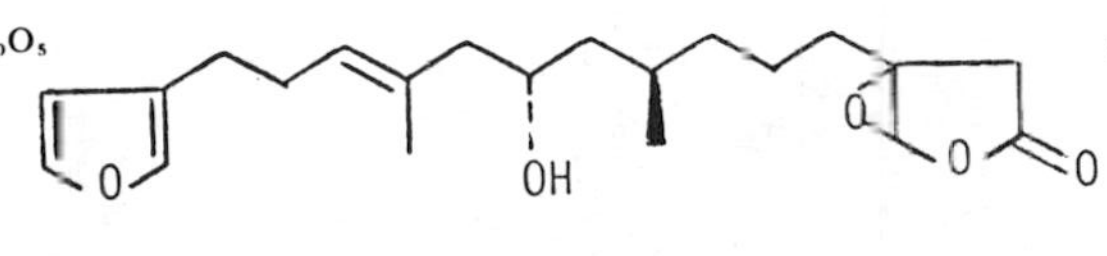

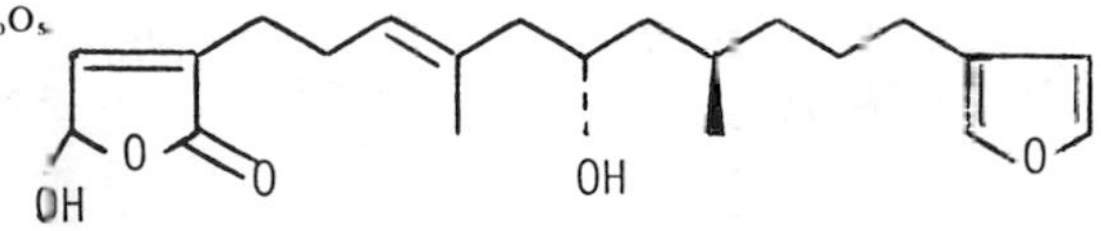

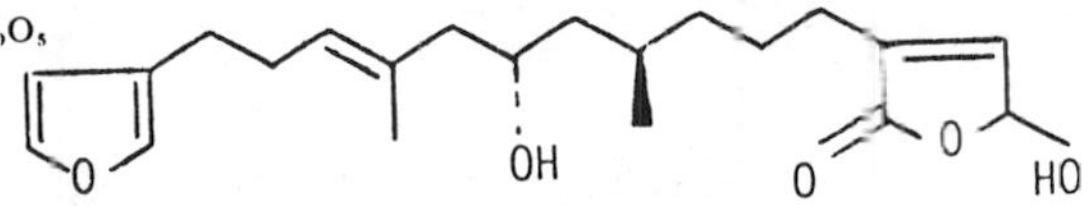

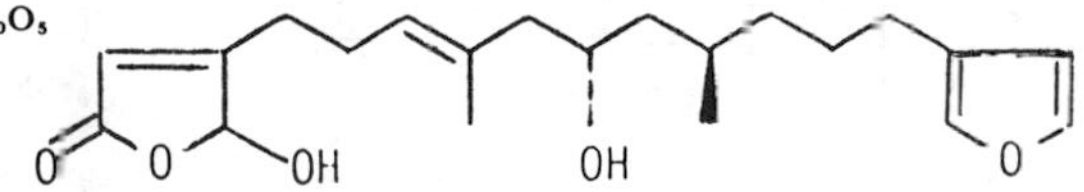

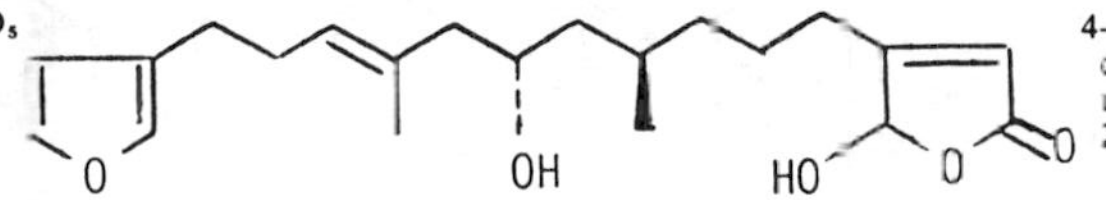

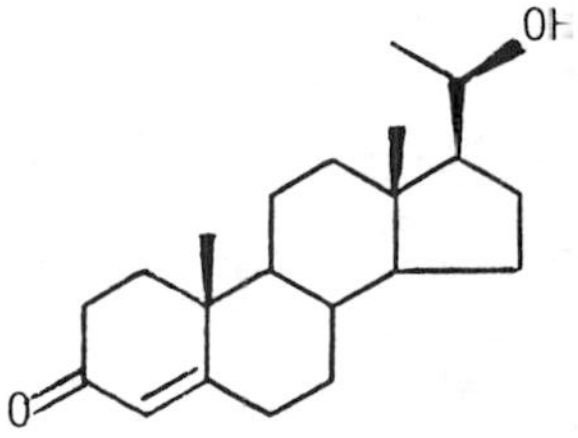

Table 2 (continued)

Molecular formula	Structural formula	Name	Source of compound	Available by synthesis	Available nonmarine sources	Activity	References
C_{21}							
241. $C_{21}H_{32}O_4$	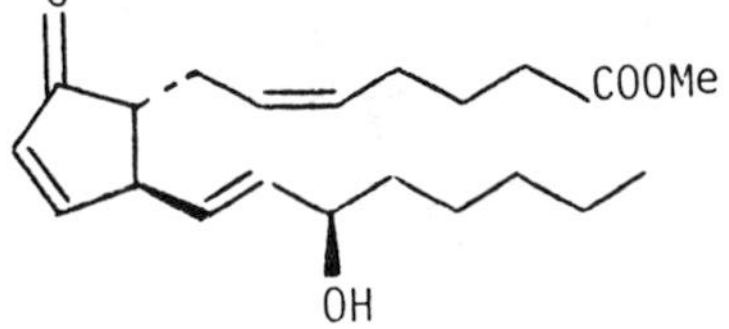	15*R*-PGA₂ methyl ester	Gorgonian: *Plexaura homomalla*				10
242. $C_{21}H_{33}N$	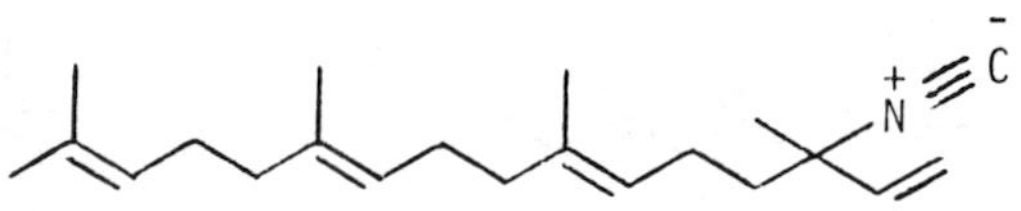	1,5,9,13-Tetramethyl-1-vinyl-4,8,12-tetradecatrienyl isocyanide	Sponge: *Halichondria* sp.				36
243. $C_{21}H_{33}NS$	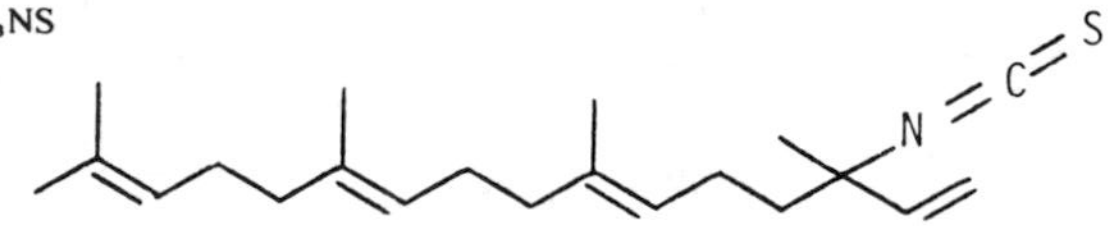	1,5,9,13-Tetramethyl-1-vinyl-4,8,12-tetradecatrienyl isothiocyanate	Sponge: *Halichondria* sp.				36
244. $C_{21}H_{34}$		*cis,cis,cis,cis,cis*-3,6,9,12,15-Heneicosapentaene	Green algae: *Monostroma* sp. *Spongomorpha arcta*				283
245. $C_{21}H_{35}NO$	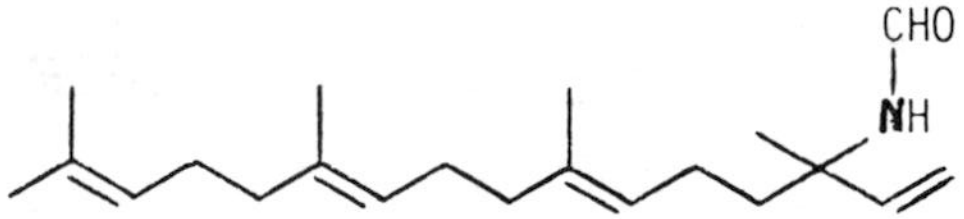	*N*-(1,5,9,13-Tetramethyl-1-vinyl-4,8,12-tetradecatrienyl)-formamide	Sponge: *Halichondria* sp.				36

Compound	Name	Source			Ref.
246. $C_{21}H_{42}O_4$	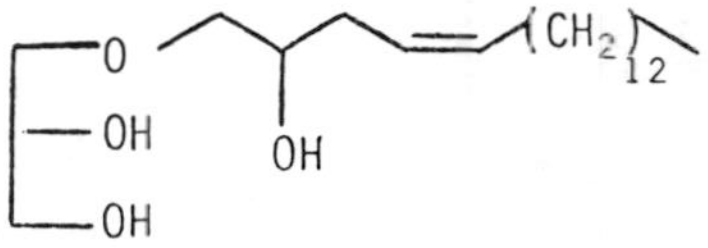 1-*O*-(2-Hydroxy-4-octadecenyl)glycerol	Shark: *Somniosus microcephalus*			113 115
247. $C_{21}H_{44}$	 Heneicosane	Algae: *Fucus distichus* *F. vesiculosus* *Laminaria saccharina* *Prasiola stipitata* *Rhodomela confervoides*			283
248. $C_{21}H_{44}O_4$	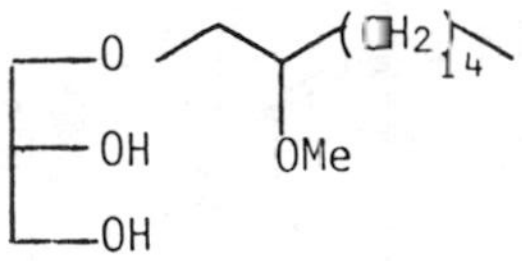 1-*O*-(2-Methoxyheptadecyl)glycerol	Cod liver oil Crayfish Herrings Mackerel Mussel Shrimp		+	114
C_{22} 249. $C_{22}H_{27}N_7O$	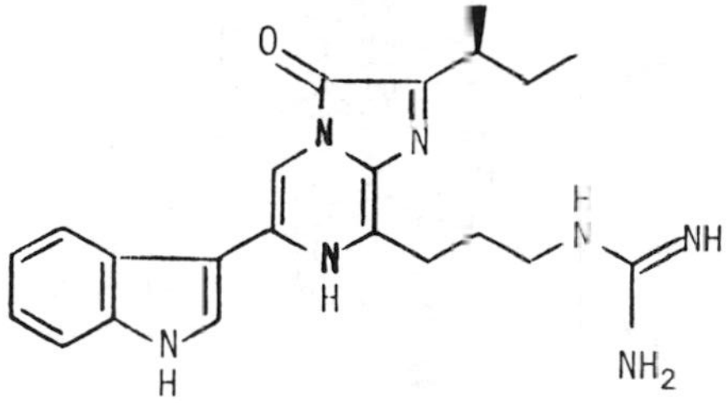 Cypridina luciferin	Crustacean: *Cypridina hilgendorfii*	+		18,55 104,105 106,107 108,149 166,184

Table 2 (continued)

Molecular formula	Structural formula	Name	Source of compound	Available by synthesis	Available nonmarine sources	Activity	References
C_{21}							
250. $C_{22}H_{28}O_4$	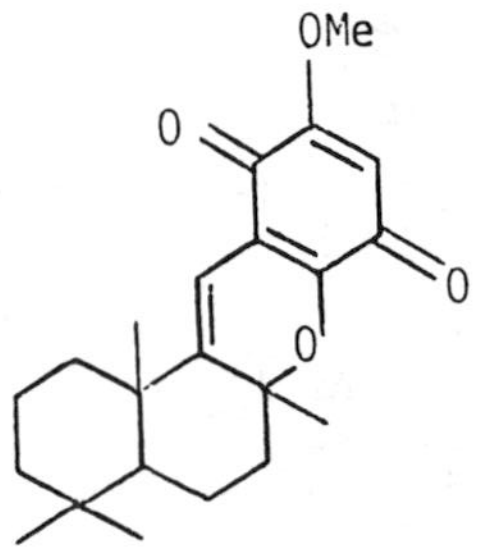 OMe, O, O, O	Didehydrocyclospongiaquinone-1	Sponge				162
251. $C_{22}H_{30}O_4$	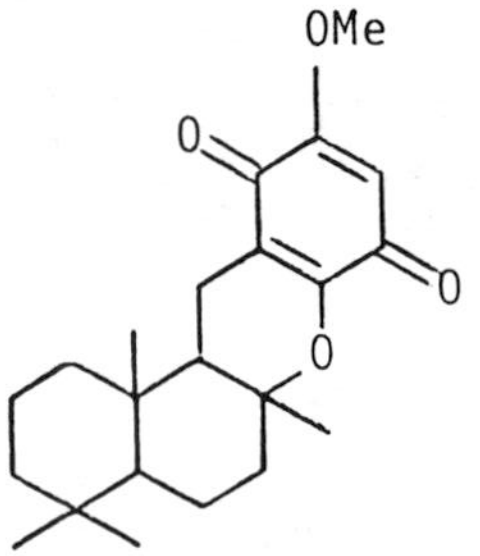 OMe, O, O, O	Cyclospongiaquinone-1	Sponge				162

No.	Formula	Name	Source	Ref.
252.	$C_{22}H_{30}O_4$	Cyclospongiaquinone-2	Sponge	162
253.	$C_{22}H_{30}O_4$	Isospongiaquinone	Sponge	162
254.	$C_{22}H_{30}O_4$	Spongiaquinone	Sponge	162

Table 2 (continued)

Molecular formula	Structural formula	Name	Source of compound	Available by synthesis	Available nonmarine sources	Activity	References
C_{22}							
255. $C_{22}H_{36}O_2$	OAc	Caulerpol acetate	Green alga: *Caulerpa brownii*				21
256. $C_{22}H_{36}O_3$	OAc, O	Epoxynephthenol acetate	Soft coral: *Nephthea* sp.				242
257. $C_{22}H_{44}O_4$	O, OH, OH, OMe, $(CH_2)_{12}$	1-*O*-(2-Methoxy-4-octadecenyl)glycerol	Shark: *Somniosus microcephalus*	+			113
258. $C_{22}H_{46}$		Docosane	Algae: *Fucus vesiculosus*, *Laminaria saccharina*, *Prasiola stipitata*, *Rhodomela confervoides*				283

Compound	Name	Source		Ref.
259. $C_{22}H_{46}O_3$	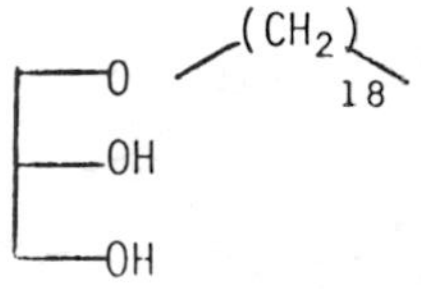 1-*O*-Nonadecylglycerol	Cod liver oil Coelenterates Crayfish Herring Mackerel Mussel Shrimp	+	114 194
260. $C_{22}H_{46}O_4$	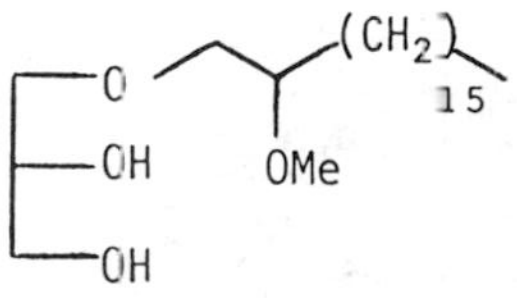 1-*O*-(2-Methoxyoctadecyl)glycerol	Cod liver oil Crayfish Herrings Mackerel Mussel Shrimp	+	114
C_{23}				
261. $C_{23}H_{32}O_2$	1-(Heptadeca-5,8,11,14-tetraenyl)-3,5-dihydroxybenzene	Brown alga: *Cystophora torulosa*		162
262. $C_{23}H_{48}$	 Tricosane	Algae: *Fucus distichus* *F. vesiculosus* *Laminaria saccharina* *Prasiola stipitata* *Rhodomela confervoides*		283
263. $C_{23}H_{48}O_3$	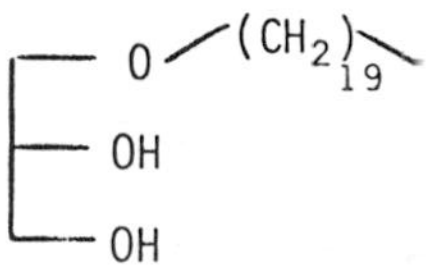 1-*O*-Eicosylglycerol	Cod liver oil Coelenterates Crayfish Herrings Mackerel Mussel Shrimp	+	114 194

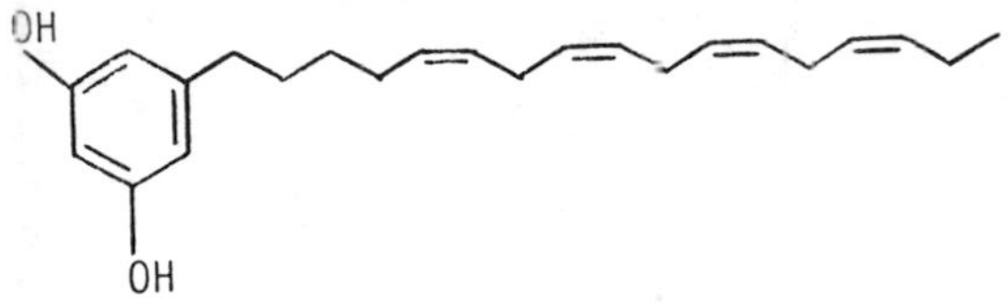

Table 2 (continued)

Molecular formula	Structural formula	Name	Source of compound	Available by synthesis	Available nonmarine sources	Activity	References
C_{22}							
264. $C_{23}H_{48}O_3$		1-*O*-Phytanylglycerol	Cod liver oil	+			114
265. $C_{23}H_{48}O_4$		1-*O*-(2-Methoxynonadecyl)glycerol	Cod liver oil Crayfish Herrings Mackerel Mussel Shrimp		+		114
C_{24}							
266. $C_{24}H_{32}O_6$		Spongiadiol diacetate	Sponges: *Spongia* sp.				162
267. $C_{24}H_{32}O_6$		*epi*-Spongiadiol diacetate	Sponges: *Spongia* sp.				162

Compound	Name	Source		Ref.
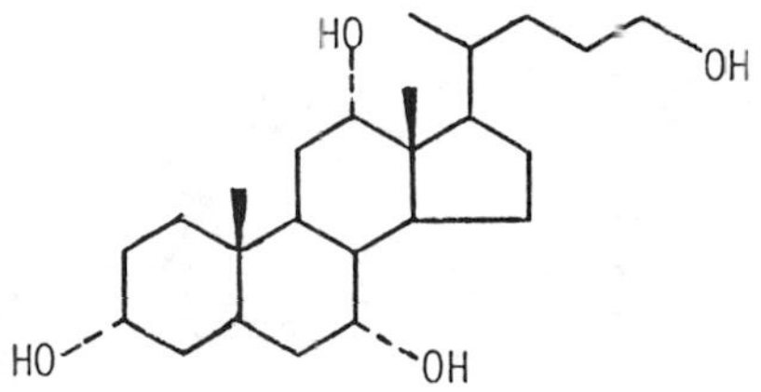 268. $C_{24}H_{42}O_4$ Occurs as 24-sulfate ester	Petromyzonol	Fish: *Petromyzon marinus*		258
269. $C_{24}H_{50}$	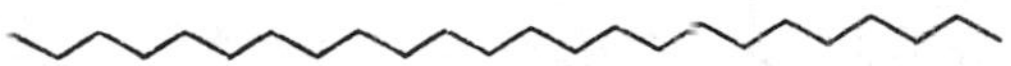 Tetracosane	Algae: *Fucus distichus* *F. vesiculosus* *Laminaria saccharina* *Prasiola stipitata* *Rhodomela confervoides*		283
270. $C_{24}H_{50}O_3$	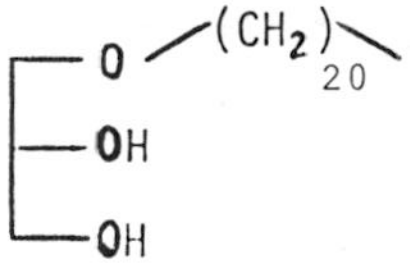 1-*O*-Heneicosylglycerol	Cod liver oil Herring: *Clupea harengus* Mussel: *Mytilus edulis*	+	114
271. $C_{24}H_{50}O$	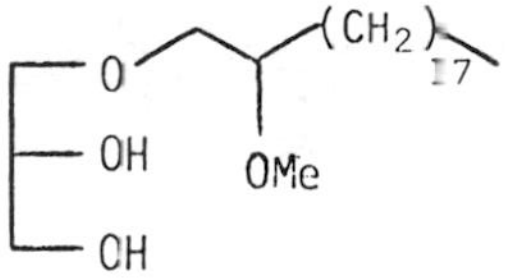 1-*O*-(2-Methoxyeicosyl) glycerol	Cod liver oil Crayfish Herrings Mussel Shrimp	+	114

Table 2 (continued)

Molecular formula	Structural formula	Name	Source of compound	Available by synthesis	Available nonmarine sources	Activity	References
C_{25}							
272. $C_{25}H_{21}N_3O_9S_2$	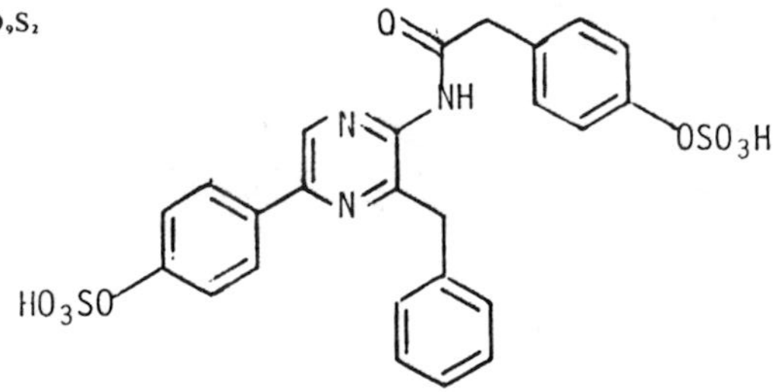	Watasenia oxyluciferin	Squid: *Watasenia scintil-lans*	+			109
273. $C_{25}H_{28}O_4$	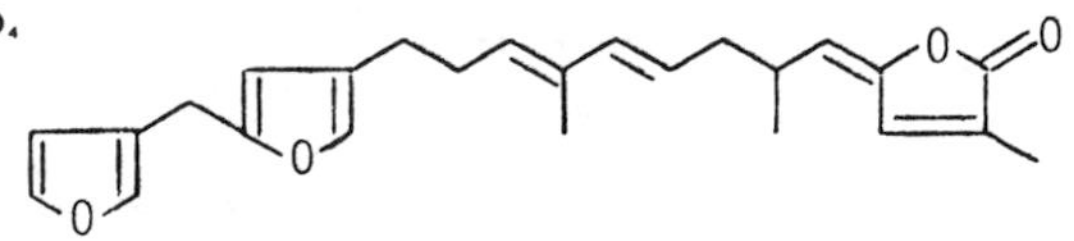	Ircinolide	Sponge: *Thorecta marginalis*				162
274. $C_{25}H_{28}O_5$	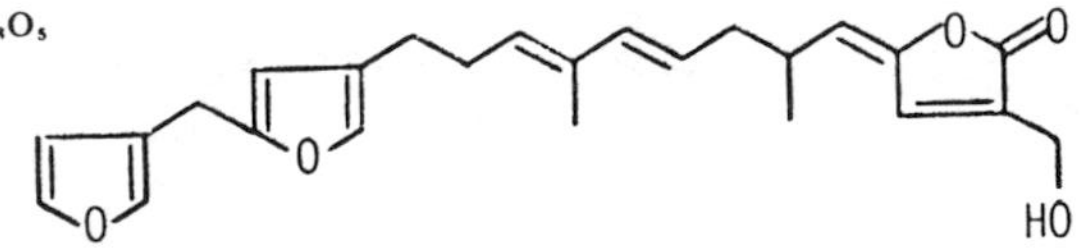	24-Hydroxyircinolide	Sponge: *Thorecta marginalis*				162
275. $C_{25}H_{32}O_4$	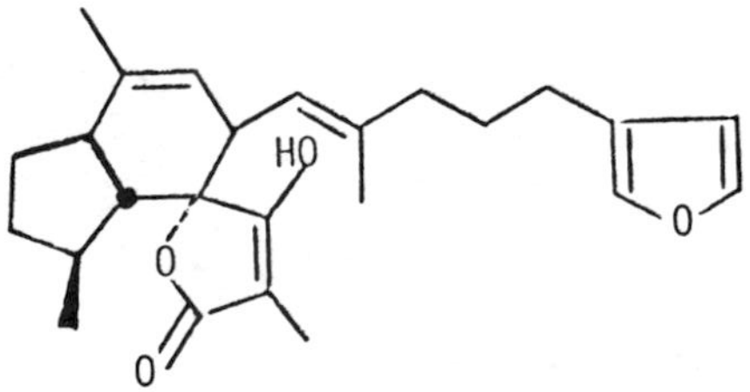	Ircinianin	Sponge: *Ircinia* sp.				72

276. $C_{25}H_{34}O_4$	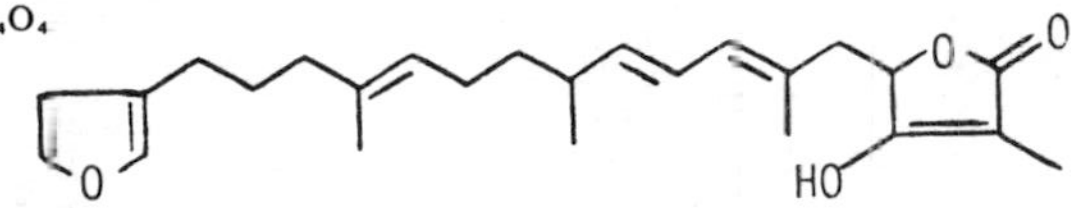 Halmiformin-1	Sponge: *Ircinia halmiformis*		162
277. $C_{25}H_{36}O_3$	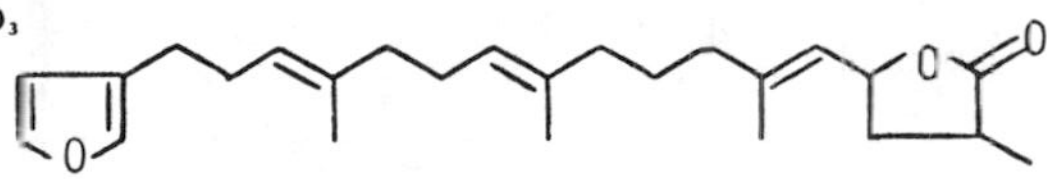 Fasciospongin	Sponge: *Fasciospongia* sp.		162
278. $C_{25}H_{42}O$	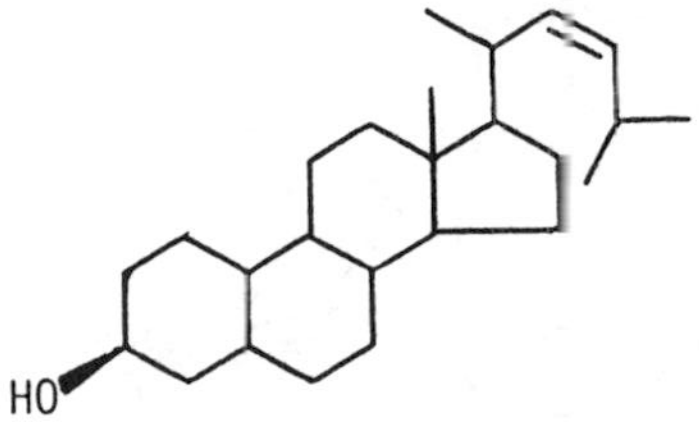 19,24-Bisnorcholest-*cis*-22-en-3β-ol	Sponge: *Axinella polypoides*		204
279. $C_{25}H_{42}O$	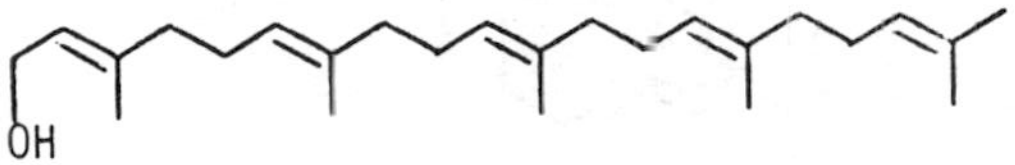 Geranylfarnesol	Sponge: *Fasciospongia fovea*		162
280. $C_{25}H_{52}O_3$	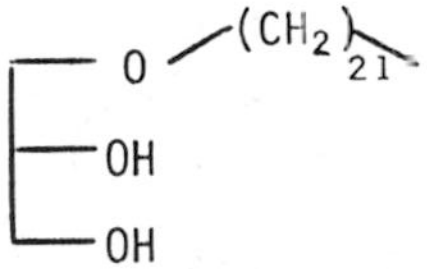 1-*O*-Docosylglycerol	Cod liver oil Herring: *Clupea harengus* Mussel: *Mytilus edulis*	+	114
281. $C_{25}H_{52}O_4$	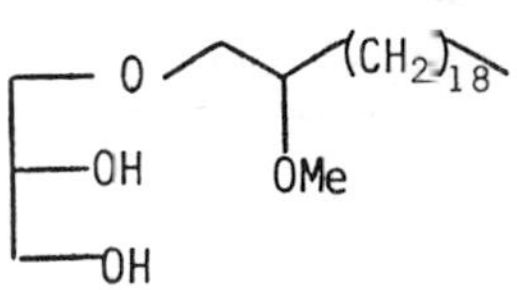 1-*O*-(2-Methoxyheneicosyl)glycerol	Herring: *Clupea harengus*	+	114

Table 2 (continued)

Molecular formula	Structural formula	Name	Source of compound	Available by synthesis	Available nonmarine sources	Activity	References
C_{26}							
282. $C_{26}H_{34}O_8$	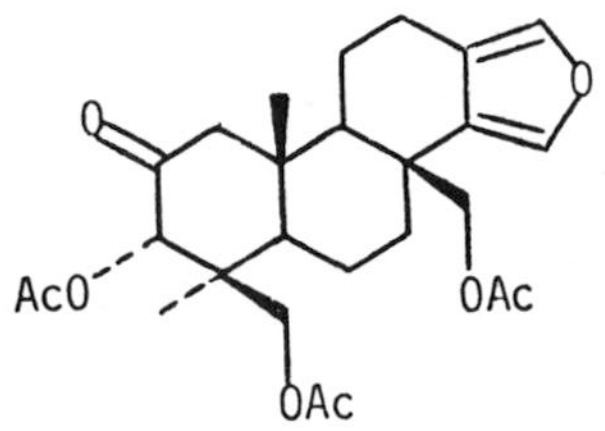	Spongiatriol triacetate	Sponges: *Spongia* sp.				162
283. $C_{26}H_{34}O_8$	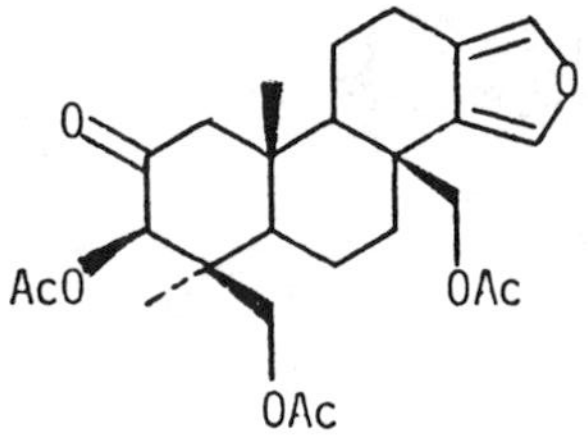	*epi*-Spongiatriol triacetate	Sponges: *Spongia* sp.				162
284. $C_{26}H_{42}NO_6S^-$	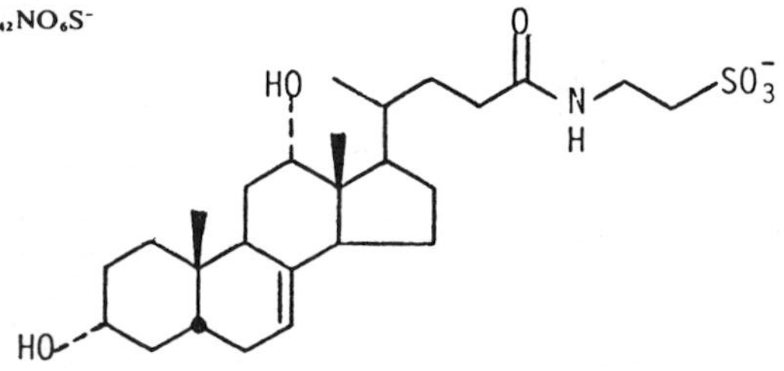	Tauro-3α,12α-dihydroxy-5β-chol-7-en-24-oate	Fish: *Myoxocephalus quadricornis*				258

Compound	Name	Source	Ref.
285. $C_{26}H_{42}O$	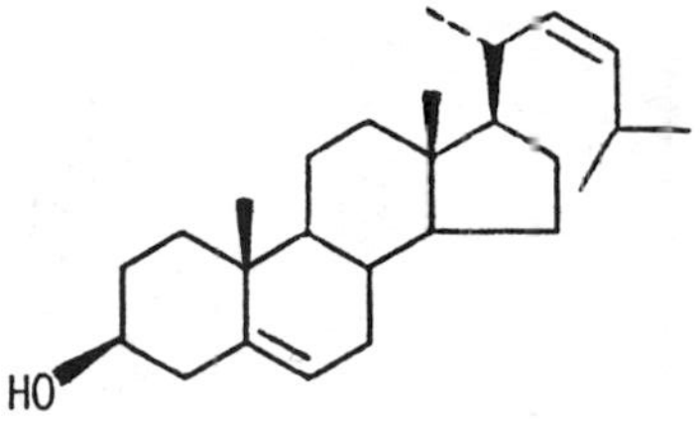 22-*cis*-24-Norcholesta-5,22-dien-3β-ol	Sponges: *Verongia aerophoba*, *V. archeri*, *V. fistularis*, *V. thiona*	68
286. $C_{26}H_{44}NO_6S^-$	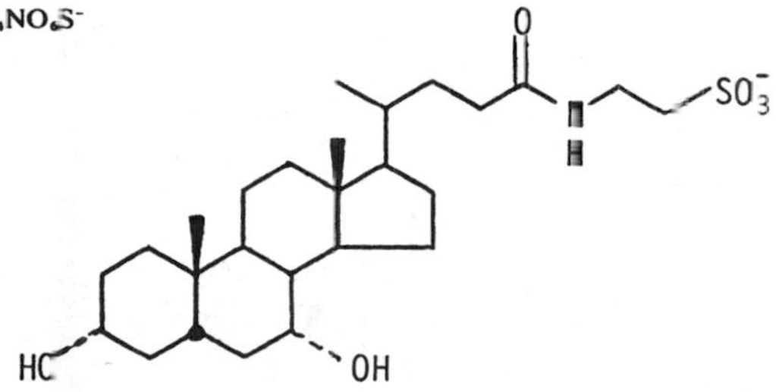 Taurochenodeoxycholate	Fish bile	258
287. $C_{26}H_{44}NO_5S^-$	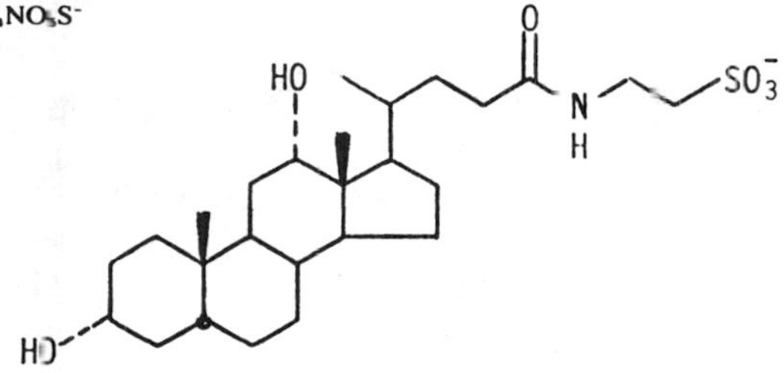 Taurodeoxycholate	Fish: *Fugu rubripes*, *Gadus callarias*, *Myoxocephalus quadricornis*	258

Table 2 (continued)

Molecular formula	Structural formula	Name	Source of compound	Available by synthesis	Available nonmarine sources	Activity	References
C_{26}							
288. $C_{26}H_{44}NO_6S^-$	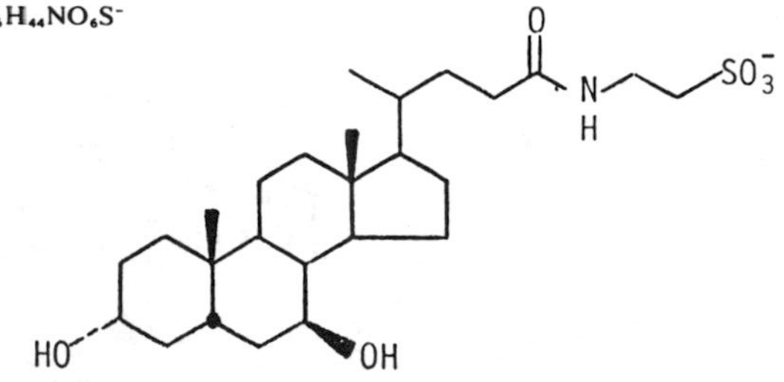	Tauro-3α,7β-dihydroxy-5β-cholan-24-oate	Fish: *Gadus callarias*				258
289. $C_{26}H_{44}NO_7S^-$	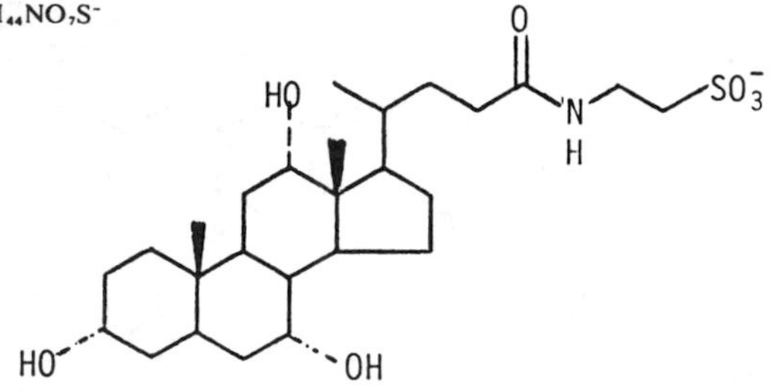	Tauroallocholate	Fish: *Fugu rubripes* *Tetrodon porphyreus*				258
290. $C_{26}H_{44}NO_7S^-$	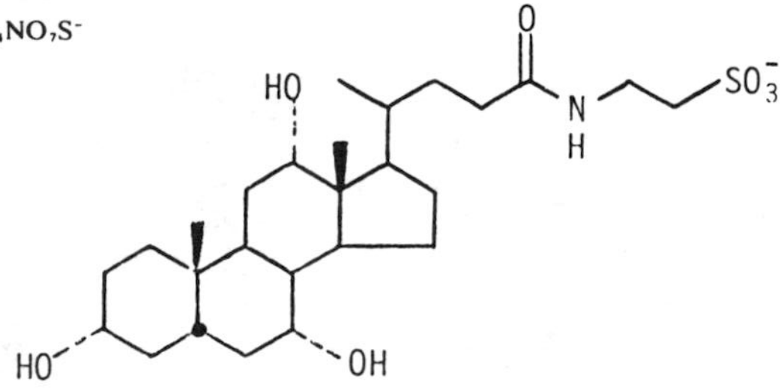	Taurocholate	Fish bile				258

No.	Formula	Name	Source	Ref.
291.	$C_{26}H_{44}NO_7S^-$	Taurohaemulcholate	Fish: *Parapristipoma trilineatum*; *Plectorhynchus cinctus*	258
292.	$C_{26}H_{44}NO_7S^-$	Tauro-3α,7β-12α-trihydroxy-5β-cholan-24-oate	Eel: *Conger myriaster*; Fish: *Gadus callarias*	258
293.	$C_{26}H_{44}O$	19-Norcholest-*trans*-22-en-3β-ol	Sponge: *Axinella polypoides*	204
294.	$C_{26}H_{46}O$	19-Norcholestan-3β-ol	Sponge: *Axinella polypoides*	204

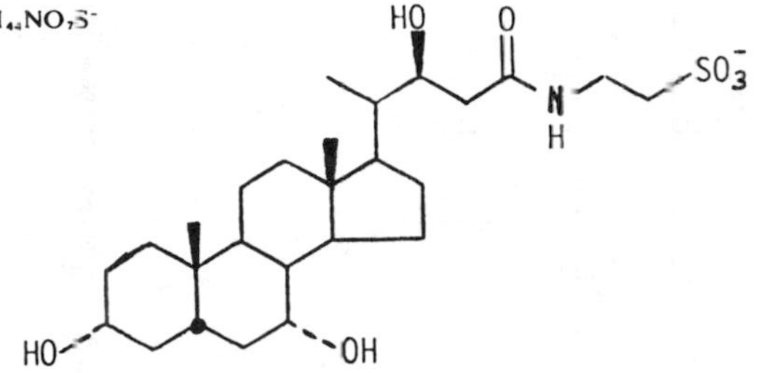

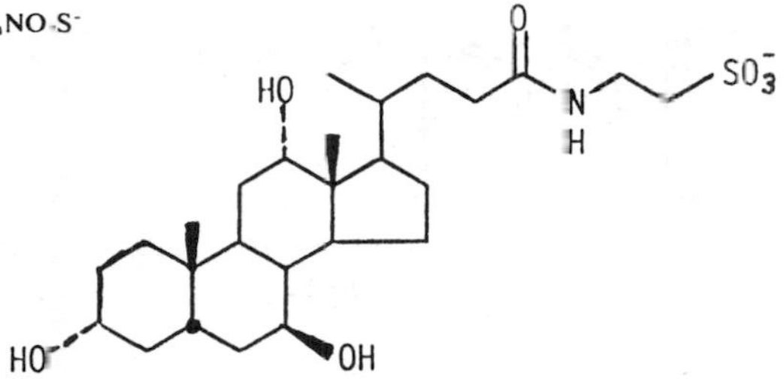

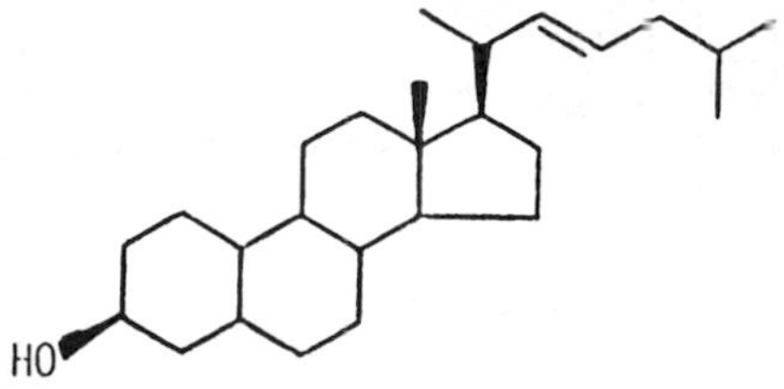

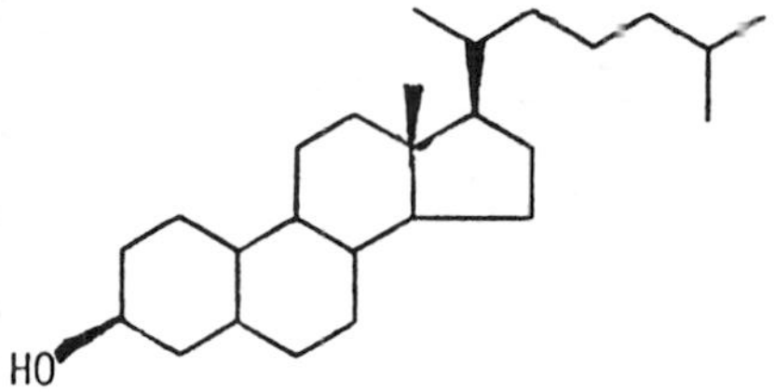

Table 2 (continued)

Molecular formula	Structural formula	Name	Source of compound	Available by synthesis	Available nonmarine sources	Activity	References
C_{26}							
295. $C_{26}H_{46}O_2$	 COOH	*cis, cis, cis*-5,9,19-Hexacosatrienoic acid	Sponge: *Microciona prolifera*				136
296. $C_{26}H_{48}O_2$	 COOH	*cis, cis*-5,9-Hexacosadienoic acid	Sponge: *Microciona prolifera*				136
297. $C_{26}H_{54}O_4$	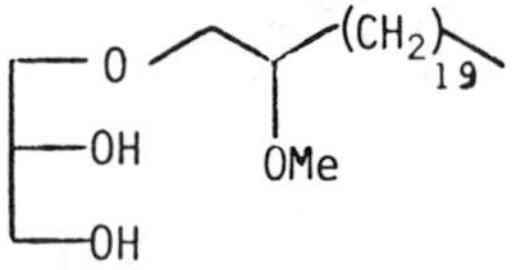 O, OH, OH, OMe, $(CH_2)_{19}$	1-*O*-(2-Methoxydocosyl) glycerol	Herring: *Clupea harengus*		+		114
C_{27}							
298. $C_{27}H_{40}O_2$	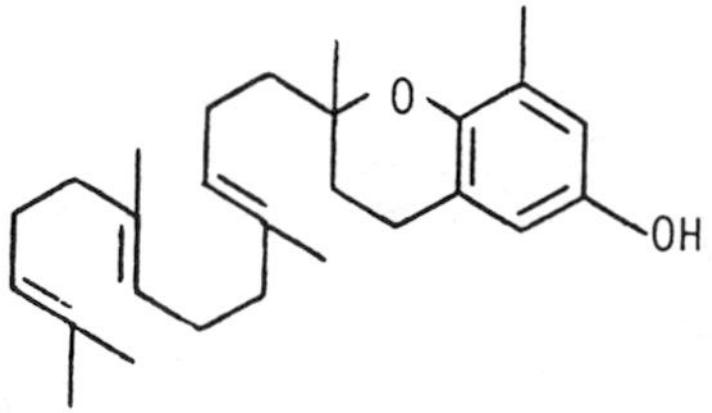 O, OH	δ-Tocotrienol	Brown alga: *Sargassum tortile*	+			176
299. $C_{27}H_{40}O_3$	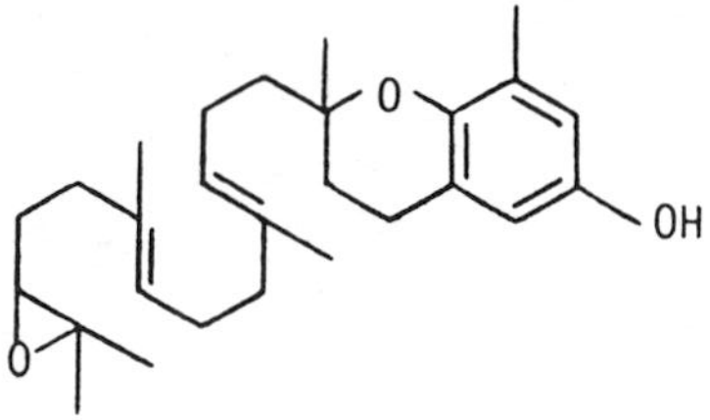 O, O, OH	δ-Tocotrienol epoxide	Brown alga: *Sargassum tortile*	+			176

No.	Name	Source		Ref.
300. $C_{27}H_{40}O_4$	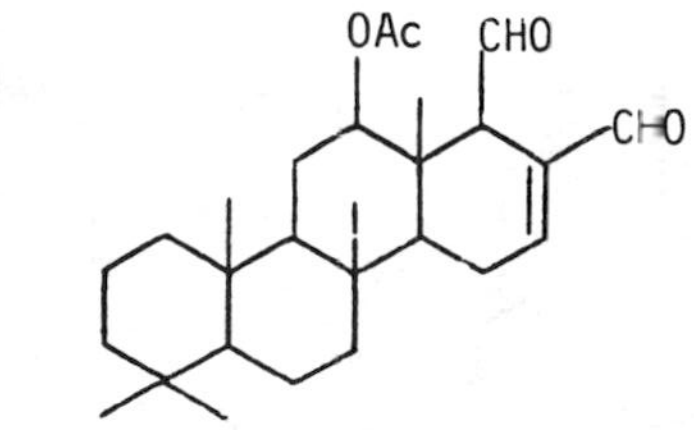 Scalaradial	Sponge: *Cacospongia mollior*		49
301. $C_{27}H_{42}O_3$	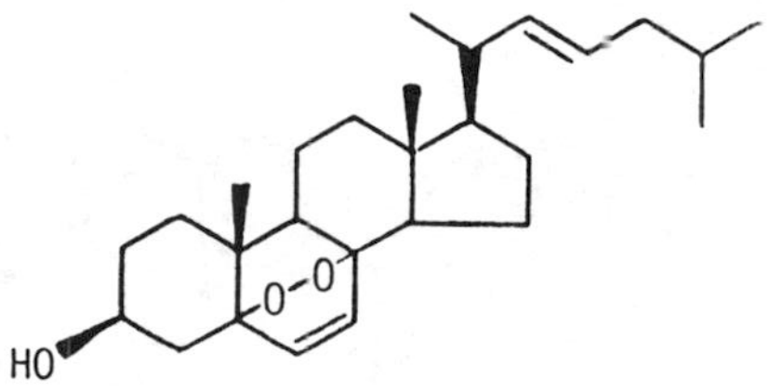 5,8-Epidioxycholesta-6,22-dien-3β-ol	Sponge: *Axinella cannabina*	+	76
302. $C_{27}H_{44}O$	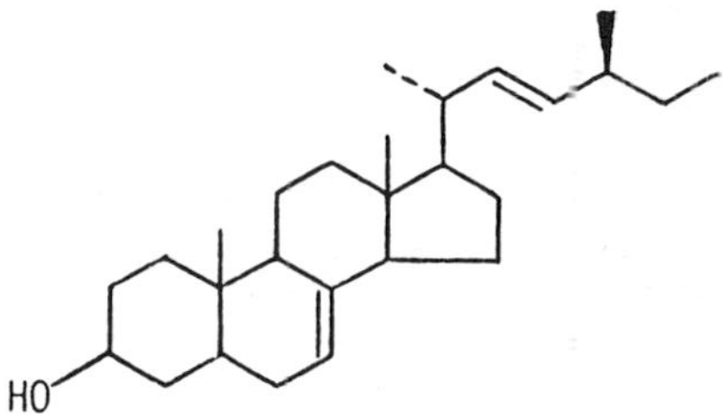 Amuresterol	Starfish	+	171
303. $C_{27}H_{44}O$	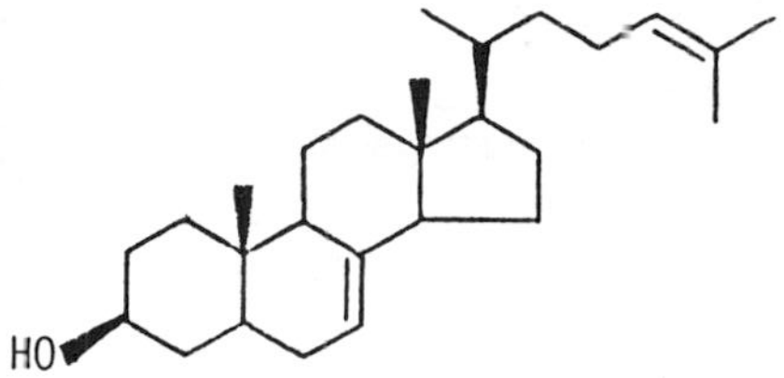 5α-Cholesta-7,24-dien-3β-ol	Starfish: *Asterias rubens*		249

Table 2 (continued)

Molecular formula	Structural formula	Name	Source of compound	Available by synthesis	Available nonmarine sources	Activity	References
C_{27}							
304. $C_{27}H_{44}O$	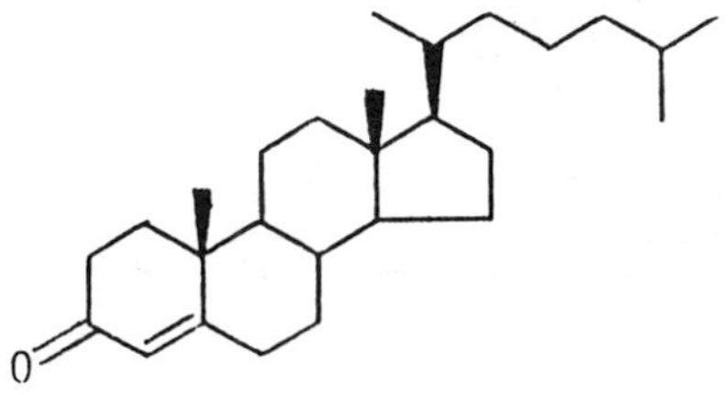	Cholest-4-en-3-one	Red algae: *Gracilaria textorii* *Meristotheca papulosa*				143 144
305. $C_{27}H_{44}O$	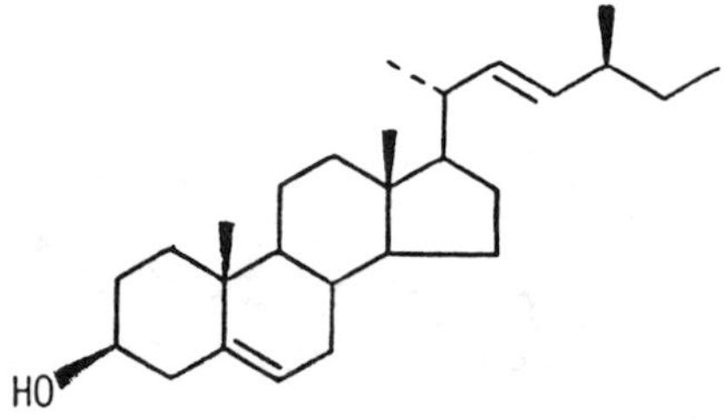	Occelasterol	Crustacean Molluscs Sea anemone Sea cucumber Sea urchin Sea worm Tunicate	+			172
306. $C_{27}H_{44}O_2$	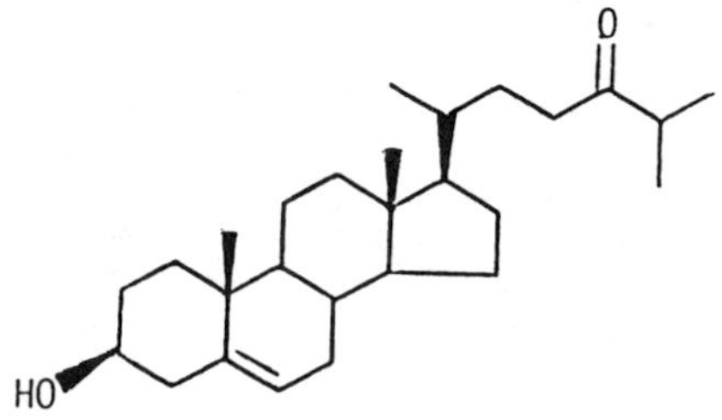	24-Ketocholesterol 24-Oxocholesterol	Brown algae: *Ascophyllum nodosum* *Laminaria saccharina* *Pelvetia canaliculata*				216 240

Compound	Name	Source			Ref.
307. $C_{27}H_{46}O$	3β-Hydroxymethyl-A-nor-5α-*cis*-cholest-22-ene	Sponge: *Axinella verrucosa*			205
308. $C_{27}H_{46}O$	24-Methylene-19-nor-cholestan-3β-ol	Sponge: *Axinella polypoides*			204
309. $C_{27}H_{46}O$	19-Nor-5α,10β-ergost-22-en-3β-ol	Sponge: *Axinella polypoides*			204
310. $C_{27}H_{46}O_2$	δ-Tocopherol	Brown algae: *Ascophyllum nodosum* *Fucus serratus* *F. spiralis* *F. vesiculosus* *Pelvetia canaliculata*	+	+	139 202

Table 2 (continued)

Molecular formula	Structural formula	Name	Source of compound	Available by synthesis	Available nonmarine sources	Activity	References
C_{27}							
311. $C_{27}H_{46}O_4S$	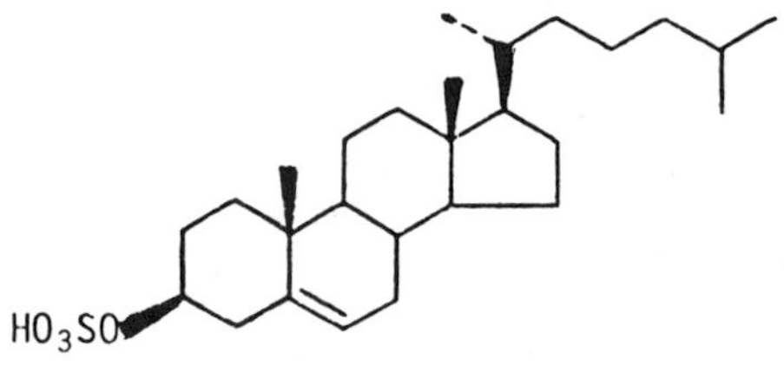	Cholesteryl sulfate	Sea urchin: *Anthocidaris crassispina* Starfish: *Asterias rubens*	+	+		20 282
312. $C_{27}H_{48}O$	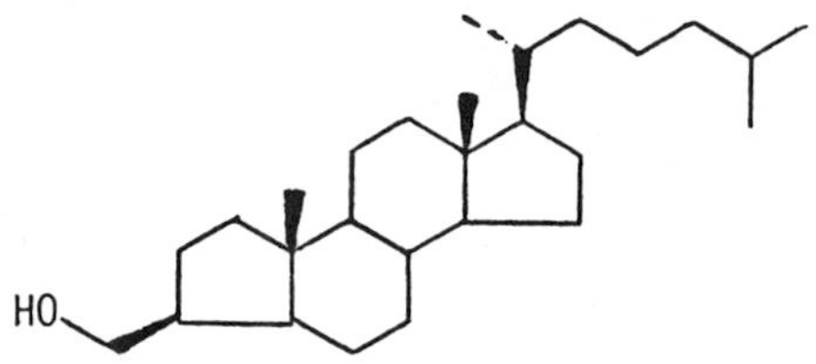	3β-Hydroxymethyl-A-nor-5α-cholestane	Sponge: *Axinella verrucosa*				205
313. $C_{27}H_{48}O$	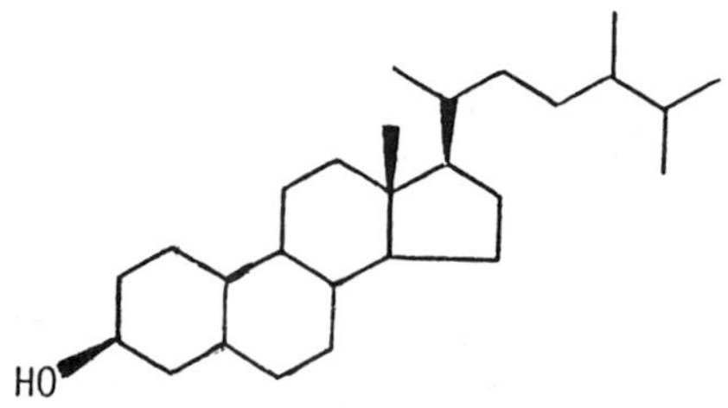	24-Methyl-19-norcholestan-3β-ol	Sponge: *Axinella polypoides*				204

No. / Formula	Name	Source	Ref.
314. $C_{27}H_{48}O_4$ 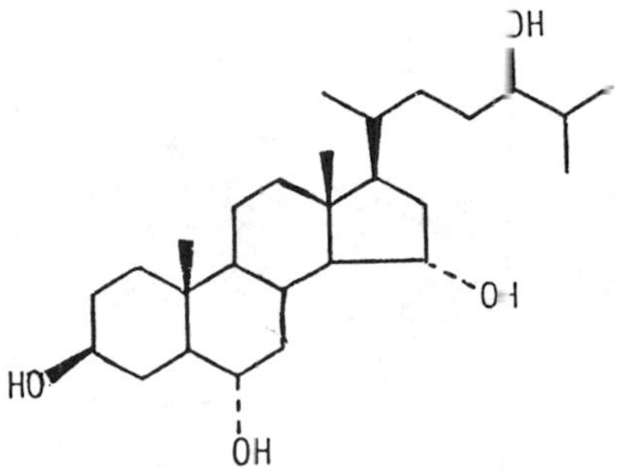	$3\beta,6\alpha,15\alpha,24\xi$-Tetrahydroxy-$5\alpha$-cholestane	Starfish: *Asterias amurensis*	142
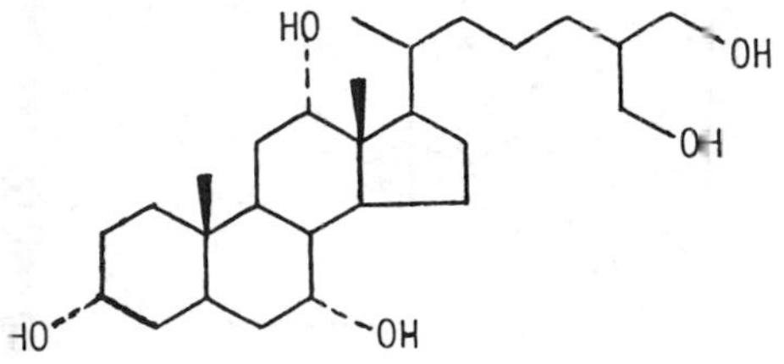 315. $C_{27}H_{48}O_5$ Occurs as 26 or 27-sulfate ester	5α-Cyprinol	Fish: *Latimeria chalumnae*	258
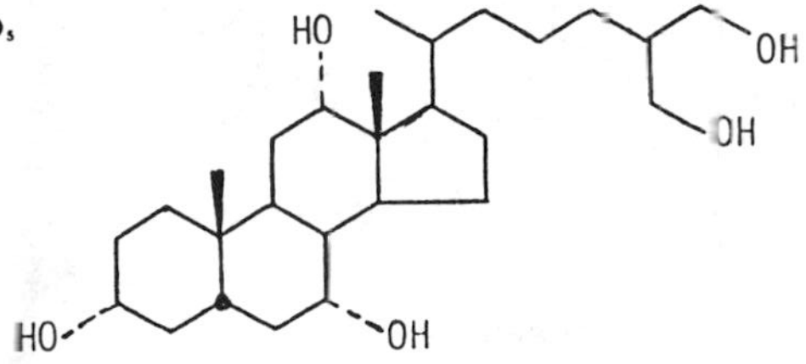 316. $C_{27}H_{48}O_5$ Occurs as 26 or 27-sulfate ester	5β-Cyprinol	Eels: *Anguilla japonica*, *Conger myriaster*, *Muraenesox cinereus*	258

Table 2 (continued)

Molecular formula	Structural formula	Name	Source of compound	Available by synthesis	Available nonmarine sources	Activity	References
C_{27}							
317. $C_{27}H_{48}O_5$	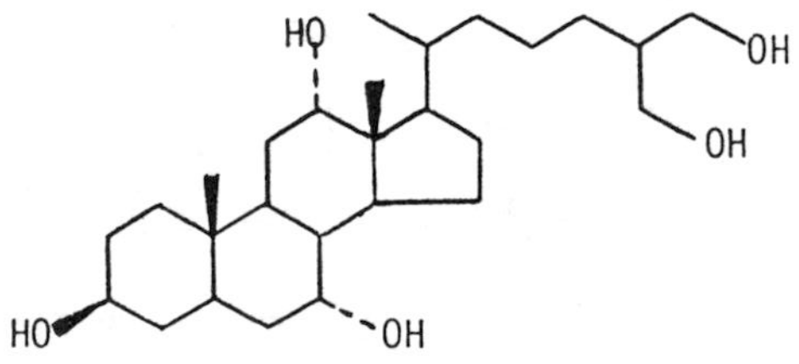 Occurs as 26 or 27-sulfate ester	Latimerol	Fish: *Latimeria chalumnae*				258
C_{28}							
318. $C_{28}H_{42}O_2$	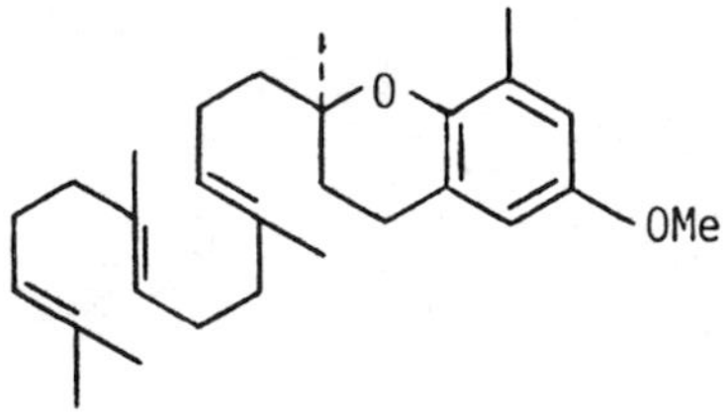	δ-Tocotrienol methyl ether	Brown alga: *Cystophora torulosa*				162
319. $C_{28}H_{42}O_4$	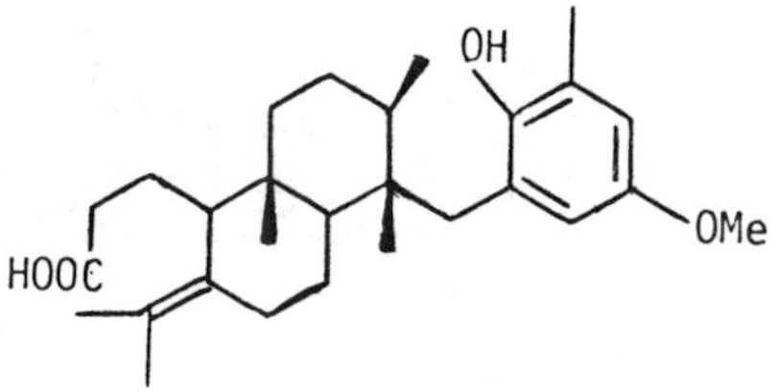	Atomaric acid	Brown alga: *Taonia atomaria*				100

320. $C_{28}H_{42}O_4$	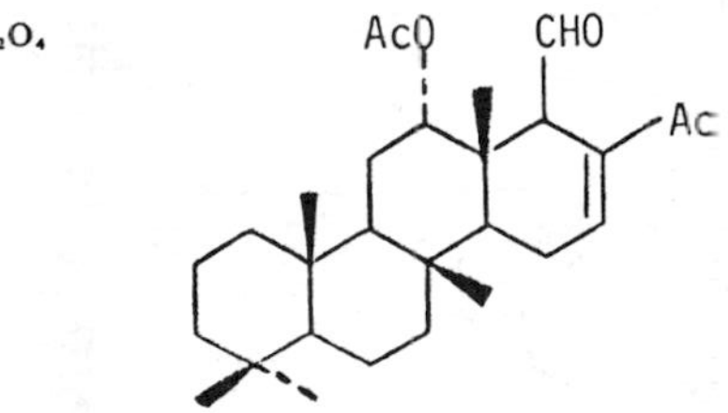 *epi*-Dendalone acetate	Sponge: *Phyllospongia dendyi*			162
321. $C_{28}H_{44}O$	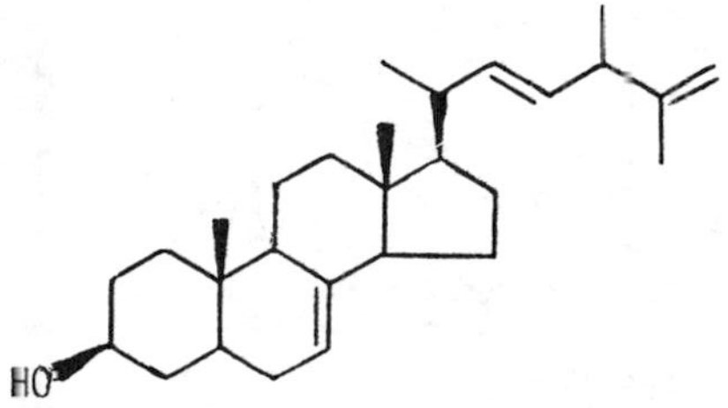 24-Methylcholesta-7,22,25-trien-3β-ol	Starfish: *Leiaster leachii*		+	261
322. $C_{28}H_{44}O_3$	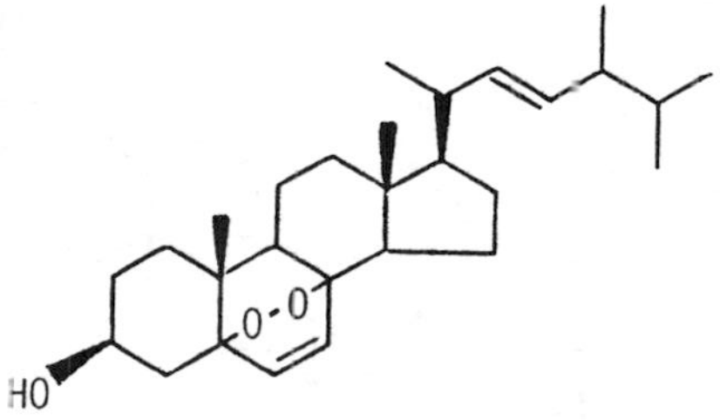 Ergosterol peroxide	Sponge: *Axinella cannabina*	+	+	13 76
323. $C_{28}H_{46}O$	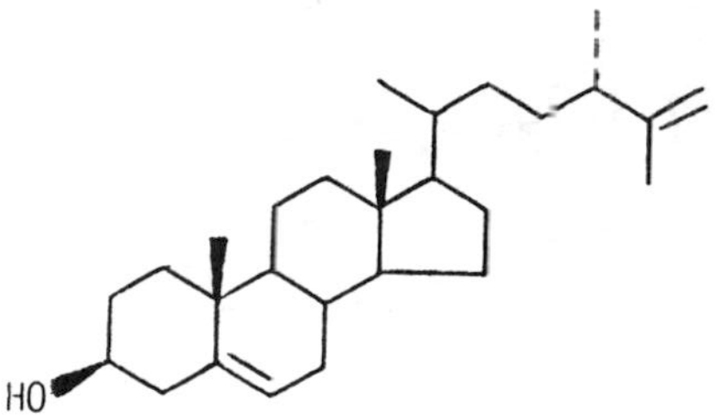 Codisterol	Green alga: *Codium fragile*			237

Table 2 (continued)

Molecular formula	Structural formula	Name	Source of compound	Available by synthesis	Available nonmarine sources	Activity	References
C_{28}							
324. $C_{28}H_{46}O$	HO	4α-Methyl-5α-cholesta-7,24-dien-3β-ol	Starfish: *Asterias rubens*				249
325. $C_{28}H_{46}O$	HO	24-Methylcholesta-5,22-dien-3β-ol	Brown alga: *Sargassum fluitans*				250
326. $C_{28}H_{46}O$	HO	24α-Methylcholesta-5,22-dien-3β-ol	Brittle star: *Ophiocomina nigra* Diatom: *Phaeodactylum tricornutum*				238

327. $C_{28}H_{46}O$	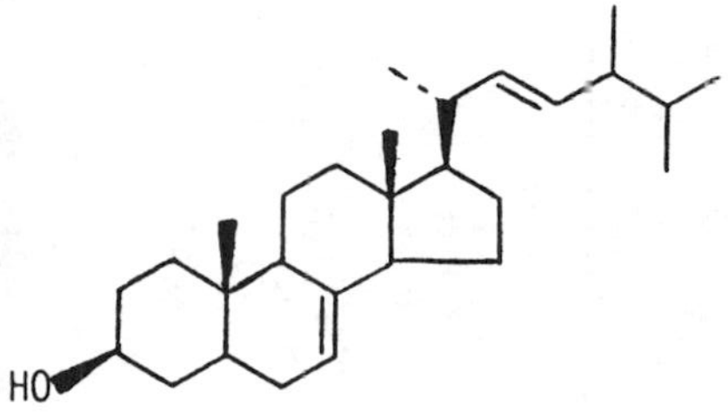 24-Methylcholesta-7,22-dien-3β-ol	Starfish: *Leiaster leachii*	261
328. $C_{28}H_{48}O$	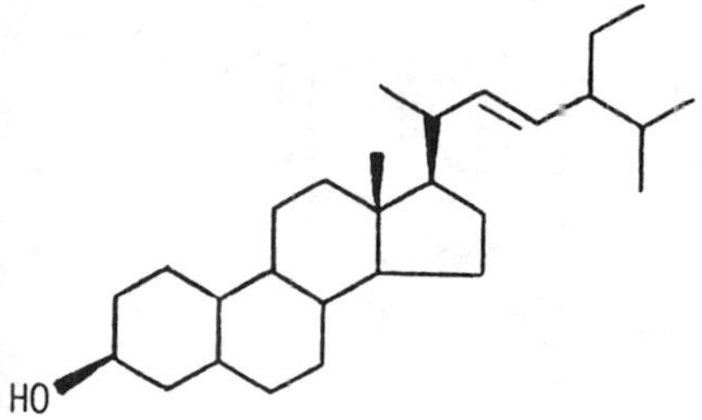 24-Ethyl-19-norcholest-*trans*-22-en-3β-ol	Sponge: *Axinella polypoides*	204
329. $C_{28}H_{48}O$	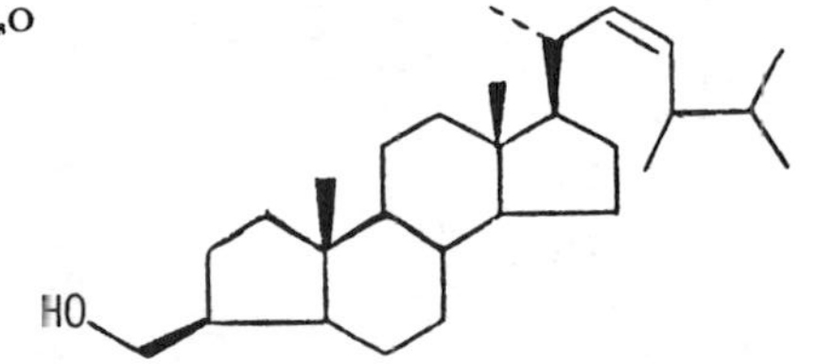 3β-Hydroxymethyl-24-methyl-A-nor-5α-*cis*-cholest-22-ene	Sponge: *Axinella verrucosa*	205
330. $C_{28}H_{48}O$	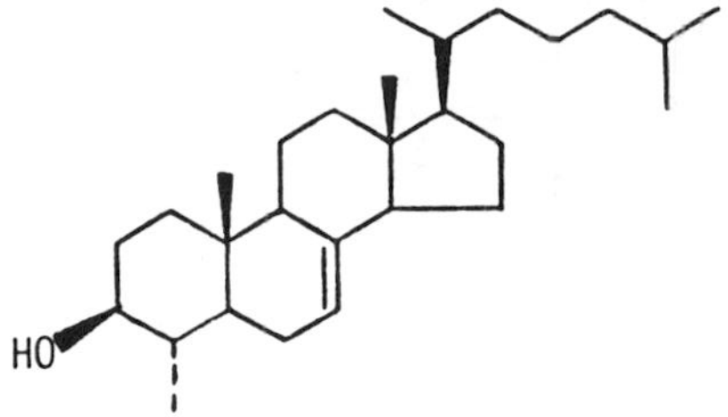 4α-Methyl-5α-cholest-7-en-3β-ol	Starfish: *Asterias rubens*	249

Table 2 (continued)

Molecular formula	Structural formula	Name	Source of compound	Available by synthesis	Available nonmarine sources	Activity	References
C_{28}							
331. $C_{28}H_{48}O$	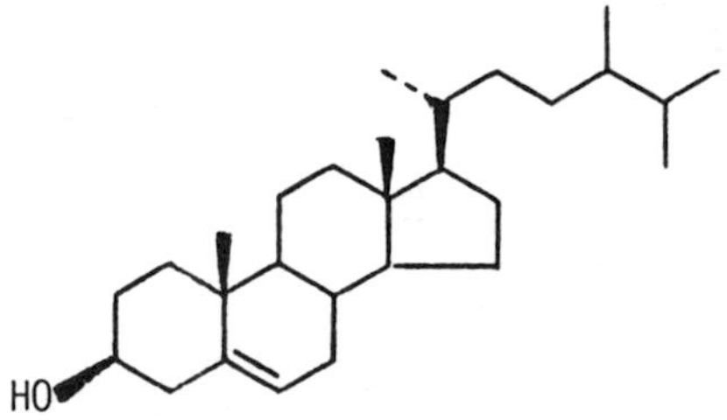	24-Methyl-5α-cholest-5-en-3β-ol	Brown alga Gorgonians Jellyfish Sea anemone Sea urchins Sea worm				22 169 172 250 275 276
332. $C_{28}H_{48}O$	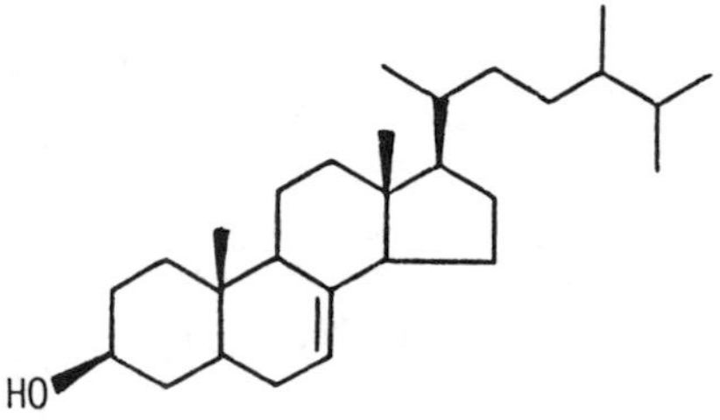	24-Methyl-5α-cholest-7-en-3β-ol	Starfish: *Asterias rubens* *Coscinasterias acutispina* *Leiaster leachii*				146 249
333. $C_{28}H_{48}O_2$	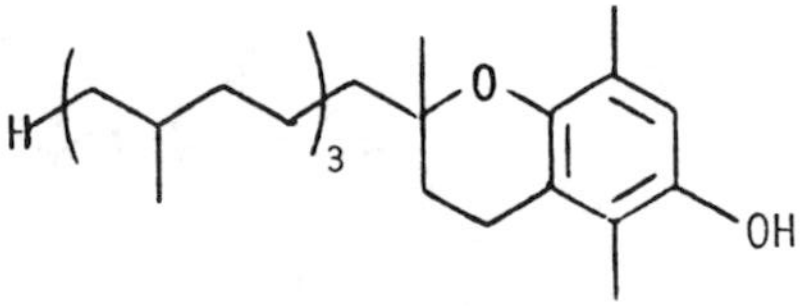	β-Tocopherol	Brown algae: *Ascophyllum nodosum* *Fucus serratus* *F. spiralis* *F. vesiculosus* *Pelvetia canaliculata*		+		139 202

C_{29}

No. / Formula	Name	Source		Ref.
334. $C_{28}H_{48}O_2$	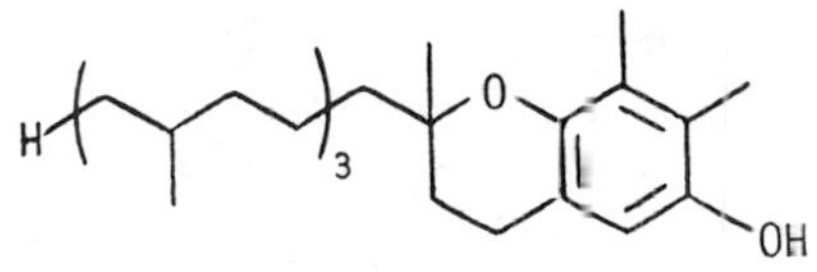 γ-Tocopherol	Brown algae: *Ascophyllum nodosum* *Fucus serratus* *F. spiralis* *F. vesiculosus* *Pelvetia canaliculata*	+	139 202
335. $C_{28}H_{50}O$	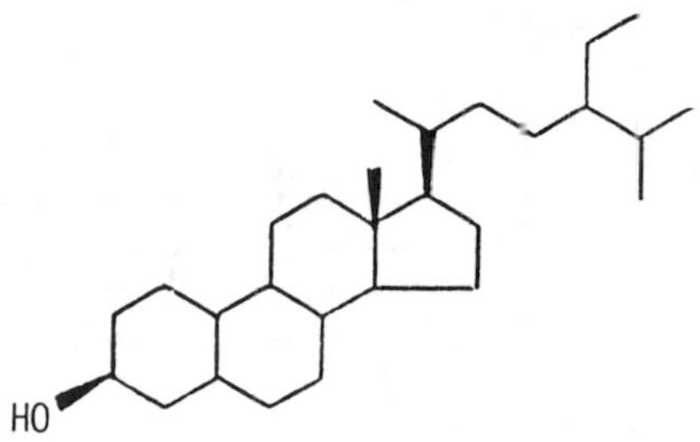 24-Ethyl-19-norcholestan-3β-ol	Sponge: *Axinella polypoides*		204
336. $C_{28}H_{50}O$	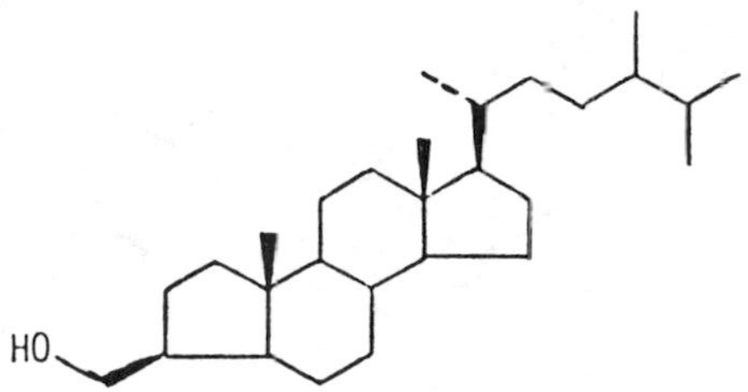 3β-Hydroxymethyl-24-methyl-A-nor-5α-cholestane	Sponge: *Axinella verrucosa*		205
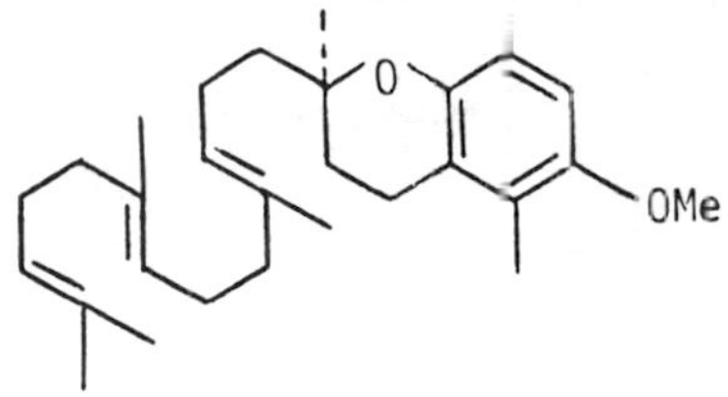 337. $C_{29}H_{44}O_2$	Methyl-δ-tocotrienol methyl ether	Brown alga: *Cystophora torulosa*		162

Table 2 (continued)

Molecular formula	Structural formula	Name	Source of compound	Available by synthesis	Available nonmarine sources	Activity	References
C_{29}							
338. $C_{25}H_{44}O_6$		Heteronemin	Sponge: *Heteronema erecta*				162
339. $C_{29}H_{49}O$		Clerosterol	Green alga: *Codium fragile*		+		237
340. $C_{29}H_{48}O$		4,4-Dimethyl-5α-cholesta-7,24-dien-3β-ol	Starfish: *Asterias rubens*				249

No. / Formula	Compound	Source	Ref.
341. $C_{29}H_{48}O$	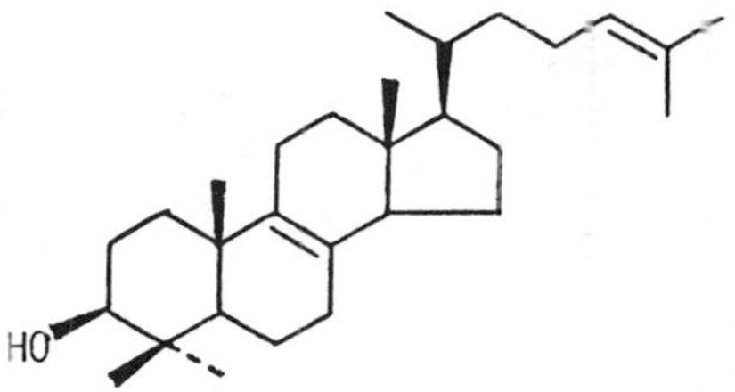 4,4-Dimethyl-5α-cholesta-8(9),24-dien-3β-ol	Starfish: *Asterias rubens*	249
342. $C_{29}H_{48}O$	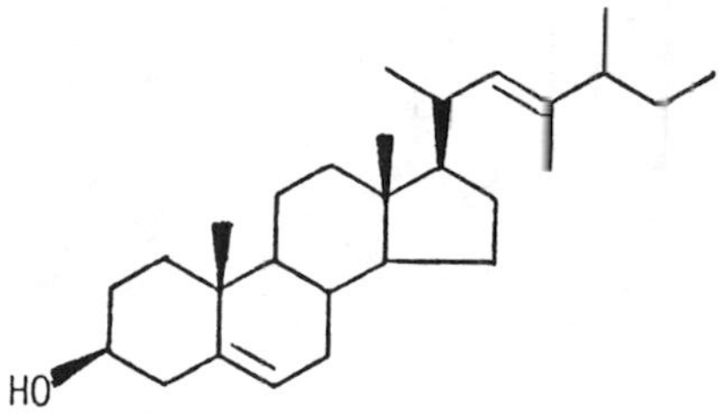 23,24-Dimethylcholesta-5,22-dien-3β-ol	Soft coral: *Sarcophyton elegans*	147
343. $C_{29}H_{48}O$	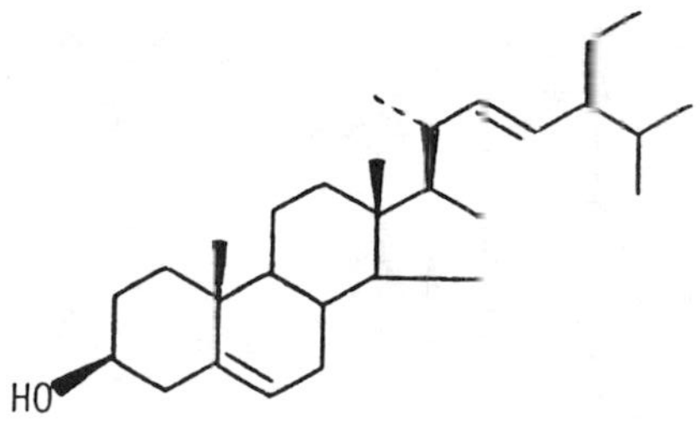 24-Ethylcholesta-5,22-dien-3β-ol	Brown alga: *Sargassum fluitans* Sea worm: *Pseudopotamilla ocelata*	169 172 250

Table 2 (continued)

Molecular formula	Structural formula	Name	Source of compound	Available by synthesis	Available nonmarine sources	Activity	References
C_{29}							
344. $C_{29}H_{48}O$	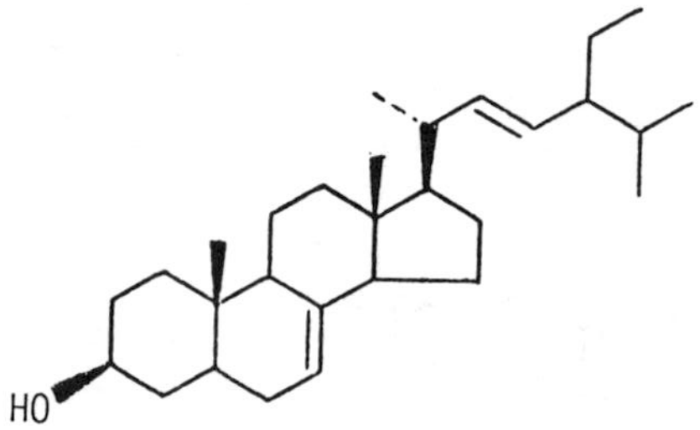	24-Ethylcholesta-7,22-dien-3β-ol	Starfish: *Asterias rubens* *Coscinasterias acutispina* *Leiaster leachii*				146 249 261
345. $C_{29}H_{48}O$	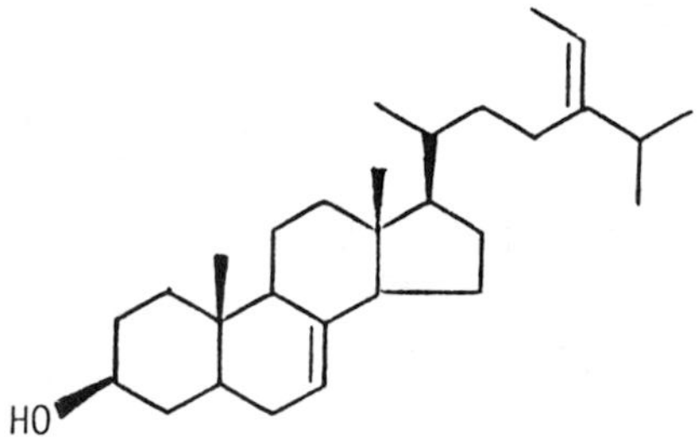	(E)-24-Ethylidene-5α-cholest-7-en-3β-ol	Starfish: *Asterias rubens* *Leiaster leachii*				146 249

No.	Formula	Name	Source	Ref.
346.	$C_{29}H_{48}O$	(Z)-24-Ethylidene-5α-cholest-7-en-3β-ol	Starfish: *Asterias rubens*	249
347.	$C_{29}H_{50}O$	4,4-Dimethyl-5α-cholest-7-en-3β-ol	Starfish: *Asterias rubens*	249
348.	$C_{29}H_{50}O$	24-Ethylcholest-5-en-3β-ol	Brown alga Jellyfish Sea anemone Seaurchins Sea worm	169 172 250 275 276
349.	$C_{29}H_{50}O$	24-Ethyl-3β-hydroxymethyl-A-nor-5α-*cis*-cholest-22-ene	Sponge: *Axinella verrucosa*	205

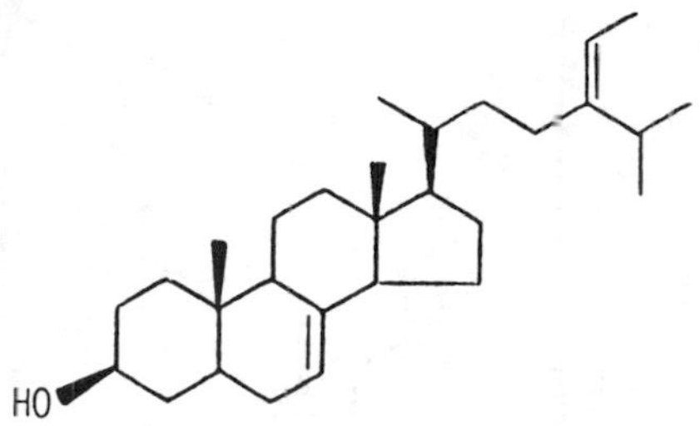

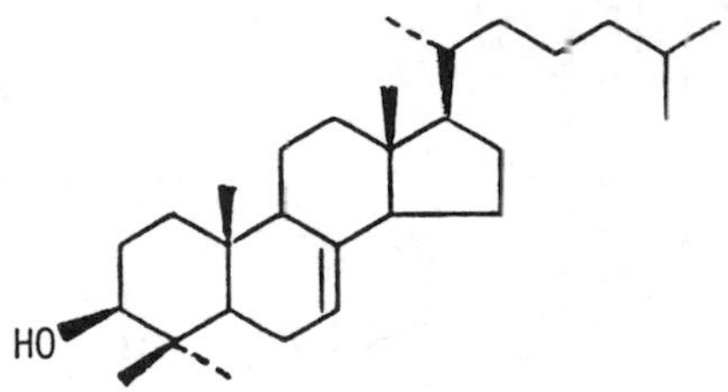

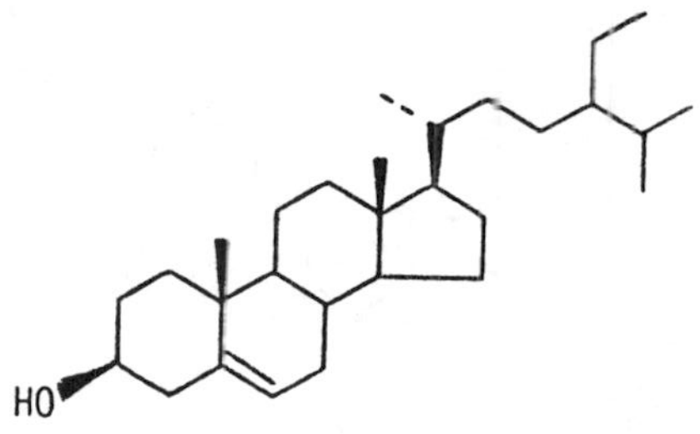

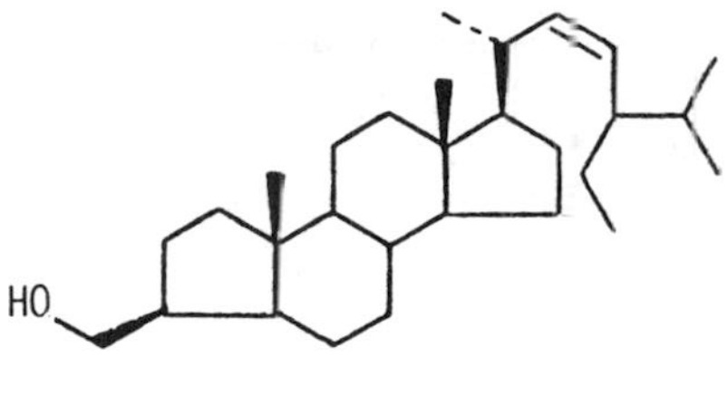

Table 2 (continued)

Molecular formula	Structural formula	Name	Source of compound	Available by synthesis	Available nonmarine sources	Activity	References
C_{29}							
350. $C_{29}H_{50}O$	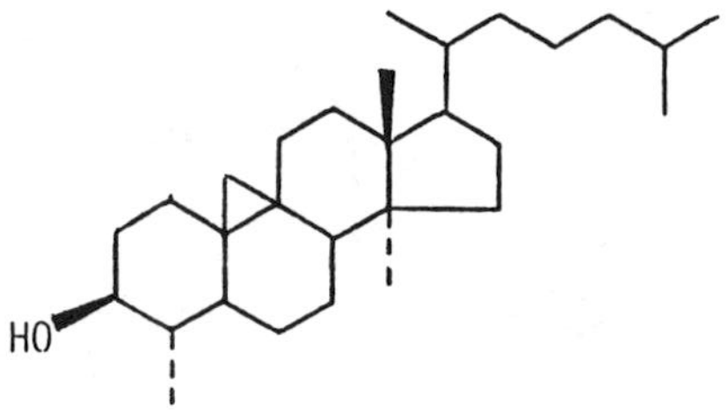	31- Norcycloartanol	Red alga: *Rhodymenia palmata* Starfish: *Asterias rubens*				84 249
351. $C_{29}H_{50}O_2$	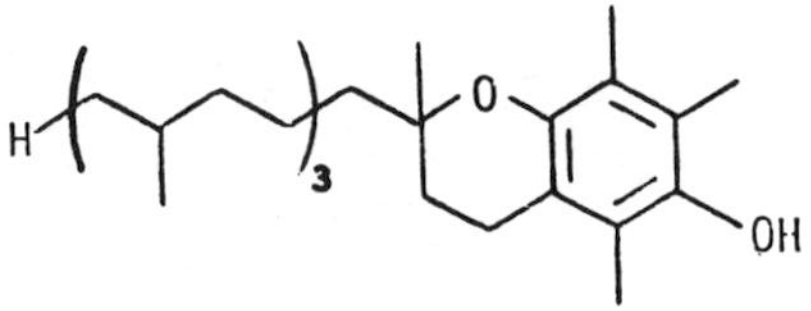	α-Tocopherol Vitamin E	Algae	+	+		139
352. $C_{29}H_{52}O$	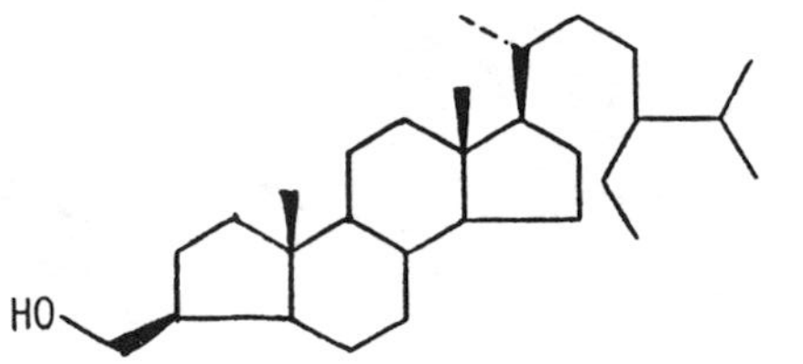	24-Ethyl-3β-hydroxymethyl-A-nor-5α-cholestane	Sponge: *Axinella verrucosa*				205

C_{30}

No. / Formula	Name	Source	Ref.
353. $C_{30}H_{46}O_5$	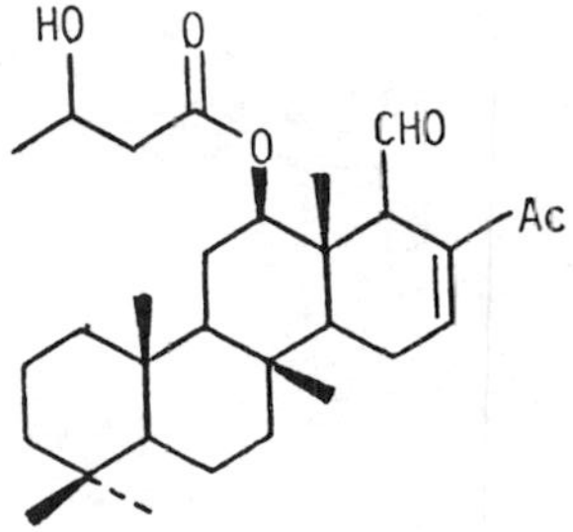 Dendalone 3-hydroxy-butyrate	Sponge: *Phyllospongia dendyi*	162
354. $C_{30}H_{46}O_5$ (replaces No. 1/431.)	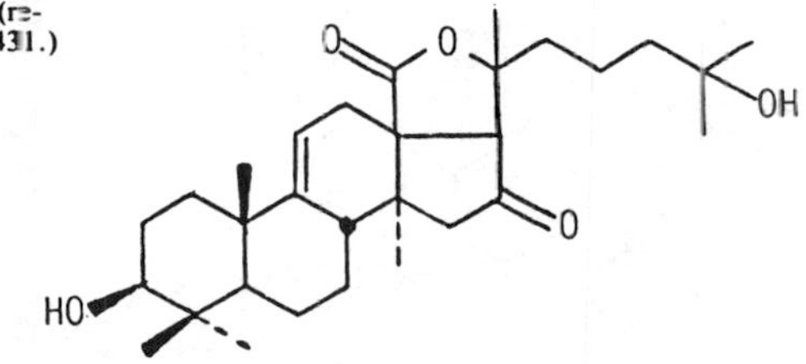 Stichopogenin A_4	Sea cucumber: *Stichopus japonicus*	168
355. $C_{30}H_{48}O_7$	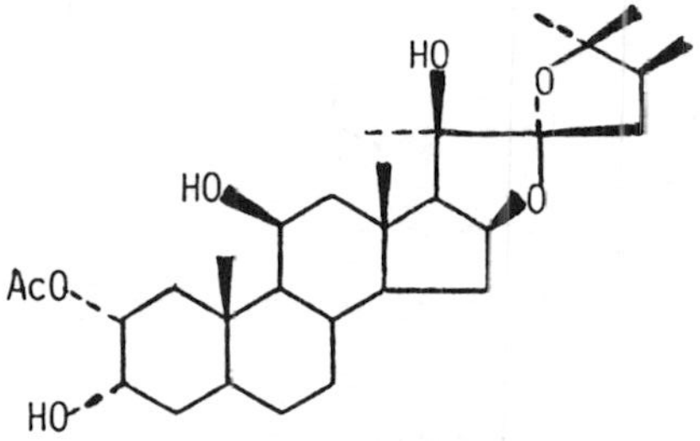 Hippurin-1	Gorgonian: *Isis hippuris*	162

Table 2 (continued)

Molecular formula	Structural formula	Name	Source of compound	Available by synthesis	Available nonmarine sources	Activity	References
C_{30}							
356. $C_{30}H_{50}O$	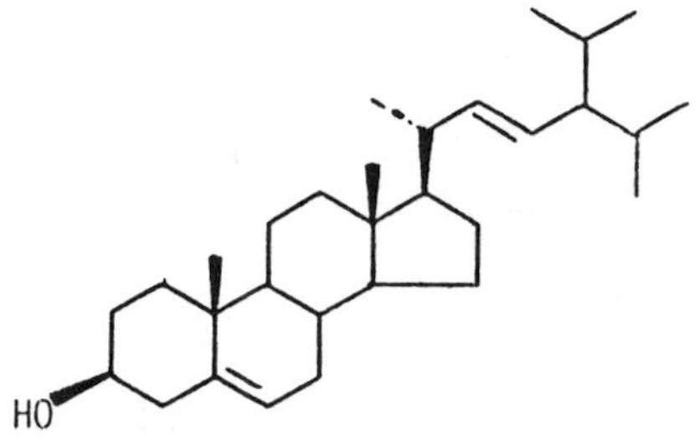	24-Isopropylcholesta-5,22-dien-3β-ol	Sponge: *Pseudaxinyssa* sp.				120
357. $C_{30}H_{52}O$	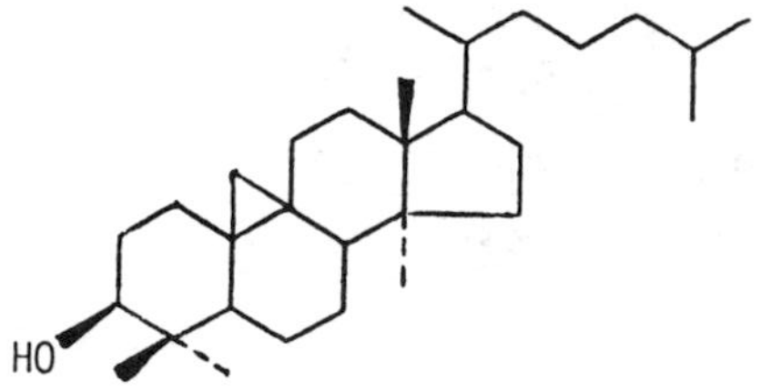	Cycloartanol	Red alga: *Rhodymenia palmata* Starfish: *Asterias rubens*				84 249
358. $C_{30}H_{52}O$	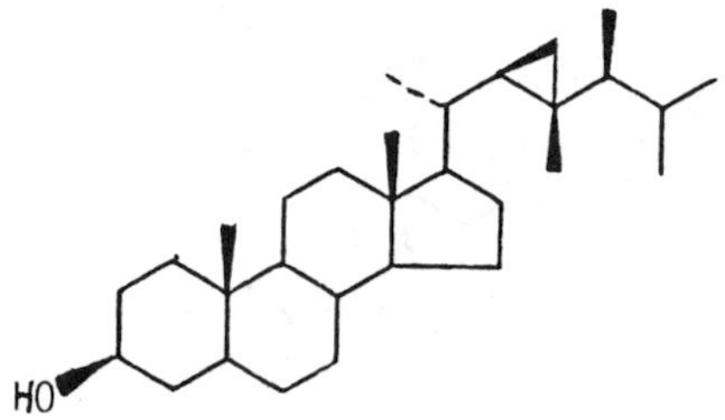	Gorgostanol	Starfish: *Acanthaster planci*				148

C_{30}					
359. $C_{30}H_{52}O$	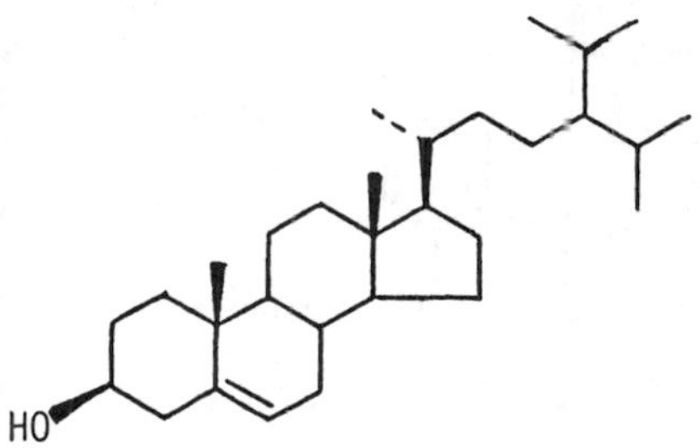 24-Isopropylcholest-5-en-3β-ol	Sponge: *Pseudaxinyssa* sp.		120
360. $C_{30}H_{52}O_5$	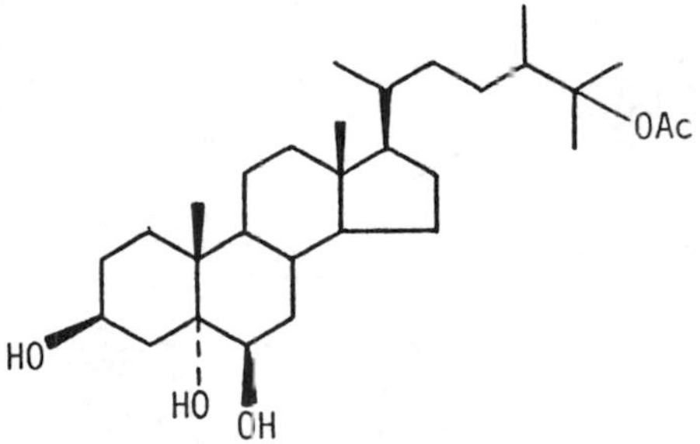 24-Metylcholestane-3β,5α,6β,25-tetrol-25-monoacetate	Soft coral: *Sarcophyton elegans*	+	209
C_{31}					
361. $C_{31}H_{46}O_2$	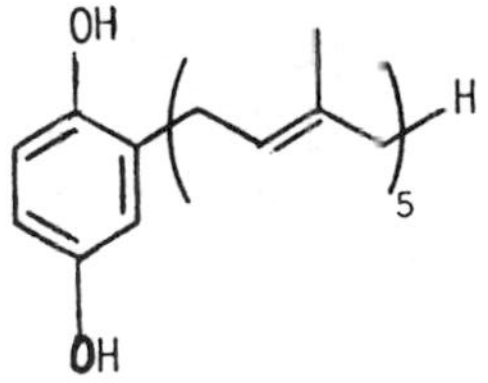 2-Pentaprenyl-1,4-benzoquinol	Sponge: *Ircinia ramosa*		162

Table 2 (continued)

Molecular formula	Structural formula	Name	Source of compound	Available by synthesis	Available nonmarine sources	Activity	References
362. $C_{31}H_{46}O_2$	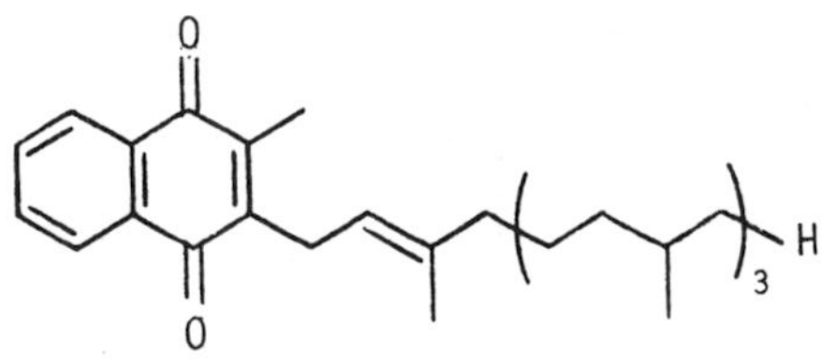	Phylloquinone Vitamin K_1	Algae	+	+	Prophylaxis + treatment of hypoproth-rombinemia	73 202 215
C_{32} 363. $C_{32}H_{47}BrO_{10}$	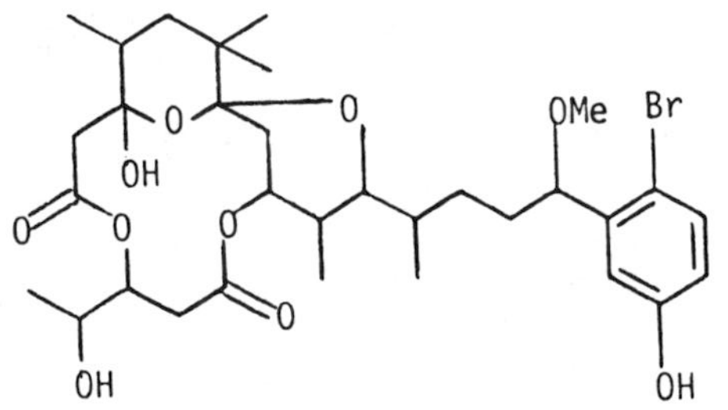	Aplysiatoxin	Sea hare: *Stylocheilus longi-cauda*				159 160
364. $C_{32}H_{48}O_{10}$	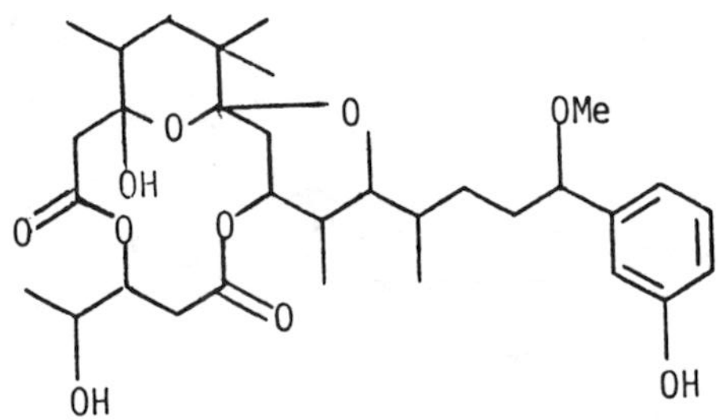	Debromoaplysiatoxin	Sea hare: *Stylocheilus longi-cauda*				159 160

C_{33}

No.	Structure	Name	Source	Ref.
365. $C_{33}H_{32}N_4O_5$	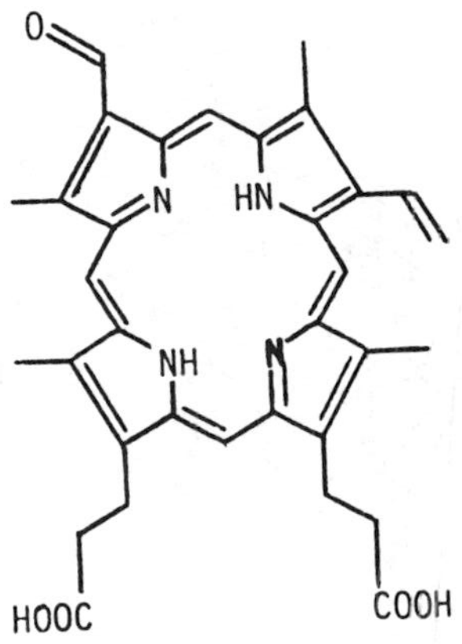 O, N, HN, NH, N, HOOC, COOH	Chlorocruoroporphyrin	Starfish: *Astropecten irregularis* *Luidia ciliaris*	231
366. $C_{33}H_{54}O_7$	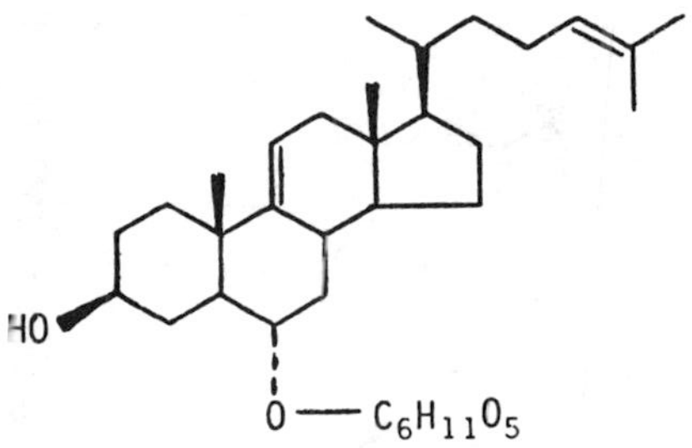 HO, O — $C_6H_{11}O_5$	3β-Hydroxy-5α-cholesta-9(11), 24-dien-6α-yl-β-D-glucoside	Starfish: *Marthasterias glacialis*	265

Table 2 (continued)

Molecular formula	Structural formula	Name	Source of compound	Available by synthesis	Available nonmarine sources	Activity	References
C_{34}							
367. $C_{34}H_{33}FeN_4O_5$	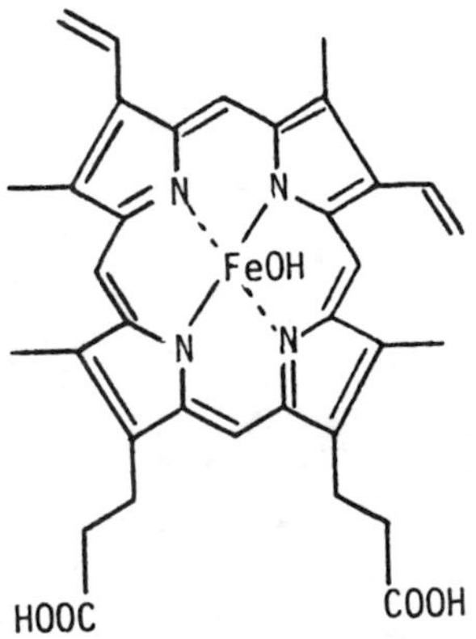 N, N, FeOH, N, N, HOOC, COOH	Hematin	Sea worms		+		231
368. $C_{34}H_{34}N_4O_4$	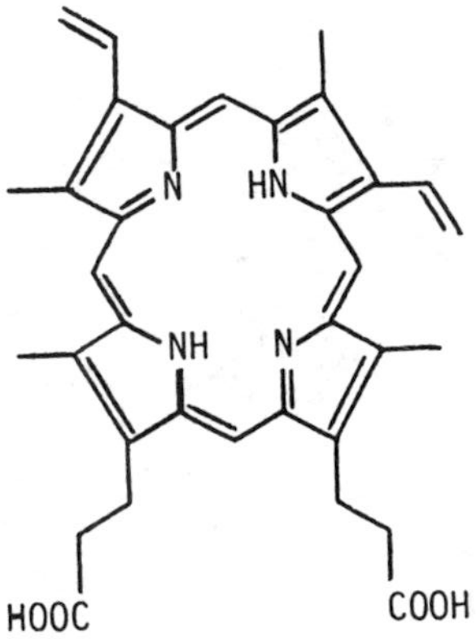 N, HN, NH, N, HOOC, COOH	Protoporphyrin IX	Mollusc: *Bankia setacea* Sea worm: *Lineus longissimus* Starfish		+		103 202 231

369. $C_{34}H_{38}N_4O_6$	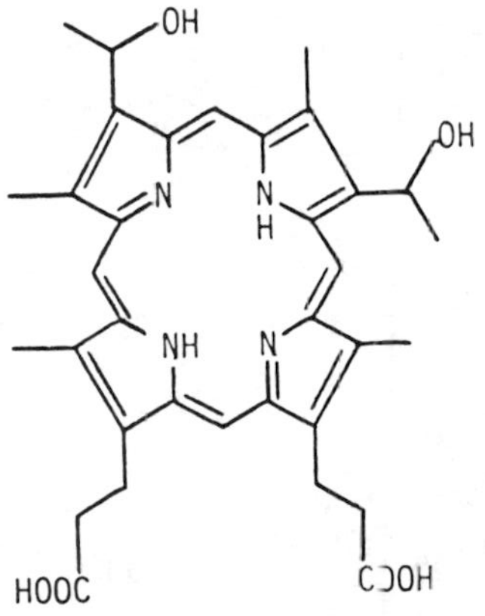	Hematoporphyrin	Corals Jellyfish Sea anemones		Antidepressant	202 231
C_{36} 370. $C_{36}H_{38}N_4O_8$	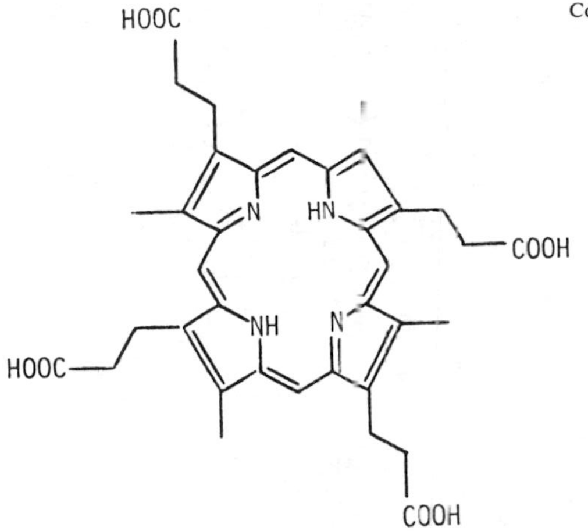	Coproporphyrin I	Sea worms: *Amphitrite johnstoni* *Arenicola marina*	+		231

Table 2 (continued)

Molecular formula	Structural formula	Name	Source of compound	Available by synthesis	Available nonmarine sources	Activity	References
371. $C_{36}H_{38}N_4O_8$	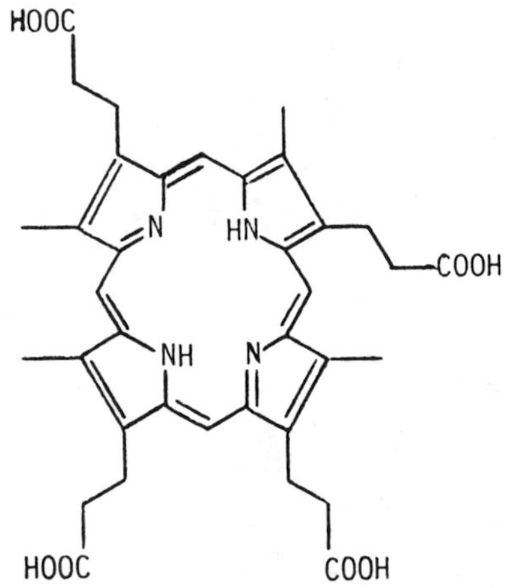 HOOC, NH, N, HN, COOH, HOOC, COOH	Coproporphyrin III	Sea worms		+		231

C_{40}

Compound	Name	Source		References
372. $C_{40}H_{38}N_4O_{16}$	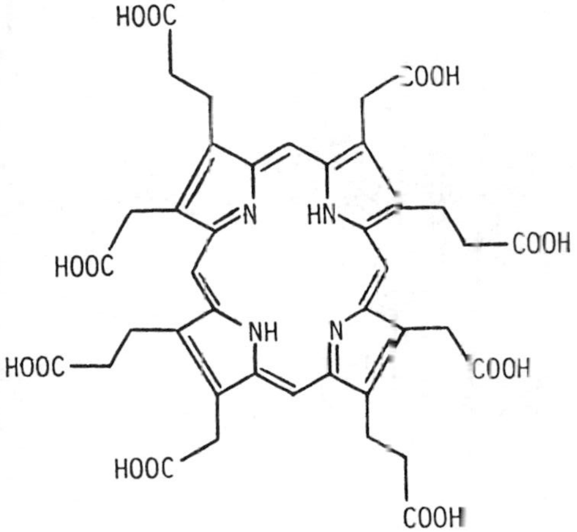 Uroporphyrin I	Molluscs	+	103 231
373. $C_{40}H_{48}O_4$	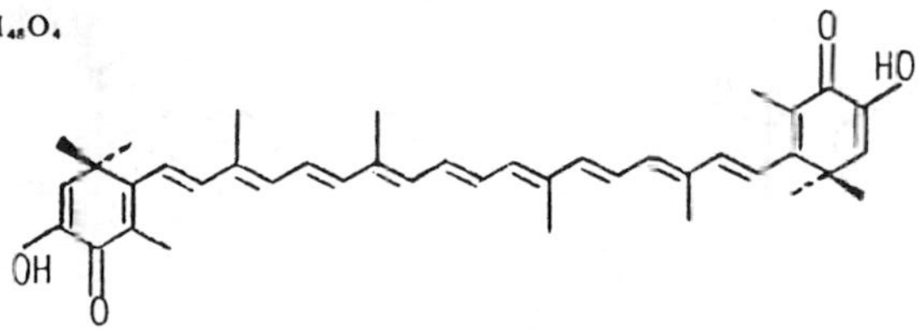 Astacene	Crustaceans Fish Sponge: *Verangia aerophoba*		61, 62, 63, 64 65,133 198,200
374. $C_{40}H_{52}O_3$	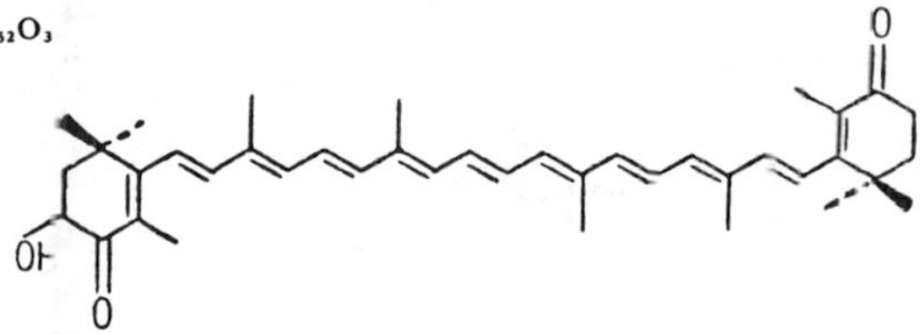 3-Hydroxycanthaxanthin Phoenicoxanthin	Crustaceans Fish Sea cucumber: *Psolus fabrichii*	+	34,133 153,156 157,158 196,199

Table 2 (continued)

Molecular formula	Structural formula	Name	Source of compound	Available by synthesis	Available nonmarine sources	Activity	References
C_{40}							
375. $C_{40}H_{52}O_3$	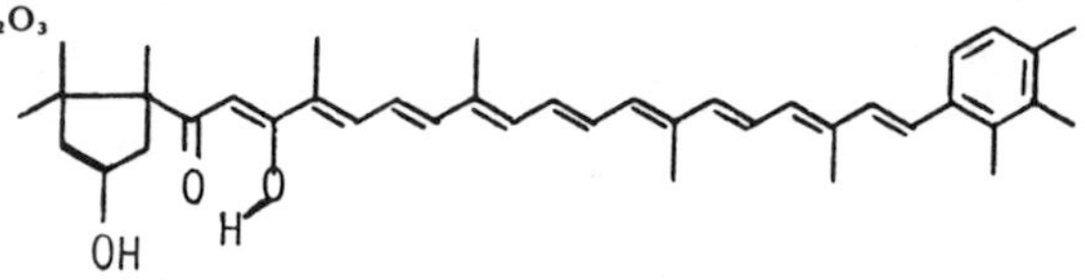	Trikentriorhodin	Sponge: *Trikentrion helium*				5
376. $C_{40}H_{54}O_2$	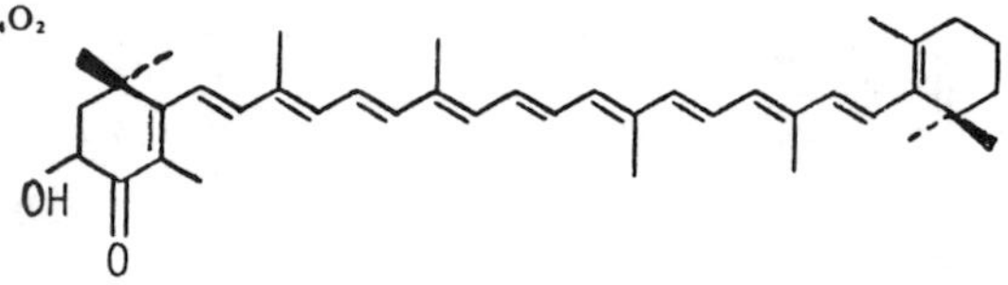	Hydroxyechinenone 3-Hydroxy-4-keto-β-carotene	Crustacean: *Arctodiaptomus salinus* Starfish: *Asterina panceri*		+		133 151
377. $C_{40}H_{54}O_2$	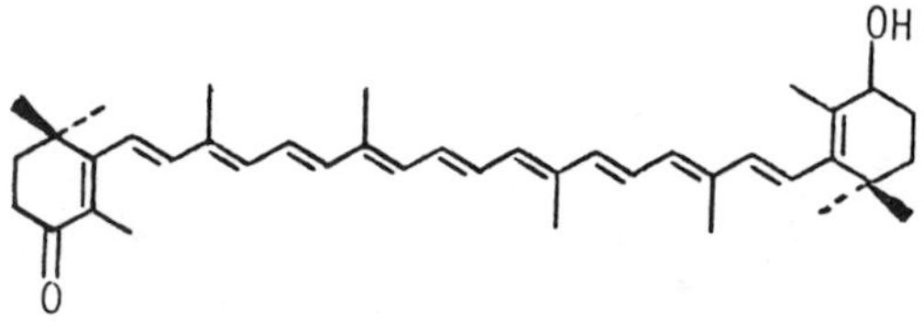	4′-Hydroxyechinenone 4-Hydroxy-4′-keto-β-carotene	Crustaceans Fish		+		64,65 133,151 156
378. $C_{40}H_{54}O_3$	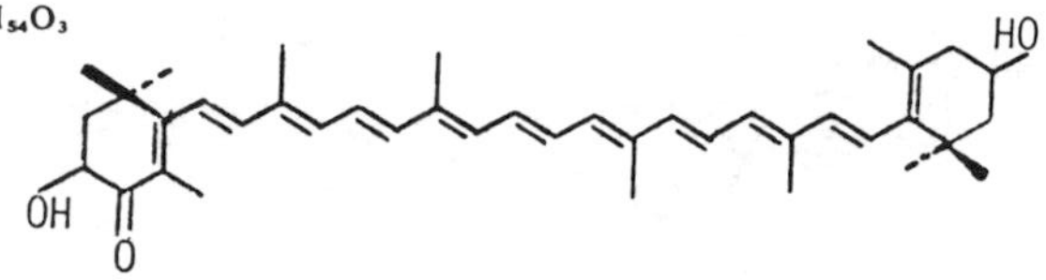	Adonixanthin 3,3′-Dihydroxy-4-keto-β-carotene β-Doradexanthin	Crustaceans Starfish: *Asterina panceri*		+		133 151 196 199

Compound	Name	Source			Remarks	References
379. $C_{40}H_{54}O_3$	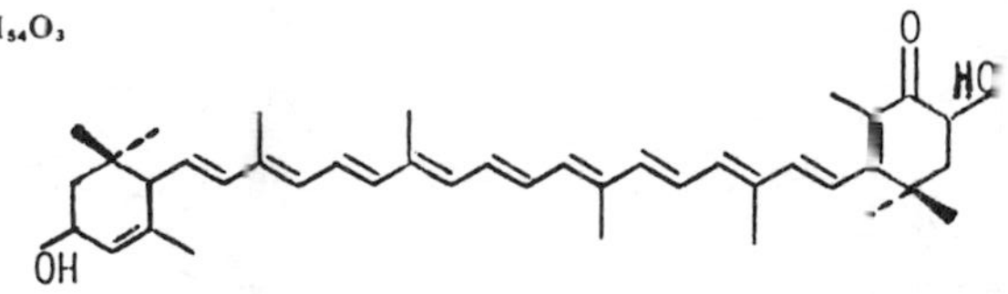 α-Doradexanthin 4-Ketolutein	Sea bream: *Chrysophrys major*				133 157
380. $C_{40}H_{54}O_4$	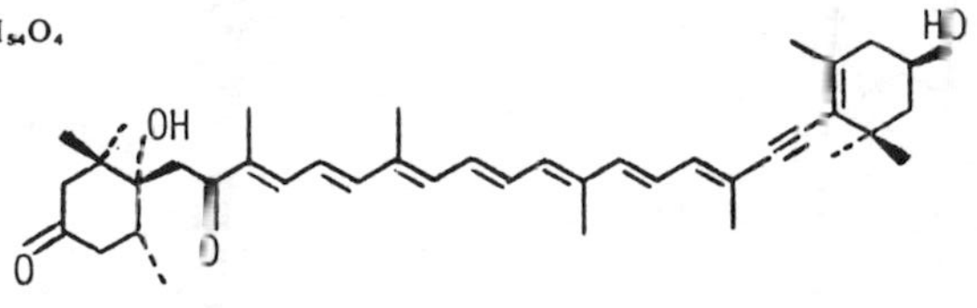 Isomytiloxanthin	Mussel: *Mytilus edulis*				165
381. $C_{40}H_{54}O_4$	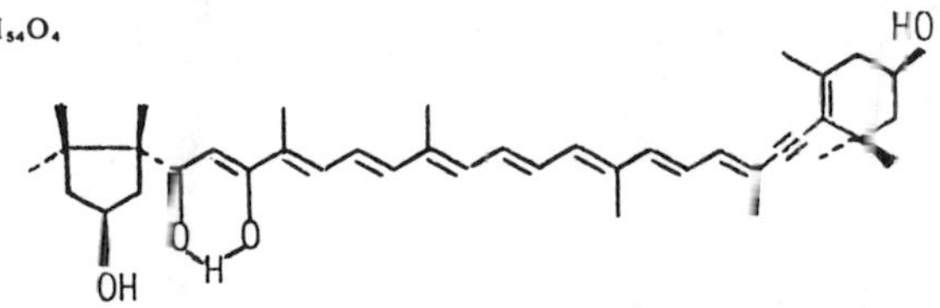 Mytiloxanthin	Mussels: *Mytilus californianus* *M. edulis*				103 165
382. $C_{40}H_{56}$	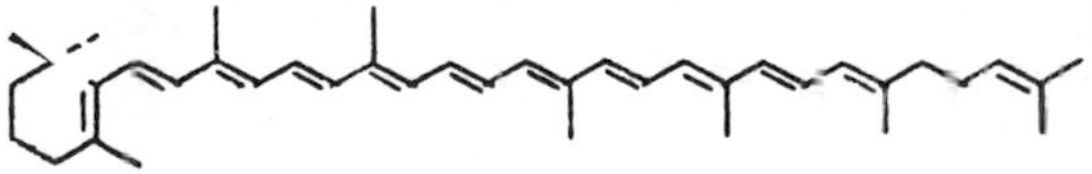 γ-Carotene	Crustaceans Sponges: *Hymeniacidon sanguinea* *Verongia aerophoba*	+	+	Vitamin A activity	61,65 102,133 153,202
383. $C_{40}H_{56}O$	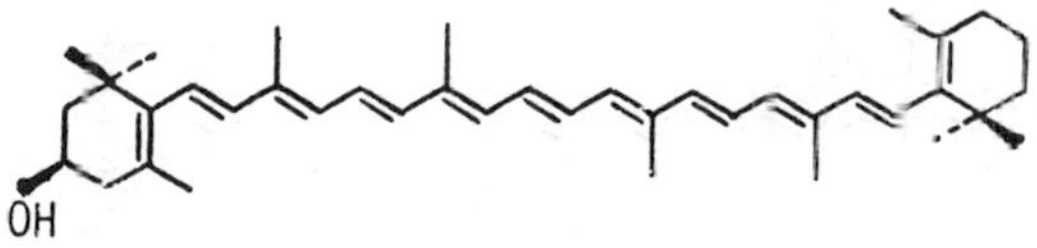 Cryptoxanthin	Crustaceans Fish Molluscs Starfish: *Asterina panceri*	+	+	Vitamin A activity	63 65 103 133 151 198 202
384. $C_{40}H_{56}O$	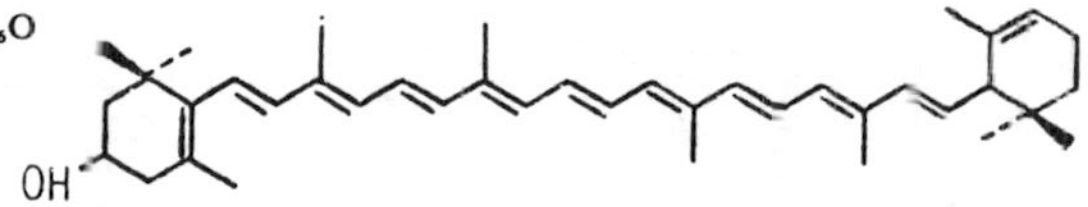 α-Cryptoxanthin	Fish				133 157 158 198

Table 2 (continued)

Molecular formula	Structural formula	Name	Source of compound	Available by synthesis	Available nonmarine sources	Activity	References
C_{40}							
385. $C_{40}H_{56}O$	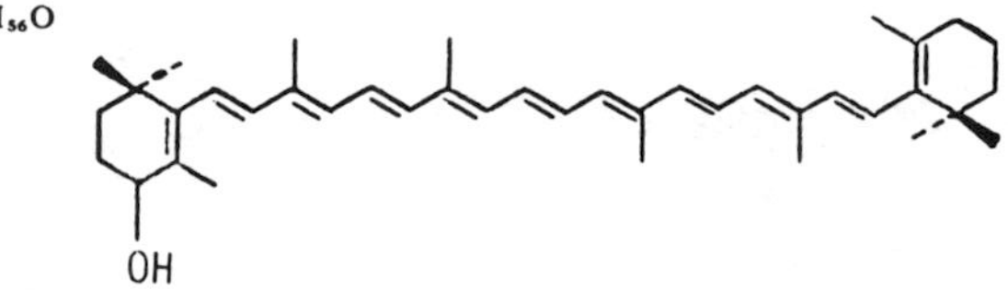	Isocryptoxanthin	Crustaceans Sea worm: *Sabella penicillis*	+	+		63 133 151 156
386. $C_{40}H_{56}O_2$	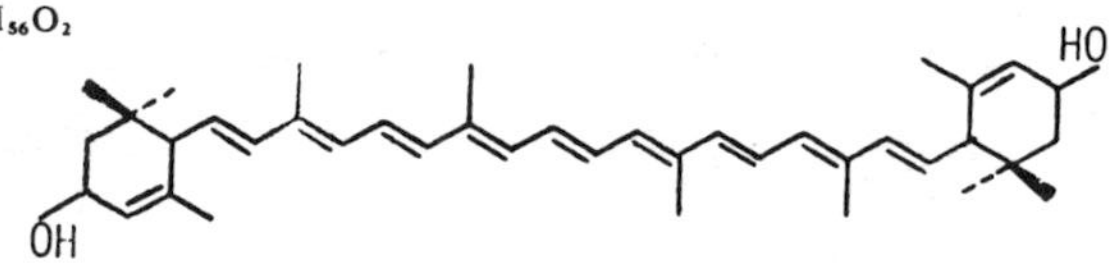	3,3′-Dihydroxy-ε-carotene Tunaxanthin	Fish Prawn: *Penaeus japonicus*				60, 62 64,133 152,154 155,157 158, 198
387. $C_{40}H_{56}O_2$	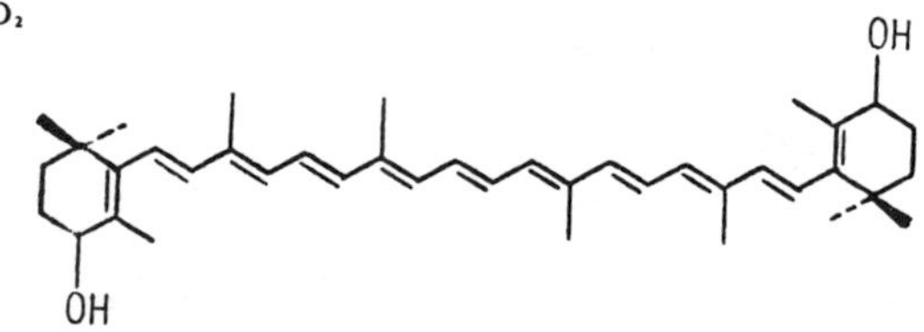	Isozeaxanthin	Crustaceans Fish Sponge: *Verongia aerophoba*	+	+		61,63 64, 65 133,151
388. $C_{40}H_{56}O_3$	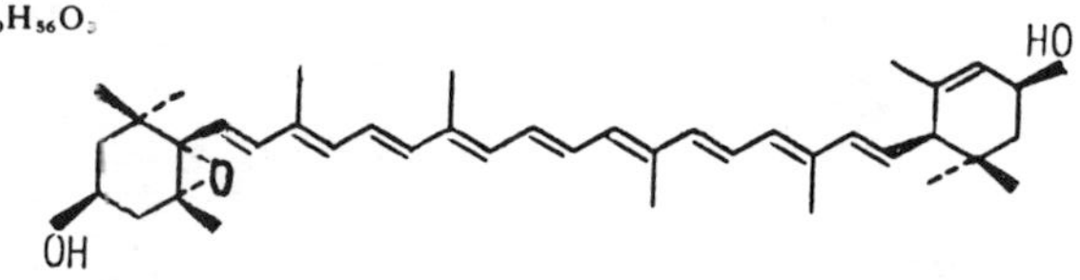	Lutein epoxide Taraxanthin	Crustaceans Fish		+		62 63 64 65 133

No.	Formula	Name	Source		References
389.	$C_{40}H_{56}O_3$	Lutein-5,8-epoxide Flavoxanthin	Crustaceans Mollusc: *Venus japonica*	+	103,133 151,200 202
390.	$C_{40}H_{56}O_4$	Crustaxanthin	Crustaceans: *Calanoides* sp. *Arctodiaptomus salinus*		133 151 153
391.	$C_{40}H_{56}O_4$	Heteroxanthin	Yellow-brown alga: *Vaucheria sessilis*		133 225
392.	$C_{40}H_{56}O_4$	Neoxanthin	Brown alga: *Fucus vesiculosus* Sponge: *Verongia aerophoba*	+	61 133 223
393.	$C_{40}H_{56}O_4$	Violaxanthin	Sponge: *Verongia aerophoba*	+	61 133 202

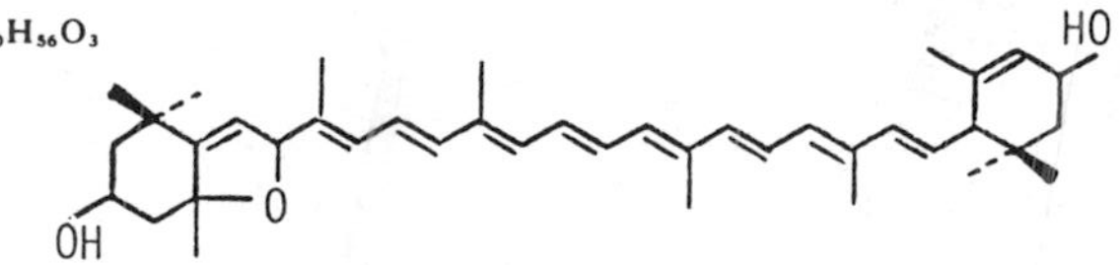

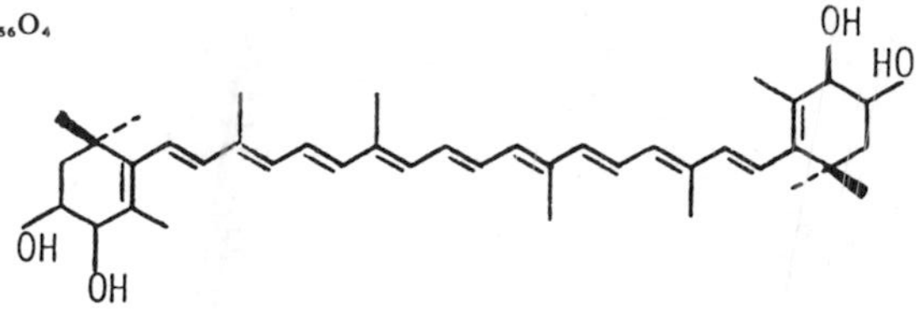

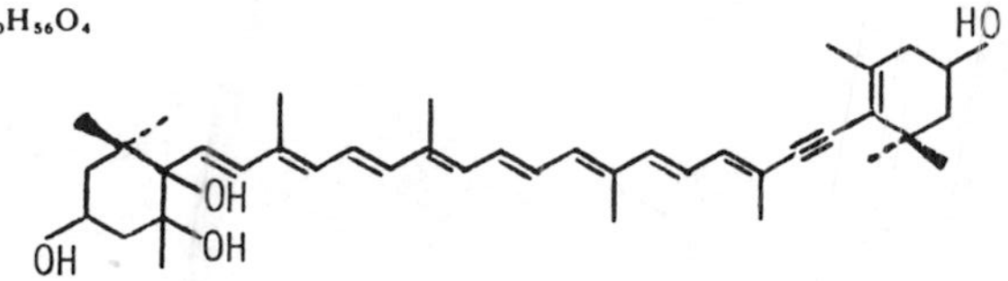

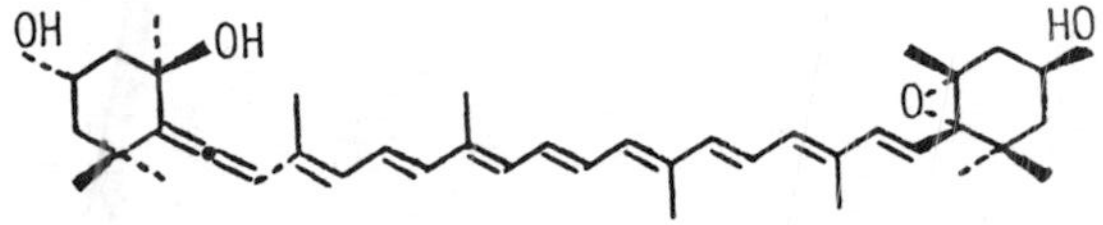

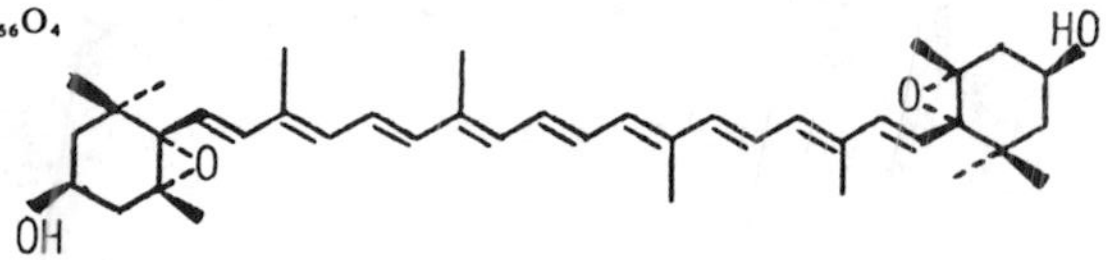

Table 2 (continued)

Molecular formula	Structural formula	Name	Source of compound	Available by synthesis	Available nonmarine sources	Activity	References
C_{42}							
394. $C_{42}H_{58}O_2$		4-Keto-4′-ethoxy-β-carotene	Crustacean: *Eupagurus bernhardus*				65
C_{53}							
395. $C_{53}H_{80}O_2$		Plastoquinone-9	Algae		+		27 73 215

C_{55} 396. $C_{55}H_{72}MgN_4O_5$ 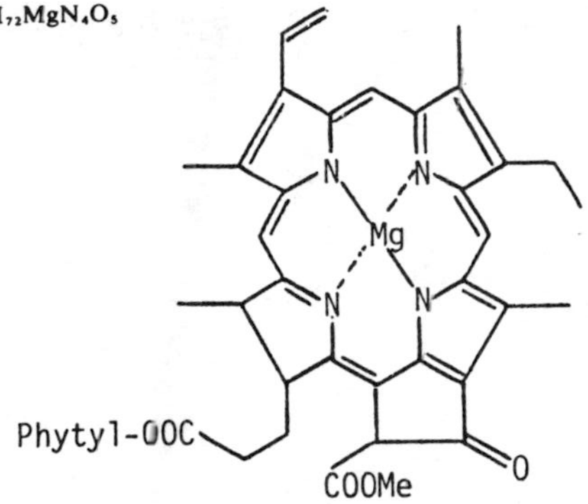	Chlorophyll a	Algae	+	+		38 73 202
C_{59} 397. $C_{59}H_{94}O_{27}$ 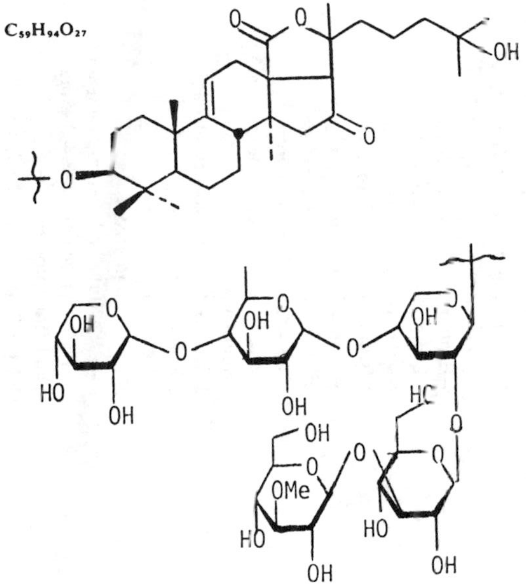	Holotoxin A	Sea cucumber: *Stichopus japonicus*			Antifungal	167 168

COMPOUND NAME INDEX FOR TABLE 2

Name	Compound number
Acanthellin-1	164
N-Acetoglycolyl-4-methyl-4,9-dideoxyneuraminic acid	128
Acetoxyfimbrolide A	102
Acetoxyfimbrolide B	103
Acetoxyfimbrolide C	101
Acetoxyfimbrolide G	100
Acetoxyintricatol	174
Acetylcholine	51
N-Acetylneuraminic acid	114
Acrylic acid	11
Adenosine diphosphate	96
Adenosine triphosphate	99
5′-Adenylic acid	91
Adonixanthin	378
AF-350 monosulfate	172
Africanol	158
Allo-laurinterol	132
α-Amino-*n*-butyric acid	22
γ-Amino-*n*-butyric acid	23
2-Aminoimidazole	14
Amuresterol	302
Anserine	98
Aplysiatoxin	363
Astacene	373
Atomaric acid	319
Aucantene	107
Avarol	229
Avarone	227
Axisonitrile-2	165
1,2,4-Benzenetriol	29
Bifuhalol	116
Bis(3,5-dibromo-4-hydroxyphenyl)methane	119
19, 24-Bisnorcholest-*cis*-22-en-3β-ol	278
Bromoacetone	12
4-Bromo-α-chamigren-8,9-epoxide	149
4-Bromo-α-chamigren-8-one	150
4-Bromo-β-chamigren-8-one	151
1-Bromo-3-chloroacetone	9
6-Bromo-2-chloromyrcene	88
(E)-9-Bromo-2-chloromyrcene	89
(Z)-9-Bromo-2-chloromyrcene	90
3-Bromo-4,5-dihydroxybenzyl alcohol	41
Bromoform	1
3-Bromo-4-hydroxybenzyl alcohol	40
2-Bromomyrcene	92
(E)-9-Bromomyrcene	93
(Z)-9-Bromomyrcene	94
(+)-*R*-4-Butylcyclohepta-2,6-dienone	109
(+)-6-Butylcyclohepta-2,4-dienone	110
Candicine chloride	113
Δ9(12)-Capnellene-3 β,8β,10α-triol	156
L-Carnitine	49
Carnosine	76
γ-Carotene	382
Cartilagineal	79
Caulerpol	216
Caulerpol acetate	255

COMPOUND NAME INDEX FOR TABLE 2 (continued)

Name	Compound number
Cetyl alcohol	171
Chimyl alcohol	194
Chlorocruoroporphyrin	365
1-Chloro-1,3-dibromoacetone	5
3-Chloro-1,1-dibromoactone	6
6-Chloro-2,(Z)-9-dibromomyrcene	85
(Z)-9-Chloro-2,6-dibromomyrcene	86
2-Chloromyrcene	95
Chlorophyll a	396
3-Chloro-1,1,3-tribromoacetone	3
5α-Cholesta-7,24-dien- 3β-ol	303
Cholest-4-en-3-one	304
Cholesteryl sulfate	311
Citrulline	35
Clerosterol	339
Codisterol	323
Coproporphyrin I	370
Coproporphyrin III	371
Creatinine	21
Crustaxanthin	390
Cryptoxanthin	383
α-Cryptoxanthin	384
(−)-β-Curcumene	153
Cycloartanol	357
Cycloeudesmol	159
Cyclospongiaquinone-1	251
Cyclospongiaquinone-2	252
Cypridina luciferin	249
5α-Cyprinol	315
5β-Cyprinol	316
Cystine	32
Dactyloxene-B	155
Dactylyne	135
Deacetyllaurencin	138
Debromoaplysiatoxin	364
1a,2,3,6,7,9,10a,13,14,14a-Decahydro-1a,5,8, 12-tetramethyloxireno[9,10]-cyclotetradeca-[1,2-b]furan	209
1a,2,3,6,7,9,10a,13,14,14a-Decahydro-1a,5,8,12-tetramethyloxireno[9,10]-cyclotetradeca[1,2-b]furan	210
Dehydroepoxynephthenol	214
epi-Dendalone acetate	320
Dendalone 3-hydroxybutyrate	353
1,3-Dibromoacetone	10
2,7-Dibromo-8-chloro-octahydro-8,10,10-trimethyl-5-methylene-6H-2,5a-methano-1-benzoxepin-3,4-diol	140
2,3-Dibromo-4,5- dihydroxybenzyl-*n*-propyl ether	80
1,3-Dibromo-2-heptanone	48
3,5-Dibromo-4-hydroxybenzaldehyde	36
3,5-Dibromo-4- hydroxyphenylacetic acid	54
3,5-Dibromo-4-hydroxyphenylpyruvic acid	66
1,1-Dibromo-3-iodoacetone	7
Didehydrocyclospongiaquinone-1	250
Dihydroflexibilide	215
Dihydrolaurene	142
7,8-Dihydroxanthopterin	31
3,3′-Dihydroxy- ε-carotene	386

COMPOUND NAME INDEX FOR TABLE 2 (continued)

Name	Compound number
2,4′-Dihydroxy-5-hydroxymethyl-3,3,′5′-tribromodiphenylmethane	124
3,3′-Dihydroxy-4-keto-β-carotene	378
3,5-Dihydroxyphenylacetic acid	57
4,4-Dimethyl-5α-cholesta-7, 24-dien-3β-ol	340
4,4-Dimethyl-5α-cholesta-8(9),24-dien-3β-ol	341
23,24-Dimethylcholesta-5,22-dien-3β-ol	342
4,4-Dimethyl-5α-cholest-7-en-3β-ol	347
[3-(Dimethylsulfonio)propyl]-trimethylammonium dichloride	64
2-Diprenyl-1,4-benzoquinol	163
Docosane	258
1-*O*-Docosylglycerol	280
1-*O*-Dodecylglycerol	162
Dopamine	60
α-Doradexanthin	379
β-Doradexanthin	378
Eicosane	222
Eicosanol	223
1-*O*-Eicosylglycerol	263
Elatol	144
5,8-Epidioxycholesta-6,22-dien-3β-ol	301
Epizoanthoxanthin A	121
Epizoanthoxanthin B	125
Epoxynephthenol acetate	256
Ergosterol peroxide	322
Erythropterin	67
β-Estradiol	183
Estrone	182
24-Ethylcholesta-5,22-dien-3β-ol	343
24-Ethylcholesta-7,22-dien-3β-ol	344
24-Ethylcholest-5-en-3β-ol	348
24-Ethyl-3β-hydroxymethyl-A-nor-5α-cholestane	352
24-Ethyl-3β-hydroxymethyl-A-nor-5α-*cis*-cholest-22-ene	349
(E)-24-Ethylidene-5α-cholest-7-en-3β-ol	345
(Z)-24-Ethylidene-5α-cholest-7-en-3β-ol	346
24-Ethyl-19-norcholestan-3β-ol	335
24-Ethyl-19-norcholest-*trans*-22-en-3β-ol	328
Ethyl urocanate	58
Fasciospongin	277
Filiformin	133
Filiforminol	134
Fimbrolide A	70
Fimbrolide B	71
Fimbrolide C	68
Flavoxanthin	389
Flexibilide	212
Floridoside	77
Fucoserratene	62
L-Fucose-4-sulfate	33
Fumaric acid	18
Furospongenol	231
Furospongenone	228
4-[11-(3-Furyl)-6-hydroxy-4,8-dimethyl-3-undecenyl]-2,6-dioxabicyclo[3.1.0]-hexan-3-one	232
4-[11-(3-Furyl)-6-hydroxy-4,8-dimethyl-8-undecenyl]-2,6-dioxabicyclo[3.1.0]-hexan-3-one	233

COMPOUND NAME INDEX FOR TABLE 2 (continued)

Name	Compound number
5-[11-(3-Furyl)-6-hydroxy-4,8-dimethyl-3-undecenyl]-2,6-dioxabicyclo[3.1.0]-hexan-3-one	234
5-[11-(3-Furyl)6-hydroxy-4,8-dimethyl-8-undecenyl]-2,6-dioxabicyclo[3.1.0]-hexan-3-one	235
3-[11-(3-Furyl)-6-hydroxy-4,8-dimethyl-3-undecenyl]-5-hydroxy-2(5H)-furanone	236
3-[11-(3-Furyl)-6-hydroxy-4,8-dimethyl-8-undecenyl]-5-hydroxy-2(5H)-furanone	237
4[11-(3-Furyl)-6-hydroxy-4,8-dimethyl-3-undecenyl]-5-hydroxy-2(5H)-furanone	238
4-[11-(3-Furyl)-6-hydroxy-4,8-dimethyl-8-undecenyl]-5-hydroxy-2(5H)-furanone	239
Geranylfarnesol	279
Glanduliferol	154
Glycerylphosphorylcholine	65
N-Glycolylneuraminic acid	115
Gorgostanol	358
Halmiformin-1	276
Hematin	367
Hematoporphyrin	369
Heneicosane	247
cis,cis,cis,cis,cis-3,6,9,12,15-Heneicosapentaene	244
1-*O*-Heneicosylglycerol	270
Heptadecane	177
Heptadecanol	178
1-(Heptadeca-5,8,11,14-tetraenyl)-3,5-dihydroxybenzene	261
1-*O*-Heptadecylglycerol	224
Heteronemin	338
Heteroxanthin	391
Hexabromo-2,2′-bipyrrole	53
cis,cis-5,9-Hexacosadienoic acid	296
cis,cis,cis 5,9,19 Hexacosatrienoic acid	295
Hexadecane	170
Hexadecanol	171
1-*O*-Hexadecylglycerol	194
Hippurin-1	355
Holotoxin A	397
Homoserine	24
Homoserine betaine	50
4-Hydroxybenzaldehyde	39
3-Hydroxycanthaxanthin	374
3*β*-Hydroxy-5*α*-cholesta-9(11),24-dien-6*α*-yl-*β*-D-glucoside	366
Hydroxyechinenone	376
4′-Hydroxyechinenone	377
Hydroxyfimbrolide A	72
Hydroxyfimbrolide B	73
Hydroxyfimbrolide C	69
1-*O*-(2-Hydroxy-4-hexadecenyl)glycerol	191
1-*O*-(2-Hydroxyhexadecyl)glycerol	195
Hydroxyhydroquinone	29
24-Hydroxyircinolide	274
3-Hydroxy-4-keto-*β*-carotene	376
4-Hydroxy-4′-keto-*β*-carotene	377
3-Hydroxy-L-kynurenine	81
(−)-*S*-Hydroxymethyl-L-homocysteine	25
3*β*-Hydroxymethyl-24-methyl-A-nor-5*α*-cholestane	336

COMPOUND NAME INDEX FOR TABLE 2 (continued)

Name	Compound number
3β-Hydroxymethyl-24-methyl-A-nor-5α-*cis*-cholest-22-ene	329
3β-Hydroxymethyl-A-nor-5α-cholestane	312
3β-Hydroxymethyl-A-nor-5α-*cis*-cholest-22-ene	307
1-*O*-(2-Hydroxy-4-octadecenyl)glycerol	246
p-Hydroxyphenylacetic acid	56
1-*O*-(2-Hydroxy-4-tetradecenyl)glycerol	176
1-*O*-(2-Hydroxytetradecyl)glycerol	180
Ichthyopterin	74
N(4-Imidazolepropionyl)histamine	106
Inosine	82
Inosinic acid	87
Iodoacetone	13
1-Iodo-3,3-dibromo-2-heptanone	43
Ircinianin	275
Ircinolide	273
Isoagatholactone	211
Isocaespitol	157
Isocryptoxanthin	385
T-Isolaureatin	136
3-*trans*-Isolaureatin	136
Isomytiloxanthin	380
24-Isopropylcholesta-5,22-dien-3β-ol	356
24-Isopropylcholest-5-en-3β-ol	359
6-Isopropyl-3,9,13-trimethyl-2,7,9,12-cyclotetradecatetraen-1-ol	213
Isospongiaquinone	253
Isoxanthopterin	28
Isozeaxanthin	387
24-Ketocholesterol	306
4-Keto-4′-ethoxy-β-carotene	394
4-Ketolutein	379
Δ^4-3-Keto-pregnen-20β-ol	240
D(−)-Lactic acid	15
Latimerol	317
T-Laureatin	137
3-*trans*-Laureatin	137
6*R*, 7*R cis*-Laurediol	145
6*S*, 7*S cis*-Laurediol	146
6*R*, 7*R trans*-Laurediol	147
6*S*, 7*S trans*-Laurediol	148
LL-PAA216	120
Lobophytolide	200
Lutein epoxide	388
Lutein-5,8-epoxide	389
Malic acid	20
L(−)-Methionine-*l*-sulfoxide	26
1-*O*-(2-Methoxydocosyl)glycerol	297
1-*O*-(2-Methoxyeicosyl)glycerol	271
1-*O*-(2-Methoxyheneicosyl)glycerol	281
1-*O*-(2-Methoxyheptadecyl)glycerol	248
1-*O*-(2-Methoxy-4-hexadecenyl)glycerol	221
1-*O*-(2-Methoxyhexadecyl)glycerol	225
1-*O*-(2-Methoxynonadecyl) glycerol	265
1-*O*-(2-Methoxy-4-octadecenyl)glycerol	257
1-*O*-(2-Methoxyoctadecyl)glycerol	260
1-*O*-(2-Methoxypentadecyl)glycerol	196
1-*O*-(2-Methoxytetradecyl)glycerol	188
4α-Methyl-5α-cholesta-7,24-dien-3β-ol	324

COMPOUND NAME INDEX FOR TABLE 2 (continued)

Name	Compound number
24-Methylcholesta-5,22-dien-3β-ol	325
24α-Methylcholesta-5,22-dien-3β-ol	326
24-Methylcholesta-7,22-dien-3β-ol	327
24-Methylcholestane-3β,5α,6β,25-tetrol-25-monoacetate	360
24-Methylcholesta-7,22,25-trien-3β-ol	321
4α-Methyl-5α-cholest-7-en-3β-ol	330
24-Methyl-5α-cholest-5-en-3β-ol	331
24-Methyl-5α-cholest-7-en-3β-ol	332
24-Methylene-19-norcholestan-3β-ol	308
1-Methylhistidine	46
3-Methylhistidine	47
N-Methylmethionine sulfoxide	34
Methyl-*trans*-monocyclofarnesate	168
N-Methylmurexine	118
24-Methyl-19-norcholestan-3β-ol	313
Methylpalmitate	175
[3-(Methylthio)propyl] trimethylammonium chloride	52
Methyl-δ-tocotrienol methyl ether	337
Multifidene	108
Myrcene	97
Myristyl alcohol	130
Mytiloxanthin	381
Neoxanthin	392
Nephthenol	217
Nidificene	152
Nidifidiene	143
Nonadecane	192
Nonadecanol	193
cis,cis,cis,cis,cis-3,6,9,12,15-Nonadecapentaene	189
cis,cis,cis,cis-4,7,10,13-Nonadecatetraene	190
1-*O*-Nonadecylglycerol	259
l-Noradrenaline	61
22-*cis*-24-Norcholesta-5,22-dien-3β-ol	285
19-Norcholestan-3β-ol	294
19-Norcholest-*trans*-22-en-3β-ol	293
31-Norcycloartanol	350
19-Nor-5α,10β-ergost-22-en- 3β-ol	309
3-Norpseudozoanthoxanthin	104
Occelasterol	305
Octadecane	185
Octadecanol	186
cis-9-Octadecen-1-ol	184
N-(1,2,3,4,4a,7,8,8a-Octahydro-4-isopropyl-1,6-dimethyl-1-naphthyl)formamide	169
1,2,3,4,4a,7,8,8a-Octahydro-4-isopropyl-1,6-dimethyl-1-naphthyl isocyanide	166
1,2,3,4,4a,7,8,8a-Octahydro-4-isopropyl-1,6-dimethyl-1-naphthyl isothiocyanate	167
2,3,6,7,10a,13,14,14a-Octahydro-1a,5,8,12-tetramethyloxireno[9,10]cyclotetradeca-[1,2,-b] furan-9(1aH)-one	201
2,3,6,7,10a,13,14,14a-Octahydro-1a,5,8,12-tetramethyloxireno[9,10]cyclotetradeca-[1,2-b]furan-9(1aH)-one	202
Oleyl alcohol	184
Ornithine	27
24-Oxocholesterol	306
Pacifenol	139

COMPOUND NAME INDEX FOR TABLE 2 (continued)

Name	Compound number
Parazoanthoxanthin A	78
Parazoanthoxanthin E	126
Parazoanthoxanthin F	127
Pentadecane	160
Pentadecanol	161
1-*O*-Pentadecylglycerol	187
2-Pentaprenyl-1,4-benzoquinol	361
Petromyzonol	268
15*R*-PGA_2 methyl ester	241
$PGF_{1\alpha}$	219
$PGF_{2\alpha}$	218
Phenylacetic acid	55
Phloroglucinol	30
Phoenicoxanthin	374
Phosphoethanolamine	2
Phylloquinone	362
1-*O*-Phytanylglycerol	264
Phytol	220
Plastoquinone-9	395
Pleraplysillin-2	199
Prepacifenol epoxide	141
Progesterone	230
n-Propyl-4-hydroxybenzoate	83
Protoporphyrin IX	368
Pseudozoanthoxanthin	117
Pterin-6-carboxylic acid	38
Renilla luciferin	198
Retinol	208
Rhodophytin	131
Riboflavine	173
Sarcophine	203
Scalaradial	300
Spongiadiol	204
epi-Spongiadiol	205
Spongiatriol	206
epi-Spongiatriol	207
Spongiadiol diacetate	266
epi-Spongiadiol diacetate	267
Spongiaquinone	254
Spongiatriol triacetate	282
epi-Spongiatriol triacetate	283
Stearyl alcohol	186
Stichopogenin A_4	354
Succinic acid	19
Taraxanthin	388
Tauroallocholate	289
Taurochenodeoxycholate	286
Taurocholate	290
Taurodeoxycholate	287
Tauro-3α,12α-dihydroxy-5β-chol-7-en-24-oate	284
Tauro-3α,7β-dihydroxy-5β-cholan-24-oate	288
Taurohaemulcholate	291
Tauro-3α,7β,12α-trihydroxy-5β-cholan-24-oate	292
1,1,3,3-Tetrabromoacetone	4
1,1,3,3-Tetrabromo-2-heptanone	42
Tetrabromopyrrole	17

COMPOUND NAME INDEX FOR TABLE 2 (continued)

Name	Compound number
Tetracosane	269
Tetradecane	129
Tetradecanol	130
1-*O*-Tetradecylglycerol	179
Tetradehydrofurospongin-1	226
3β,6α,15α,24ξ-Tetrahydroxy- 5α-cholestane	314
N-(1,5,9,13-Tetramethyl-1-vinyl-4,8,12-tetradecatrienyl)-formamide	245
1,5,9,13-Tetramethyl-1- vinyl-4,8,12-tetradecatrienyl isocyanide	242
1,5,9,13-Tetramethyl-1-vinyl-4,8,12-tetradecatrienyl isothiocyanate	243
Thelepin	123
α-Tocopherol	351
β-Tocopherol	333
γ-Tocopherol	334
δ-Tocopherol	310
δ-Tocotrienol	298
δ-Tocotrienol epoxide	299
δ-Tocotrienol methyl ether	318
1,1,3-Tribromoacetone	8
2,3,6-Tribromo-4,5-dihydroxybenzyl alcohol	37
1,1,3-Tribromo-2-heptanone	44
1,3,3-Tribromo-2-heptanone	45
Tricosane	262
Tridecane	122
Trifuhalol	181
Trikentriorhodin	375
Trimethylamine oxide	16
Tunaxanthin	386
Tyramine	59
cis,trans-1,3,5-Undecatriene	111
trans,trans,trans-2,4,6-Undecatriene	112
5′-Uridylic acid	75
N-Urocanylhistamine	105
Uroporphyrin I	372
L-Valine betaine	63
Violacein	197
Violacene	84
Violaxanthin	393
Vitamin A	208
Vitamin B_T	49
Vitamin E	351
Vitamin K_1	362
Watasenia oxyluciferin	272

REFERENCES

1. **Abe, H., Uchiyama, M., and Sato, R.,** Isolation of phenylacetic acid and its *p*-hydroxy derivative as auxin-like substances from *Undaria pinnatifida, Agric. Biol. Chem.,* 38, 897, 1974.
2. **Abe, S. and Kaneda, T.,** Occurrence of homoserine betaine in the hydrolyzate of an unknown base isolated from a green alga, *Monostroma nitidum, Bull. Jpn. Soc. Sci. Fish.,* 40, 1199, 1974.
3. **Ackermann, D.,** Über das Vorkommen von Homarin, Taurocyamin, Cholin, Lysin und anderen Aminosäuren sowie Bernsteinsaure in dem Meereswurm-*Arenicola marina, Hoppe-Seyler's Z. Physiol. Chem.,* 302, 80, 1955.
4. **Adachi, K. and Tanaka, J.,** Synthesis of cadalene from bromobenzene, *J. Syn. Org. Chem. (Japan),* 31, 322, 1973.
5. **Aguilar-Martinez, M. and Liaaen-Jensen, S.,** Animal carotenoids. IX, Trikentriorhodin, *Acta Chem. Scand. Ser. B,* 28, 1247, 1974.
6. **Andersen, R. J., Wolfe, M. S., and Faulkner, D. J.,** Autotoxic antibiotic production by a marine *Chromobacterium, Mar. Biol.,* 27, 281, 1974.
7. **Andersen, R. J., and Faulkner, D.J.,** The synthesis of (±) aeroplysinin-1 and related compounds, *Proc. Food-Drugs from the Sea,* 1974, Marine Technology Society, Washington, D.C., 1976, 263.
8. **Andrewes, A. G., Borch, G., Liaaen-Jensen, S., and Snatzke, G.,** Animal carotenoids. 9. On the absolute configuration of astaxanthin and actinioerythrin, *Acta Chem. Scand. Ser. B,* 28, 730, 1974.
9. **ApSimon, J. W. and Eenkhoorn, J. A.,** Marine organic chemistry. II. Synthesis of 3β,6α-dihydroxy-5α-pregn-9(11)-en-20-one, the major sapogenin of the starfish *Asterias forbesi, Can. J. Chem.,* 52, 4113, 1974.
10. **Baker, J. L.,** U.S. Patent 3,778, 469 (Cl. 260-499), 1972.
11. **Baker, J. T. and Murphy, V.,** *Handbook of Marine Science., Compounds from Marine Organisms,* Vol I, CRC Press, Cleveland, Ohio, 1976.
12. **Banville, J. and Brassard, P.,** The synthesis of some naturally occurring anthraquinones, 9th. I.U.P.A.C. Symp. Natural Products,(Abstr.) Ottawa, 1974, 55A.
13. **Barton, D. H. R., Leclerc, G., Magnus, P. D., and Menzies, I. D.,** An unusual synthesis of ergosterol acetate peroxide, *Chem. Commun.,* p. 447, 1972.
14. **Baxter, J. G.,** Vitamin A, *Comp. Biochem.,* 9, 169, 1963.
15. **Bender, J. A., DeRiemer, K., Roberts, T. E., Rushton, R., Boothe, P., Mosher, H. S., and Fuhrman, F. A.,** Choline esters in the marine gastropods *Nucella emarginata* and *Acanthina spirata;* a new choline ester, tentatively identified as *N*- methylmurexine, *Comp. Gen. Pharmac.,* 5, 191, 1974.
16. **Bergmann, W. and Stempien Jr., M. F.,** Contributions to the study of marine products. XLIII. The nucleosides of sponges. V. The synthesis of spongosine. *J. Org. Chem.,* 22, 1575, 1957.
17. **Bernstein, J., Shmeuli, U., Zadock, E., Kashman, Y., and Neeman, I.,** Sarcophine, a new epoxy cembranolide from marine origin, *Tetrahedron,* 30, 2817, 1974.
18. **Bersis, D.,** Beitrag zur Chemilumineszenz und Biolumineszenz, *Folia Biochim. et Biolog. Graeca,* 11, 30, 1974.
19. **Billups, W. E., Chow, W. Y., and Cross, J. H.,** Synthesis of (±)-dictyopterene A and (±)-dictyopterene C′, *Chem. Commun.,* p. 252, 1974.
20. **Björkman, L. R., Karlsson, K.-A., Pascher, I., and Samuelsson, B. E.,** The identification of large amounts of cerebroside and cholesterol sulfate in the sea star *Asterias rubens, Biochim. Biophys. Acta,* 270, 260, 1972.
21. **Blackman, A. J. and Wells, R. J.,** personal communication.
22. **Block, J. H.,** Marine sterols from some gorgonians, *Steroids,* 23, 421, 1974.
23. **Blumer, M.,** Fossile Kohlenwasserstoffe und Farbstoffe in Kalksteinen. Geochemische Untersuchungen III. *Mikrochemie,* 36/ 37, 1048, 1951.
24. **Boar, R. B. and Widdowson, D. A.,** Biosynthesis, *Annu. Rep. Progr. Chem. Sect. B,* 71, 455, 1974.
25. **Boeryd, B., Hallgren, B., and Ställberg, G.,** Studies on the effect of methoxy-substituted glycerol ethers on tumour growth and metastasis formation, *Brit. J. Exp. Pathol.,* 52, 221, 1971.
26. **Boeryd, B., Hallgen, B., and Ställberg, G.,** Antitumor activity of methoxy-substituted glycerol ethers, *11th Int. Cancer Congr.* p. 139, 1974.
27. **Boger, E. A. and Johansen, H. W.,** Plastoquinones in coralline red algae (Corallinaceae) *Phyton,* 32, 129, 1974.
28. **Boll, P. M.,** Synthesis of asterosterol, a novel C_{26} marine sterol, *Acta Chem. Scand. Ser. B,* 28, 270, 1974.
29. **Bordes, D. B., Morton, G. O., and Wetzel, E. R.,** Structure of a novel bromine compound isolated from a sponge, *Tetrahedron Lett.,* p. 2709, 1974.
30. **Borys, H. K., Weinreich, D., and McCaman, R. E.,** Determination of glutamate and glutamine in individual neurons of *Aplysia californica, J. Neurochem.,* 21, 1349, 1973.

31. **Botticelli, C. R., Hisaw, F. L., Jr., and Wotiz, H. H.,** Estradiol-17β and progesterone in ovaries of starfish *(Pisaster ochraceous), Proc. Soc Exp. Biol. Med.,* 103, 875, 1960.
32. **Botticelli, C. R., Hisaw, F. L., Jr., and Wotiz, H. H.,** Estrogens and progesterone in the sea urchin *Strongylocentrotus franciscanus* and pecten *(Pecten hericius), Proc. Soc. Exp. Biol. Med.,* 106, 887, 1961.
33. **Brooks, D. E., Mann, T., and Martin, A. W.,** The occurrence of carnitine and glycerylphosphorylcholine in the octopus spermatophore, *Proc. R. Soc. London Ser. B.* 186, 79, 1974.
34. **Bullock, E. and Dawson, C. J.,** Carotenoid pigments of the holothurian *Psolus favrichii* Düben and Koren (the scarlet psolus), *Comp. Biochem. Physiol.,* 34, 799, 1970.
35. **Burkholder, P. R.,** The ecology of marine antibiotics and coral reefs, in *Biology and geology of coral reefs: Vol.II, Biology (Part 1),* Jones, O. A. and Endean, R., Eds., Academic Press, New York, 1973, 117.
36. **Burreson, B. J. and Scheuer, P. J.,** Isolation of a diterpenoid isonitrile from a marine sponge, *Chem. Commun.,* p. 1035, 1974.
37. **Burreson, B. J., Christophersen, C., and Scheuer, P. J.,** Cooccurrence of a terpenoid isocyanide-formamide pair in the marine sponge *Halichondria* sp., *J. Amer. Chem. Soc.,* 97, 201, 1975.
38. **Calabrese, G. and Felicini, G. P.,** Research on red algal pigments. V. The effect of the intensity of white and green light on the rate of photosynthesis and its relationship to pigment components in *Gracilaria compressa* (C. Ag.) Grev. (Rhodophyceae, Gigartinales), *Phycologia,* 12, 195, 1973.
39. **Cariello, L. and Prota, G.,** Occurrence of 3-hydroxy-L-kynurenine in gorgonians, *Comp. Biochem. Physiol. B,* 41, 195, 1972.
40. **Cariello, L., Crescenzi, S., Prota, G., and Zanetti, L.,** New zoanthoxanthins from the Mediterranean zoanthid *Parazoanthus axinellae, Experientia,* 30, 849, 1974.
41. **Cariello, L., Crescenzi, S., Prota, G., and Zanetti, L.,** Methylation of zoanthoxanthins, *Tetrahedron,* 30, 3611, 1974.
42. **Cariello, L., Crescenzi, S., Prota, G., and Zanetti, L.,** Zoanthoxanthins of a new structural type from *Epizoanthus arenaceus* (Zoantharia), *Tetrahedron,* 30, 4191, 1974.
43. **Chantraine, J.-M., Combaut, G., and Teste, J.,** Phenols bromes d'une algue rouge, *Halopytis incurvus:* acides carboxyliques, *Phytochemistry,* 12, 1793, 1973.
44. **Chapman, D. J. and Fox, D. L.,** Bile pigment metabolism in the sea-hare, *Aplysia, J. Exp. Mar. Biol. Ecol.,* 4, 71, 1969.
45. **Cimino, G., De Stefano, S., and Minale, L.,** Methyl *trans*-monocyclofarnesate from the sponge, *Halichondria panicea, Experientia,* 29, 1063, 1973.
46. **Cimino, G., De Stafano, S., and Minale, L.,** Occurrence of hydroxyhydroquinone and 2-aminoimidazole in sponges, *Comp. Biochem. Physiol. B,* 47, 895, 1974.
47. **Cimino, G., De Stefano, S., and Minale, L.,** Oxidized furanoterpenes from the sponge *Spongia officinalis, Experientia,* 30, 18, 1974.
48. **Cimino, G., De Stefano, S., and Minale, L.,** Pleraplysillin-2, a further furanosesquiterpenoid from the sponge *Pleraplysilla spinifera, Experientia,* 30, 846, 1974.
49. **Cimino, G., De Stefano, S., and Minale, L.,** Scalaradial, a third sesterterpene with the tetracarbocyclic skeleton of scalarin, from the sponge *Cacospongia mollior, Experientia,* 30, 846, 1974.
50. **Cimino, G., De Rosa, D., De Stefano, S., and Minale, L.,** Isoagatholactone, a diterpene of a new structural type from the sponge *Spongia officinalis, Tetrahedron,* 30, 645, 1974.
51. **Coll, J. C., Kazlauskas, R., Murphy, P. T., Wells, R. J., and Hawes, G. B.,** personal communication.
52. **Corey, E. J. and Washburn, W. N.,** The role of the symbiotic algae of *Plexaura homomalla* in prostaglandin biosynthesis, *J. Amer. Chem. Soc.,* 96, 934, 1974.
53. **Cormier, M. J. and Hori, K.,** Studies on the bioluminescence of *Renilla reniformis,* IV. Non-enzymatic activation of renilla luciferin, *Biochim. Biophys. Acta,* 88, 99, 1964.
54. **Cormier, M. J., Hori, K., and Karkhanis, Y.D.,** Studies on the bioluminescence of *Renilla reniformis.* VII. Conversion of luciferin into luciferyl sulfate by luciferin sulfokinase, *Biochemistry,* 9, 1184, 1970.
55. **Cormier, M. J., Wampler, J. E., and Hori, K.,** Bioluminescence: chemical aspects, *Fortschr. Chem. Org. Naturst.,* 30, 1, 1973.
56. **Cormier, M. J., Hori, K., Karkhanis, Y. D., Anderson, J. M., Wampler, J. E., Morin, J. G., and Hastings, J. W.,** Evidence for similar biochemical requirements for bioluminescence among the coelenterates, *J. Cell. Physiol.,* 81, 291, 1973.
57. **Cormier, M. J., Hori, K., and Anderson, J. M.,** Bioluminescence in coelenterates, *Biochim. Biophys. Acta,* 346, 137, 1974.
58. **Cottrell, G. A. and Laverack, M. S.,** Invertebrate pharmacology., *Annu. Rev. Pharmacol.,* 8, 273, 1968.
59. **Crews, P. and Kho, E.,** Cartilagineal. An unusual monoterpene aldehyde from marine alga, *J. Org. Chem.,* 39, 3303, 1974.
60. **Crozier, G. F.,** Pigments of fishes, *Chem. Zool.,* 8, 509, 1974.

61. **Czeczuga, B.,** Investigations of carotenoids in some fauna of the Adriatic Sea. I. *Verongia aerophoba* (Porifera : Spongiidae), *Mar. Biol.*, 10, 254, 1971.
62. **Czeczuga, B.,** Carotenoids in fish. II. Carotenoids and vitamin A in some fishes from the coastal region of the Black Sea, *Hydrobiologia*, 41, 113, 1973.
63. **Czeczuga, B.,** Investigations of carotenoids in some fauna of the Adriatic Sea. III. *Leander (Palaemon) serratus* and *Nephrops norvegicus* (Crustacea : Decapoda). *Mar. Biol.*, 21, 139, 1973.
64. **Czeczuga, B.,** Carotenoids in the fish milt, *Bull. Acad. Pol. Sci.*, 22, 211, 1974.
65. **Czeczuga, B.,** Comparative studies of carotenoids in the fauna of the Gullmar Fjord (Bohuslän, Sweden). II. Crustacea: *Eupagurus bernhardus, Hyas coarctatus* and *Upogebia deltaura, Mar. Biol.*, 28, 95, 1974.
66. **Das, N. P., Lim, H. S., and Teh, Y. F.,** Histamine and histamine-like substances in the marine sponge *Suberites inconstans, Comp. Gen. Pharmacol.*, 2, 473, 1971.
67. **Dasgupta, S.K., Crump, D. R., and Gut, M.,** New preparation of desmosterol, *J. Org. Chem.*, 39, 1658, 1974.
68. **De Rosa, M., Minale, L., and Sodano, G.,** Metabolism in porifera II. Distribution of sterols, *Comp. Biochem. Physiol. B*, 46, 823, 1973.
69. *Dictionary of Organic Compounds*, 4th ed., Eyre & Spottiswoode Publishers, London, 1965.
70. **Doig, M. T. III, and Martin, D. F.,** Anticoagulant properties of a red tide toxin, *Toxicon*, 11, 351, 1973.
71. **Doughterty, R. C, Strain, H. H., Svec, W. A, Uphaus, R. A., and Katz, J J.,** Structure of chlorophyll c. *J. Amer. Chem. Soc.*, 88, 5037, 1966.
72. **Dunstan, P. J., Hofheinz, W., and Oberhansli, W. E.,** personal communication.
73. **Egger, K.,** Die Verbreitung von Vitamin K_1 und Plastochinon in Pflanzen, *Planta*, 64, 41, 1965.
74. **Enwall, E. L. and van der Helm, D.,** The crystal structure and absolute configuration of the 3-*p*-iodobenzoate-11-acetate of secogorgosterol, *Rec. Trav. Chim. Pays-Bas*, 93, 53, 1974.
75. **Fagerlumd, U. H. M. and Idler, D. R.,** Marine sterols. VI. Sterol biosynthesis in molluscs and echinoderms, *Can. J. Biochem. Physiol.*, 38, 997, 1960.
76. **Fattorusso, E., Magno, S., Santacroce, C., and Sica, D.,** Sterol peroxides from the sponge *Axinella cannabina, Gazz. Chim. Ital.*, 104, 409, 1974.
77. **Fattorusso, E., Magno, S., Mayol, L., Santacroce, C., and Sica, D.,** Isolation and structure of axisonitrile-2. A new sesquiterpenoid isonitrile from the sponge *Axinella cannabina, Tetrahedron*, 30, 3911, 1974.
78. **Faulkner, D. J., Stallard, M. O., and Ireland, C.,** Prepacifenol epoxide, a halogenated sesquiterpene diepoxide, *Tetrahedron Lett.*, p. 3571, 1974.
79. **Faulkner, D. J.,** personal communication.
80. **Fenical, W.,** Rhodophytin, a halogenated vinyl peroxide of marine origin, *J. Amer. Chem. Soc.*, 96, 5580, 1974.
81. **Fenical, W. and Sims, J. J.,** Cycloeudesmol, an antibiotic cyclopropane containing sesquiterpene from the marine alga *Chondria oppositiclada* Dawson, *Tetrahedron Lett.*, p. 1137, 1974.
82. **Fenical, W.,** Polyhaloketones from the red seaweed *Asparagopsis taxiformis, Tetrahedron Lett.*, p. 4463, 1974.
83. **Fenical, W.,** Geranyl hydroquinone, a cancer-protective agent from the tunicate *Aplidium* sp., Proc. Food-Drugs from the Sea 1974, Marine Technology Society, Washington, D. C., 1976.
84. **Ferezou, J. P., Devys, M., Allais, J. P., and Barbier, M.,** Sur le sterol a 26 atomes de carbone de l'algue rouge *Rhodymenia palmata, Phytochemistry*, 13, 593, 1974.
85. **Firnhaber, H. J. and Wells, R. J.,** personal communication.
86. **Fisher, L. R., Kon, S. K., and Thompson, S. Y.** Vitamin A and carotenoids in certain invertebrates. II. Studies of seasonal variations in some marine crustacea. *J. Mar. Biol. Assoc. U. K.*, 33, 589, 1954.
87. **Fisher, L. R., Kon, S. K., and Thompson, S. Y.,** Vitamin A and carotenoids in certain invertebrates. III. Euphausiacea. *J. Mar. Biol. Assoc. U. K.*, 34, 81, 1955.
88. **Fornasiero, U., Antonello, C., and Guiotto, A.,** Naphthaquinone pigments of *Psammechinus microtuberculatus* Blainville, *Ann. Chim. (Roma)*, 63, 387, 1973.
89. **Fraenkel, G.,** The distribution of vitamin B_T (Carnitine) throughout the animal kingdom, *Arch. Biochem. Biophys.*, 50, 486, 1954.
90. **Fujita, Y.,** Seitai busshitsu no toriatsukaiho, in *Sorui-Jikkenho*, Tamiya, H. and Watanabe, T., Eds., 274, 1965.
91. **Fujita, Y. and Shimura, S.,** Phycoerythrin of the marine blue-green alga *Trichodesmium thiebautii, Plant Cell Physiol.*, 15, 939, 1974.
92. **Fujiwara-Arasaki, T. and Mino, N.,** The distribution of trimethylamine and trimethylamine oxide in marine algae, Proc. 7th Int. Seaweed Symp., University of Tokyo Press, 1971, 506.
93. **Gilchrist, B. M. and Welton, L. L.,** Carotenoid pigments and their possible role in reproduction in the sand crab *Emerita analoga* (Stimpson, 1857), *Comp. Biochem. Physiol, B*, 42, 263, 1972.

94. **Glombitza, K.-W., Rosener, H.-U., Vilter, H., and Rauwald, W.,** Antibiotica aus Algen. 8. Phloroglucin aus Braunalgen, *Planta Med.*, 24, 301, 1973.
95. **Glombitza, K.-W., and Sattler, E.,** Trifuhalol, ein neuer Triphenyldiäther aus *Halidrys siliquosa, Tetrahedron Lett.*, p. 4277, 1973.
96. **Glombitza, K.-W. and Rösener, H.-U.,** Bifuhalol: ein Diphenyläther aus *Bifurcaria bifurcata, Phytochemistry*, 13, 1245, 1974.
97. **Glombitza, K.-W., Stoffelen, H., Murawski, U., Bielaczek, J., and Egge, H.,** Antibiotica aus Algen. 9. Bromphenole aus Rhodomelaceen. *Planta Med.*, 25, 105, 1974.
98. **Goad, L. J., Rubinstein, I., and Smith, A. G.,** The sterols of echinoderms, *Proc. R. Soc. London Ser. B.*, 180, 223, 1972.
99. **González, A. G., Darias, J., Martin, J. D., and Peréz, C.,** Revised structure of caespitol and its correlation with isocaespitol, *Tetrahedron Lett.*, p. 1249, 1974.
100. **González, A. G., Darias, J., Martin, J. D., and Norte, M.,** Atomaric acid, a new component from *Taonia atomaria, Tetrahedron Lett.*, p. 3951, 1974.
101. **Gonzáles, A. G., Darias, J., Martin, J. D., Peréz, C., Sims, J. J., Lin, G. H. Y., and Wing, R. M.,** Isocaespitol, a new halogenated sesquiterpene from *Laurencia caespitosa, Tetrahedron*, 31, 2449, 1975.
102. **Goodwin, T. W.,** Pigments of porifera, *Chem. Zool.*, 2, 37, 1968.
103. **Goodwin, T. W.,** Pigments of mollusca, *Chem. Zool.*, 7, 187, 1972.
104. **Goto, T. and Kishi, Y.,** Luciferins, bioluminescent substances, *Angew. Chem. Int. Ed. Engl.*, 7, 407, 1968.
105. **Goto, T.,** Chemistry of bioluminescence, *Pure Appl. Chem.*, 17, 421, 1968.
106. **Goto, T.,** Cypridina bioluminescence IV. Synthesis and chemiluminescence of 3,7-dihydroimidazo(1,2-a)pyrazin-3-one and its 2-methyl derivative, *Tetrahedron Lett.*, p. 3873, 1968.
107. **Goto, T., Inoue, S., Sugiura, S., Nishikawa, K., Isobe, M., and Abe, Y.,** Cypridina bioluminescence, V. Structure of emitting species in the luminescence of cypridina luciferin and its related compounds, *Tetrahedron Lett.*, p. 4035, 1968.
108. **Goto, T., Isobe, M., Coviello, D. A., Kishi, Y., and Inoue, S.,** Cypridina bioluminescence. VIII. The bioluminescence of cypridina luciferin analogs, *Tetrahedron*, 29, 2035, 1973.
109. **Goto, T., Iio, H., Inoue, S., and Kakoi, H.,** Squid bioluminescence. I. Structure of watasenia oxyluciferin, a possible light-emitter in the bioluminescence of *Watasenia scintillans, Tetrahedron Lett.*, p. 2321, 1974.
110. **Gough, J. and Sutherland, M. D.,** The structure of spinochrome B, *Tetrahedron Lett.*, p. 269, 1964.
111. **Gough, J. H. and Sutherland, M. D.,** Marine pigments. VII. 3-Acetyl-2,5,6,7-tetrahydroxy-1,4-naphthoquinone, a new spinochrome from *Salmacis sphaeroides* (Lovén). *Aust. J. Chem.*, 20, 1693, 1967.
112. **Gough, J. H. and Sutherland, M. D.,** Pigments of marine animals. IX. A synthesis of 6-acetyl-2,3,5,7-tetrahydroxy-1,4-naphthoquinone, its status as an echinoid pigment, *Aust. J. Chem.*, 23, 1839, 1970.
113. **Hallgren, B. and Ställberg, G.,** Methoxy-substituted glycerol ethers isolated from Greenland shark liver oil, *Acta Chem. Scand. Ser. B*, 21, 1519, 1967.
114. **Hallgren B., Niklasson, A., Ställberg, G., and Thorin, H.,** On the occurrence of 1-0-(2-methoxyalkyl)glycerols and 1-0- phytanylglycerol in marine animals, *Acta Chem. Scand.*, 28, 1035, 1974.
115. **Hallgren, B. and Ställberg, G.,** 1-0-(2-Hydroxyalkyl)glycerols isolated from Greenland shark liver oil, *Acta Chem. Scand. Ser. B*, 28, 1074, 1974.
116. **Harada, K. and Yamada, K.,** Distribution of trimethylamine oxide in fishes and other aquatic animals, V. Teleosts and elasmobranchs, *Fish. Inst. Res. Rep. (Japan)*, 22, 77, 1973.
117. **Hashimoto, Y., Konosu, S., Fusetani, N., and Nose, T.,** Attractants for eels in the extracts of short-necked clam. I. Survey of constituents eliciting feeding behaviour by the omission test, *Bull. Jpn. Soc. Sci. Fish.*, 34, 78, 1968.
118. **Herring, P. J.,** Depth distribution of the carotenoid pigments and lipids of some oceanic animals. 2. Decapod crustaceans, *J. Mar. Biol. Assoc. U. K.*, 53, 539, 1973.
119. **Higa, T. and Scheuer, P. J.,** Thelepin, a new metabolite from the marine annelid *Thelepus setosus, J. Amer. Chem. Soc.*, 96, 2246, 1974.
120. **Hofheinz, W.,** personal communication.
121. **Hori, K., Wampler, J. E., Matthews, J. C., and Cormier, M. J.,** Identification of the product excited states during the chemiluminescent and bioluminescent oxidation of *Renilla* (sea pansy) luciferin and certain of its analogs, *Biochemistry*, 12, 4463, 1973.
122. **Hori, K., Wampler, J. E., and Cormier, M. J.,** Chemiluminescence of *Renilla* (sea pansy) luciferin and its analogues, *Chem. Commun.*, p. 492, 1973.
123. **Hori, K, and Cormier, M. J.,** Structure and chemical synthesis of a biologically active form of *Renilla* (sea pansy) luciferin, *Proc. Natl. Acad. Sci. U.S.A.*, 70, 120, 1973.

124. **Hotta, K., Kurokawa, M., and Isaka, S.**, Comparative studies of sialic acids from the jelly coat of sea urchin eggs, *J. Jpn. Biochem. Soc.*, 45, 911, 1973.
125. **Howden, M.E.H. and Williams, P. A.**, Occurrence of amines in the posterior salivary glands of the octopus *Hapalochlaena maculosa* (Cephalopoda), *Toxicon*, 12, 317, 1974.
126. **Ichikawa, N., Naya, Y., and Enomoto, S.**, New halogenated monoterpenes from *Desmia (Chondrococcus) hornemanni*, *Chem. Lett.*, p. 1333, 1974.
127. **Idler, D. R., Wiseman, P. M., and Safe, L. M.**, A new marine sterol, 22-*trans*-24-norcholesta-5,22-dien-3β-ol, *Steroids*, 16, 451, 1970.
128. **Idler, D. R. and Wiseman, P.**, Identification of 22-*cis*-cholesta-5,22-dien-3β-ol and other scallop sterols by gas-liquid chromatography and mass spectrometry, *Comp. Biochem. Physiol. A*, 38, 581, 1971.
129. **Iguchi, M., Niwa, M., and Yamamura, S.**, Stereostructure of zonarene, *Bull. Chem. Soc. Jpn.*, 46, 2920, 1973.
130. **Ike, T., Inanaga, J., Nakano, A., Okukado, N., and Yamaguchi, M.**, Total synthesis of natural acetylenic analogues of isorenieratene and renieratene, *Bull. Chem. Soc. Jpn.*, 47, 350, 1974.
131. **Inoue, N., Hosokawa, Y., and Akiba, M.**, Pristane in salmon muscle lipid, *Hokkaido Univ. Bull. Fac. Fish.*, 23, 209, 1973.
132. **Ishihara, K., Oguri, K., and Taniguchi, H.**, Isolation and characterization of fucose sulfate from jelly coat glycoprotein of sea urchin egg, *Biochim. Biophys. Acta*, 320, 628, 1973.
133. **Isler, O., Ed.**, Carotenoids, *Birkhäuser Verlag, Basel, 1971.*
134. **Jaenicke, L., Müller, D. G. , and Moore, R. E.**, Multifidene and aucantene, C_{11} hydrocarbons in the male-attracting essential oil from the gynogametes of *Cutleria multifida* (Smith) Grev. (Phaeophyta), *J. Amer. Chem. Soc.*, 96, 3324, 1974.
135. **Jaenicke, L. and Seferiadis, K.**, Die Stereochemie von Fucoserraten, dem Gametenlockstoff der Braunalge *Fucus serratus* L., *Chem. Ber.*, 108, 225, 1975.
136. **Jefferts, E., Morales, R. W., and Litchfield, C.**, Occurrence of *cis*-5,*cis*-9-hexacosadienoic and *cis*-5,*cis*-9,*cis*-19 -hexacosatrienoic acids in the marine sponge *Microciona prolifera*, *Lipids*, 9,244, 1974.
137. **Jeffrey, S. W.**, Purification and properties of chlorophyll c from *Sargassum flavicans*, *Biochem. J.*, 86, 313, 1963.
138. **Jeffs, P. W. and Lytle, L. T.**, Isolation of (−)-α-curcumene, (−)-β-curcumene and (+)- β-bisabolene from gorgonian corals. Absolute configuration of (−)-β-curcumene, *Lloydia*, 37, 315, 1974.
139. **Jensen, A.**, Tocopherol content of seaweed and seaweed meal. I. Analytical methods and distribution of tocopherols in benthic algae, *J. Sci. Food Agric.*, 20, 449, 1969.
140. **Jones, O. A. and Endean, R., Eds.**, *Biology and geology of coral reefs.* Vol. II: *Biology*, Part 1., Academic Press, New York, 1973.
141. **Kaisin, M., Sheikh, Y. M., Durham, L. J., Djerassi, C., Tursch, B., Dalo ze, D., Braekman, J. C., Losman, D., and Karlsson, R.**, Capnellane — a new tricyclic sesquiterpene skeleton from the soft coral *Capnella imbricata*, *Tetrahedron Lett.*, p. 2239, 1974.
142. **Kamiya, Y., Ikegami, S., and Tamura, S.**, A novel steroid, 3β, 6α, 15α,24ξ-tetrahydroxy-5α-cholestane from asterosaponins, *Tetrahedron Lett.*, p. 655, 1974.
143. **Kanasawa, A. and Yoshioka, M.** Occurrence of cholest-4-en-3-one in red alga *Meristotheca papulosa*, *Bull. Jpn. Soc. Sci. Fish.*, 37, 397, 1971.
144. **Kanazawa, A. and Yoshioka, M.**, The occurrence of cholest-4-en-3-one in the red alga *Gracilaria textorii*, *Proc. 7th Int. Seaweed Symp.*, University of Tokyo Press, Japan, 1971, 502.
145. **Kanazawa, A., Yoshioka, M., and Teshima, S.**, Sterols in some red algae, *Mem. Fac. Fish. Kagoshima Univ.*, 21, 103, 1972.
146. **Kanazawa, A., Teshima, S., and Ando, T.**, (E)-24-Ethylidene-cholest-7-en-3β-ol and other sterols in asteroids, *Mem. Fac. Fish. Kagoshima Univ.*, 22, 21, 1973.
147. **Kanazawa, A., Teshima, S., Ando, T., and Tomita, S.**, Occurrence of 23,24-dimethylcholesta-5,22-dien-3β-ol in a soft coral, *Sarcophyta elegans*, *Bull. Jpn. Soc. Sci. Fish.*, 40, 729, 1974.
148. **Kanazawa, A., Teshima, S., Tomita, S., and Ando, T.**, Gorgostanol, a novel C_{30} sterol from an asteroid, *Acanthaster planci*, *Bull. Jpn. Soc. Sci. Fish.*, 40, 1077, 1974.
149. **Karpetsky, T. P. and White, E. H.**, The synthesis of *Cypridina* etioluciferamine and the proof of structure of *Cypridina* luciferin, *Tetrahedron*, 29, 3761, 1973.
150. **Kashman, Y., Zadock, E., and Neeman, I.**, Some new cembrane derivatives of marine origin, *Tetrahedron*, 30, 3615, 1974.
151. **Katayama, T., Yokoyama, H., and Chichester, C. O.**, The biosynthesis of astaxanthin. II. The carotenoids in benibuna *Carassius auratus*, especially the existence of new keto carotenoids, α-doradecin and α-doradexanthin, *Bull. Jpn. Soc. Sci. Fish*, 36, 702, 1970.
152. **Katayama, T., Hirata, K., Yokoyama, H., and Chichester, C. O.**, The biosynthesis of astaxanthin. III. The carotenoids in sea breams, *Bull. Jpn. Soc. Sci. Fish.*, 36, 709, 1970.
153. **Katayama, T., Hirata, K., and Chichester, C. O.**, The biosynthesis of astaxanthin. IV. The carotenoids in the prawn *Penaeus japonicus* Bate (Part I), *Bull. Jpn. Soc. Sci. Fish.*, 37, 614, 1971.

154. **Katayama, T., Katama, T., and Chichester, C. O.**, The biosynthesis of astaxanthin. VI. The carotenoids in the prawn *Penaeus japonicus* Bate (Part II), *Int. J. Biochem.*, 3, 363, 1972.
155. **Katayama, T., Shintani, K., and Chichester, C. O.**, The biosynthesis of astaxanthin. VII. The carotenoids in sea bream *Chrysophrys major* Temminck and Schlegel, *Comp. Biochem. Physiol. B*, 44, 253, 1973.
156. **Katayama, T., Kunisaki, Y., Shimaya, M., Sameshima, M., and Chichester, C. O.**, The biosynthesis of astaxanthin. XIII. The carotenoids in the crab *Portunus trituberculatus*, *Bull. Jpn. Soc. Sci. Fish.*, 39, 283, 1973.
157. **Katayama, T., Miyahara, T., Kunisaki, Y., Tanaka, Y., and Imai, S.**, Carotenoids in the sea bream *Chrysophrys major* Temminck and Schlegel. II. *Mem. Fac. Fish. Kagoshima Univ.*, 22, 63, 1973.
158. **Katayama, T., Miyahara, T., Tanaka, Y., Sameshima, M., Simpson, K. L. , and Chichester, C. O.**, The biosynthesis of astaxanthin. XV. The carotenoids in Chidai, red sea bream, *Evynnis japonica* Tanaka and the incorporation of labelled astaxanthin from the diet of the red sea bream to their body astaxanthin, *Bull. Jpn. Soc. Sci. Fish.*, 40, 97, 1974.
159. **Kato, Y. and Scheuer, P. J.** Aplysiatoxin and debromoaplysiatoxin, constituents of the marine mollusk *Stylocheilus longicauda* (Quoy and Gaimard, 1824), *J. Amer. Chem. Soc.*, 96, 2245, 1974.
160. **Kato, Y. and Scheuer, P. J.**, The aplysiatoxins, *Pure Appl. Chem.*, 41, 1, 1975.
161. **Kazlauskas, R., Murphy, P. T., and Wells, R. J.**, personal communication.
162. **Kazlauskas, R., Murphy, P. T., Quinn, R. J., and Wells, R. J.**, personal communication.
163. **Kerkut, G. A.**, Catecholamines in invertebrates, *Br. Med. Bull.*, 29, 100, 1973.
164. **Keyl, M. J., Michaelson, I. A., and Whittaker, V. P.**, Physiologically active choline esters in certain marine gastropods and other invertebrates, *J. Physiol.*, 139, 434, 1957.
165. **Khare, A., Moss, G. P., and Weedon, B. C. L.**, Mytiloxanthin and isomytiloxanthin, two novel acetylenic carotenoids, *Tetrahedron Lett.*, p. 3921, 1973.
166. **Kishi, Y., Goto, T., Hirata, Y., Shimomura, O., and Johnson, F. H.**, Cypridina bioluminescence. I. Structure of *Cypridina* luciferin, *Tetrahedron Lett.*, p. 3427, 1966.
167. **Kitagawa, I., Sugawara, T., and Yosioka, I.**, Structure of holotoxin A, a major antifungal glycoside of *Stichopus japonicus* Selenka, *Tetrahedron Lett.*, p. 4111, 1974.
168. **Kitagawa, I., Sugawara, T., Yosioka, I., and Kuriyama, K.**, Structure of stichopogenin A4, the genuine aglycone of holotoxin A, isolated from *Stichopus japonicus* Selenka, *Tetrahedron Lett.*, p. 963, 1975.
169. **Kobayashi, M., Nishizawa, M., Todo, K., and Mitsuhashi, H.**, Marine sterols. I. Sterols of annelida, *Pseudopotamilla occelata* Moore, *Chem. Pharm. Bull. Tokyo*, 21, 323, 1973.
170. **Kobayashi, M., Todo, K., and Mitsuhashi, H.**, Marine sterols. III. Synthesis of asterosterol, a novel C_{26} sterol from asteroids, *Chem. Pharm. Bull. Tokyo*, 22, 236, 1974.
171. **Kobayashi, M. and Mitsuhashi, H.**, Marine sterols. IV. Structure and synthesis of amuresterol, a new marine sterol with unprecedented side chain, from *Asterias amurensis* Lütken, *Tetrahedron*, 30, 2147, 1974.
172. **Kobayashi, M. and Mitsuhashi, H.**, Marine sterols. V. Isolation and structure of occelasterol, a new 27-norergostane-type sterol, from an annelida, *Pseudopotamilla occelata*, *Steroids*, 24, 399, 1974.
173. **Komura, T., Wada, S., and Nagayama, H.**, The identification of fucosterol in the marine brown alga *Hizikia fusiformis*, *Agr. Biol. Chem.*, 38, 2275, 1974.
174. **Komura, T., Nagayama, H., and Wada, S.**, Studies on the lipase activator in marine algae. IV. Isolation of phytol from the Hiziki unsaponifiable matter and its activating effect on pancreatic liapse activity. *J. Agr. Chem. Soc. Japan*, 48, 459, 1974.
175. **Konosu, S., Watanabe, K., and Shimizu, T.**, Distribution of nitrogenous constituents in the muscle extracts of eight species of fish, *Bull. Jpn. Soc. Sci. Fish.*, 40, 909, 1974.
176. **Kumanireng, A. S., Kato, T., and Kitahara, Y.**, Structure and synthesis of diterpenephenols isolated from marine plants, *17th I.U.P.A.C. Symp. Chem. Natural Products*, Tokyo, p. 94, 1973.
177. **Kurosawa, E., Fukuzawa, A., and Irie, T.**, *trans*-and *cis*- laurediol, unsaturated glycols from *Laurencia nipponica* Yamada, *Tetrahedron Lett.*, p. 2121, 1972.
178. **Lam, J. K.K. and Sargent, M. V.**, Synthesis of methyl tri-0-methylptilometrate (methyl 1,6,8-trimethoxy-3-propylanthraquinone-2-carboxylate), *J. Chem. Soc. Perkin Trans. I*, p. 1417, 1974.
179. **Lee, A. S. K., Vanstone, W. E., Markert, J. R., and Antia, N. J.**, UV-absorbing and UV-fluorescing substances in the belly skin of fry of coho salmon *(Oncorhynchus kisutch)*, *J. Fish. Res. Board, Canada*, 26, 1185, 1969.
180. **Liaaen-Jensen, S.**, Selected examples of structure determination of natural carotenoids, *Pure Appl. Chem.*, 20, 421, 1969.
181. **Lowe, M. E. and Horn, D. H. S.**, Bioassay of the red chromatophore concentrating hormone of the crayfish, *Nature*, 213, 408, 1967.
182. **Lowe, M. E., Horn, D. H. S., and Galbraith, M. N.**, The role of crustecdysone in the moulting crayfish, *Experientia*, 24, 518, 1968.

183. **McBride, W. J., Freeman, A. R., Graham, L. T., Jr., and Aprison, M. H.**, The content of several amino acids in the external cell sheath and four giant axons of a nerve bundle from the CNS of the lobster, *Brain Res.*, 59, 440, 1973.
184. **McCapra, F. and Manning, M. J.**, Bioluminescence of coelenterates: chemiluminescent model compounds, *Chem. Commun.*, p. 467, 1973.
185. **McDonald, F. J., Campbell, D. C., Vanderah, D. J., Schmitz, F. J., Washecheck, D., M., Burks, J. E., and van der Helm, D.**, Marine natural products. Dactylyne, an acetylenic dibromochloro ether from the sea hare *Aplysia dactylomela*, *J. Org. Chem.*, 40, 665, 1975.
186. **McLachlan, J., Craigie, J. S., Chen, L. C.-M., and Ogata, E.**, *Porphyra linearis* Grev. An edible species of nori from Nova Scotia, *Proc. 7th Int. Seaweed Symp.*, University of Tokyo Press, Japan, 1971, 473.
187. **McMillan, J. A., Paul, I. C., White, R. H., and Hager, L. P.**, Molecular structure of acetoxyintricatol: a new bromo compound from *Laurencia intricata*, *Tetrahedron Lett.*, p. 2039, 1974.
188. **McMurry, J. E. and von Beroldingen, L. A.**, Ketone methylenation without epimerization: total synthesis of (±) laurene, *Tetrahedron*, 30, 2027, 1974.
189. **Madgwick, J. C., Ralph, B. J., Shannon, J. S., and Simes, J. J.**, Non-protein amino acids in Australian seaweeds, *Arch. Biochem. Biophys.*, 141, 766, 1970.
190. **Madgwick, J. C. and Ralph, B. J.**, Free amino acids in Australian marine algae, *Bot. Mar.*, 15, 205, 1972.
191. **Mann, T., Martin, A. W., Thiersch, J. B., Lutwak-Mann, C., Brooks, D. E., and Jones, R.**, D(−)-Lactic acid and D(−)- lactate dehydrogenase in octopus spermatozoa, *Science*, 185, 453, 1974.
192. **Manning, W. M. and Strain, H. H.**, Chlorophyll d, a green pigment of red algae, *J. Biol. Chem.*, 151, 1, 1943.
193. **Marderosian, A. D.**, Marine pharmaceuticals, *J. Pharm. Sci.* 58, 1, 1969.
194. **Marsh, M. E. and Ciereszko, L. S.**, Alpha-glyceryl ethers in coelenterates, *Proc. Oklahoma Acad. Sci.*, 53, 53, 1973.
195. **Martin, D. F. and Padilla, G. M., Eds.**, *Marine Pharmacognosy*, Academic Press, New York, 1973.
196. **Matsuno, T., Kusumoto, T., Watanabe, T., and Ishihara, Y.**, Carotenoid pigments of spiny lobster, *Bull. Jpn. Soc. Sci. Fish.*, 39, 43, 1973.
197. **Matsuno, T., Akita, T., and Hara, M.**, Carotenoid pigments of Japanese anchovy, *Bull. Jpn. Soc. Sci. Fish.*, 39, 51, 1973.
198. **Matsuno, T., Nagata, S., Sato, Y., and Watanabe, T.**, Comparative biochemical studies of carotenoids in fishes. II. Carotenoids of horse mackerel, swellfishes, porcupine fishes and striped mullet, *Bull. Jpn. Soc. Sci. Fish.*, 40, 579, 1974.
199. **Matsuno, T., Watanabe, T., and Nagata, S.**, Carotenoid pigments of crustacea. II. The carotenoid pigments of *Scyllarides squamosus (= Scyllarus sieboldi)* and *Parribacus antarcticus (= Parribacus ursus-major)*, *Bull. Jpn. Soc. Sci. Fish.*, 40, 619, 1974.
200. **Matsuno, T. and Watanabe, T.** Carotenoid pigments of crustacea. III. The carotenoid pigments of *Sesarma (Holometopus) haematocheir* and *Sesarma (Sesarma) intermedia*, *Bull. Jpn. Soc. Sci. Fish.*, 40, 767, 1974.
201. **Mazzarella, L. and Puliti, R.**, Crystal structure and absolute configuration of aeroplysinin-1, *Gazz. Chim. Ital.*, 102, 391, 1972.
202. *Merck Index*, 8th Ed., Merck & Co., Rahway, New Jersey, 1968.
203. **Mettrick, D. F. and Telford, J. M.**, The histamine content and histidine decarboxylase activity of some marine and terrestrial animals from the West Indies, *Comp. Biochem. Physiol.*, 16, 547, 1965.
204. **Minale, L. and Sodano, G.**, Marine sterols, 19-nor-stanols from the sponge *Axinella polypoides*, *J. Chem. Soc. Perkin Trans. I*, p.1888, 1974.
205. **Minale, L. and Sodano, G.**, Marine sterols: unique 3β-hydroxymethyl-A-nor-5α-steranes from the sponge *Axinella verrucosa*, *J. Chem. Soc. Perkin Trans. I*, 2380, 1974.
206. **Minale, L., Riccio, R., and Sodano, G.**, Acanthellin-1, an unique isonitrile sesquiterpene from the sponge *Acanthella acuta*, *Tetrahedron*, 30, 1341, 1974.
207. **Minale, L., Riccio, R., and Sodano, G.**, Avarol, a novel sesquiterpenoid hydroquinone with a rearranged drimane skeleton from the sponge *Disidea avara*, *Tetrahedron Lett.*, p. 3401, 1974.
208. **Miyazawa, K. and Ito, K.**, Isolation of L-methionine-I-sulfoxide and *N*-methylmethionine sulfoxide from a red alga *Grateloupia turuturu*, *Bull. Jpn. Soc. Sci. Fish.*, 40, 655, 1974.,
209. **Moldowan, J. M., Tursch, B. M., and Djerassi, C.**, 24ξ-Methylcholestane-3β,5α,6β,25-tetrol 25-monoacetate, a novel polyhydroxylated steroid from an alcyonarian, *Steroids*, 24, 387, 1974.
210. **Momzikoff, A.**, Mise en évidence d'une excretion de ptérines par une population naturelle de copépodes planctoniques marins, *Cah. Biol. Mar.*, 14, 323, 1973.
211. **Momzikoff, A. and Gaill, F.**, Mise en évidence d'une émission d'isoxanthoptérine et de riboflavine par différentes espèces d'ascidies, *Experientia*, 29, 1438, 1973.
212. **Moore, R. E. and Scheuer, P. J.**, Palytoxin: a new marine toxin from a coelenterate, *Science*, 172, 495, 1971.

213. **Moore, R. E. and Yost, G.,** Dihydrotropones from *Dictyopteris, Chem. Commun.*, p. 937, 1973.

214. **Moore, R. E., Pettus, J. A., Jr., and Mistysyn, J.,** Odoriferous C_{11} hydrocarbons from Hawaiian *Dictyopteris, J. Org. Chem.*, 39, 2201, 1974.

215. **Morton, R. A.,** Ubiquinones, plastoquinones and vitamins K, *Biol. Rev.*, 46, 47, 1971.

216. **Motzfeldt, A.-M.,** Isolation of 24-oxocholesterol from the marine brown alga *Pelvetia canaliculata* (Phaeophyceae), *Acta Chem. Scand.*, 24, 1846, 1970.

217. **Mulder, J. W. and Wells, R. J.,** personal communication.

218. **Müller, D. G. and Jaenicke, L.,** Fucoserraten, the female sex attractant of Fucus serratus L. (Phaeophyta), *FEBS Lett.*, 30, 137, 1973.

219. **Müller, D. G.,** Sexual reproduction and isolation of a sex attractant in *Cutleria multifida* (Smith) Grev. (Phaeophyta), *Biochem. Physiol. Pflanzen*, 165, 212, 1974.

220. **Mynderse, J. S. and Faulkner, D. J.,** Violacene, a polyhalogenated monocyclic monoterpene from the red alga *Plocamium violaceum, J. Amer. Chem. Soc.*, 96, 6771, 1974.

221. **Nakayama, T. O. M.,** Carotenoids, in *Physiology and Biochemistry of Algae,* Lewin, R. A., Ed., Academic Press, New York, 1962, 409.

222. **Neeman, I., Fishelson, L., and Kashman, Y.,** Sarcophine — a new toxin from the soft coral-*Sarcophyton glaucum* (Alcyonaria), *Toxicon*, 12, 593, 1974.

223. **Nitsche, H.,** Neoxanthin and fucoxanthinol in *Fucus vesiculosus, Biochim. Biophys. Acta,* 338, 572, 1974.

224. **Nomura, T., Ogata, H., and Ito, M.,** Occurrence of prostaglandins in fish testis, *Tohoku J. Agr. Res.*, 24, 138, 1973.

225. **Norgård, S., Svec, W. A., Liaaen-Jensen, S., Jensen, A., and Guillard, R. R. I.,** Algal carotenoids and chemotaxonomy, *Biochem. Syst. Ecol.*, 2, 7, 1974.

226. **Ogura, Y.,** Recent studies on Fugu toxin, *Setai no Kagaku,* 9, 281, 1958.

227. **Osborne, N. N.,** Occurrence of glycine and glutamic acid in the nervous system of two fish species and some invertebrates, *Comp. Biochem. Physiol. B,* 43, 579, 1972.

228. **Patil, V. D., Nayak, U. R., and Dev, S.,** Chemistry of ayurvedic crude drugs. II. Guggulu (resin from *Commiphora mukul*) 2. Diterpenoid constituents, *Tetrahedron,* 29, 341, 1973.

229. **Pedersén, M., Saenger, P., and Fries, L.,** Simple brominated phenols in red algae, *Phytochemistry,* 13, 2273, 1974.

230. **Pettit, G. R., Day, J. F., Hartwell, J. L., and Wood, H. B.,** Antineoplastic components of marine animals, *Nature,* 227, 962, 1970.

231. **Rimington, C. and Kennedy, G. Y.,** Porphyrins: structure, distribution and metabolism, *Comp. Biochem.*, 4, 557, 1962.

232. **Rinehart, K. L., Jr., Johnson, R. D., Paul, I. C., MacMillan, J. A., Siuda, J. F., and Krejcarek, G. A.,** Identification of compounds in selected marine organisms by gas chromatography-mass spectrometry, field desorption mass spectrometry and other physical methods, *Proc. Food- Drugs from the Sea 1974,* Marine Technology Society, Washington, D. C., 1976, 434.

233. **Roseghini, M., Erspamer, V., Ramorino M. L., and Gutierrez, J. E.,** Choline esters, their precursors and metabolites in the hypobranchial gland of prosobranchiate molluscs *Concholepas concholepas* and *Thais chocolata, Eur. J. Biochem.*, 12, 468, 1970.

234. **Roseghini, M. and Fichman, M.,** Choline esters and imidazole acids in extracts of the hypobranchial gland of *Thais haemastoma, Comp. Gen. Pharmacol.*, 4, 251, 1973.

235. **Roseghini, M., Alcala, A. C., and Vitali, T.,** Occurrence of *N*-imidazolepropionyl-histamine in the soft tissues of the Philippine gastropod *Drupa concatenata* Lam., *Experientia,* 29, 940, 1973.

236. **Roseghini, M. and Alcala, A. C.,** Occurrence of *N*-urocanylhistamine in the soft tissues of the gastropod mollusc *Drupa concatenata* Lam., *Biochem. Pharmacol.*, 23, 1431, 1974.

237. **Rubinstein, I. and Goad, L. J.,** Sterols of the siphonous marine alga *Codium fragile, Phytochemistry,* 13, 481, 1974.

238. **Rubinstein, I. and Goad, L. J.,** Occurrence of (24S)-24-methylcholesta-5, 22E-dien-3β-ol in the diatom *Phaeodactylum tricornutum, Phytochemistry,* 13, 485, 1974.

239. **Saavedra, J. M., Brownstein, M. J., Carpenter, D. O., and Axelrod, J.,** Octopamine: presence in single neurons of *Aplysia* suggests neurotransmitter function, *Science,* 185, 364, 1974.

240. **Safe, L. M., Wong, C. J., and Chandler, R. F.,** Sterols of marine algae, *J. Pharm. Sci.*, 63, 464, 1974.

241. **Schantz, E. J.,** Some toxins occurring naturally in marine organisms, in *Microbial Safety of Fishery Products,* Chichester, C. O., Ed., Academic Press, New York, 1973, 151.,

242. **Schmitz, F. J., Vanderah, D. J., and Ciereszko, L. S.,** Marine natural products: nephthenol and epoxynephthenol acetate, cembrene derivatives from a soft coral, *Chem. Commun.*, p. 407, 1974.

243. **Schmitz, F. J. and McDonald, F. J.,** Marine natural products: Dactyloxene-B, a sesquiterpene ether from the sea hare *Aplysia dactylomela, Tetrahedron Lett.*, p. 2541, 1974.

244. **Schmitz, F. J., Campbell, D. C., and McDonald, F. J.,** Dactylyne, a halogenated acetylenic oxetane isolated from the sea hare *Aplysia dactylomela, 9th I.U.P.A.C. Symp. Chem. Natural Products, Abstr.* Ottawa, 1974. 12E.

245. **Sieburth, J. M.,** Acrylic acid, an antibiotic principle in *Phaeocystis* blooms in antarctic waters, *Science,* 132, 676, 1960.
246. **Sims, J. J., Fenical, W., Wing, R. M., and Radlick, P.,** Marine natural products. IV. Prepacifenol, a halogenated epoxy sesquiterpene and precursor to pacifenol from the red alga *Laurencia filiformis, J. Amer. Chem. Soc.,* 95, 972, 1973.
247. **Sims, J. J., Lin, G. H. Y., and Wing, R. M.,** Marine natural products. X. Elatol, a halogenated sesquiterpene alcohol from the red alga *Laurencia elata, Tetrahedron Lett.,* p. 3487, 1974.
248. **Siuda, J. F., VanBlaricom, G. R., Shaw, P. D., Johnson, R. D., White, R. H., Hager, L. P., and Rinehart Jr., K. L.,** 1-lodo-3,3-dibromo-2-heptanone, 1,1,3,3-tetrabromo-2-heptanone, and related compounds from the red alga *Bonnemaisonia hamifera, J. Amer. Chem. Soc.,* 97, 937, 1975.
249. **Smith, A. G., Rubinstein, I., and Goad, L. J.,** The sterols of the echinoderm *Asterias rubens, Biochem. J.,*135, 443, 1973.
250. **Smith. L. L., Dhar, A. K., Gilchrist, J. L., and Lin, Y. Y.,** Sterols of the brown alga *Sargassum fluitans, Phytochemistry,* 12, 2727, 1973.
251. **Stallard, M. O. and Faulkner, D. J.,** Chemical constituents of the digestive gland of the sea hare *Aplysia californica.* I. Importance of diet, *Comp. Biochem. Physiol. B,* 49, 25, 1974.
252. **Stoffelen, H., Glombitza, K.-W., Murawski, U., Bielaczek, J., and Egge, H.,** Bromphenole aus-*Polysiphonia lanosa* (L.) Tandy, *Planta Med.,* 22, 396, 1972.
253. **Strain, H. H., Manning, W. M. and Hardin, G.,** Xanthophylls and carotenes of diatoms, brown algae, dinoflagellates and sea anemones, *Biol. Bull.,* 86, 169, 1944.
254. **Suyama, M. and Yoshizawa, Y.,** Free amino acid composition of the skeletal muscle of migratory fish, *Bull. Jpn. Soc. Sci. Fish.,* 39, 1339, 1973.
255. **Suzuki, M., Kurosawa, E., and Irie, T.,** Three new sesquiterpenoids containing bromine minor constituents of *Laurencia glandulifera* Kützing, *Tetrahedron Lett.,* p. 821, 1974.
256. **Suzuki, M., Kurosawa, E., and Irie, T.,** Glanduliferol, a new halogenated sesquiterpenoid from-*Laurencia glandulifera* Kützing, *Tetrahedron Lett.,* p. 1807, 1974.
257. **Takagi, M. and Okumura, A.,** On a new amino acid, S-hydroxymethyl-L-homocysteine isolated from *Chondrus ocellatus, Bull. Jpn. Soc. Sci. Fish.,* 30, 837, 1964.
258. **Tammar, A. R.,** Bile salts in fishes, *Chem. Zool.,* 8, 595, 1974.
259. **Tanaka, Y. and Katayama, T.,** Carotenoids in the sea bream, *Chrysophrys major* Temmick and Schlegel. III. The carotenoids in mysis and the internal organs of squid as the food for sea bream, *Mem. Fac. Fish. Kagoshima Univ.,* 23, 117, 1974.
260. **Taylor, F. R., Ikawa, M., Sasner Jr., J. J., Thurberg, F. P., and Andersen, K. K.,** Occurrence of choline esters in the marine dinoflagellate *Amphidinium carteri, J. Phycol.,* 10, 279, 1974.
261. **Teshima, S., Kanazawa, A., and Ando, T.,** Isolation of a novel C_{28}-sterol, 24-methycholesta-7,22,25-trien-3β-ol from a starfish, *Leiaster leachii, Bull. Jpn. Soc. Sci. Fish.,* 40, 631, 1974.
262. **Torres Pombo, J.,** Contribución al conocimiento quimico del alga *"Gelidium sesquipedale* (Clem.)Thuret" y a la estructura de su agar, *Acta Cient. Compostelana,* 9, 53, 1972.
263. **Toyama, Y. and Takagi, T.,** Sterols and other unsaponifiable substances in the lipids of shell fishes, crustacea and echinoderms. XV. Occurrence of $\Delta 7$: 8-cholestenol as a sterol component of starfish *Asterias amurensis* Lütken, *Bull. Chem. Soc. Japan,*27, 421, 1954.
264. **Tsukuda, N. and Kitahara, T.,**Composition of the esterified fatty acids of astaxanthin diester in the skin of seven red fishes, *Bull. Tokai Reg. Fish. Res. Lab.,* No. 77, 89, 1974.
265. **Turner, A. B.,** Starfish saponins, *9th I.U.P.A.C. Symp. Chem. Natural Products* Abstr., Ottawa, 1974, 8E.
266. **Tursch, B., Braekman, J. C., Daloze, D., Fritz, P., Kelecom, A., Karlsson, R., and Losman. D.,** Chemical studies of marine invertebrates. VIII. Africanol, an unusual sesquiterpene from *Lemnalia africana* (Coelenterata, Octocorallia, Alcyonacea), *Tetrahedron Lett.,* p. 747, 1974.
267. **Tursch, B., Braekman, J. C., Daloze, D., Herin, M., and Karlsson R.,** Chemical studies of marine invertebrates. X. Lobophytolide, a new cembranolide diterpene from the soft coral *Lobophytum cristagalli* (Coelenterata, Octocorallia, Alcyonacea), *Tetrahedron Lett.,* p. 3769, 1974.
268. **Van der Helm, D., Ealick, S. E., and Weinheimer, A. J.,**15(R)-Acetoxy-6(S),10(S)-dibromo-3a(S),4,7,8,11,12,13,14(S),15,15a(R)-decahydro-6,10-14-trimethyl-3-methylene-5(R),9(R)-epoxycyclotetradeca[b]2-furanone,$C_{22}H_{32}Br_2O_5$, *Cryst. Struct. Commun.,* 3, 167, 1974.
269. **Viala, J., Devys, M., and Barbier, M.,**Sur la structure des stérols à 26 atomes de carbone du tunicier *Halocynthia roretzi, Bull. Soc. Chim. Fr.,* p. 3626, 1972.
270. **Waraszkiewicz, S. M. and Erickson, K. L.,**Halogenated sesquiterpenoids from the Hawaiian marine alga *Laurencia nidifica*: nidificene and nidifidiene, *Tetrahedron Lett.,* p. 2003, 1974.
271. **Weinheimer, A. J.,** The discovery of 15-*epi* PGA_2 in *Plexaura homomalla, Stud. Trop. Oceanogr.,* 12, 17, 1974.
272. **Wilkie, D. W.,** The carotenoid pigmentation of *Pleuroncodes planipes* Stimpson (Crustacea: Decapoda: Galatheidae), *Comp. Biochem. Physiol. B,* 42, 731, 1972.
273. **Wotiz, H. H., Botticelli, C. R., Hisaw, F. L., Jr., and Olsen, A. G.,** Estradiol-17β, estrone and progesterone in the ovaries of dogfish *(Squalus suckleyi), Proc. Natl. Acad. Sci. U.S.A.,* 46, 580, 1960.

274. **Yamada, Y., Kim, J.-S., Iguchi, K., and Suzuki, M.,** An effective synthesis of a bromine-containing antibacterial compound from marine sponges, *Chem. Lett.*, p. 1399, 1974.

275. **Yasuda, S.,** Sterol compositions of jelly fish (Medusae), *Comp. Biochem. Physiol. B.*, 48, 225, 1974.

276. **Yasuda, S.,** Sterol compositions of echinoids (sea urchin, sand dollar and heart urchin), *Comp. Biochem, Physiol. B*, 49, 361, 1974.

277. **Yasumoto, T. and Endo, M.,** Toxicity study on a marine snail, *Turbo argyrostoma* I. Presence of two sulfur- containing amines in the acetone-soluble fraction, *Bull. Jpn. Soc. Sci. Fish.*, 39, 1055, 1973.

278. **Yasumoto, T. and Endo, M.,** Toxicity study on a marine snail, *Turbo argyrostoma* II. Identification of (3-methylthiopropyl)trimethylammonium chloride, *Bull. Jpn. Soc. Sci. Fish.*, 40, 217, 1974.

279. **Yasumoto, T. and Endo, M.,** Toxicity study on a marine snail, *Turbo argyrostoma* III. Occurrence of candicine, *Bull. Jpn. Soc. Sci. Fish.*, 40, 841, 1974.

280. **Yasumoto, T. and Sano, F.,** Occurrence of homoserine betaine and valine betaine in the ovary of shellfish *Callista brevishiphonata*, *Bull. Jpn. Soc. Sci. Fish.*, 40, 1163, 1974.

281. **Yasumoto, T.,** Toxicity study on a marine snail, *Turbo argyrostoma*. IV. Occurrence of [3-(dimethylsulfonio)propyl]trimethylammonium dichloride, *Bull. Jpn. Soc. Sci. Fish.*, 40, 1169, 1974.

282. **Yoshizawa, T. and Nagai, Y.,** Occurrence of cholesteryl sulfate in eggs of the sea urchin *Anthocidaris crassispina*, *Jpn. J. Exp. Med.*, 44, 465, 1974.

283. **Youngblood, W. W. and Blumer, M.,** Alkanes and alkenes in marine benthic algae, *Mar. Biol.*, 21, 163, 1973.

Index

INDEX

A

B

C

D

E

F

G

H

I

J

K

L

M

N

O

P

R

S

U

V

W

X

Y